Y0-AHR-839

THE SHELFBREAK: CRITICAL INTERFACE ON CONTINENTAL MARGINS

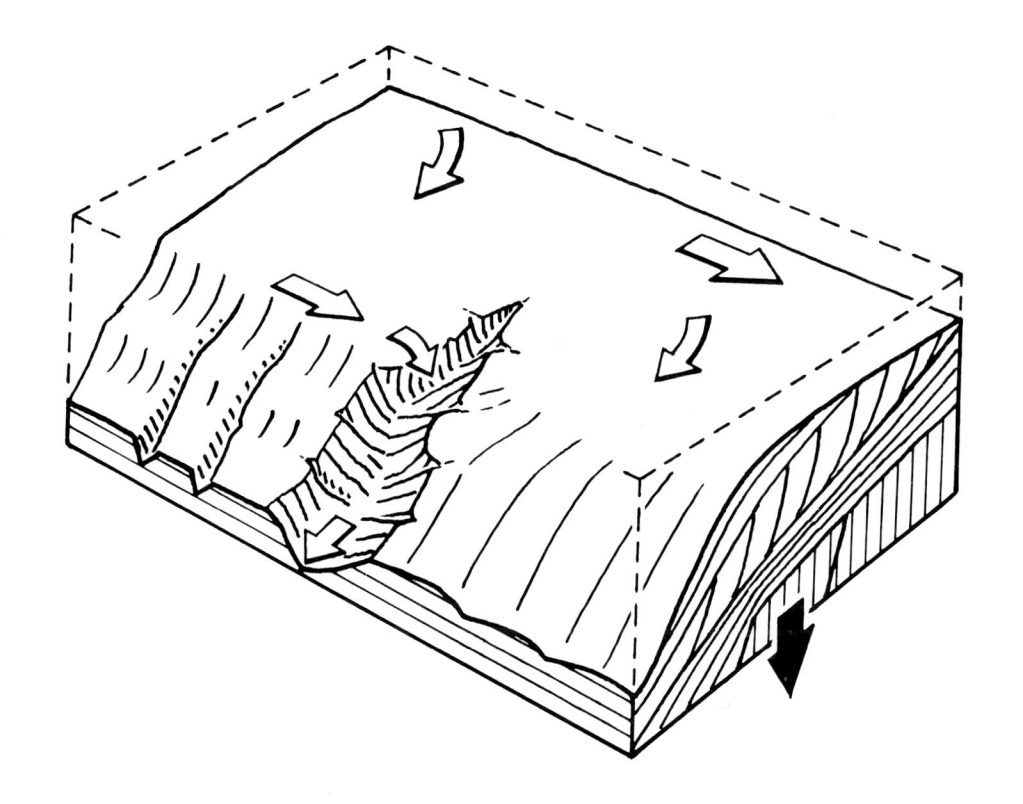

Edited by
Daniel Jean Stanley
Division of Sedimentology
Smithsonian Institution
and
George T. Moore
Chevron Oil Field Research Company
La Habra, California

Copyright 1983 *by*

SOCIETY OF ECONOMIC PALEONTOLOGISTS AND MINERALOGISTS

Special Publication No. 33

Tulsa, Oklahoma, U.S.A.   June, 1983

PREFACE

The shelfbreak is that point where the first major change in gradient occurs on the outermost edge of the continental shelf. The break, or shelfedge as it is also termed, can be traced for more than 300,000 km along the world's continental margins, plateaus and islands. Although this environment delimits the boundary between two principal and well-defined provinces, the continental shelf and slope—and thus is of the first order importance on continental margins—it has received surprizingly little specific attention in either modern oceans or in the rock record. The break clearly is a significant boundary mapped by, among others: structural geologists investigating the origin of the outer margin, physical oceanographers studying distinct water mass regimes on the shelf and slope, biologists recording faunal distributions, and sedimentologists interpreting the dispersal paths and depositional facies between the inner margins and deep basins.

The foregoing considerations and recent economic exploitation near and beyond the shelfbreak prompted the organization of a Research Symposium on this topic under the sponsorship of the Society of Economic Paleontologists and Mineralologists. Twenty-two presentations were made in three sessions convened at the joint Annual Meeting of the SEPM and American Association of Petroleum Geologists in San Francisco on June 2, 1981. These, and an additional six invited chapters, constitute the first compendium dedicated specifically to the shelfbreak.

The material is organized in a manner to illustrate examples of the shelfbreak in both modern oceans and the rock record. The organization of the volume also takes into account a series of basic questions that need to be addressed. For example, how diverse are the shelfedge configurations on different world margins? What changes are noted from the landward to oceanward side of the break? What morphologic and genetic relations exist, if any, between slope features such as canyons and the outermost shelf and break? To what extent does sealevel—and, more specifically, eustatic—fluctuations affect the shelfedge and its position? These and related points are discussed in Part I.

Questions pertaining to larger-scale aspects of the structural and stratigraphic framework are emphasized in Part II. In addition to eustatic controls, what are the responses in the subsurface (transitory position and configuration of break) to different tectonic settings in modern oceans? To what extent can these criteria be used to distinguish passive and active margins? Are comparable paleo-shelfbreaks recognized in the rock record, and what are their more obvious criteria? Does the geological evolution of a margin (divergent, transform or convergent motion; rate of subsidence) affect the preservation of the break in the ancient record?

What, for example, are the responses at the break to large volumes of terrigenous material transported onto and across this environment? Part III focuses on the modifications induced by large fluvial input and the presence of deltas which can affect not only the shelf proper but the break as well. How does the break on carbonate margins in warm seas and at lower latitudes differ from that in terrestrial clastic-dominated sectors? Chapters in Part IV compare the salient characters of carbonate shelf-to-slope breaks in the modern ocean and ancient record, and also focus on diagenetic effects mapped on some platform margins.

Information on the flow of water masses at and near the break, still a developing field of research, is critical for understanding how sediments move from the shelf to slope—and, possibly, vice-versa. It is recognized that the modern shelfbreak is not a static area insofar as sediment transfer is concerned. What are some of the essential factors that influence flow and particle motion at the shelfedge? How constant are these physical processes? To what extent is the plexus of processes similar on different margins? How are the short and long-term flow pattern variations to be explained? These aspects of dynamics are highlighted in Part V, while the sedimentary responses to diverse flow processes are summarized in Part VI. In this respect what is, for example, the long-term temporal and spacial configuration of surficial sediment facies resulting from the almost continuous movement of water above the seafloor? To what extent do variations in grain-size—generally, sand to mud—coincide with the shelfbreak? Can these changes in sediment facies be recognized by seismic techniques? Is offshelf spill-over concentrated in canyons, or are sediments also transported across the shelfbreak onto the slope away from canyon heads? To what extent are geotechnical properties of sediments at the shelfbreak distinct from shelf or deeper water deposits? What physical properties provide clues to mass failure of sediments near and below the shelfbreak?

Chapters in Part VII focus on faunas and organic matter. Is there evidence of a marked faunal change at or near the shelfbreak? Are such changes recorded by planktonic and infaunal, as well as benthonic, assemblages? To what extent does the content of organic matter reflect the marked change from shelf to slope? If correlations between these and other parameters exist, such could have important implications in interpreting the paleogeography of ancient margin deposits and delimiting paleo-shelfbreaks in the rock record.

All these questions bear on the economic prospects and legal aspects at the shelfbreak. Are any economic deposits unique to the region of the shelfbreak? Has the break acted as a locus for oil and gas? What appear to be the conditions favoring mineral concentrations and where would these likely occur? Both economic and political pressures affecting the offshore have increased substantially in recent years: what are the legal ramifications? How will maritime law as presently interpreted affect research and exploration? These and related points are raised in Part VIII.

Knowledge of the shelfbreak remains rudimentary and it would be pretentious to assert that we have progressed much beyond a preliminary understanding of this environment. Many of the questions we have posed in this preface have yet to be answered satisfactorily. Thus we hope that this volume highlights the extent to which the shelfbreak is a real "frontier." It is the most critical interface at the outer shelf-to-slope part of continental margins. Moreover, it is one of the most important submarine evironments requiring study. Because of its complexity, the systematic multidisciplinary approach using refined technological advances is needed. The collected contributions presented in this volume provide a major step in this direction.

We thank the many referees who assisted us in the review process. In addition to the large number of authors who served as an editorial pool, we acknowledge the following who, among others, generously gave of their time: W. H. Abbott, G. W. Andrews, M. A. Arthur, S. Bachman, G. Bellaiche, A. H. Bouma, W. R. Bryant, R. T. Buffler, H. E. Cook, R. L. Ellison, R. W. Embley, W. K. Gealey, A. R. Green, A. Guilcher, R. B. Halley, D. G. Howell, G. H. Keller, J. S. Kelley, R. Knecht, G. P. Lohmann, I. N. McCave, F. T. Manheim, W. J. Meyers, D. G. Moore, J. R. Moore, J. W. Pierce, R. W. Pfeil, Jr., D. Schnitker, D. W. Scholl, P. D. Snavely III, E. Suess, E. Uchupi, P. R. Vail, M. Wimbush and R. A. Young. Appreciation is also expressed to L. Kemp and V. E. Welsh for their valuable assistance in the editorial process. Funding from Smithsonian Scholarly Studies grant 1233S201 to one of us (DJS) helped defray some costs involved with the preparation of this volume.

Daniel Jean Stanley
and
George T. Moore
Co-Editors

THE SHELFBREAK: CRITICAL INTERFACE ON CONTINENTAL MARGINS

Edited by Daniel Jean Stanley and George T. Moore

CONTENTS

PART I. MORPHOLOGY AND SEALEVEL CHANGE

Shelfbreak Physiography: An Overview .... *Jean-René Vanney and Daniel Jean Stanley* ... 1

Breaching the Shelfbreak: Passage from Youthful to Mature Phase in Canyon Evolution .... *John A. Farre, Bonnie A. McGregor, William B. F. Ryan, and James M. Robb* ... 25

The Effect of Sealevel Change on the Shelfedge and Slope of Passive Margins .... *Walter C. Pitman III and Xenia Golovchenko* ... 41

PART II. STRUCTURAL AND STRATIGRAPHIC FRAMEWORK

Prograding Shelfbreak Types on Passive Continental Margins: Some European Examples .... *Denis Mougenot, Gilbert Boillot, and Jean-Pierre Rehault* ... 61

Mississippian Continental Margins of the Conterminous United States .... *Raymond C. Gutschick and Charles A. Sandberg* ... 79

Structural Dynamics of the Shelf-Slope Boundary at Active Subduction Zones .... *George W. Moore* ... 97

Recognition of the Shelf-Slope Break Along Ancient, Tectonically Active Continental Margins .... *Raymond V. Ingersoll and Stephan A. Graham* ... 107

PART III. SEDIMENTATION AND DELTAIC INFLUENCE

Deltaic Influences on Shelfedge Instability Processes .... *J. M. Coleman, David B. Prior, and John F. Lindsay* ... 121

Unstable Progradational Clastic Shelf Margins .... *Charles D. Winker and Marc B. Edwards* ... 139

Topography and Sedimentary Processes in an Epicontinental Sea .... *Donald L. Woodrow* ... 159

PART IV. CARBONATE MARGINS

Modern Carbonate Shelf-Slope Breaks .... *Albert C. Hine and Henry T. Mullins* ... 169

Shelf-Slope Break in Fossil Carbonate Platforms: An Overview .... *Noel P. James and Eric W. Mountjoy* ... 189

Carbonate Internal Breccias: A Source of Mass Flows at Early Geosynclinal Platform Margins in Greece .... *Hans Fuchtbauer and Detlev K. Richter* ... 207

PART V. PROCESSES AND SEDIMENT DYNAMICS

Factors that Influence Sediment Transport at the Shelfbreak .... *Herman A. Karl, Paul R. Carlson, and David A. Cacchione* ... 219

Shelfbreak Circulation, Fronts and Physical Oceanography: East and West Coast Perspectives .... *Leonard J. Pietrafesa* ... 233

Shelfedge Dynamics and the Nepheloid Layer in the Northwestern Gulf of Mexico .... *David W. McGrail and Michael Carnes* ... 251

MODERN SEDIMENT DYNAMICS AT THE SHELF-SLOPE BOUNDARY OFF
 NOVA SCOTIA .... *Philip R. Hill and Anthony J. Bowen* 265

PART VI. SEDIMENTARY RESPONSES TO PROCESSES

THE MUDLINE: VARIABILITY OF ITS POSITION RELATIVE TO
 SHELFBREAK .... *Daniel J. Stanley, Sunit K. Addy and E. William Behrens* 279

SEISMIC FACIES OF SHELFEDGE DEPOSITS, U.S. PACIFIC
 CONTINENTAL MARGIN ...... *Michael E. Field, Paul R. Carlson, and Robert K. Hall* 299

ROLE OF SUBMARINE CANYONS ON SHELFBREAK EROSION AND SEDIMENTATION:
 MODERN AND ANCIENT EXAMPLES .... *Jeffrey A. May,
 John E. Warme, and Richard A. Slater* 315

SEAFLOOR CHARACTERISTICS AND DYNAMICS AFFECTING GEOTECHNICAL PROPERTIES
 AT SHELFBREAKS .... *Richard H. Bennett and Terry A. Nelsen* 333

PART VII. FAUNAL AND ORGANIC MATTER DISTRIBUTION

BENTHIC FORAMINIFERA AT THE SHELFBREAK: NORTH AMERICAN ATLANTIC
 AND GULF MARGINS .... *Stephen J. Culver and Martin A. Buzas* 359

DIATOMS IN SEDIMENTS AS INDICATORS OF THE SHELF-SLOPE
 BREAK .... *Constance A. Sancetta* 373

INFAUNAL-SEDIMENT RELATIONSHIPS AT THE SHELF-SLOPE BREAK ...... *Norman J. Blake
 and Larry J. Doyle* 381

ORGANIC MATTER CHARACTERISTICS NEAR THE SHELF-SLOPE
 BOUNDARY .... *Robert W. Jones* 391

PART VIII. ECONOMIC PROSPECTS AND LEGAL ASPECTS

OCCURRENCES OF OIL AND GAS IN ASSOCIATION WITH THE
 PALEO-SHELFBREAK .... *William C. Krueger and F. K. North* 409

MINERAL DEPOSITS AT THE SHELFBREAK .... *Michael J. Cruickshank
 and T. John Rowland Jr.* 429

THE SHELFBREAK: SOME LEGAL ASPECTS .... *David A. Ross and K. O. Emery* 437

SUBJECT INDEX 443

PART I
MORPHOLOGY AND SEALEVEL CHANGE

SHELFBREAK PHYSIOGRAPHY: AN OVERVIEW

JEAN-RENÉ VANNEY
Département de Géologie Dynamique, Pierre et Marie Curie University,
and Institut de Géographie, Sorbonne, Paris, France 75005

DANIEL JEAN STANLEY
Division of Sedimentology, Smithsonian Institution, Washington, D.C. 20560

ABSTRACT

The shelfbreak (SB) is a distinct, critical interface of continental margins which delineates the major physiographic boundary between two major submarine provinces, shelf and slope. The shelfbreak is defined as that point of the first major change in gradient at the outermost edge of the continental shelf, and its depth, distance from shore and configuration are highly variable. Although structural framework is a dominant controlling factor, depositional regime and consequent progradational and regradational development generally modifies, substantially, the shelf-to-slope configuration. These depositional considerations include, among others, sediment supplied by rivers, carbonate reef build-up, influence of ice transport at high latitudes, and the interplay of fluid-driven and gravitative processes active in environments at and adjacent to the break. Moreover, the imprint (relict) of earlier eustatic oscillations, particularly low stands when the SB was at or close to the coastline, is still in evidence.

This overview, which incorporates observations in diverse geological and geographic settings, focuses on geomorphological aspects and is an attempt to synthesize shelfbreak type by means of a descriptive-genetic classification which takes into consideration the interaction of the dominant controlling factors. These are: (a) structural framework and rate of substrate motion which are functions of the larger-scale geological evolution of a margin, (b) the overprint of earlier (largely Quaternary) climatic and eustatic events, and (c) sediment supply and processes (reviewed in terms of climatic belts). The interplay of these three large-scale parameters has, of course, varied considerably in time and space, giving rise to a diverse suite of "end-member" break and transitional variants. In summary, we view the shelfbreak as a reworked palimpsest feature which has not yet attained complete equilibrium with presently active processes.

INTRODUCTION

The shelfbreak, long recognized as a well-defined submarine feature (de Marsilli, 1725), is an environment of first order importance on the earth's surface (Heezen et al., 1959). It is a ubiquitous and integral part of continental margins, and as such may be traced almost continuously for more than 300,000 km. The shelfbreak generally denotes that part of the outer edge of the shelf where there is a marked increase in gradient. Thus it defines the limit between two major and distinct provinces, the continental shelf and slope (Dietz and Menard, 1951). In terms of physiography and sedimentation, the shelfbreak is a boundary comparable in importance to the coastline and base-of-slope (Southard and Stanley, 1976). Throughout this chapter the terms shelfbreak, break, shelfedge and outermost edge of the shelf are used interchangeably and are abbreviated SB.[1]

Until recently, most information about the SB was derived from irregularly spaced echo-sounding and seismic transects and systematic hydrographic surveys. Factors most commonly taken into consideration where profiles have been made along restricted sectors of the outer margin include distance from coast, depth below sealevel, and shape. Most cross-section profiles reveal the marked transition between continental shelf and slope, and the SB denotes the point of inflection beyond which slope gradient increases. The position and depth of this point may be determined mechanically in several ways (Fig. 1, A–C) as shown by Wear et al. (1974) and Zembruscki (1979). In a very general way, the SB is a boundary landward of which the near-horizontal seafloor and surficial sediment cover (most commonly clastic) are affected by currents and waves which interface with, and are, largely affected by atmospheric conditions and tides. Below the SB, colder water masses (lying beneath the pycnocline) prevail on the slope. Here the planktonic fraction and particles settling from suspension become progressively more important in a seaward direction.

The shelfbreak has been defined more recently as that point where the first major change occurs in the gradient at the outermost edge of the continental shelf (Fig. 1A; see also Wear et al., 1974). The average relief between the shelfbreak and base

---

[1] Commonly used terms in other languages: French = rebord de la plate-forme; German = Gefälleknick; Russian = kraj čelfa; Spanish = (re)borde de la plataforma; Italian = limite della piattaforma; Portuguese (Brazil) = Quebra (bordo) da plataforma.

Copyright © 1983, The Society of Economic Paleontologists and Mineralogists

SB shapes as a response to different origins. Although the shelf-to-slope transition is commonly depicted as a sharp zone of inflection forming an obtuse angle it may, in some instances, also present a very gentle convex-up configuration on which it is difficult to identify the SB. Analyses of a series of transects across the outer shelf-to-slope sectors reveal several configurations. In the simplest case, the profile appears as a convex-up arc (Fig. 1A). Four sub-types are recognized: gentle, gradual, sharp and abrupt (Southard and Stanley, 1976, their Fig. 2). Commonly, the configuration is more complex, since the shelf-to-slope sector is often broken by a series of terraces (Fig. 1, D–F). It is important to recall that profiles made with echo-sounding and seismic systems generally reveal an artificial shape that results from high vertical exaggeration ranging from ×10 to ×40. The typical upper slope surface rarely exceeds 10°, and most often is less than 5°.

The diversity of the morphology, subbottom configuration, and sediment thickness and type in the transition zone between the outer shelf and uppermost slope record an evolution that involves the interplay of several controlling factors that vary in time and space (Fig. 2). This chapter serves as a review in which we outline a series of SB types, taking into account some of the major factors which control their physiography. Although our approach is largely a geomorphological one, a simplified classification of shelfbreak types is developed which involves the interplay of substrate stability and motion (Figs. 3, 4), and sedimentation processes as related to geographic region and major climatic belts (Fig. 5).

Not every type of shelfbreak is discussed in this review, and in consequence we recognize that the classification of SB types, outlined in Table 1 and illustrated in Figs. 6 to 10 is likely an oversimplification. Moreover, lack of data on some margins or of SB forms necessarily leads to some speculation. Nevertheless, it is the purpose of this overview to present a synthesis of the outer shelf-to-slope physiography which may be applicable to the study of still poorly-known margins in different oceans.

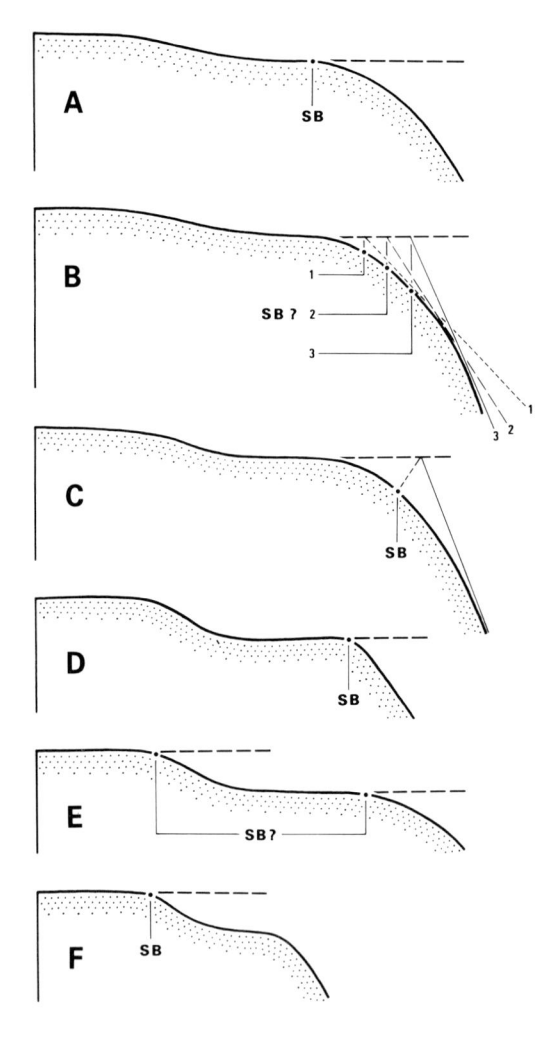

FIG. 1.—A, method used to define the shelfbreak (SB) in this study. B, C, two other possible methods to define the SB. D–F, profiles showing more complex configuration due to terraces at and near the SB. Vertical scale is highly exaggerated. Modified from Wear et al. (1974) and Southard and Stanley (1976).

of the slope is almost 4 km. According to Shepard (1973), the shelfbreak lies at an average distance of about 75 km (40 nautical miles) from shore at an average depth of 132 m (72 fathoms). There is considerable deviation, however, and breaks at depths of less than 20 to over 200 m and at distances from a few kilometers to more than 300 km from land are recorded (Emery, 1981).

In plan view, the SB rarely extends as a straight line for any distance. Systematic surveys indicate that it is generally gently sinuous to highly crenulated, and locally is poorly defined. Both echo-sounding and seismic profiles reveal a diversity of

BASIS FOR SHELFBREAK CLASSIFICATION

Continental margin studies abound but few investigations have focused on the break. As noted by Wear et al. (1974), it is surprising that a physiographic feature of such widespread importance remains so poorly known. This is in part because the SB lies between two distinct zones of interest in oceanographic research: for the specialist of coastal and shelf environments, the SB represents the seaward limit, while for those who focus on deeper realms, the uppermost slope represents the shoaler, landward boundary. Thus, the shelfbreak

TABLE 1.—CLASSIFICATION OF MAJOR SHELFBREAK (SB) TYPES AND VARIANTS DEFINED AND ILLUSTRATED IN THIS CHAPTER, BASED PRIMARILY UPON STRUCTURAL CONTROL (FIGS. 3, 4) AND SEDIMENTATION PATTERNS AS RELATED TO GEOGRAPHIC REGIONS AND CLIMATIC BELTS (FIG. 5)

Sedimentation as Related to Major Climatic Belts (Fig. 5)	Depositional Influence	Structural Control ⇒	Relatively Stable Margin (Fig. 4A)	Unstable Convergent Margin (Fig. 4B)	Unstable Margin (Fig. 4C)
	Polar Region	Glacial	Fig. 6A [a]: Embanked variant [b]: Chamfer variant	Fig. 6B	Fig. 6C
		Not-Glaciated	Fig. 7	—	—
	Temperate Region		Fig. 10A [a]: Convex, arcuate variant [b]: Terraced variant [c]: Crestal variant	Fig. 10B [a]: Depositional dominated variant [b]: Substrate displaced variant	Fig. 10C [a]: Abrasional variant [b]: Prograded, out-built variant [c]: Prograded, up-built variant
	Tropical Region	Carbonate Reef Build-Up (Non-Fluvial)	Fig. 8A (both attached and isolated platforms)	Fig. 8B	Fig. 8C
		Fluvial (deltaic)	Fig. 9A	Fig. 9B [a]: Fore-Arc variant [b]: Outer-Arc ridges variant	Fig. 9C [a]: Broken, rectilinear variant [b]: Oblique, festooned variant

has remained, research-wise, a scientific "no man's land" for which there is a dearth of systematically collected data (Vanney, 1976). The objective of this chapter is to consider some of the major controlling factors, and their interaction in space and time, which determine the shelfbreak morphology as presently preserved.

In earlier reviews Guilcher (1963) and Shepard (1973), among others, have called attention to the remarkable variability of sediment distribution on continental shelves as well as of the geographic configuration of the outer continental margin. Reasons for this are rarely clear-cut. For example, there does not appear to be a direct relation between depth of shelf margin and degree of wave exposure, with the exception of narrow shelves which are commonly shallower in regions of active coral growth. However, most classifications of shelves (for example, Thompson, 1961; Il'in, 1962; Vanney, 1976) or margins in the broader sense (Guilcher, 1963; Emery, 1968, 1969, 1980; Hedberg, 1970; Shepard, 1973) recognize the primary role of the underlying structural framework, and the modifying effects by sedimentation (Fig. 2).

In any case, the shelf-to-slope transition does not always conform to the uniform and simple scheme as often depicted but, in fact, comprises a highly variable series of forms. The classification of shelfbreak types introduced here attempts to take these variants into account and systematically group them in some coherent fashion. Each of the SB types cited is discussed in terms of major generalized morphological traits and thus each represents a composite. Our text is illustrated by block diagrams which synthesize numerous actual examples, rather than by selected echo-sounding or seismic profiles, many of which are reproduced in the following chapters of this volume. SB physiography is the resultant of several controlling factors, two of which are dominant (see Table 1) and briefly reviewed in the two following sections.

Influence of Structural Mobility and Substrate Displacement

The primary control of SB type is geological structure (Emery, 1968), and in many instances the marked change in gradient between shelf and slope corresponds to a major tectonic discontinuity (Fig. 2). Surveys in the different oceans show that there are essentially three types which may be defined by style and rate of tectonic activity (Boillot, 1979; Mougenot et al., this volume). The underlying structural control is schematically depicted in Figure 3 and the SB response in Figure 4; the three types are briefly summarized below.

SB on Relatively Stable Margins (Fig. 3, Type I; Fig. 4A).—This type occurs on the outer edge of the shelf on passive continental margins originally formed by rifting. Typically, long-term struc-

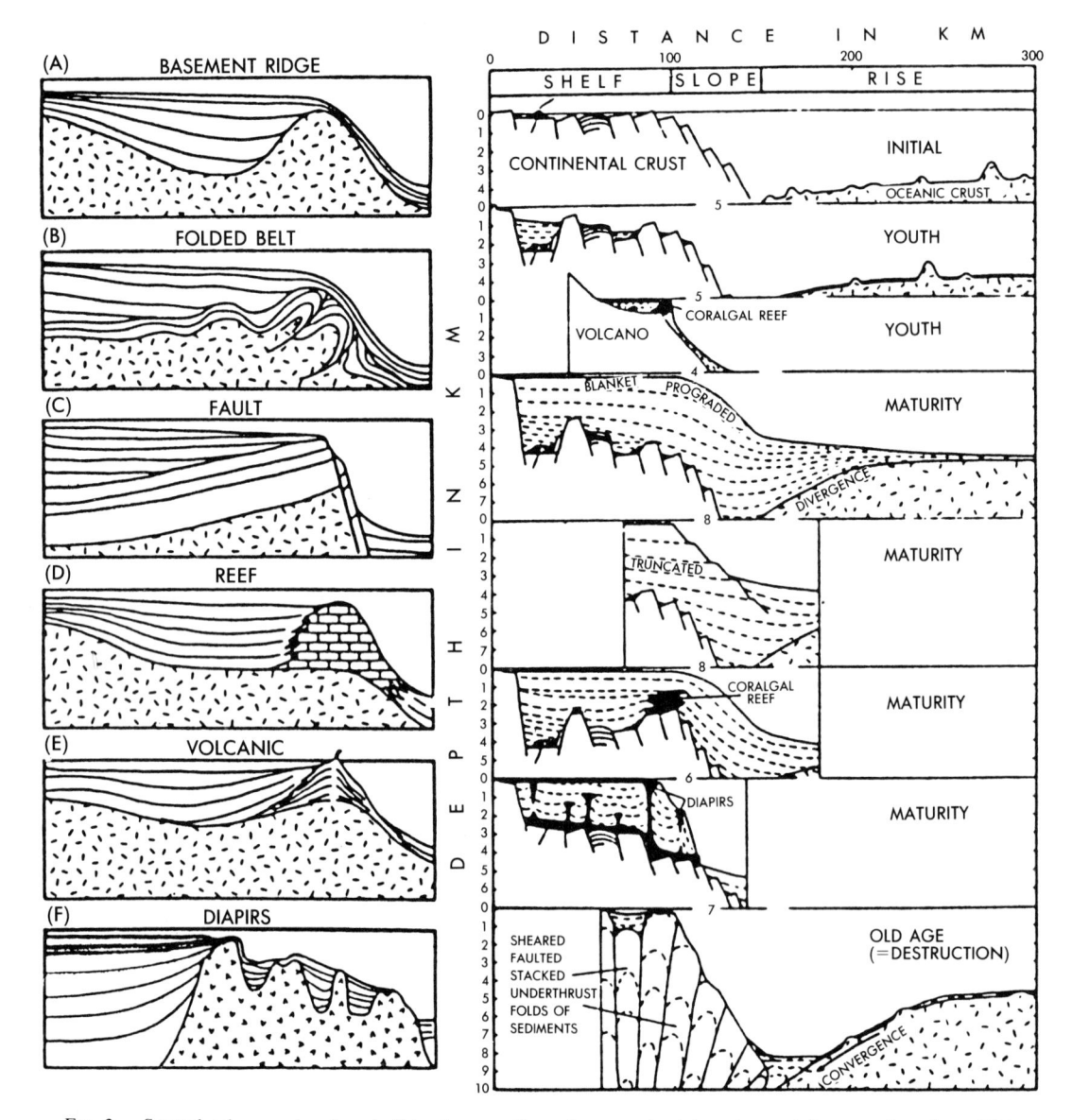

FIG. 2.—Several schemes showing shelf-to-slope configurations emphasizing structural framework and modifying effects of sedimentary processes. The diagrams in the two sequences (left, after Hedberg, 1970; right, after Emery, 1980) show the role of different types of dams and their effect on the SB.

tural stability and slow subsidence have inevitably affected the configuration of a thick sedimentary cover (Mougenot et al., this volume). The SB may be envisioned as attaining a stage of relative maturity (Mougenot and Vanney, 1980).

SB on Unstable Convergent (Subducted) Margin (Fig. 3, Type II; Fig. 4B).—This SB type occurs on margins that are being actively deformed in compressive settings, as recently indicated by several authors (cf. Karig, 1979). Detailed morphological data from this type are generally more limited, in part because of an inadequate number of systematic surveys and the generally poor resolution of geophysical soundings across the narrow shelf-to-slope zone. In some cases, the SB corresponds to an inflection point on an upper allochthonous slab (1)[1] or imbricate wedge of the accretion-

---

[1] Numbers in parentheses in this and following sections refer to features depicted on the block diagrams in Figures 4 and 6 to 10.

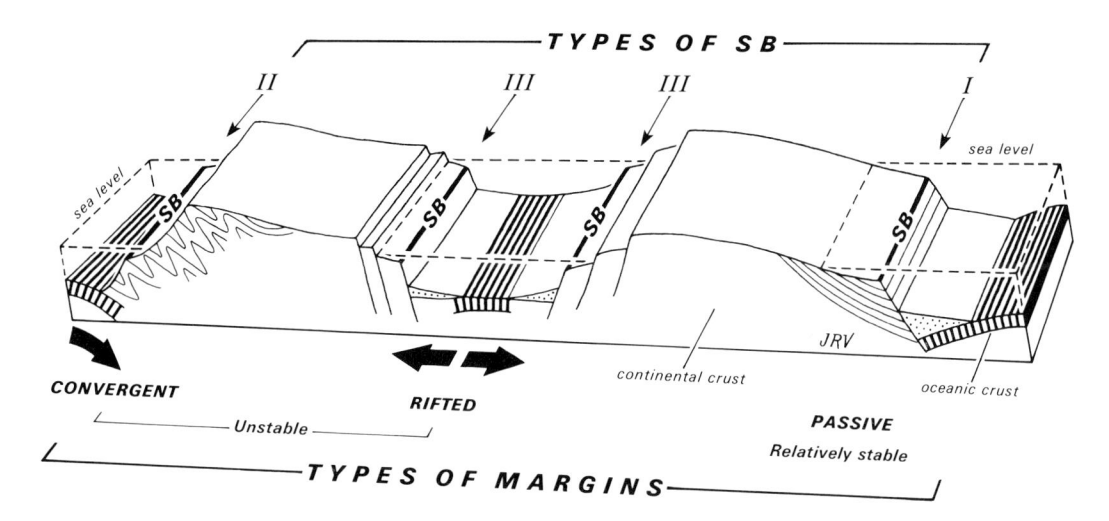

FIG. 3.—Three distinct SB types based on structural control as discussed in this chapter. Large dark arrows show crustal movement.

ary prism (2). The uppermost slope (3) is an unstable surface prone to failure. This convergent type results in the development of two types of depressions that cut across the SB: short and narrow gullies (4), discontinuous as a result of active displacement; and larger, more continuous submarine canyons (5) which are kept open by transportation of sediment from land to deeper basins.

SB on Unstable Rifted Margins (Fig. 3, Type III; Fig. 4C).—This type is exemplified by young margins formed by recent rifting. The SB may occur close to a coast that has been structurally elevated or depressed and, in some cases, strongly deformed. Moreover, structural deformation results in the formation of an SB close to the edge of a block that is faulted and tilted (Fig. 4C-a). In other cases, an older relict SB (1) may be deeply buried beneath a deformed sedimentary cover (2), and the modern SB occurs at the seaward edge of a Holocene sediment prism (3) which is narrow and shallow (Fig. 4C-b).

Influence of Sedimentary Processes

In addition to substrate motion, interpretation of SB morphology must also take into account sedimentary evolution, including progradation or regradation (Mougenot et al., this volume) at the outer edge of a continental shelf (Fig. 4A). The position and shape of the break are controlled by the volume of clastic sediment supply and seafloor displacement (Pitman and Golovchenko; Mougenot et al., and other chapters, this volume). It has long been recognized that the seaward extent of the coastline coincided with, or occupied positions close to, the SB during eustatic low stands (Veatch and Smith, 1939; Emery, 1969; Shepard, 1973). In consequence, development of the SB is a long-term response to diverse processes (fluvial and glacial transport and littoral processes) when sealevel stands were low and to fluid-driven and gravitative marine processes at times of higher sealevel stands. The SB may thus be envisioned as the seaward limit on continental margins where subaerial processes (wind, river, ice) have left a direct imprint, along with marine depositional events (Vanney, 1976; Southard and Stanley, 1976).

Since the beginning of the last rise in sealevel, modern processes have not yet had sufficient time (~17,000 years) to completely modify the shape of the SB. In view of its present distance from shore and depth (often 100 m or greater) below present sealevel, the SB is neither fully in equilibrium with present processes nor is it a Pleistocene (or older relict) feature. The shelf-to-slope sector, however, remains a high energy zone as a result of the interplay of tides, storm waves, breaking internal waves and bottom turbulence produced by fronts separating shelf and slope waters (such as on the northwest Atlantic margin). These processes, summarized by Southard and Stanley (1976) and Stanley and Wear (1978), are discussed in the chapters in Part V of this volume. In addition to physical oceanographic parameters and associated fluid-driven processes, the SB approximates the upper limit of progradational mechanisms (slides, sediment flows) that prevail on the slope. Discussions of sediment instability and failure and responses to such processes are presented in the chapters in Part VI of this volume.

In summary, the SB is a unique sector of the continental margin which records events that have shaped the seafloor during both subaerial and sub-

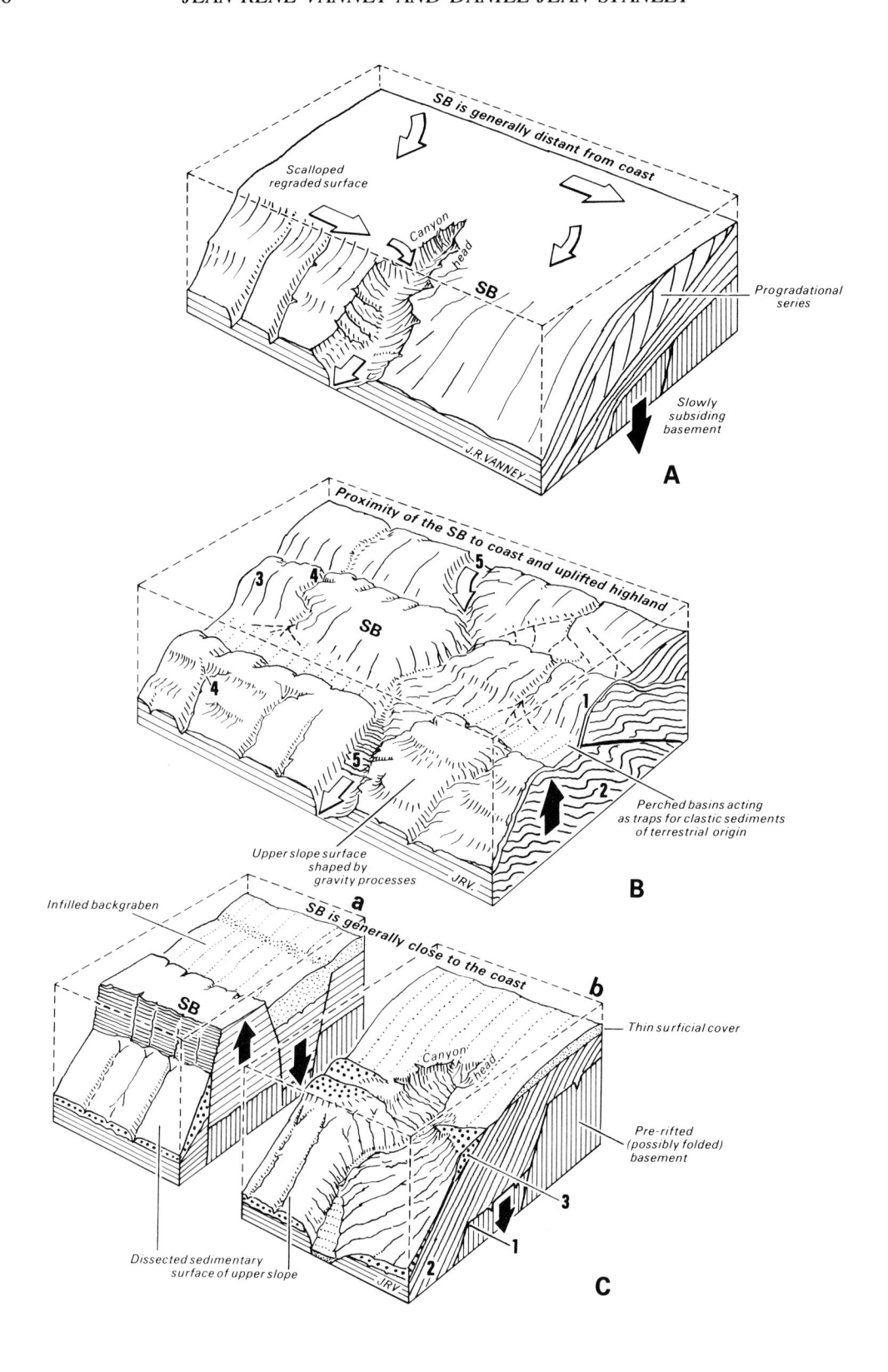

A

SB is generally distant from coast

Scalloped regraded surface

Canyon head

SB

Progradational series

Slowly subsiding basement

J.R. VANNEY

B

Proximity of the SB to coast and uplifted highland

SB

Perched basins acting as traps for clastic sediments of terrestrial origin

Upper slope surface shaped by gravity processes

J.R.V.

C

Infilled backgraben

SB is generally close to the coast

Thin surficial cover

Canyon head

Pre-rifted (possibly folded) basement

Dissected sedimentary surface of upper slope

J.R.V.

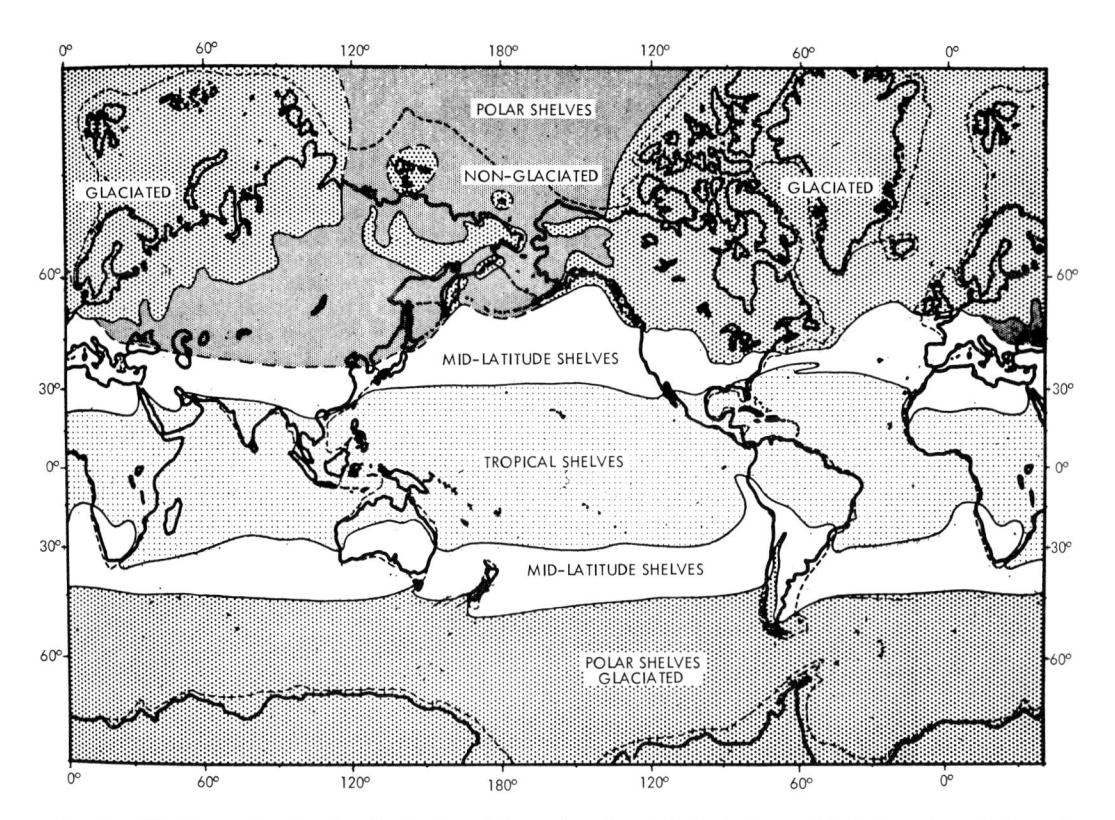

FIG. 5.—World map showing the distribution of the major climatic belts (polar, mid-latitude and tropical) as discussed in this chapter. Poorly-defined boundaries are shown by dashed lines. Short dashed lines depict outer shelf limit. After Thompson (1961).

merged phases in the Quaternary. That the shelf-to-slope transition is a palimpsest feature is evident from the present interplay of processes that are reworking relict sediments and modifying the shape inherited from prior geological, primarily tectonic, events. The classification of major SB shapes presented herein takes into account factors of sedimentation in the different geographic-climatic zones (Fig. 5) as well as the role of structural mobility cited earlier (Figs. 3, 4).

Discussion of SB types in terms of major climatic belts is useful in that each region is characterized by a distinct association of sedimentary processes. For example, in the high-latitude polar regions, glacial and sea ice and frost action are major agents affecting sedimentation on the coast and beyond, on the continental margin. In contrast, low latitude tropical regions are more often characterized by higher fluvial input and/or the formation of reefs and development of carbonate platforms. As might be expected, sedimentary conditions at mid-latitudes are transitional between the former two.

SHELFBREAK IN POLAR GLACIAL REGIONS (FIG. 6)

Of all shelf-to-slope transitions, the configuration of the SB in polar glacial regions is perhaps the most distinctive. It is in large part a response to the dominant agent, sheet and sea ice, which extends onto and across the shelves. The shelfbreak is locally affected by glacial fronts and grounding of ice which, during the Pleistocene (and locally at present), extended to or, at least, near the shelfedge. In response to the movement of ice, there is marked erosion of the seafloor, deposition of characteristic ice-transported materials, and the effects of iso-

FIG. 4.—Schematic block diagrams showing the effects of geological framework and structure on shelfbreak (SB) development. A, relatively stable margin; B, unstable convergent (subducted) margin; C, unstable rifted margin. Large dark arrows show substrate motion; open arrows depict sediment dispersal. [a] and [b] in diagram C represent variants (variants are also shown in Figs. 6A, 8A, 9B, C, and 10 A–C). Numbers are explained in text. Position of sealevel shown relative to SB in these diagrams and those in Figs. 6–10.

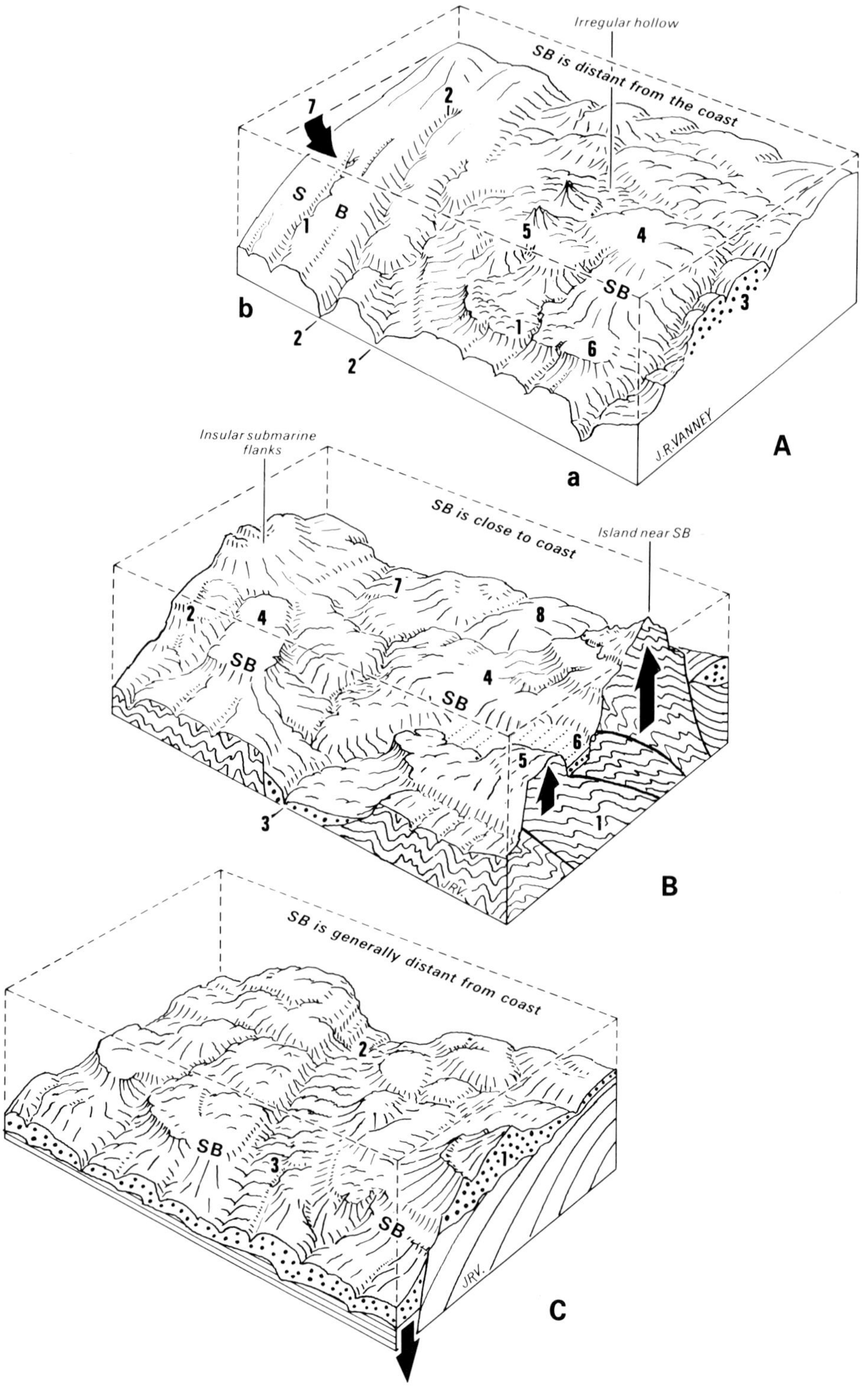
Irregular hollow
SB is distant from the coast
7
2
S
B
1
5
4
SB
3
b
2
2
1
6
J.R.VANNEY
A
a
Insular submarine flanks
SB is close to coast
Island near SB
7
2
4
8
SB
4
SB
6
5
3
1
JRV.
B
SB is generally distant from coast
2
SB
3
1
SB
JRV.
C

static rebound, all of which affect the SB shape and depth.

SB on Relatively Stable Margins (Fig. 6A)

Examples include the Grand Banks off Newfoundland, and the Weddell and Ross seas in the Antarctic. In these cases the SB is generally quite distant from the coast and may be slightly incised by gullies (1) and heads of canyons (2). These V-shaped valleys were cut by subglacial sediment failure, including mass flow processes. Moraine deposition is, or has been, dominant and wide-spread; resulting deposits are (3). Two variants are distinguished.

Embanked SB Variant (Fig. 6A-a).—The most common type, embanked, is characterized by broad, discontinuous and undulatory banks (4), the seaward edge of which defines the SB. The SB tends to be shallow and festooned, characterized by associated local highs (5) in the form of pinnacles and ridges. Moraine deposits are of the frontal or terminal type in the case where the ice-front coincides (or coincided) with the SB. Deeper terraces (6) below the SB are probably moraines of the basal type deposited at times when the leading edge of the ice extended seaward of the SB; these terraces thus record an older pre- or interglacial SB. Examples include those on the Antarctic margin (Zhivago and Evteev, 1970; Vanney et al., 1981; Johnson et al., 1982).

Chamfer SB Variant (Fig. 6A-b).—Another variant is the poorly-defined break (or chamfer-type); it occurs on the leading edge of a gently inclined plane and generally lies well below normal SB depths. This SB displays a very gentle convex-up profile, which sometimes is incised by gullies (1). Its configuration may be related to the effects of recent tilting or incomplete isostatic rebound (shown by arrow, 7). Examples include sectors off the Grand Banks of Newfoundland and the outer Ross Shelf of Antarctica.

SB on Unstable Convergent Margins (Fig. 6B)

Examples include the shelfedge of the Aleutians, Southern Alaska (Burk, 1965; von Huene, 1972; Fisher, 1979; Carlson et al., 1982), Antarctic peninsula (Vanney and Johnson, 1976; Johnson et al., 1982), and probably, the southern Chilean margins. This SB type is often located close to a peninsular, high-relief coast, and is more incised than variants shown in Fig. 6A. The SB represents the crestal part of the accretionary prism (1) which has been sufficiently uplifted to serve, during most of the Quaternary, as a dam blocking offshelf sediment spill-over. Glacial morphological and depositional features are less obvious and less well preserved than in the former type (Fig. 6A) due to the smaller volumes and less extensive distribution of the ice derived from these insular regions, and to the concurrent displacement of the substrate. This SB is generally shallow, possibly in response to isostatic readjustement. It is scalloped and incised by gullies (2) and heads of canyons (3). The SB is also bounded by a series of discontinuous banks (4), sometimes isolated as advanced highs and outliers seaward of the shelfedge (5) behind which occur perched fore-arc basins (6). These external banks, probably uplifted by rebound, are usually incised and eroded by glaciers; they often enclose marginal channels (7). These latter, formed by scour and ice erosion, are broad and have been subsequently filled by glacial till (8).

SB on Unstable Rifted Margins (Fig. 6C)

Examples include shelfedges off Norway, Greenland, and East Antarctica. This SB type is well known (cf. Nansen, 1904; Holtedahl, 1958, 1970; Holtedahl and Holtedahl, 1961; Sommerhoff, 1973; Shepard, 1973; Zhivago, 1978, Vanney and Johnson, 1976, 1979a, b). It generally lies at a greater depth than the preceding type and, in plan view, appears straighter and more regular in response to movement along faults which have been active during the advances and retreats of the ice sheet. Along most of its extent, the SB is bounded by banks of irregular relief corresponding to the outer limit of widespread morainal deposits (1). Large reentrants are due to large channels (2) oriented perpendicularly to the margin (the "djupet" of Norway, for example); these are the terminal portions of submarine glaciated valleys, generally separated from the fjord by sills situated approximately at the coastline. These large channels are several tens of kilometers in width and have a U-shaped cross-sectional profile. The axial profile shows a shallowing seaward, toward the shelfedge, in response to terminal moraine deposits (3) released by glacial tongues.

SHELFBREAK IN POLAR NON-GLACIATED REGIONS (FIG. 7)

This SB type occurs primarily on relatively stable margins and examples include Siberia (Holmes and Creager, 1974; Naugler et al., 1974), the Arctic Alaskan margin (Barnes and Reimnitz, 1974) and Bering Sea, and perhaps Patagonia (Leonardi and Ewing, 1971). Topographic details of this SB type, however, are poorly known (with the exception of the Bering and Beaufort Seas). General character-

---

FIG. 6.—Schematic block diagrams depicting the shelfbreak and adjacent margin in polar glacial regions. The SB is shown (A) on relatively stable margins (a = embanked variant; b = chamfer variant); on unstable (B) convergent and (C) rifted margins. Large arrows show substrate displacement. Numbers refer to features described in text.

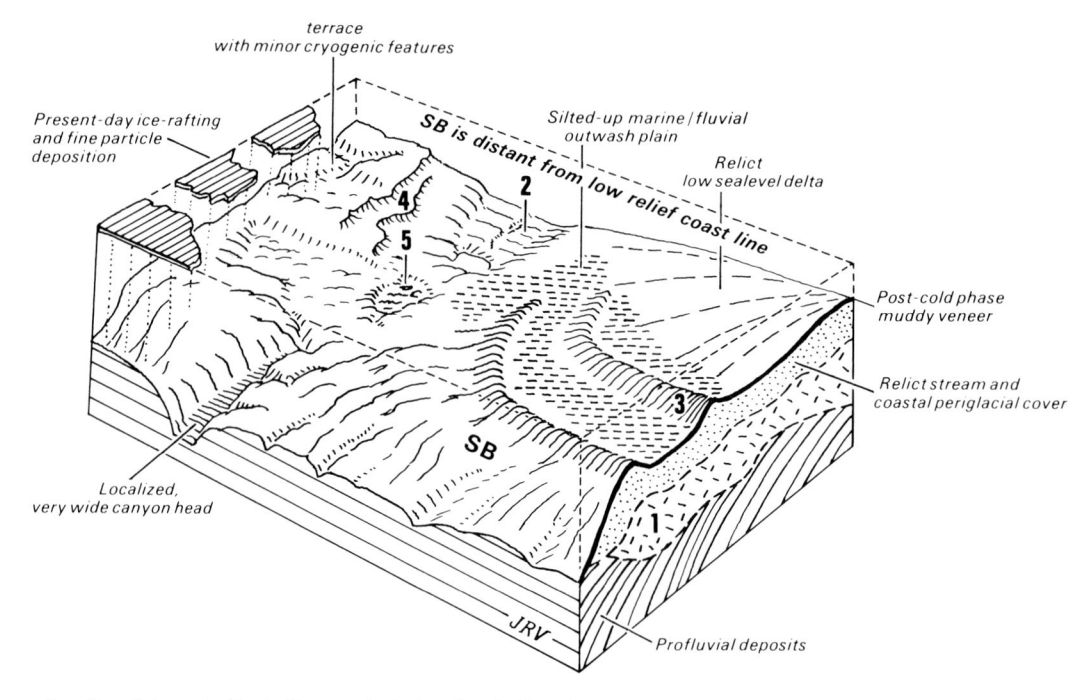

FIG. 7.—Schematic block diagram depicting the shelfbreak and adjacent margin in polar non-glacial regions. Important substrate displacement is assumed. Numbers refer to features described in text.

istics include shaping by periglacial fluvial and coastal processes effective during lower stands of sealevel. Local modifications were, and continue to be, produced by ice rafting and abrasion during and since the Holocene rise in sealevel. In general, the SB is shallow (50–70 m), slightly sinuous and distant from low-relief coastlines.

Typically, the SB occurs on the outer shelf that appears as a uniform plane, with only minor undulations and incisions. In the Beaufort Sea, there is evidence of persistent relict (Pleistocene) permafrost (1) beneath the seafloor. The following features are recognized and interpreted: (a) U-shaped, flat-floored valleys of low relief which are discontinuous and, in some sectors, do not extend seaward all the way across to the shelfbreak (2); these valleys trace the paths of periglacial rivers which were eroded by drift ice at times of eustatic low stands. (b) Low-relief and extensive undulations (3) represent beach-dune ridge complexes, offshore bars and delta levees, or the surface expressions from ice-scraping and a piling up of sediment by the leading-edge of sea ice. The SB is bounded by an irregular distribution of low-relief escarpments (4) and small depressions (5) or kettles, which are perhaps relict lakes of thermokarstic origin.

SHELFBREAK IN TROPICAL NON-FLUVIAL REGIONS
(FIG. 8)

The development of this SB type occurs in areas devoid of runoff and is closely related to biogenic carbonate sedimentation, particularly in the form of coral reefs and bioherms and an absence of or decreased clastic sediment supply. Reefs and bioherms formed near the outer edge of the shelf during low eustatic sealevel stands. During the rise in sealevel (and since its stabilization) reef development was either inhibited or continued to develop upward, but only in selected areas near the break. The most distinct traits of the SB are its abrupt configuration and shallow depth. Moreover, isolated and often well-defined, high-relief mounds commonly occur seaward of the SB. Along most carbonate-rich shelfedges, there is a remarkable diversity of SB shapes resulting from modifications of coral reef build-up, which has varied with time and space. Reefs are affected by the trade wind patterns on western sides of continents. We note that in very arid tropical and sub-tropical regions devoid of runoff, coral reefs are absent as a result of cold waters due to upwelling (Namibia, Peru, Chile, western Sahara).

FIG. 8.—Schematic block diagrams depicting the shelfbreak and adjacent margin in tropical, non-fluvial regions. Large arrows show substrate displacement. Numbers refer to features described in text.

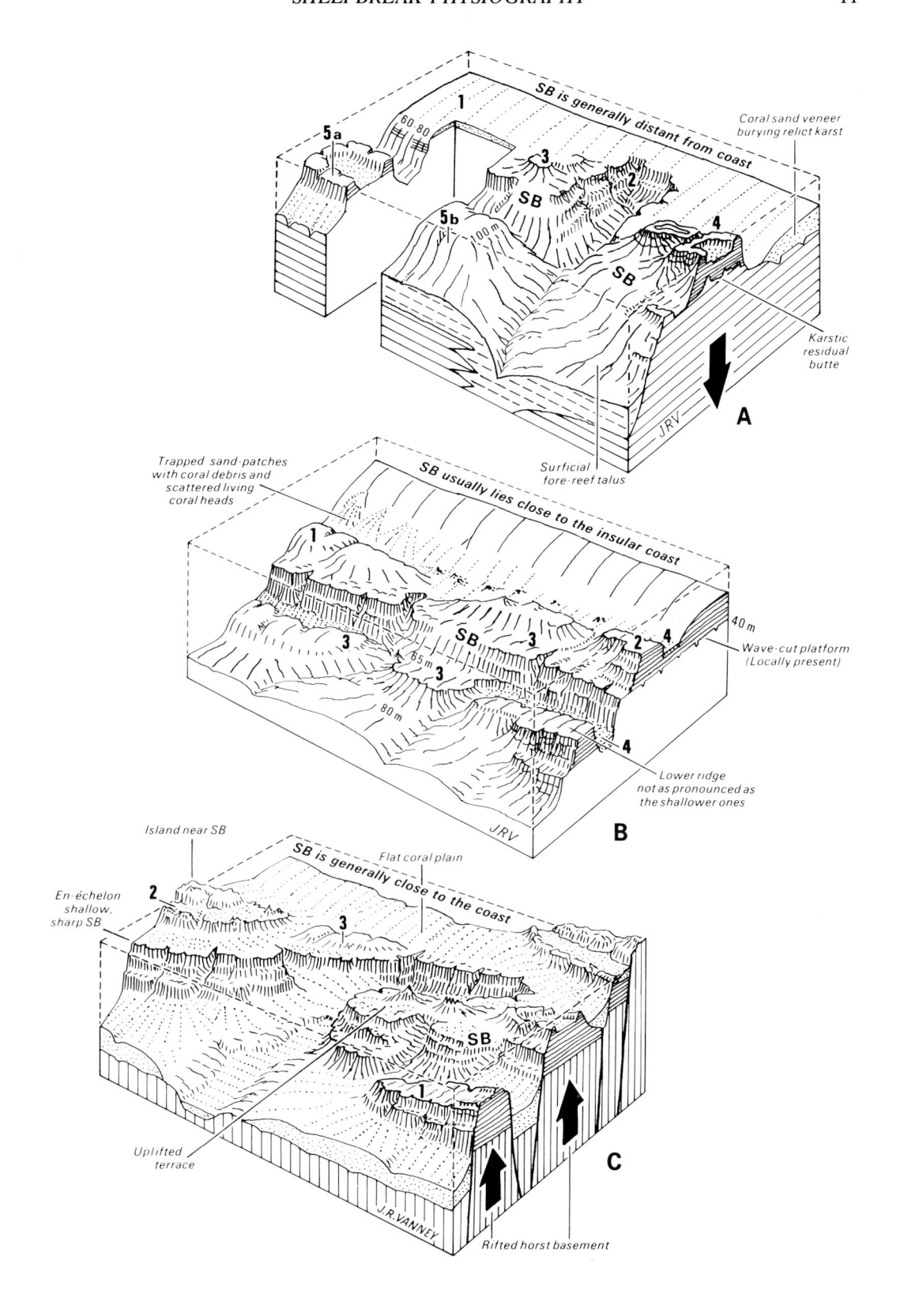

SHELFBREAK PHYSIOGRAPHY

SB on Relatively Stable Margins (Fig. 8A)

This SB type, generally distant from the coast, typically results from the almost continuous formation of a thick carbonate build-up, under subtropical to tropical warm water conditions and accumulation on a long-term subsiding substrate. Examples include those off Yucatán, Florida, Bahamas (Logan et al., 1969; Uchupi, 1969; Neumann et al., 1977; Lighty, 1977; Lighty et al., 1978) and off the Sahul shelf, northern Australia (Van Andel, 1965; Van Andel and Veevers, 1967; and possibly the Espirito Santo shelf off Brazil (Zembruscki et al., 1972). The SB records not only long-term reef development, but also the effects of the interplay of subsidence and eustatic oscillations. The SB on the outermost sector of the continental platform (1) (cf. Kendall and Schlager, 1981) is well-defined and abrupt, and locally incised by large re-entrants (2). The break, lying at a depth of about 60 to 90 meters or less, is bordered locally by small mounds (3) interpreted as drowned reefs; carbonate reef development in this case has been slowed or terminated by high rates of subsidence and/or a rapid rise in sealevel. The SB is also bordered locally by larger reef build-ups (4) which have continued to develop on the shallow outer portion of the attached platform (perhaps on residual karstic forms) during and since the last eustatic low stand. On many margins, the outer edge of the attached platform is separated from a series of isolated carbonate mounds which display a tabular mesa-like form (5). These mesa-like features are similar in shape to the SB at the outer edge of the attached platform: their profile may be sharp (5a) or rounded (5b). The sharp type results from either a rapid rate of sealevel rise, or rapid subsidence, matched by reef development. The rounded type results from erosion and occurs where either sealevel rise or subsidence is not matched by coral growth and reef build-up.

SB on Unstable Convergent Margins (Fig. 8B)

Although still inadequately studied, this SB type occurs in areas such as Barbados, the Lesser Antilles, and some island arc sectors in the tropical Pacific (Fink, 1972; Adey and Burke, 1976; Bouysse and Martin, 1979). The SB parallels and is usually close to the insular coast. Shelf-to-slope profiles usually reveal a series of mounds (1), wall-like ridge (2), and terraces (3) which, off Barbados, are concentrated at depths ranging from 20 to 80 meters (Macintyre, 1967, 1972). Some step-like terraces cut into, and below, the shelfedge are interpreted as submerged barrier reefs. In plan view, mounds near the SB are separated by linear depressions (4) which are filled by coralline debris and sand dispersed among living coral heads. The discontinuous series of mounds and ridges parallel to, and seaward of, the SB records the direct relation between reef construction and climatic condition. It appears that the spatial distribution and development of reefs has been interrupted periodically by displacement of the substrate and/or acceleration of coastal erosion and increased rates of sedimentation.

SB on Rifted Substrates (Fig. 8C)

This SB type is perhaps less common and has been less well studied systematically than the previous type. It occurs in the Red Sea and Gulf of Aden (Guilcher, 1955; Nestéroff, 1955; Laughton, 1966; Laughton et al., 1970), and on the eastern Baja California margin in the Gulf of California (Normark and Curray, 1968). Its shape is similar to that of the two previously cited types (Fig. 8A, B). In profile, there is a sharp and abrupt transition from shelf to slope; associated isolated platforms (1) occur in front of the SB. The break differs from the previous two in that, in plan view, it is highly irregular, has significant offsets and is often oblique with respect to the trend of the margin as a result of displacement by rift faults. Some reef-surrounded islands close to the SB occur on uplifted, tilted horsts (2). On some charts, the outermost shelf morphology adjacent to the SB appears uniform and relatively flat; the SB in such settings has been influenced by a masking effect by coral debris sedimentation. Some drowned reefs (3) also occur near the break.

SHELFBREAKS IN TROPICAL FLUVIAL-DOMINATED REGIONS (FIG. 9)

This group of SB types includes those whose morphological evolution is dominated by the sediment input transported by very large tropical rivers. The role of carbonate reef and bioherm build-up and carbonate platform construction is marginal to nonexistent or subordinate to clastic deposition.

SB on Relatively Stable Margins (Fig. 9A).

Examples of this SB type include those off Guinea, west Africa (Allen, 1964 a, b, 1965; Allen and Wells, 1962; Martin, 1971, 1973; McMaster et al., 1971), off the Bengal coast (Ganges-Brahmaputra fluvial systems), and off the Amazon-Guiana-Orinoco shelf (Nota, 1958; Kowsmann et al., 1977; França, 1979; Palma, 1979; Zembruscki, 1979). Moreover, comparable SB types occur in the subtropical Gulf of Mexico (thoroughly documented by Coleman, 1981). In the case of the best-studied example, the Mississippi River, mapping reveals an earlier Quaternary SB, buried by the modern prodeltaic wedge, lying at shallow depths. In most of the above-cited examples, the SB in plan

FIG. 9.—Schematic block diagrams depicting the shelfbreak and adjacent margin in tropical fluvial regions. Large arrows show substrate displacement. Numbers refer to features described in text.

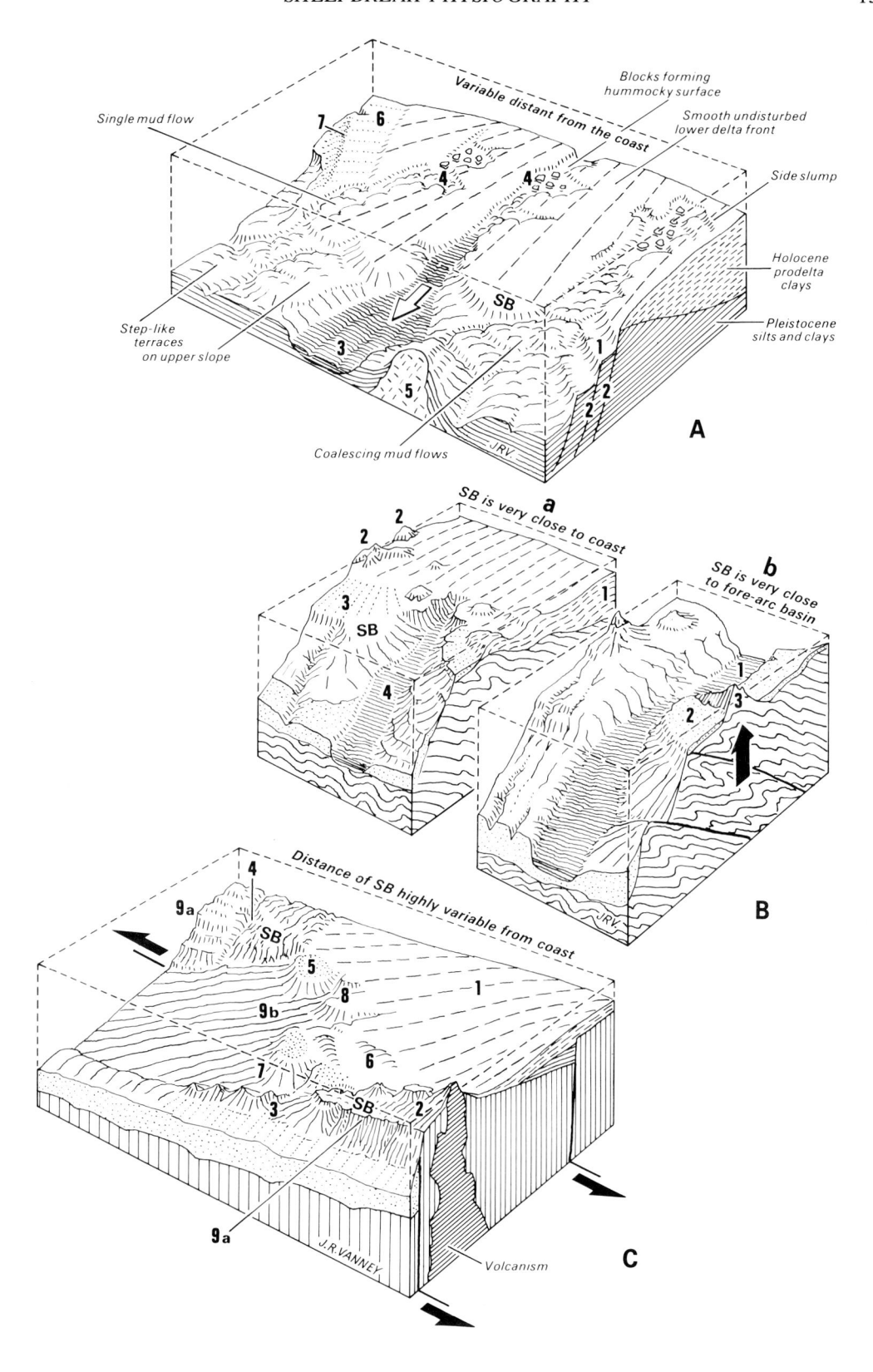

SHELFBREAK PHYSIOGRAPHY

view, as defined by the outer edge of a prodelta, appears as a seaward-protruding bulge. The SB shape results from the interplay of sediment supply and lowering of the seafloor surface, due to compaction of the thick sediment wedge and subsidence. The shelf-to-slope profile is generally a gentle and relatively smooth convex-up arc resulting from seaward progradation by rapidly accumulating fluvial muds. Detailed morphologic analysis at and near the SB records the effects of gravitational processes; a series of concave-up and scalloped surfaces (1) record the marked effects of sediment instability and failure in response to very high sedimentation rates and growth-fault systems (2). The SB in such settings also may be incised by the heads of canyons (3) and gullies (4) that are partially filled by mud-flow lobes (Coleman et al., this volume). Locally, the SB may be modified by small, high-relief sedimentary features of diapiric origin (5), comparable to the mud lump islands seaward of Mississippi River distributaries. In areas that are not affected by active modern prodeltaic sedimentation, the SB may represent a relict paleodelta edge (6) where sedimentary sequences of early Holocene age and older are preserved. On these relict surfaces are mapped narrow ridges (7) consisting of coarse sands. These ridges, lying at depths ranging from 100 to 130 meters, parallel the SB along extensive distances in the Gulf of Guinea and Gulf of Mexico, as well as off Guiana and the Amazon, where they are incised by canyon heads.

SB on Unstable Convergent Margins (Fig. 9B)

This SB type is present on the outer margin of relatively narrow and seismically active platforms where deposition has been affected by considerable structural displacement and deformation of the substrate, and by associated gravitative processes. These movements have fostered active spill-over and bypassing of sediments from the outer shelf onto the slope and into submarine canyons, some of which serve as active channels for sediment transported to deep basins. Although morphological details of this SB type are still poorly known, two variants are recognized.

Simple Variant: Fore-Arc SB (Fig. 9B-a).—Breaks displaying this configuration are present off Ecuador (Lonsdale, 1978), Central America (Gierloff-Emden, 1958; Underwood and Karig, 1980), some sectors off Indonesia (Rodolfo, 1969; Hamilton, 1979 a, b; Kieckhefer et al., 1981), and off northern Columbia. The shelfedge is part of a prograding sedimentary prism (1) which is locally bordered by large insular promontories (2) which may represent arcuate-shaped segments of recently uplifted prodeltaic wedges. A discontinuous Pleistocene wedge consisting of fine-grained terrigenous and carbonate sediment crops out seaward of these small islands at the outer shelfedge (3). A probable example may be Preparis Island off the Irrawaddy River in the Indian Ocean. The irregular, markedly offset configuration, usually depicted by this type of shelfedge, may result from a combination of the large supply of sediment transported by rivers to the adjacent margin, and subsequent mass movements, triggered by seismic activity and substrate instability. This failure induces a landward erosion concentrated in canyon heads (4). This SB type most closely resembles that illustrated in the model recently proposed by Underwood and Karig (1980).

Complex Variant: Outer Arc-Ridge SB (Fig. 9B-b).—This SB variant differs from the preceding by its location, usually on tabular and insular blocks isolated in front of fore-arc basins. Examples include the discontinuous SB found between Burma and Indonesia (Weeks et al., 1967; Rodolfo, 1969; Hamilton, 1979 a, b; Moore and Curray, 1980; Moore et al., 1980; Kieckhefer et al., 1981). One may also include in this category the SB on the margin of islands off Venezuela, which form insular blocks in a foreward position north of the Bonaire Basin. A diagnostic characteristic associated with this SB type is the presence of a fore-arc basin which has acted as a trap for fluvial-derived muds and fine-grained turbidites. Progressive uplift and, at times, concurrent tilting, coupled with a small supply of sediment, have resulted in a SB shape that is in large part a response to geological structure, rather than sedimentation. The outer edge of the shelf and the SB is dissected by very large interinsular transverse channels (1) extending from the fore-arc basin seaward to the slope, such as occurs on the margin of Nias Island off Sumatra. Here, the SB constitutes part of a cuesta-like form, with a steep gradient oriented seaward, and sub-bottom acoustic reflectors dipping landward; the break is covered by a thin, discontinuous sediment cover (2) and is punctuated by a series of small islands (3) that are, in part, of coral origin.

SB on Unstable Rifted Margins (Fig. 9C)

Examples of this SB type include those on the eastern margin of the Gulf of California (Curray and Moore, 1964; Rusnak et al., 1964; Moore, 1973), the Andaman Sea (Weeks et al., 1967; Rodolfo, 1969), and the Caribbean in areas such as off northeast Venezuela and in the Gulf of Honduras (Kornicker and Bryant, 1969; Pinet, 1971, 1972, 1975). In plan view, the break in these areas presents an irregular *en-échelon* configuration resulting from rifting (lateral as well as vertical) movement of the substrate on which delta progradation has occurred (1) prior to and during tectonic displacement. The shelfedge includes two morphologic variants.

Broken, Rectilinear SB Variant (Fig. 9C-a).—The first type, rectilinear, is distinct and well defined; it occurs on, or along, transform faults. The series of prograding (prodelta) sediment (1) deposited prior to rifting (such as those which accumulated during the early phases of opening of the Gulf of California), are truncated and/or offset by faults. Thus, the SB is the upper part of a fault-related escarpment. The adjacent outer shelf sector tends to be broken into low, parallel ridges (2) resulting from initial uplift and subsequent subsidence. The outer shelf sector of structurally, more deformed rifted margins are bordered and extended by linear spurs (3). These spurs are tectonically displaced tabular ridges of reef origin. Examples include tilted, insular blocks such Islas Tres Marias and de La Bahia located, respectively, in the Gulfs of California and Honduras. In some instances, the SB is incised by canyons (4) located along faults that are oriented perpendicularly to the SB.

Oblique, Festooned SB Variant (Fig. 9C-b).— The second variant differs by its oblique orientation relative to [9C-a], and its highly irregular configuration. Its origin is related to post-rift progradation by deltaic (Pleistocene) sedimentation. Examples include those on the shelfedge seaward of submarine deltas located between Mazatlan and Guaymas (Gulf of California). In general, the cross-sectional profile is gentle convex-up in those cases where the pre-rifted topography is largely buried by sedimentary series. The uniform topography in the vicinity of the SB is related to a Pleistocene erosional surface formed during eustatic oscillations and subsequent burial under a series of detrital sands or carbonate-rich sediments (5). The modern deltaic accumulation on the inner shelf is largely responsible for the formation of dome-like (6) structures (diameter 0.2–2.5 km) such as those in the Gulf of Honduras interpreted as diapiric folds (Pinet, 1972). Locally, the steep slope (7) seaward of the SB records the effects of sediment instability, and may be incised by the heads of gullies (8); these features resulted when deltaic sediments spilled over the break during the late Quaternary low sea-level stands.

SHELFBREAKS IN TEMPERATE REGIONS (FIG. 10)

The SB that forms in mid-latitudes, unlike those described earlier, is often characterized by an absence of three factors which dominate in the polar and tropical regions: obvious glacial features, dominant reef and carbonate (bioherm) sedimentation, and important fluvial deposition. Of course there are exceptions. Morphologic attributes of the temperate region SB have been acquired during the past several millions of years, in large part as a result of the marked effects of climatic cooling events. This cooling has been severe enough to reduce or prevent reef development except at some subtropical latitudes; climatic cooling, however, has not been so rigorous as to produce glaciers, which are important in the shaping of the SB at higher latitudes. Since Pliocene and Quaternary time, the outer shelves and SB in temperate regions have been affected by periodic subaerial exposure and development of periglacial conditions. Two factors in particular are important: erosion resulting from rivers which migrated across shelves and extended to the break, and littoral processes at and near the SB which, at such times, was the coastline. Moreover, coastal processes were intensified at such times: higher wave energy may have resulted from marked changes in meteorological patterns and displacement of polar fronts toward lower latitudes. Surveys on different margins show that these fluvial and littoral effects played a major role in shaping the SB, regardless of the nature of the substrate on which it formed.

SB on Relatively Stable Margins (Fig. 10A)

The dominant erosive regime is clearly recorded on the outer edge of structurally stable and inactive shelves by the relative thinning of the prograding Plio-Pleistocene sediment prism. The SB is moderate to variable in distance from the coast on the three variants recognized (Fig. 10A-a, B, C). The effects of erosion also are apparent, i.e. by the highly complex configuration of seismic reflectors which suggest sedimentation affected by conditions of high energy, and gravitational processes (slumping, sediment gravity flows) on the uppermost slope. These features are characteristic on many margins which have been studied and are relatively well-distributed regionally in the temperate mid-latitude belt.

Convex Arcuate Profile SB Variant (Fig. 10A-a).—This variant, commonly recorded on both bathymetric and seismic transects, differs genetically from its tropical counterpart. That is, the arcuate profile is not an equilibrium profile that results simply from progressive Quaternary progradation (1) on a subsiding substrate. Rather, the shelf-to-slope profile records the long-term effects of alternating deposition and erosion, the latter produced by the spill-over of material seaward across the SB. This spill-over triggers sediment failure and, in turn, a landward retreat of the SB and the formation of gullies on the uppermost slope (2) and erosion in the head of canyons (3). It is noted that even at the SB located seaward of the mouths of moderate to large rivers such as the Garonne, Sebou, and Tejo, shapes typical of deltaic progradation are not observed (Vanney, 1977, 1982; Vanney and Mougenot, 1981).

Terraced SB Variant (Fig. 10A-b).—This second variant is particularly well-developed along sectors off the eastern U.S. (Fig. 1, D–F; see also among others, Stanley et al., 1968; Uchupi, 1968; Swift et al., 1972; Wear et al., 1974). In this region, the SB lies seaward and below relict, fluvial deltas pre-

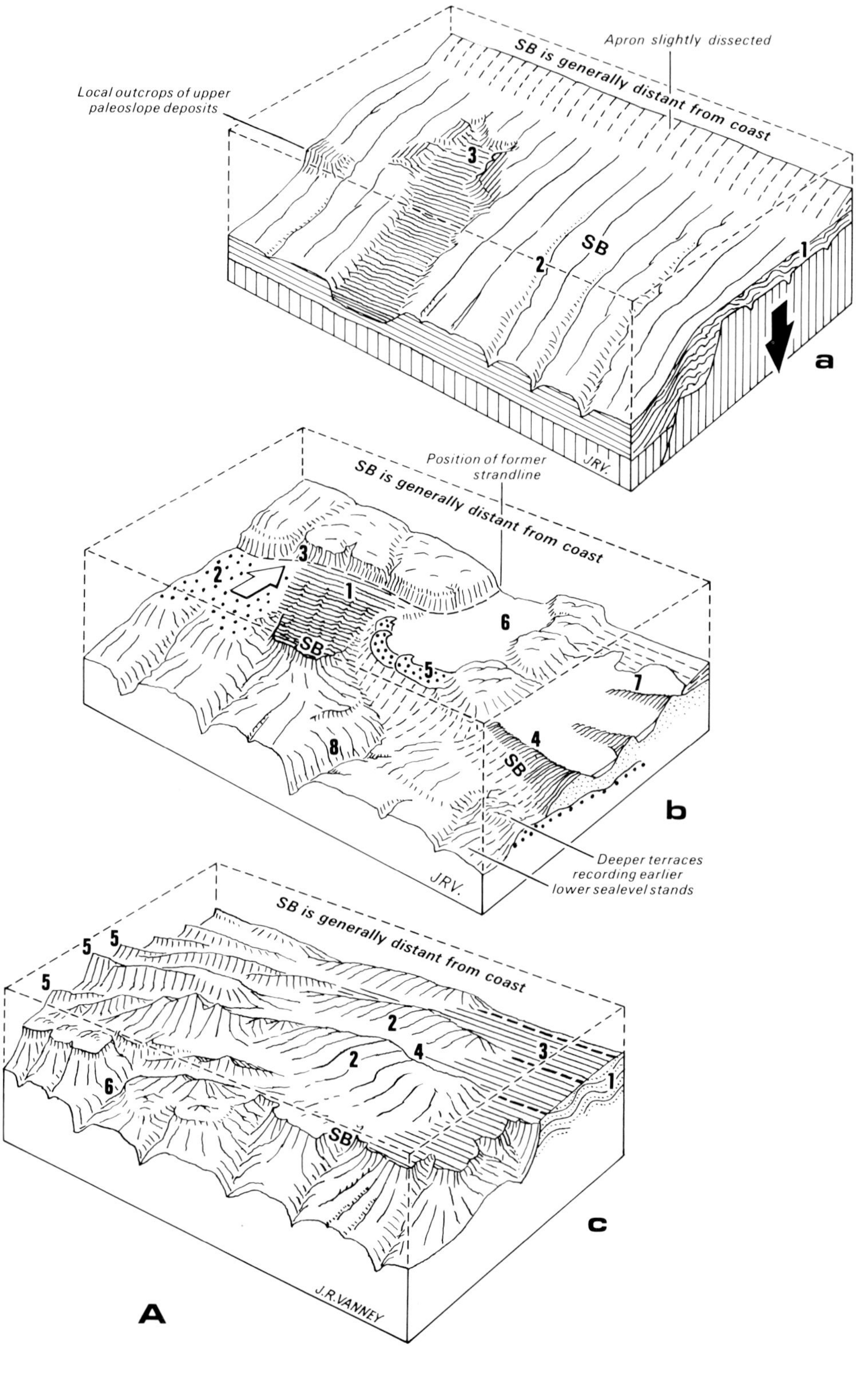
Apron slightly dissected
SB is generally distant from coast
Local outcrops of upper paleoslope deposits
3
SB
2
1
a
JRV.
Position of former strandline
SB is generally distant from coast
3
2
1
SB
6
5
7
4
SB
8
b
Deeper terraces recording earlier lower sealevel stands
JRV.
SB is generally distant from coast
5
5
5
2
2
4
3
1
6
SB
c
J.R.VANNEY
A

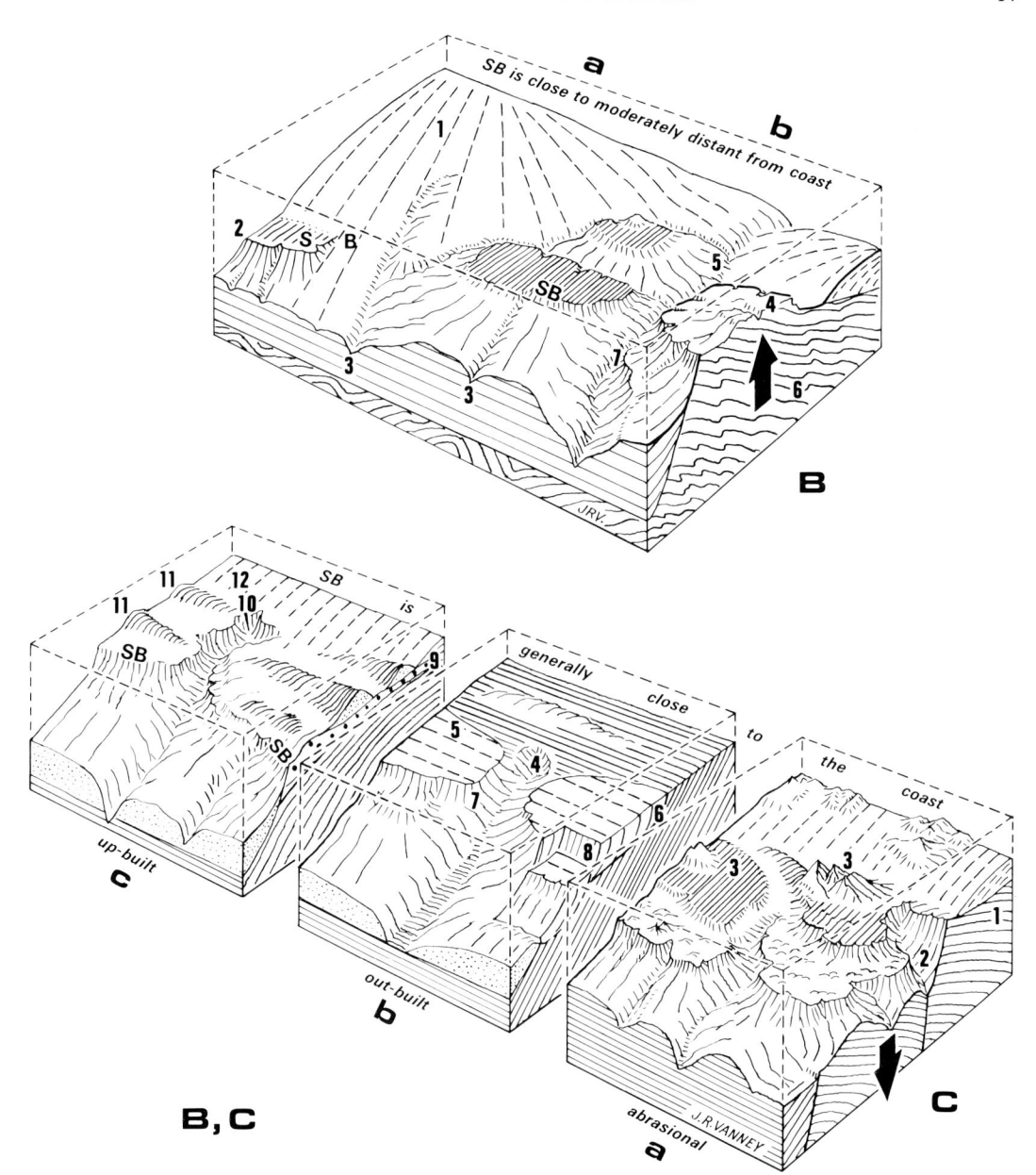

FIG. 10.—Schematic block diagrams depicting the shelfbreak and adjacent margin in temperate regions. The SB is shown (A) on relatively stable margins (a = convex arcuate profiles SB variant; b = terraced SB variant; c = crestal SB variant); (B) unstable convergent margin; and (C) unstable rifted margin (three SB variants include: a = abrasional, b = prograded, out-built, and c = prograded, up-built). Large arrows show substrate displacement. Numbers refer to features described in text.

served on the shelf. On margins on both sides of the Atlantic, profiles reveal single or a series of eroded terraces; those off the Mid-Atlantic States commonly occur between about 80 and 150 meters. Terraces are cut in coarse sediment (gravelly sand and muddy sandstone) and these display a generally undulatory surface. Terraces of this type have been interpreted differently according to their shape and position. Small, often distinct, step-like features may represent submerged wave-cut (1) or wave-built (2) littoral features associated with erosion during low sealevel stands; examples include the

Franklin, Fortune and Nichols shores as recorded by low cliffs (3) (detailed by Veatch and Smith, 1939; Garrison and McMaster, 1966; Wear et al., 1974). In the Golfe de Gascogne (Bay of Biscay), elongate mounds (4, 5) border the SB in a discontinuous manner; their seaward gradient is steeper than that of the landward gradient. These features are interpreted as offshore barrier island (4) or baymouth bars (5) related to deposition in depressions (6, 7) which are relict estuaries (6) or lagoons (7) formerly located near the outermost shelf-uppermost slope. In this region, there is evidence of slump-related dissection (8) of the upper reaches of canyons which head near the SB (Pinot, 1974; Vanney, 1977). In various regions, differences in depth, angularity and abruptness of the SB and associated submerged terraces below the SB are probably related, in part, also to subsidence or warping of the outer margin and to incomplete hydrostatic rebound since the end of the Pleistocene.

Crestal SB Variant (Fig. 10A-c).—This variant is perhaps the most distinctive SB type on stable margins. The shelf-to-slope sector is commonly formed by relict outer shelf sand bodies (1) which have been markedly re-shaped into multi-crested banks during the most recent eustatic low stand. Each bank is topped by a series of ridges which display diverse spacing and relief. For example, on many parts of the outer Moroccan shelf, we find a series of widely-spaced, low relief banks (2) which appear as gentle undulations bordering the SB (Vanney, 1980, 1983). These banks are formed in partially lithified sands, which originated as carbonate build-ups (Macintyre and Milliman, 1970) or as submerged coastal dunes; longitudinal depressions between the crests are either truncated dunes (3) or interdune deflation plains (4). Another type of bank, commonly located near the SB in shallow sectors of the Celtic Sea, is characterized by more pronounced relief and high, closely-spaced crests (5). A well-documented example in this region is the La Chapelle Bank which lies several tens of meters above the shelfedge proper; this SB is incised by multiple relict interfluvial channels (6) between the Black Mud and Ouessant canyons (see, among others, Cartwright and Stride, 1958; Hinschberger, 1970; Bouysse et al., 1976). It is likely that the crests represent relict offshore bars or perhaps submerged dunes. Bottom currents induced by cascading, tides or strong wave action, during and following the most recent rise in sea-level, have contributed to the development and reworking of these crests.

SB on Unstable Convergent Margins (Fig. 10B)

The diversity of shape and depth of this type of SB results from the interplay of displacement by compression and variations in amounts of sediment trapped by structural damming. Examples include those on the margin off Japan (Mogi, 1979), off New Zealand (Pantin, 1963; Lewis, 1973), on the outer edge of the shelf off Washington and Oregon (Byrne, 1962, 1963 a, b; Bales and Kulm, 1969; Kulm and Fowler, 1974), and in the Hellenic Arc (Le Quellec, 1979; Leite, 1980) and the Calabrian Arc (Rossi and Gabbianelli, 1978) in the Mediterranean.

Deposional Dominated SB Variant (Fig. 10B-a).—On some margins, the morphological configuration of the outer shelf is a response largely to an abundant terrigenous supply and, in such cases, the modern depositional prism (1) buries a relict shelfbreak. This relict SB, formed during the last eustatic low stand, crops out locally and forms small terraces (2) below late Quaternary deposits. The modern SB usually displays a gentle convex-up profile, which, in plan view, appears highly sinuous as a result of erosion (gullies, 3) resulting from the shelf-to-slope sediment bypassing.

Substrate Displaced SB Variant (Fig. 10B-b).— On margins where SB shape is largely a response to substrate displacement rather than sedimentation, many sectors of the outermost shelf are bordered by banks (4) resulting from uplift and tilting. The substrate is of Neogene age off Oregon, and Cretaceous to mid-Quaternary age off New Zealand; some banks bordering the SB in these regions are formed by this series. Examples include the Heceta and Coquille banks off Oregon (Kulm and Fowler, 1974), and Ariel, Lachlan, Adams and Whareama banks off the North Island of New Zealand (chart in Pantin, 1963; seismic profiles in Lewis, 1973). The banks may, in some instances, form a continuous ridge (the Lachlan Ridge off New Zealand), or they may be offset as insular blocks (Hellenic Arc) and separated by marked depressions (5), or they may occur off the prolongation of some headlands. The banks off Oregon, for example, have a relief in excess of several tens of meters, and thus are shallower than the SB. Truncated landward, dipping strata (6) at and near the SB (probable Pleistocene erosional surface off Oregon) form a cuesta-like relief. The evolution of SB shape is not only a response to tectonic effects, but also to the modifying influence of concomitant headward erosion, particularly at the heads of gullies (3) and submarine valleys (7). This latter mechanism has produced an SB which is locally highly festooned and steepened.

SB on Unstable Rifted Margins (Fig. 10C)

The main attribute of this SB type is its presence on an outer shelf that has recently been separated by extension and bordered by a depressed marginal basin. Depending upon the imprint of post-rifting sedimentation, the SB may display several variants including those well-defined in the Mediterranean and off California (Emery, 1960; Moore, 1969).

Abrasional SB Variant (Fig. 10C-a).—This variant occurs on the scarps of faulted, tilted blocks in the pre-rifted series where the effects of post-rifting sedimentation have been slight to nil. The SB configuration is generally sharp. Erosion of consolidated series (1) has occurred in the heads of canyons (2) whose location relative to the SB is often controlled by faults oriented perpendicularly to the margin. The outermost shelf surface is generally either smooth or gently undulatory, and may be covered by a carbonate-rich sediment veneer; some topographic features (3) landward of the SB are horst blocks or erosional remnants. Examples include the narrow, step-like margin of some sectors off western Corsica (Gennesseaux, 1972; Gennesseaux and Rehault, 1975) and Provence (Vanney, 1972; Froget, 1974); and ridges (or spurs), frequently topped by islands, in the Aegean Sea (Stanley and Perissoratis, 1977) and Tyrrhenian Sea (Selli, 1970; Selli and Fabbri, 1971; Viaris de Lesegno, 1978).

Prograded, Out-Built SB Variant (Fig. 10C-b).—This variant has been mapped on sectors of the outer edge of the narrow Ligurian shelf in the western Mediterranean (Fierro et al., 1973). Dominant characteristics are a relatively featureless shelf-to-slope profile, and moderate indentation by canyon heads (4). The SB has developed on the distal part of a progradational prism (5) which comprises features of coastal origin, including elongate bars (6) located on the frontal sector of the prism; these bars have been truncated and progressively levelled by wave erosion. The seaward face of the prism (7) is generally steep and the sediment cover thereon prone to failure. Slump scars (8) and other features produced by failure and associated gravity-driven processes abound on the upper slope, and these parallel the SB. A well accentuated example of this SB progradational variant has been described off Israel (Almagor, 1976; Neev et al., 1976). Here, alternating well- and poorly-cemented carbonate (Pleistocene) strata form the outer edge of the shelf and appear below the SB. Outcrops of the harder layers tend to retain modern surficial sediments which move downslope by creep. The outcrops thus form an irregular topography consisting of highs and lows (with a relief ranging from 5 to 25 meters) oriented parallel to the SB.

Prograded, Up-built SB Variant (Fig. 10C-c).— In contrast with the preceding type, this third SB variant occurs where the effects of important post-rift foundering are gradually compensated by the thick accumulation of fluvial sediment. As shown by seismic surveys, a combination of foreset and topset prograded strata form a wedge (9). In the western Mediterranean, the best examples occur on shelves of moderate to extensive width located seaward of gulfs and large bays (gulfs of Lion, and of Asinara west of Sardinia; Monaco, 1970; Got, 1973; Got et al., 1979; Aloïsi et al., 1975; Fanucci et al., 1976). The SB displays a highly crenulated and, in profile, scalloped configuration. The crenulation results from important cirque-like headward erosion forming large recesses of canyon heads (10). In the Gulf of Lion off southern France, the SB and adjacent margin record the presence of a coarse, fluvial wedge (9, consisting of pebbles, sand and silty mud) deposited during the last eustatic lowering of sealevel. Active rejuvenation of the upper slope by pre-Holocene stream erosion resulted in pronounced landward-oriented dissection. Just landward of the SB the slightly seaward-inclined, outer shelf surface displays a series of distinct undulations (11) which are interpreted as sand bars. These features, unconformable on the outer shelf surface, originated during the last low sealevel stand. Landward of these, we commonly find a subdued depression (12) whose sediment surface slopes gently seaward. This type of depression denotes the position of an ancient lagoon subsequently covered by fine-grained material that accumulated during the rise in sealevel.

CONCLUSIONS

The shelfbreak, the point of the first major change in gradient at the outermost edge of the continental shelf, is highly variable in depth and distance from shore. This overview calls attention to the diversity of shelf-to-slope configurations, and synthesizes shelfbreak types by means of a genetic-descriptive classification. Although the emphasis is largely a geomorphological one, we recognize, as have all earlier workers concerned with continental margins, that variations in shelfbreak configuration result from several controlling factors. The dominant factors are geological framework and substrate mobility (cf. reviews in Hedberg, 1970, and Emery, 1980). Progradation and erosion related to the sedimentation regime further modify, often substantially, the break configuration (Southard and Stanley, 1976). Depositional considerations include, among others, sediment supplied by rivers, formation of carbonates and varying rates of reef build-up, influence of ice transport in higher latitudes, and the interplay of fluid-driven and gravitative processes active in outer shelf-to-slope environments. Moreover, physical, biological and chemical processes which modified the outer edge of the shelf were particularly important during lower eustatic sealevel stands when the break was located at or close to the coast. In most instances this relict imprint is still clearly evident. Thus, we view the shelfbreak as a reworked relict, or palimpsest, feature which has not reached complete equilibrium with processes active at present.

An assessment of the different margins that have been well studied reveals the presence of a remarkable diversity of shelfbreak types and variants. It is important to recall, however, that only a few shelfbreak sectors have been systematically surveyed and, in consequence, our classification most likely outlines a rather incomplete series of shelfbreak varieties rather than all possible types. The classification resulting from our didactic approach tends to focus on the more diagnostic and/or better studied end-members, and we are aware that a sequence of transitional shelfbreak varieties is overlooked. These undoubtedly will become better known as studies of this critical interface on world margins are intensified. In summary, the present classification takes into consideration the interaction among (a) the rate of substrate motion and style of displacement, which are a function of the large-scale evolution of a particular margin, (b) the overprint of earlier (largely Quaternary) climatic and eustatic events, and (c) sediment supply and processes (reviewed in terms of climatic belts). The interplay of these three larger-scale parameters has varied considerably, giving rise to the suite of possible SB types illustrated in Figures 6 to 10.

In response to the interplay of the above three major controlling factors, a similar (at least superficially) shelfbreak configuration can develop under a set of highly different conditions. For example, an abrupt break on the edge of a narrow shelf may occur in a rifted, carbonate reef setting in the tropical zone, or in a relatively stable setting in a high latitude ice-affected margin, or in a highly unstable convergent margin in a mid-latitude temperate belt. Thorough interpretation thus requires an integration of several methods, including precise echo-sounding profiling and close-grid seismic surveys, deep-sea drilling, and technologies that involve bottom sampling and long-term measurements of water mass movement. These aspects and their ramifications are discussed at length in the following chapters of this volume.

Definition and interpretation of the shelfbreak are no longer theoretical excercises, but have become crucial in view of the ever-increasing pressures by man's activities on the outer margin. Hydrocarbon and mineral exploration has intensified, as has fishing and the dumping of wastes. Even as this manuscript is being written, we are aware of newsworthy legal arbitration (Libyan-Tunisian and Canada-USA offshore boundary delimitation by the International Court of Justice), and the military confrontation of the Falkland-Malvinas Islands and their margin by Argentina and Great Britain. Moreover, it is disturbing that international organizations are using arbitrary definitions based on artificial considerations that extend the edge of the shelf to "the outer edge of the continental margin" or to "a distance of 200 nautical miles from the baselines from which the breadth of the territorial sea is measured." The role of marine scientists thus clearly becomes all the more important in presenting a precise and accurate view of the shelfbreak—one which will stand the test of time.

ACKNOWLEDGMENTS

We thank the following reviewers for their comments which have improved the manuscript: R. Embley (Washington, D.C.), A. Guilcher (Brest, France), G. T. Moore (La Habra, California) and J. W. Pierce (Washington, D.C.). Appreciation is expressed to Mlle. J. Leuridan (CNRS-Paris) for assistance with drafting. Funding for travel was provided to D. J. Stanley by Smithsonian Scholarly Studies grant 1233S201.

REFERENCES

ADEY, W. H., AND BURKE, R., 1976, Holocene bioherms (algal ridges and bank-barrier reefs) of the eastern Caribbean: Geol. Soc. America Bull., v. 87, p. 95–109.

ALLEN, J. R. L., 1964a, Sedimentation in the modern delta in the River Niger, West Africa, in Van Straaten, L. M. J. U., ed., Deltaic and Shallow Deposits: Amsterdam, Elsevier, p. 26–34.

———, 1964b, The Nigerian continental margin: Bottom sediments, submarine morphology and geological evolution: Mar. Geology, v. 1, p. 289–332.

———, 1965, Late Quaternary Niger Delta and adjacent areas: Sedimentary environments and lithofacies: Am. Assoc. Petroleum Geologists Bull., v. 49, p. 547–600.

———, AND WELLS, J. W., 1962, Holocene coral banks and subsidence in the Niger Delta: Jour. Geology, v. 70, p. 381–397.

ALMAGOR, G., 1976, The bathymetric chart of the Israeli continental shelf and slope off the Ashqelon-Tel Aviv coast: Israeli Geol. Survey Rept. MG/7/76, 20 p.

ALOÏSI, J. C., MONACO, A., THOMMERET, J., AND THOMMERET, Y., 1975, Evolution paléogéographique du plateau continental languedocien dans le cadre du Golfe du Lion. Analyse comparée des données sismiques, sédimentologiques et radiométriques concernant le Quaternaire récent: Rev. Géogr. phys. Géol. dyn., v. 17, p. 13–22.

BALES, W. E., AND KULM, L. D., 1969, Structure of the continental shelf off southern Oregon: Am. Assoc. Petroleum Geologists Bull., v. 53, p. 471.

BARNES, P. W., AND REIMNITZ, E., 1974, Sedimentary processes on Arctic shelves off the northern coast of Alaska, in Reed, J. C., and Sater, J. E., eds., The coast and shelf of the Beaufort Sea: Arlington, Virginia, Arctic Inst. North America, p. 439–476.

BOILLOT, G., 1979, Géologie des Marges Continentales (2nd Edition, 1983): Paris, Masson, 140 p.

BOUYSSE, P., HORN, R., LAPIERRE, F., AND LE LANN, F., 1976, Etude des grands bancs de sable du Sud-Est de la Mer Celtique: Mar. Geology, v. 20, p. 251–275.

———, AND MARTIN, P., 1979, Caractères morphostructuraux et évolution géodynamique de l'arc insulaire des Petites Antilles (Campagne Arcante I): Bur. Rech. Géol. Miner. Bull., v. 4, p. 185–210.

BURK, C. A., 1965, Geology of the Alaska Peninsula—Island arc and continental margin: Geol. Soc. America Memoir 99 (Part I), p. ix–250.

BYRNE, J. V., 1962, Geomorphology of the continental terrace off the central coast of Oregon: The Ore Bin, v. 24, p. 65–74.

———, 1963a, Geomorphology of the continental terrace off the northern coast of Oregon: The Ore Bin, v. 25, p. 201–209.

———, 1963b, Geomorphology of the continental terrace of Coos Bay: The Ore Bin, v. 25, p. 149–157.

CARLSON, P. R., BRUNS, T. R., MOLNIA, B. F., AND SCHWAB, W. C., 1982, Submarine valleys in the northeastern Gulf of Alaska: Characteristics and probable origin: Mar. Geology, v. 47, p. 217–242.

CARTWRIGHT, D. E., AND STRIDE, A. H., 1958, Large sand wave near the edge of the continental shelf: Nature, v. 180, p. 41.

COLEMAN, J. M., 1981, Deltas, Processes and Models of Deposition for Exploration (2nd Edition): Minneapolis, Minnesota, Burgess Publishing Co., 124 p.

———, PRIOR, D. B., AND LINDSAY, J. F., 1983, Deltaic influences on shelfedge instability processes, in Stanley, D. J., and Moore, G. T., eds., The Shelfbreak: Critical interface on continental margins: Soc. Econ. Paleontologists Mineralogists Spec. Pub. 33, p. 121–137.

CURRAY, J. R., AND MOORE, D. G., 1964, Pleistocene deltaic progradation of continental terrace, Costa de Nayarit, Mexico, in Van Andel, T. H., and Shor, G. C., eds., Marine Geology of the Gulf of California, A Symposium: Am. Assoc. Petroleum Geologists Memoir 3, p. 193–215.

DIETZ, R. S., AND MENARD, H., 1951, Origin of abrupt change in slope at continental shelf margin: Am. Assoc. Petroleum Geologists Bull., v. 35, p. 1194–1216.

EMERY, K. O., 1960, The Sea off Southern California: A Modern Habitat of Petroleum: New York, John Wiley and Sons, p. xi–366.

———, 1968, Shallow structure of continental shelves and slopes: Southeastern Geology, v. 9, p. 173–194.

———, 1969, The continental shelves: Scientific American, v. 221, p. 107–122.

———, 1980, Continental margins—Classification and petroleum prospects: Am. Assoc. Petroleum Geologists Bull., v. 64, p. 297–315.

———, 1981, Geological limits of the "continental shelf": Ocean Div. Internat. Law Journal, v. 10, p. 1–11.

FANUCCI, F., FIERRO, G., ULZEGA, A., GENNESSEAUX, M., REHAULT, J. P., AND VIARIS DE LESEGNO, L., 1976, The continental shelf of Sardinia: Structure and sedimentary characteristics: Bol. Soc. Geol. Ital., v. 95, p. 1201–1217.

FIERRO, G., GENNESSEAUX, M., AND REHAULT, J. P., 1973, Caractères structuraux et sédimentaires du plateau continental de Nice à Gênes (Méditerranée nord-occidentale): Bur. Rech. Géol. Miner. Bull., v. 4, p. 193–208.

FINK, L. K., 1972, Bathymetric and geologic studies of the Guadeloupe region, Lesser Antilles island arc: Mar. Geology, v. 12, p. 267–288.

FISHER, M. A., 1979, Structure and tectonic setting of continental shelf southwest of Kodiak Island, Alaska: Am. Assoc. Petroleum Geologists Bull., v. 63, p. 301–310.

FRANÇA, A. M. C., 1979, Geomorfologia da margem continental leste-brasileira e da bacia oceânica adjacente, in Chaves, H. A. F., ed., Geomorfologia da margem continental Brasileira e das areas oceânicas adjacentes: Projeto REMAC, Rio de Janeiro, v. 7, p. 89–128.

FROGET, C., 1974, Essai sur la géologie du précontinent de la Provence occidentale: Univ. Aix-Marseille [thesis], 219 p.

GARRISON, L. E., AND MCMASTER, R. L., 1966, Sediments and geomorphology of the continental shelf off southern New England; Mar. Geology, v. 4, p. 273–289.

GENNESSEAUX, M., 1972, La structure du plateau continental des Bouches de Bonifacio (Corse): C. R. Acad. Sci. Paris, v. 275, p. 2295–2297.

———, AND REHAULT, J. P., 1975, La marge continentale Corse; Soc. Geol. France Bull., v. 17, p. 505–518.

GIERLOFF-EMDEN, H. G., 1958, Die Küsten-Schelf von El Salvador in Zusammenhang mit der Morphologie und Geologie des Festlandes: Dtsch. Hydr. Ztschr., v. 11, p. 240–246.

GOT, H., 1973, Etude des corrélations tectonique-sédimentation au cours de l'histoire quaternaire du précontinent pyrénéo-catalan: Univ. Sci. Tech. Montpellier [thesis], p. xxxiii–294.

———, ALOÏSI, J. C., LEENHARDT, O., MONACO, A., SERRA-RAVENTOS, J., AND THEILEN, F., 1979, Structures sédimentaires sur les marges du Golfe du Lion et de Catalogne: Rev. Géol. dyn. Géogr. phys., v. 21, p. 281–293.

GUILCHER, A., 1955, Géomorphologie de l'extrémité septentrionale du Banc Farsan (Mer Rouge), in Résultats scientifiques des campagnes de la "Calypso", I, Campagne en Mer Rouge (1951–1952): Ann. Inst. Océanogr., v. 30, p. 55–100.

———, 1963, Continental shelf and slope (continental margin), in Hill, M. N., ed., The Sea: New York, Interscience, v. 3, p. 281–311.

HAMILTON, W., 1979a, Subduction in the Indonesian region, in Talwani, M., and Pitman III, W. C., eds., Island Arcs, Deep Sea Trenches, and Back-arc Basins: Am. Geoph. Union, Maurice Ewing Series, v. 1, p. 15–31.

———, 1979b, Tectonics of the Indonesian region: U.S. Geol. Survey Professional Paper 1078, 345 p.
HEDBERG, H. D., 1970, Continental margins from viewpoint of the petroleum geologist: Am. Assoc. Petroleum Geologists Bull., v. 54, p. 3–43.
HEEZEN, B. C., THARP, M., AND EWING, M., 1959, The floors of the oceans. 1. The North Atlantic: Geol. Soc. America Spec. Paper 65, 122 p.
HINSCHBERGER, F., 1970, L'Iroise et les abords d'Ouessant et de Sein. Etude de morphologie et de sédimentologie sous-marines: Caen, Fac. Lettres Sci. Humaines, 309 p.
HOLMES, M. L., AND CREAGER, J. S., 1974, Holocene history of the Laptev Sea continental shelf, in Herman, Y., ed., Marine Geology and Oceanography of the Arctic Seas: New York, Springer, p. 211–229.
HOLTEDAHL, O., 1958, Some remarks on geomorphology of continental shelves off Norway, Labrador and southeast Alaska: Jour. Geology, v. 66, p. 461–471.
———, 1970, On the geomorphology of the West Greenland shelf, with general remarks on the "marginal channel" problem: Mar. Geology, v. 8, p. 155–172.
———, AND HOLTEDAHL, H., 1961, On "marginal channels" along continental borders and the problems of their origins: Geol. Inst. Uppsala Bull., v. 40, p. 183–187.
HUENE, R. VON, 1972, Structure of the continental margin and tectonism of the eastern Aleutian Trench: Geol. Soc. America Bull., v. 83, p. 3613–3629.
IL'IN, A. V., 1962, Geomorphology of the continental shelf in the northern Atlantic Ocean [in Russian]: Moscow, Trudy Inst. Okeanol. Akad. Nauk SSSR, v. 56, p. 3–14.
JOHNSON, G. L., VANNEY, J. R., AND HAYES, D., 1982, The Antarctic continental shelf, in Craddock, C., ed., Antarctic Geoscience (Symposium on Antarctic Geology and Geophysics): Madison, Univ. Wisconsin Press, p. 995–1002.
KARIG, D. E., 1979, Growth patterns on the upper trench slope, in Talwani, M., and Pitman III, W. C., eds., Island arcs, deep sea trenches, and back-arc basins: Am. Geoph. Union, Maurice Ewing Series, v. 1, p. 175–181.
KENDALL, C. ST. G., AND SCHLAGER, W., 1981, Carbonate and relative changes in sea level: Mar. Geology, v. 44, p. 181–212.
KIECKHEFER, R. M., MOORE, G. F., AND EMMEL, F. S., 1981, Crustal structure of the Sunda forearc region west of central Sumatra from gravity data: Jour. Geoph. Res., v. 86, p. 7003–7012.
KORNICKER, L. S., AND BRYANT, W. E., 1969, Sedimentation on continental shelf of Guatemala and Honduras: Am. Assoc. Petroleum Geologists Memoir 11, p. 244–257.
KOWSMANN, R. O., COSTA, M. P. A., VICALVI, M. A., COUTIHNO, M. G. N., AND CAMBOA, L. A. P., 1977, Modelo de sedimentacâo holocênica na plataforma continental sul brasileira, in Evoluçâo sedimentar holocênica da plataforma continental e do talude do Sul do Brasil: Projeto REMAC, Rio de Janeiro, v. 2, p. 7–26.
KULM, L. D., AND FOWLER G. A., 1974, Oregon continental margin structure and stratigraphy: A test of the imbricate thrust model, in Burk, C. A., and Drake, C. L., eds., The Geology of Continental Margins: Berlin, Springer, p. 261–283.
LAUGHTON, A. S., 1966, The Gulf of Aden: Phil. Trans. Roy. Soc. London, v. A-259, p. 150–171.
———, WHITMARSH, R. D., AND JONES, M. T., 1970, The evolution of the Gulf of Aden: Phil. Trans. Roy. Soc. London, v. A-267, p. 227–266.
LEITE, O., 1980, La marge continentale sud-crétoise. Géologie et structure: Univ. Paris VI [thesis], 210 p.
LE QUELLEC, F., 1979, La marge continentale ionienne du Péloponnèse, Géologie et structure: Univ. Paris VI [thesis], 210 p.
LEWIS, K. B., 1973, Erosion and deposition on a tilting continental shelf during Quaternary oscillations of sea level: New Zealand Jour. Geology Geophys., v. 16, p. 281–301.
LIGHTY, R. G., 1977, Relict shelf-edge Holocene coral reef, southeast coast of Florida, in Proc. Third Internat. Coral Reef Symposium, v. 2, p. 215–222.
———, MACINTYRE, I. G., AND STUCKENRATH, R., 1978, Submerged early Holocene barrier reef southeast Florida shelf: Nature, v. 276, p. 59–60.
LOGAN, B. W., HARDING, J. L., AHR, W. M., WILLIAMS, J. C., AND SNEATH, R. G., 1969, Late Quaternary carbonate sediment in Yucatan Shelf, in Logan, B. W. et al., eds., Carbonate Sediments and Reefs, Yucatan Shelf, Mexico: Am. Assoc. Petroleum Geologists Memoir 11, p. 5–128.
LONARDI, A. G., AND EWING, M., 1971, Sediment transport and distribution in the Argentine Basin, IV. Bathymetry of the continental margin, Argentine Basin, and other related provinces. Canyons and sources of sediment: Phys. Chem. Earth, v. 8, p. 19–121.
LONSDALE, P., 1978, Ecuadorian subduction system: Am. Assoc. Petroleum Geologists Bull., v. 62, p. 2454–2477.
MACINTYRE, I. G., 1967, Submerged coral reefs, west coast of Barbados, West Indies: Canadian Jour. Earth Sci., v. 4, p. 461–474.
———, 1972, Submerged reefs of eastern Caribbean: Am. Assoc. Petroleum Geologists Bull., v. 56, p. 720–738.
———, AND MILLIMAN, J. D., 1970, Physiographic features on the outer shelf and upper slope, Atlantic continental margin, southeastern United States: Geol. Soc. America Bull., v. 81, p. 2577–2598.
MARSILLI, L. F., COMTE de, 1725, Histoire Physique de la Mer. Ouvrage Enrichi de Figures Dessinées d'Après le Naturel: Amsterdam, Académie des Sciences de Paris, 173 p.
MARTIN, L., 1971, The continental margin from Cape Palmas to Lagos: Bottom sediments and submarine morphology, in Delaney, F. M., ed., The Geology of the East Atlantic Continental Margin: London Inst. Geol. Sci. Rept. 76/16, p. 79–95.
———, 1973, Morphologie, sédimentologie et paléogéographie au Quaternaire récent du plateau continental ivoirien: Univ. Paris VI [thesis], 340 p.

McMaster, R. L., Milliman, J. D., and Ashraf, A., 1971, Continental shelf and upper slope sediments off Portugese Guinea, Guinea and Sierra Leone, west Africa: Jour. Sed. Petrology, v. 41, p. 150–158.

Mogi, A., 1979, An Atlas of the Sea Floor Around Japan. Aspects of Submarine Geomorphology: Tokyo, Univ. Tokyo Press, 96 p.

Monaco, A., 1970, Contribution à l'étude géologique et sédimentologique du plateau continental du Roussillon (Golfe du Lion): Univ. Sci. Tech. Languedoc Montpellier, unpublished, 295 p.

Moore, D. G., 1969, Reflection profiling studies of the California continental borderland: Structure and Quaternary turbidite basins: Geol. Soc. America Spec. Paper 117, 142 p.

———, 1973, Plate-edge deformation and crustal growth, Gulf of California structural province: Geol. Soc. America Bull., v. 84, p. 1883–1906.

Moore, G. F., and Curray, J. R., 1980, Structure of the Sunda Trench lower slope off Sumatra from multichannel seismic reflection data: Marine Geophys. Res., v. 4, p. 319–340.

———, ———, Moore, D. G., and Karig, D. E., 1980, Variations in geologic structure along the Sunda Forearc, northeastern Indian Ocean, in The Tectonic and Geologic Evolution of Southeast Asia Seas and Islands: Am. Geophys. Union Geophys. Monograph 23, p. 145–160.

Mougenot, D., Boillot, G., and Rehault, J. P., 1983, Prograding shelfbreak types on passive continental margins: Some European examples, in Stanley, D. J., and Moore, G. T., eds., The shelfbreak: Critical Interface on Continental Margins: Soc. Econ. Paleontologists Mineralogists Spec. Pub. 33, p. 61–77.

———, and Vanney, J. R., 1980, Géomorphologie et profils de réflexion sismique: Interpretation de surfaces remarquables d'une plate-forme continentale, in Problèmes Géomorphologiques de la Marge Continentale Européenne: Ann. Inst. Océanogr., v. 56, p. 85–100.

Nansen, F., 1904, The bathymetrical features of the North Polar seas with a discussion of the continental shelves and previous oscillations of shoreline, in The Norwegian North Polar Expedition Scientific Results: Kristiania (Oslo), v. 4, 231 p.

Naugler, F. P., Silverberg, N., and Creager, J. S., 1974, Recent sediments of the East Siberian Sea, in Herman, Y., ed., Marine Geology and Oceanography of the Arctic Seas: Berlin, Springer, p. 191–210.

Neev, D., Almagor, G., Arad, A., Ginzburg, A., and Hall, J. K., 1976, The geology of the southeastern Mediterranean Sea: Israel Geol. Survey Bull., v. 68, p. 1–51.

Nesteroff, W. D., 1955, Les récifs coralliens du Banc Farsan Nord (Mer Rouge), in Résultats scientifiques des campagnes de la "Calypso", I, Campagne en Mer Rouge (1951–1952): Ann. Inst. Océanogr., v. 30, p. 7–53.

Neumann, A. C., Kofoed, J. W., and Keller, G. H., 1977, Lithoherms in the Straits of Florida: Geology, v. 5, p. 4–10.

Normark, W. R., and Curray, J. R., 1968, Geology and structure of the tip of Baja California, Mexico: Geol. Soc. America Bull., v. 69, p. 1589–1600.

Nota, D. J. G., 1958, Sediments of the western Guiana shelf, in Reports of the Orinoco Shelf Expeditions, v. 2, p. 1–98.

Palma, J. J. C., 1979, Geomorfologia da plataforma continental norte Brasileira, in Chaves, H. A. F., ed., Geomorfologia da margem continental Brasileira e das areas oceânicas adjacentes: Projecto REMAC, Rio de Janeiro, v. 7, p. 25–52.

Pantin, H. M., 1963, Submarine morphology east of the North Island, New Zealand: New Zealand Oceanogr. Inst. Memoir 14, 44 p.

Pitman III, W. C., and Golovchenko, X., 1983, The effect of sealevel change on the shelfedge and slope of passive margins, in Stanley, D. J., and Moore, G. T., eds., The Shelfbreak: Critical Interface on Continental Margins: Soc. Econ. Paleontologists Mineralogists Spec. Pub. 33, p. 41–58.

Pinet, P. R., 1971, Structural configuration of the northwestern Caribbean plate boundary: Geol. Soc. America Bull., v. 82, p. 2027–2032.

———, 1972, Diapirlike features offshore Honduras: Implications regarding tectonic evolution of Cayman Trough and Central America: Geol. Soc. America Bull, v. 83, p. 1911–1922.

———, 1975, Structural evolution of the Honduras continental margin and the sea floor south of the western Cayman Trough: Geol. Soc. America Bull., v. 86, p. 830–836.

Pinot, J. P., 1974, Le précontinent breton, entre Penmarc'h, Belle-Ile et l'escarpement continental, étude géomorphologique: Lannion, Impram, 245 p.

Rossi, S., and Gabbianelli, G., 1978, Geomorphologia del Golfo di Tarento: Bol. Soc. Geol. Ital., v. 97, p. 423–2437.

Rodolfo, K. S., 1969, Bathymetry and marine geology of the Andaman Basin, and tectonic implications for southeast Asia: Geol. Soc. America Bull., v. 80, p. 1203–1230.

Rusnak, G. A., Fisher, R. L., and Shepard, F. P., 1964, Bathymetry and faults of the Gulf of California, in Van Andel, T. H., and Shor, G. C., eds., Marine Geology of the Gulf of California, A Symposium: Am. Assoc. Petroleum Geologists Memoir 3, p. 59–75.

Selli, R., 1970, Richerche geologiche preliminari nel mare Tirreno: Giornale di Geologia, v. 37, p. 4–24.

———, and Fabbri, A., 1971, Tyrrhenian: A Pliocene deep sea: Acc. Naz. Lincei, Rend. Sc. Fis. Mat. Nat., v. 50, p. 104–116.

Shepard, F. P., 1973, Submarine geology: New York, Harper and Row, 517 p.

Sommerhoff, G., 1973, Formenschatz und morphologische Gliederung des Südostgrönländischen Schelfgebietes und Kontinentalabhanges, in "Meteor" Forsch. Erg: Berlin-Stuggart, Gebrüder Borntraeger, v. 15, p. 1–54.

Southard, J. B., and Stanley, D. J., 1976, Shelf-break processes and sedimentation, in Stanley, D. J., and Swift,

D. J. P., eds., Marine Sediment Transport and Environmental Management: New York, Wiley-Interscience, p. 351–377.

STANLEY, D. J., DRAPEAU, G., AND COK, K. E., 1968, Submerged terraces on the Nova Scotia Shelf: Ztschr. Geomorphol. suppl. 5, p. 85–94.

——— AND PERISSORATIS, C., 1977, Aegean Sea ridge barrier and basin sedimentation pattern: Mar. Geology, v. 24, p. 97–107.

———, AND WEAR, C. M., 1978, The "mud-line": An erosion-deposition boundary on the upper continental slope: Mar. Geology, v. 28, p. M19–M29.

SWIFT, D. J. P., DUANE, D. B., AND PILKEY, O. H., eds., 1972, Shelf Sediment Transport: Process and Pattern: Stroudsburg, Pennsylvania, Dowden, Hutchinson and Ross, 656 p.

THOMPSON, W. C., 1961, A genetic classification of continental shelves: Proc. 9th Pacific Sci. Congress, v. 12, p. 30–39.

UCHUPI, E., 1968, Atlantic continental shelf and slope of the United States: Physiography: U.S. Geol. Survey Professional Paper 529-C, 30 p.

———, 1969, Morphology of the continental margin of southeastern Florida: Southeastern Geology, v. 11, p. 129–134.

UNDERWOOD, M. B., AND KARIG, D. E., 1980, Role of submarine canyons in trench and trench-slope sedimentation: Geology, v. 8, p. 432–436.

VAN ANDEL, T. H., 1965, Morphology and sediments of the Sahul Shelf: Trans. New York Acad. Sci., v. 28, p. 81–89.

———, AND VEEVERS, J. J., 1965, Submarine morphology of the Sahul Shelf, Northwestern Australia: Geol. Soc. America Bull., v. 76, p. 695–700.

——— AND ———, 1967, Morphology and sediments of the Timor Sea: Australia Bur. Min. Res., Geol. Geophys., v. 83, 173 p.

VANNEY, J. R., 1972, Cartes bathymétriques de la plate-forme continentale de la Provence au 1/50 000. Feuille de Toulon: Paris, CNEXO data file.

———, 1976, Géomorphologie des Plates-formes Continentales: Paris, Doin, 300 p.

———, 1977, Géomorphologie de la Marge Continentale Sud-amoricaine: Paris, SEDES, 472 p.

———, 1980, Cartes bathymétriques de la marge continentale atlantique du Maroc au 1/100 000: Feuilles de Rabat, Safi, unpublished.

———, 1983 in press, La plate-forme continentale du Gharb et de la Meseta Marocaine. Régions et problèmes morphologiques, in Vanney, J. R., ed., La Plate-forme Continentale du Gharb et de la Meseta Marocaine: Serv. Géol. Maroc., Mém. 327.

———, AND JOHNSON, G. L., 1976, Geomorphology of the Pacific continental margin of the Antarctic Peninsula, in Hollister, C. D., Craddock, C. et al., eds., Initial Reports of the Deep Sea Drilling Project I, v. 35, p. 279–289.

——— AND ———, 1979a, The sea floor morphology seaward of Terre Adélie (Antarctica): Deutsch. Hydrogr. Ztschr., v. 32, p. 77–87.

———AND ———, 1979b, Wilkes Land continental margin physiography, east Antarctica: Polarforschung, v. 49, p. 20–29.

———, FALCONER, R. K., AND JOHNSON, G. L., 1981, Geomorphology of the Ross Sea and adjacent oceanic provinces: Mar. Geology, v. 41, p. 73–102.

——— AND MOUGENOT, D., 1981, La plate-forme continentale du Portugal et les provinces adjacentes: Analyse géomorphologique: Lisboa, Serv. Geol. Portugal Memoir 28, 145 p.

VIARIS DE LESEGNO, L., 1978, Etude structurale de la Mer Tyrrhénienne septentrionale: Univ Paris VI [thesis], 170 p.

VEATCH, C. M., AND SMITH, P. A., 1939, Atlantic submarine valleys: Geol. Soc. America Spec. Paper 7, 101 p.

WEAR, C. M., STANLEY, D. J., AND BOULA, J. E., 1974, Shelfbreak physiography between Wilmington and Norfolk Canyons: Marine Tech. Soc. Jour., v. 8, p. 37–48.

WEEKS, L. A., HARBISON, R. N., AND PETER, G., 1967, Island arc system in Andaman Sea: Am. Assoc. Petroleum Geologists Bull., v. 51, p. 1803–1815.

ZEMBRUSCKI, S. G., 1979, Geomorfologia da margem continental sul Brasileira e das bacias oceânicas adjacentes, in Chaves, H. A. F., ed., Gemorfol gia da margem continental Brasileira e das areas Oceânicas adjacentes: Projeto REMAC, Rio de Janeiro, v. 7, p. 129–177.

———, BARRETO, H. G., PALMA, J. C., AND MILLIMAN, J. D., 1972, Estudo preliminar das provincias geomorfológicas da margem continental brasileira: Anais XXVI Congr. Bas. Geologia, Soc. Bras. Geol., v. 2, p. 187–209.

ZHIVAGO, A. V., 1978, Morphostructure of the Antarctic continental shelf [in Russian], in Geomorfologija i Paleogeografija čelfa: Moscow, Nauka, p. 57–97.

——— AND EVTEEV, S. A., 1970, Shelf and marine terraces of Antarctica: Quaternaria, v. 12, p. 89–114.

BREACHING THE SHELFBREAK: PASSAGE FROM YOUTHFUL TO MATURE PHASE IN SUBMARINE CANYON EVOLUTION

JOHN A. FARRE,[1] BONNIE A. McGREGOR,[2]
WILLIAM B. F. RYAN,[1] AND JAMES M. ROBB[3]

Lamont-Doherty Geological Observatory, Columbia University, Palisades, New York 10964;[1]
U.S. Geological Survey, Fisher Island Station, Miami, Florida 33139;[2]
and U.S. Geological Survey, Woods Hole, Massachusetts 02543[3]

ABSTRACT

Mid-range side-scan sonar images of the U.S. Middle Atlantic continental margin show the presence of a variety of erosional features. Amphitheatre-shaped scars are present on the continental slope near Carteret Canyon. Several slope canyons, whose heads do not appreciably breach the shelfbreak, have a pinnate drainage pattern on the upper, sediment-draped slope. The thalwegs of these canyons follow nearly straight paths down the slope. In contrast, Wilmington, a shelf-indenting canyon, follows a curved to meandering path down the slope.

On the basis of these and other data, we propose a preliminary explanation of evolution of canyons on the middle Atlantic slope. In this model, localized slope failure begins the process of canyon formation. By headward erosion, these depressions extend upslope and form linear sediment chutes. These slope canyons represent the youthful phase in canyon evolution. Slope canyons begin the transition to a mature phase when the canyon heads breach the shelfbreak. Access to the continental shelf leads to the transport of shelf-derived materials through the canyon.

During the youthful phase, the dominant mechanism of canyon erosion is the failure of the slope itself. In the mature phase, entrenchment is augmented by the episodic cutting by turbulent sediment suspensions enroute from the shelf to the continental rise and abyssal plain. The shelfbreak, in this model, is an important factor in the evolution of a passive continental margin.

INTRODUCTION

The major submarine canyons of the U.S. Middle Atlantic continental margin (Fig. 1), begin on the continental shelf, cross the shelfbreak, and continue down the continental slope to the continental rise (Veatch and Smith, 1939; Heezen et al., 1959; Kelling and Stanley, 1970). Many of these canyons having second and third order tributaries in their upper reaches have been considered to be the seaward continuation of terrestrial drainage systems that crossed the shelf during low stands of sealevel in the Pleistocene (Spencer, 1903; Stetson, 1936). The similarities between the submarine dendritic drainage system and fluvial systems on land seem to imply that the same or similar processes could be responsible for their formation (Shepard, 1933; Shepard and Dill, 1966; Shepard, 1981).

There is little doubt that subaerial erosion was effective in creating some presently buried shelf valleys (McMaster and Ashraf, 1973; Knebel et al., 1979). Pleistocene rivers delivered to the outer shelf suspended and bedload sediment that subsequently entered the submerged canyon heads (Twichell et al., 1977). Questions do exist about whether canyon initiation and excavation are the products of the erosive forces of shelf sediments that travel across the slope.

The glacially-derived sediment brought to the outer shelf during the last low stand of the late Pleistocene (Wisconsinan Stage) sea has yet to be buried by Holocene sediments (Emery, 1968). Landward of the shelfbreak, this relict surficial sediment is principally coarse- to medium-grained, shelly terrigenous sand (Trumbull, 1972; Schlee, 1973) reworked by nearshore currents during the Holocene transgressions (Swift et al., 1980). The presence of similar sandy sediments containing displaced shallow-water fauna on the western Atlantic abyssal plains led Ericson et al. (1961) to propose that shelf-indenting canyons act as major conduits for sediment crossing the shelfbreak and slope. The abundance of Pleistocene sand with autochthonous cold-water fauna in cores on the outer continental rise supports the association of submarine canyon activity with low stands of sealevel, when the canyon heads were in the near-shore region of moving sand (Middleton and Hampton, 1973).

With the recognition that the submarine canyons were the conduits for this long distance transport and that much of the sand which resides in the continental rise and abyssal plain came from the shelf, canyon formation and deepening was attributed to cascading sand flows (Dill, 1964) and to powerful, episodic, erosive turbidity currents (Daly, 1936; Heezen and Ewing, 1957). Fault control has been suggested to explain canyons that follow straight paths down the slope at an angle to the regional slope (Belderson and Kenyon, 1976) and canyons that show abrupt changes in downslope trend (Kelling and Stanley, 1970).

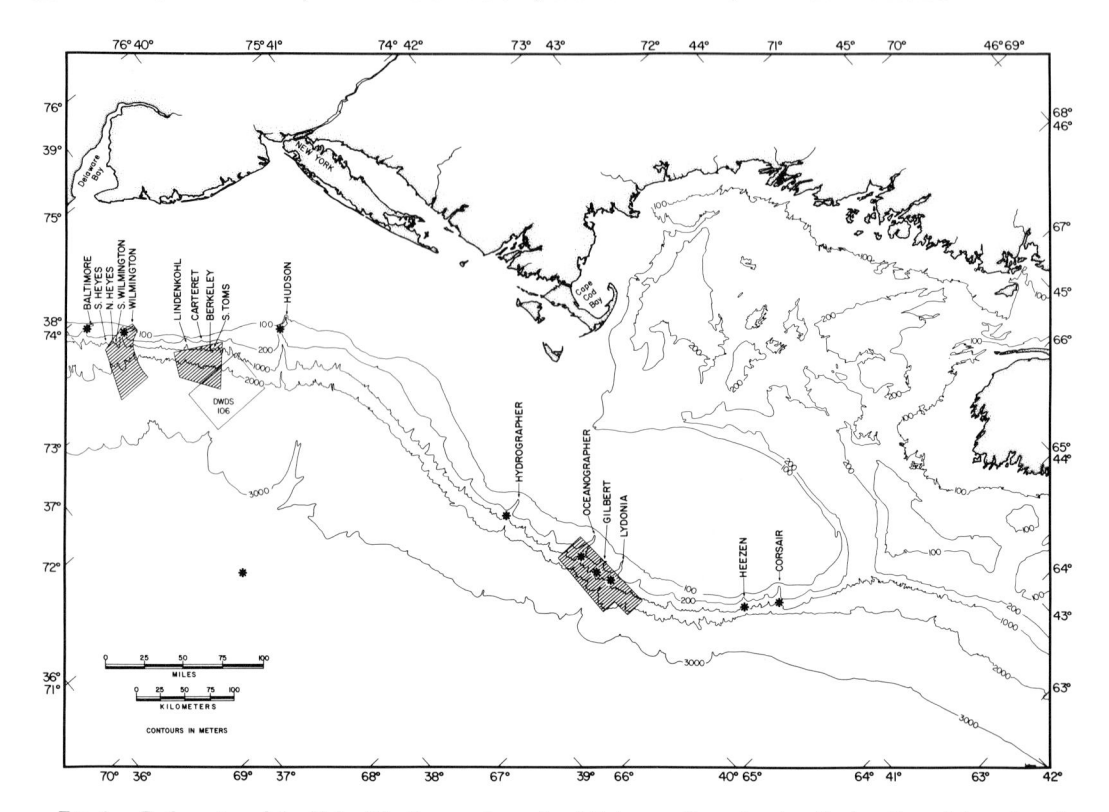

FIG. 1.—Bathymetry of the U.S. Atlantic margin north of Delaware Bay, showing the location of the submarine canyons cited in the text. Shaded areas were surveyed with Sea MARC I during the USGS L-DGO 1980 field program. Asterisks denote canyons visited by the authors or associates in submersible vessels.

The U.S. Middle Atlantic slope is draped mainly by silt- and clay-sized material with only thin laminae of sand (Doyle et al., 1979; Stanley et al., 1981). This drape ranges in thickness from nearly 0 to over 300 m (Robb et al., 1981; McGregor and Bennett, 1981). Beneath the drape lie subhorizontal, semi-lithified Tertiary paleoshelf strata which have been truncated by erosion (Schlee et al., 1976; Uchupi et al., 1977).

Besides the regional erosion surfaces, there are numerous localized incisions into the continental slope (Uchupi, 1968; McGregor, 1977). Many incisions begin at the middle (800–1500 m) and lower (1500–2100 m) slope depths. The largest of these are canyons (Carteret and Lindenkohl, for example) with their heads very near the shelfbreak.

The presence of both localized and regional erosional features suggests that it is unlikely that erosion of the continental slope is exclusively the result of sediment bypass from the shelf to the abyss. Much of our research has focused on the areas of the slope between the major submarine canyons. Data collected by new types of exploration tools indicate that pervasive mass wasting phenomenon exist in addition to processes by which shelf sediment bypasses the slope.

METHODS

Various Survey Programs

Our field program in 1980 utilized a new mid-range side-scan sonar. The field surveys extend from the outermost continental shelf across the continental slope to the uppermost continental rise. We explored three selected segments of the Middle Atlantic continental margin (Fig. 1) that had been targeted by the U.S. Department of the Interior for future lease sales to the oil and gas industry in 1981 and 1982.

The side-scan sonar data presented in this report were acquired jointly by the United States Geological Survey (USGS) and Lamont-Doherty Geological Observatory (L-DGO). The field program was part of a long-term study, originated and planned by the USGS, of slope erosional processes related to environmental aspects of petroleum exploration and development (Robb et al., 1981).

L-DGO collected visual observations and photographs of the seabed and bathymetry of parts of the Middle Atlantic continental slope between 1975 and 1980 during geological and biological assessment studies funded by the U.S. Environmental Protection Agency (Hanselman and Ryan, 1979) and the U.S. Bureau of Land Management (Hecker et al., 1980). Sampling and bathymetric mapping in Wilmington Canyon have been undertaken by the U.S. National Oceanic and Atmospheric Administration (Stubblefield et al., 1982).

Instruments and Techniques

International Submarine Technology, using specifications and assistance provided by L-DGO scientists and engineers, designed and constructed the side-scan sonar system called Sea MARC I. Side-scan sonar measures the variation in low-angle acoustic reflectivity of the sea floor. The backscattering energy is influenced by the attitude and orientation of the sea floor with respect to the survey vehicle and by the local acoustic impedance and roughness of the sediment/water interface on a scale from tens of meters to centimeters. The dark areas of the side-scan images illustrated in this paper represent high backscatter energy whereas pure white areas are acoustic shadows.

Sea MARC I, operating at frequencies of 27 and 30 kHz, provides sufficient resolution to define topographic features having dimensions of a few meters. Real-time correction of slant range to horizontal range and adjustment of paper advance proportional to vehicle speed results in sonar images suitable for the construction of plan view mosaics. A 4.5-kHz subbottom profiler having a beam width of approximately 40° is included in the sonar vehicle.

Figure 2 illustrates the underway operation of the Sea MARC I sonar vehicle and the associated side-scan and subbottom sonar records produced. A neutrally buoyant vehicle is towed between 100 and 400 meters above the sea floor. All data are transmitted up a 1.73 cm-diameter armored coaxial cable to the survey ship where several electro-sensitive dry paper graphic recorders produce the kind of grey-scale images that have been photographed for inclusion in this chapter. Typical towing speeds fluctuate from 3.3 to 4.6 km/hour depending upon wire length and weather conditions. The swath width of sea floor surveyed is selectable among 1, 2, and 5 km. The slope surveys were undertaken predominantly in the 5-km swath mode, and more than 400 km^2 of sea floor were insonified each survey day.

Navigation of the sonar vehicle is accomplished by a combination of ship positioning (Loran-C, Transit Satellite, etc.), the acoustically determined slant-range distance between the ship and the sonar vehicle, and measurements of depth, heading, and

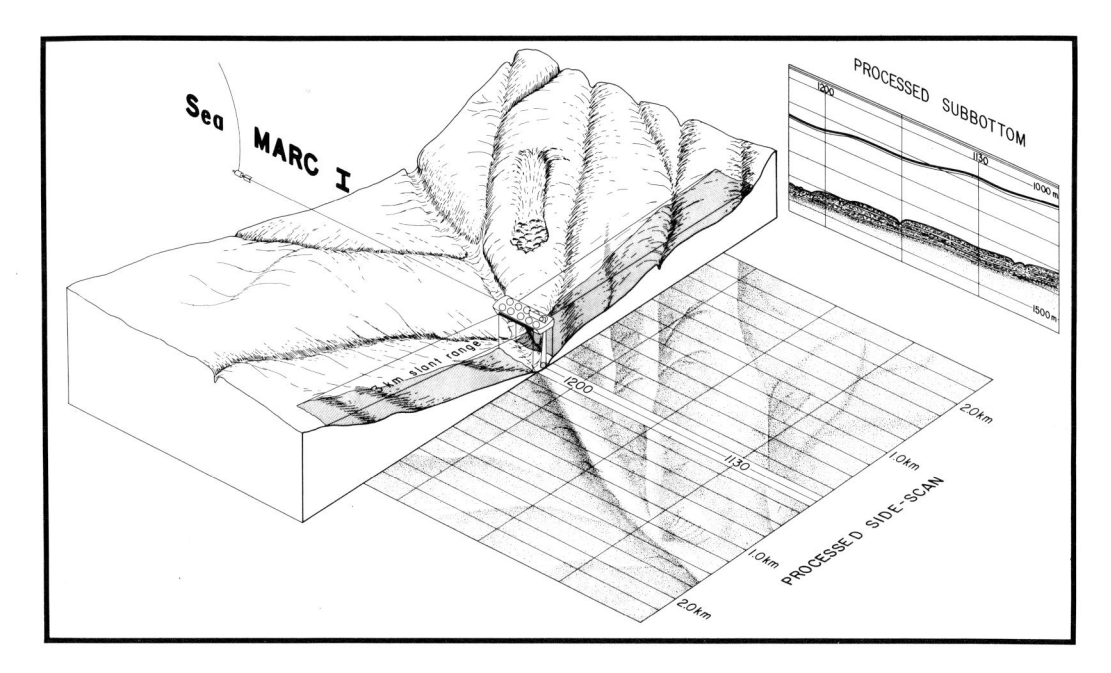

FIG. 2.—Diagram illustrating how side-scan sonar data discussed in text are collected. Sea MARC I is a neutrally buoyant sonar mapping vehicle towed within a few hundred meters of the seafloor. The side-scan sonar, operating at 27 and 30 kHz, insonifies a strip of the seafloor and the data are processed into an orthorectified, 5-km wide plan-view image. The down-looking 4.5-kHz sonar produces a subbottom reflection profile.

speed of the vehicle. The 5 km-swath images of Sea MARC I span the gap between the high-resolution DEEP TOWED INSTRUMENTATION SYSTEM (Spiess and Tyce, 1973) and the long-range GLORIA (Laughton, 1981).

Considerable "ground truth" visual observations and stratigraphic sampling have been obtained for the interpretation of the sonar images by using manned submersible vessels (DSRV ALVIN and NR-1) in Deep-Water Dumpsite 106 and in Corsair, Lydonia, Gilbert, Oceanographer, Hydrographer, Hudson, Wilmington and Baltimore canyons (Fig. 1). Data is obtained from various sources, some published (Heezen, 1975; Ryan et al., 1978; Rawson and Ryan, 1978; Hanselman and Ryan, 1979; Stubblefield et al., 1982; Heezen, unpublished data, 1975–1977; R. Freeman-Lynde and M. Rawson, personal communications in 1979; Ryan and Farre, in press).

Dives of submersible vessels have been concentrated in canyon thalwegs in order to take advantage of local outcrops of bedrock. Several dives in Deep Water Dumpsite 106 near the mouth of South Toms Canyon, explored erosional scars and debris aprons near the continental slope-rise boundary.

Data Presentations

In most of the side-scan surveys, vehicle tracks were parallel to each other and oblique to the regional contours and were spaced approximately 6 km apart. Plan-view side-scan mosaics have been made of each survey area. Bathymetric profiles have been computed by adding vehicle pressure depth (calibrated by acoustic ranging to the sea surface) to vehicle altitude.

The bathymetric contours of several canyons have incorporated the seabed observations and measurements as well as a network of surface-ship echo-sounding lines at an average track spacing of <1 km. A contour map of the Middle Atlantic slope between Lindenkohl and South Toms canyons was prepared from surface ship soundings alone (Robb et al., 1981). That map was modified using the side-scan images to interpolate features between echo sounding tracks (Kirby et al., in press). A further refinement awaits an additional side-scan data set collected by L-DGO which is proprietary and has not yet been published.

OBSERVATIONS

Bathymetry

The heavily dissected nature of the U.S. Middle Atlantic continental slope is depicted in the regional bathymetric maps published by USGS (Uchupi, 1970) and the U.S. National Oceanic and Atmospheric Administration, National Ocean Survey (1975–1978) and detailed maps by Fefe (*in* Malahoff et al., 1980), Bennett et al. (1978), Ryan et al. (1978), McGregor et al. (1979), Robb et al. (1981), and Kirby et al. (in press).

Canyons that indent the shelf, such as Oceanographer and Baltimore canyons, have side tributaries that are apparent by contours in Figure 3A and B. The continental slope has an overall seaward gradient that averages between 4° and 8°. Valleys on the slope have similar gradients along their thalwegs, but canyons that indent the shelfbreak have gradients as low as 2.5°.

Many valleys commence on the continental slope at depths well below the shelfbreak. The available detailed contouring of the slope indicates that such valleys are downslope trending chutes without obvious tributaries. Surface-ship echo-sounding, even with narrow beam transducers, often fails to accurately resolve the width and shape of valley floors on the slope.

The deployment of manned submersible vessels in shelf-indenting canyons (see Trumbull and McCamis, 1967; Stanley, 1974; Ryan et al., 1978; Cacchione et al., 1978; Stubblefield et al., 1982) has allowed the canyon floors to be explored. The observations from submersible vessels indicate that the central thalweg of a shelf-indenting canyon generally follows a curved to sinuous path down the slope and commonly contains a surficial carpet of rippled sand of shelf origin. Hanging valleys occur in Baltimore Canyon where the main valley is more deeply entrenched into the slope than the side tributaries. R. Freeman Lynde and M. Rawson explored the hanging valleys of Baltimore Canyon with the submarine NR-1 in 1979.

Stratigraphic Relationships

The gently seaward dipping sedimentary layers beneath the continental shelf continue out beneath the slope. These pre-Pleistocene strata have been truncated by erosion (Grow and Sheridan, 1976; Robb et al., 1981). Paleoshelf strata, as ancient as Early Cretaceous, outcrop at the base of locally steep walls inside shelf-indenting canyons (Ryan et al., 1978; Valentine et al., 1980). Former shelf strata as old as Eocene underly a thin Pleistocene veneer in inter-canyon settings on the lower slope (Stetson, 1949; Northrup and Heezen, 1956; Hollister et al., 1972; Robb et al., 1981). The Pleistocene veneer is generally conformable to the seafloor and is thickest (>100 m) on the upper slope (~200–800 m) as is shown for the Carteret Canyon area in Figure 4. Complex stratigraphic relationships in the canyons indicate that erosion and deposition have alternated with multiple phases of cut and fill (Ryan and Miller, 1981).

MORPHOLOGIES REVEALED BY SIDE-SCAN SONAR

Shelf-Indenting Canyon

Wilmington Canyon, the largest canyon covered in our side-scan surveys, indents the shelf approx-

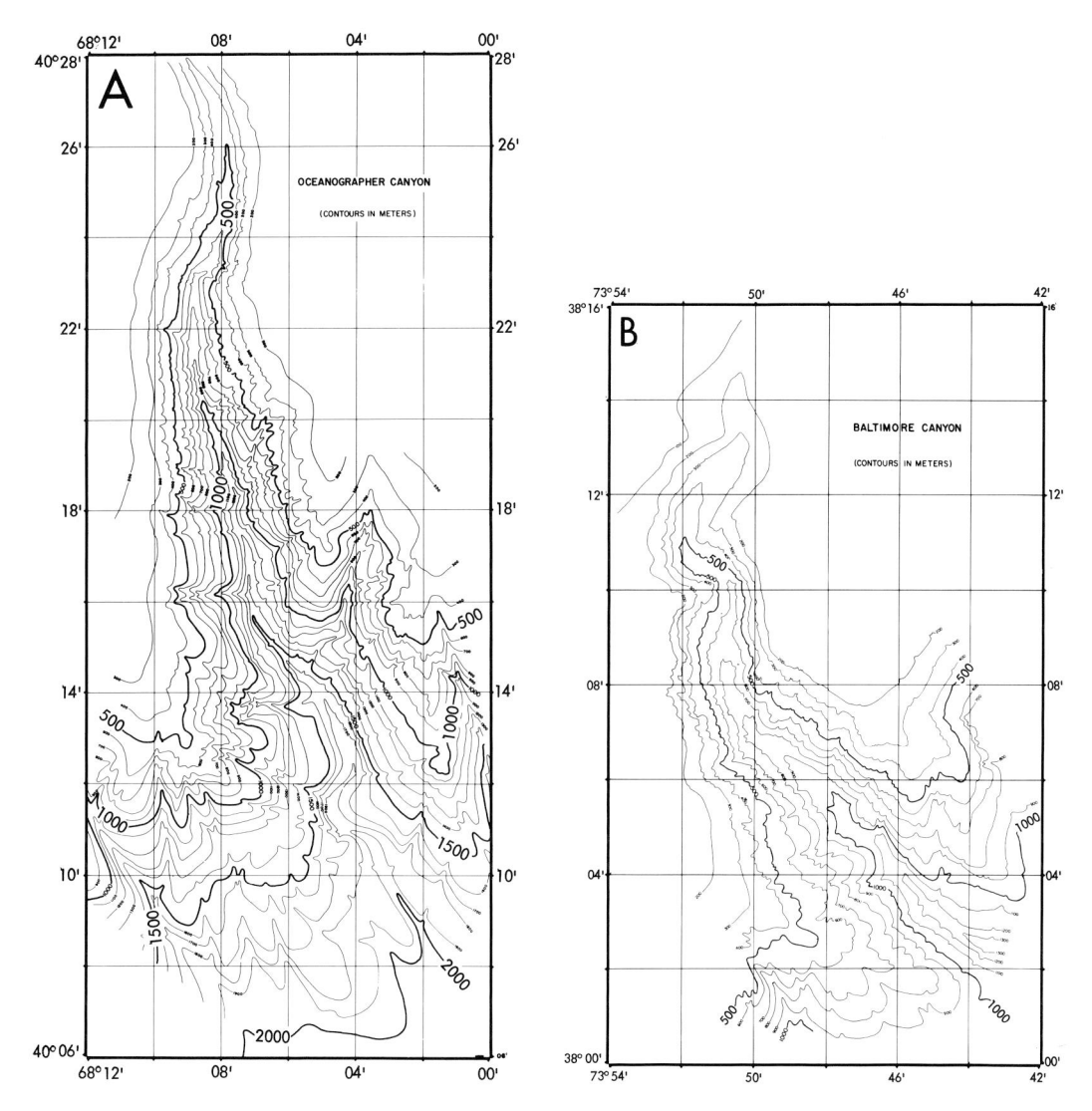

FIG. 3.—Bathymetric maps of Oceanographer (A) and Baltimore (B) canyons. Both these shelf-indenting canyons have a dendritic drainage pattern in which the central thalweg is deeply entrenched, especially at the shelfbreak, and follows curved paths down the continental slope. Tracks of surface ship surveys used to construct these maps are less than 1 km apart. Contour interval is 100 m. From Hecker et al. (1980).

imately 19 km and follows a curved to meandering path across the slope and rise. Figure 5 shows the sinuous path of Wilmington Canyon between approximately 1600 and 1800 m that is remarkably similar to the paths of entrenched meandering rivers on land. Here the thalweg is approximately 200 m wide and has incised down about 200 m into the slope, as shown by the contours published by McGregor et al. (1982) and modified by Stubblefield et al. (1982). The steep canyon walls are cut by short (~200 m length) side gullies. Observations from the DSRV ALVIN (Stubblefield et al., 1982) show that the meanders in Wilmington Canyon have steep outer walls, while the inner walls have gentler slopes.

In Wilmington Canyon there are few segments where the thalweg is directed straight downslope. In some places, the walls are cut by crescent-shaped niches or scallops concave towards the thalweg (Fig. 6). Adjacent to some scallops are lobate-shaped debris aprons (~10,000 m^2 in area) on the channel floor. The uniformly high reflectivity of the channel floor indicates that it is a reasonably smooth surface having a superimposed small-scale rough-

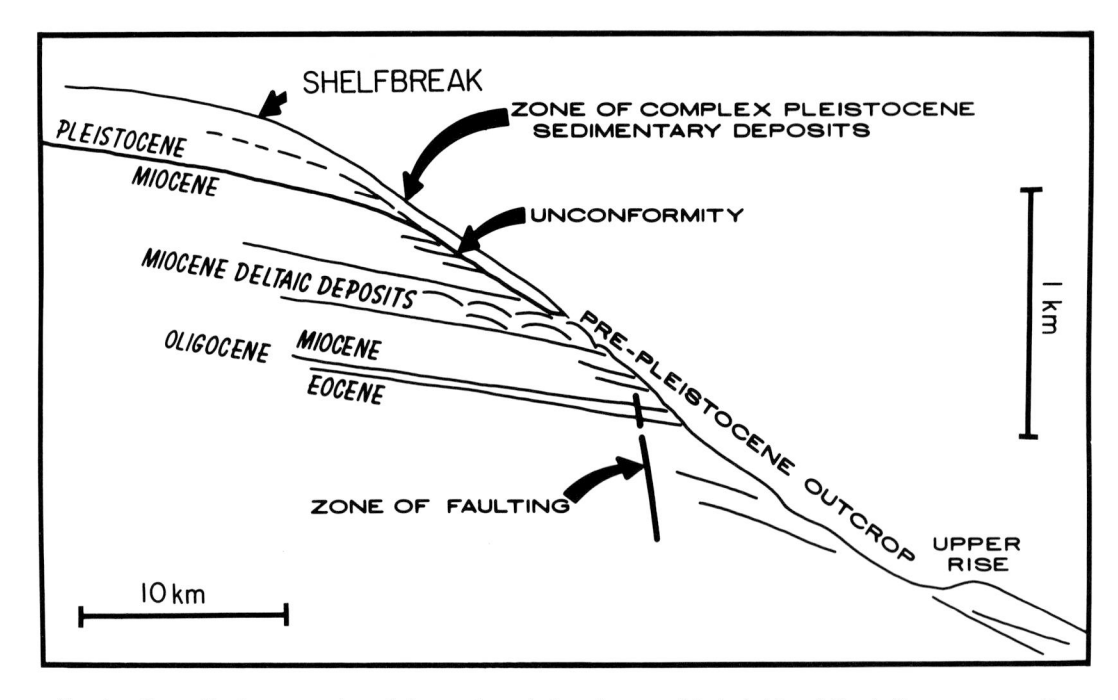

FIG. 4.—Generalized cross section of the continental slope between Lindenkohl and South Toms canyons. From Robb et al. (1981).

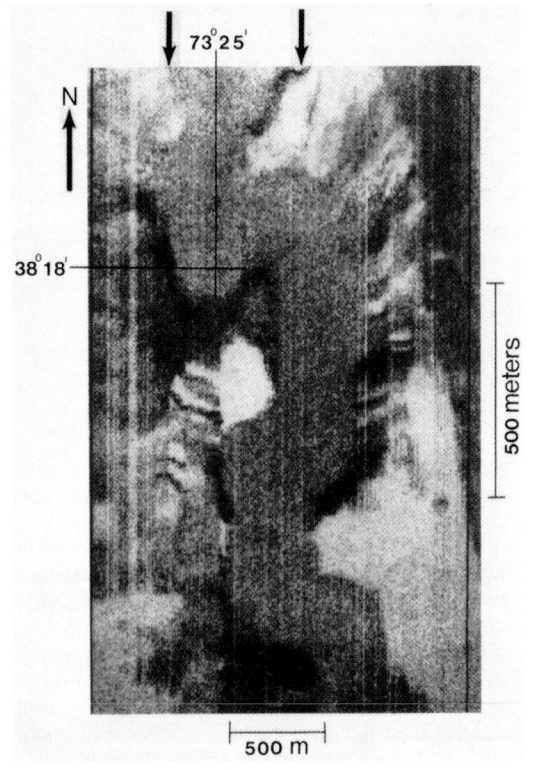

ness, possibly representing the extensively bioturbated surface seen from the submersible vessel (Stubblefield et al., 1982).

Slope Canyons

Pleistocene sediment is thickly draped above the erosion surface in the Wilmington Canyon area (McGregor and Bennett, 1981). Here, much of the slope is dissected by valleys that run the entire length of the slope. South Wilmington, North Heyes, and South Heyes are examples of slope canyons that follow relatively direct paths down the slope and rise but do not appreciably indent the shelfbreak (McGregor et al., 1982). Canyons in the Carteret Canyon area that fit this description include Berkely, Carteret, South Toms, and Lindenkohl canyons.

The upper reaches of these slope canyons have a drainage pattern in which the main valley receives many closely spaced tributaries (~250 m separa-

FIG. 5.—Unprocessed side-scan sonar image of Wilmington Canyon between approximately 1600 and 1800 m on the continental slope. Sand, derived from the shelf, is believed to be the principal agent cutting the observed meanders. Downslope is towards the bottom of the figure. Heavy double arrows denote the direction of incoming sound.

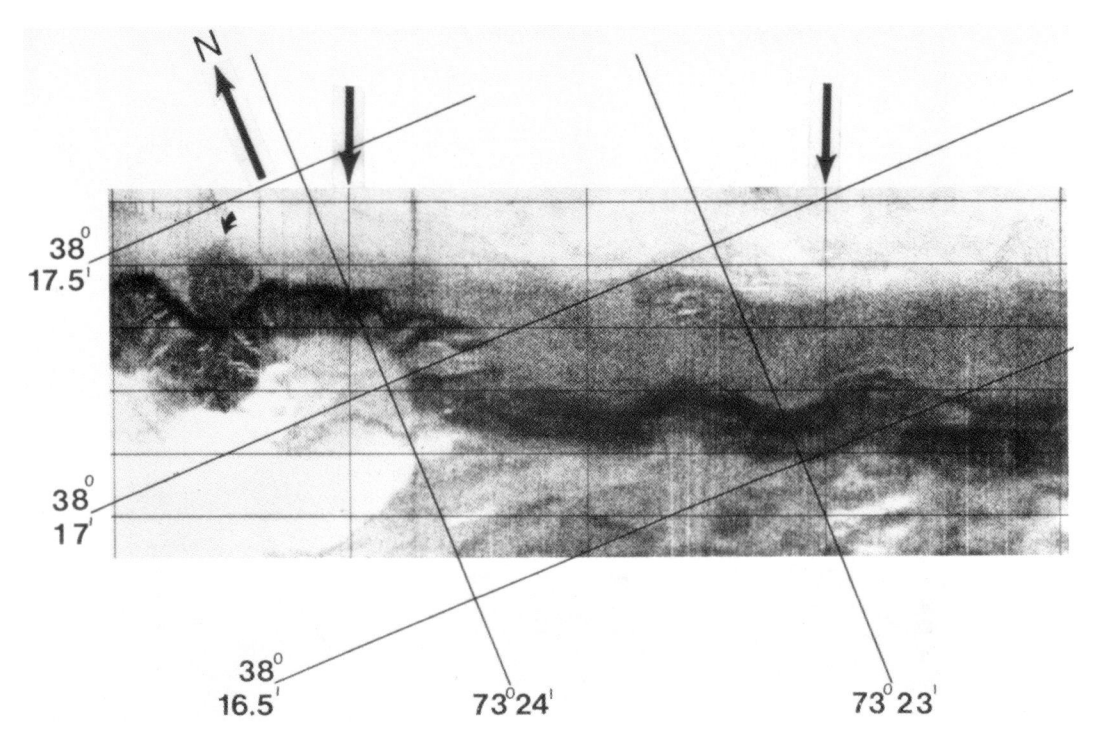

FIG. 6.—Processed side-scan sonar image of the southern wall of Wilmington Canyon at approximately 1900 m (center of figure) on the lower continental slope. Small crescent-shaped scars apparently result from slumping of the canyon walls. The curved arrow points to a debris-flow deposit that has not yet been removed, smoothed, and/or buried by the canyon bedload. Downslope is towards the right of the figure. Heavy double arrows denote the direction of incoming sound.

tion) that join it at acute angles (67° ± 15°). This drainage pattern resembles a fern leaf in plan view. The heads of the second order tributaries commonly begin at the crest of knife-edged, steep-sided spurs that form the divides between the drainage systems of adjacent submarine canyons of the upper and middle continental slope (Fig. 7). Such spurs have also been observed on the northern margin of the Bay of Biscay and on the continental slope between southwest Ireland and Spain, by using the GLORIA side-scan system (Belderson and Kenyon, 1976; Kenyon et al., 1978). Secondary tributaries also occur along the middle and upper reaches of Lindenkohl Canyon. They are present, but are not as fully developed, on the middle and upper reaches of Berkely, Carteret and South Toms canyons.

On the lower slope, the main thalwegs of slope canyons widen. Where present, the side gullies have a more orthogonal intersection with the main valley and do not always extend up to the crest of slope spurs (Fig. 8).

The thalweg of South Heyes Canyon contains downslope-trending lineations (Fig. 8). If the lineations have relief, then it cannot be resolved with our subbottom profiler, which has a rather large acoustic footprint at normal operating altitudes. Some of the lineations seem to emanate from the mouths of second order tributaries.

Amphitheatre-Shaped Depressions

Sonar images generated by Sea MARC I on the middle and lower continental slope near Carteret Canyon, a region where the Pleistocene drape is thin to nonexistent, show large amphitheatre-shaped depressions that open downslope into chutes previously mapped as slope valleys (Fig. 9). The depressions have steep (up to vertical) bounding walls. Some of the headwall scarps have a relief of more than 200 m. Large cracklike features several hundred meters in length radiate laterally and up-slope from the headwalls of many of the depressions into the thin sediment drape of the slope. A third of the amphitheatre shown in the lower part of Figure 9 is in shadow. This feature has been insonified from a different perspective as shown in Figure 10. This alternate view reveals that several progressively deeper, subparallel surfaces, separated by steep walls, make up the depression floor.

Amphitheatre-shaped depressions are common on the lower and middle slope in the Carteret Can-

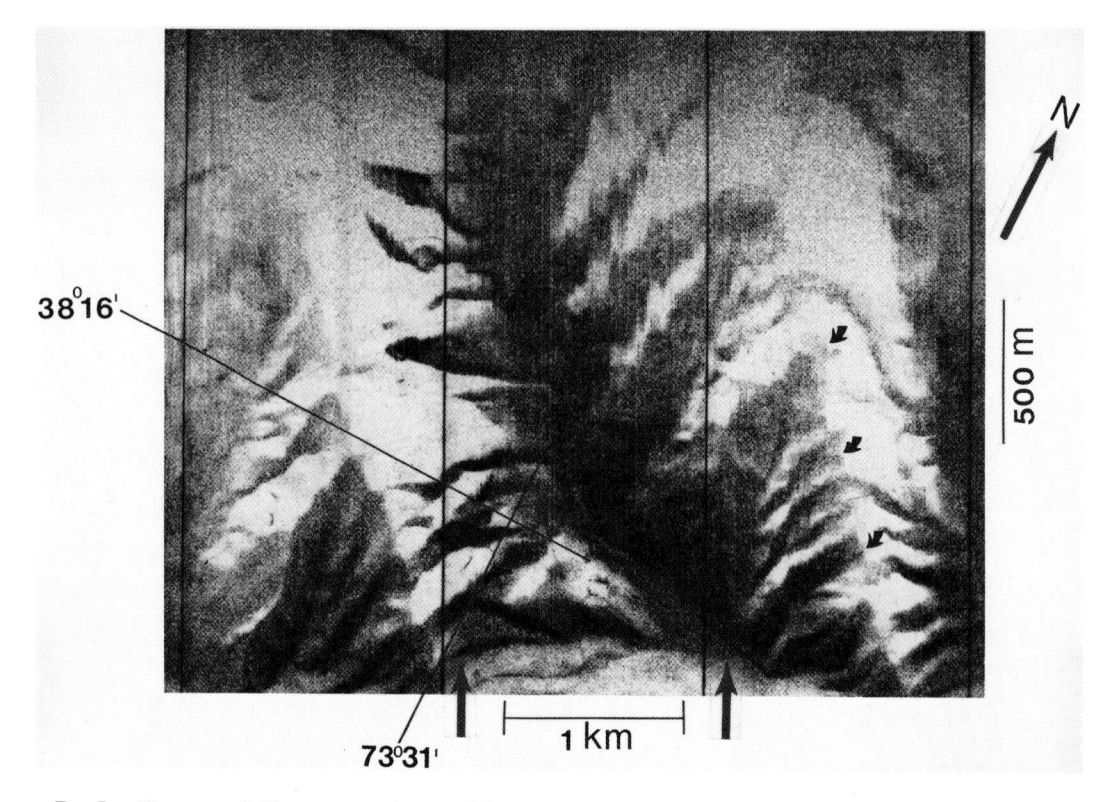

FIG. 7.—Unprocessed side-scan sonar image of the upper reaches of South Wilmington Canyon. A pinnate drainage system is apparent. South Wilmington's drainage network is separated from an unnamed slope canyon to the northeast by the knife-edged spur in the vicinity of the curved arrows. Water depth is approximately 1300 m at the deepest part of the canyon axis. Relief of the canyon walls is approximately 200 m. The heavy double arrows denote the direction of incoming sound.

yon area. Some scars appear fresh and young, and have high reflectivity on their floors compared with that of the adjacent area. Others have outlines that are more diffuse, and have more subtle changes in reflectivity, suggesting that they are older features that have been partially buried by hemipelagic sedimentation.

DISCUSSION

The variety in morphology present on the Middle Atlantic continental margin implies that several processes and variables are involved in slope evolution. In some places, we see possible morphological analogues to the submarine features on land.

Amphitheatre-shaped depressions on the middle and lower slope near Carteret Canyon have similarities in appearance to snow-slab failures in mountainous terrain. The steep headwall in the tensile region, the striated basal fracture plane, and the common crevasses on the adjacent snow sheet all have counterparts in the submarine environment. Through a variety of mechanisms (Perla and La Chapelle, 1970), a horizon of relatively low shear strength forms in the snow and ice unit. This horizon, which may be within granular ice, becomes the detachment surface upon which slabs of snow and ice slide down the slope.

On the continental slope, the stripping of materials is commonly observed to be along surfaces that continue as subbottom reflections (i.e. bedding planes) beneath the adjacent undisturbed area (Summerhayes et al., 1979). The association of strong reflectors with the basal slip surface suggests that detachment takes place at the contact between different lithologies or between layers having contrasting induration. The steplike floor of the scar shown in Figure 10 may represent the outcrops of several bedding planes that have acted as detachment surfaces.

Amphitheatre-shaped scars have been found on the middle and lower slope of the Carteret Canyon area where semilithified Tertiary rocks are very close to the seafloor. Possibly, a structural fabric present in these materials controls the failure of older strata in this area. The transport of coherent blocks to deeper environments results in the formation of

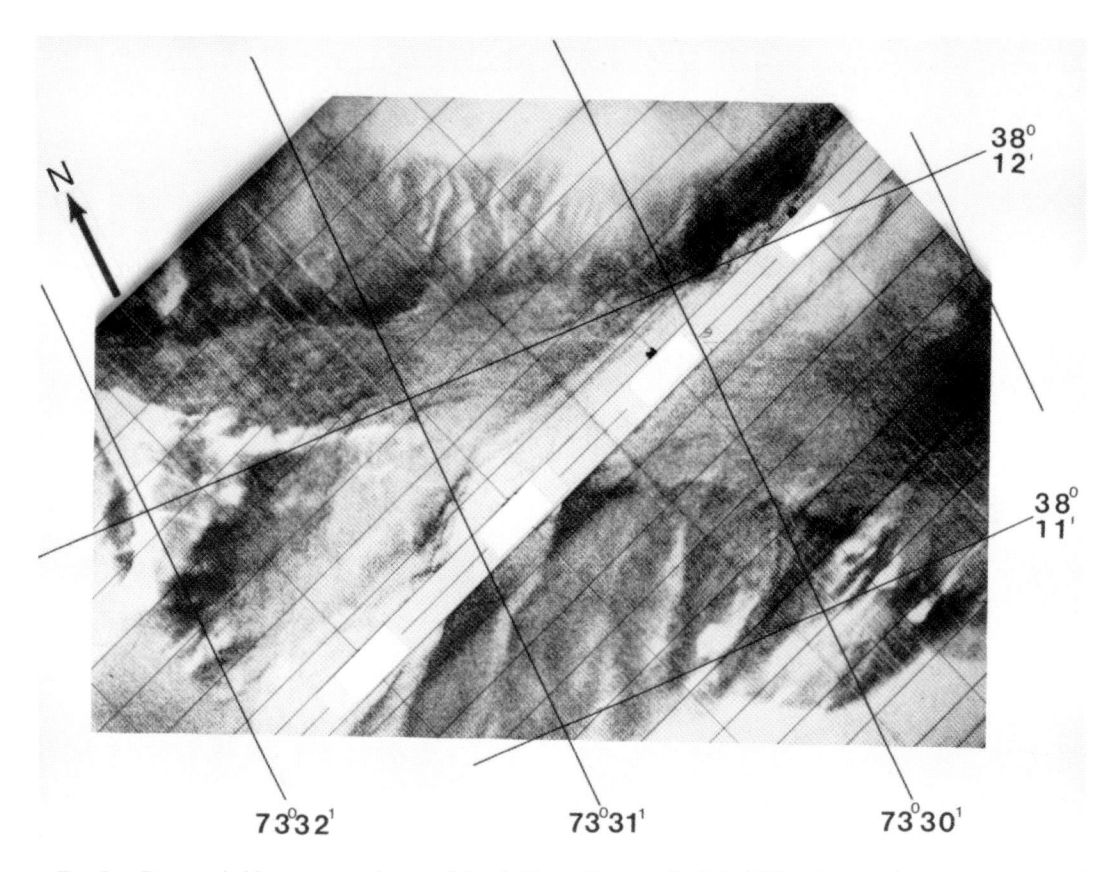

FIG. 8.—Processed side-scan sonar image of South Heyes Canyon; depth is 1500 m in area shown at the center of the figure. The downslope-trending lineations, resembling lateral moraines in glaciated valleys, suggest that creep may be an important sediment transport process in slope canyons. Alternatively, the lineations may be slide marks produced during mass sediment flows. Downslope is toward the right of the figure. The sound source was towed through the area shown as a blank strip passing through the center of the figure.

olistostromes (Kelling and Stanley, 1976).

We see a range in the freshness of these scars on the slope near Carteret Canyon, which implies that erosion is intermittent and that conditions suitable for failure may come and pass. We think, however, that the headwall of a scar should be the preferred location for the next failure (either large or small scale) because of the steepness of the bounding wall.

On land, the fernlike drainage patterns form on relatively steep slopes of homogeneous substrate, and this morphology is referred to as "pinnate drainage" (D. O'Leary, personal comm., 1981). The upper and middle continental slope having such pinnate drainage has a general dip of about 8° and is covered by fine-grained sediment.

Probably several processes are responsible for the pinnate drainage pattern. Biological borings of erosional escarpments and canyon walls (Dillon and Zimmerman, 1970; Warme et al., 1978; Malahoff et al., in press) and possible groundwater sapping (Johnson, 1939) could lead to weakening and undermining of cliffs, which eventually would collapse into talus. This type of biological, chemical, and/or mechanical defacement proceeds perpendicularly to the local contours. Like failures in terrestrial drainages or snow-covered terrains, the failure advances upslope (i.e. is directed headward). Upslope retreat of erosional surfaces by mass wasting is a retrogressive process that might propagate for a considerable distance in a single event or might be intermittent with long periods of gradual weathering. The fern pattern signifies that the erosion is not directed by structural lineaments but instead is guided by the shape of the local slope and takes place in a physically homogeneous cover. The knife-edged spurs that separate the drainage systems of adjacent canyons indicate that erosion is widespread and has affected the entire upper slope between the canyon thalwegs (Fig. 7).

The knife-edged spurs are especially significant because they show that erosion is not caused by the

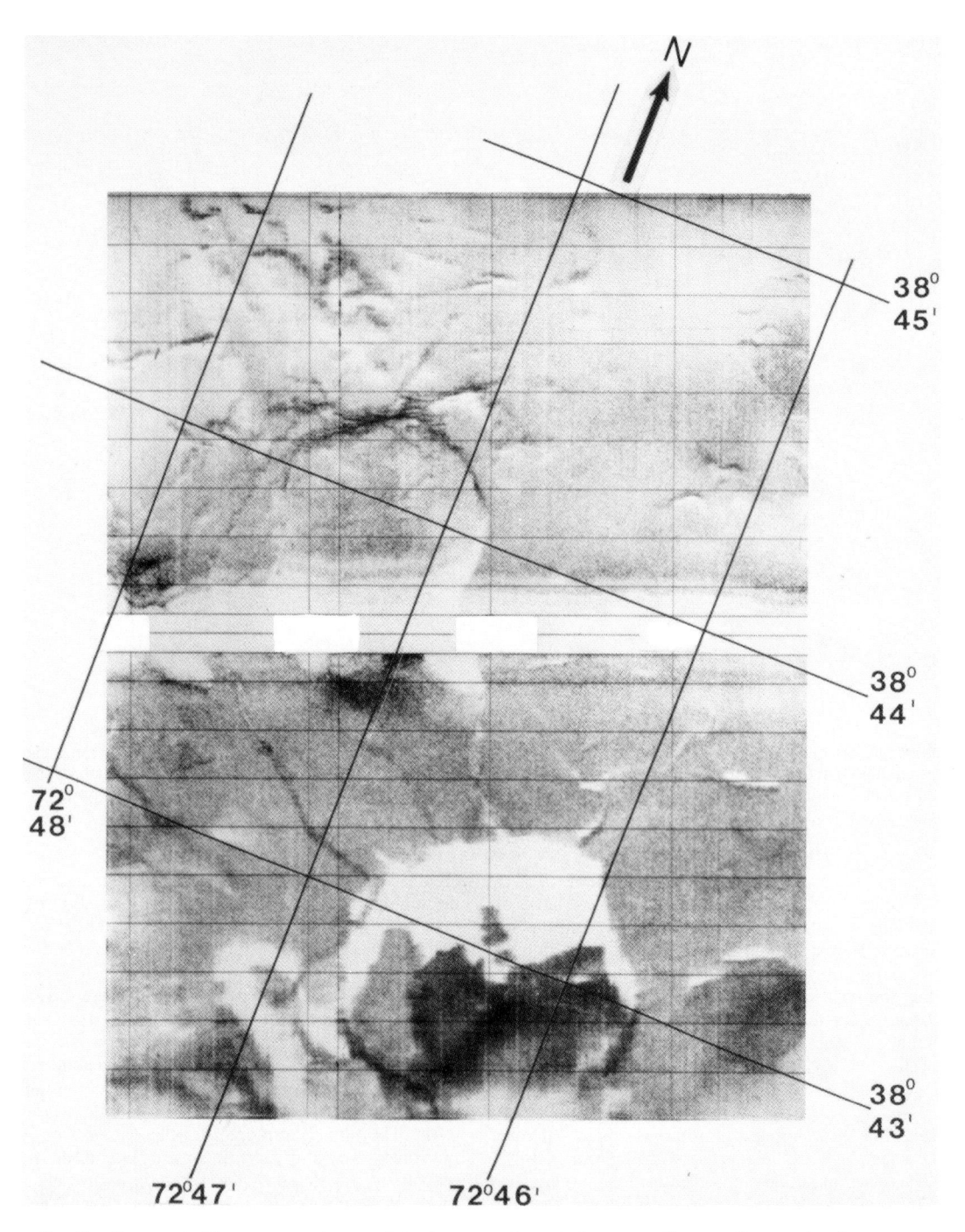

FIG. 9.—Processed side-scan sonar image of the lower continental slope near Carteret Canyon showing two amphitheatre-shaped scars. Note the high reflectivity of the floor and the cracklike features radiating from the headwall of the lower scar. Water depth is approximately 2050 m in the area shown at the center of the figure. Sound source was towed through the area shown as a blank strip passing through the center of the figure.

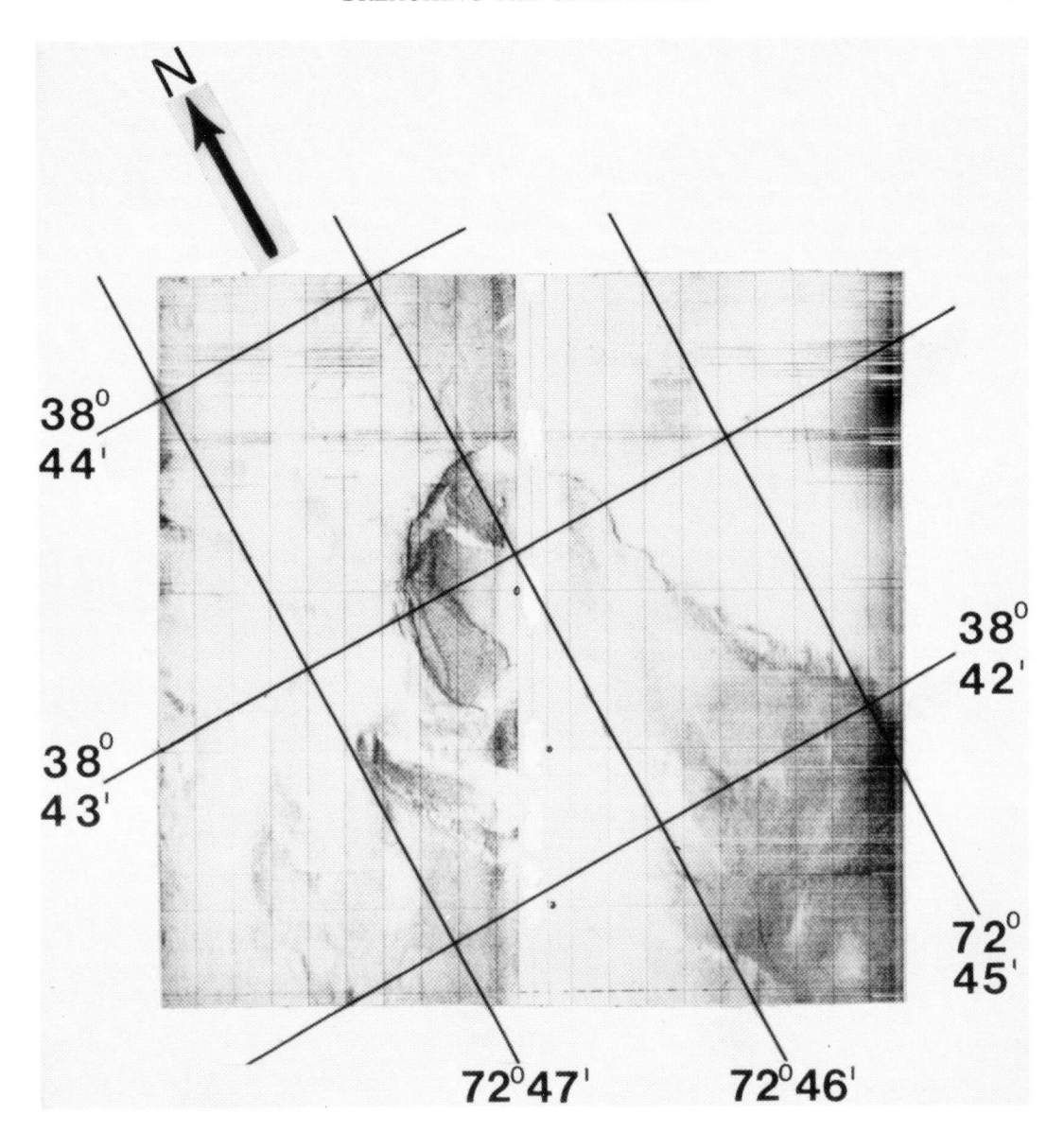

FIG. 10.—Processed side-scan sonar image showing a different view of the scar shown in the lower part of Fig. 9. The floor of the scar consists of several subparallel surfaces, offset by steep walls. These may be bedding planes that have acted as detachment surfaces. Sound source was towed through the area shown as a blank strip passing through the center of the figure.

introduction of materials from a point source (i.e. shelf sands entering a canyon head) or by the abrasive power of sediments crossing the slope. The observed pattern of erosion can be explained as the scars where sediment has been removed by mass failure.

The canyon thalweg is the route for the eroded slope materials that are redeposited at greater depths. The high reflectivity of the thalwegs is produced by a small scale roughness whose most likely cause is chaotic debris. The lineations on the floor of South Heyes Canyon could be the result of one or several sediment transport processes. Creeping of the canyon fill could result in the construction of lateral moraines similar to those in glaciated valleys. Alternatively, the lineations may be slide marks (low relief, parallel grooves or striations) similar to those produced on channel floors during subaqueous debris flow experiments (Hampton, 1972).

The thalwegs of shelf-indenting canyons are deeply entrenched, especially on the upper slope and near the shelfbreak. The occurrence of hanging side-tributaries in Baltimore Canyon indicates that the main valley has entrenched at a rate greater than that of the side-tributaries. The crescent-shaped scars and debris aprons on the floor of Wilmington Canyon may attest to undermining and oversteepening of canyon walls by erosive currents traveling through the thalweg. Downcanyon-traveling currents may have a tendency to cut meanders in the canyon walls just as terrestrial rivers do. The crescent-shaped scars in the walls of the canyon may become smoothed and enlarged by the currents into meanders.

WORKING HYPOTHESIS

We have summarized the variety of morphology present near the shelfbreak and on the U.S. Middle Atlantic continental slope. In the process of integrating these data into a workable model, we have constructed a flow chart (Fig. 11) that places the different morphologies into a preliminary scheme of evolution of canyons on the slope.

In this model, localized slope failure begins the process of canyon formation. By a variety of mechanisms, the erosional scars grow headward. Because of changing physical or environmental conditions, erosion may cease and normal hemipelagic sedimentation may prevail. Side-tributaries form by the local failure of steep walls. While the valley is confined to the slope, no significant slope bypassing processes can operate; that is, the amount of sediment having traveled through the valley is roughly equal to the amount of sediment that formerly occupied the valley and tributaries.

Through continued headward erosion, the slope valley may eventually breach the shelfbreak. Then, a new mechanism for canyon erosion becomes possible. Sand transported along the continental shelf, especially during times of low sealevel, can more easily enter the canyon heads. Either in a steady state or through catastrophic processes which flush the canyon head, the canyon becomes the route for shelf materials to the continental rise and abyssal plain. The erosion associated with slope bypassing processes is concentrated in the main thalweg, thereby leaving the side tributaries hanging above. Continued indentation into the shelf will be intensified if land-based drainage systems are captured

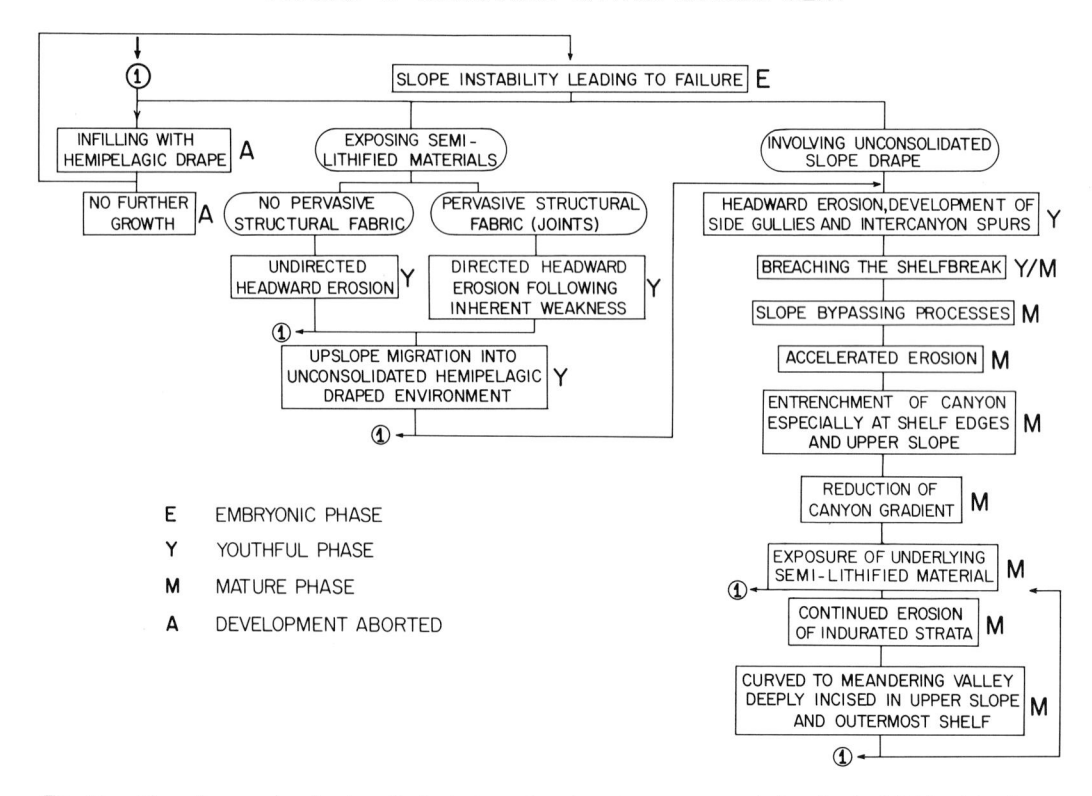

FIG. 11.—Flow diagram showing hypothetical phases in submarine canyon evolution for the Middle Atlantic continental margin.

during low stands of sealevel. Shelfedge deltas formed during the regressive phase provide additional sources of sand to accelerate canyon-head entrenchment.

We wish to emphasize that, because of changing conditions, erosion can stop at any time, and canyon morphologies can be subsequently buried beneath sediment cover. During the Holocene transgression, for example, the upper reaches of several canyons on the U.S. Atlantic shelf were infilled (Twichell et al., 1977). Episodes of cut and fill explain both the fresh and subdued erosional scars seen in many of the slope canyons surveyed in our field program.

CONCLUSIONS

A study of mid-range side-scan sonar images indicates that the embryonic phase of dissection on the continental margin begins on the continental slope. The predominant mass-wasting process appears to be slumping and sliding. During this phase, material is delivered to the rise by debris flows. The erosional regime apparently can abort and normal sedimentation can return.

In the youthful phase, as the unconsolidated cover is removed, a pinnate drainage network appears. Once the tributaries advance headward and meet at knife-edged spurs, the entire open slope becomes available to erosion. As more strata are exhumed, the wasting process involves indurated materials, which fissure and collapse in slides and avalanches. The observed amphitheatre-shaped depressions are the headwall scars of rock slides. Lower slope gullies are avalanche chutes excavated into older formations by the debris flows activated at the time of mass-wasting.

A mature phase is reached once upslope-directed erosion breaches the shelfbreak, and the canyon head can tap a reservoir of shelf sands. In the mature phase, the canyon head indents the shelfedge, the canyon gradient decreases, and the axial thalweg entrenches deeply into the upper slope. Density currents enroute from the shelf to deeper environments use the mature canyons as transport routes. This bypassing tends to cut and smooth canyon walls to a curving or meandering path.

According to the above model, the shelfbreak, marking the boundary between the shelf and slope, plays a major role in influencing the ultimate morphology of the mature submarine canyon system. The morphology of the shelfbreak, in turn, becomes modified by the submarine canyon in a complex inter-relationship.

ACKNOWLEDGMENTS

Special thanks to Jack Grimm and members of TITANIC '80, Inc., who funded the design, construction and initial field testing of Sea MARC I. Technical assistance from D. Chayes, J. Kosalos, L. Robinson, I. Bjorkhein, K. Scanlon, C. Parmenter, J. Dodd, L. Goodman, K. Parolski, I. Bitte, A. Brosius, J. DiBernardo, A. Hagan, D. Edwards, M. A. Luckman and J. Wollin is appreciated. Appreciation to the officers and crew of the R. V. GYRE who made the cruise both successful and enjoyable. Finally, we thank T. Aldrich and R. Tirey, U.S.G.S. at Woods Hole, Mass. who offered effective logistical support.

Funding for the fieldwork for this study was provided by the U.S. Bureau of Land Management to the U.S. Geological Survey under Memorandum of Understanding AA 851-MUO-18 and Interagency Agreement AA 851-IA1-17. Funding for technical assistance was provided by the U.S. Geological Survey to Lamont-Doherty Geological Observatory through award OCE-77-23274B from the National Science Foundation. The first author received support in the form of a fellowship from the Phillips Petroleum Company. Some of the ideas presented here were developed during discussions with D. Twichell and D. O'Leary U.S.G.S.; they, and D. J. Stanley, critically read the manuscript resulting in many improvements.

This is L-DGO Contribution No. 3438.

REFERENCES

BELDERSON, R. H., AND KENYON, N. H., 1976, Long-range sonar views of submarine canyons: Mar. Geology, v. 22, p. M69–M74.
BENNETT, R. H., LAMBERT, D. N., MCGREGOR, B. A., FORDE, E. B., AND MERRILL, G. F., 1978, Slope map: A major submarine slide on the U.S. Atlantic continental slope east of Cape May: U.S. Dept. Commerce, NOAA, A-5787, USCOMM-NOAA-DC [chart].
CACCHIONE, D. A., ROWE, G. T., AND MALAHOFF, A., 1978, Submersible investigations of outer Hudson Submarine Canyon, in Stanley, D. J., and Kelling, G., eds., Sedimentation in Submarine Canyons, Fans and Trenches: Stroudsburg, Pa., Dowden, Hutchinson and Ross, p. 42–50.
DALY, R. A., 1936, Origin of submarine canyons: Am. Jour. Sci., v. 31, p. 401–420.
DILL, R. F., 1964, Contemporary submarine erosion in Scripps Submarine Canyon [Ph.D. thesis]: La Jolla, Calif., Univ. California, Scripps Inst. Oceanogr., 248 p.
DILLON, W. P., AND ZIMMERMAN, H. B., 1970, Erosion by biological activity in two New England submarine canyons: Jour. Sed. Petrology, v. 40, p. 524–547.
DOYLE, L. J., PILKEY, O. H., AND WOO, C. C., 1979, Sedimentation on the eastern United States continental slope, in Doyle, L. J., and Pilkey, O. H., eds., Geology of Continental Slopes: Soc. Econ. Paleontologists Mineralogists Spec. Paper 27, p. 119–129.

EMERY, K. O., 1968, Relict sediments on continental shelves of the world: Am. Assoc. Petroleum Geologists Bull., v. 52, p. 445–464.

ERICSON, D., EWING, M., WOLLIN, G., AND HEEZEN, B. C., 1961, Atlantic deep-sea sediment cores: Geol. Soc. America Bull., v. 72, p. 193–286.

GROW, J. A., AND SHERIDAN, R. E., 1976, High velocity sedimentary horizons beneath the outer continental shelf off New Jersey: EOS, v. 57, 265 p.

HAMPTON, M. A., 1972, The role of subaqueous debris flow in generating turbidity currents: Jour. Sed. Petrology, v. 42, p. 775–793.

HANSELMAN, D. H., AND RYAN, W. B. F., 1979, 1978 Atlantic 3800 meter radioactive waste disposal site survey—Sedimentary, micromorphologic and geophysical analysis: Washington, D.C., Environ. Protection Agency Tech. Rept., p. 1–41.

HECKER, B., BLECHSCHMIDT, G., AND GIBSON, P., 1980, Epifaunal zonation and community structure in 3 mid- and north-Atlantic canyons: Canyon Assessment Study Final Rept., BLM contract AA551CT849, Appendix F, 73 p.

HEEZEN, B. C., 1975 Photographic reconnaissance of continental slope and upper continental rise: May 1974 baseline investigation of deep-water dumpsite 104: NOAA Dumpsite Evaluation Rept. 75-1, p. 27–104.

———, AND EWING, M., 1957, Turbidity currents and submarine slumps and the Grand Banks earthquakes: Am. Jour. Sci., v. 250, p. 849–873.

———, THARP, M., AND EWING, M., 1959, The Floors of the Oceans. 1. The North Atlantic: Geol. Soc. America Spec. Paper 65, 122 p.

HOLLISTER, C. D., EWING, J. I., HABIB, J., HATHAWAY, J. C., LANCELOT, Y., LUTERBACHER, H., PAULUS, F. J., POUG, C. W., WILCON, J. A., AND WORSTELL, P., 1972, Site 107 upper continental rise and Site 108 continental slope: Initial Reports of the Deep Sea Drilling Project, v. 11, p. 351–363.

JOHNSON, D. W., 1939, The Origin of Submarine Canyons, a Critical Review of Hypothesis: New York, Columbia Univ. Press, 126 p.

KELLING, G., AND STANLEY, D. J., 1970, Morphology and structure of Wilmington and Baltimore submarine canyons, eastern U.S.: Jour. Geology, v. 78, p. 637–660.

——— AND ———, 1976, Sedimentation in canyon, slope and base of slope environments, in Stanley, D. J., and Swift, D. J. P., eds., Marine Sediment Transport and Environmental Management: New York, Wiley-Interscience, p. 379–435.

KENYON, N. H., BELDERSON, R. H., AND STRIDE, A. H., 1978, Channels, canyons and slump folds on the continental slope between south-west Ireland and Spain: Oceanol. Acta, v. 1, p. 369–380.

KIRBY, J. R., ROBB, J. M., AND HAMPSON, J. C., in press, Detailed bathymetry of the U.S. continental slope between Lindenkohl Canyon and South Toms Canyon, offshore middle Atlantic U.S.: U.S. Geol. Survey Misc. Field Inv. Map MF 1443.

KNEBEL, H. J., WOOD, S. A., AND SPIKER, E., 1979, Hudson River: Evidence for extensive migration on the exposed continental shelf during the Pleistocene: Geology, v. 7, p. 254–258.

LAUGHTON, A. S., 1981, The first decade of GLORIA: Jour. Geophys. Res., v. 86, B12, p. 11, 511–11, 534.

MALAHOFF, A., EMBLEY, R. W., PERRY, R., AND FEFE, C., 1980, Submarine mass-wasting of sediments on the continental slope and upper rise south of Baltimore Canyon: Earth and Planetary Sci. Letters, v. 49, p. 1–7.

———, ———, AND FORNARI, D., 1982 in press, Geomorphology of Norfolk and Washington canyons and the surrounding continental slope and upper rise as observed from DSRV ALVIN, in Scrutton, R., ed., Bruce Heezen Memorial Volume.

MCGREGOR, B. A., 1977, Geophysical assessment of submarine slide northeast of Wilmington Canyon: Mar. Geotechnology, v. 2, p. 229–243.

———, AND BENNETT, R. H., 1981, Sediment failure and sedimentary framework of the Wilmington Geotechnical Corridor, U.S. Atlantic margin: Sed. Geology, v. 30, p. 213–234.

———, ———, AND LAMBERT, D. N., 1979, Bottom processes, morphology, and geotechnical properties of the continental slope south of Baltimore Canyon: Applied Ocean Res., v. 1, p. 177–187.

———, STUBBLEFIELD, W. L., RYAN W. B. F., AND TWICHELL, D. C., 1982, Wilmington Submarine Canyon: A marine fluvial-like system: Geology, v. 10, p. 27–30.

MCMASTER, R. L., AND ASHRAF, A., 1973, Extent and formation of deeply buried channels on the continental shelf off southern New England: Jour. Geology, v. 81, p. 374–379.

MIDDLETON, G. V., AND HAMPTON, M. A., 1973, Sediment gravity flows: Mechanisms of flow and deposition, in Middleton, G. V., and Bouma, A. H., eds., Turbidites and Deep Water Sedimentation: Soc. Econ. Paleontologists Mineralogists Pacific Sect., Short Course, Anaheim, p. 1–38.

NATIONAL OCEAN SURVEY BATHYMETRIC MAPS, 1975–1978, Wilmington Canyon, NJ 18-6, Hudson Canyon, NJ 483, Atlantis Canyon, NJ 19-1, Hydrographer Canyon, NK 19-11.

NORTHRUP, J., AND HEEZEN, B. C., 1956, An outcrop of Eocene sediment on the continental slope: Jour. Geology, v. 59, p. 396–399.

PERLA, R. I., AND LA CHAPELLE, E. R., 1970, A theory of snow slab failure: Jour. Geophys. Res., v. 75, p. 7619–7627.

RAWSON, M. D., AND RYAN, W. B. F., 1978, Geological observation of Deepwater Radioactive Waste Dumpsite 106: Washington, D.C., Environ. Protection Agency Tech. Rept., p. 1–79.

ROBB, J. M., HAMPSON, J. C., KIRBY, J. R., AND TWICHELL, D. C., 1981, Geology and potential hazards of the continental slope between Lindenkohl and South Toms canyons, offshore mid-Atlantic states: U.S. Geol. Survey Open-file Rept. 81-600, p. 1–33.

Ryan, W. B. F., Cita, M. B., Miller, E. L., Hanselman, D., Nesteroff, W. D., Hecker, B., and Nibbelink, M., 1978, Bedrock geology in New England submarine canyons: Oceanol. Acta, v. 1, p. 233–254.

———, and Farre, J. A., 1982, Potential of waste disposal on the continental margin by natural dispersal processes, in Duedall, I. W., ed., Wastes in the Ocean: New York, John Wiley and Sons, Inc., in press.

———, and Miller, E. L., 1981, Evidence of a carbonate platform beneath Georges Bank: Mar. Geology, v. 44, p. 213–228.

Schlee, J., 1973, Atlantic continental shelf and slope of the United States—Sediment texture of the northeastern part: U.S. Geol. Survey Professional Paper 529-L, 64 p.

———, Behrendt, J. C., Grow, J. A., Robb, J. M., Mattick, R. E., Taylor, P. T., and Lawson, B. J., 1976, Regional framework off northeastern United States: Am. Assoc. Petroleum Geologists Bull., v. 60, p. 926–951.

Shepard, F. P., 1933, Canyons beneath the seas: Scientific Monthly, v. 37, p. 31–39.

———, 1981, Submarine canyons: Multiple causes and long-time persistence: Am. Assoc. Petroleum Geologists Bull., v. 65, p. 1062–1077.

———, and Dill, R. F., 1966, Submarine Canyons and Other Sea Valleys: Chicago, and McNally, 397 p.

Spencer, J. W., 1903, Submarine valleys off the American coasts and in the north Atlantic: Geol. Soc. America Bull., v. 14, p. 207–226.

Spiess, F. N., and Tyce, R. C., 1973, Marine physical laboratory deep-tow instrumentation system: Scripps Inst. Oceanogr. Ref., 73-4, p. 1–37.

Stanley, D. J., 1974, Pebbly mud transport in the head of Wilmington Canyon: Mar. Geology, v. 16, p. M1–M8.

———, Sheng, H., Lambert, D. N., Rona, P. A., McGrail, D. W., and Jenkyns, J. S., 1981, Current-influenced depositional provinces, continental margin off Cape Hatteras, identified by petrologic method: Mar. Geology, v. 40, p. 215–235.

Stetson, H. C., 1936, Geology and paleontology of the Georges Bank canyons: Geol. Soc. America Bull., v. 47, p. 339–366.

———, 1949, The sediments and stratigraphy of the east coast continental margin: Massachusetts Inst. Tech. Papers Phys. Oceanogr. Meteorol., v. 11, 60 p.

Stubblefield, W. L., McGregor, B. A., Forde, E. B., Lambert, D. N., and Merrill, G. F., 1982, Reconnaissance in DSRV ALVIN of a "fluvial-like" meander system in Wilmington Canyon and slump features in south Wilmington Canyon: Geology, v. 10, p. 31–36.

Summerhayes, C. P., Bornhold, B. P., and Embley, R. W., 1979, Surficial slides and slumps on the continental slope and rise of South West Africa: A reconnaissance study: Mar. Geology, v. 31, p. 265–277.

Swift, D. J. P., Moir, R., and Freeland, G. L., 1980, Quaternary rivers on the New Jersey shelf: Relation of seafloor to buried valleys: Geology, v. 8, p. 276–280.

Trumbull, J. V. A., 1972, Atlantic continental shelf and slope of the United States—Sand sized fraction of bottom sediments, New Jersey to Nova Scotia: U.S. Geol. Survey Professional Paper 529-K, 45 p.

———, and MacCamis, M. J., 1967, Geological exploration in an east coast submarine canyon from a research submersible: Science, v. 158, p. 370–372.

Twichell, D. C., Knebel, H. J., and Folger, D. W., 1977, Delaware River: Evidence for its former extension to Wilmington Submarine Canyon: Science, v. 195, p. 483–484.

Uchupi, E., 1968, Atlantic continental shelf and slope of the United States—Physiography: U.S. Geol. Survey Professional Paper 529-C, 30 p.

———, 1970, Atlantic continental shelf and slope of the United States—Shallow structure: U.S. Geological Survey Professional Paper 529-I, 44 p.

———, Ballard, R. D., and Ellis, J. P., 1977, Continental slope and upper rise off western Nova Scotia and Georges Bank: Am. Assoc. Petroleum Geologists Bull., v. 60, p. 1483–1492.

Valentine, P. C., Uzmann, J. R., and Cooper, R. A., 1980, Geology and biology of the Oceanographer Submarine Canyon: Mar. Geology, v. 38, p. 283–312.

Veatch, A. C., and Smith, P. A., 1939, Atlantic submarine valleys of the United States and the Congo Submarine Valley: Geol. Soc. America Spec. Paper 7, 101 p.

Warme, J. E., Slater, R. A., and Cooper, R. A., 1978, Bioerosion in submarine canyons, in Stanley, D. J., and Kelling, G., eds., Sedimentation in Submarine Canyons, Fans, and Trenches: Stroudsburg, Pa., Dowden, Hutchinson and Ross, p. 65–70.

THE EFFECT OF SEALEVEL CHANGE ON THE SHELFEDGE AND SLOPE OF PASSIVE MARGINS[1]

WALTER C. PITMAN III AND XENIA GOLOVCHENKO[2]
Lamont-Doherty Geological Observatory of Columbia University, Palisades, New York 10964

ABSTRACT

Subsidence of passive margins appears to be of thermal origin, and increases from near zero at a landward hinge zone to a maximum value at the ocean-continent boundary. At all points on the attenuated margin the driving subsidence is at a maximum immediately subsequent to rifting, and decreases thereafter as a function of time. The present driving subsidence at the ocean-continent boundary of the U.S. East Coast may be greater than 1 cm/1000 years. Rifting at this margin ceased at least 160 m.a. It is concluded that glacial fluctuation is the only known mechanism that can cause world-wide sealevel to change at rates in excess of 1 cm/1000 yrs and with a magnitude greater than 100 m. We do not preclude the possibility that mechanisms as yet unknown may be sufficient to change sealevel by rates and with magnitudes in excess of the above. Considered here are geological periods during which glacial fluctuation was minimal to non-existent. We have assumed that at most passive margins, the shelfedge lies at (or even seaward of) the ocean-continent boundary. Under these conditions it is not likely that the shoreline can move out over the shelfedge because the driving subsidence at the shelfedge is greater than the rate at which sealevel may fall. We have also shown that if by chance the rate of sealevel fall is several times greater than the rate of subsidence at the shelfedge it may take several million years to displace the shoreline seaward to the shelfedge.

Alternatively, at a starved margin, the shelfedge may lie well landward of the ocean-continent boundary (well up toward the hinge zone). In this case, rate of subsidence may be much less than 1 cm/1000 yrs and rate of sealevel fall may be much greater than rate of subsidence at the shelfedge. In this case the coastline may move seaward over the shelfedge, but this will still require a time interval of a million years or greater. If sealevel drops sufficiently to allow the coastline to move seaward over the shelfedge and if at the same time sealevel ceases falling, then within a time interval of less than two million years the combined effects of both erosion and subsidence will generally cause the coastline to retreat onto the shelf and transgress toward the hinge zone.

INTRODUCTION

A variety of stratification patterns mapped on passive margins at or near the shelfedge, and on the slope and rise, indicate that the shoreline, at times, was positioned seaward of the shelfedge. Depositional patterns include strata that onlap the slope (Fig. 1A, B). On the basis of geometry of these stratal patterns it has been concluded that during the Cretaceous and Tertiary sealevel frequently dropped in excess of 100 m in less than 1 million years (my), (Vail et al., 1977b). Because of the rapidity and magnitude of these changes it has been concluded that they must have been produced by glacial fluctuations. It is recognized that since the mid-Miocene, glacial fluctuations of sufficient amplitude have occurred but there is no geological evidence that such events took place during the Mesozoic and Cenozoic prior to mid-Miocene.

We can show that the conditions under which the coastline may be displaced seaward over the shelfedge during non-glacial periods are unusual and this phenomenon only occurs rarely. On the other hand the conditions under which the coastline may move near the shelfedge are much more probable. It seems likely that the sequences of strata that appear to onlap the slope are deposited during periods when the coastline has retreated to a position near the shelfedge. In this chapter we define the conditions under which the beach can migrate to, or over, the shelfedge during non-glacial episodes of sealevel change. Not only is the rate of sealevel change important, but also the rate at which the margin subsides. We can show that, considering various possible mechanisms, the rate of sealevel lowering is usually less than the rate of subsidence at the ocean-continent boundary. Under these conditions it is not likely that the coastline can move seaward of the shelfedge, if the shelfedge is at or near the ocean-continent boundary.

DEVELOPMENT OF PASSIVE MARGINS

Passive margins are formed by rifting and separation of continents (see Part II, this volume). Sediments at the shelfedge may reach thicknesses as great as 14 km and may comprise shallow water deposits from the base to top of a section.

By a technique called backstripping, Watts and Ryan (1976), Steckler and Watts (1978), and others have calculated layer-by-layer the amount of subsidence caused by sediment load. The difference

---

[1]Lamont-Doherty Geological Observatory Contribution No. 3438.
[2]Present address: Marathon Oil Co., Denver Research Center, P.O. Box 269, Littleton, Colorado 80160.

Copyright © 1983, The Society of Economic Paleontologists and Mineralogists

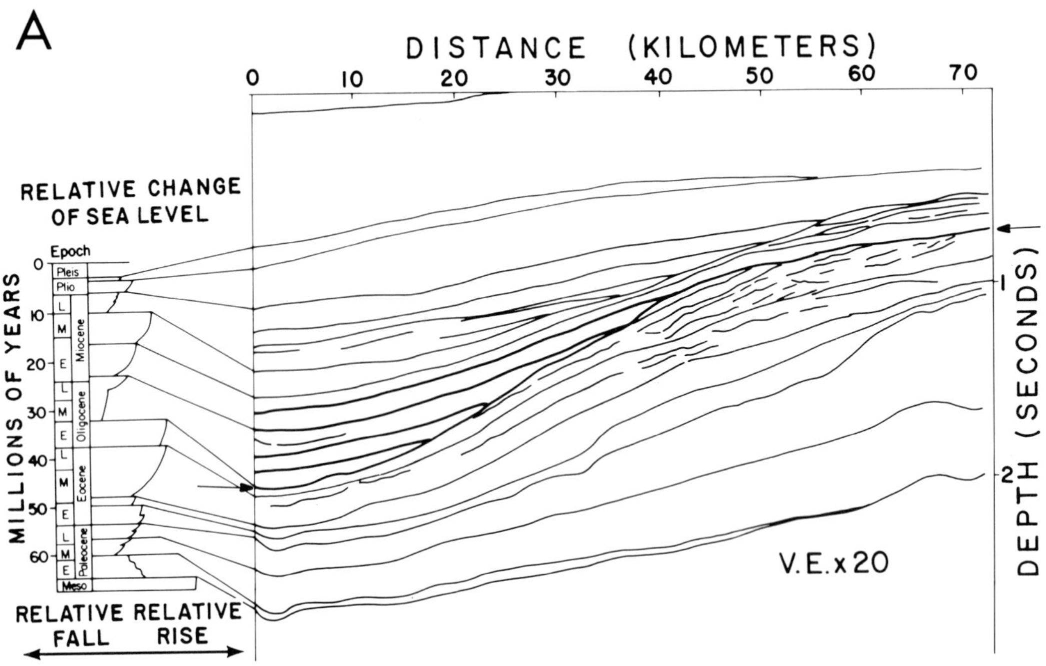
A
DISTANCE (KILOMETERS)
0 10 20 30 40 50 60 70
RELATIVE CHANGE OF SEA LEVEL
Epoch
Pleis
Plio
Miocene
Oligocene
Eocene
Paleocene
Meso
MILLIONS OF YEARS
DEPTH (SECONDS)
V.E.x20
RELATIVE FALL
RELATIVE RISE

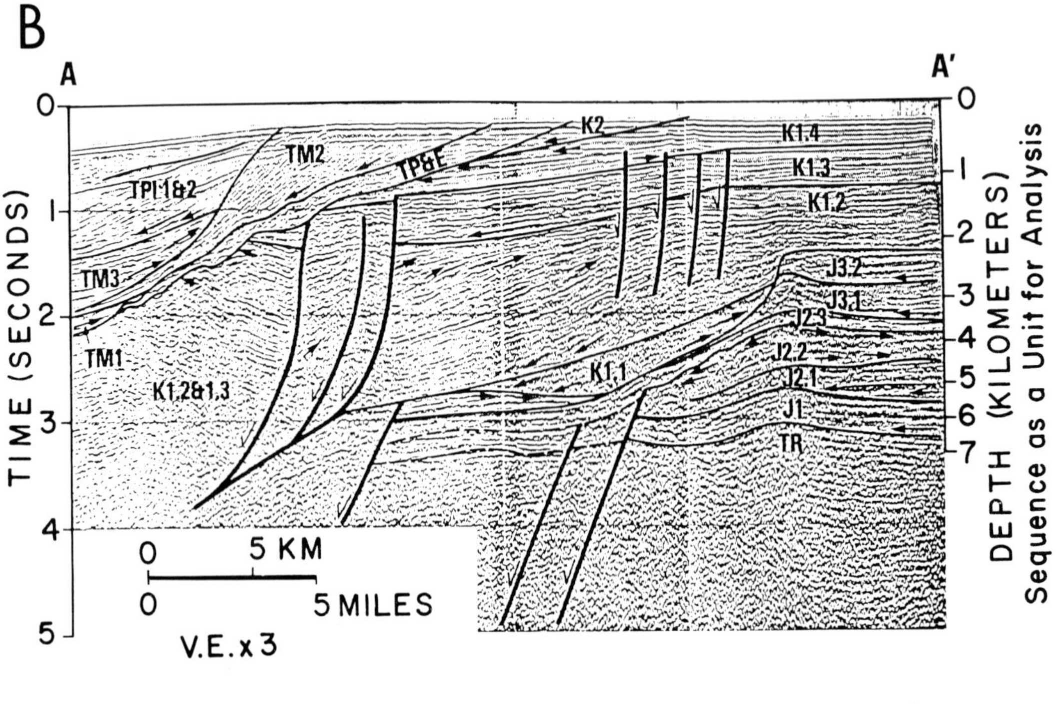
B
A
A′
TIME (SECONDS)
DEPTH (KILOMETERS)
Sequence as a Unit for Analysis
TM2
TP&E
K2
K1.4
K1.3
K1.2
TPl 1&2
TM3
TM1
K1.2&1.3
K1.1
J3.2
J3.1
J2.3
J2.2
J2.1
J1
TR
0 5 KM
0 5 MILES
V.E.x3

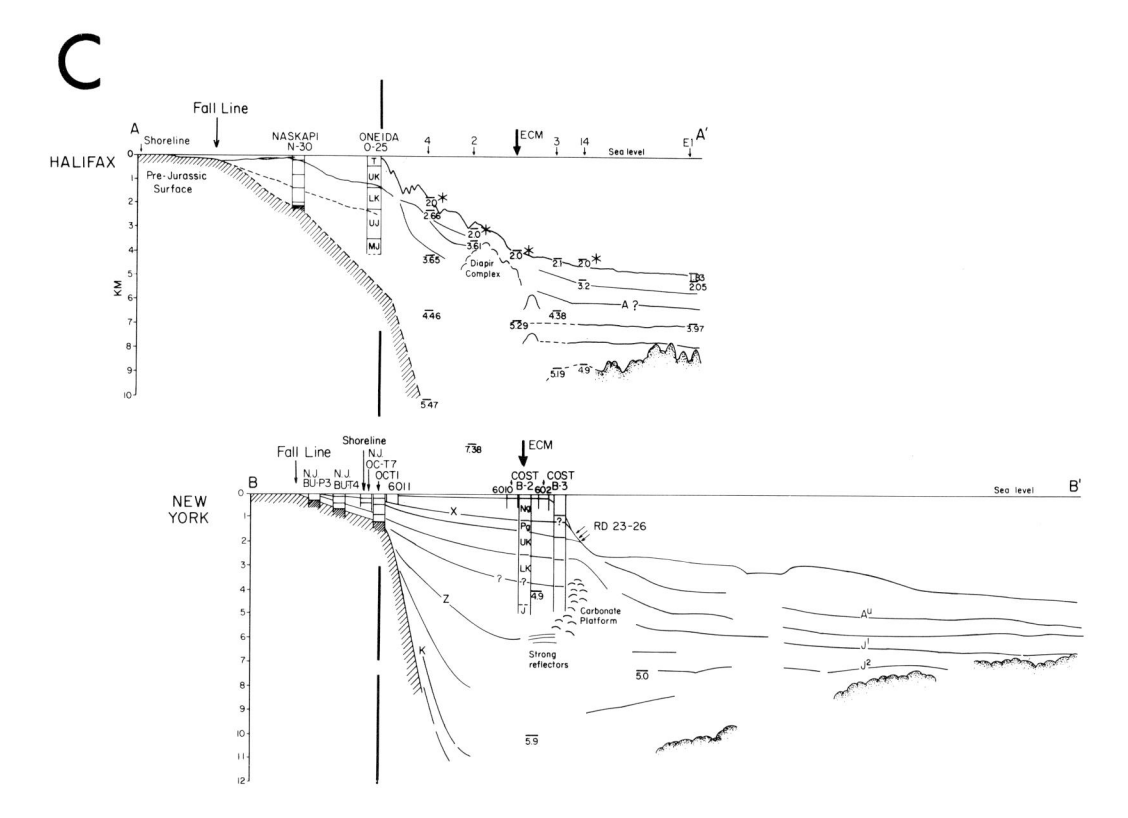

FIG. 1.—A, An interpreted sedimentary section from the North Sea (Vail et al., 1977a). The surface marked by the arrows at the end points is regarded as a regressive surface representing an outer shelf to the right and grading to an upper slope toward the left. The strata immediately above this show progressive onlap in the landward direction. This section was interpreted as indicating that during a regressive episode the beach had traveled from the shelf on the right out over the slope and that sealevel may have dropped at least as far as the deepest point of onlap (Vail et al., 1977a). B, An interpreted section from the northwestern margin of Africa (Vail et al., 1977a). The arrows pointing seaward indicate offlap. Those directed landward indicate onlap. TR = Triassic, J = Jurassic, K = Cretaceous, TP and E = Paleocene and Eocene, TM = Miocene, and TPl = Pliocene. Tertiary strata appear as prograding and onlapping sequences over the shelfedge. C, Interpreted section from several points along the east coast of North America. Note that the shelfedge on the section off Nova Scotia appears to lie well landward of the ocean-continent boundary, whereas on the section through the COST B-2 well off New Jersey the shelfedge lies above or even seaward of the ocean-continent boundary (from Watts, 1981).

between the total subsidence and that caused by the sediment load is termed *driving subsidence*. When this is plotted as a function of time—beginning at the end of the rifting phase—it is apparent that the rate of driving subsidence decays with time and the shape of the subsidence curve is similar to that of mid-ocean ridges. The shape of the driving subsidence curve suggests a thermal cooling mechanism. The driving subsidence creates the depression that the sediments infill, in turn causing further subsidence. There also appears to be significant subsidence and sediment accumulation during the rifting stage (Falvey, 1974).

McKenzie (1978) proposed a two-phase model that could account for all of these observations as follows (Fig. 2A):

1. First, there is a stretching (rifting) phase during which both the crust and lithosphere are thinned. Under most conditions subsidence occurs during the stretching phase as isostatic equilibrium is maintained. This phase lasts only a few million years (McKenzie, 1978; Royden and Keen, 1980). On an Atlantic-type margin the stretching phase is terminated by separation of the attenuated continental lithosphere to form an ocean.

2. The stretching phase is followed by a long, protracted phase of subsidence caused by cooling. Initially, the lithosphere-asthenosphere boundary, defined by the 1330°C isotherm, will lie at a shallow depth and heat flow will be high. As cooling takes place, the 1330° isotherm descends gradually and the margin subsides. The rate of subsidence

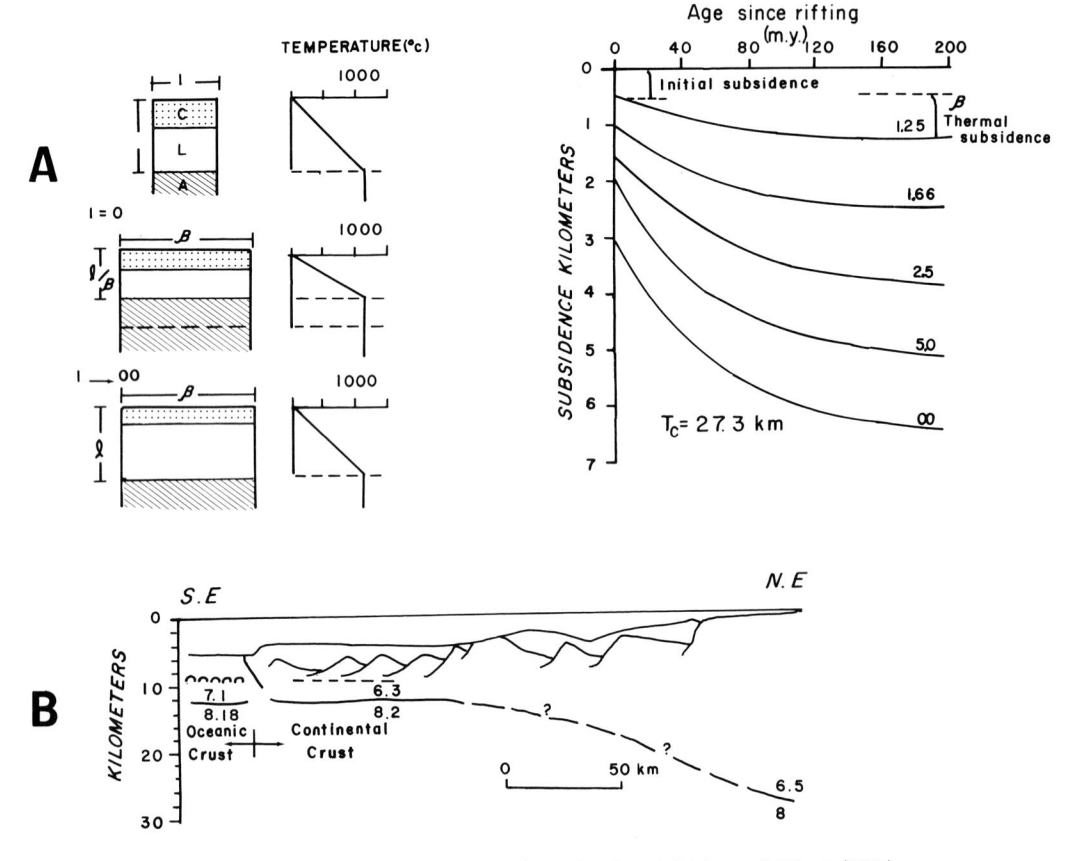

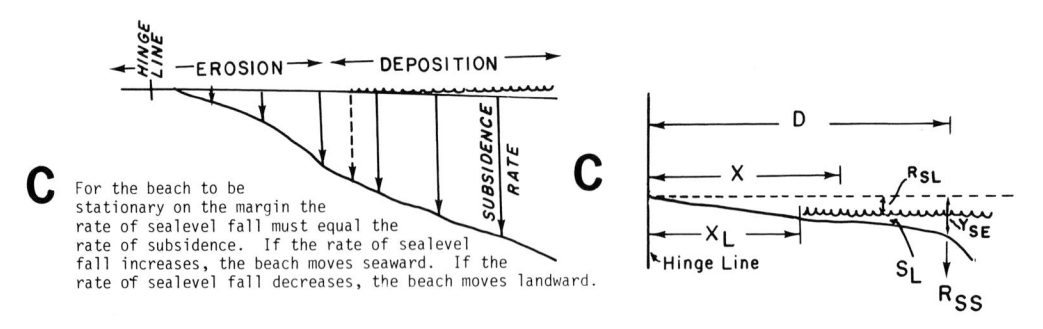

FIG. 2.—*A* (left), McKenzie model is shown schematically. A normal crustal-lithospheric section is stretched by a factor β and thinned by a factor 1/β. During the stretching (presumed to take place in just a few million years), isostatic subsidence (or uplift) takes place as adjustment to the new density profile. After stretching (rifting), the 1350°C isotherm will be at a depth of ℓ/β. This causes high heat flow, cooling and subsidence. *A* (right), Subsidence for any factor of β is time dependent and lasts at least 200 my. *B*, Sedimentary structures show that the sedimentary wedge thickens seaward, and thus the rate of subsidence increases seaward. The rate of thermal subsidence is a function of the amount of stretching, and this indicates that the stretching factor β increases from near zero at the hinge zone to a maximum at the ocean-continent boundary. *C*, A simplified profile across a passive margin showing the hinge

decays as a function of time. The time-dependent subsidence for various stretching factors is shown in Figure 2A. $\beta = \infty$ is applicable in the case where final separation has taken place and an ocean is being formed. The amount and rate of driving subsidence is a maximum for oceanic lithosphere ($\beta = \infty$) and decreases for decreasing β.

3. In calculating the effects of post-rift cooling, McKenzie (1978) assumed the 1330°C isotherm gradually descends to a maximum depth of 125 km. Oceanic lithosphere is an end-point of the McKenzie model ($\beta = \infty$). Two different empirical subsidence curves were found by Parsons and Sclater (1977) to describe subsidence of oceanic lithosphere. These are given in Table I (in Fig. 3). It is assumed that at the landward-edge of a passive margin, the amount of crustal and lithosphere attenuation that has taken place is near zero and, hence, the amount of subsidence is small. We also assume that the attenuation increases seaward to a maximum value at the ocean-continent boundary. We will consider data from the COST B2 well (Fig. 1C) which is at, or very near, the ocean-continent boundary; the subsidence history thus will be nearly the same as for oceanic lithosphere (b = ∞).

Parsons and Sclater (1977) showed that the exponential form of the oceanic subsidence curve gives the best fit for ages greater than 80 my. But Heestand and Crough (1981) have argued that many areas of the ocean are anomalously elevated because of hot spot activity. They found that when appropriate corrections are made to the oceanic depth data, a $\sqrt{T}$ form for the subsidence curve fits best (d(t) depth = 2700 + 295 $\sqrt{T}$).

The rate of subsidence has been plotted as a function of time for each of the subsidence formulas given in Table I (in Fig. 3). We calculated the rate of subsidence at the COST B2 well for the period 65 my to present. At the COST B2 well, 1500 m of sediment have accumulated since the Cretaceous. Using the method of Steckler and Watts (1978) to backstrip, a minimum average post-Cretaceous driving subsidence of 1.32 cm/1000 yrs is estimated. We consider the present record to be about 170 my after the end of rifting, and therefore the end of the Cretaceous occurred 105 my after the end of rifting. The theoretical driving subsidence rates calculated for 105 my and 170 my from the three different subsidence formulas are given in Table I (in Fig. 3). As can be seen, the theoretical subsidence rates given by the exponential form of the equation are, on the average, low by a factor of 2. In our calculations we have ignored rigidity and assumed pure local loading. If rigidity had been considered, then the theoretical exponential subsidence rates would probably be low by a factor of 3. It appears that the $\sqrt{T}$ function gives a more reasonable value for subsidence at the shelf. We will assume that this relationship best approximates the subsidence at a passive margin.

SUBSIDENCE AND SEALEVEL CHANGES

Thermal cooling causes the margin to continuously subside, creating a space which the sediment infills, causing further subsidence. Because the driving subsidence is deep-seated and most likely thermal in origin it is probably not episodic but varies smoothly with time (Fig. 2A–C). Also, it should vary smoothly across, and along, the margin (Fig. 2C). On the other hand, because of differential compaction and processes such as salt flow, the subsidence caused by the sediment loading might not be a smooth function of time and may vary abruptly spatially (Fig. 1C). However, multichannel seismic data show that the strata are relatively undisturbed, and affected by minor growth faults only (Fig. 1C). The general picture is that of a platform subsiding about a hinge zone, which is the region of zero subsidence. The rate of subsidence increases seaward and varies smoothly with distance and descreases smoothly with time.

Thus, an essential condition to get the shoreline displaced seaward over the shelfedge is the rate of sealevel lowering which must at least be greater than the rate of driving subsidence at the shelfedge. Not only must the rate of sealevel lowering exceed the rate of subsidence at the shelfedge, but it must persist long enough to lower the sealevel sufficiently. It should also be immediately obvious that the only condition which will allow the coastline to remain on the shelf is a persistent sealevel fall (Figs. 1B, 2C).

SEALEVEL CHANGES—RATES AND MAGNITUDE

The rate, direction and magnitude of sealevel changes control the pattern of deposition at passive margins (Pitman, 1978, 1979, see also chapters in parts II and III, this volume).

There are a variety of mechanisms likely to cause sealevel change. For each of these we will estimate

---

line, shelf, shelfedge, and slope. For simplicity, it is assumed that the shoreline is the dividing line between erosion in the direction landward and deposition seaward, such that erosional and depositional processes maintain a constant slope. It is also assumed that the rate of thermal subsidence at any point on the shelf is proportional to the distance from the hinge zone. R_{SL} = rate of sealevel change (positive upwards) as measured with respect to the stable craton. Y_{SE} = elevation of the shelfedge as measured with respect to the hinge zone. R_{SS} = rate of subsidence of the shelfedge. SL = slope of the shelf. D = distance from the hinge zone to the shelfedge. X = distance from the hinge zone to any point on the shelf. X_L = distance from the hinge zone to the shoreline.

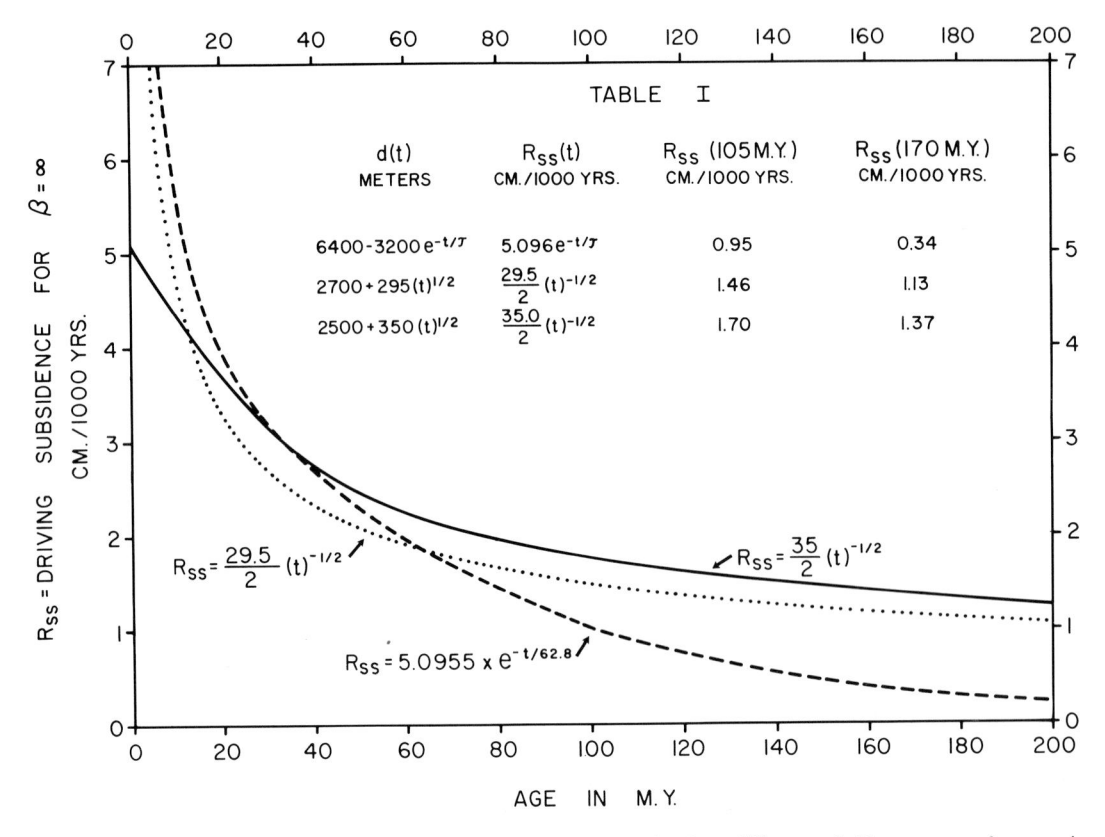

FIG. 3.—Plots of subsidence rate as a function of time are shown for three different subsidence curves for oceanic crust. These give subsidence rate with a water load for oceanic lithosphere and should approximate the driving subsidence for the region just seaward of the ocean-continent boundary. Time is in my after the end of the rifting stage. For the East Coast of North America, at least 170 my have elapsed since the end of the rifting stage. The table shows driving subsidence calculated from each of the three curves for 105 my and 170 my. Note that subsidence rates calculated using the exponential form of the subsidence curve are much too low.

the rate and magnitude of the potential or possible change.

Glacial Functions

If the present-day continental ice sheets melted, the depth of oceans would increase sufficiently to raise sealevel by about 60 m. During the past 20,000 to 10,000 years, because of glacial melt, sealevel may have been raised by 100 m (or perhaps as much as 150 m) at rates of 1000 to 1500 cm/1000 yrs. Probably a number of glacial oscillations of this magnitude have occurred since the mid-Miocene. But there is no geologic evidence to suggest glacial events of this magnitude at any time in the Mesozoic or Cenozoic prior to the mid-Miocene. We might expect post-Cretaceous glacial events to take place in Antarctica with a maximum sealevel change of 25 to 50 m at most. Isotopic data (Shackleton and Kennett, 1975; Shackleton and Boersma, 1981) indicate that Late Cretaceous seas were significantly warmer than at present; however, by early Tertiary time they began to cool. The ^{18}O record provides no indication of significant glacial events prior to the mid-Miocene.

Change in Volume of Mid-Ocean Ridge System

The mid-ocean ridge system occupies a very large volume in the world ocean basins. The elevation of the ridge is of thermal origin. There are several ways to change the volume of the mid-ocean ridge. One is by changing the spreading rate: an increase in the spreading rate will increase the volume, a decrease will reduce the volume. If spreading were to cease and the ridge system were allowed to cool for 70 my, sealevel would be lowered by 500 m. A second way to change volume is to subduct segments of existing ridges, and a third way is to create new ridge sectors by rifting. Pitman (1978) has calculated that in about 70 my, sealevel has dropped approximately 350 m because of a decrease in volume of the mid-ocean ridge system.

Collisional Type Orogeny

When two continents collide at a convergent zone (such as the Indus suture), continental shortening takes place, continental area is reduced and oceanic area is increased causing a sealevel drop. Pitman (in prep.) has estimated the effect of Himalayan and Alpine orogenies as shown in the table, Figure 4.

Sediment Influx and Removal

The vast amount of sediments contributed to the oceans is by rivers, with minor additions from shoreline processes. Some of the sediment accumulations are complexly deformed or destroyed at subduction zones and at the margins of colliding plates. The net balance between influx and removal will cause sealevel to rise or fall accordingly. Pitman and Golovchenko (in prep.) estimated that there has been a sealevel rise of about 85 m since the Late Cretaceous as a result of net influx of sediments.

Hot Spots

Heestand and Crough (1981) have estimated that hot spots may cause up to 100 m rise or fall in sealevel. These are thermal events and have approximately the same thermal behavior as the mid-ocean ridge system. Estimated normal and maximum values for each of the above causal mechanisms are given in the table, in Figure 4.

Collisional type events cause sealevel to drop but also cause mountain building which will serve to increase sediment influx into the oceans. Most importantly, except for glacial fluctuation, the maximum sustained rate of sealevel fall that can be

CAUSES; MAGNITUDES; RATES OF SEALEVEL CHANGE

MECHANISM	PROBABLE MAXIMUM MAG. M.	PROBABLE MAXIMUM RATE CM./1000Y.	MAXIMUM MAXIMUM MAG. M.	MAXIMUM MAXIMUM RATE CM./1000Y.	TIME INTERVAL (M.Y.)
GLACIATION ↓↑	150	1000	250	1000	0.1
RIDGE VOLUME ↓↑	350	0.75	500	1.2	70
OROGENY ↓	70	0.10	150	0.20	70
SEDIMENT ↓↑	60	0.11	85	0.25	70
HOT SPOTS ↓↑	50	0.08	100	0.14	70
FLOODING OF OCEAN BASINS ↓ (MED.)	15		INSTANTANEOUS		

PITMAN (1978)

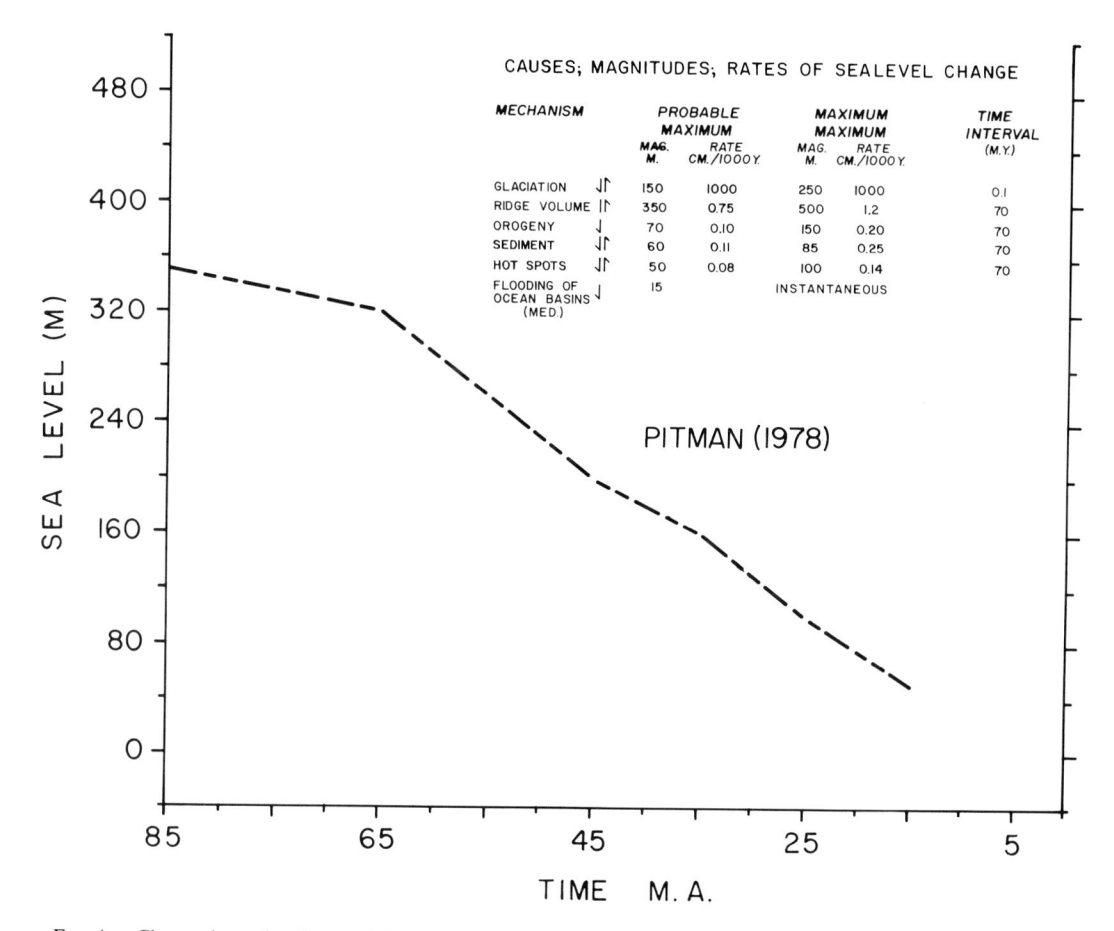

FIG. 4.—Change in sealevel caused by change in volume of the mid-ocean ridge system plus post mid-Miocene buildup of ice in Greenland and Antarctica (Pitman, 1978). Calculations do not include the large fluctuation caused by major glacial fluctuations from Pleistocene to Recent. We presume that there were probably no glacially-driven sealevel changes greater than 25 m during the Mesozoic and Cenozoic prior to mid-Miocene. Other mechanisms such as the Himalayan and Alpine orogenies and sediment flux seem to have canceled each other. The table in this figure gives a listing of various mechanisms that may cause sealevel change. For each mechanism, an estimate of the normal maximum, the maximum-maximum magnitude, and rate of changes are given.

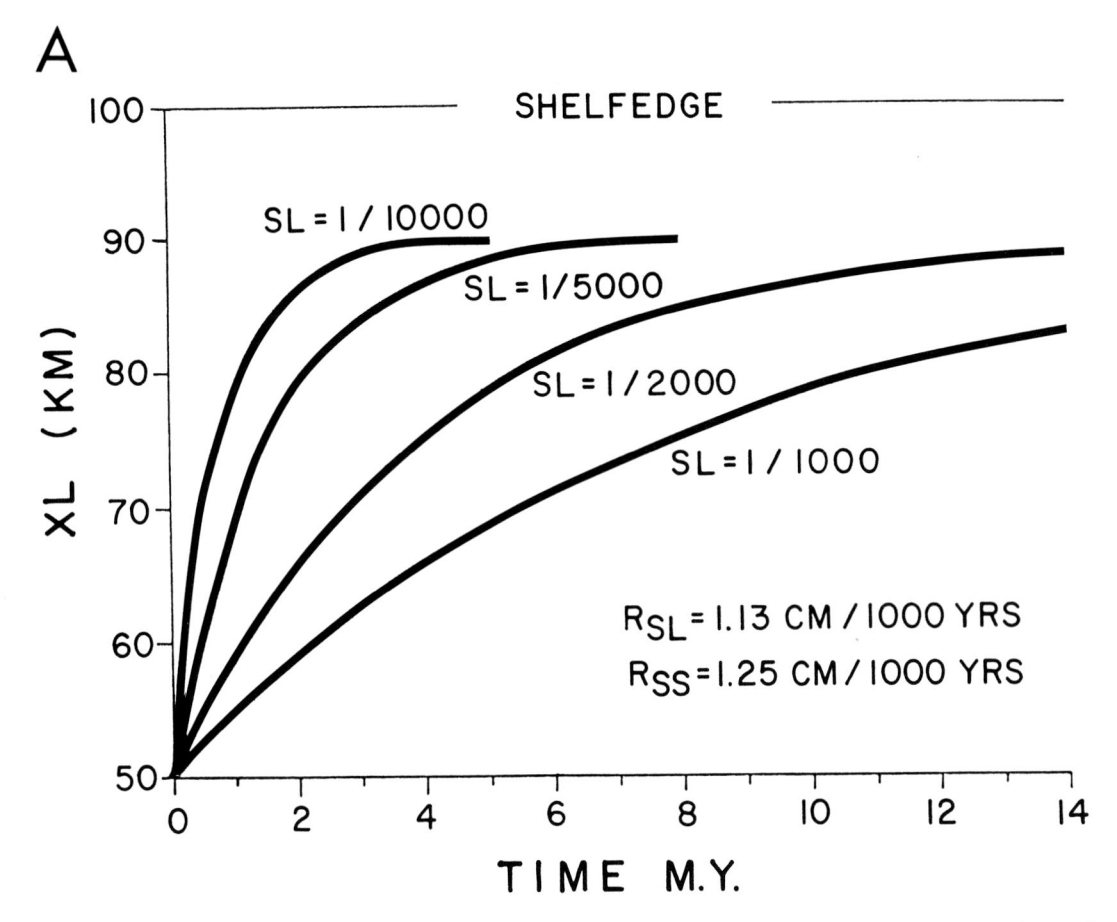

FIG. 5.—A, Position of a beach as a function of time after an increase in the rate of sealevel fall from 0.625 cm/1000 yrs to 1.13 cm/1000 yrs has been calculated for various shelf-slope sectors. Note that in all cases the shoreline gradually approaches the point where the rate of thermal subsidence is equal to the rate of sealevel fall ($X_L/D \cdot R_{SS} = R_{SL}$; $X_L = D \cdot R_{SL}/R_{SS}$). The shallower the slope, the more rapid the regression. B, Case where the rate of sealevel fall decreases from 1.13 cm/1000 yrs to 0.0. A transgression occurs even though sealevel is not rising. The shallower slopes exhibit the most rapid transgressions.

achieved by combining mechanisms appears to be about 1.7 cm/1000 yrs and by combining what we call more probable rates appears to be 1.01 cm/1000 yrs, that is, rates that can be sustained for several millions of years. This latter rate is less than the estimated rate of subsidence at the shelfedge today in the vicinity of the Baltimore Trough. Thus, without significant glacial build-up and/or other catastrophic events there does not appear to be a way for the shoreline to move out to the shelfedge where the shelfedge lies over the ocean-continent boundary.

THE ANCIENT SHELF

We assume that under normal conditions of slow but persistent sealevel change, sedimentary and erosional processes are able to maintain a dynamic graded slope across the exposed and subaqueous portions of the shelf (Fig. 2C). This means that erosional and depositional processes are able to keep pace with subsidence so as to maintain a graded slope. During the Pleistocene, and probably as far back as mid-Miocene, there have been very rapid and large glacially-driven sealevel changes. During the most recent of these episodes sealevel fell by perhaps 150 m in several tens of thousands of years, and then rose by the same amount during the past 20,000 to 10,000 years. At the time of each eustatic lowering the shoreline was drawn out over the shelfedge, exposing the shelf to erosion. The subsequent sealevel rise flooded river valleys and the shelf. In each case, and in particular during the most recent sealevel rise, the rate and magnitude of sealevel change exceeded the capability of the depositional processes to keep pace. The present-

B

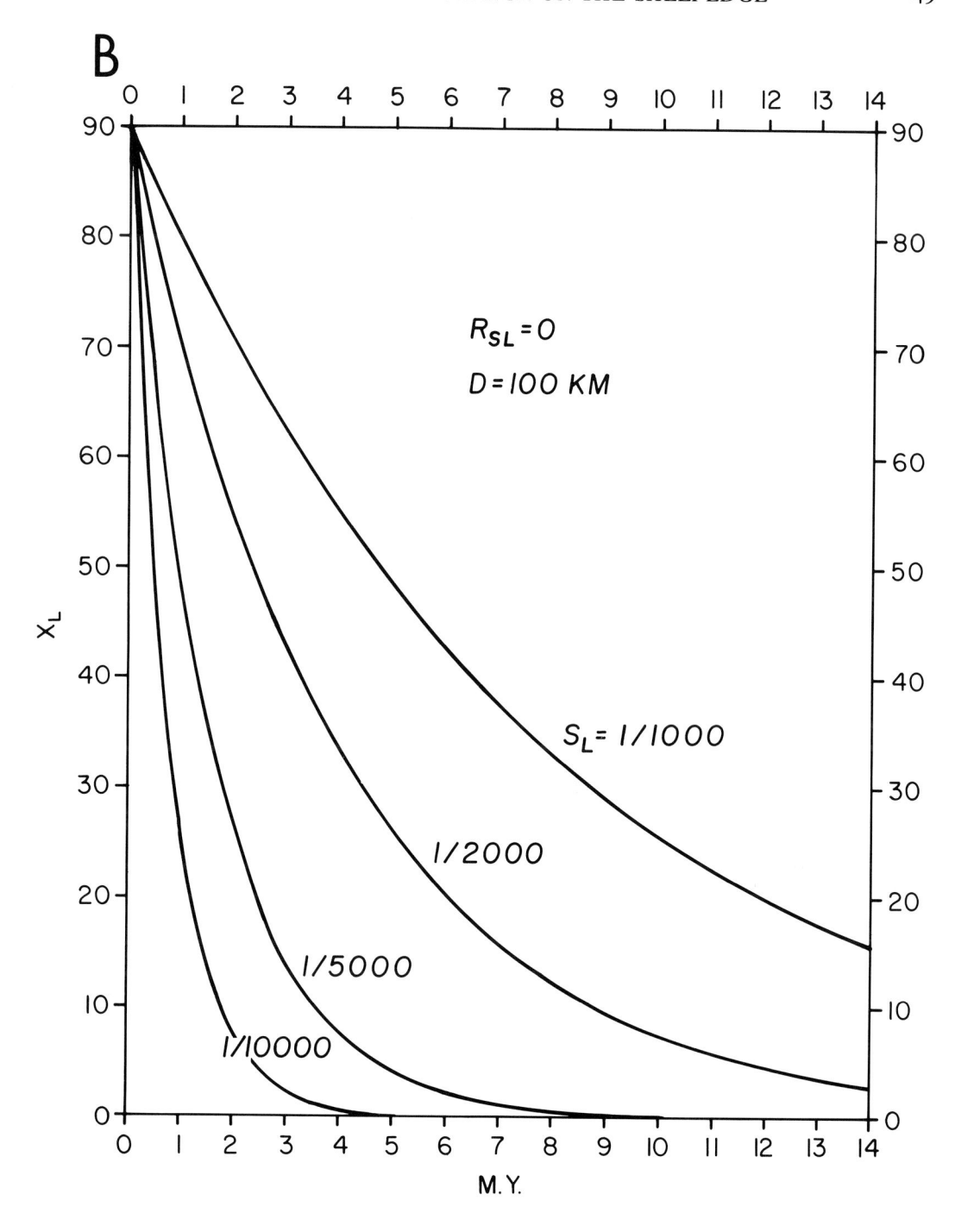

$R_{SL} = 0$

$D = 100$ KM

$S_L = 1/1000$

$1/2000$

$1/5000$

$1/10000$

M.Y.

day shelf areas, with the possible exception of those lying adjacent to active deltaic regions, are thus not in dynamic equilibrium. The slope of the shelf immediately to the west of the Mississippi Delta is about 1:5000 and the slope of the delta itself is 1:8500. We suggest that the slope of shelves as low as 1:5000 may have been quite common during non-glacial periods.

SEALEVEL CHANGE AND POSITION OF COASTLINE

We will now develop quantitative models showing the relationship of the shoreline to the shelf and the shelfedge as a function of time and various other parameters such as the shelf-slope configuration, shelfedge, subsidence rate, rate of sealevel change. As stated above it appears that the rate of driving subsidence increases smoothly from near zero at the hinge zone to a maximum at the shelfedge. We make the simplifying assumption that the rate of subsidence R_{SX} at point x is proportional to the distance x from the hinge line (Fig. 2C).

$$R_{SX} = \frac{x}{D} R_{SS}$$

We also assume that the shelf and coastal plain are dynamically-graded slopes, that deposition takes place seaward of the shoreline and erosion landward of the shoreline. The depositional and erosional processes are such as to maintain a constant grade. The grade of the depositional region will probably be different from the grade of the erosional region, although each will tend to maintain a constant grade. We assume that the grade of the shelf and that of the coastal plain are equivalent. This assumption does not affect the arguments to follow. From Figures 2 and 3 it can be seen that the only way in which the coastline can be sustained on the shelf is when sealevel is persistently falling. The position at which the beach will remain stationary (a stillstand) is where the rate of driving subsidence at X_L is equal to the rate of sealevel fall (R_{SL} is positive upwards).

$$R_{SL} = R_{SXL} = \frac{X_L}{D} R_{SS}$$

If the rate of sealevel lowering increases, the beach will migrate seaward toward a new stable position. If the rate of sealevel fall is greater than the rate of subsidence at the shelfedge the shoreline will eventually migrate over the shelfedge. If the rate of sealevel fall decreases, the shoreline will migrate landward toward a new equilibrium position. The new position of the shoreline is given by:

$$X_L = \frac{R_{SL}}{R_{SS}} D + X_{LI} e^{\frac{-T R_{SS}}{D S L} \frac{-R_{SL}}{R_{SS}} D} \quad (1)$$

where X_L = the position of the beach at the end of the time interval T and X_{LI} is the position of the beach at the beginning of the time interval T.

Examples of the regressive and transgressive events that may take place for shelf regions with varying slopes are shown in Figures 5A, B and 6. In Figure 5A prior to T = 0 my, the rate of sealevel fall had been 0.625 cm/1000 yrs for sufficiently long that the beach was at the $X_{LI} = R_{SL}/R_{SS}$. D = 50 km point. At T = 0 the rate of sealevel fall is increased to 1.13 cm/1000 yrs and a regression occurs as the shoreline moves toward the new equi-

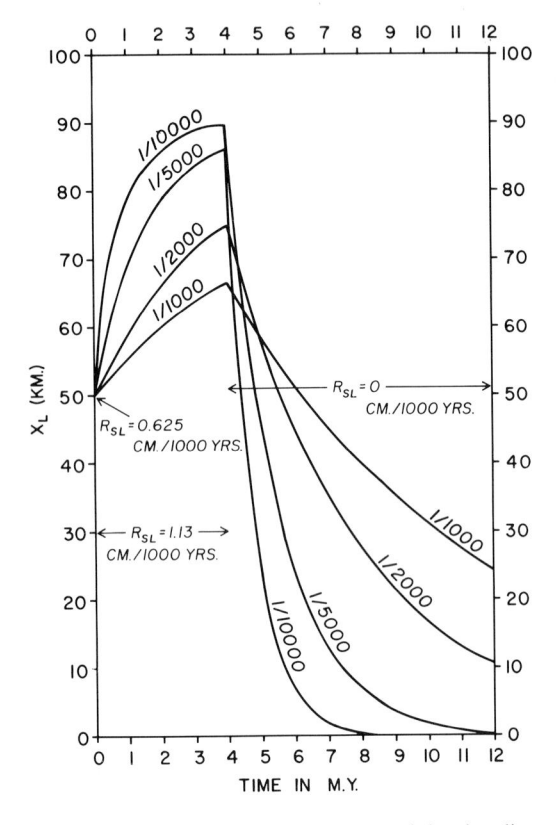

FIG. 6.—Example where the position of the shoreline is plotted as a function of time for a sequence of two changes in R_{SL}. First R_{SL} is increased from 0.625 cm/1000 yrs to 1.13 cm/1000 yrs for 4 my. At 4 my the rate of sealevel change is reduced to 0.0. Regressions and transgressions are most rapid on the shelf with the more gentle slope.

librium point at 90 km. On a shelf with a very gentle shallow slope of 1:10,000, the regression is quite rapid but nevertheless takes 4 my. On the other hand, on a shelf with a steep slope of 1:1000, after 10 my the shoreline is still only 78 km from the hingeline and not near its equilibrium position. During those 10 my, sealevel will have dropped 113 m, which is a significant change.

In Figure 5B we assume that, prior to T = 0, the rates of sealevel fall had been 1.13 cm/1000 yrs for sufficiently long such that the shoreline in each case was stationary at $X_L = R_{SL}/R_{SS}$. D = 90 km. At T = 0, R_{SL} is reduced to zero so a transgression occurs as the shoreline migrates toward its new equilibrium position at X_L = 0.0 km. Again the beach on the plain with the very shallow slope proceeds rapidly (4 my) to the new equilibrium position. At the end of 10 my the beach on a platform with a slope of 1:1000 has transgressed only to X_L = 20 km.

In Figure 6 these events are combined. At T = 0 the rate of sealevel lowering is increased from 0 to 1.13 cm/1000 yrs. At 4 my it is decreased to 0.0 cm/1000 yrs. Although the onset of the regressive and transgressive events is synchronous, the transgression virtually ends on a very gently dipping platform at 8 my while the shoreline is still rapidly transgressing on a shelf with a slope of 1:1000.

We have also calculated the position of the coastline on a shelf as a function of time for various rates of shelfedge subsidence as well as for various rates of sealevel change. In these calculations, equation (1) has been slightly modified:

$$\frac{X_L}{D} = \frac{R_{SL}}{R_{SS}} + \frac{X_{LI}}{D} e^{\frac{-T T_{SS}}{Y_{SE}}} - \frac{R_{SL}}{R_{SS}} \quad (2)$$

where $Y_{SE} = D \cdot S_L$ is the elevation of the shelfedge as measured from a horizontal plane drawn through the intersection of the coastal plain with the hinge zone (positive downward) (Fig. 2C). In the calculations we set $Y_{SE} = 50$ meters, and $X_{LI} = 0.0$ km. In Figure 7A, graphs of X_L/D as a function of time are shown for the case in which the rate of sealevel lowering is $R_{SL} = 0.5$ cm/1000 yrs. The range of values for R_{SS} is 0.5 to 4.0 cm/1000 yrs. A value of $R_{SS} = 0.5$ cm/1000 yrs at the ocean-continent boundary would require that the margin be over 1 billion years old in the case of a $\sqrt{T}$ subsidence curve. On the other hand, if we assume subsidence follows an exponential form, then an ocean-continent boundary 144 my old would be a sufficient cooling time to reduce the driving subsidence to 0.5 cm/1000 yrs. But from the arguments presented previouxly, a $\sqrt{T}$ dependence seems preferable. Alternatively (see Fig. 9A, 9B), if the shelf has not prograded out to the vicinity of the ocean-continent boundary but, because of a paucity of sediments, the shelfedge sits well back on the continental platform toward the hinge zone, then the subsidence at the shelfedge may be less than 0.5 cm/1000 yrs. This may have been the case for the West African margin during the Cenozoic. Sealevel, since the Late Cretaceous, has fallen about 350 m (see Fig. 4) at an average rate of 0.5 cm/1000 yrs. This may have been sufficient to move the coastline out to or near to the shelfedge of the Northwest African margin causing deposition to take place on the slope and upper part of the rise (Fig. 1B).

The rate of margin subsidence used is 4 cm/1000 yrs (high end of the range) and this rate would occur on oceanic lithosphere 12 to 20 my into the cooling phase (Fig. 6), depending on the formula chosen for subsidence. What is most obvious is that even when the rate of sealevel fall exceeds the rate of subsidence at the shelfedge it takes considerable time for the shoreline to migrate to a position close to the shelfedge. In Figure 7A, the curve for $R_{SS} = 0.5$ cm/1000 yrs at the end of 10 my still places the shoreline only 62.5% of the way to the shelfedge ($X_L/D = 0.625$).

In Figure 7B, where $R_{SL} = 1.0$ cm/1000 yrs, which is close to the normal maximum sustained rate of the level change for nonglacial periods (Pitman, 1978), it takes 7 my for the shoreline to reach the shelfedge when the shelfedge subsidence rate is 0.5 cm/1000 years. In this case the amount of sealevel fall is 70 m and the depth Y_{SE} to the shelfedge is 50 m as defined in Figure 2C. $R_{SL} = 1.11$ cm/1000 yrs can be regarded as the maximum reasonable combined rate of sustained sealevel fall exclusive of glacial fluctuation (table, in Fig. 4). For the curve in which the rate of subsidence is 1 cm/1000 yrs (Fig. 7B) at the end of 10 my, during which time sealevel would have dropped 100 m, $X_L/D = 0.86$. In each case where the rate of shelfedge subsidence equals the rate of sealevel fall it takes millions of years before the shoreline even approaches the shelfedge. Figure 7C shows the regression curves for a rate of sealevel fall of 2.0 cm/1000 yrs. This might be the maximum-maximum sustainable rate allowable by the mechanisms considered here (see table in Fig. 4). Even for margins at which the subsidence rate at the shelfedge is 0.5 cm/1000 yrs and 1.0 cm/1000 yrs it takes about 3 my (a drop of 60 m in sealevel) for the coastline to reach the shelfedge. Again, even with what are considered high rates of sealevel lowering it takes several millions of years to get the beach to the shelfedge. Figures 7D and 7E show calculations for rates of sealevel lowering of 3 and 5 cm/1000 yrs, respectively. These are both well in excess of what we believe to be maximum sustainable rate of sealevel fall. In the case in which $R_{SL} = 5$ cm/1000 yrs, it still takes about 1 my for the coastline to reach the shelfedge. In each case in which the shoreline eventually migrates to the shelfedge the amount of sealevel drop that has occurred is greater than the depth to the shelfedge (Y_{SE}). This is due to thermal subsidence and erosional and depositional processes which maintain a constant vertical distance Y_{SE} between the shelfedge and a horizontal plane drawn through the hinge zone.

SEALEVEL BELOW THE SHELFEDGE

We will now consider the consequences of very rapid (geologically instantaneous) and large sealevel change, after which the shoreline may lie on the continental slope 50 to 100 m below the former shelfedge. If sealevel stops falling, the shoreline will eventually transgress up and over the shelfedge because of subsidence and erosion. If R_{SL} remains at zero, the shoreline will eventually transgress to the vicinity of the hinge zone.

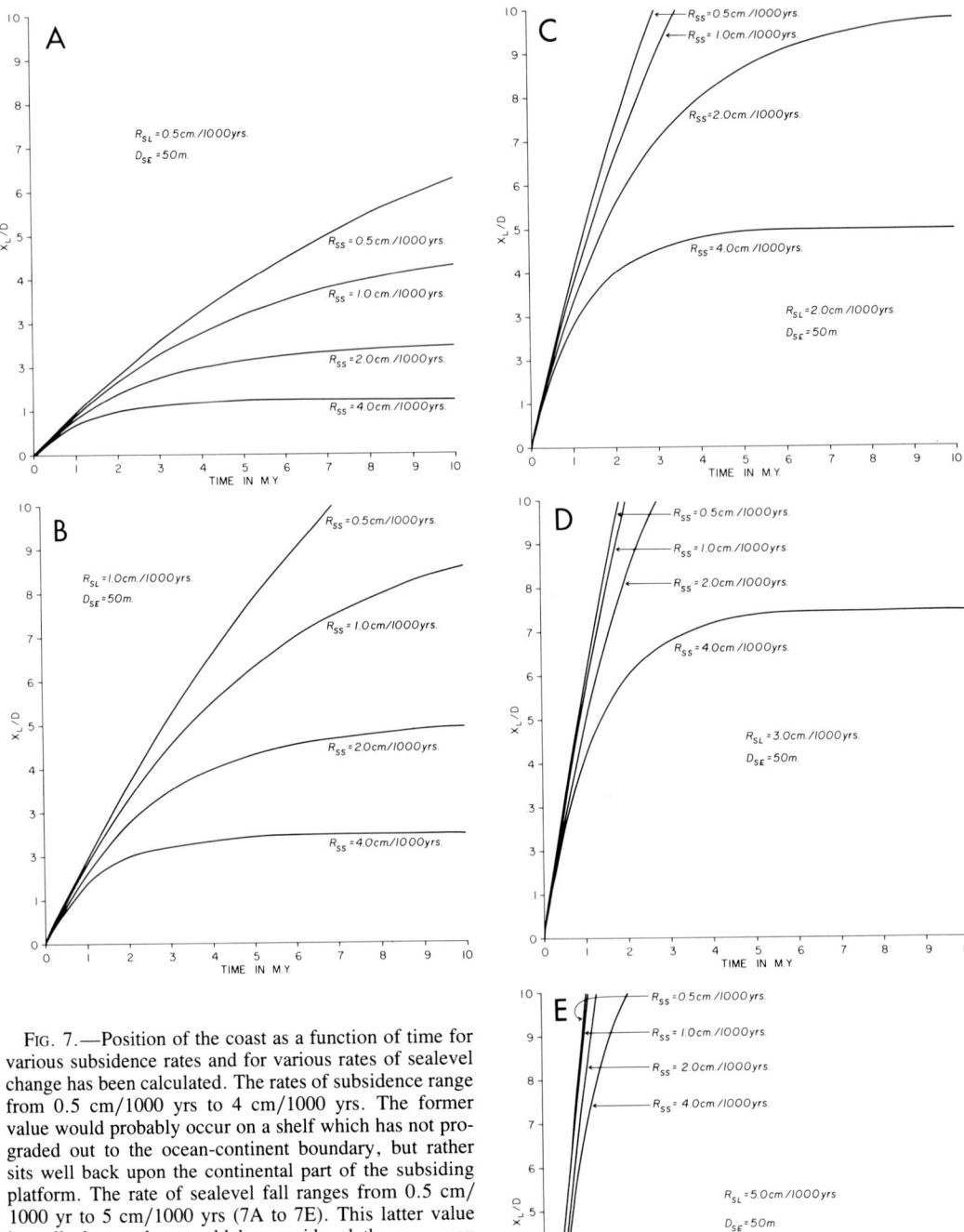

FIG. 7.—Position of the coast as a function of time for various subsidence rates and for various rates of sealevel change has been calculated. The rates of subsidence range from 0.5 cm/1000 yrs to 4 cm/1000 yrs. The former value would probably occur on a shelf which has not prograded out to the ocean-continent boundary, but rather sits well back upon the continental part of the subsiding platform. The rate of sealevel fall ranges from 0.5 cm/1000 yr to 5 cm/1000 yrs (7A to 7E). This latter value is well above what would be considered the max-max sustainable rate.

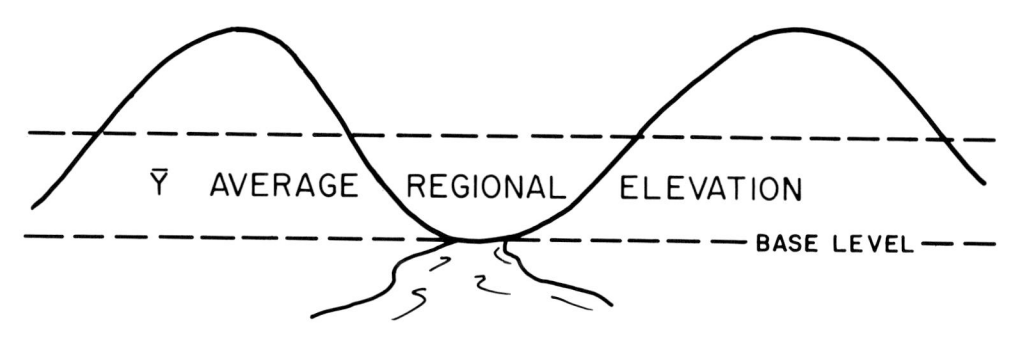

$$\frac{d\bar{Y}_D}{dt} = \text{REGIONAL DENUDATION RATE}$$

WE ASSUME THAT $\frac{d\bar{Y}_D}{dt} = K \cdot \bar{Y}$

$$\frac{\rho_c}{\rho_m} \cdot \frac{d\bar{Y}_D}{dt} = \text{UPLIFT DUE TO DENUDATION} = \frac{\rho_s}{\rho_m} \cdot K\bar{Y}$$

R_{SL} = RATE OF SEALEVEL (BASE LEVEL) CHANGE

$\frac{d\bar{Y}}{dt}$ = RATE OF DEGRADATION = RATE AT WHICH REGIONAL ELEVATION CHANGES WITH RESPECT TO BASE LEVEL

$$\frac{d\bar{Y}}{dt} = -\left[\left(K \cdot \bar{Y} - \frac{\rho_c}{\rho_m} \cdot K \cdot \bar{Y}\right) + R_{SL}\right]$$

$$Y = Y_0\, e^{-(1-\frac{\rho_c}{\rho_m})\cdot K\cdot T} - (1-e^{-(1-\frac{\rho_c}{\rho_m})\cdot K\cdot T}) \cdot \frac{R_{SL}}{(1-\frac{\rho_c}{\rho_m}) \cdot K}$$

Fig. 8.—An idealized profile across a dissected landscape. This is a mature landscape, and the river shown is at base level (sealevel).

In order to attain some estimate of the timing of these events we must first make some estimate of the rate at which the exposed shelfedge can be degraded to base level (Fig. 8). The development of these arguments is discussed in detail elsewhere by Pitman and Golovchenko (in prep.). We consider here a dissected landscape in which the rivers have been cut to base level (sealevel). The landscape need not be sinusoidal as shown in Figure 8. $\bar{Y}$ is the average regional elevation above base level and $d\bar{Y}/dt$ is the degradation rate or rate at which the regional elevation approaches base-level. $d\bar{Y}_D/dt$ is the average regional denudation rate. Following Ahnert (1970) we assume that the regional denudation rate $d\bar{Y}_D/dt$ is proportional to the average regional elevation, $\bar{Y}$

$$\frac{d\bar{Y}_D}{dt} = K\bar{Y}$$

where K is of the *order* of $10^{-4}/1000$ yrs. The rate of regional uplift caused by denudation is

$$\frac{\rho c}{\rho m} \frac{d\bar{Y}_D}{dt} = \frac{\rho c}{\rho m} K \bar{Y}$$

where ρc is the density of the material being eroded and ρm is the density at the depth of compensation. If R_{BL} is the rate of base level change (positive upwards), then

$$\frac{d\tilde{Y}}{dt} = -\left[\left(KY - \frac{\rho c}{\rho m}KY\right) + R_{SL}\right] \quad (3)$$

$$\tilde{Y} = \tilde{Y}_e e^{-(1-\rho c/\rho m)K \cdot T}$$

$$- (1 - e^{-(1-\rho c/\rho m)K \cdot T}) \cdot \frac{R_{SL}}{\left(1 - \frac{\rho c}{\rho m}\right)K} \quad (4)$$

$\tilde{Y}$ is the regional elevation at the end of the time interval T, $\tilde{Y}o$ is the regional elevation at the beginning of the time interval T, and R_{BL} is the rate of sealevel change during the time interval T.

It is of note that base level may be changed by raising or lowering sealevel or by raising or lowering the regional landscape. In our case, once the shoreline is positioned on the slope and the shelf is entirely exposed, and sealevel ceases falling, there will still be a base level change produced by thermal subsidence of the shelf. This effect will be equivalent to a sealevel rise.

The erosional constant, K, is of the order of 10^{-4}/1000 yrs. For regions such as the high Himalayas where the average regional elevation is 5 km, a value of $K = 1 \times 10^{-4}$/1000 yrs gives a regional denudation rate of 50 cm/1000 yrs. The same value of K seems appropriate for the Appalachian Mountains and gives a denudation rate of about 5 cm/1000 yrs. In our case we are concerned with unconsolidated sediments so that a value of $K = 10 \times 10^{-4}$/1000 yrs is used. This gives a denudation rate of 10 cm/1000 yrs on an exposed shelf 100 meters above sealevel.

We now consider the case of two different locations on the same margin (Figs. 9A, B, and 10). It is assumed that everywhere on this margin the rate of driving subsidence at the ocean-continent boundary is 2 cm/1000 yrs. At one location the shelf has prograded out to the ocean-continent boundary so the rate of driving subsidence at the shelfedge is 2 cm/1000 yrs and $Y_{SE} = 100$ m. However, the shelf at the other location has prograded only half-way from the hinge zone to the ocean-continent boundary. In this case the rate of driving subsidence at the shelfedge is 1.0 cm/1000 yrs and we assume that the depth to the shelf-edge $Y_{SE} = 50$ m. The rate of sealevel fall is 5 cm/1000 yrs. After 1.1 my (Fig. 10) the coastline will have regressed to the shelfedge on the margin at which $R_{SS} = 1$ cm/1000 yrs. This shelf will gradually emerge because sealevel continues to fall at 5 cm/100 yrs. The rate of emergence is given by equation 3. The total emergence for a given time interval is given by equation 4. In the equation, R_{SL} is equal to the rate of subsidence minus the rate of sealevel fall ($R_{SL} = 4$ cm/1000 yrs). The regional elevation (Yo) is equal to 0 at the time, To, when

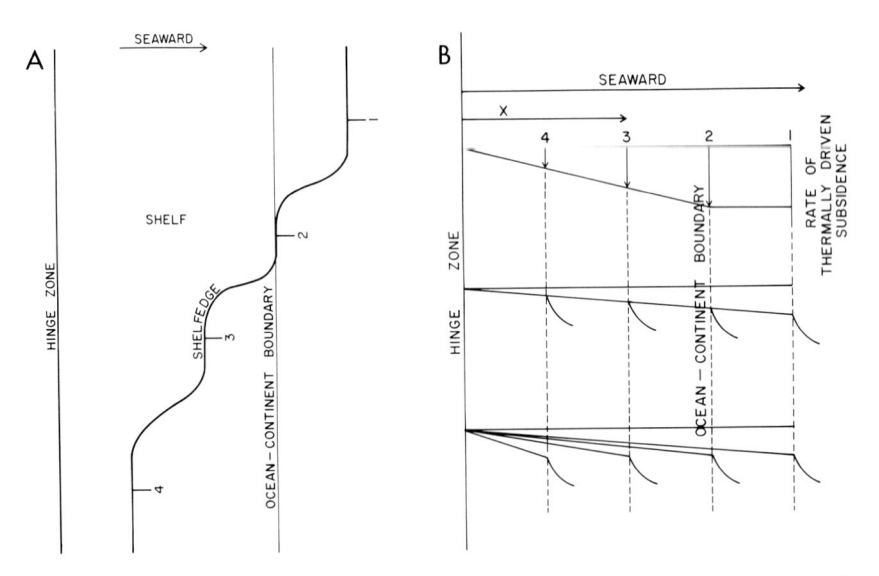

FIG. 9.—Map view of continental margin from hinge zone oceanward showing four different positions of the shelf-edge with respect to the ocean-continent boundary. The extent to which the shelf progrades out over the attenuated continental platform and onto the oceanic lithosphere is variable and depends on the width of the attenuated zone and sediment supply. In cases 4 and 3 as shown in A and B, the shelf has not prograded to the ocean-continent boundary; in these cases the subsidence rate is proportionally lower than at the ocean-continent boundary. We are assuming that at the ocean-continent boundary the driving subsidence is approximately equal to that of purely oceanic lithosphere. In case 2 the shelfedge sits at the ocean-continent boundary, and in case 1 the shelf has prograded out onto oceanic lithosphere.

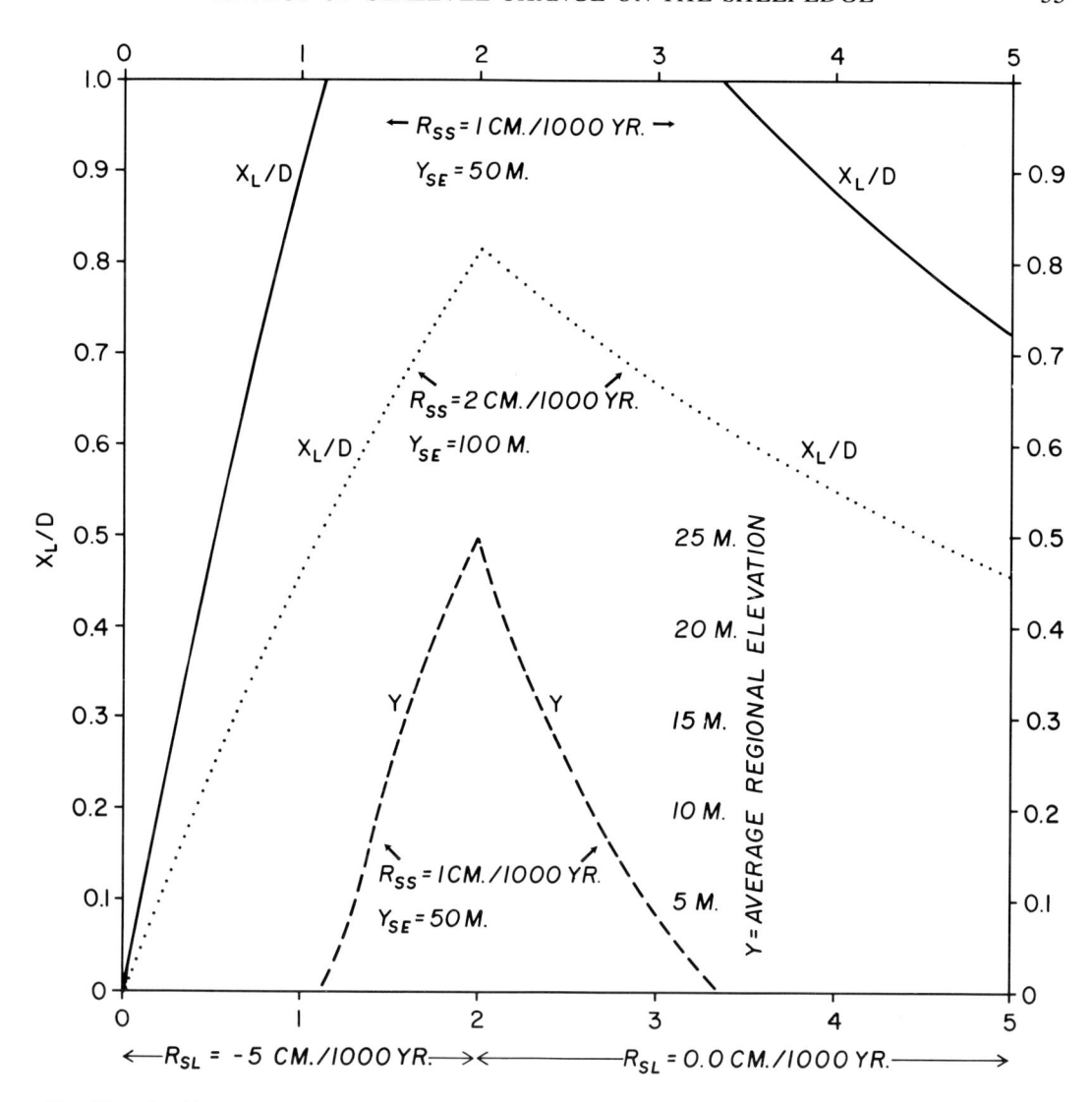

FIG. 10.—Combination degradation and regression curves for various values of driving subsidence at the shelfedge. Where the shelfedge is well back on the attenuated continental part of the platform, such as off of Nova Scotia, a sealevel fall of appropriate rate and duration may be sufficient to bring the coast out over the shelfedge on the less prograded part of the margin but not on the more prograded section. The shelfedge will continue to emerge as the shoreline retreats over the shelfedge and sealevel continues to drop. In this case the rate of sealevel drop is greater than the rate of degradation caused by driving subsidence and erosion. Sealevel at the end of 2 my has dropped 100 m and the rate of sealevel fall is decreased to 0.0 cm/1000 yrs, but it will still take another 1.37 my before the shoreline can transgress landward of the shelfedge.

the shoreline has reached the shelfedge (at 1.1 my).

The curve at the bottom of Figure 10 shows the emergence of the shelfedge that takes place between 1.1 my, when the shoreline reached the shelfedge, and 2 my, when the rate of sealevel fall decreased to zero. During this same time interval the shoreline at the other location on the shelf (R_{SS} = 2 cm/1000 yrs) has continued to move seaward. By 2 my it has traveled 0.8 of the way to the shelfedge. Sealevel has dropped 100 m, equal to the depth at the shelfedge of this margin (D_{SE} = 100 m), but this is still not sufficient to displace the shoreline to the shelfedge.

At 2 my, the rate of sealevel fall decreases to zero. The shoreline on the shelf, at which R_{SS} = 2 cm/1000 yrs, immediately begins to move landward as the shelf subsides beneath it. On the other shelf, however, the transgression cannot begin until the shelfedge has been degraded to sealevel. This takes place by a combination of subsidence and denudation. As can be seen in Figure 10, it is not until 3.35 my that the degradation is sufficient to allow a transgression to begin. In the meantime, the transgression will have proceeded on margin 2 such that the shoreline will have moved landward from $X_L/D = 0.82$ to $X_L/D = 0.46$. Thus, synchronous eustatic sealevel changes have caused non-synchronous transgressive and regressive events on the margin.

The final example (Fig. 11) is one in which we will assume that there has been a precipitous drop of sealevel of 150 m in 10,000 years, after which sealevel does not change ($R_{SL} = 0.0$). The depth (Y_{SE}) to all the shelfedges in this example is 50 m. The rate of subsidence of the shelfedge varies from $R_{SS} = 0.5$ cm/1000 yrs to $R_{SS} = 6.0$ cm/1000 yrs. This latter could only occur at or near the axis of a brand new ridge. Because the sealevel drop is so rapid, the shelfedge in each case has emerged to an elevation of 100 m ($Yo = 50$ m). For each value of subsidence rate we have calculated the degradation curve and the amount of time required to degrade the emerged shelf to sealevel. In each case, K, the erosional constant $= 10 \times 10^{-4}/1000$ yrs and $\rho_c = \rho_s = 2.0$ gm/cm^3. The transgression then begins at that point in time when the shelf has been degraded to sealevel. Again, it can be seen that synchronous sealevel changes cause apparently non-synchronous events on the shelf. In comparing the case of $R_{SS} = 4$ cm/1000 yr shelf with that of the $R_{SS} = 0.5$ cm/1000 yr shelf it is apparent that by the time the transgression begins on the latter shelf the shoreline on the 4 cm/1000 yr shelf has transgressed to within 5 km of the hinge zone.

SUMMARY

1. It has been demonstrated by other authors that driving subsidence on Atlantic-type margins is caused by cooling. Even after 170 my, the rate of

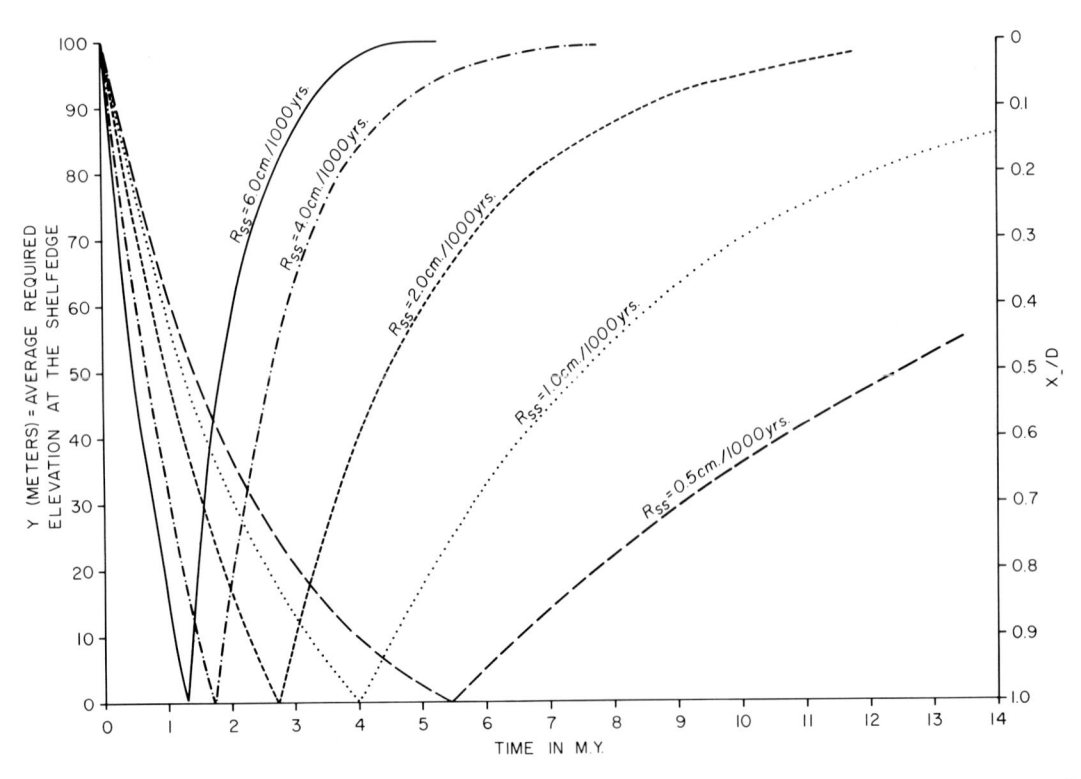

FIG. 11.—Combined degradation and transgression curves have been calculated for margins with various values of driving subsidence at the shelfedge. All cases assume $T = 0$ and $Y_{SE} = 50$ m, and also that just prior to $T = 0$ my there was an extremely rapid drop in sealevel of 150 m in less than 10,000 yrs. At $T = 0$ the rate of sealevel fall decreases to 0.0 cm/1000 yrs; at $T = 0$ the elevation of the shelfedge is 100 m ($Y = 100$ m). In each case transgression cannot begin until the shelf has been degraded to base level. This takes place by a combination of denudation and subsidence. In each case it is assumed that: $K = 10 \times 10^{-4}/1000$ yrs, $\rho_c = \rho_s = 2.0$ gm/cm^3, and $R_{BL} = R_{SS}$. The scale on left gives average elevation of the shelfedge as a function of time and is applicable only for the interval during which each shelf degrades to $Y = 0$. Note that shelves with the fastest subsidence rates degrade most rapidly. In each case the transgression begins only at the time the shelf has degraded to zero elevation. The scale on the right gives the position of the coast relative to the shelfedge.

subsidence at the ocean-continent boundary may exceed 1 cm/1000 yr. In general, the rate of driving subsidence is a maximum at the ocean-continent boundary and decreases progressively landward to zero at the hinge zone. At an old margin, if the shelfedge lies well landward of the ocean-continent boundary, the subsidence rate may be much less than 1 cm/1000 yrs and it is then possible for the shoreline to regress beyond the shelfedge. However, if the shelf has prograded out to the vicinity of the ocean-continent boundary, it is unlikely that the shoreline will regress out over the shelfedge. This results because, except during periods of major glacial fluctuation (>50 m of sealevel change), it is unusual to have periods of sustained sealevel fall in excess of 1 cm/1000 yrs.

2. We note that the water locked up in present day ice sheets is equivalent to a drop in sealevel of 50 to 60 meters. There is no evidence that indicates the presence of ice sheets larger than the present ones during the Mesozoic and Cenozoic up to the mid-Miocene. In fact, the only known glaciation for this time period took place in Antarctica (Hayes and Frakes, 1975) and may have caused sealevel fluctuations equivalent to the building and melting of 2/3 the present day ice sheet (~40 m).

3. Two conditions must be fulfilled in order that the shoreline migrate over the shelfedge: (a) First, the rate of sealevel fall must be greater than the rate of subsidence at the shelfedge. At any lesser rate of sealevel fall the coastline will migrate to that point on the subsiding platform where the rate of sealevel fall is equal to the local rate of subsidence; (b) Second, the lowering in sealevel must be of sufficient magnitude to drop sealevel down to the shelfedge. A drop of sealevel of high rate but insufficient magnitude will only displace the shoreline part way out on the shelf.

4. In order for the shoreline to remain on the subsiding part of the platform, sealevel must be falling. If sealevel is falling at a constant rate, the beach will migrate to that part of the platform which is subsiding at the same rate and stay there as long as the rate of sealevel fall does not change. This would result in what is called a stillstand, usually equated in the past with periods of unchanging sealevel. As shown here, however, such conditions must be equated with a steady drop in sealevel.

5. Several models are proposed to show what can happen if the shoreline migrated over the shelfedge and sealevel fall ceases (R_{SL} goes to zero), i.e. the shoreline will not transgress the platform until the shelf has degraded to sealevel. At that time the shoreline will begin to migrate slowly toward the hinge zone.

6. Even if changes in sealevel are synchronous worldwide, they cause non-synchronous transgressive and regressive events on various margins depending on local subsidence rates and specific geometry.

7. It has been concluded by Vail et al. (1977b) and Pitman (1978) that since the Late Cretaceous there has been a sealevel lowering of perhaps 350 m, at an average rate of about 0.5 cm/1000 yrs. Under these conditions the coastline would have generally been located well out on the shelf, particularly on old margins such as the East Coast off North America and West Africa or on slowly subsiding margins such as the North Sea. In this case a rapid build-up of glacial ice in the Antarctic, sufficient to lower sealevel by 30 m in mid-Oligocene, may have been sufficient to displace the coastline rapidly seaward to the shelfedge on many margins. If this ice did not melt and sealevel did not rise, or in other words, the rate of sealevel change was zero, it would take several millions of years for a transgression of significant magnitude to take place.

8. It must be emphasized that our inability at present to propose a mechanism that will cause lowerings in sealevel of the rate and magnitude suggested by Vail and others does not mean that such events did not occur. We can only conclude that either there is a sufficient, but as yet unknown, mechanism or that the interpretation of the sedimentary patterns in terms of sealevel changes is incorrect.

ACKNOWLEDGMENTS

The reviewer and editors have provided many suggestions which have improved the manuscript. This research was supported by National Science Foundation grants OCE-79-26308 and OCE-79-22884.

REFERENCES

Ahnert, F., 1970, Functional relationships between denudation, relief, and uplift in large, mid-latitude drainage basins: Am. Jour. Sci., v. 268, p. 243–263.

Falvey, D. A., 1974, The development of continental margins in plate tectonic theory: Aust. Petroleum Explor. Assoc. Jour., p. 95–106.

Hayes, D. E., and Frakes, L. A., 1975, General synthesis, deep sea drilling project Leg 28, in Hayes, D. E., Frakes, L. A., et al., eds., Initial Reports of Deep Sea Drilling Project, Volume 28: Washington, U.S. Govt. Printing Office, p. 919–9042.

Heestand, R. L., and Crough, S. T., 1981, The effect of Hot Spots on the oceanic depth relation: Jour. Geophys. Res., v. 86, p. 6107–6114.

Le Pichon, S., and Sibuet, J. C., 1981, Passive margins: a model of formation: Jour. Geophys. Res., v. 86, p. 3708–3720.

McKenzie, D., 1978, Some remarks on the development of sedimentary basins: Earth and Planetary Sci. Letters, v. 40, p. 25–32.

Parsons, B., and Sclater, J. G., 1977, An analysis of the variations of ocean floor bathymetry and heat flow with age: Jour. Geophys. Res., v. 82, p. 803–827.

Pitman, III, W. C., 1978, Relationship between eustacy and stratigraphic sequences of passive margins: Geol. Soc. America Bull., v. 89, p. 1389–1403.

――――, 1979, The effect of eustatic sea level changes on stratigraphic sequences at Atlantic margins: Am. Assoc. Petroleum Geologists Memoir 29, p. 453–460.

――――, and Golovchenko, X., in prep., Quantitative analysis of landscape development.

Royden, L., and Keen, C. E., 1980, Rifting process and thermal evolution of the continental margin of Eastern Canada determined from subsidence curves: Earth and Planetary Sci. Letters, v. 51, p. 343–361.

Shackleton, N. J., and Kennett, J. P., 1975, Paleotemperature history of the Cenozoic and the initiation of Antarctic glaciation: oxygen and carbon isotope analyses in DSDP sites 277, 279, and 281, in Kennett, J. P., Houtz, R. E., et al., eds., Initial Reports of Deep Sea Drilling Project, Volume 29: Washington, U.S. Govt. Printing Office, p. 743–756.

――――, and Boersma, A., 1981, The climate of the Eocene Ocean: Jour. Geol. Soc. London, v. 138, p. 153–157.

Sclater, J. G., Anderson, R. N., and Bell, M. L., 1971, Elevation of ridges and evolution of the central Eastern Pacific: Jour. Geophys. Res., v. 76, p. 7888–7915.

Steckler, M., and Watts, A. B., 1978, Subsidence of the Atlantic-type continental margin off New York: Earth and Planetary Sci. Letters, v. 41, p. 1–13.

Vail, P. R., Mitchum, R. M., Jr., and Thompson, S., III, 1977a, Seismic stratigraphy and global changes of sea level, part 3: relative changes of sea level from coastal onlap: Am. Assoc. Petroleum Geologists Memoir 26, p. 63–82.

――――, ――――, and ――――, 1977b, Seismic stratigraphy and global changes of sea level, part 4: global cycles of relative changes of sea level: Am. Assoc. Petroleum Geologists Memoir 26, p. 83–97.

Watts, A. B., and Ryan, W. B. F., 1976, Flexure of the lithosphere and continental margin basins: Tectonophysics, v. 36, p. 25–44.

――――, and Steckler, M. S., 1979, Subsidence and eustacy at the continental margin of eastern North America, in Talwani, M., Hay, W., and Ryan, W. B. F., eds., Deep Drilling Results in the Atlantic Ocean: Continental Margins and Paleoenvironment: Maurice Ewing Symposium, Series 3: Washington, D.C., Am. Geoph. Union, p. 218–234.

――――, 1981, The U.S. Atlantic continental margin: subsidence history, crustal structure and thermal evolution, in Bally, A. B., and Schreiber, B. C., eds., Geology of Passive Continental Margins: History, Structure and Sedimentologic Record: Tulsa, Oklahoma, Am. Assoc. Petroleum Geologists, p. 2-1–2-75.

PART II
STRUCTURAL AND STRATIGRAPHIC FRAMEWORK

PROGRADING SHELFBREAK TYPES ON PASSIVE CONTINENTAL MARGINS: SOME EUROPEAN EXAMPLES[1]

DENIS MOUGENOT, GILBERT BOILLOT, AND JEAN-PIERRE REHAULT
Université Pierre et Marie Curie de Paris, Laboratoire de Géodynamique sous-marine,
06230 Villefranche-sur-mer, France

ABSTRACT

This study focuses on the origin of prograding shelfbreaks on passive margins, using as examples selected seismic sections recorded on Iberian Atlantic and Western Mediterranean margins. Five main factors appear to control shelf progradation: (1) the amount and nature of sediment contributed to the outer shelf; (2) the equilibrium depth H at which sedimentary particles come to rest, a factor depending on grain-size distribution and specific hydrodynamic conditions at the depositional site (according to the models presented herein, H is the shelfbreak depth); (3) the morphology of the margin, i.e. the shelfbreak is significantly prograding only where the shelf forms on a slightly inclined slope or a marginal plateau; (4) the geological activity of the margin, i.e. prograding shelves have a sigmoid configuration on young subsiding margins, and an oblique configuration on mature, slowly subsiding margins; (5) the eustatic sealevel changes, i.e. the shelfbreak is eroded during periods of low sealevel and is built-up and progrades during periods of high sealevel. In sum, the prograding outeredge of shelves provides a fairly reliable record of Quaternary sealevel changes and of the geological evolution of margins, of which they are an integral part.

INTRODUCTION

The origin of the shelfbreak on passive margins (or stable margins, i.e. margins resulting from plate divergence) is the result of the interplay of various factors which have been summarized by many workers (see for instance Emery, 1977, or Vanney, 1977). These factors include, among others: (1) tectonic collapse of the outer shelf by normal faulting and retrogressive erosion of the continental slope (Fig. 1A), giving rise to "regrading" (landward migrating) type shelfbreaks; (2) buildup of reefs on the outer shelf resulting in barriers (Fig. 1B), whose seaward edges constitute a constructed shelfbreak; the position of this shelfbreak type remains almost constant as long as the reefs continue to grow and maintain their position relative to sealevel as the shelf subsides; and (3) sedimentary progradation of the shelf that results in seaward migrating type shelfbreaks (Fig. 1C). The potential for the above phenomena acting successively results in shelfbreaks having a complex origin. For instance, surveys of the Portuguese Shelf off Lisbon reveal a prograding outer shelf and shelfbreak built on the edge of a normal fault scarp (Fig. 2).

We are concerned in this paper primarily with the progradational type. During the course of structural studies of continental margins of western Iberia and the western Mediterranean, we have observed relevant examples of prograding shelfbreaks. The examples presented (Fig. 3) include interpretation of the seismic profiles and also take into account dominant hydrodynamic parameters, tilting and subsidence of the margin, and eustatic changes of sealevel. Nevertheless, the lack of drill holes on the studied prograding shelves precludes time-scale refinement of the interpretation and, admittedly, our models, inspired in part from Dietz and Menard (1951) and recent investigations by Dailly (1975), Sangree et al. (1978), Mitchum et al. (1977) and Boillot (1979), remain partly speculative.

OBLIQUE PROGRADATION

In the simplest case, a continental shelf is either an area of bypassing or of erosion, and the deposition of sedimentary particles occurs on the upper continental slope, seaward of the shelfbreak.

Conceptual Model (Fig. 4, Model 1)

The shelfbreak depth (H), in this model, is assumed to be controlled by local hydrodynamic and sedimentary conditions. H compares with the level of "depositional equilibrium," as defined by Moore and Curray (1964), and this latter provides a substitute for the "wave-base concept" reviewed and rejected by Dietz (1963, 1964). According to this scheme, the energy above level H prevents the deposition of the sedimentary particles, which migrate towards the shelfbreak. Below H (i.e. beyond the shelfbreak), the amount of energy decreases such that sedimentary particles settle and form a sediment talus on the upper part of the slope. Here, the prograding layers are termed "oblique" (Sangree et al., 1978).

According to the above model, three parameters must be taken into consideration:

1) The hydrodynamic conditions of water masses

---

[1]Contribution n°168 from the "Groupe d'Etude de la Marge Continentale" (ERA 605).

Copyright © 1983, The Society of
Economic Paleontologists and Mineralogists

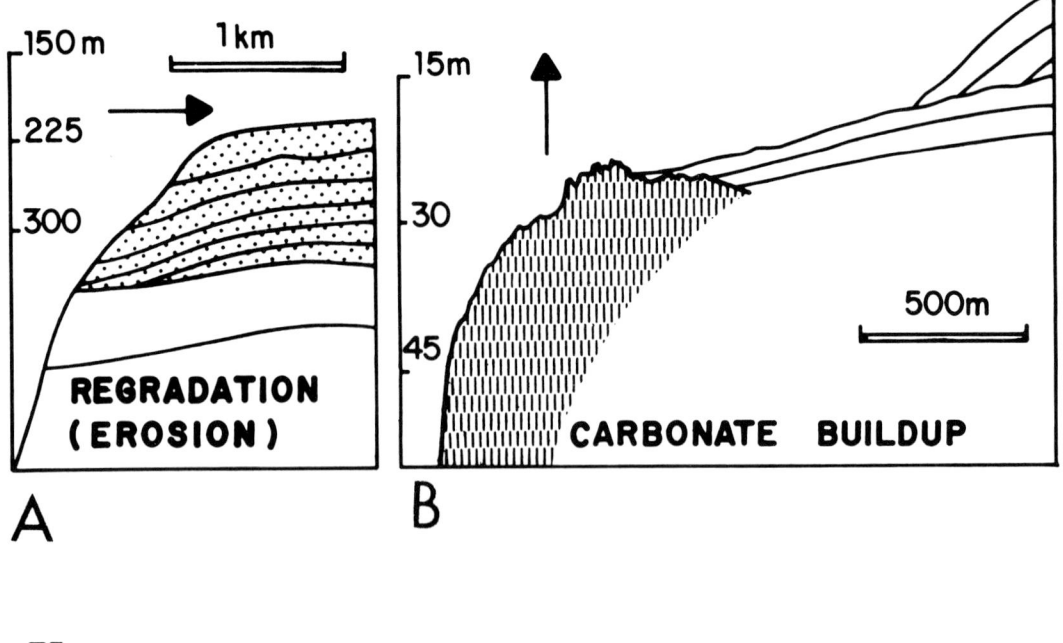

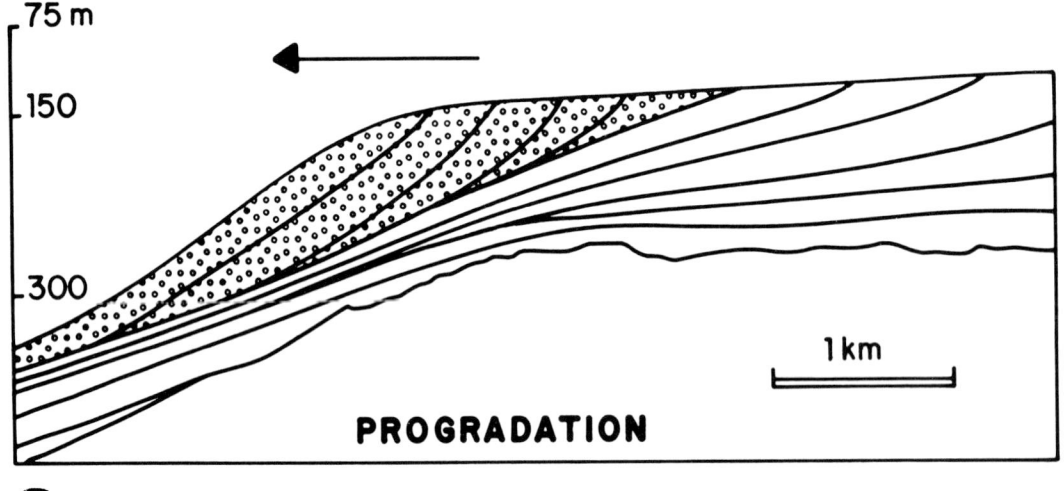

FIG. 1.—Various types of shelfbreaks on passive continental margins, based on seismic information. (A) Regrading type shelfbreak due to erosion concentrated on the upper continental slope (profile off northern Portugal, after Vanney and Mougenot, 1981); (B) Shelfbreak resulting from carbonate build-up, or reef barrier type (profile off Bahamas, after Hine et al., 1981); (C) Prograding type shelfbreak induced by horizontal transport of sediment (profile off southern Portugal, after Vanney and Mougenot, 1981). The arrows show the direction of shelfbreak migration.

near the shelfbreak, which vary from margin to margin and with time. Generally, the outer shelf to upper slope is an environment affected by high energy, where tidal currents, internal waves and other factors are active (McCave, 1972; Southard and Stanley, 1976; and chapters in Parts V and VI of this volume).

2) The amount of sedimentary input, which depends on climatic and tectonic factors, and on the distance from the shoreline where the main sources of terrigenous particles are river mouths and coastal erosion.

3) The size and density of particles, which decrease proportionally with a decrease in energy,

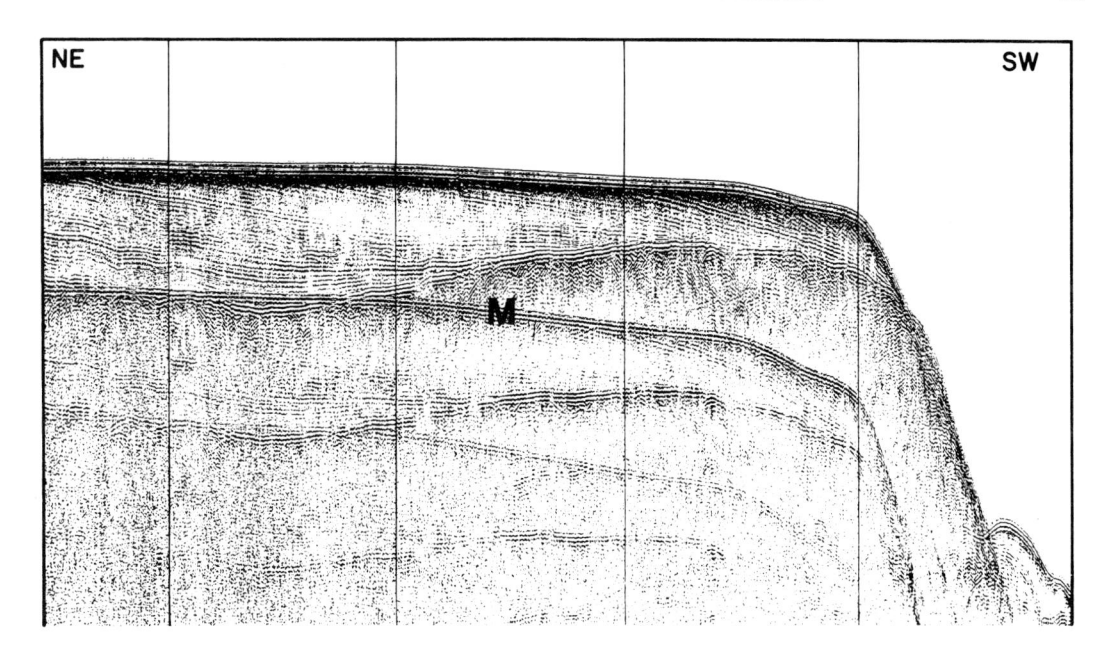

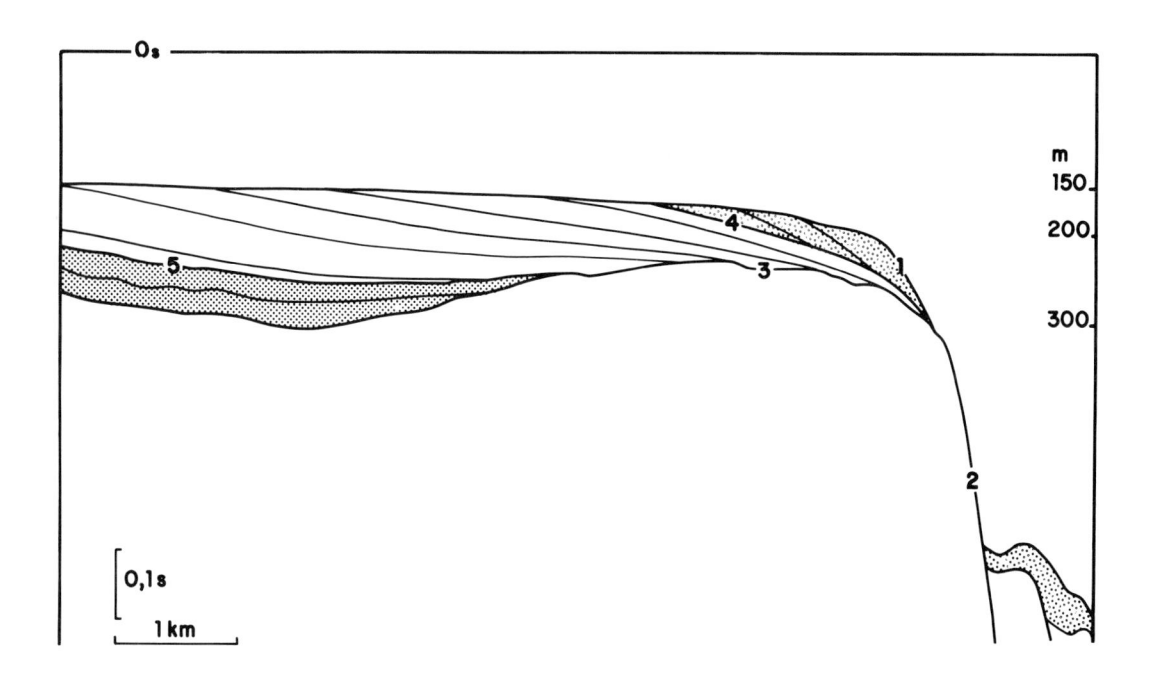

FIG. 2.—Shelfbreak having a multiple origin (Sparker seismic profile recorded on the Portuguese Shelf off Lisbon). The seaward prograding shelfbreak is parallel to a fault scarp which limits its offshore progression. Interpretation: (1) prograding sediment talus; (2) normal fault scarp; (3) Oligocene erosional surface truncating the acoustic basement; (4) discontinuity between Neogene and Quaternary prograding layers; (5) discontinuity between Middle Miocene prograding layers and Lower Miocene series; M: multiple. The position of the profile is shown in Fig. 3. Approximate vertical scale exaggeration: ×7.

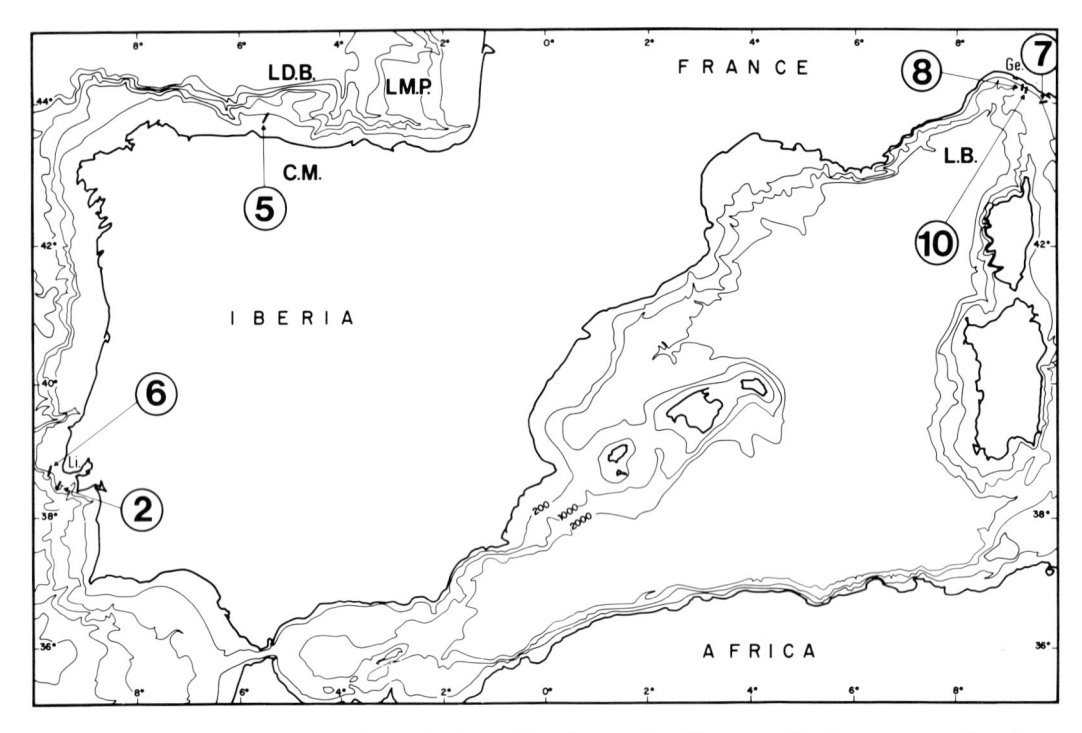

FIG. 3.—Chart showing position of the seismic profiles discussed in this paper. Numbers correspond to figures. LMP: Landes Marginal Plateau; LDB: Le Danois Bank; CM: Cantabrian Mounts: Li: Lisbon; Ge: Genova; LB: Ligurian Basin. Depths in meters.

usually with an increase in depth. The changes in equilibrium slope of talus thus depend on the sediment texture and composition.

North-Spanish Continental Shelf

The seismic profile depicted in Figure 5 shows the superposition of a slightly deformed sedimentary cover on a diffracting substratum. This substratum is composed of Mesozoic deposits folded during the early Paleogene (Boillot et al., 1971). Sediment above the unconformity contains, at its base, Late Eocene layers. The prograding wedge, 3700 m wide, is located on the outer part of the present shelf and comprises Neogene and Pleistocene units. The seaward prograding phase of the shelf has lasted, at the most, 20 million years, and the shelfbreak has migrated offshore at a rate of roughly 185 m per million years. The configuration of prograding layers is typically "oblique," and this example can be compared to the scheme outlined in the preceding section (Fig. 4, model 1).

The surficial deposits of the North-Spanish Shelf are composed of green sands mixed with a small amount of lutite (Lamboy, 1976). At present, these sediments appear to be transported periodically on the shelf, particularly near the shelfbreak where the effects of tidal currents are generally greater than on the inner shelf (Maze, 1980). On the other hand, the seismic records show that the inner continental shelf is an erosional surface which cuts the folded Mesozoic layers (Fig. 5). This surface represents a zone of sedimentary transit and bypassing and/or erosion. These observations would conform with conditions ascribed to model 1 in Fig. 4, i.e. conditions where shelfbreak depth may coincide with the "depositional equilibrium" level.

We recall, however, that present conditions were initiated less than 7000 years ago, at the end of the Flandrian transgression, when sealevel stand stabilized at about its present level. Even if the shelf is in equilibrium with the present environment, its progradation during the Holocene has been negligible (3 meters, at the most, assuming an average rate of 185 m per million years). It would appear that the shelfbreak records a relict phase that developed during an earlier Quaternary high level stand (the result of Pleistocene eustatic regressions is discussed at greater length in a following section).

Large scale progradation is not generally observed on the northern Spanish margin. It is observed only where the continental shelf extends above a marginal plateau (Le Danois Bank, north of Asturia, and the Landes Marginal Plateau, north

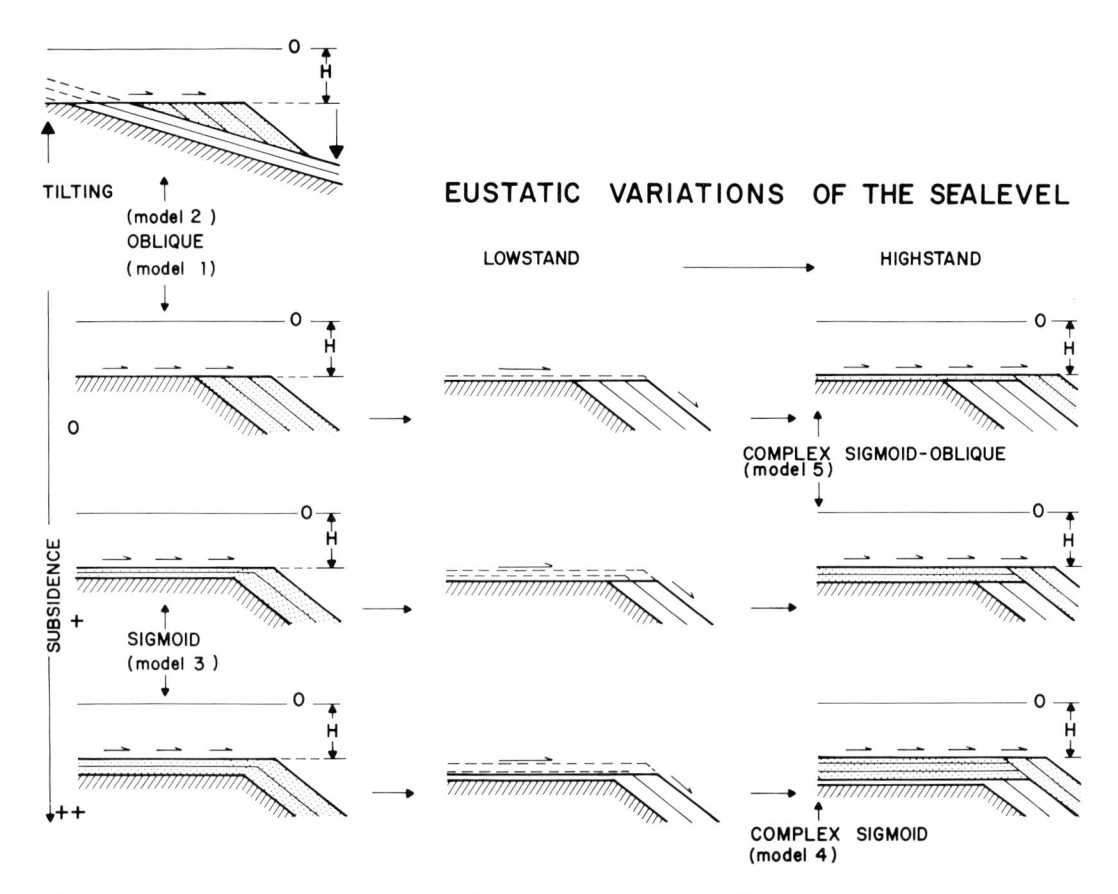

FIG. 4.—Progradational models and variants discussed in text. H is the shelfbreak depth below sealevel. H is also the assumed equilibrium level at which most sedimentary particles come to rest. The strata which have accumulated during a specific phase are stippled. Arrows indicate the direction of sedimentary by-passing and dashed lines show eroded layers.

of the coastal Basque region; Fig. 3). Elsewhere, in regions where the shelfbreak bounds a steep continental slope, the prograding wedge of the outer shelf is either small or totally missing. In that case, oblique strata developed at the top of the continental slope are probably unstable, and are likely to collapse after a period of seaward development. Thus we suggest that oblique progradation is unlikely to foster an unlimited seaward prolongation of the continental shelf.

West-Portuguese Continental Shelf

A seismic profile obtained off Lisbon (Fig. 6) shows a prograding continental shelf formed by parallel southward-dipping layers. These layers, of early Miocene age, overlie an Oligocene erosional surface which cuts Cretaceous and Paleogene formations (Baldy et al., 1977). The observed configuration is a response, in part, to tectonic tilting which occurred during Middle Miocene time and which resulted in deepening of the margin toward the south and in uplift toward the north.

In this example, oblique strata progradation results largely from tectonic movement (Fig. 4, model 2). It appears that, after shelf downwarp, the shelfbreak depth H remained constant as a result of the development of a sedimentary wedge. At the same time, erosion occurred above, in the uplifted area, where the tilted Miocene layers and underlying strata are truncated. This erosion probably resulted in formation of the prograding prism. On the same seismic record (Fig. 6), most of the shelf is covered by a thin sedimentary layer which pinches out on the outer shelf. This deposit accumulated recently (i.e. during the Holocene), and lies on a Quaternary erosional surface (Vanney and Mougenot, 1981). Near the shelfbreak, only a thin relict unconsolidated cover composed of glauconite-rich calcarenites crops out (Monteiro and Moita, 1971). Thus, the shelfbreak, like that of the northern Spanish Shelf described earlier, is most likely a Pleistocene relict feature.

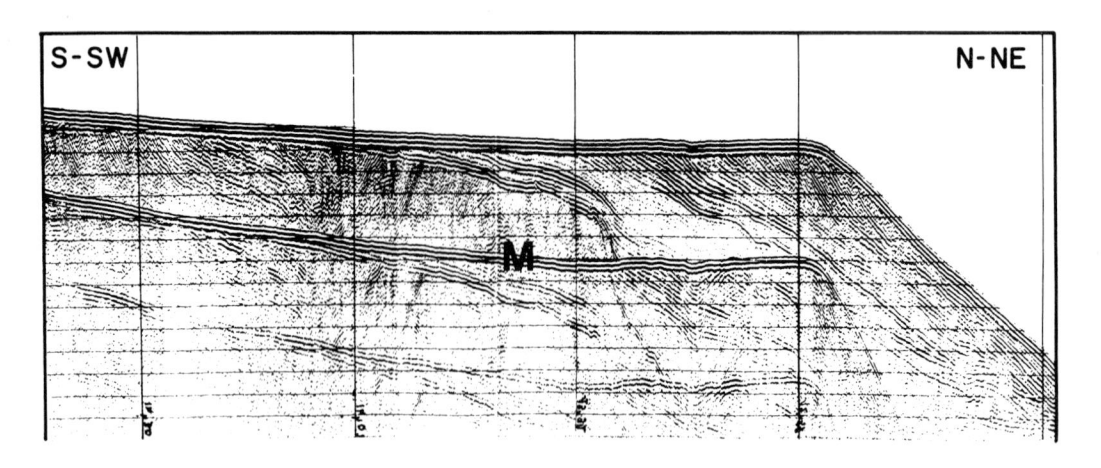

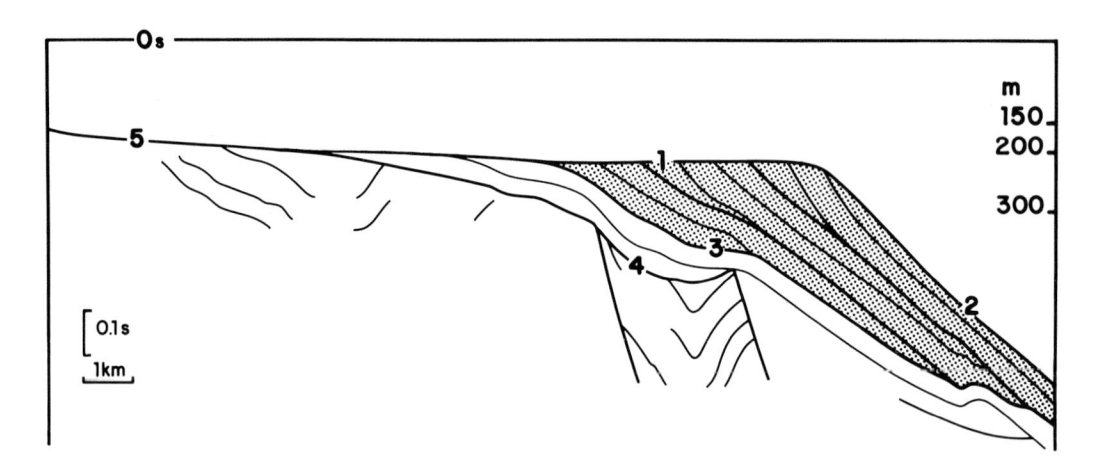

FIG. 5.—Example of an oblique-type prograding shelf (see Fig. 4, model 1); Sparker seismic profile recorded on the northern-Spanish continental shelf. Interpretation: (1) erosional or non-depositional surface truncating prograding Neogene layers; (2) prograding upper slope surface; (3) top of tilted Upper Eocene sediment strata; (4) Middle Eocene erosional surface cutting older deformed layers; (5) Quaternary erosional surface. M = multiple. The position of the profile is shown in Fig. 3. Approximate vertical scale exaggeration: ×11.

SIGMOID PROGRADATION

Sigmoid-type progradation (Sangree et al., 1978) refers to contemporaneous deposition of a sedimentary layer on the continental shelf and on the slope seaward of the shelfbreak. This configuration is most common in the case of subsiding passive margins.

Conceptual Model (Fig. 4, Model 3)

The shelfbreak depth H is assumed to be the same as in the development of oblique progradation. According to this hypothesis, sediment particles are in motion above H, and settle out below H as a result of a decrease in transport energy. Thus, as the shelf subsides, depth H remains constant as deposition continues on the shelf proper. If the volume of sediment supplied is sufficient, deposition occurs also at the break and uppermost continental slope. The major difference with the oblique progradation model is the enhanced role of margin subsidence, a significant parameter.

Gulf of Genova Continental Shelf, Example 1

Along the entire Gulf of Genova (Western Mediterranean) a thick Plio-Quaternary sequence cov-

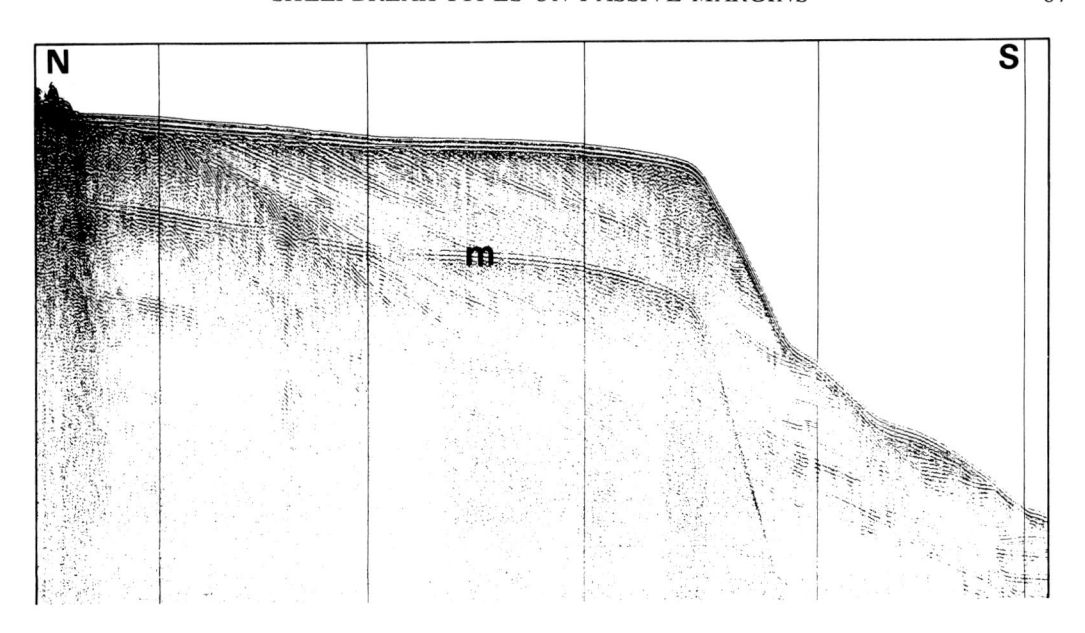

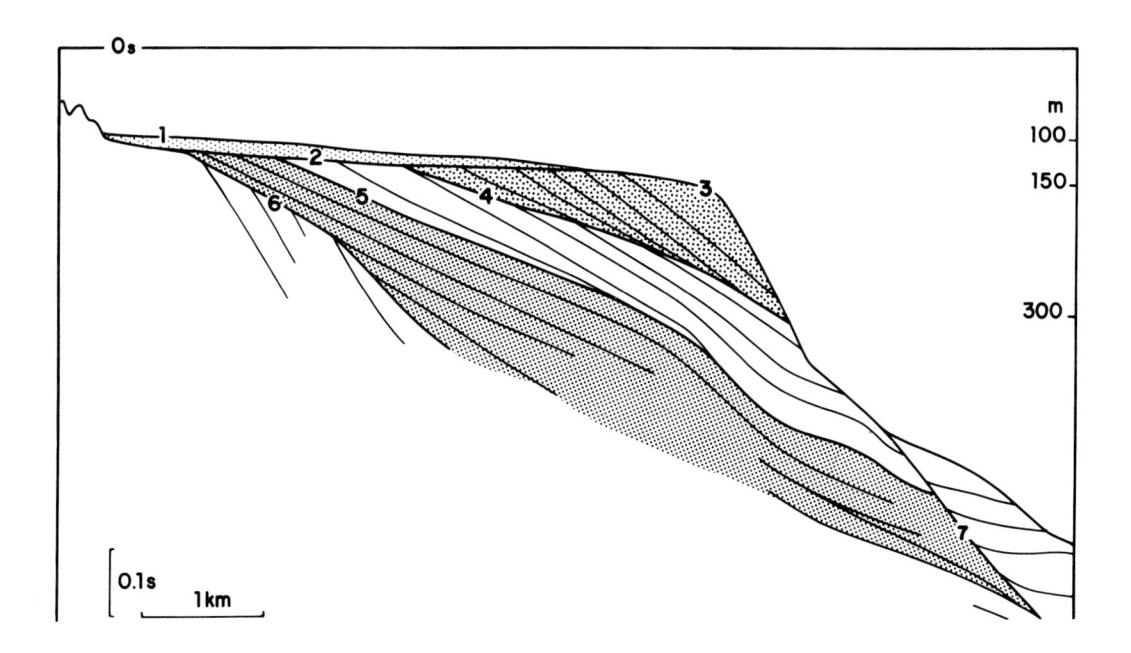

FIG. 6.—Example of oblique-type prograding shelf, resulting from tectonic tilt (see Fig. 4, model 2); Sparker profile recorded on the Portuguese continental shelf off Lisbon. Interpretation: (1) Holocene deposits forming a narrow sediment prism that pinches out near the break; (2) Pleistocene erosional surface; (3) shelfbreak formed by oblique progradation; (4) discontinuity between Middle and Upper Miocene prograding layers; (5) top of the tilted Lower Miocene formations; (6) Oligocene erosional surface truncating underlying deformed Cretaceous and Paleogene layers; (7) normal gravity fault. m = multiple. The position of the profile is shown in Fig. 3. Approximate vertical scale exaggeration: ×7.

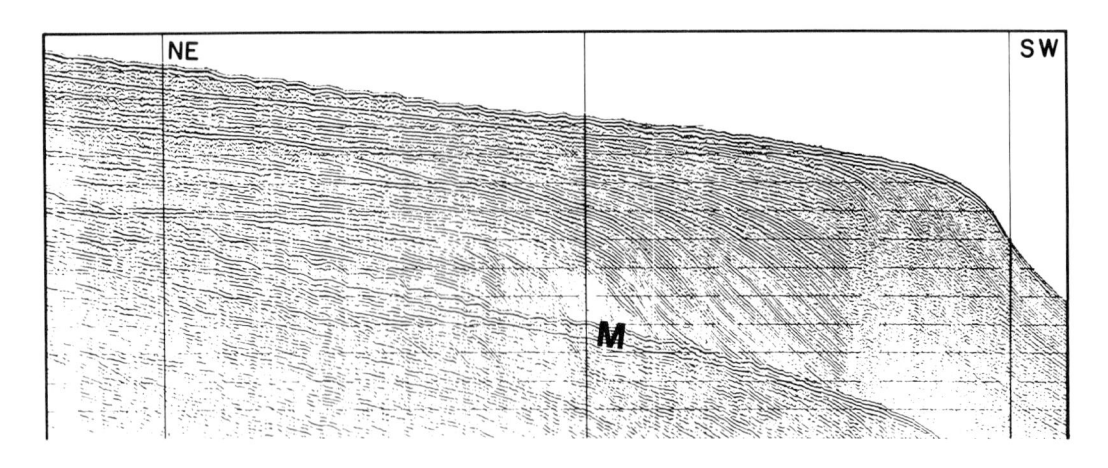

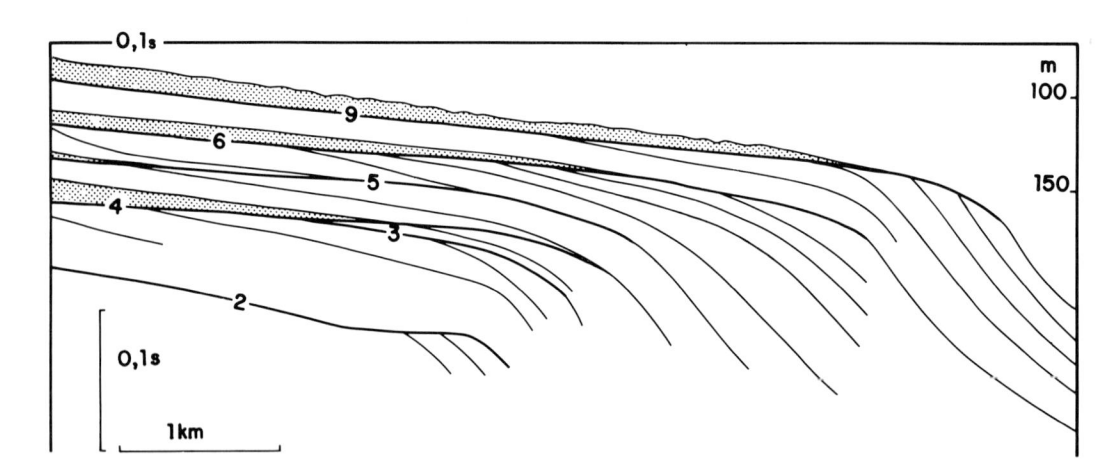

FIG. 7.—Example of sigmoid-type prograding shelf (see Fig. 4, model 3); air-gun seismic profile recorded on the Gulf of Genova continental shelf. The sigmoid-shaped discontinuities recorded by the sedimentary deposits are numbered sequentially, as the erosional surfaces in Fig. 8. The basal stratum of each sigmoid unit has been stippled in order to emphasize similarity with the Holocene layer (above 9) that covers most of the shelf. Undulations of the seafloor are sand waves produced by bottom currents. M = multiple. The position of the profile is shown in Fig. 3. Approximate vertical scale exaggeration: ×11.

ering the Messinian erosional surface forms a typical prograding shelf (Fanucci et al., 1974). Bottom sampling of surficial deposits indicates that this shelf is an area of active sedimentation (Fierro et al., 1973). Samples were recovered from the thin Holocene layer (at the most, a few meters thick) which covers the inner and middle shelf. This layer pinches out toward the shelfbreak, where outcrops of older formations are mapped.

The seaward pinch-out of the Holocene layer may be explained in two ways: intensified current erosion at the shelfedge results in reduced modern sediment accumulation; moreover, the amount of sediment input since the Holocene transgression has been insufficient to bury the entire shelf. In response to the first case, sediment particles remain in motion on the shelf as far as the break, and perhaps farther seaward. In the second case, sediment remains mobile as far as the pinch-out of the Holocene layer, as suggested by the presence of large current ripple bedforms observed on the seafloor (Fig. 7). In as much as the pinch out is located at a depth close to that of the shelfbreak, it would appear that conditions assumed by model 3 (Fig. 4) prevail, i.e. shelfbreak depth H corresponds approximately with the depositional equilibrium horizon at which sedimentary particles come to rest.

Detailed observation of prograding Plio-Quaternary deposits on the Gulf of Genova continental shelf shows an extensive suite of erosional (or nondepositional) surfaces that separate the sigmoid layers. This configuration suggests that prograding processes have not been continuous. Seismic reflectors indicate that phases of active sedimentation on the shelf were followed by phases of non-deposition, or erosion, on the shelf. This phenomenon almost certainly is a response to Quaternary eustatic sealevel changes. The effects of these eustatic changes, however, are only partially superposed on those produced by almost continuous subsidence of the shelf. Since the end of Messinian time (about 5 million years ago), the basal Pliocene horizon has subsided by about 125 m below the present seafloor, and deposition has occurred at a rate of about 25 m per million years. During the same time, the shelfbreak has prograded 2500 m seaward, or at about a rate of 500 m per million years.

The above observations may be interpreted in terms of regional geology. It is believed that oceanic accretion in the Ligurian Sea ended about 18 million years ago (Edel, 1980; Rehault, 1981), and subsidence of the Gulf of Genova shelf is a probable consequence of rapid lithospheric cooling which followed. Moreover, progradation of the shelf in the Gulf of Genova has been rapid during the past 5 million years. This is due partly to uplift of the adjoining Alps and Apennines, and to the narrow width of the shelf which facilitated the transport of large amounts of terrigenous material across the shelf and onto the slope beyond. Another aspect should also be considered: the northeastern Gulf of Genova prograding shelf is preserved only because it formed on a slope that dips slightly toward the Ligurian Basin plain (Segre, 1960; Rehault et al., 1974). Thus conditions of shelfbreak preservation are comparable to those of the northern Spanish Shelf (Fig. 5), particularly those on the Asturian marginal plateau.

EFFECTS OF EUSTATIC SEALEVEL CHANGES

Conceptual Models (Fig. 4, Models 4 and 5)

During eustatic lowering of sealevel, increased sediment transport and erosion (or non-deposition) are expected on a continental shelf surface. At such times, the shoreline approaches and may actually reach the shelfbreak, with resulting increased deposition on the continental slope and rise. In contrast, when sealevel rises and the coast is located near a position occupied prior to regression, prograding conditions at and near the shelfbreak progressively become reestablished. The amount of sediment eroded during a low stand is replaced either by an equivalent amount of sediment (Fig. 4, model 5, without subsidence), or by a larger volume of sediment (Fig. 4, models 4 and 5, with subsidence). In either case, shelfbreak depth H is closely related to sedimentary processes. In the case of a fast eustatic rise, the model highlights a two-phase sedimentary build-up: first, deposition of a horizontal layer on the shelf which reestablished depth H by replacing sediments eroded during the regressive phase and by compensating any subsidence which may have occurred during the low stand; then, deposition of oblique prograding layers beyond the shelfbreak. Thus, the deposition of oblique layers occurs only when the H datum is reached on the shelf.

Gulf of Genova Continental Shelf, Example 2

The seismic profile shown in Figure 8 was recorded on the Gulf of Genova Shelf off Sestri-Levante, about 50 km west of the section shown in Figure 7. The shelf is formed by thick prograding sediment sequences of Pliocene to Quaternary age displaying a complex sigmoid structure; these cover a Messinian erosional surface that dips gently toward the Ligurian Basin. The thickness of sedimentary build-up ranges to about 100 m, which corresponds to an average accumulation rate of 20 m per million years. The seaward extension of the shelfedge exceeds 2000 m, indicating an offshore migration averaging 400 m per million years. These calculations indicate that the Holocene progradation for the last 7000 years is minimal (less than 3 m), and thus the present shelfbreak is essentially relict, of late Pleistocene age.

Seismic reflection profiles reveal the complexity of the subsurface shelf configuration (Fig. 8). A series of discontinuities, numbered from 1 to 9, separates a rhythmic depositional sequence, numbered from H1 to H9. Each sigmoid unit consists of two parts: an upper, generally horizontal, and well-stratified layer; and a deeper, oblique talus-like deposit characterized by an acoustically transparent facies. The near-horizontal series represent outermost shelf, and oblique layers uppermost continental slope deposits. The oblique layers, in each depositional unit, partially cover the horizontal series. Deposition occurs as follows: (1) accumulation on the outer shelf of a well-stratified horizontal layer which progrades only slightly, if at all; (2) non-deposition on the outer shelf and rapid progradation of the break; (3) erosion or non-deposition on the shelf, without evidence of contemporaneous progradation. Phases (1) and (2) record essentially high sealevel stands (eustatic transgression), while phase (3) correlates with low stands (eustatic regressions).

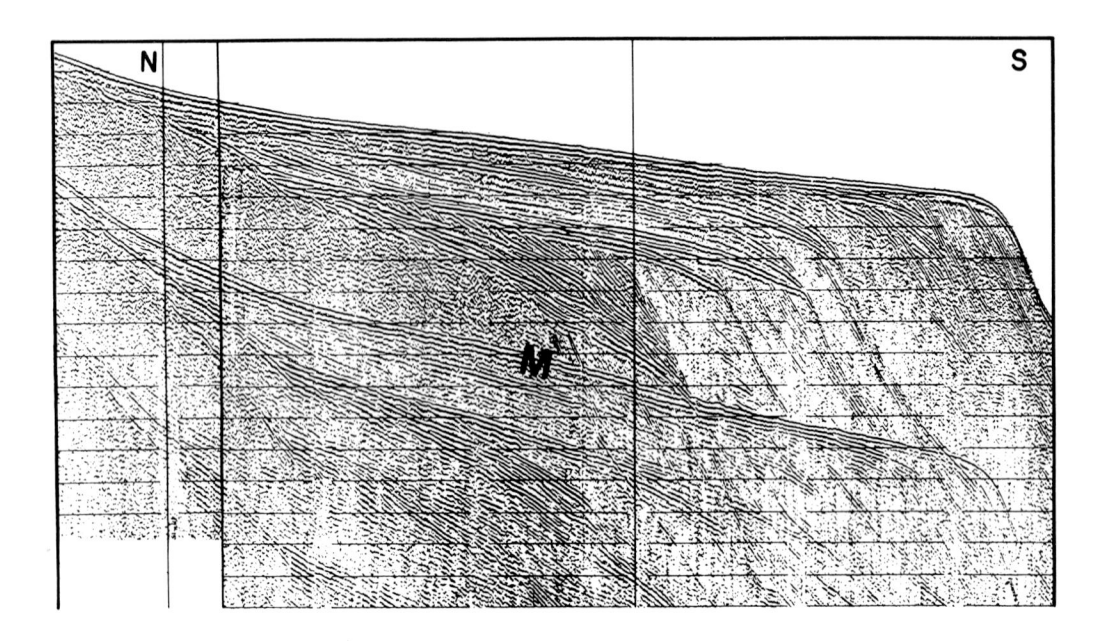

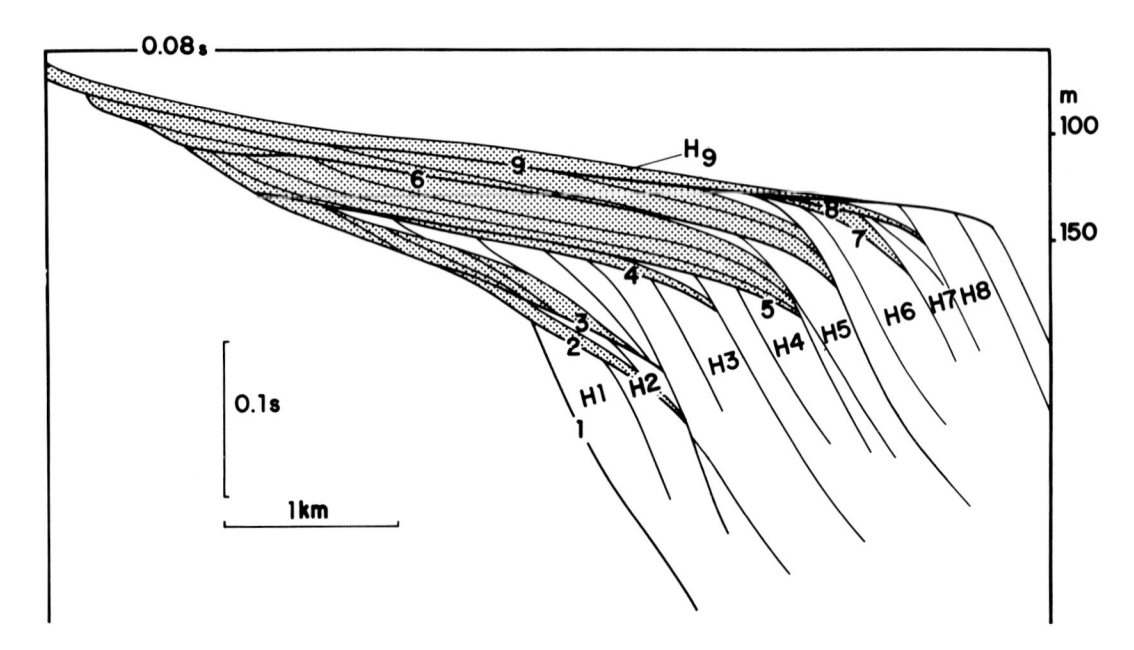

FIG. 8.—Example of complex sigmoid-type prograding shelf (see Fig. 4, model 4); air-gun seismic profile recorded on the Gulf of Genova continental shelf. Each sedimentary unit, numbered H1 to H9 (Holocene) corresponds to a high sealevel stand. The erosional surfaces are numbered 1 (probable Messinian surface) to 9 (uppermost Pleistocene). Horizontal, well-stratified layers within each sigmoid unit are stippled. M = multiple. The position of the profile is shown in Fig. 3. Approximate vertical scale exaggeration: ×11.

Figure 9, depicting the detailed evolution of the shelf with time, serves to illustrate the above interpretation. On the Gulf of Genova margin, at least nine transgressive phases, each separated by a regressive episode, have occurred since Messinian time. Of all the transgressive phases, the most recent one has left the least important imprint. This is not surprising since the Holocene sealevel rise lasted only about 12,000 years (Labeyrie et al., 1976), or an almost instantaneous time-span when considering the Neogene record. Consequently, the recent subbottom configuration provides a valuable insight on conditions that occurred at the beginning of previous eustatic high stands. A horizontal layer is deposited on the shelf, and the shelfbreak proper is a relict feature formed during preceding cycles; it is nevertheless almost in equilibrium with the present environment. Thus, shelfbreak depth will fluctuate only slightly as a new progradation cycle is initiated.

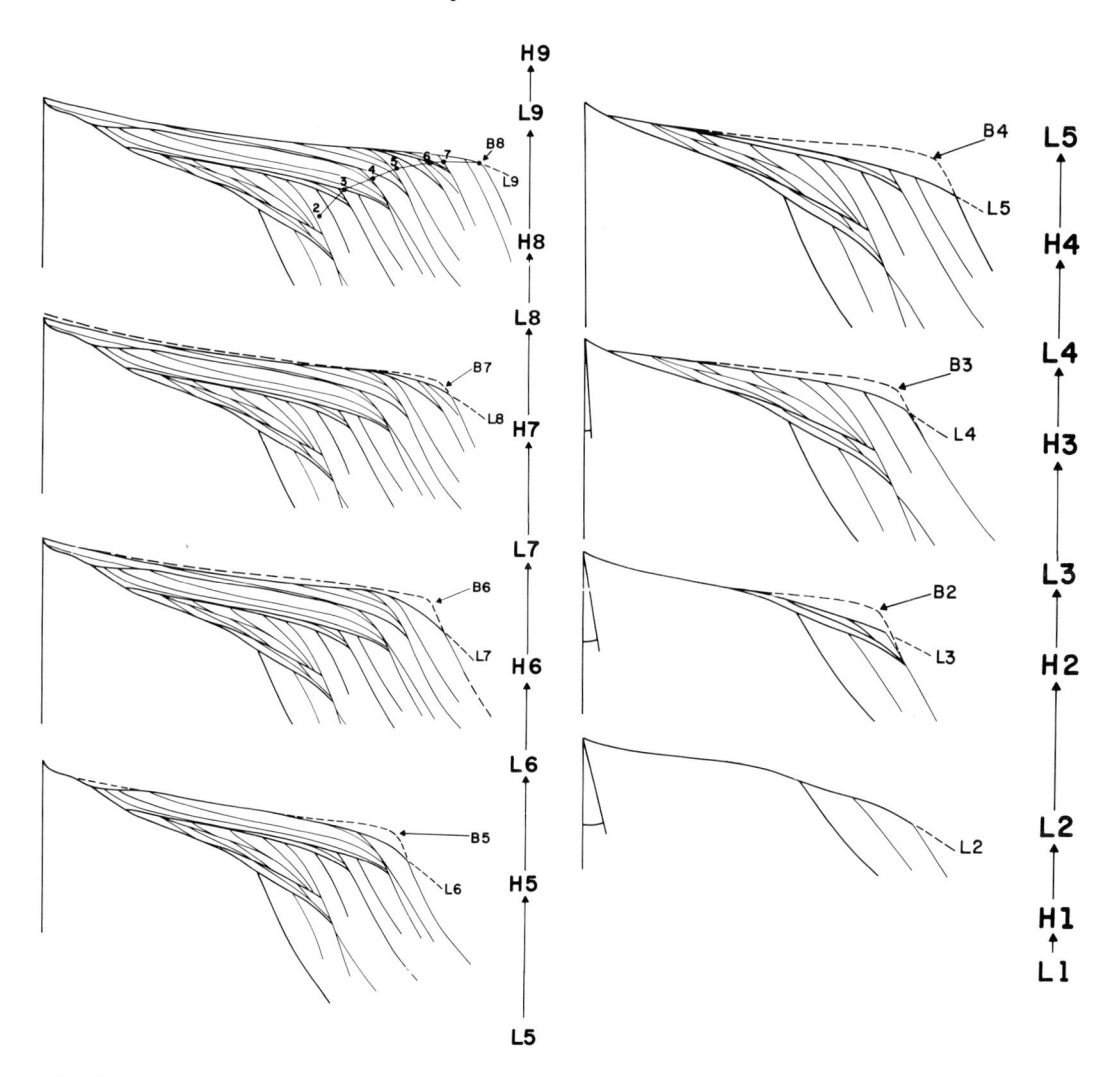

FIG. 9.—Profiles showing the subbottom configuration of the Gulf of Genova outer shelf and upper slope. Reflectors record the Plio-Quaternary evolution of the shelf shown in Fig. 8. The shelf equilibrium profile is assumed to be similar to H8 which formed prior to the deposition of the Holocene layer and prior to the weak erosion recording the previous eustatic low sealevel (L9). By superposing the configuration of profile H8 on the upper part of each sedimentary unit, one can determine the shelfbreak position (B2 to B8) during each high stand (H2 to H7), and also the amount of erosion during each successive low stand (L3 to L8). Shelfbreak development occurs in two stages. From B2 to B3, the shelf appears to record a progressive tectonic tilt as depicted on the right. This tilt likely occurred during the Pliocene. Later subsidence noted on profiles (from B4 to B8) would likely correspond to thermal subsidence affecting the margin during the Quaternary. B = shelfbreak; H = high sealevel stand; L = low sealevel stand.

Figure 9 also indicates that erosion of the shelf during eustatic regressive phases is weak, i.e. less than 20 m near the shelfbreak. Consequently, the thickness of a horizontal layer deposited during a transgressive phase serves as a base to estimate the amount of subsidence that occurred during this event and the previous regressive phase. According to this interpretation, two successive stages are associated with the Gulf of Genova subsidence: at first, margin downwarp seems to result from tectonic tilting (H1 to H3), as best highlighted by model 2 (Fig. 4). The margin then is affected by regional subsidence (from H4 to H9), and configuration of layers corresponds to that depicted by the models 3 and 4 (Fig. 4). In the Western Mediterranean, the first stage (tilting) has been well defined. Structural displacement was particularly important in the Middle and Upper Pliocene, when layers were markedly tilted in the lower reaches of the Var and Roya rivers (Bourcart, 1963; Lorenz, 1971). It is thus possible to identify the stratigraphic interval between the Pliocene and Quaternary (after the H3 sequence) on the section shown in Figures 8 and 9. In this region, the Pleistocene reveals five eustatic cycles (H4 to H8), which may be correlated with a comparable Quaternary sequence observed on the Gulf of Lion shelf (Monaco, 1971), and perhaps with five major glaciations and deglaciations, i.e. Donau, Gunz, Mindel, Riss and Würm. This interpretation, however, remains hypothetical due to lack of drilling, and we hesitate to establish a close correlation with Quaternary events described by various workers (cf. Kukla, 1977).

The profile in Figure 10, recorded near the one shown in Figure 8, shows a somewhat different structural configuration. In the former, the Messinian surface dips slightly basinward, and the sedimentary sequence of probable Pliocene age (between 2 and 4) reveals primarily oblique progradation. On the other hand, units of Quaternary age (between 4 and 9) display a sigmoid configuration. This configuration (Fig. 4, model 5) is generally termed "complex sigmoid-oblique" (Mitchum et al., 1977). The units in Figure 8 however are clearly observed in Figure 10, and the phenomena described earlier in this section of our paper have played a major role on the whole Gulf of Genova margin.

The record of glacio-eustatic fluctuations would not be preserved in a prograding type shelfedge were it not for the influence of important subsidence. The development of both northern Spanish (Fig. 5) and western Portuguese (Fig. 6) shelves has occurred in a manner comparable to that of the Ligurian Shelf (Figs. 7 and 8). In the Bay of Biscay and adjacent Atlantic margin, however, subsidence has occurred more slowly. Horizontal Plio-Pleistocene layers on these shelves are absent and the successive erosional surfaces (from L2 to L9) are not distinguishable. Only prograding layers are preserved and these do not clearly reveal the effects of eustatic variation. However, on seismic profiles, variations on slope from one oblique layer to another are observed (Figs. 2, 5 and 6), and these may correspond to effects of eustatic and/or tectonic-controlled depth variations.

DISCUSSION

From examples selected on several margins, we have attempted to synthetize the dominant factors resulting in progradational processes at and near the shelf-to-slope break, and to summarize the various sedimentary configurations resulting from the interplay of these factors. The major parameters likely to induce progradation are briefly considered.

Sedimentary Input

The nature of prograding shelfedge sedimentary deposits remains ill-defined, due to a lack of drill-holes. It is believed, however, that these are essentially detrital in origin. The outer prograding portions of shelves of the Gulf of Genova and northern Spain lie seaward of high relief (Alps-Apennines and Cantabric Mountains), land areas which have been subjected to intense erosion; the Portuguese Shelf is maintained by fluvial deposits (derived largely from the Tejo and Sado rivers).

All the examples cited herein occur in regions where the shelf is moderately to very narrow (between 10 and 30 km). In such cases, the distance between the main source of detrital input (river mouths and coastal erosion) and the shelfbreak is reduced. This attribute reduces the amount of sediments trapped on the shelf, and favors shelfbreak progradation.

Local Hydrodynamic Conditions

Bottom current energy generally decreases with depth, and at a critical level, defined as H, deposition of sediment particles occurs. For a given grain size and density, H represents the depth at which the average accumulation rate becomes equivalent to erosional rate. According to this concept, sediment particles remain mobile landward of the shelfbreak, while below it, particles come to rest. This results in a stable layer and progradation at the break. Our examples of prograding shelfbreaks record responses to a previous high sealevel. However, the sum of observations suggests that present conditions are generally comparable to those existing during the preceding late Quaternary high stand. Moreover, variations in the H equilibrium level (-150 m in the western Mediterranean, and to -200 m in the Atlantic seaward of the Iberian Peninsula) record differences in hydrodynamic conditions affecting diverse margins in different seas.

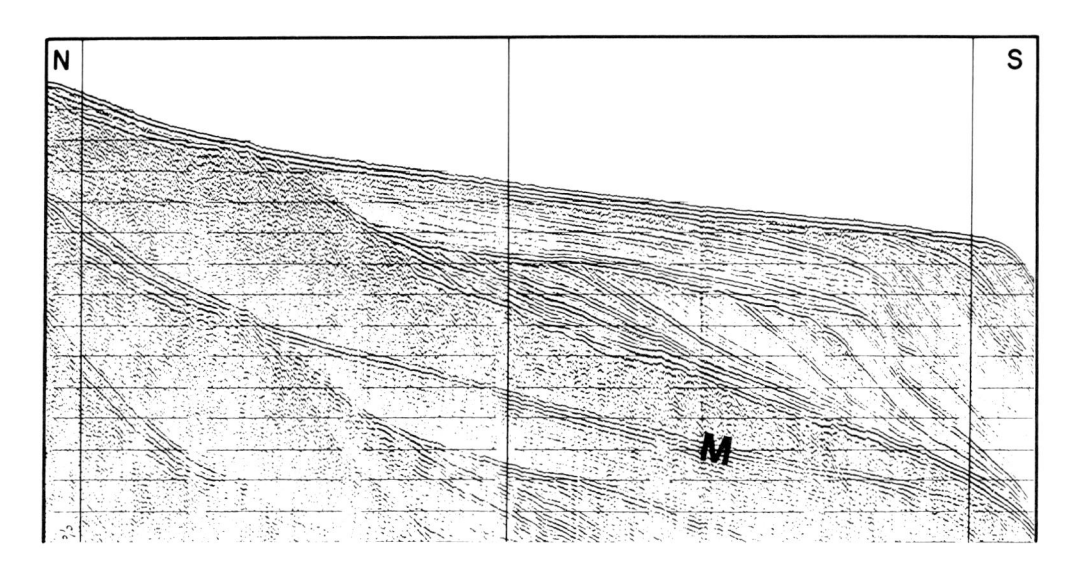

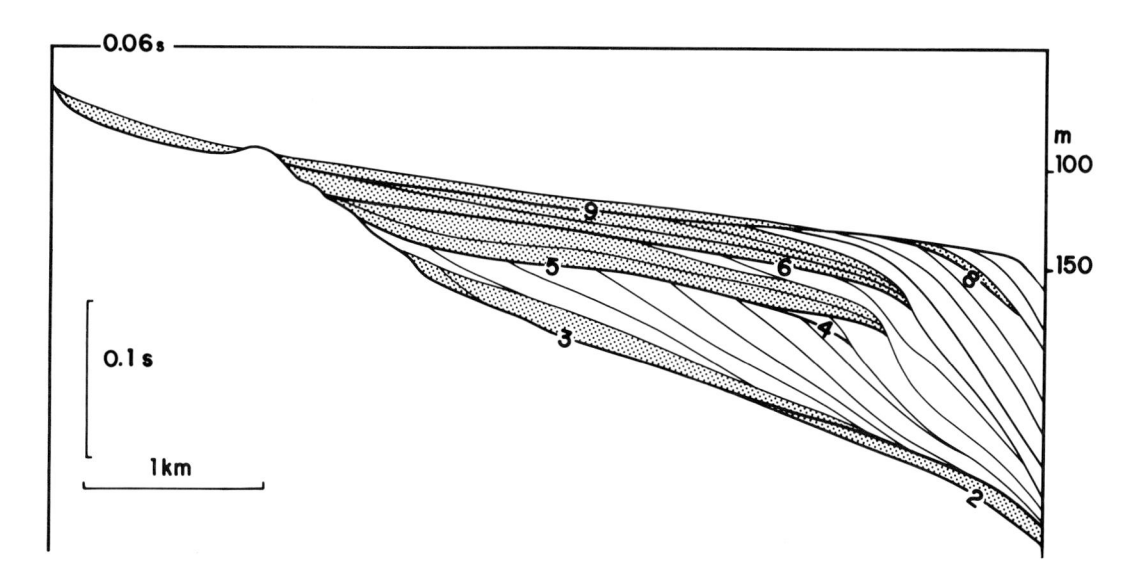

FIG. 10.—Example of complex sigmoid-oblique-type prograding shelf (see Fig. 4, model 5); air-gun seismic profile recorded on the Gulf of Genova continental shelf. The erosional surfaces separating each sigmoid or oblique sedimentary unit are numbered sequentially, as in Fig. 8. Subhorizontal, well-stratified layers are stippled. Oblique layers between discontinuities 3 and 4 resulted from Pliocene tilting (see Fig. 4, model 2); tilting also may have induced the variable dip of older discontinuities (2 and 3) Approximate vertical exaggeration: ×11.

Grain-Size Distribution of Prograding Sediments

Seismic surveys in the Gulf of Genova suggest that subbottom sediment series forming the shelf-edge in this region are composed primarily of silts and clays, whereas the superficial samples and a few drillholes on the Portuguese Shelf (Mougenot, 1979) record an enhanced proportion of detrital sands. It is obvious that such grain-size differences, plus variations of particle density (terrigenous versus carbonate components) influence the H equilibrium level. This, in turn, affects shelfbreak depth and also the equilibrium slope on the upper part of a prograding talus.

Margin Physiography

Active progradation occurs on slopes that are gently inclined seaward (case of the Gulf of Genova Shelf) or on marginal plateaus (case of the Iberian shelves). In contrast, a prograding shelfedge is unstable and subject to failure (slumping) when it is positioned above a steeply inclined continental slope such as off Lisbon (Fig. 2). Continental shelf progradation is thus extensive only in certain settings. For example, on steep slopes (Fig. 1A) progradation is limited in size and thickness, and thus the break results from the interplay of other processes.

Tectonic Movement

Tilting of crustal blocks forming the margin may give rise to some conditions described in the preceding paragraph (development of a marginal plateau or of a slightly inclined surface) which would favor the formation of a prograding shelf. For instance, west of the Iberian Peninsula, increased sediment flux and formation of a gently inclined slope, two conditions suitable for the deposition of prograding wedges, are related to the tilt of underlying Lower Miocene layers (Fig. 6). Similarly, offset of the Messinian surface has played a major role in the Plio-Quaternary progradation recorded at the shelfedge throughout the entire Gulf of Genova region (Figs. 7, 8, 10).

Subsidence

Subsidence is the major controlling factor which results in different types of prograding shelfbreaks formed during the same time period. In the case of pronounced subsidence, sediments accumulate both horizontally on the shelf proper and obliquely on the upper part of the continental slope. The resulting depositional configuration is sigmoid, and examples are likely to be observed on young margins (Fig. 11) that are affected by rapid thermal cooling (case of Ligurian Basin margin). If subsidence is minor, very slow or non-existent, sediments are likely to settle primarily below the shelfbreak on the continental slope, producing a slight dip as observed on seismic profiles. The configuration of shelfbreak deposits is thus oblique, and the upper part of each stratum appears truncated by an erosional or non-depositional surface. The latter case is commonly associated with mature, slowly subsiding margins (Fig. 11). An example is the Atlantic continental margin off the Iberian Peninsula. The sigmoid-oblique configuration (superposition of both the above-cited configurations) is also a common one, and often results from the additional influence of eustatism.

Eustatic Sealevel Changes

In our interpretation of acoustic-sedimentary configurations recorded on seismic profiles, we assume that Plio-Quaternary changes of sealevel occurred rapidly (instantaneously when considered in light of the geological time-scale). In contrast, it is probable that sealevel had occupied its highest and lowest stands for extended periods of time. This agrees with eustatic curves determined by means of other data and methods (see, for instance, Pastouret et al., 1980). The various types of progradational wedges are summarized in terms of a natural succession of sedimentary cycles, each composed of four terms:

(1) Low Sealevel Stand and Erosion on the Shelf.—A small amount of erosion is generally associated with this phase. On the Gulf of Genova margin (Fig. 8), maximum erosion is concentrated near the shelfbreak, but the eroded layer is thin (less than 20 m on Fig. 9). Because of rapid compaction of the sediment sequence, erosion plays a relatively reduced role, except perhaps in the head of submarine canyons. During such lowstand phases, sediment transport is in large part diverted towards the rise via the continental slope and submarine canyons; this shelfbreak-slope bypassing results in the development of the rise wedge and deep-sea fans. On slightly subsiding margins, this erosional phase can result in the complete removal of the thin horizontal stratum deposited on the shelf during the previous high-level stand. In this case, an oblique configuration follows the earlier sigmoid configuration (Fig. 4).

(2) Return to a High Sealevel Stand.—This rise results in the deposition of a new horizontal stratum on the shelf. This sediment accumulation replaces the sediment removed during the previous erosional stage and possibly also compensates for the subsidence that may have occurred during the low sealevel stand. According to this scheme, sediments are deposited on the continental shelf and are no longer bypassed to the deep margin (Gibbs, 1981). This phase is the one prevailing at present (Aloisi et al., 1977).

(3) High Sealevel Stand; Equilibrium Level Attained.—When the equilibrium level H is attained, sediments form a horizontal layer across the shelf, and particles also bypass the shelfbreak, thus accumulating on the upper continental slope. An oblique progradational pattern prevails during this phase.

(4) Return to a Low Sealevel Stand.—At the beginning of a eustatic regression, progradation may in some cases continue for a short time on the upper continental slope. Erosion, however, prevails on the shelf proper.

Events related to phases (2) and (3) cited above result in the development of a sigmoid-shaped layer which may form upon an oblique type progradational layer. In this manner, the complex sigmoid-oblique configuration (Fig. 4, model 5), commonly observed on Plio-Quaternary prograding shelves, develops.

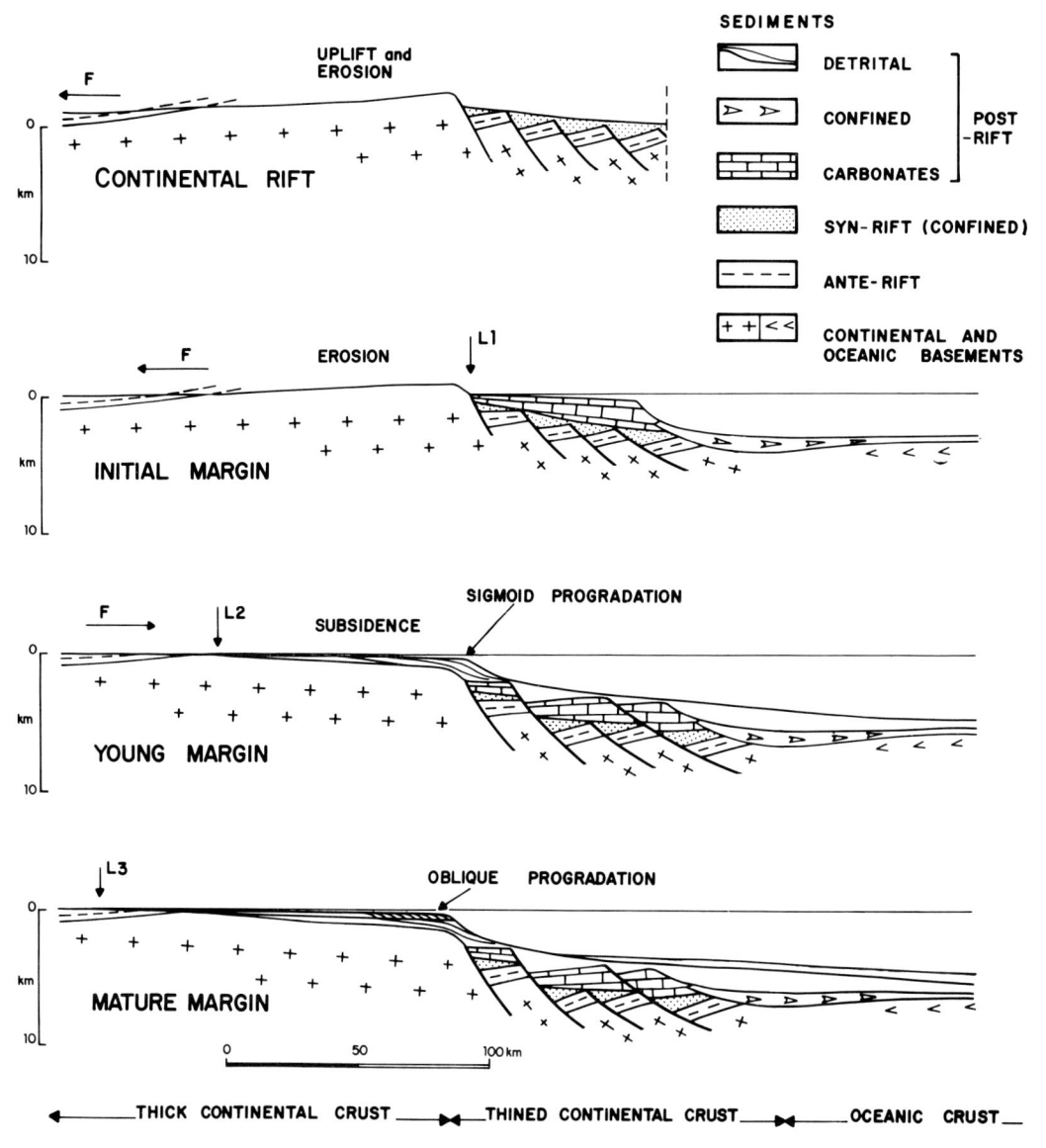

FIG. 11.—Interpretation of the progressive evolution of a passive continental margin, starting from continental rift phase and ending at mature phase as exemplified by present Atlantic margins (modified after Boillot, 1979). The sediments involved in this evolution are differentiated by their lithofacies and by their stratigraphic position in the development of the margin. The sigmoid progradation shelfbreak type is associated with subsidence of young margins (e.g. Western Mediterranean margins). The oblique progradation shelfbreak type is more typically associated with the slightly subsiding mature margins (e.g. Iberian Atlantic margins). L1 to L3 = successive shoreline position. F = direction of major sediment transport.

SUMMARY

Seismic profiles of prograding shelves and associated shelf-to-slope breaks record the importance of Quaternary eustatic oscillations. The dominant factor controlling progradational processes is hydrodynamics. We assume a "depositional equilibrium" level (H) for transport of sediment particles, H being the shelfbreak depth. Physiography also is an important factor: prograding wedges are preserved only if they are formed on a gently inclined slope or a marginal plateau. A third controlling factor is margin subsidence. Sigmoid progradation occurs on young, rapidly subsiding margins; oblique progradation occurs on mature, slowly subsiding margins. Detailing the configuration of prograding shelves, as observed on seismic profiles, is most useful for interpreting the geological evolution of passive margins.

ACKNOWLEDGMENTS

We gratefully acknowledge useful suggestions provided by D. J. Stanley and his help in translating the French manuscript. We also acknowledge M. Gennesseaux, J. A. Malod, A. Monaco, J. R. Vanney, and reviewers, for valuable comments and discussion.

REFERENCES

ALOISI, J. C., AUFFRET, G. A., AUFFRET, J. P., BARUSSEAU, J. P., HOMMERIL, P., LARSONNEUR, C., AND MONACO, A., 1977, Essai de modélisation de la sédimentation actuelle sur les plateaux continentaux français: Soc. Géol. France Bull. 7, v. 19, p. 183–195.
BALDY, P., BOILLOT, G., DUPEUBLE, P. A., MALOD, J. A., MOÏTA, I., AND MOUGENOT, D., 1977, Carte géologique du plateau continental sud-portugais et sud-espagnol (Golfe de Cadix): Soc. Géol. France Bull. 7, v. 19, p. 703–724.
BOILLOT, G., 1979, Géologie des Marges Continentales: Paris, Masson, 150 p. [English edition translated by A. Scarth, 1981, London, Longman, 140 p.]
———, DUPEUBLE, P. A., LAMBOY, M., D'OZOUVILLE, L., AND SIBUET, J. C., 1971, Structure et histoire géologique de la marge continentale au nord de l'Espagne (entre 4° et 9°W), in Histoire Structurale du Golfe de Gascogne: Paris, Technip, p. V.6-1–V.6-52.
BOURCART, J., 1963, La Méditerranée et la révolution du Pliocène, in Livre à la Mémoire du Professeur Paul Fallot: Soc. Géol. France Mém. H.-Ser., v. 1, p. 103–116.
DAILLY, G., 1975, Some remarks on regression and transgression in deltaic sediments, in Canada's continental margins and offshore petroleum exploration: Canada Soc. Petroleum Geologists Memoir 4, p. 791–820.
DIETZ, R. S., 1963, Wave-base, marine profile of equilibrium, and wave-built terraces: A critical appraisal: Geol. Soc. America Bull., v. 74, p. 971–990.
———, 1964, Wave-base, marine profile of equilibrium, and wave-built terrace: reply: Geol. Soc. America Bull., v. 75, p. 1275–1282.
———, AND MENARD, H. W., 1951, Origin of abrupt change in slope at continental shelf margin: Am. Assoc. Petroleum Geologists Bull., v. 35, p. 1994–2016.
EDEL, J. B., 1980, Etude paléomagnétique en Sardaigne. Conséquences pour la géodynamique de la Méditerranée occidentale: [Thesis], Institut Physique du Globe Strasbourg, Univ. of Strasbourg, 310 p.
EMERY, K. O., 1977, Stratigraphy and structure of pull-apart margins, in Geology of Continental Margins: Am. Assoc. Petroleum Geologists Continuing Education Course 5, p. B1–B20.
FANUCCI, F., FIERRO, G., REHAULT, J. P., AND TERRANOVA, T., 1974, Le plateau continental de la mer Ligure de Portofino à La Spezia: Étude structurale et évolution plio-quaternaire: C. R. Acad. Sci. Paris D., v. 279, p. 1151–1154.
FIERRO, G., GENNESSEAUX, M., AND REHAULT, J. P., 1973, Caractères structuraux et sédimentaires du plateau continental de Nice à Gênes (Méditerranée nord-occidentale): [Thesis], Bur. Rech. Géol. Miner. Bull., 2è ser., sect. 4, no. 4, p. 193–208.
GIBBS, R. J., 1981, Sites of river-derived sedimentation in the ocean: Geology, v. 9, p. 77–80.
HINE, A. C., WILBER, R. J., AND NEUMANN, A. C., 1981, Carbonate sand bodies along contrasting shallow bank margins facing open seaways in northern Bahamas: Am. Assoc. Petroleum Geologists Bull., v. 65, p. 261–290.
KUKLA, G. J., 1977, Pleistocene land-sea correlations. I. Europe: Earth-Science Rev., v. 13, p. 307–374.
LABEYRIE, J., LALOU, C., MONACO, A., AND THOMMERET, J., 1976, Chronologie des niveaux eustatiques sur la côte du Roussillon de −33.000 ans B.P. à nos jours: C. R. Acad. Sci., Paris D, v. 282, p. 349–352.
LAMBOY, M., 1976, Géologie marine et sous-marine du plateau continental au nord-ouest de l'Espagne. Genèse des glauconies et des phosphorites. [Thèse d'Etat]: Rouen, Univ. of Rouen, 285 p.
LORENZ, C., 1971, Observations sur la stratigraphie du Pliocène ligure. La phase tectonique du Pliocène moyen: C. R. Somm. Soc. Géol. France, no. 8, p. 441–445.
MAZE, M. R., 1980, Formation d'ondes internes stationnaires sur le talus continental. Application au Golfe de Gascogne: Ann. Hydrogr. 5è ser., v. 8, no. 754, p. 45–58.
McCAVE, I. N., 1972, Transport and escape of fine-grained sediment from shelf areas, in Swift, D. J. P., Duane, D. B., and Pilkey, O. H., eds., Shelf Sediment Transport: Process and Pattern: Stroudsburg, Pa., Dowden, Hutchinson and Ross, p. 225–248.
MITCHUM, R. M., VAIL, P. R., AND SANGREE, J. B., 1977, Seismic stratigraphy and global changes of sealevel; Part 6: Stratigraphic interpretation of seismic reflection patterns in depositional sequences, in Seismic stratigraphy—Applications to hydrocarbon exploration: Am. Assoc. Petroleum Geologists Memoir 26, p. 117–133.
MONACO, A., 1971, Contribution à l'étude géologique et sédimentologique du plateau continental du Roussillon (Golfe du Lion) [Thèse d'Etat]: Univ. Sci. Techn. Languedoc, Montpellier, 295 p.
MONTEIRO, J. H., AND MOÏTA, I., 1971, Morphologia e sedimentos da plataforma continental e vertente continental superior a largo da peninsula de Setubal: 1er Congres Luso-Ispano-Amer. Geol. Econ., sec. 6, p. 301–330.
MOORE, D. G., AND CURRAY, J. R., 1964, Wave-base, marine profile of equilibrium, and wave-built terraces: Discussion: Geol. Soc. America Bull., v. 75, p. 1267–1274.

Mougenot, D., Monteiro, J. H., Dupeuble, P. A., and Malod, J. A., 1979, La marge continentale sud-portugaise; Evolution structurale et sédimentaire; Ciências da Terra, v. 5, Univ. Nov. Lisbon, p. 223–246.

Pastouret, L., Auzende, J. M., Le Lann, A., and Olivet, J. L., 1980, Témoins des variations glacio-eustatiques du niveau marin et des mouvements tectoniques sur le banc de Gorringe (Atlantique du nord-est): Palaeogeogr. Palaeoclimatol. Palaeoecol., v. 32, p. 99–118.

Rehault, J. P., 1981, Evolution tectonique et sédimentaire du Bassin Ligure (Méditerranée occidentale) [Thèse d'Etat]: Paris, Université Paris VI, 2 vols., 132 p.

———, Olivet, J. L., and Auzende, J. M., 1974, Le bassin nord-occidental méditerranéen; Structure et évolution: Soc. Géol. France Bull. 7, v. 16, p. 281–294.

Sangree, J. B., Waylett, D. C., Frazier D. E., Amery, G. B., and Fennessy, W. J., 1978, Recognition of continental-slope seismic facies, offshore Texas-Louisiana, in Bouma, A. H., Moore, G. T., and Coleman, J. M., eds., Framework, facies, and oil-trapping characteristics of the upper continental margin: Am. Assoc. Petroleum Geologists Studies in Geology no. 7, p. 87–116.

Segre, A., 1960, Carta batimetrica del Mediterraneo centrale; Mari Ligure e Tirreno settentrionale. Carte no. 1250: Genova, Istituto Idrografico della Marina (chart).

Southard, J. B., and Stanley, D. J., 1976, Shelfbreak processes and sedimentation, in Stanley, D. J., and Swift, D. J. P., eds., Marine Sediment Transport and Environmental Management: New York, Wiley, p. 351–377.

Vanney, J. R., 1977, Géomorphologie des Plates-formes Continentales: Paris, Doin, 300 p.

———, and Mougenot, D., 1981, La plate-forme continentale du Portugal et les provinces adjacentes; Analyse géomorphologique: Mem. Serv. Geol. Portugal, Lisbon, v. 28, 150 p.

MISSISSIPPIAN CONTINENTAL MARGINS OF THE CONTERMINOUS UNITED STATES

RAYMOND C. GUTSCHICK
U.S. Geological Survey and Department of Earth Sciences,
University of Notre Dame, Notre Dame, Indiana 46556-1020

CHARLES A. SANDBERG
U.S. Geological Survey, Federal Center, Denver, Colorado 80225

ABSTRACT

The paleogeography, paleotectonics, and paleoceanography of continental margins and shelfedges around the present western, southern, and eastern sides of the conterminous United States are reconstructed for a brief span (about 1.5 m.y.) of Mississippian time. The time is that of the middle Osagean *anchoralis-latus* conodont Zone (latest Tournaisian, Mamet foram zone 9). At this time, a shallow tropical sea covered most of the southern North American continent and was the site of a broad carbonate platform. Bordering this platform were three elongate foreland troughs, each containing several bathymetrically distinct starved basins on their inner (continentward) sides. The foreland troughs were bordered on their outer sides by orogenic highlands or a welt that formed in response to successive collisions or convergences with North America by Africa and Europe to the east, by an oceanic plate to the west, and by South America to the south.

During a eustatic rise of sealevel that accompanied the orogenies and culminated during the *anchoralis-latus* Zone, the carbonate platform prograded seaward while the troughs subsided and carbonate sediments were transported over the passive shelfedges to intertongue with thin carbonate foreslope deposits and thin (~10 m) phosphatic basinal sediments. Simultaneously, thick (~500 m) flysch and deltaic terrigenous sediments, such as the Antler flysch on the west and the Borden deltaic deposits on the east, were shed into the outer parts of the foreland basins from active margins along orogenic highlands. This Mississippian reconstruction provides a unique opportunity to compare and contrast passive and active shelfedges of a Paleozoic continent during a high stand of sealevel. The passive shelfedges can be recognized and mapped by application of a six-part sedimentation and paleoecologic model developed for the shelfedge of the Deseret starved basin in Utah, Idaho, and Nevada.

INTRODUCTION

Reconstruction of the paleogeography of the United States during a brief span of Paleozoic time enables us to recognize and evaluate the patterns and styles of more than 8,000 km of passive and active shelfedges. This brief time span, which is based on conodont zonation, is that of the *anchoralis-latus* Zone of about 1.5 m.y. duration. In other terms, this zone is in the middle of the Mississippian, middle Osagean, latest Tournaisian, or Mamet foram zone 9 (Gutschick et al., 1980; Sandberg and Gutschick, 1980; Lane et al., 1980). Reasonably reliable contemporaneous time-rock control between the conodonts and other fossil groups has been established for this zone. This time span, which was chosen because it shows a maximum eustatic rise of sealevel, was characterized by: (1) progradation away from the craton of an extensive carbonate platform with passive shelfedges, (2) maximum development of peripheral foreland troughs and starved basins that subsided as sealevel rose, (3) active molasse and deltaic development on the east, and (4) active flysch trough development on the west. Our study provides a six-part model and establishes criteria for recognition of passive shelfedges that can be applied not only to the Mississippian but also to high stands of sealevel throughout the Paleozoic, when the craton was a site for widespread carbonate platform development.

Our interpretation of the paleogeography, paleotectonics, and paleoceanography of the *anchoralis-latus* Zone in the conterminous United Stated evolved from a sedimentational and conodont biofacies model of the Deseret starved basin of Utah, Nevada, and Idaho (Sandberg and Gutschick, 1976, 1979, 1980). Other Mississippian regional studies of parts of the United States have recognized similar sediment-starved conditions peripheral to a carbonate platform (Lineback, 1969; Sando et al., 1976; Rose, 1976; Yurewicz, 1977; and Lane, 1978). Impetus to our study was given by six talks presented at a symposium on Mississippian carbonate shelf margins at the Ninth International Carboniferous Congress in Urbana, Illinois, May 25, 1979. Formal papers resulting from three of the talks were published in the SEPM Paleozoic Paleogeography of West-Central United States Symposium in Denver, Colorado (June 10, 1980): Gutschick et al. (1980); Lane and De Keyser (1980); and Sandberg and Gutschick (1980). The other three talks have, to date, appeared only as abstracts: Bamber et al. (1979); Manger and Thompson (1979); and Whitehead (1979).

This paper concentrates on the *anchoralis-latus* Zone passive shelfedges (shelf-foreslope breaks) as shown in Figure 1, and the geologic characteristics that facilitate their recognition. The continental framework that formed these shelfedges is portrayed by an east-west cross section (Fig. 2). A map of the United States shows representative stratigraphic units used for our reconstruction (Fig. 3). Paleogeographic-lithofacies and paleoceanographic maps of the United States at the time of the *anchoralis-latus* Zone are used to demonstrate the geologic history of the shelfedges.

OROGENIC HISTORY

A triad of orogenic events—Acadian, Antler, and proto-Ouachita—were focused on the eastern, western, and southern margins of the North American craton and molded the structural pattern for Mississippian time (Fig. 1). In succession, continental plates collided on the east, an oceanic plate collided on the west, and a continental plate converged on the south. The Appalachian and Antler Highlands were formed by these compressions. A stable cratonic platform emerged along the Transcontinental arch, which was subparallel to the paleoequator, and a series of structural troughs—the Eastern Interior, Ouachita, and Antler foreland troughs—formed around the platform in a horseshoe-like pattern. Smaller, bathymetrically distinct depositional basins occur within the troughs (e.g., the Deseret, Illinois, Michigan, Anadarko, and Marathon basins).

Acadian Orogeny

The Acadian orogeny (Middle Devonian to Early Mississippian) in the northeastern United States was the first major deformation to directly affect the pattern of middle Osagean paleogeography and lithofacies. This effect is reflected in the history of the Eastern Interior trough into which a succession of superposed alluvial clastic wedges (the Catskill and Pocono) prograded westward toward the trough from the Appalachian Highlands (Fig. 2). Distal muds from the Catskill wedge were deposited in the trough early in Late Devonian time. The Eastern Interior foreland trough, which was silled at its south end, became anaerobic when sealevel lowered and prevented the circulation of oxygenated surface water from the Ouachita trough into the Eastern Interior trough. The black New Albany, Chattanooga, and Antrim Shales then accumulated in the trough in relatively deep water, about 150 m (Cluff, 1980). The end of Devonian time was marked by a eustatic rise in sealevel that allowed oxygenated waters to flow into the Eastern Interior trough. This oxygen-

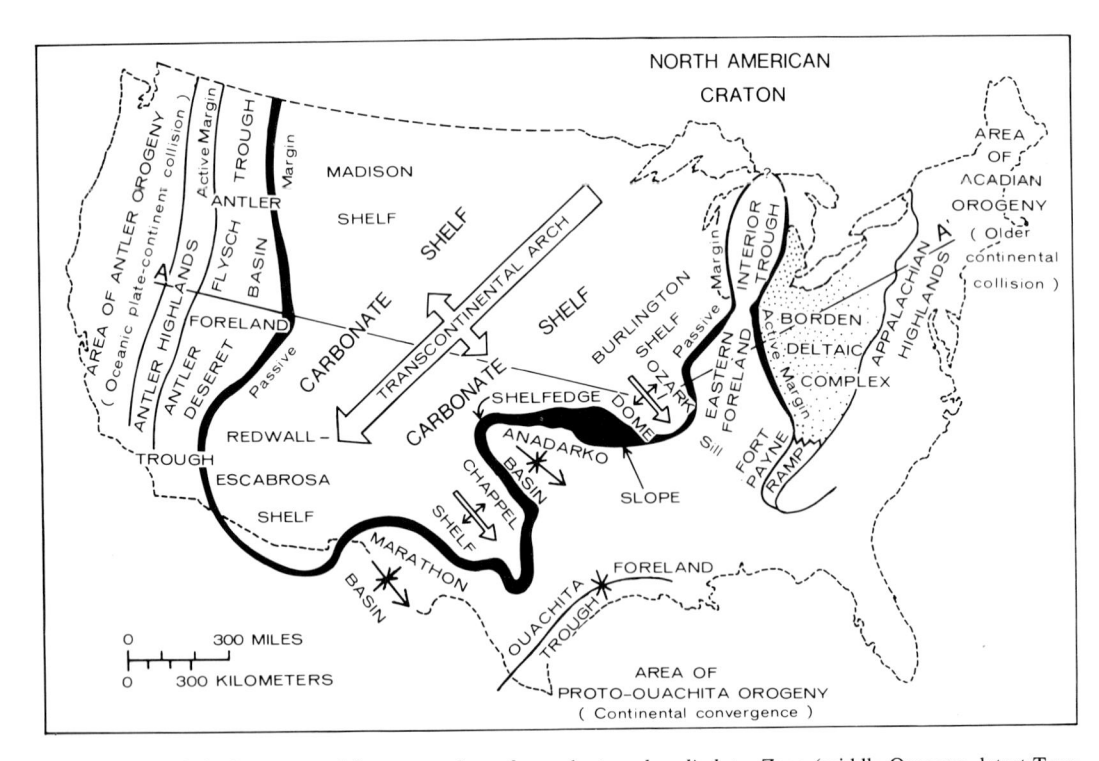

Fig. 1.—Mississippian structural features at time of conodont *anchoralis-latus* Zone (middle Osagean, latest Tournaisian, Mamet foram zone 9). Shows location of cross-section A-A' detailed in Figure 2.

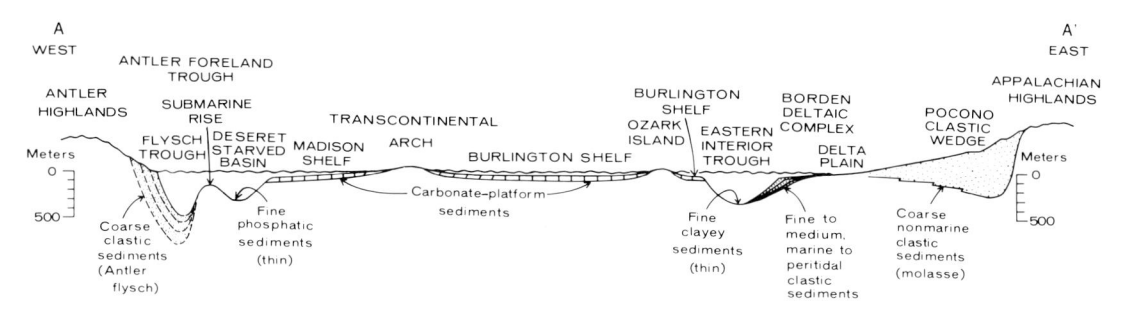

FIG. 2.—Cross-section of conterminous United States showing Mississippian structural and geographic features above and generalized Mississippian sediment types below. Location of section A-A' is shown in Fig. 1.

ation is reflected by the abrupt change in sediment colors from black or dark gray to gray, green, and blue in the Hannibal Shale and equivalents. Thereafter, subsidence in the trough coupled with rising sealevel and decrease in sediment supply caused the trough to become starved. During the time of the *anchoralis-latus* Zone, the carbonate platform prograded into the trough from the west and the Borden deltaic complex prograded into the trough from the east (Fig. 2). A broad delta plain formed by the Borden Formation represents the distal, marine part of a westward-prograding clastic wedge (Kepferle, 1977, 1978). The proximal, nonmarine part of this wedge was formed by the Pocono, Burgoon, and Grainger Formations (Fig. 3), which are coarse, clastic molasse deposits that filled the Appalachian basin during the final stages of the Acadian orogeny (Edmunds et al., 1979).

Antler Orogeny

The Antler orogeny (latest Middle Devonian to Early Mississippian) in the western United States

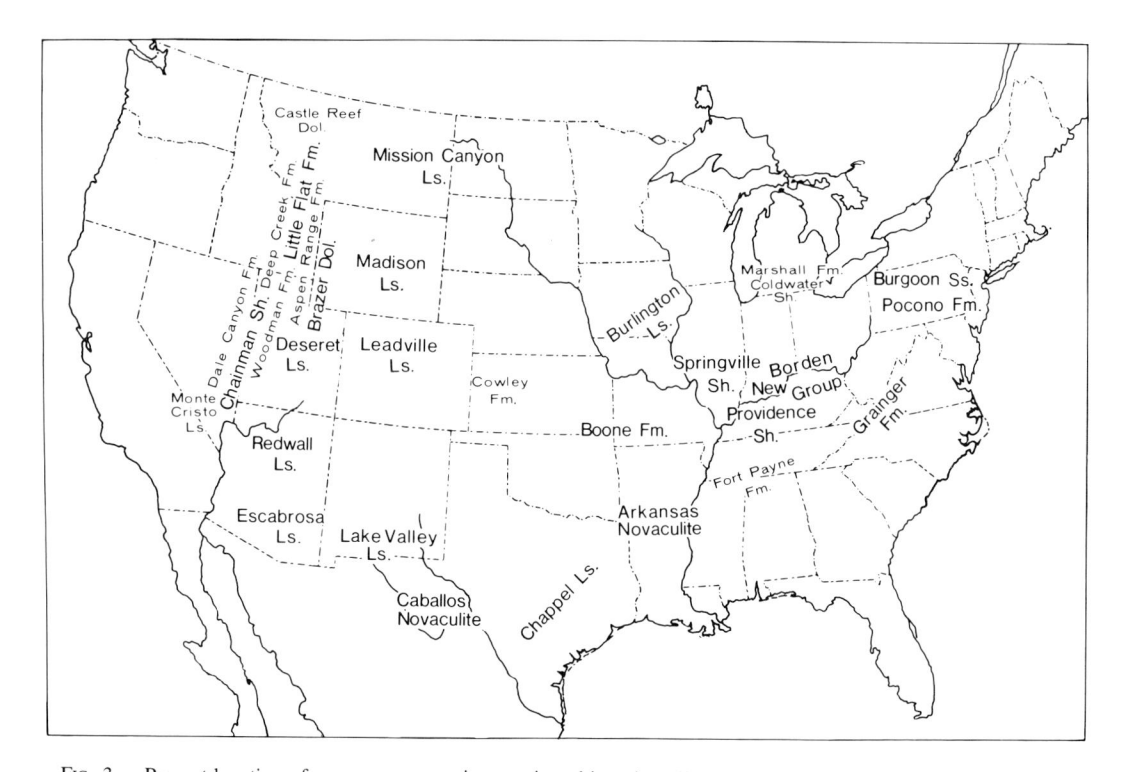

FIG. 3.—Present location of some representative stratigraphic units, all or part of which were deposited during time of *anchoralis-latus* conodont Zone. Because of diachronism, some formations may represent different zones or ages in locations other than those shown.

greatly influenced the configuration and development of the Antler foreland trough (Fig. 2) during the Early Mississippian (Gutschick et al., 1980; Sandberg and Gutschick, 1980). The Antler was the second major orogeny to affect Mississippian North America, and it climaxed while the Acadian orogeny was waning. Deformation produced a medial welt or submarine rise, similar to the Schwellen in the Variscan geosyncline of West Germany, that divided the foreland trough into two parts, a rapidly subsiding flysch trough adjacent to the highlands and the Deseret starved basin along the carbonate platform (Fig. 2; Sandberg and Gutschick, 1980). This largely submarine rise had an island near its north end, and the northernmost part of the rise acted as a sill to restrict circulation at the north end of the Deseret basin (Figs. 4, 5). The carbonate platform prograded into the starved basin as sealevel rose. The foreslope along the shelf margin increased from about 3° to 5° inclination. North of the Deseret starved basin, the trough was fed by sediments from the Antler Highlands on the west and the carbonate platform on the east.

Proto-Ouachita Orogeny

The southern margin of the Mississippian North America was affected by a convergence of the South American plate that culminated in a continental collision by latest Pennsylvanian time (Walper and Rowett, 1972; Ross, 1979). The earlier movement and resulting early orogenic activity are evidenced by the absence of marine Upper Devonian and Lower Mississippian rocks from the northern part of the South American craton. This absence was recognized as a result of our reevaluation of described Devonian and Carboniferous fossil collections from that area. The Ouachita foreland trough, situated beween converging North and South America was a throughgoing deep-water trough having extensive blooms of pelagic siliceous radiolarian plankton that contributed to the formation of novaculite and chert in Devonian and Early Mississippian time. Buckling of the Ouachita trough produced a medial welt similar to the submarine rise in the Antler foreland trough. Because parts of the welt within the Ouachita trough apparently were emergent, the feature is herein named the Caballos-Arkansas Island Chain. The concept of an elongate welt with low islands explains the juxtaposition of shallow-water and deep-water novaculite and chert lithofacies that have been so well documented by Folk and McBride (McBride and Thomson, 1970; Folk, 1973; Folk and McBride, 1976; McBride and Folk, 1977). Neither the carbonate platform nor the low island chain contributed much sediment to the inner part of the trough adjacent to the craton, however, and that part of the trough became starved.

PALEOGEOGRAPHY AND LITHOFACIES

The paleogeography of the conterminous United States at the time of the *anchoralis-latus* Zone is shown by Figure 5. Wide distribution of limestone and dolomite marks the extensive shallow-water carbonate platform. The axis of the Transcontinental arch and shelves on both sides are dotted by a few large islands. Around the platform, carbonate-shelf rocks grade laterally through a narrow belt of foreslope clinoform micritic limestones into a thin

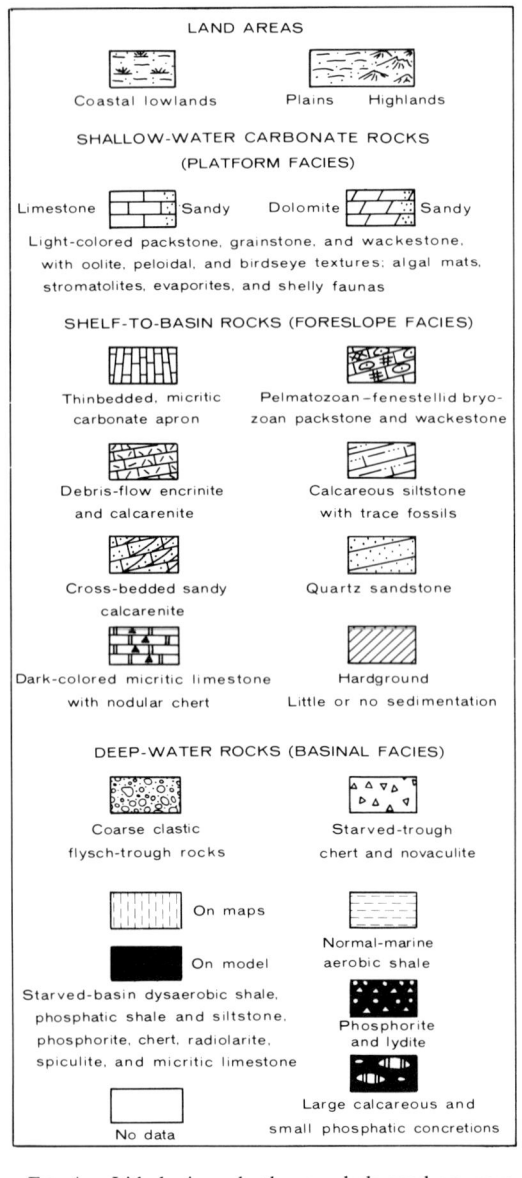

FIG. 4.—Lithologic and other symbols used on maps and model in Figures 5, 6 and 7.

deep-water condensed suite of phosphatic shales and bedded cherts, interbedded with distal tongues of micritic limestone. This basinal suite, in turn, grades into novaculite and chert in the Ouachita trough and around the Caballos-Arkansas Island Chain.

The active east margin of the Eastern Interior foreland trough is dominated by fine clastic deposits along the Borden delta front. Clastic sedimentation diminishes toward the south end of the Appalachian Highlands, farthest from the area of maximum uplift, and gives way to carbonate sedimentation on the Fort Payne Ramp. In the west, the Antler flysch trough received interbedded sandstone and shale turbidite deposits along with coarse deep-sea fan deposits (Nilsen, 1977).

PALEOCEANOGRAPHY AND PALEOBATHYMETRY

The paleoceanography of the United States (Fig. 6) during the time of the *anchoralis-latus* Zone is constructed to be compatible with and to reinforce the interpretation of the paleogeographic and lithofacies map (Fig. 5). Our interpretive paleobathymetric map encourages studies of directional properties of the sediments that may correct or improve the depicted water circulation pattern in Mississippian time. It also serves as a guide and stimulus to help document and evaluate the paleobathymetric interpretation.

Interpreting paleobathymetry in terms of absolute depth is difficult. In addition to the references cited in this section, the following studies have been particularly helpful: Hallam (1967), Eicher (1969); Heckel (1972); Mamet (1972); Byers (1977); Kepferle (1977); Benedict and Walker (1978); and Shanmugam and Walker (1980).

The eustatic rise of sealevel reached a maximum for the Mississippian at the time of the *anchoralis-latus* Zone. The shallow, clear epeiric sea covering the cratonic carbonate platform exceeded 2,330,000 km^2 within the conterminous United States and continued into western Canada (Bamber and others, 1979). The general characters of depth, extent, substrate, sediments, and probable energy distribution of this Mississippian example closely resemble models by Edie (1958), Shaw (1964), and Irwin (1965). Schopf (1980), however, doubts the former existence of an epeiric sea such as envisaged by Shaw and by Irwin. Circulation of water on this shelf would be greatly restricted or eliminated owing to bottom friction, and tidal exchange would be limited to the seaward edges of the shelf.

We place the outer limit of the shelf at an average depth of 50 m, although the depth must have varied slightly from place to place. We interpret the shelf seas and upper foreslope down to 200 m to be within the neritic zone and the lower slope and basins and troughs, except for the Michigan basin, to be within the upper bathyal zone between 200 m and 600 m. The Illinois starved basin contained water as deep as 300 m (Lineback, 1969). On the east side of the Illinois basin, the toe of the slope of the Borden delta front was placed at a depth of about 180 m, and the carbonate platform margin of the Muldraugh Member of the Borden Formation was placed at about 50 m by Kepferle (1977). The maximum water depth in the Deseret starved basin in western Utah was interpreted to be at least 300 m by Sandberg and Gutschick (1979, 1980). The Antler flysch trough in Nevada reached bathyal depths in excess of 500 m and perhaps approaching 1,000 m (Harbaugh and Dickinson, 1981). Water depth in the Ouachita trough is thought to have exceeded 300 m (McBride and Thomson, 1970).

Surface sea currents in most of the Eastern Interior Sea are interpreted to be controlled by southeast paleo-trade winds and to conform to the counterclockwise Coriolis effect of the southern hemisphere; those in the Madison Sea north of the paleoequator are interpreted to fit the pattern of northeast paleo-trade winds and the clockwise Coriolis effect. Using this pattern, westward-moving surface currents in the Ouachita trough probably divided and some entered the Illinois basin. Upwelling was an important factor along the margins of much of the carbonate shelf for the nourishment of benthic faunas, especially echinoderms and bryozoans and for the development of build-ups, bioherms, banks, and Waulsortian mounds.

SHELFEDGE MODELS FOR EAST SIDE
OF DESERET STARVED BASIN

The passive carbonate shelfedge between the Deseret starved basin and the Madison Shelf has received the most concentrated biostratigraphic study of any of the carbonate shelfedges in the United States for *anchoralis-latus* Zone time (Sandberg and Gutschick, 1976, 1979, 1980; Gutschick and others, 1980). Based on our observations, a six-part graphic model (Fig. 7A–F) for recognition of the shelfedge has been developed. This model integrates the distribution and depth criteria of lithofacies, rock colors and organic carbon, corals and algae, conodonts, foraminifera and radiolaria, and trace fossils. Similar models previously have been proposed for other carbonate shelfedges (Mamet, 1972; Wilson, 1974, 1975a; Gutschick and others, 1976; Armstrong and Mamet, 1976; Lewis and Potter, 1978). The results of our shelf-foreslope-basin analysis, as shown in the model, have been used to interpret the paleogeography and paleoceanography of other regions of the conterminous United States, much as if we were assembling pieces of a jigsaw puzzle, to produce a reconstruc-

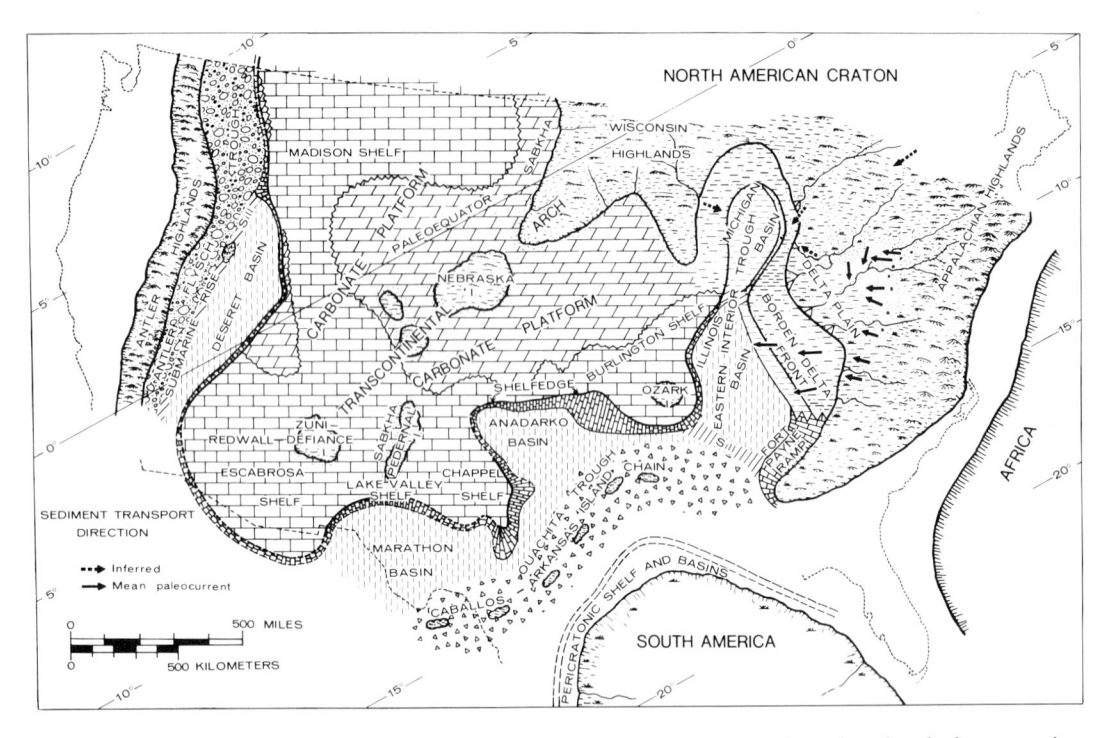

FIG. 5.—Paleogeographic and lithofacies map of conterminous United States at time of *anchoralis-latus* conodont Zone. Data unavailable in blank areas.

tion that is internally and externally consistent in terms of paleotectonics, paleoecology and sedimentology. Application of the six-part model provides much sharper resolution of the passive carbonate shelfedges that has been possible heretofore.

Lithofacies Model

In the lithofacies model (Fig. 4, 7A), three suites of rocks are recognized: deep-water rocks or basinal facies (Sandberg and Gutschick, 1979, p. 128, their Fig. 14), shelf-to-basin rocks or foreslope facies (Gutschick and others, 1980, p. 118; Sandberg and Gutschick, 1980, p. 139–140), and shallow-water carbonate rocks or platform facies (Gutschick et al. 1980). Crinoid-bryozoan packstones and grainstones that occur along the shelfedge commonly are dolomitized, as in the Brazer Dolomite of Utah and Wyoming. The shelfedge carbonate rocks commonly are thicker bedded and more massive than their foreslope carbonate equivalents.

Rock-Color and Organic-Carbon Model

Rock colors are conspicuous in the field and differences are closely related to rock composition, depositional site, and subsequent weathering. Rock colors (Goddard et al., 1948) for the spectrum of rocks in this study (Fig. 7B) are arranged to match the lithofacies model. The greatest control of colors is the oxidation-reduction state of the organic carbon and the iron compounds. Another important control is the dominance of clays, phosphorite, and glauconite, which occur together in the starved basin. Platform carbonate rocks are light colored and basinal rocks are dark colored and commonly are petroleum source rocks (Hosterman and Whitlow, 1981). The color model is idealized, but it expresses the general correlation between rock color and position in the basin-foreslope-platform model.

Coral and Algae Biofacies Model

Recent study of the paleoecology of Mississippian corals in the western United States resulted in recognition of deep-water and shallow-water biofacies (Sando, 1981). A summary of the paleoecologic assignments, as they pertain to the time-span discussed in this paper, is presented in Figure 7C. Corals found on the slope tend to be small, simple, solitary, nondissepimented forms. Colonial rugose corals were restricted to shallow-water carbonate-platform environments. There appears to be a sharp break between these biofacies. Green and red benthic calcareous algae were associated within the euphotic zone on the carbonate platform, but only red algae lived in the disphotic zone (Mamet, 1972).

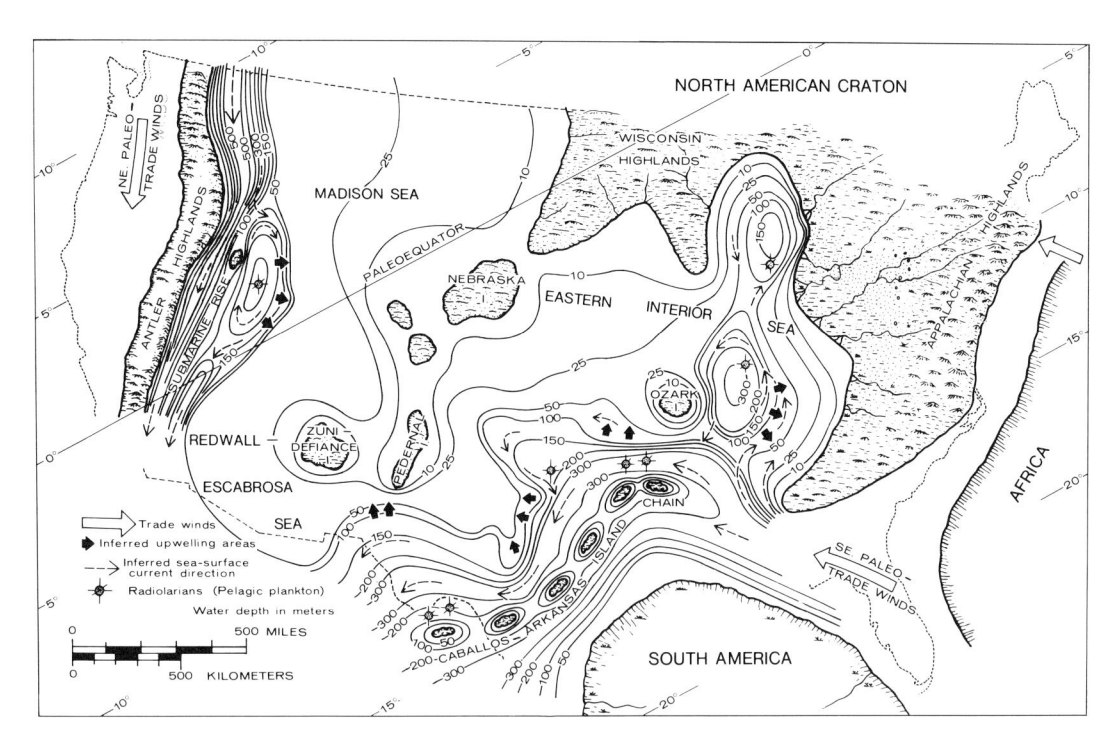

FIG. 6.—Paleoceanographic and paleobathymetric map of conterminous United States at time of *anchoralis-latus* conodont Zone. Circulation patterns of surface sea currents and directions of upwelling are based on inferred directions of paleo-trade winds.

Conodont Biofacies Model

Great strides have been made in the use of conodonts from the Deseret starved basin for zonation and paleoecologic interpretation (Sandberg, 1979; Sandberg and Gutschick, 1979, p. 130, their fig. 16; Sandberg *in* Lane et al., 1980, p. 123–126, their table 3). This information is summarized in Figure 7D. The individual elements of the phosphatic conodont apparatus are scattered in bottom sediments, because the conodont animal (the soft parts of which are unknown,) is interpreted to have been nektonic, with a wide range of niches in benthic and pelagic environments. Numbers of conodont elements and diversity of genera and species are greatest on the foreslope where upwelling currents provided maximum aeration and food supply. *Eotaphrus* is perhaps the most useful genus for shelfedge recognition, because it occurs mainly in shallow upper-foreslope and shelfedge settings. The association of the conodont *Eotaphrus*, which ranges above the *anchoralis-latus* Zone, with the solitary coral *Ankhelasma* in the western United States is being studied by C. A. Sandberg and W. J. Sando. The occurrence of either or both of these taxa apparently provides a marker for the shelfedge or upper foreslope in rocks slightly younger than those of the *anchoralis-latus* Zone. The greatest diversity of platform-type or pectiniform conodont elements, such as *Pseudopolygnathus, Polygnathus, Gnathodus, Bactrognathus, Doliognathus,* and *Scaliognathus,* is on the platform margin and upper foreslope.

Foraminifera and Radiolaria Biofacies Model

The calcareous, secreted foraminifera and siliceous, agglutinated foraminifera are exclusively benthic and appear to be virtually antithetic in their distributions (Fig. 7E). Planktonic foraminifera had not yet appeared in the Paleozoic, so that calcareous forms are generally absent on the foreslope below the platform. Rarely, transported tests of calcareous foraminifera that lived on the platform are found in debris flow deposits in the foreslope facies. Paleozoic smaller calcareous foraminifera (tournayellids, forschiids, endothyrids, and earlandiids) were restricted to shallow waters of the platform and usually were associated with euphotic algae down to a water depth of about 30 m (Mamet, 1977). On the other hand, agglutinate foraminifera were adapted to the slope and basin floor in aerobic and dysaerobic habitats. Study of agglutinate foraminiferan faunas and their paleoecology in the Deseret starved basin is still in progress, but to date three intergradational biofacies have been recognized: sacca-

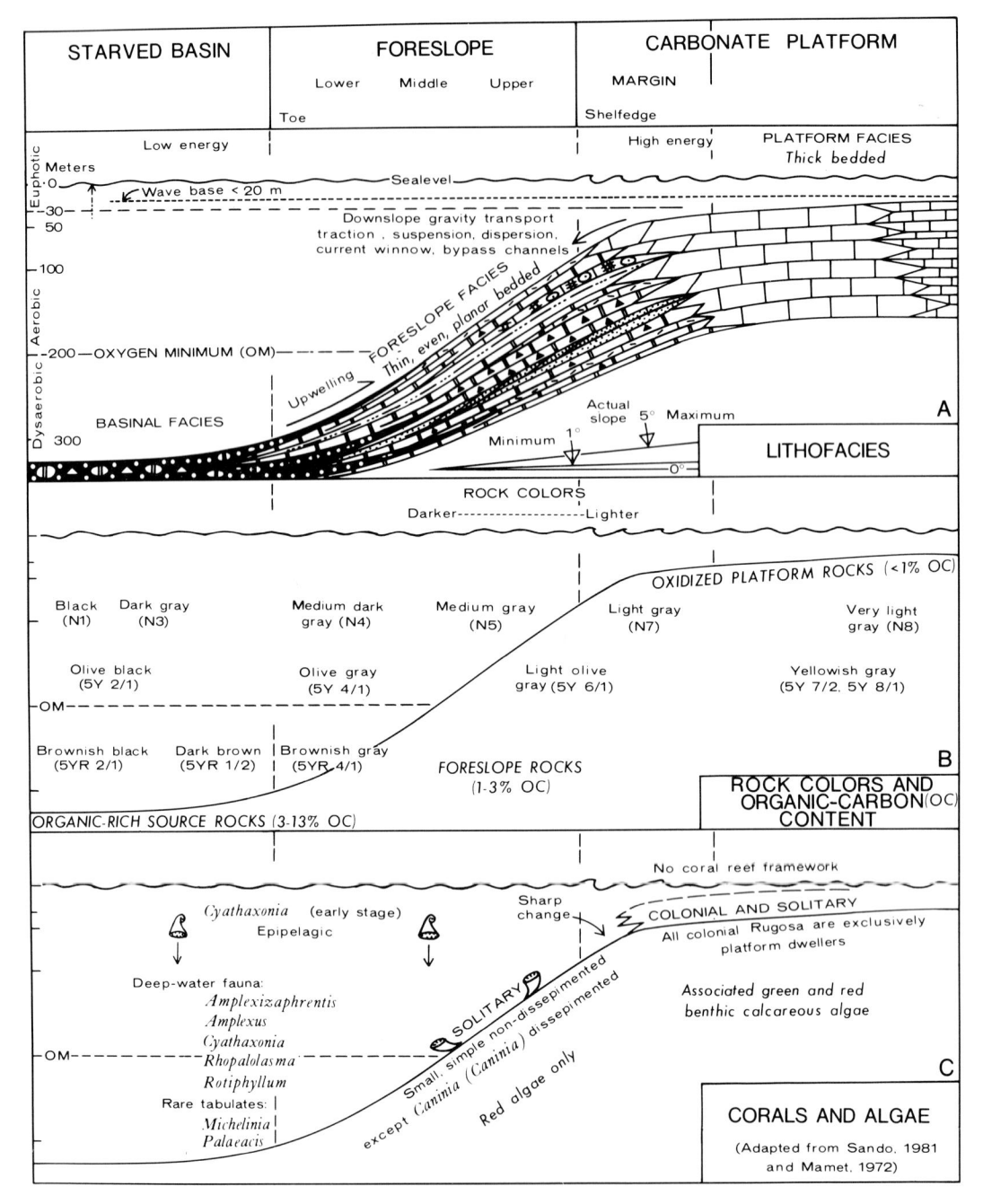

FIG. 7.—Six-part sedimentational and paleoecologic model for recognition of *anchoralis-latus* conodont Zone shelf-edge on east side of Deseret starved basin, Utah, southeastern Idaho, and eastern Nevada.

minid biofacies, upper foreslope; *Hyperammina* biofacies, middle and lower foreslope; and *Reophax* biofacies, deep basin. The foraminiferans are highly useful in recognizing the shelfedge (Fig. 7E), because, in general, calcareous foraminifera represent the platform and agglutinate foraminifera represent the foreslope.

Pelagic planktonic radiolarian faunas, including one representative fauna from the *anchoralis-latus* Zone, are present in rocks deposited in the Deseret basin. However, the paleoecology of these faunas is not yet understood. Radiolarians have been recovered from the basinal facies, where they are abundant and diverse, and from foreslope facies, but not from the platform facies. Study of the radiolarians is in progress, but only by their scarcity on the platform do they offer a general guide to shelfedge recognition at present.

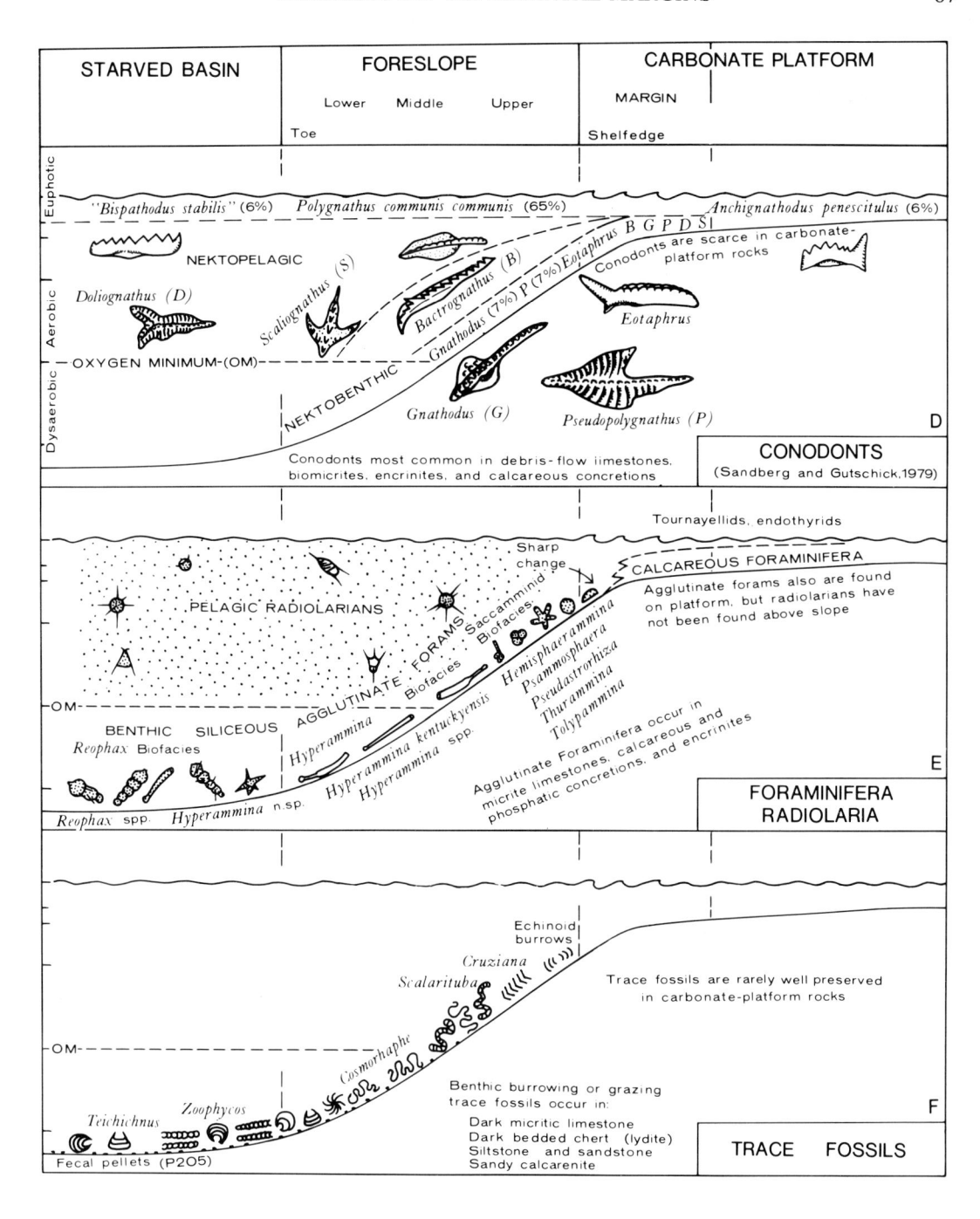

Trace Fossil Biofacies Model

Trace fossils have been found in the basin and on the foreslope in a conventional deep-water pattern (Seilacher, 1967), but not on the carbonate platform (Fig. 7F). The trace-fossil assemblage, which closely resembles the forms and distribution described by Kepferle (1977) and Chaplin (1980) from the Borden Formation of Kentucky, is currently under study. Deep-water assemblages of *Zoophycos* and *Teichichnus* occur in dark micritic limestones, bedded cherts, and siltstones. *Cosmorhaphe* (*Helminthoida* and *Phycosiphon* of some authors) and *Scalarituba* occur characteristically along the foreslope; the former predominates lower on the slope and *Scalarituba* higher, but often the two forms are found together. Single finds of trilobite(?) trails (*Cruziana*) in sandy crinoidal calcarenite and echinoid(?) burrows in silty limestone were also made in foreslope facies. The abundance of trace fossils in foreslope-carbonate facies and their scarcity in platform-carbonate facies can be used to approximate the position of the shelfedge. Trace fossils are more commonly observed in shallow-water clastic rocks, such as in the Borden Formation of Kentucky, so that they may be more useful for defining the shelfedge in clastic facies.

PASSIVE SHELFEDGES

Burlington Shelfedge

The Burlington Shelf (Figs. 1, 2, 5) occupies the carbonate platform east of the Transcontinental arch and north of the Anadarko basin (Lane, 1978). It is north, northwest, and west of Ozark Island in the Eastern Interior Sea (Fig. 6). The foreslope between the Burlington shelfedge and the deep, gradually subsiding Illinois starved basin is abrupt (Lineback, 1981); but the descent into the Ouachita trough and Anadarko basin to the south is more gradual.

The carbonate sheet across the stable Burlington Shelf represents the rapid, uninterrupted, progradational advance of carbonate sediments during a rising sealevel to a high-stand position. Shelf sedimentation took place throughout Osagean time, but only the Burlington Limestone and time-equivalent strata represent the *anchoralis-latus* Zone. The chronology of this prograding wedge is illustrated by Lane (1978), who used the boundaries between conodont zones as time planes for lithogenetic units.

The general nature and character of the Burlington Limestone and equivalent strata of the Burlington carbonate platform have been studied by Van Tuyl (1922) and many other authors referenced by Lane (1978). Thickness and distribution have been treated by Lane (1978), Willman and others (1975), and Lane and De Keyser (1980). Outer shelf rocks are essentially light-colored, coarsely crystalline, cherty, crinoidal, bioclastic limestone and some dolomite. Inner shelf rocks have been almost entirely dolomitized. The Burlington Limestone is well known for the profusion and excellent preservation of its echinoderm faunas.

The outer Burlington Shelf can be regarded as an extensive biogenic carbonate-bank complex dominated by pelmatozoan skeletons and bryozoan-frond debris, with some brachiopods and horn corals. Two dominant facies, a crinoid-arenite bank suite and crinoid-wacke inertia flow suite, were recognized by Carozzi and Gerber (1979) in Missouri, Illinois, and Iowa. A similar three-fold depositional facies pattern, build-up core, build-up flank, and resedimented intermound unit, was presented by King (1980) for central Missouri.

The shelf margin along the west side of the Illinois basin was a long, linear, prograding crinoidal carbonate-bank complex. The shelfedge is marked by thick grain-supported limestones and cherty lime mudstones, which wedge out down the slope (Lane, 1978), or by distal fine-grained crinoid-bryozoan calcisiltites deposited by inertia flows (Carozzi and Gerber, 1979). The toe of the slope is a sharp change from carbonate rocks to a greatly condensed section of Springville Shale comprising greenish-gray shale with phosphatic nodules (Collinson and Scott, 1958; Lineback, 1969; Lane, 1978). The Burlington shelfedge and steep foreslope into the Illinois basin do not crop out, but they were delineated from subsurface well data (Lineback, 1966, 1981).

Depositional conditions along the Burlington shelfedge south and west of Ozark Island were somewhat different. Carbonate buildups and bioherms were numerous along the outer shelf in Oklahoma, Missouri, and Arkansas (Harbaugh, 1957; Troell, 1962; and King, 1980). A broad gentle foreslope extended into the sediment-starved Ouachita trough. Moderate upwelling onto the slope from the trough helped to nourish organisms that produced Waulsortian-type bioherms. Because of the gentle foreslope, the shelfedge is less well-defined than on the eastern part of the Burlington Shelf. Condensed conodont faunas occur at the toe of the slope in clay muds (State Pond Member of Springville Shale in Illinois) and in lime muds and sands (e.g., Walls Ferry Limestone Bed of the St. Joe Limestone Member of the Boone Formation).

Chappel Shelfedge

The Chappel Shelf is similar to, and perhaps an extension of, the Burlington Shelf, from which it is separated by the Anadarko basin (Figs. 5, 6). The Chappel Shelf terminates seaward with a crinoidal bank complex that wedges out at the shelfedge above the foreslope into the starved Ouachita trough. Outcrop sections of the Chappel Limestone on the Llano uplift give an east-west cross section of this abrupt facies change, whereas subsurface data farther north reveal carbonate crinoidal-bryozoan bioherms and banks of the shelf margin. The Chappel Limestone is a thin, discontinuous, dark-gray crinoidal packstone in scattered outcrops around the northern and eastern side of the Llano uplift in central Texas (Cloud and Barnes, 1949; Plummer, 1950; Kier and others, 1979). The Chappel is thicker light-gray to pink fossiliferous crinoidal packstone and grainstone on the western side of the Llano uplift. The Chappel thickens into the subsurface north of the uplift (Turner, 1957; Pray, 1958; Asquith, 1980).

The Chappel shelfedge is the irregular margin of a crinoidal-bryozoan bank complex facing a deepwater starved basin, and it must have resembled the Burlington shelfedge. The bank complex prograded seaward as sealevel rose. Offshore currents caused upwelling up the slope onto the bank front. Supply of sediment into the basin was extremely limited and sediment-starved conditions prevailed. Bank cores in the Chappel Limestone are composed of crinoid-bryozoan lime wackestone, often dolomitized, and their flanks are crinoid-bryozoan grainstone and packstone (Asquith, 1980). Interbank deposits of argillaceous crinoid-bryozoan wackestone are thinner than bank cores. Brachiopods described by Carter (1967) are abundant and diverse along the shelfedge and upper foreslope.

Basinal rocks of the Chappel Limestone are much thinner and colors are much darker than equivalent shelf rocks (Plummer, 1950; Cloud et al., 1957). In the basinal sequence, pelmatozoan detritus is common, but shelly organisms such as brachiopods are greatly diminished in numbers and size (Carter, 1967). Conodonts of three successive faunal zones are abundant in the condensed basinal sections; although discernible physical breaks do not exist between the zones, some older Devonian conodonts are reworked in each zone (Hass, 1959; Hoyle, 1978). Dark organic-rich shale of the Upper Mississippian Barnett Formation rests conformably on the Chappel Limestone in the basin. The lower part of the Barnett and its fauna resemble the deepwater, starved-basin facies of the Deseret Limestone and equivalent strata in western Utah (Sandberg and Gutschick, 1979, their Fig. 14).

Lake Valley Shelfedge

The narrow Lake Valley Shelf of southern New Mexico borders the Marathon basin between the Chappel Shelf to the east in Texas and the Redwall-Escabrosa Shelf to the west in Arizona (Fig. 5). The stratigraphy, microfacies, petrography, and paleogeography of the Mississippian Lake Valley Limestone carbonate-shelf rocks and the Rancheria Formation basinal rocks were described by Laudon and Bowsher (1941, 1949); Pray (1958, 1961); Armstrong (1962); Lane (1974); Wilson (1975b); Yurewicz (1977); Armstrong et al. (1980); and Lane and De Keyser (1980).

The Lake Valley Limestone consists of a lower cherty limestone with crinoid bioherms, a middle siltstone containing fenestellid bryozoans, and an upper cherty crinoidal limestone (Laudon and Bowsher, 1941). The core facies of the bioherms is dark-colored lime mudstone with fenestrate bryozoans and scattered pelmatozoan segments reminiscent of Waulsortian bioherms (Pray, 1958).

The Lake Valley Shelf in the Sacramento Mountains of New Mexico represents a Waulsortian-type carbonate-mound sea bottom with coalescing flank beds and younger intermound fill (De Keyser, 1978; Lane and De Keyser, 1980). The carbonate rocks of this shelf wedge out to the south and probable time-equivalent rocks of the deep-water starved basin are dark-gray, cherty, sparsely fossiliferous, fine-grained limestone of the Las Cruces and younger Rancheria Formations (Laudon and Bowsher, 1949; Lane, 1974; Yurewicz, 1977).

Redwall-Escabrosa Shelfedge

The Escabrosa Limestone in the south (Purves, 1978; Peirce, 1979; Armstrong et al., 1980), and the Redwall Limestone in the north (McKee and Gutschick, 1969; Kent and Rawson, 1980), make up the Redwall-Escabrosa Shelf in Arizona. This shelf, at the south end of the Transcontinental arch (Fig. 5), extends into southwestern Colorado where the rock names change to Leadville Limestone (Armstrong and Mamet, 1976) and into Utah, where it merges northward into the Madison Shelf. The shallow-water limestones and dolomites of this broad shelf are characterized by peloids, ooids, algae, calcareous foraminiferans, bryozoans, and especially crinoids.

The northwestern extension of the Redwall Limestone is the Monte Cristo Limestone (Hewett, 1931; Pierce and Langenheim, 1974), the carbonate rocks of which make up the shelf margin in southeastern Nevada. The shelfedge is in the Bullion Member of the Monte Cristo.

Little is known about the southwestern Redwall-Escabrosa shelfedge. The open-marine carbonate environment of the Redwall-Escabrosa Shelf continues from the Lake Valley Shelf southward into northernmost Sonora, Mexico, as far as Cananea and Chihuahua (Wilson, 1975b; Bahlburg and Silver, 1976). The southernmost part of this shelf has intershelf to shelfedge patch-reef coarse grainstones and massive coralline(?) boundstones and oolites facing a deeper water marine environment with finely laminated peloidal mudstones and wackestones (Bahlburg and Silver, 1976). To the west in the vicinity of Caborca, Sonora, Mexico, a Mississippian exposure has revealed lithologies and fossils (Easton and others, 1958) that suggest a carbonate shelf in proximity to deeper water. The small outcrops are complicated structurally and the rocks have been altered, so some uncertainty exists concerning the stratigraphic succession. Outcrops and well data in southwestern Arizona, west of the Cordilleran overthrust belt, are scarce (McKee, 1947, 1951; Miller, 1970), so that opinions vary about the Mississippian paleogeography of this area.

There is an apparent symmetry of the Madison, Redwall-Escabrosa, and Burlington shelves relative to the axis of the Transcontinental arch (Fig. 1). The paleogreographic pattern and mirror-image symmetry across the Transcontinental arch explain the strikingly similar stratigraphy and lithofacies of Upper Devonian and Mississippian rocks between the Ouachita trough sequence from Arkansas to west Texas and the Antler trough sequence in Nevada and Utah (Gutschick and Moreman, 1967; Ketner, 1980). We envision that this pattern is maintained by the Ouachita and Antler foreland troughs, and therefore we connect them around the southwest end of the Redwall-Escabrosa Shelf. This interpretation of the paleogeography is highly speculative in view of the paucity of information in northern Mexico and the complexity of its post-Mississippian history.

Madison Shelfedge North of Deseret Basin

North of the Deseret basin, the Madison Shelf is juxtaposed to the Antler flysch trough (Fig. 5), so that the trough received sediments both from the carbonate platform and the Antler Highlands. The Mission Canyon Limestone makes up the carbonate platform and the upper part of the McGowan Creek Formation on the east side of the trough contains western-derived flysch sediments (Sandberg, 1975). The shelfedge lies along the Montana-Idaho border, but details are lacking because this area is in the structurally complex Overthrust belt. Lithologic and sedimentational differences between the carbonate platform and flysch trough were documented by Huh (1967); Sandberg (1975); Gutschick et al. (1980); and Skipp and Hall (1980).

Information on only the carbonate platform is known in the Sun River Canyon area and other parts of the Sawtooth Range in northern Montana. There, the Castle Reef Dolomite (Fig. 3) makes up the carbonate platform (Mudge et al. 1962; Nichols, 1980). Farther west in northern Montana, where foreslope rocks would be expected to occur, most Paleozoic rocks have been eroded. Waulsortian-type bioherms, which are present in Kinderhookian rocks of the Sun River area (Haines, 1977), have not been recognized in the Castle Reef.

ACTIVE SHELFEDGES

The active *anchoralis-latus* Zone shelfedges bordered the western side of the Appalachian Highlands, which had formed earlier during the Acadian orogeny. We consider these shelfedges to be active, in contrast to the passive shelfedges that bordered the craton, even though the predominant clastic sediments are of slightly post-orogenic, or molasse, type. In a mirror-image position on the eastern side of the younger Antler Highlands, the foreland trough was so close to the highlands that an active shelf or shelfedge could not be developed.

Borden Deltaic Complex

The northern part of the eastern side of the Eastern Interior foreland trough during *anchoralis-latus* Zone time was bordered by the Borden deltaic complex, which prograded westward down the paleoslope from the Appalachian Highlands (Fig. 1). The stratigraphy, sedimentology, and history of the Borden Formation have been discussed by Stockdale (1931, 1939); Swann et al. (1965); Lineback (1966, 1968); Weir et al. (1966); Peterson and Kepferle (1970); Suttner and Hattin (1973); Kepferle (1977, 1978); Whitehead (1976, 1978); Sable (1979); Chaplin (1980); and many other authors.

The deltaic complex consisted of a series of delta lobes that were built outward from laterally shifting streams and distributary systems along the delta front. The Borden Formation encompasses rocks ranging from prodelta greenish-gray shales through slope siltstones to platform sandstones, siltstones, and shales (Kepferle, 1977; Gray, 1979).

Two separate marine basins, the Michigan and Illinois basins, received Borden deltaic sediments along their eastern sides. The feature that separated them may have been a delta lobe. In the Michigan basin, the depositional site was closer to the source of sediments, and the basin subsided to accommodate the active accumulation (Cohee, 1979). In the Illinois basin, progradation of the Borden delta front was across a gently inclined ramp into a subsiding starved basin. This depositional site was farther from the sediment source than the Michigan basin, so that the rate and amount of sedimentation were less and the sediments were finer grained.

Borden deposition in the Illinois basin started with the New Providence Shale at the end of Kinderhookian time and continued with steady westward progradation through the Osagean. The starved basin continued to receive distal, bottomset laminated muds, which accumulated slowly during the Osagean. Basinal clay-rich muds on the east side of the Illinois basin are called New Providence Shale, whereas equivalent muds on the west side of the basin are called Springville Shale. Late Osagean (Keokuk Limestone) carbonate sediments were deposited directly on the New Providence and Springville Shales in the deep part of the basin, where Borden sediments were very thin or absent due to basin starvation.

Although post-Borden erosion around the Cincinnati arch has removed much of the shelfedge of *anchoralis-latus* Zone time across Indiana and part of northern Kentucky, this margin can be reconstructed from the detailed sedimentologic analysis of Kepferle (1977, 1978). The shelfedge was between virtually flat topset beds and inclined foreset beds, which dipped from 0°16' to 1°17' (Kepferle, 1977). The water depth at the shelfedge was about 45 m or less. The bottom was silty mud becoming more muddy downslope, and water was too turbid for limestone-producing organisms. Infaunal traces of burrowing and grazing organisms are common, and bioturbation has obscured many sedimentary structures. The outer shelf margin was populated by communities of brachiopods, bryozoans, gastropods, and epifaunal trilobites. Low-energy currents from waves and tides swept the bottom. This shelfedge environmental pattern characterized the entire delta front along the east side of the Illinois basin.

Fort Payne Ramp

The Fort Payne Ramp (Fig. 5), which was the depositional site for the Fort Payne Formation, adjoins the south end of the Appalachian Highlands, far distant from the locus of the Acadian orogeny and well beyond the distribution of Borden delta clastics. Relief of the land along the carbonate shore was low, and terrigenous clastic influx was minimal. The Fort Payne Ramp bordered the southeastern margin of the Illinois basin and occupied a critical position between the Borden deltaic complex to the north and the Ouachita trough to the west (Thomas, 1972; Milici et al., 1979). In the Ouachita trough, siliceous sedimentation predominated (Fig. 5). The divergence of surface waters (Fig. 6) caused deep silica-rich waters to well up from the Ouachita trough over the sill and enter the Illinois basin beside the Fort Payne Ramp. This upwelling apparently enriched the dominantly carbonate sediments with silica to form chert.

Deposition of the Fort Payne Formation involved Waulsortian-type mounds formed on a shallow slowly subsiding ramp during a eustatic rise of sealevel (Macquown and Perkins, 1982). Crinoidal bioherms in the Fort Payne have also been described for Tennessee and Kentucky (Thadens and others, 1961). Bioherms in the basal part of the Fort Payne Chert in Tennessee rest directly on the Chattanooga Shale. Their cores are fine-grained, argillaceous, dolomitic crinoidal limestones (Marcher, 1962) or bryozoan grainstones with mud-supported microfacies (Macquowan and Perkins, 1982). The geometry, lithologies, microfacies, and paleoenvironmental relations of the mounds are similar to those described for the Burlington Shelf on the opposite side of the Illinois basin. Both occurrences have in common subsiding slopes, rising sealevel, and upwelling. These conditions require troughward progradation of the Fort Payne mounds to form banks as on the Burlington shelfedge. Continued progradation of lime muds and skeletal material down the ramp, in concert with the water circulation pattern, caused a lobe of dark siliceous lime mud to form eventually in the deep basin and to be spread in counter-clockwise direction by Coriolis currents all the way to southern Illinois (Lineback, 1966, 1969).

CONCLUSIONS

The patterns and styles of Mississippian *anchoralis-latus* Zone shelfedges around the western, southern, and eastern sides of the conterminous United States can be interpreted from paleotectonic-paleogeographic and paleoceanographic reconstructions of the southern part of the North American continent:

1. Passive shelfedges faced away from the craton. They represent almost exclusively the slope-break between a broad carbonate platform covered by shallow seas and carbonate foreslopes that descended into elongate foreland trough on three sides of the craton. Because of the high stand of sea level during *anchoralis-latus* Zone time, most of the craton was submerged and little clastic sediment was shed seaward from the few remaining land areas. The troughs contained several bathymetrically distinct, deep to moderately deep, sediment-starved basins, including the Anadarko and Marathon basins. The passive shelfedges can be recognized and mapped by application of a six-part model developed for the shelfedge of the Deseret starved basin in Utah, Nevada, and Idaho.

2. Active shelfedges faced the craton and bordered orogenic highlands that resulted from an older continent-continent collision (Appalachian Highlands) or from a younger oceanic plate-continent collision (Antler Highlands). Bordering the Appalachian Highlands, the active shelfedge was formed partly by a deltaic complex and partly by a carbonate ramp. Bordering the Antler Highlands, a shelfedge was not developed, because coarse clastic sediments fed directly into a deep foreland trough. Bordering the outer side of the Ouachita foreland trough, only a welt with a few low islands was formed, because continental convergence was just beginning during *anchoralis-latus* Zone time. This welt, which is herein named the Caballos-Arkansas Island Chain, explains the juxtaposition of shallow-water and deep-water lithofacies that have been documented by Folk and McBride. Later in the Paleozoic, the welt may have evolved into orogenic highlands analogous to the Appalachian and Antler Highlands.

ACKNOWLEDGMENTS

We are grateful to W. J. Sando, who reviewed both this paper and an earlier version, to E. D.

McKee and J. L. Wilson, who reviewed the earlier version, and to D. L. Woodrow and an anonymous reviewer, who provided many helpful suggestions for improving this paper. The authors bear sole responsibility for the ideas expressed herein. Preprints of formal papers were generously furnished by A. V. Carozzi and M. S. Gerber, D. W. Harbaugh and W. R. Dickinson, and O. T. King, Jr.

REFERENCES

ARMSTRONG, A. K., 1962, Stratigraphy and paleontology of the Mississippian System in southwestern New Mexico and adjacent southeastern Arizona: New Mexico Bur. Mines and Mineral Resources Memoir 8, 99 p.

——, AND MAMET, B. L., 1976, Biostratigraphy and regional relations of the Mississippian Leadville Limestone in the San Juan Mountains, southwestern Colorado: U.S. Geol. Survey Professional Paper 985, 25 p.

——, ——, AND REPETSKI, J. E., 1980, The Mississippian System of New Mexico and southern Arizona: in Fouch, T. D., and Magathan, E. R., eds., Paleozoic Paleogeography of West-Central United States: Soc. Econ. Paleontologists Mineralogists Rocky Mountain Sec., West-Central United States Paleogeography Symposium 1, p. 82–99.

ASQUITH, G. B., 1980, Subsurface carbonate depositional models; The application of petrophysical logs, lithologies, and geometry to carbonate reservoir exploration: Rocky Mountain Assoc. Geologists Short Course Notes, Denver, Colo., Nov. 20, 1980.

BAHLBURG, W., AND SILVER, B. A., 1976, Depositional environments of Tamaroa Sequence (Upper Devonian and Mississippian), Pedregosa basin, southeastern Arizona, southwestern New Mexico, northeastern Sonora, and northwestern Chihuahua [abs.]: Am. Assoc. Petroleum Geologists Bull., v. 60, p. 646–647.

BAMBER, E. W., MACQUEEN, R. W., AND RICHARDS, B. C., 1979, Facies relations at the Mississippian carbonate platform margin, Western Canada, in Abstracts of Papers: Urbana, Illinois, 9th Internat. Congress of Carboniferous Strat. and Geol., p. 10.

BENEDICT, G. L., III, AND WALKER, K. R., 1978, Paleobathymetric analysis in Paleozoic sequences and its geodynamic significance: Am. Jour. Sci., v. 278, p. 579–607.

BYERS, C. W., 1977, Biofacies patterns in euxinic basins; A general model, in Cook, H. E., and Enos, P., eds., Deep-Water Carbonate Environments: Soc. Econ. Paleontologists Mineralogists Spec. Pub. 25, p. 5–17.

CAROZZI, A. V., AND GERBER, M. S., 1979, Crinoid arenite banks and crinoid wacke inertia flows; A depositional model for the Burlington Limestone (Middle Mississippian), Illinois, Iowa, and Missouri, USA, in Abstracts of Papers: Urbana, Illinois, 9th Internat. Congress of Carboniferous Strat. and Geol., p. 29.

CARTER, J. L., 1967, Mississippian brachiopods from the Chappel Limestone of central Texas: Am. Paleontology Bull., v. 53, p. 253–488.

CHAPLIN, J. R. 1980, Stratigraphy, trace fossil associations, and depositional environments in the Borden Formation (Mississippian), northeastern Kentucky: Geol. Soc. Kentucky, Kentucky Geol. Survey Ann. Field Conf., 114 p.

CLOUD, P. E., JR., AND BARNES, V. E., 1949, The Ellenburger Group of central Texas: Univ. Texas Bull. 4621, 473 p.

——, ——, AND HASS, W. H., 1957, Devonian-Mississippian transition in central Texas: Geol. Soc. America Bull., v. 68, p. 807–816.

CLUFF, R. M., 1980, Paleoenvironment of the New Albany Shale Group (Devonian-Mississippian) of Illinois: Jour. Sed. Petrology, v. 50, p. 767–780.

COHEE, G. V., 1979, Michigan Basin Region, in Craig, L. C., and Connor, C. W., eds., Paleotectonic Investigations of the Mississippian System in the United States: U.S. Geol. Survey Professional Paper 1010, Part 1, p. 49–57.

COLLINSON, C., AND SCOTT, A. J., 1958, Age of the Springville Shale (Mississippian) of southern Illinois: Illinois State Geol. Survey Circ. 254, 12 p.

DE KEYSER, T. L., 1978, The Early Mississippian of the Sacramento Mountains, New Mexico—An ecofacies model for carbonate shelf margin deposition [Ph.D. Dissert.]: Corvallis, Oregon, Oregon State Univ., 304 p.

EASTON, W. H., SANDERS, J. E., KNIGHT J. B., AND MILLER, A. K., 1958, Mississippian fauna in northwestern Sonora Mexico: Smithsonian Misc. Collections, v. 119, no. 3, 87 p.

EDIE, R. W., 1958, Mississippian sedimentation and oil fields in southeastern Saskatchewan, in J. A. Goodman, ed., Jurassic and Carboniferous of western Canada: Am. Assoc. Petroleum Geologists, John Andrew Allen Memorial Vol., p. 331–363.

EDMUNDS, W. E., BERG, T. M., SEVON, W. D., PIOTROWSKI R. C., HAYMAN, L., AND RICKARD L. V., 1979, The Mississippian and Pennsylvanian (Carboniferous) Systems in the United States—Pennsylvania and New York: U.S. Geol. Survey Professional Paper 1110-B, 33 p.

EICHER, D. L., 1969, Paleobathymetry of Cretaceous Greenhorn Sea in eastern Colorado: Am. Assoc. Petroleum Geologists Bull., v. 53, p. 1075–1090.

FOLK, R. L., 1973, Evidence for peritidal deposition of Devonian Caballos Novaculite, Marathon Basin, Texas: Am. Assoc. Petroleum Geologists Bull., v. 57, p. 702–725.

——, AND MCBRIDE, E. F., 1976, The Caballos Novaculite revisited; Part 1, Origin of novaculite members: Jour. Sed. Petrology, v. 42, p. 659–669.

GODDARD, E. N., AND OTHERS, 1948, Rock-color chart: Geol. Soc. America, 6 p.

GRAY, H. H., 1979, The Mississippian and Pennsylvanian (Carboniferous) Systems in the United States—Indiana: U.S. Geol. Survey Professional Paper 1110-K, 20 p.

GUTSCHICK, R. C., MCLANE, M. J., AND RODRIGUEZ, J., 1976, Summary of Late Devonian-Early Mississippian biostratigraphic framework in western Montana, in Tobacco Root Geol. Soc. Guidebook, 1976 Field Conf.: Montana Bur. Mines Geol. Spec. Pub. 73, p. 91–124.

———, AND MOREMAN, W. L., 1967, Devonian-Mississippian boundary relations along the cratonic margin of the United States, in Oswald, D. H., ed., Internat. Symposium on Devonian System, Calgary, Alberta, Sept., 1967: Alberta Soc. Petroleum Geologists, v. 1, p. 1009–1023.

———, SANDBERG, C. A., AND SANDO, W. J., 1980, Mississippian shelf margin and carbonate platform from Montana to Nevada, in Fouch, T. D., and Magathan, E. R., eds., Paleozoic Paleogeography of West-Central United States: Soc. Econ. Paleontologists Mineralogists Rocky Mountain Sec., West-Central United States Paleogeography Symposium 1, p. 111–128.

HAINES, F. E., 1977, Lower Mississippian sedimentation in northwestern Montana [Ph.D. Dissert.]: Rolla, Missouri, Univ. Missouri, 116 p.

HALLAM, A., ed., 1967, Depth indicators in marine sedimentary environments: Mar. Geology Spec. Issue, v. 5, p. 327–555.

HARBAUGH, D. W., AND DICKINSON, W. R., 1981, Depositional facies of Mississippian clastics, Antler Foreland Basin, central Diamond Mountains, Nevada: Jour. Sed. Petrology, v. 51, p. 1223–1234.

HARBAUGH, J. W., 1957, Mississippian bioherms in northeast Oklahoma: Am. Assoc. Petroleum Geologists Bull., v. 41, p. 2530–2544.

HASS, W. H., 1959, Conodonts from the Chappel Limestone of Texas: U.S. Geol. Survey Professional Paper 294-J, p. 365–399.

HECKEL, P. H., 1972, Ancient shallow marine environments, in Rigby, J. K., and Hamblin, W. K., eds., Recognition of Ancient Sedimentary Environments: Soc. Econ. Paleontologists Mineralogists Spec. Pub. 16, p. 226–286.

HEWETT, D. F., 1931, Geology and ore deposits of the Goodsprings quadrangle, Nevada: U.S. Geol. Survey Professional Paper 162, 172 p.

HOSTERMAN, J. W., AND WHITLOW, S. I., 1981, Munsell color value as related to organic carbon in Devonian shale of Appalachian Basin: Am. Assoc. Petroleum Geologists Bull., v. 65, p. 333–335.

HOYLE, B. L., 1978, Lower Mississippian conodont biostratigraphy of the Chappel Limestone, central Texas, and Welden Limestone, southern Oklahoma [M.A. Thesis]: Austin, Texas, Univ. Texas, 290 p.

HUH, O. K., 1967, The Mississippian System across the Wasatch line, east-central Idaho, extreme southwestern Montana: in Montana Geol. Soc. 18th Annual Field Conf. Guidebook, p. 31–62.

IRWIN, M. L., 1965, General theory of epeiric clear water sedimentation: Am. Assoc. Petroleum Geologists Bull., v. 49, p. 445–459.

KENT, W. N., AND RAWSON, R. R., 1980, Depositional environments of the Mississippian Redwall Limestone in northeastern Arizona in Fouch, T. D., and Magathan, E. R., eds., Paleozoic Paleogeography of West-Central United States: Soc. Econ. Paleontologists and Mineralogists Rocky Mountain Sec., West-Central United States Paleogeography Symposium 1, p. 101–109.

KEPFERLE, R. C., 1977, Stratigraphy, petrology, and depositional environment of the Kenwood Siltstone Member, Borden Formation (Mississippian), Kentucky and Indiana: U.S. Geol. Survey Professional Paper 1007, 49 p.

———, 1978, Prodelta turbidite fan apron in Borden Formation (Mississippian) Kentucky and Indiana, in Stanley, D. J. and Kelling, G., eds., Sedimentation in Submarine Canyons, Fans, and Trenches: Stroudsburg, Pa., Dowden, Hutchinson & Ross, Inc., p. 224–238.

KETNER, K. B., 1980, Stratigraphic and tectonic parallels between Paleozoic geosynclinal siliceous sequences in northern Nevada and those of the Marathon Uplift, Texas, and Ouachita Mountains, Arkansas and Oklahoma, in Fouch, T. D., and Magathan, E. R., eds., Paleozoic Paleogeography of West-Central United States: Soc. Econ. Paleontologists Mineralogists Rocky Mountain Sec., West-Central United States Paleogeography Symposium 1, p. 363–369.

KIER, R. S., BROWN, L. F., JR., AND MCBRIDE, E. F., 1979, The Mississippian and Pennsylvanian (Carboniferous) Systems in the United States—Texas: U.S. Geol. Survey Professional Paper 1110-S, 45 p.

KING, O. T., JR., 1980, Depositional environments and genetic stratigraphy of the Burlington Limestone (Osagean) in central Missouri [abs.]: Geol. Soc. America Abstracts with Programs, v. 12, p. 463.

LANE, H. R., 1974, The Mississippian of southeastern New Mexico and west Texas—A wedge-on-wedge relation: Am. Assoc. Petroleum Geologists Bull., v. 58, p. 269–282.

———, 1978, The Burlington Shelf (Mississippian, north-central United States): Geologica et Palaeontologica, v. 12, p. 165–176.

———, AND DE KEYSER, T. L., 1980, Paleogeography of the Late Mississippian (Tournaisian 3) in the central and southwestern United Sates, in Fouch, T. D., and Magathan, E. R., eds., Paleozoic Paleogeography of West-Central United States: Soc. Econ. Paleontologists Mineralogists Rocky Mountain Sec., West-Central United States Paleogeography Symposium 1, p. 149–162.

———, SANDBERG, C. A., AND ZIEGLER, W., 1980, Taxonomy and phylogeny of some Lower Carboniferous conodonts and preliminary standard post-*Siphonodella* zonation: Geologica et Palaeontologica, v. 14, p. 117–164.

LAUDON, L. R., AND BOWSHER, A. L., 1941, Mississippian formations of Sacramento Mountains, New Mexico: Am. Assoc. Petroleum Geologists Bull., v. 25, p. 2107–2160.

——, AND ——, 1949, Mississippian formations of southwestern New Mexico: Geol. Soc. America Bull., v. 60, p. 1–87.
LEWIS, R. Q., SR., AND POTTER, P. E., 1978, Surface rocks in the western Lake Cumberland area, Clinton, Russell, and Wayne Counties, Kentucky: Geol. Soc. Kentucky Ann. Field Conf. Guidebook, p. 1–41.
LINEBACK, J. A., 1966, Deep-water sediments adjacent to the Borden Siltstone (Mississippian) delta in southern Illinois: Illinois State Geol. Survey Circ. 401, 48 p.
——, 1968, Turbidites and other sandstone bodies in the Borden Siltstone (Mississippian) in Illinois: Illinois State Geol. Survey Circ. 425, 29 p.
——, 1969, Illinois basin—Sediment-starved during Mississippian: Am. Assoc. Petroleum Geologists Bull., v. 53, p. 112–126.
——, 1981, The eastern margin of the Burlington-Keokuk (Valmeyeram) carbonate bank: Illinois State Geol. Survey Circ. 520, 24 p.
MACQUOWN, W. C., AND PERKINS, J. H., 1982, Stratigraphy and petrology of petroleum-producing Waulsortian-type carbonate mounds in Fort Payne Formation (Lower Mississippian) of north-central Tennessee: Am. Assoc. Petroleum Geologists Bull., v. 66, p. 1055–1075.
MAMET, B. L., 1972, Essai de reconstruction paléoclimatique basé sur les microflores algaires du Viséen: 24th Internat. Geol. Congress, Sec. 7, Montreal, Quebec, p. 282–291.
——, 1977, Foraminiferal zonation of the Lower Carboniferous: Methods and stratigraphic implications, *in* Kauffman, E. G., and Hazel, J. E., eds., Concepts and Methods of Biostratigraphy: Stroudsburg, Pa., Dowden, Hutchinson & Ross, Inc., p. 445–462.
MANGER, W. L., AND THOMPSON, T. L., 1979, Lower Mississippian carbonate shelf, southern Missouri, eastern Oklahoma, and northern Arkansas, *in* Abstracts of Papers: Urbana, Illinois, 9th Internat. Congress of Carboniferous Strat. and Geol., p. 125–126.
MARCHER, M. V., 1962, Crinoidal bioherms in the Fort Payne Chert (Mississippian) along the Caney Fork River, Tennessee: U.S. Geol. Survey Professional Paper 450-E, p. 43–45.
MCBRIDE, E. F., AND THOMSON, A., 1970, The Caballos Novaculite, Marathon region, Texas: Geol. Soc. America Spec. Paper 122, 129 p.
——, AND FOLK, R. L., 1977, The Caballos Novaculite revisited; Part II, Chert and shale members and synthesis: Jour. Sed. Petrology, v. 47, p. 1261–1286.
MCKEE, E. D., 1947, Paleozoic seaways in western Arizona: Am. Assoc. Petroleum Geologists Bull., v. 31, p. 282–292.
——, 1951, Sedimentary basins of Arizona and adjoining areas: Geol. Soc. America Bull., v. 62, p. 481–505.
——, AND GUTSCHICK, R. C., eds., 1969, History of the Redwall Limestone of Northern Arizona: Geol. Soc. America Memoir 114, 726 p.
MILICI, R. C., BRIGGS, G., KNOX, L. M., SITTERLY, P. D., AND STATLER, A. T., 1979, The Mississippian and Pennsylvanian Systems in the United States—Tennessee: U.S. Geol. Survey Professional Paper 1110-G, 38 p.
MILLER, F. K., 1970, Geologic map of the Quartzsite quadrangle, Yuma County, Arizona: U.S. Geol. Survey Geologic Quadrangle Map GQ-841.
MUDGE, M. R., SANDO, W. J., AND DUTRO, J. T., JR., 1962, Mississippian rocks of Sun River Canyon area, Sawtooth Range, Montana: Am. Assoc. Petroleum Geologists Bull., v. 46, p. 2003–2018.
NICHOLS, K. M., 1980, Depositional and diagenetic history of porous dolomitized grainstones at the top of the Madison Group, Disturbed belt, Montana, *in* Fouch, T. D., and Magathan, E. R., eds., Paleozoic Paleogeography of West-Central United States: Soc. Econ. Paleontologists Mineralogists Rocky Mountain Sec., West-Central United States Paleogeography Symposium 1, p. 163–173.
NILSEN, T. H., 1977, Paleogeography of Mississippian turbidites in south-central Idaho, *in* Stewart, J. H., Stevens, C. H., and Fritsche, A. E. eds., Paleozoic Paleogeography of the West-Central United States: Soc. Econ. Paleontologists Mineralogists Pacific Sec., Pacific Coast Paleogeography Symposium 1, p. 275–299.
PEIRCE, H. W., 1979, The Mississippian and Pennsylvanian (Carboniferous) Systems in the United States—Arizona: U.S. Geol. Survey Professional Paper 1110-Z, 20 p.
PETERSON, W. L., AND KEPFERLE, R. C., 1970, Deltaic deposits of the Borden Formation in central Kentucky: U.S. Geol. Survey Professional Paper 700-D, p. 49–54.
PIERCE, R. W., AND LANGENHEIM, R. L., JR., 1974, Platform conodonts of the Monte Cristo Group, Mississippian, Arrow Canyon Range, Clark Country, Nevada: Jour. Paleontology, v. 48, p. 149–169.
PLUMMER, F. B., 1950, The Carboniferous rocks of the Llano region of central Texas: Univ. Texas Bull. 4329, 170 p.
PRAY, L. C., 1958, Fenestrate bryozoan core facies, Mississippian bioherms, southwestern United States: Jour. Sed. Petrology, v. 28, p. 261–273.
——, 1961, Geology of the Sacramento Mountains escarpment, Otero County, New Mexico: New Mexico Bur. Mines and Mineral Resources Bull. 35, 144 p.
PRYOR, W. A., AND SABLE, E. G., 1974, Carboniferous of the Eastern Interior Basin: Geol. Soc. America Spec. Paper 148, p. 281–313.
PURVES, W. J., 1978, Paleoenvironmental evaluation of Mississippian age carbonate rocks in central and southeastern Arizona [Ph.D. Dissert.]: Tuscon, Arizona, Univ. Arizona, 672 p.
ROSE, P. R., 1976, Mississippian carbonate shelf margins, Western United States: U.S. Geol. Survey Jour. Res., v. 4, p. 449–466.

Ross, C. A., 1979, Late Paleozoic collision of North and South America: Geology, v. 7, p. 41–44.
Sable, E. G., 1979, Eastern Interior Basin region, in Craig, L. C., and Connor, C. W., eds., Paleotectonic Investigations of the Mississippian System in the United States, Part 1: U.S. Geol. Survey Professional Paper 1010, p. 57–106.
Sandberg, C. A., 1975, McGowan Creek Formation, new name for Lower Mississippian flysch sequence in east-central Idaho: U.S. Geol. Survey Bull. 1405-E, 11 p.
———, 1979, Devonian and Lower Mississippian conodont zonation of the Great Basin and Rocky Mountains, in Sandberg, C. A., and Clark, D. L., eds., Conodont Biostratigraphy of the Great Basin and Rocky Mountains: Brigham Young Univ. Geology Studies, v. 26, pt. 3, p. 87–106.
———, and Gutschick, R. C., 1976, Paleotectonic, biostratigraphic, and economic significance of Osagean starved basin in Utah [abs.]: Geol. Soc. America Abstracts with Programs, v. 8, p. 1083, 1085.
———, and ———, 1979, Guide to conodont stratigraphy of Upper Devonian and Mississippian rocks along the Wasatch Front and Cordilleran Hingeline, Utah, in Sandberg, C. A., and Clark, D. L., eds., Conodont Biostratigraphy of the Great Basin and Rocky Mountains: Brigham Young Univ. Geology Studies, v. 26, pt. 3, p. 107–134.
———, and ———, 1980, Sedimentation and biostratigraphy of Osagean and Meramecian starved basin and foreslope, Western United States, in Fouch, T. D., and Magathan, E. R., eds., Paleozoic Paleogeography of West-Central United States: Soc. Econ. Paleontologists Mineralogists Rocky Mountain Sec., West-Central United States Paleogeography Symposium 1, p. 129–147.
Sando, W. J., 1981, The paleoecology of Mississippian corals in western conterminous United States: Acta Palaeontologica Polonica, v. 25, p. 619–631.
———, Dutro, J. T., Jr., Sandberg, C. A., and Mamet, B. L., 1976, Revision of Mississippian stratigraphy, eastern Idaho and northeastern Utah: U.S. Geol. Survey Jour. Res., v. 4, p. 467–479.
Schopf, T. J. M., 1980, Paleoceanography: Cambridge, Massachusetts, Harvard Univ. Press, 341 p.
Seilacher, A., 1967, Bathymetry of trace fossils: Mar. Geology, v. 5, p. 413–428.
Shanmugam, G., and Walker, K. R., 1980, Sedimentation, subsidence, and evolution of a foredeep basin in the middle Ordovician, southern Appalachians: Am. Jour. Sci., v. 280, p. 479–496.
Shaw, A. B., 1964, Time in Stratigraphy: New York, McGraw-Hill, 365 p.
Skipp, B., and Hall, W. E., 1980, Upper Paleozoic paleotectonics and paleogeography of Idaho, in Fouch, T. D., and Magathan, E. R., eds., Paleozoic Paleogeography of West-Central United States: Soc. Econ. Paleontologists Mineralogists Rocky Mountain Sec., West-Central United States Paleogeography Symposium 1, p. 387–422.
Stockdale, P. B., 1931, The Borden (Knobstone) rocks of southern Indiana: Indiana Dept. Conserv., Div. Geology Pub. 98, 330 p.
———, 1939, Lower Mississippian rocks of east-central interior: Geol. Soc. America Spec. Paper 22, 248 p.
Suttner, L. J., and Hattin, D. E., eds., 1973, Field conference on Borden Group and overlying limestone units, south-central Indiana: Soc. Econ. Paleontologists Mineralogists, Great Lakes Sec., 3rd Ann. Mtg., Bloomington, Indiana, 113 p.
Swann, D. H., Lineback, J. A., and Frund, E., 1965, The Borden Siltstone (Mississippian) delta in southwestern Illinois: Illinois State Geol. Survey Circ. 386, 20 p.
Thadens, R. E., Lewis, R. Q., Cattermole, J. M., and Taylor A. R., 1961, Reefs in the Fort Payne Formation of Mississippian age, south-central Kentucky: U.S. Geol. Survey Professional Paper 424-B, p. 88–90.
Thomas, W. A., 1972, Regional Paleozoic stratigraphy in Mississippi between Ouachita and Appalachian Mountains: Am. Assoc. Petroleum Geologists Bull., v. 56, p. 81–106.
Troell, A. R., 1962, Lower Mississippian bioherms of southwestern Missouri and northwestern Arkansas: Jour. Sed. Petrology, v. 32, p. 629–644.
Turner, G. L., 1957, Paleozoic stratigraphy of the Fort Worth Basin, in Study of Lower Pennsylvanian and Mississippian rocks of the Northeast Llano Uplift: Abilene and Fort Worth Geol. Societies Joint Field Trip Guidebook, p. 57–77.
Van Tuyl, F. M., 1922, The stratigraphy of the Mississippian formations of Iowa: Iowa Geol. Survey Ann. Rept. 30, p. 33–349.
Walper, J. L., and Rowett, C. L., 1972, Plate tectonics and the origin of the Caribbean Sea and the Gulf of Mexico: Trans. Gulf Coast Assoc. Geol. Soc., v. 22, p. 105–116.
Weir, G. W., Gualtieri, J. L., and Schlanger, S. O., 1966, Borden Formation (Mississippian) in south- and southeast-central Kentucky: U.S. Geol. Survey Bull. 1224-F, 38 p.
Whitehead, N. H., III, 1976, The stratigraphy, sedimentology, and conodont paleontology of the Floyds Knob Bed and Edwardsville Member of the Muldraugh Formation (Valmeyeran), southern Indiana and north-central Kentucky [M.S. Thesis]: Urbana, Illinois, Univ. Illinois, 443 p.
———, 1978, Lithostratigraphy, depositional environments, and conodont biostratigraphy of the Muldraugh Formation (Mississippian) in southern Indiana and north-central Kentucky: Southeastern Geology, v. 19, p. 83–109.
———, 1979, Paleogeography and depositional environments of the Lower Mississippian of the East-central United States, in Abstracts of Papers: Urbana, Illinois, 9th Internat. Congress of Carboniferous Strat. and Geol., p. 231.
Willman, H. B., and others, 1975, Handbook of Illinois Stratigraphy: Illinois State Geol. Survey Bull. 95, 261 p.
Wilson, J. L., 1974, Characteristics of carbonate-platform margins: Am. Assoc. Petroleum Geologists Bull., v. 58, p. 810–824.
———, 1975a, Carbonate Facies in Geologic History: New York, Springer-Verlag, 471 p.

———, 1975b, Regional Mississippian facies and thickness in southern New Mexico and Chihuahua, *in* Pray, L. C., ed., A Guidebook to the Mississippian Shelf-Edge and Basin Facies Carbonates, Sacramento Mountains and Southern New Mexico: Dallas, Texas, Dallas Geol. Soc., p. 124–128.

YUREWICZ, D. A., 1977, Sedimentology of Mississippian basin-facies carbonates, New Mexico and West Texas—The Rancheria Formation, *in* Cook, H. E., and Enos, P., eds., Deep-Water Carbonate Environments: Soc. Econ. Paleontologists Mineralogists Spec. Paper 25, p. 203–219.

STRUCTURAL DYNAMICS OF THE SHELF-SLOPE BOUNDARY AT ACTIVE SUBDUCTION ZONES

GEORGE W. MOORE
U.S. Geological Survey, Menlo Park, California 94025

ABSTRACT

About 40 subduction-zone segments have been identified worldwide on the basis of intermediate-focus earthquakes, calc-alkalic volcanic arcs, and lines of rapid tectonic uplift. The total length of these actively convergent plate boundaries is 57,000 km. Of this length, 42% is of the Japan type, in which the upper plate is relatively stable with respect to the subduction zone; 37% is of the Andes type, in which the upper plate actively overrides the trench; and 21% is of the Himalaya type, in which continental plates or microplates collide with other continental bodies.

Subduction zones of both the Japan and Andes types are marked by basement highs at the trench-slope break. Uplift of the crust and upper mantle at the edge of the upper plate causes these basement highs where a relatively low-density prism of accreted oceanic material is emplaced below. The accretionary prism for each cycle of subduction forms within 5 m.y. after a new subduction zone is established, while the megathrust is evolving from an initial dip of about 30° near the zone's seafloor outcrop to a steady-state dip of about 10°. Except during this relatively brief period of accretion, most oceanic sediment at subduction zones is believed to be carried deeply into the lithosphere.

Elongate sedimentary basins form on both sides of the uplift at the trench-slope break: forearc basins toward the arc, and trench-slope basins toward the trench. Depending on the balance between sedimentation and tectonic displacement, the topographic shelf-slope boundary may be located anywhere in the forearc basins, approximately to the upper edge of the trench-slope basins. At the lower trench slope, compression usually removes the seawater involved in primary oil migration before thermal maturation of oil precursors can occur. Elsewhere at the active margin, although a low geothermal gradient caused by the subduction of cool oceanic crust delays hydrocarbon maturation, such thermal maturation can nevertheless resume when a normal geothermal gradient is reestablished after continental collision or after the subduction-zone alignment moves to a new position.

INTRODUCTION

The smoothly curved volcanic arcs and deep-sea trenches of subduction zones, many of them spanning thousands of kilometers, are among the Earth's longest and most regular features (Fig. 1). The shelf-slope boundary at active subduction zones is controlled both by tectonic deformation and by local conditions of sedimentation. At mature steady-state subduction zones, the shelf-slope boundary commonly coincides with the trench-slope break (Fig. 2); but some trench-slope ridges, such as the one at Java, enclose deep bathymetric basins, and some shelf-like surfaces, such as the one at Japan, lie deeper than typical shelf depths. Ingersoll and Graham (this volume) describe conditions of sedimentation at the shelf-slope break at convergent continental margins. This paper deals with the tectonic development of such depositional sites, and particularly with the initial stages of that development, which profoundly affect the preserved rock record.

The characteristics of active steady-state subduction zones vary considerably, and these variations can be correlated to help us understand operative causal relations. The depth to which earthquakes occur in subduction zones is clearly connected with both the rate of convergence and the age of the oceanic crust being subducted (Ruff and Kanamori, 1980). Because the coolness, and hence the brittleness, of the crust is related to the seafloor-spreading ages of the slabs being subducted, the correlation of the focal depth of earthquakes with crustal age and with rate of subduction shows that the plasticity of the slabs determines the focal depth. Young hot slabs, and those that move slowly, soon acquire sufficient additional heat from the hot mantle which they are penetrating to become too soft and plastic to sustain earthquakes. For example, in the young 9-my-old slab below Mount St. Helens, Washington, which slips at only about 3 cm/yr, subcrustal earthquakes are virtually absent; whereas in the 150-my-old slab under the Marianas Islands, which moves at the very high rate of 11 cm/yr, the foci extend to a (world-maximum) depth of 700 km (Moore, 1981; Golovchenko et al., 1981).

This effect is seen most clearly along the coast of South America north of the Chile Rise, where the relative convergence proceeds uniformly at 8 to 9 cm/yr, but where the subducting Nazca plate is segmented by fracture zones into seafloor tracts of differing ages. The hot young tracts have few deep-focus earthquakes and abundant volcanoes, whereas the cold old tracts have abundant deep-focus earthquakes and few volcanoes.

The occurrence of many now-inactive subduction zones of various ages, identifiable by onshore belts of the mantle-derived ultramafic rocks that are as-

Copyright © 1983, The Society of Economic Paleontologists and Mineralogists

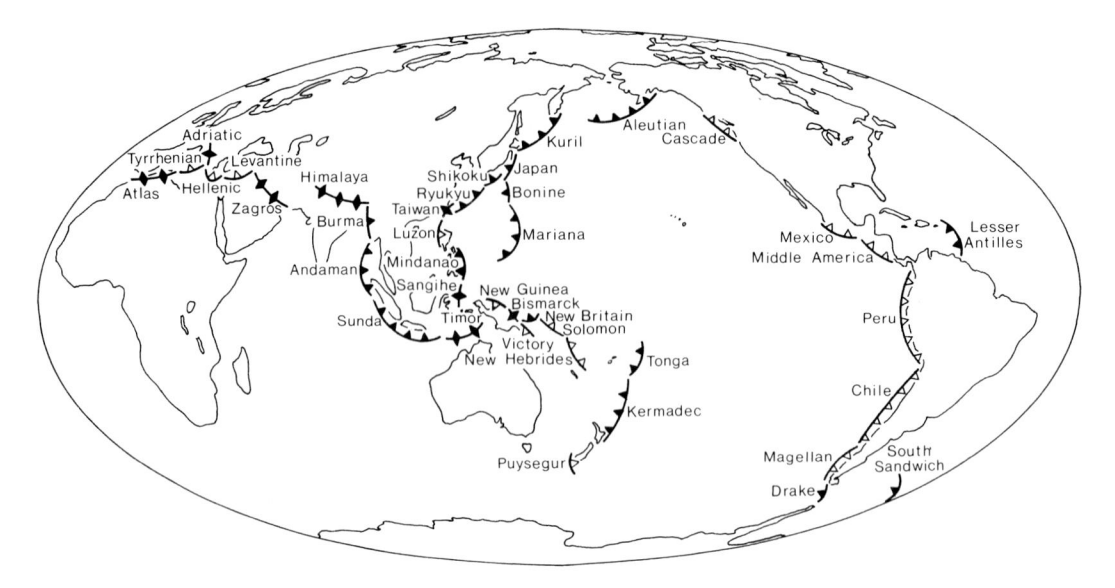

FIG. 1.—The world's major subduction-zone segments. Solid single barbs = fixed subduction zone; open single barbs = overstepped zone; solid double barbs = collided zone.

sociated with deep-ocean radiolarian chert and pillow lava, shows that individual subduction zones eventually become inactive, and hence, by implication, shows that new subduction zones originate along new alignments. Reconstructions of past plate relations suggest that subduction zones pass through the stages of birth, a nearly steady state, and final inactivation.

About 20% of all the subduction zones active today are of the collision type, in which the zone of convergence will eventually become inactive, and subduction will move to a new alignment (Table 1). This relatively high percentage indicates that the process by which a weakening subduction zone is incorporated into a continent and is subsequently fossilized is a slow one. Low-density continental crust is too buoyant to be deeply subducted, and so folding and uplift occur at the old convergent margin, the flanking terrane is forced to contract, and a new generally parallel subduction zone is established commonly nearby, but in some cases thousands of kilometers away.

Inactivation by collision has been inferred for many ancient foldbelts around the world; a fre-

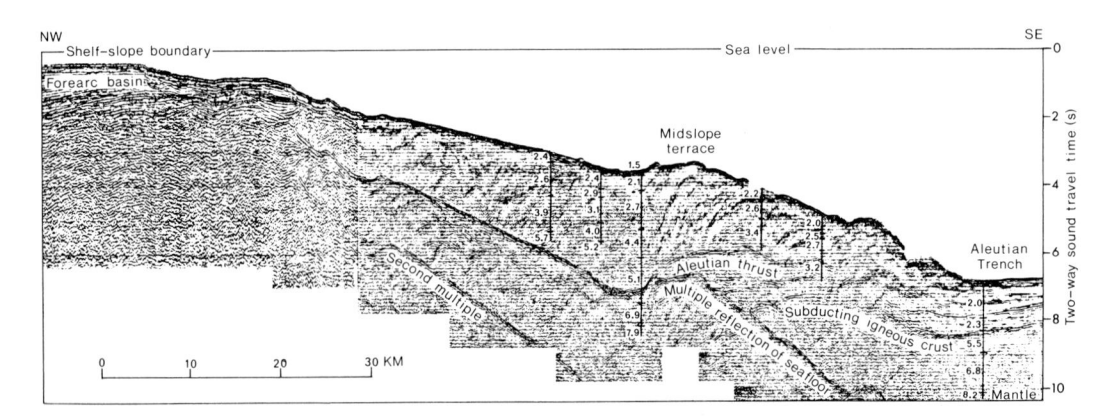

FIG. 2.—Subbottom profile at the Aleutian subduction zone 150 km east of Kodiak Island from the continental shelf to the trench (after von Huene et al., 1979). At this subduction zone, the shelf-slope boundary coincides with the trench-slope break. The Aleutian thrust and the oceanic crust seem to rise under the trench slope because the sound velocity is slower in the water than in the strata, so the subbottom reflections arrive early where the water is shallow. Sound-velocity values in km/s along vertical lines are from seismic-refraction analysis.

TABLE 1.—LENGTH (IN KM) OF THE WORLD'S MAJOR SUBDUCTION-ZONE SEGMENTS, CLASSIFIED AS *FIXED* WHERE THE EDGE OF THE UPPER PLATE DOES NOT MOVE ACROSS THE DEEP SEISMIC ZONE (BUT MARGINAL BASINS MAY OPEN); *OVERSTEPPED*, WHERE THE UPPER PLATE MOVES TOWARD THE TRENCH; AND *COLLIDED*, WHERE CONTINENTAL CRUST ON BOTH PLATES COMES TOGETHER

Segment	Fixed (Japan Type)	Overstepped (Andes Type)	Collided (Himalaya Type)
Atlas			2000
Tyrrhenian		800	
Adriatic			800
Hellenic		1300	
Levantine		1400	
Zagros			2700
Himalaya			3000
Burma	700		
Andaman	1300		
Sunda	2900		
Timor			1700
New Guinea		1200	
Bismarck			600
Victory	700		
New Britain	700		
Solomon		1500	
New Hebrides		1700	
Tonga	1100		
Kermadec	2200		
Puysegur		700	
Sangihe			700
Mindanao	800		
Luzon		900	
Taiwan			400
Ryukyu	1400		
Shikoku	700		
Mariana	1600		
Bonine	1000		
Japan	800		
Kuril	2100		
Aleutian	3000		
Cascade	1300		
Mexico		1100	
Middle America		1700	
Lesser Antilles	1100		
Peru		3100	
Chile		2900	
Magellan		1400	
Drake	800		
South Sandwich	1100		
Total	24000	21000	11900
Percent	42	37	21

quently cited example is the Carboniferous to Permian collision of North America and Africa at the Appalachian foldbelt (Wilson, 1966). Abandonment can also cause the inactivation of a subduction zone. Such an abandonment took place when the Mesozoic subduction zone along the south edge of the Bering Shelf was replaced during the early Tertiary by the present Aleutian subduction zone, up to 1000 km farther seaward (Scholl and Buffington, 1970).

The range of variability near the shelf-slope boundary is demonstrated by the topography and structure in the area between present-day volcanic arcs and deep-sea trenches. These variations record the properties of steady-state subduction zones and show the influence on them of rate of shelf sedimentation, rate of subduction, age of crust, dip of subduction zone, presence of subducting sediment and oceanic plateaus, and movement of the upper plate with respect to deep parts of the subduction zone (Dickinson and Seely, 1979).

Events during the earliest stages of a new subduction zone dominate the geologic history of the shelf-slope boundary. Broad convergent margins generally have had complex and multicycled histories that include two or more separate subduction-zone alignments.

TRENCH-SLOPE BREAK AT NEW SUBDUCTION ZONES

Several new subduction zones of the world are in various stages of formation, parallel with collision zones. Taiwan is a dramatic incipient example (Fig. 3A). The island is now riding up, flaplike, on the east edge of China's continental margin (Page, 1978), while China wedges under it at the north end of the east-dipping Luzon subduction zone. The west-dipping Ryukyu subduction zone to the north is likely soon to extend around the seaward side of Taiwan, and the island will then become coupled with China on the upper plate of the Ryukyu thrust.

An example of a more advanced stage, one of the youngest subduction zones that has developed a prominent trench, but not as yet a deep Benioff earthquake zone, lies beyond the south end of the Mindanao thrust, east of the island of Halmahera, Indonesia (Fig. 3B). The north end of Halmahera has collided with the west-dipping Sangihe subduction zone, which still has earthquake foci to a depth of 650 km (Silver and Moore, 1978). About 300 km to the east, the Mindanao thrust, which also dips to the west, is taking over from the Sangihe along an alignment that includes Halmahera on the upper plate. When the process is completed, a new subducting slab will have been emplaced in the mantle 300 km east of the old one, and a new active continental margin will flank Halmahera.

These and similar examples of incipient subduction zones that are separated from older zones, when taken together with long-established examples such as the Aleutian, show that tectonic plates can insert themselves into the mantle in previously unpenetrated places. The process generally progresses by propagation from one end at an existing plate boundary; the new alignment, however, can differ greatly from the old, so that when the process is complete, the lower mantle is cut along an entirely new line.

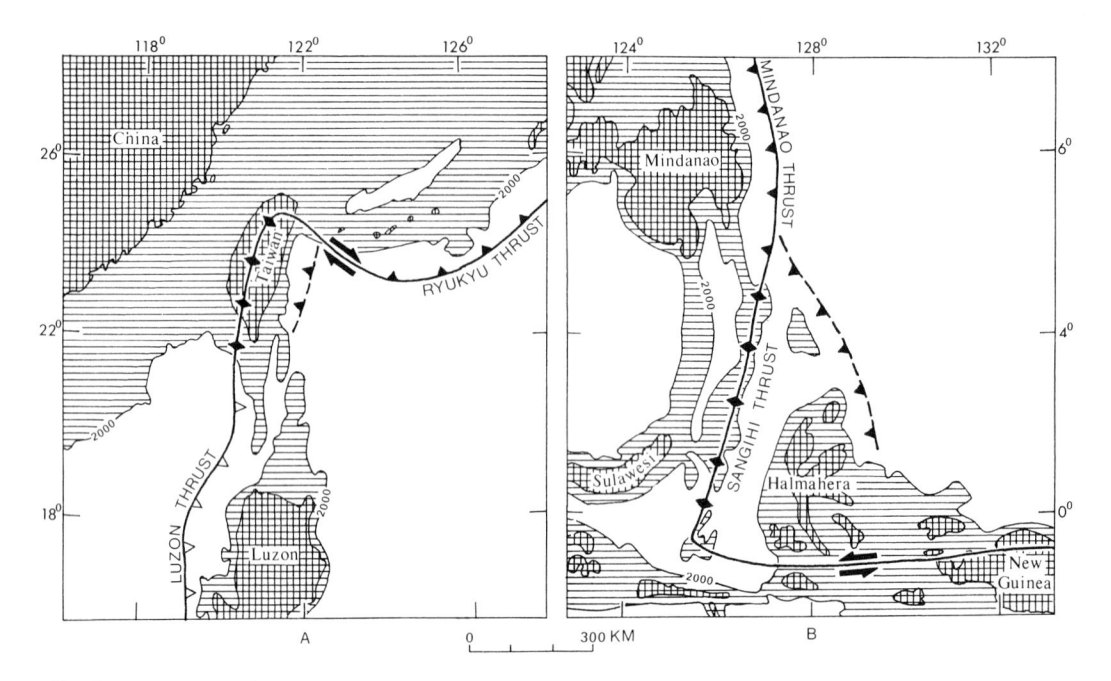

FIG. 3.—*A*, Probable future subduction-zone alignment (dashed) east of Taiwan; *B*, incipient subduction (dashed) east of Halmahera. Dark pattern, land; light pattern, shelf and upper slope; symbols same as for Fig. 1.

The initial stages of these new subduction zones differ in important ways from the later steady-state stages. Whereas established megathrusts pass through a smooth, gentle curve from the horizontal crust of the seafloor to the average dip of the subduction zone, new subduction zones are believed to obey the rules of compressional failure in shear. Consequently, the initial rupture surface should intersect the Earth's surface at a dip of about 30° (Fig. 4A).

Subsequent subduction of the angular edge between the rupture surface and the crustal surface sets into action a chain of transient processes that often leave traces in the geologic record. The margin of the upper plate is uplifted, usually resulting in a major unconformity (Fig. 4B). The unsupported edge of the upper plate subsequently collapses, resulting first in surficial slides and ultimately in a mélange, as both the slide material and tectonically brecciated material move into the subduction zone (Fig. 4C).

The top of the crust of the lower plate at the angular edge is sheared off in a series of sheets. Initially the cuts go deeply into the igneous lithosphere. Later, as the curve of the subduction zone becomes more gentle, only the sediment is cut off. Still later, either a surface layer of sediment is cut off or, where the sediment is thin, no surficial layer at all.

The sheets accumulate in the form of an imbricated stack (the accretionary prism) in the stress shadow that remains when the angular edge between the rupture surface and the crustal surface begins to move downward. This accretionary prism structurally underlies the mélange, which in turn underlies the upper mantle and crust of the upper plate.

At typical rates of subduction (5 to 10 cm/yr), the leading edge of the new lower plate would traverse the slope distance to the base of the upper plate (approximately 200 km) in 2 to 4 m.y. At the end of this time, volcanism would be expected to begin along a new volcanic arc. At approximately the same time, the megathrust would have acquired the smoothly curved profile that is characteristic of steady-state subduction zones.

Subsequently, as the lithospheric plate with its sediment and mafic-igneous crust moves along the smooth curve of the subduction zone, the average position of the surface of maximum shear intersects the seafloor at almost precisely the base of the trench slope. Igneous crust is rarely accreted during this stage, in which the dominant process is one of sediment subduction (Scholl et al., 1980). Where the oceanic sediment is especially thick, the line of maximum shear may lie a short distance below the floor of the trench, as is the case off Barbados, where about 80 m of oceanic sediment are accreted at the same time that 1200 m are subducted (Peter and Westbrook, 1976). In the more usual case for established subduction zones, however, the surface of maximum shear is believed to crop out a short distance above the trench floor, so that the previously formed accretionary prism begins to be removed by a slow tectonic plucking of the sole of the upper plate, called subduction-erosion (Scholl et al., 1980).

Although oceanic crust usually is not accreted from the lower plate during the steady-state phase of subduction, ophiolitic material including serpentinite can enter the system if the upper plate overrides the subduction zone. At the Peru megathrust, for example, the Nazca plate moves eastward at 6 cm/yr relative to the lower mantle, and the South America plate westward at 3 cm/yr (Minster and Jordan, 1978). Most of the resulting 9 cm/yr of convergence is taken up on the surface of maximum shear that crops out near the trench. The overriding requires either that the zone be pressed downward and trail drape-like under the continent (Dewey, 1980) or, more likely, that it step westward from time to time. In either case, part of the strain caused by the overriding is taken up on landward-dipping thrust faults that lie above the accretionary prism. This adjustment permits additional crust or upper-mantle material to be added near the shelfbreak as mélange (Fig. 4C). In the case of South America, the mélange might be suspected to be composed of continental crust, whereas during the steady-state stage of subduction in a more oceanic setting, serpentinite and other mantle rocks can be brought upward and exposed (Fisher and Engel, 1968), while overlying materials are eroded off or slide down the trench slope. This same process near the shelf-slope boundary can carry blueschist on the base of the upper plate to the surface from deep-seated parts of the subduction zone.

SEISMIC SOUNDING, DEEP-SEA DRILLING, AND FOREARC ISLANDS

Dickinson and Seely (1979) and Seely (1979) have written excellent reviews describing the evolution

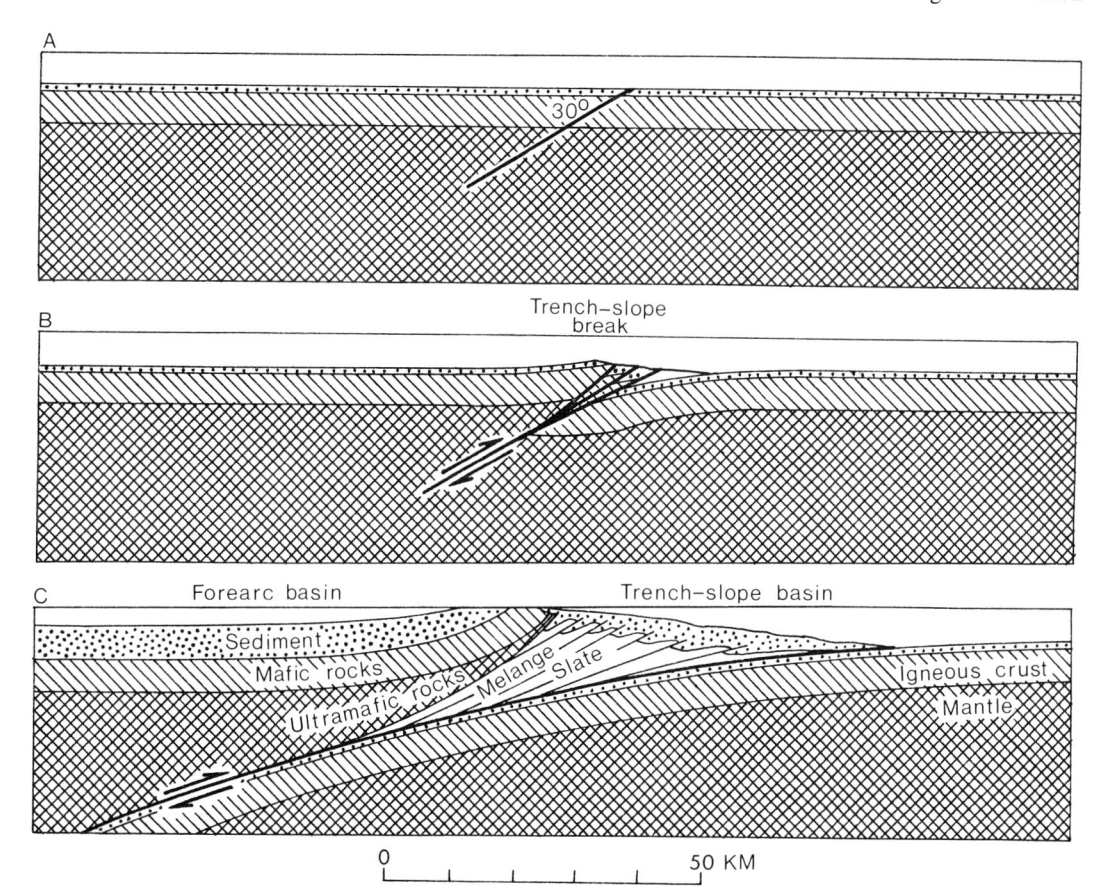

FIG. 4.—A, Incipient subduction breaks the seafloor at about 30°; B, Compression imbricates lower-plate crustal slabs and together with isostacy elevates the trench-slope break as the wedge-shaped leading edge of the lower plate is subducted; C, Later steady-state subduction causes little disruption of the lower plate; the shelf-slope boundary migrates across the forearc basin, controlled by a balance between sediment supply and tectonic undulation; the boundary commonly coincides with the trench-slope break.

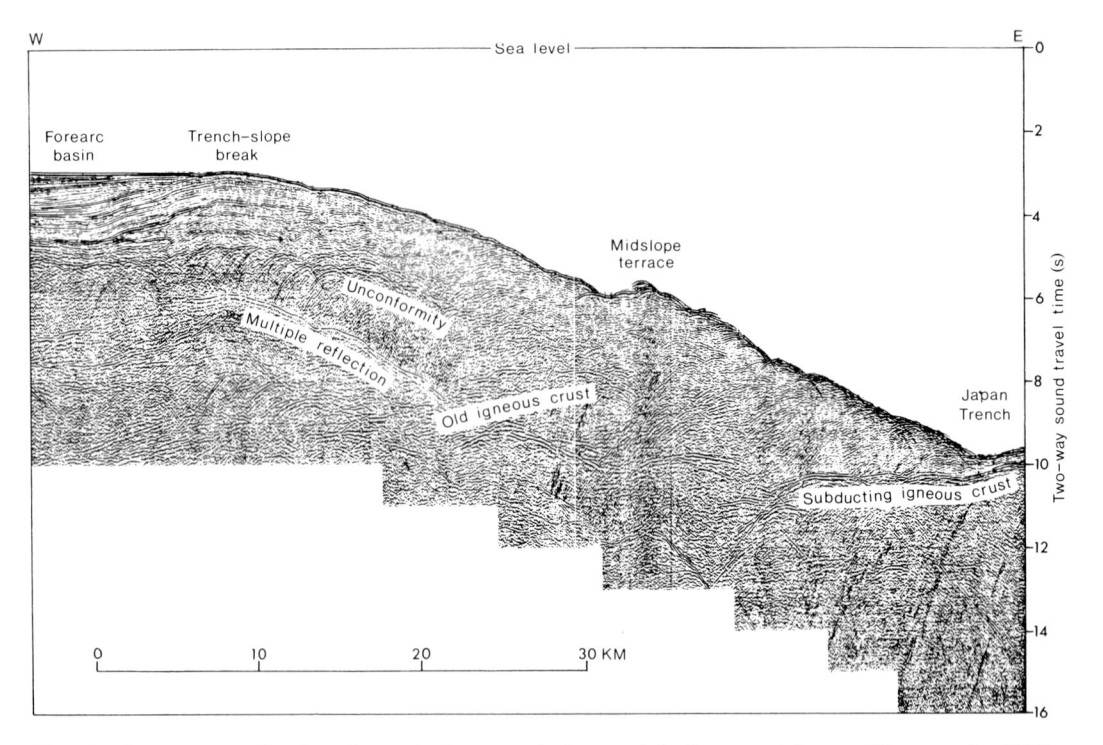

FIG. 5.—Subbottom profile, from forearc basin to trench, Japan subduction zone (after von Huene et al., 1980).

of the structural highs that border forearc basins. Important new information is now available from seismic profiles and deep-sea drilling near the Japan Trench (von Huene et al., 1980). In common with some other structural highs at the trench-slope break, such as at the Java Ridge (Hamilton, 1979), off Japan the break is depressed to a water depth of about 1500 meters. This favorable circumstance met the water-depth drilling requirements of the *Glomar Challenger,* so that the drilling vessel could core at the seaward margin of the forearc basin.

The most notable discovery resulting from the drilling off Japan was that Cretaceous rocks extend to the trench-slope break, and seismic profiling provides evidence that they probably extend to the midslope terrace, about halfway down the trench slope (Fig. 5). These Cretaceous rocks are overlain by an Oligocene subaerial unconformity. On seismic profiles they are underlain by a prominent reflector that dips gently landward about 3000 meters above the clearly recorded oceanic crust of the Pacific plate. This gently dipping reflector is believed to be the top of the old oceanic crust and upper mantle that was cut off when the present alignment of the Japan subduction zone was established during the Miocene (Fig. 6).

Several outer-arc islands at various subduction zones around the world provide peepholes into forearc structural highs and reveal details that are not discernible in drill cores or subbottom profiles. Some manifest only shelf sediment, but others (such as Nias Island, Barbados, and Kodiak Island) contribute significantly to our understanding of convergent-margin processes. Detailed studies at these places suggest that individual subduction zones have shorter lifetimes than usually has been postulated.

Nias Island, off Sumatra, consists of Miocene slabs of mélange interleaved with reasonably coherent sedimentary rocks (Karig et al., 1980). The strike is parallel with the adjacent Sunda Trench, and the rocks on Nias may represent a single fairly steady subduction-zone cycle. The initial break probably lay landward from the island, approximately under the seaward edge of the deep Mentawei forearc basin.

At Barbados, two subduction cycles are suggested by shelf rocks which cover an Oligocene to Miocene subduction mélange that strikes approximately east, at right angles to the adjacent Lesser Antilles subduction zone (Speed, 1981). This same eastward trend is repeated in the volcanic arc at Granada and at Trinidad and Tobago to the south. Hence, two subduction-zone alignments and stages of the geologic history seem to be recorded.

Kodiak Island has had an even more complicated history (von Huene et al., 1979; Nilsen and Moore, 1979), an evolution that involves at least three in-

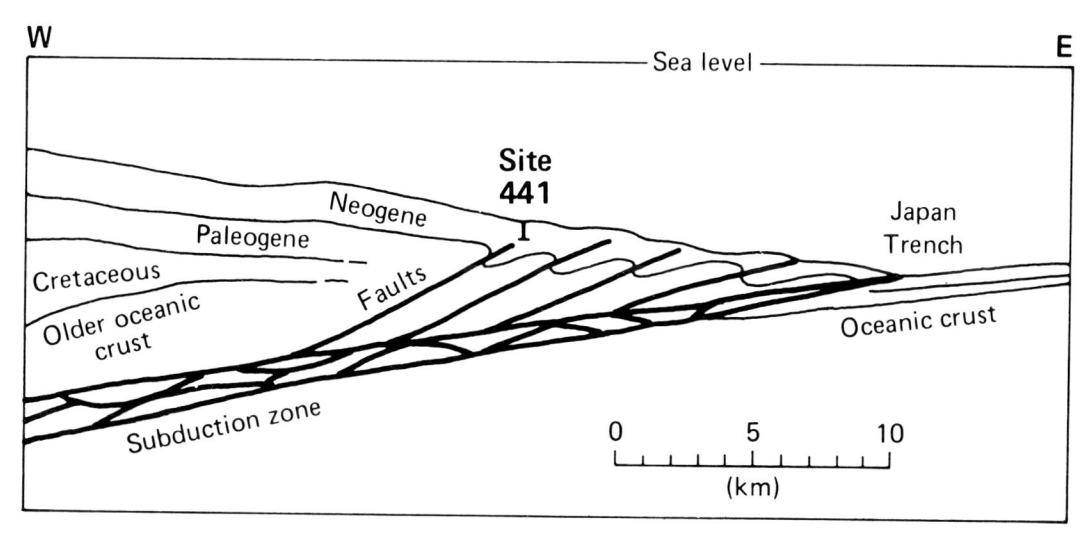

FIG. 6.—Interpretation of the Japan subduction zone (after Moore, 1980).

dependent subduction cycles. The island has sometimes been considered a typical example of a long-lived accretionary margin, because of successive seaward belts of Triassic volcanic rocks, Jurassic blueschist, serpentinite, Cretaceous mélange, Late Cretaceous flysch, Paleogene trench-slope deposits, and Neogene shelf rocks. But paleomagnetic and paleoecologic studies in Alaska have shown that the Triassic and Jurassic materials were accreted from an equatorial source as a coherent microplate during the Cretaceous (Coney et al., 1980). The distribution and alignment of Cretaceous magnetic lineations in the Pacific Basin indicate that the Mesozoic volcanic rocks and blueschist on Kodiak Island are related not to a forerunner of the nearby Aleutian subduction zone, but to a subduction zone that probably rifted apart from New Guinea or Australia during the Jurassic.

The Cretaceous mélange and flysch do indeed seem to be a single-cycle subduction complex that backed up against the exotic microplate after it had been added to the continental margin. But where lay the initial point of contact with North America? Paleomagnetic data on early Tertiary igneous rocks intrusive into and interbedded with Alaskan sedimentary rocks show that they were emplaced about 25° to the south, or approximately where southern California is now located (Grommé and Hillhouse, 1981; Plumley and others, 1981). Hence the "continuous" accretionary margin at Kodiak Island is believed to have had a stage of subduction in the Southern Hemisphere, then moved across the Pacific to central North America (probably on the Kula plate), then moved northward along the coast of North America (probably on the Pacific plate), collided with Alaska along an early eastern Aleutian Trench during the late Tertiary, and only then began the present cycle of subduction along the present Aleutian Trench.

The history of the Gulf of Alaska has had a counterpart in California, where Blake and Jones (1981) have shown that the emplacement of Franciscan rocks during the Mesozoic and Cenozoic was interrupted at least four times by periods of strike-slip motion or by the accretion of microplates large enough to cause the subduction zone to step out to a new alignment. But, like the Cook Inlet Basin of Alaska, the Great Valley of California was able to continue as a depositional basin through the various subduction cycles. It survives today as a local depocenter, even though subduction ceased in central California in the Miocene.

The evidence from seismic sounding, deep-sea drilling, and forearc islands indicates that the lifetimes of single-cycle subduction zones are usually no longer than about 25 million years. Continuous deposition in forearc basins directly landward from the trench-slope break normally endures for about the same length of time. Geologic structures within the arc-trench gap that lie landward from the forearc basin should be studied with care; most have a geologic history that began before the present subduction cycle.

CONCLUSIONS

The major conclusion of this paper is that the trench-slope break at convergent margins undergoes very rapid changes during the first five million years after a new subduction-zone alignment is established. During this brief period, a massive body of oceanic crust, upper mantle material, and imbricated sediment is accreted. The succeeding evolution of the convergent margin as a steady-state subduction zone ordinarily includes much less (or no further) accretion by the incorporation of ocean-floor sediment. Indeed, subduction-erosion may slowly reduce the volume of the accreted prism.

Where broad continental margins do consist of belts of subduction-related rocks that are progressively younger seaward and which encompass a relatively long age range, data from intervening accreted microplates, multiple magmatic arcs, and alternating sequences of ocean-floor and trench-slope sediments show that the margins resulted from a succession of several independent subduction zones.

During the evolution of a single subduction zone, the trench-slope break forms above the quickly emplaced and imbricated accretionary prism, and a midslope terrace commonly marks the original toe of the upper plate. The trench-slope break constitutes a structural high, underlain (in downward sequence) by strata that were on the crust when the subduction began, by the old igneous crust and upper mantle, by mélange, by the more coherent parts of the accretionary prism, and by the young subducting sediment and igneous crust.

Passage of the wedge-shaped leading edge of the new lower plate uplifts the shelf-slope boundary strongly, and this uplift is succeeded by subsidence, when the tabular part of the plate follows along. Subduction-erosion and cooling of the upper plate cause part of the subsidence (Howell and von Huene, 1981); cooling of newly emplaced igneous slabs in the accretionary prism is particularly effective. The main cause of the subsidence, however, is the total subduction of the angular edge of the lower plate.

Continental-margin sedimentation marks the subsequent steady-state evolution of the shelf-slope boundary. The margin is compressed and elevated, forming the islands of trench-slope ridges. The trench-slope break is almost everywhere antiformal, because of consolidation and compression of the accretionary prism below.

During the steady-state stage, imbricated slabs of trench sediment emplaced during the accretionary stage are succeeded on the slope by submarine-fan deposits (Nilsen and Moore, 1979). Where the surface of maximum shear is below the floor of the trench, folds of trench sediment are added to the lower trench slope (Moore and Karig, 1976). This effect is most prominent near junctions between subduction zones and transform faults where sediment subduction must end, for example at North Island, New Zealand, and in the Gulf of Oman (Katz and Wood, 1980; White, 1977).

A deep sedimentary basin forms landward from the trench-slope break, and an effectively unlimited sink for any bypassed sediment also exists on the trench slope. During the steady-state stage of the subduction zone, because of the constant subduction of cool oceanic crust below, the strata in these two depocenters usually remain too cool for petroleum maturation. Extrapolation of temperature logs in holes along the trench slope of the Japan Trench shows that the temperature near the subducting igneous crust below is only about 50°C, so that rocks within the upper plate are below the temperature generally recognized as the minimum temperature for generating petroleum (von Huene et al., 1980). But when, whether by abandonment or by collision, the subduction zone becomes inactive, a normal geothermal gradient is established, and petroleum precursors may then proceed to thermal maturity.

ACKNOWLEDGMENTS

For helpful reviews of the manuscript for this paper, I am indebted to William K. Gealey, David G. Howell, Raymond V. Ingersoll, and David W. Scholl.

REFERENCES

BLAKE, M. C., JR., AND JONES, D. L., 1981, The Franciscan assemblage and related rocks in northern California: a reinterpretation, in Ernst, W. G., ed., The Geologic Development of California: Englewood Cliffs, New Jersey, Prentice-Hall, p. 307–328.

CONEY, P. J., JONES, D. L., AND MONGER, J. W. H., 1980, Cordilleran suspect terranes: Nature, v. 288, p. 329–333.

DEWEY, J. F., 1980, Episodicity, sequence, and style at convergent plate boundaries: Geol. Assoc. Canada Spec. Paper 20, p. 553–573.

DICKINSON, W. R., AND SEELY, D. R., 1979, Structure and stratigraphy of forearc regions: Am. Assoc. Petroleum Geologists Bull., v. 63, p. 2–31.

FISHER, R. L., AND ENGEL, C. G., 1968, Peridotite and dunite dredged from the nearshore flank of Tonga Trench on Expedition Nova, 1967: Trans. Am. Geophys. Union., v. 49, p. 217.

GOLOVCHENKO, X., LARSON, R. L., AND PITMAN, W. C., III, 1981, Magnetic lineations, scale 1:10,000,000: Am. Assoc. Petroleum Geologists Circum-Pacific Plate-Tectonic Map, 5 sheets.

GROMMÉ, S., AND HILLHOUSE, J. W., 1981, Paleomagnetic evidence for northward movement of the Chugach terrane, southern and southeastern Alaska: U.S. Geol. Survey Circ. 823-B, p. 70–72.

HAMILTON, W., 1979, Tectonics of the Indonesian region: U.S. Geol. Survey Professional Paper 1078, 345 p.

HOWELL, D. G., AND VON HUENE, R., 1981, Tectonics and sediment, along active continental margins, in Douglas, R. G., Colburn, I. P., and Gorsline, D. S., eds., Depositional Systems of Active Continental Margin Basins: Soc. Econ. Paleontologists Mineralogists Pacific Sec., p. 1–13.

INGERSOLL, R. V., AND GRAHAM, S. A., 1983, Recognition of shelf-slope break along tectonically active, ancient, tectonically active continental margins, *in* Stanley, D. J., and Moore, G. T., eds., The Shelfbreak: Critical Interface on Continental Margins: Soc. Econ. Paleontologists Mineralogists Spec. Pub. No. 33, p. 107–117.

KARIG, D. E., LAWRENCE, M. B., MOORE, G. F., AND CURRAY, J. R., 1980, Structural framework of the forearc basin, N. W. Sumatra: Geol. Soc. London Jour., v. 137, p. 77–91.

KATZ, H. R., AND WOOD, R. A., 1980, Submerged margin east of the North Island, New Zealand, and its petroleum potential: United Nations CCOP/SOPAC Tech. Bull. No. 3, p. 221–235.

MINSTER, J. B., AND, JORDAN, T. H., 1978, Present-day plate motions: Jour. Geophys. Res., v. 83, p. 5331–5345.

MOORE, G. W., 1980, Slickensides in deep sea cores near the Japan Trench, Leg 58, Deep Sea Drilling Project: Deep Sea Drilling Project Initial Repts., v. 56–57, p. 1107–1115.

———, 1981, Plate perimeters and motion vectors, scale 1:1:10,000,000: Am. Assoc. Petroleum Geologists Circum-Pacific Plate-Tectonic Map, 5 sheets.

MOORE, J. C., AND KARIG, D. E., 1976, Sedimentology, structural geology, and tectonics of the Shikoku subduction zone, southwestern Japan: Geol. Soc. America Bull., v. 87, p. 1259–1268.

NILSEN, T. H., AND MOORE, G. W., 1979, Reconnaissance study of Upper Cretaceous to Miocene stratigraphic units and sedimentary facies, Kodiak and adjacent islands, Alaska: U.S. Geol. Survey Professional Paper 1093, 34 p.

PAGE, B. M., 1978, Franciscan melanges compared with olistostromes of Taiwan and Italy: Tectonophysics, v. 47, p. 223–246.

PETER, G., AND WESTBROOK, G. K., 1976, Tectonics of the southwestern North Atlantic and Barbados Ridge complex: Am. Assoc. Petroleum Geologists Bull., v. 60, p. 1078–1106.

PLUMLEY, P. W., BYRNE, T. B., COE, R. S., AND MOORE, J. C., 1981, Paleomagnetism of the Ghost Rocks volcanics on Kodiak indicates 24° northward displacement: EOS, v. 62, p. 854.

RUFF, L., AND KANAMORI, H., 1980, Seismicity and the subduction process: Physics Earth Planet. Interiors, v. 23, p. 240–252.

SCHOLL, D. W., AND BUFFINGTON, E. C., 1970, Structural evolution of Bering continental margin: Cretaceous to Holocene (abstract): Am. Assoc. Petroleum Geologists Bull., v. 54, p. 2503.

———, VON HUENE, R., VALLIER, T. L., AND HOWELL, D. G., 1980, Sedimentary masses and concepts about tectonic processes at underthrust ocean margins: Geology, v. 8, p. 564–568.

SEELY, D. R., 1979, The evolution of structural highs bordering major forearc basins: Am. Assoc. Petroleum Geologists Memoir 29, p. 245–260.

SILVER, E. A., AND MOORE, J. C., 1978, The Molucca Sea collision zone, Indonesia: Jour. Geophys. Res., v. 83, p. 1681–1691.

SPEED, R. C., 1981, Geology of Barbados: implications for an accretionary origin: Oceanol. Acta, v. 4, supp. C3, p. 259–265.

VON HUENE, R., MOORE, G. W., AND MOORE, J. C., 1979, Cross section, Alaska Peninsula—Kodiak Island—Aleutian Trench, scale 1:250,000: Geol. Soc. America Map and Chart Ser. MC-28A, 2 sheets.

———, NASU, N., ARTHUR, M. A., BARRON, J. A., BELL, G. D., CADET, J. P., CARSON, B., FUJIOKA, K., HONZA, E., KELLER, G., MOORE, G. W., REYNOLDS, R., SATO, S., AND SCHAFFER, B. L., 1980, Sites 438 to 441: Deep Sea Drilling Project Initial Repts., v. 56–57, p. 23–191, 225–354.

WHITE, R. S., 1977, Recent fold development in the Gulf of Oman: Earth Planet. Sci. Letters, v. 36, p. 85–91.

WILSON, J. T., 1966, Did the Atlantic close and then re-open?: Nature, v. 211, p. 676–681.

RECOGNITION OF THE SHELF-SLOPE BREAK ALONG ANCIENT, TECTONICALLY ACTIVE CONTINENTAL MARGINS

RAYMOND V. INGERSOLL[1]
Department of Geology, University of New Mexico, Albuquerque, New Mexico 87131

STEPHAN A. GRAHAM
School of Earth Sciences, Stanford University, Stanford, California 94305

ABSTRACT

Tectonically active continental margins include transform, protoceanic and convergent settings. In transform settings, numerous small basins develop on oceanic, transitional and continental crust. Protoceanic gulfs may be formed by orthogonal or oblique rifting; both types are characterized by segmented crust with consequent juxtaposition of deep basins and active uplifts. Convergent margins include the following types of basins: intramassif basins, major forearc basins and accretionary basins (trench-slope basins). Extensional backarc margins have structural styles and histories similar to protoceanic gulfs.

Shelf-slope breaks in these settings tend to be transient in time and space because of rapid vertical movements; abrupt facies changes are the result. Two types of shelf-slope breaks are common: clastic-starved and progradational. The former type is characterized by unconformities or biostratigraphically compressed intervals separating shallow-marine/nonmarine from slope/basinal deposits. Commonly, glauconitic and phosphatic lithologies mark the clastic-starved shelf-slope break. Progradational shelf-slope breaks are characterized by deltaic outbuilding, which results in coarse shallow-marine, shoreline and fluvial deposits prograding over slope mudrocks. Clastic-starved shelf-slope breaks tend to predominate in transformal and protoceanic settings due to the segmented nature of crust and the structural control on shelfedges. Most detritus bypasses outer-shelf and upper-slope environments. Deltaic progradation is the dominant process in forearc basins, with abundant detritus supplied by neighboring magmatic arcs. During the latest stages of filling of forearc basins, shelf-slope breaks may correspond with the structural boundary between forearc basins and subduction complexes. Shelf-slope break deposits involved in continental collision in general are destined to be destroyed during continental suturing.

INTRODUCTION

This article reviews general characteristics of ancient shelf-slope breaks in tectonically active settings. We are not comprehensive in our coverage of ancient examples. Rather, our intent is to illustrate with examples, primarily from California, features commonly found at the stratigraphic boundaries between outer-shelf and slope deposits. The application of Walther's Law (e.g., Middleton, 1973) to the study of vertical successions of facies allows for the reconstruction of paleoenvironments in three dimensions, which in turn, provides data necessary for paleotectonic interpretations. In addition, certain structural styles and petrologic characteristics are indicative of settings that generally correspond to shelf or slope paleoenvironments along ancient active margins. Recognition of these features and application of these techniques allow for definition of a key paleogeographic element in basin reconstruction (the shelf-slope break).

TYPES OF MARGINS

Tectonically active continental margins include transform, protoceanic and convergent settings (Dickinson, 1974). Transform settings include both transtensional and transpressional types, depending on relations between plate motions and fault orientations, as well as local heterogeneities in crustal features. The best understood examples of both modern and ancient transformal continental margins come from the Cenozoic of California (Fig. 1). Since approximately 30 my BP, the San Andreas fault system has lengthened until the entire California coast south of Cape Mendocino is now a complex transform margin (Atwater, 1970; Graham, 1978). Within this setting, numerous small basins, developed on oceanic, transitional and continental crust, display almost every type of depositional environment. Crowell (1974 a, b) has discussed the role of vertical movements in basin formation within the overall framework of transform tectonics (Fig. 2). The fill of Oligocene-Holocene transform-related basins exposed within the California Coast Ranges provides the majority of our examples of transform-related paleoenvironments.

Protoceanic gulfs may be formed either by or-

---

[1]Present address: Department of Earth and Space Sciences, University of California, Los Angeles, California 90024.

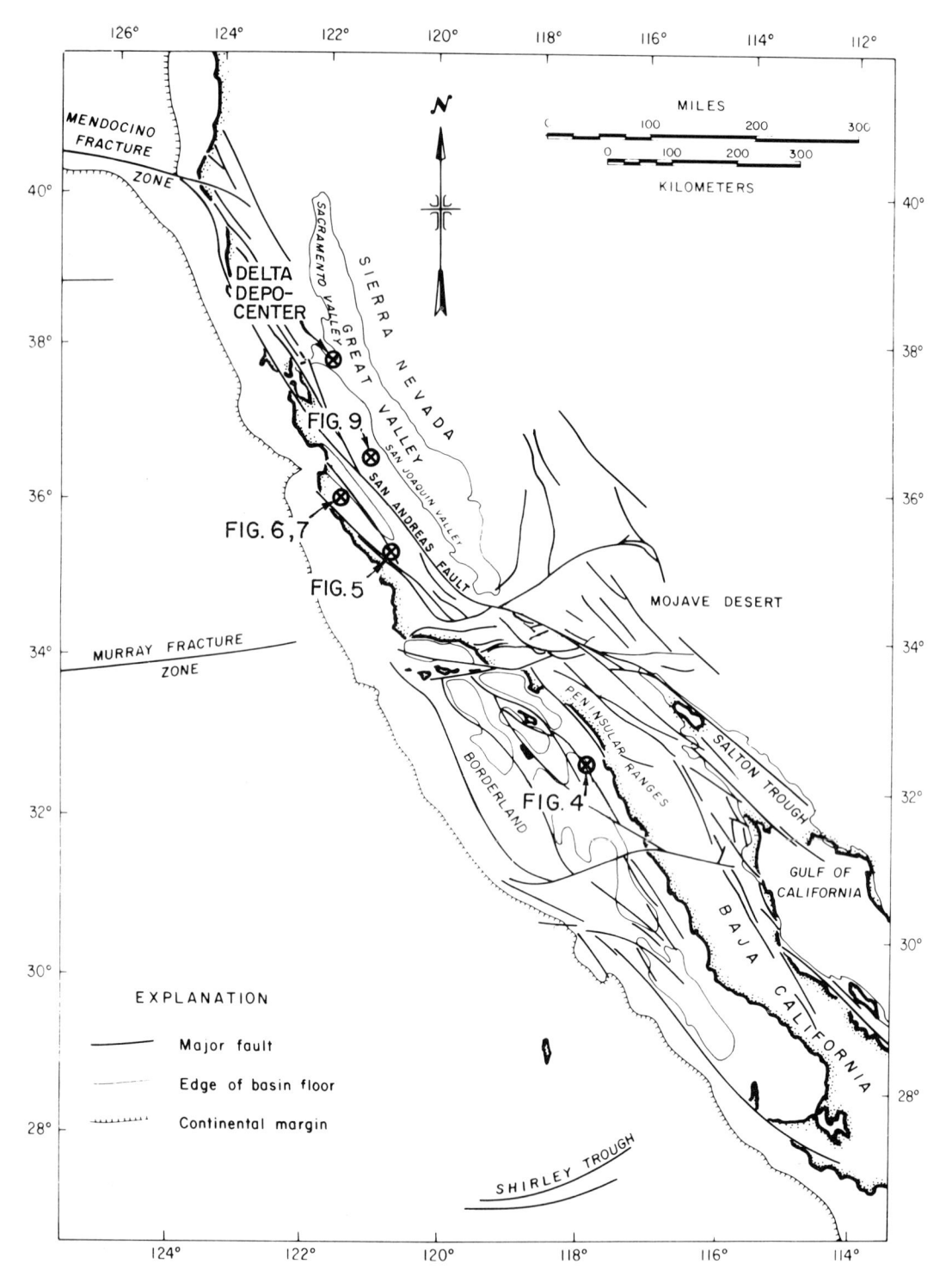

FIG. 1.—Map of California showing major faults and locations discussed in text (modified slightly from Crowell, 1974a).

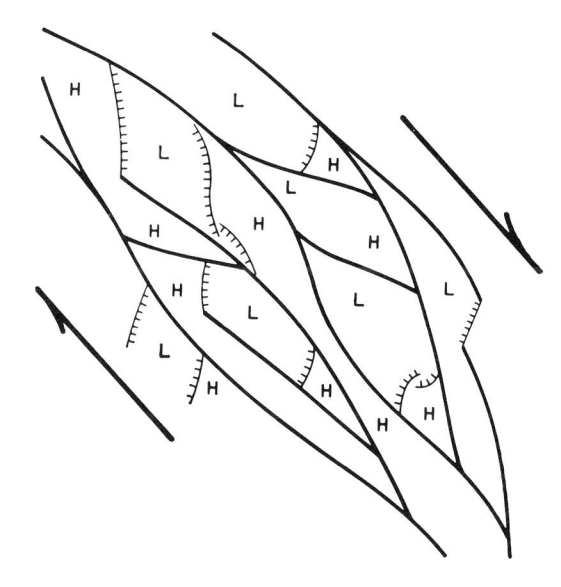

FIG. 2.—Sketch map across a hypothetical strike-slip fault zone illustrating typical anastomozing fault pattern and juxtaposed regions of uplift/erosion (H) and subsidence/sedimentation (L) (from Crowell, 1974b). Compare with Fig. 1.

thogonal (e.g., Red Sea) or by oblique (e.g., Gulf of California, Fig. 1) rifting (e.g., Lowell and Genik, 1972; Moore, 1973). Although plate-tectonic controls and structural development may differ in primarily transformal and primarily extensional settings, resulting basin evolution and paleoenvironments may be similar. Both settings are characterized by attenuated and block-faulted crust, resulting in close juxtaposition of deep basins and active uplifts. Because of these similarities, the two settings are treated together in later discussion of ancient examples. Sedimentologic, stratigraphic and paleoenvironmental data by themselves seldom distinguish these two settings. Additional information regarding structural style and tectonic setting usually is necessary.

Forearc areas of convergent continental margins include several closely related but distinct types of basins (Dickinson and Seely, 1979; Seely, 1979). Of most importance are the following three types: intramassif basins, major forearc basins and accretionary basins (trench-slope basins) (Fig. 3). Intramassif and backarc basins display structural styles similar to protoceanic basins in that they are extensional and normal-fault bounded; the predominance of arc volcanics and volcaniclastics distinguishes arc-related basins from the latter. Forearc and trench-slope basins have distinctive structural styles as a result of the subduction-accretionary process. Accretion of oceanic materials at the base of the trench slope commonly results in vertical growth of the subduction complex (Karig and Sharman, 1975). For a review of variations in this simple model, see Scholl et al. (1980). The subduction complex commonly provides a barrier to sediment dispersal, resulting in the accumulation of thick forearc sediments. Trench-slope basins are formed on top of growing subduction complexes between imbricate slices of deformed trench and oceanic sediments (Moore and Karig, 1976). A variety of

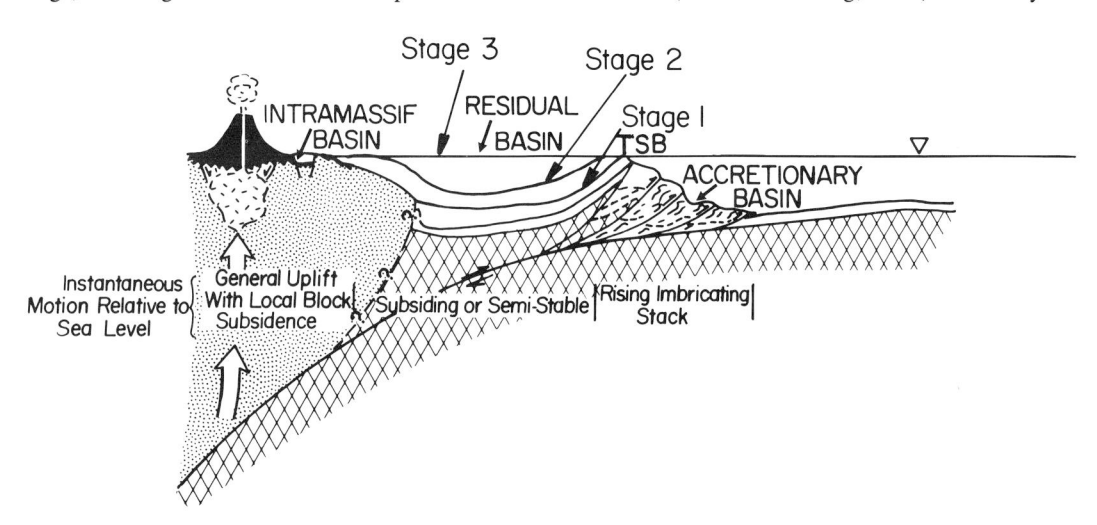

FIG. 3.—Possible types of basins and tectonic processes in forearc regions. During early history (Stage 1), deep-marine deposits accumulate along-floor of the residual forearc basin. During Stage 2, basin begins to fill by the progradation of deltaic/slope deposits derived from magmatic arc, until Stage 3, when the forearc basin is filled to near sealevel. During this last stage, the shelf-slope break may correspond to the trench-slope break (TSB) (see text for discussion). This figure is modified from Dickinson and Seely (1979) and Seely (1979).

geometries and paleoenvironments typifies both modern and ancient forearcs (Dickinson and Seely, 1979; Seely, 1979). At some stage in the evolution of a mature forearc region, the forearc basin commonly is filled to shelf depths and shelf sediments may prograde directly over the subduction complex (stage 3 of Figure 3). In this situation, the subduction complex represents predominantly slope environments, and the forearc-basin strata represent predominantly shelf environments. Thus, the shelf-slope break represents both an environmental and a tectonic transition. During earlier stages of development (e.g., stage 2 of Figure 3), shelf environments prograde over slope environments within the forearc basin in a structurally uncomplicated manner. Sedimentary processes are similar to those occurring along many mature rifted continental margins with rapid sediment input [deltaic settings along continental embankments (e.g., Dickinson, 1974) such as the modern Gulf of Mexico and Niger Delta areas]. The overall setting and the volcaniclastic nature of the detritus in forearc basins distinguishes these contrasting tectonic settings. The most thoroughly studied ancient forearc area displaying examples of all of these types of shelf-slope transitions is the late Mesozoic and Paleogene forearc basin of northern and central California (Dickinson and Seely, 1979; Dickinson et al., 1979; Ingersoll, 1978, 1979).

SHELF-SLOPE CHARACTERISTICS IN ACTIVE SETTINGS

Tectonically active continental margins have certain characteristics in common regardless of whether transform, divergent or convergent tectonics dominates. Shelf-slope breaks tend to be transient in both time and space, as a result of rapid vertical movements. Abrupt facies changes thus typify the ancient record. Due to the nondepositional character of narrow, tectonically controlled shelves, shelf-slope breaks commonly are expressed in the stratigraphic record as unconformities separating shallow-marine/nonmarine and slope/basinal deposits. Alternatively, shelf and basinal deposits may be separated by clastic-starved intervals, including glauconite, phosphorite and biogenic deposits, which represent the shelf-slope break. Due to higher sedimentation rates both upsection and downsection from shelf-slope break deposits, biostratigraphic zones usually are compressed at this transition.

Shelf-slope break deposits of active continental margins commonly are obscured by rapid rates and shifting patterns of sedimentation in adjacent areas. Extensive sediment bypassing of outer shelves and slopes occurs, with feeding of sediment from nearshore environments directly into the heads of submarine canyons (Underwood and Karig, 1980); most of this sediment ultimately is deposited within submarine fan complexes (Gorsline and Emery, 1959). Thus, both shelf and slope environments may contain finer and thinner detrital accumulations than neighboring nearshore and basinal deposits. Detrital grain size tends to decrease from shoreline to shelfedge, although authigenic and organic sediments may be coarse-grained on outer shelves (Emery, 1952). Identification of bracketing deeper- and shallower-marine facies is the usual key to locating ancient shelf-slope breaks.

In areas of rapid deltaic outbuilding, shelf-slope breaks commonly correspond to delta-top/delta-front transitions (continental embankments). In this setting, the shelf-slope break is expressed by slope mudrocks overlain by voluminous shallow-marine, shoreline and fluvial deposits. Detrital sediments may be coarse-grained just above the shelf-slope break.

SHELF-SLOPE BREAKS IN WRENCH
AND FAULT-BLOCK SETTINGS

Broad zones of anastomozing faults characterize many strike-slip dominated continental margins (Fig. 2). These complexities give rise to numerous sites for sedimentation and provide rapidly uplifting source areas. Such basins typically are small but deep, steeply fault-bounded, rhomb-shaped (note offshore borderland in Fig. 1) and destined for extensive deformation. Subsidence and uplift rates may be extremely rapid due to the constant jostling of neighboring blocks during continued strike-slip movement. Marine environments tend to dominate over nonmarine environments due to the fragmentation and attenuation of continental crust by multiple splay faults and lower-crustal extension. As a result, marine basins tend to have narrow, unstable sediment-starved shelves and steep slopes, with juxtaposition of disparate thicknesses of sediment across nearly vertical strike-slip faults. Seismic lines of the modern southern California margin (Fig. 4) illustrate such typical features.

Shelf-slope breaks of strike-slip margins commonly are notable for their lack of record within sedimentary sequences; unconformities separate contrasting shallow-marine and deep-marine deposits (Fig. 5). Where outcrops are good and/or well data complete, one may find a thin but critical record of the shelf-slope break in the form of sediment-starved shelfedge lithologies such as phosphorite, glauconite and/or biogenic deposits (Figs. 5, 6), similar to deposits at the shelfedge off modern continental margins (Birch, 1979; Odin and Matter, 1981; Parker, 1975). Biostratigraphic compression of this interval is typical due to more rapid sedimentation rates in underlying and overlying shallower-marine and deeper-marine deposits (Fig. 7).

Locally, where the proper balance of structural longevity, sedimentation and subsidence occurs, the

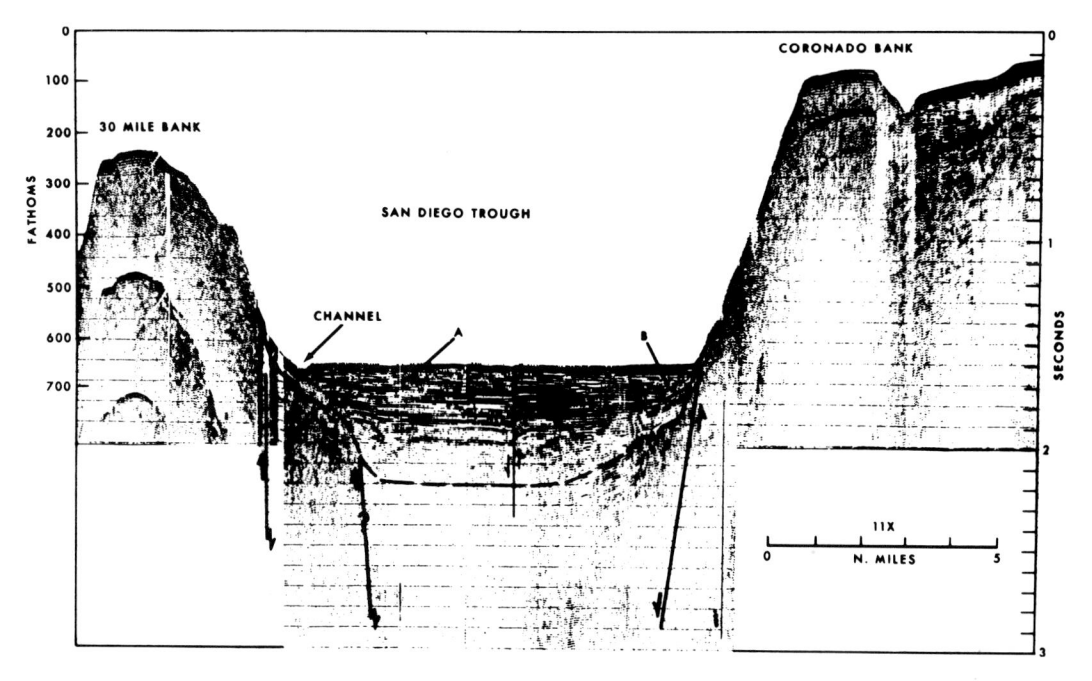

FIG. 4.—Seismic profile from Coronado Bank on the east to 30 Mile Bank on the west (from Moore, 1969). See Fig. 1 for location. Although faults display marked dip slip, most of them likely have much larger components of strike slip. Note the isolated, bank-like character of the shelf-slope break on Coronado Bank. This clastic-starved setting lies between the San Diego shelf at the right margin of the profile and the ponded turbidites of the San Diego Trough in the center.

shelfedge section is thick enough to be mappable as a distinct unit. Figure 7 shows lower Miocene shallow-marine bar sands overlain by 50 meters of middle Miocene alternating shale and phosphorite (upper bathyal), which in turn, is overlain by a thick sequence of upper Miocene basinal laminated siliceous mudrocks. The middle Miocene section is stratigraphically compressed relative to nearby coeval basinal sections of approximately 2000 m thickness.

Commonly, recognition of ancient shelf-slope breaks in wrench-tectonic regimes depends on bracketing from above and below by diagnostic deeper and shallower facies. The typical basinal facies of many wrench basins are, in particular, most distinctive. Two types of basinal facies are especially common. First, ponded submarine fans may form within small fault-bounded basins that experience syndepositional structural deformation along their margins (Fig. 4). Within this setting, there commonly is little opportunity for extensive proximal-to-distal differentiation of fan facies. Second, where structural barricades preclude major input of coarse clastics and where submerged sills with thresholds within oceanographic oxygen-minimum layers create anoxic basinal conditions, deep-marine facies may consist of laminated organic-rich, fine-grained biogenous sediments (e.g., Mon-

terey Formation of California) (Fig. 7). An intermediate mud-dominated regime is illustrated by the Rio Dell Formation of northern California (Piper et al., 1976), which was deposited in a convergent setting that is now becoming a transform setting due to the northward movement of the Mendocino triple junction (Atwater, 1970).

Divergent active margins (protoceanic) contain the same elements as wrench settings, except that juxtaposition of drastically different rock types and stratal thicknesses (Fig. 4) is less common. Deformation takes the form of continued subsidence and rotation of crustal blocks, usually without significant concurrent uplift of neighboring basement (Lowell and Genik, 1972). As a result, sedimentation rates are slower than in wrench regimes, and evaporites and/or carbonates may dominate, assuming appropriate climatic regimes. Recognition of the shelf-slope break in such a nondetrital regime would be based on the bracketing of shallow-marine and deep-marine paleoenvironments and the recognition of biostratigraphically compressed zones and unconformities. Paleoecologically determined paleobathymetry might be especially useful in carbonate deposits formed along sediment-starved outer shelves and upper slopes. In contrast to mature divergent ("passive") margins where vertical accumulation of shelfedge reef complexes is typical,

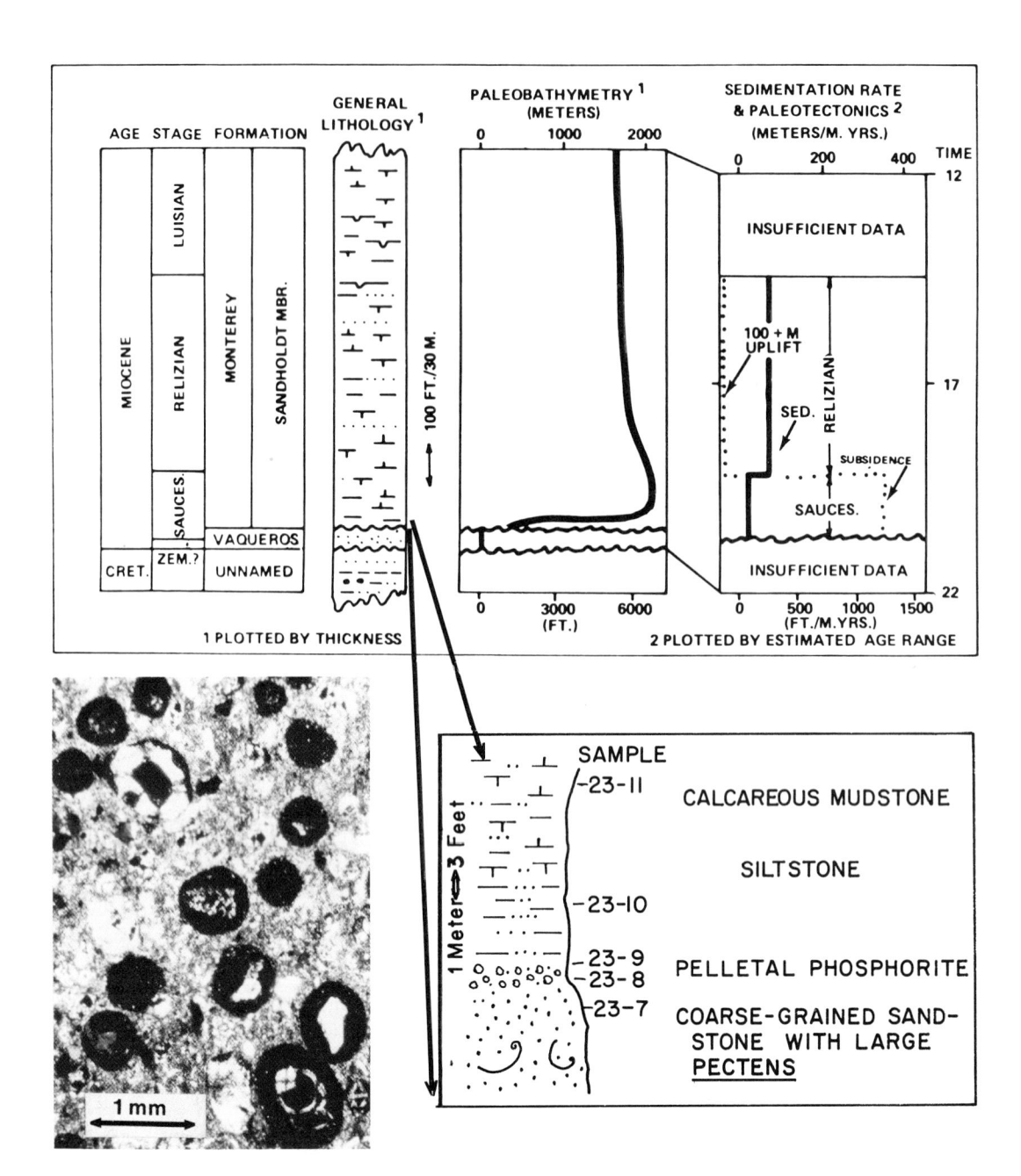

FIG. 5.—Unconformity representing the shelf-slope break in a transgressive Miocene section at Atascadero (see Fig. 1) (modified from Graham, 1980). Column at top shows abrupt paleobathymetric change (and high subsidence rates) from early Miocene shelf environments to middle Miocene basinal environments in an active wrench-tectonic regime. Blowup of the contact (lower right) reveals a very thin horizon of phosphorite of the foundered shelf-slope break separating deposits of the drastically different environments. Photomicrograph (crossed nicols) of pelletal phosphorite of sample 23-8 (lower left) shows foraminifera forming nuclei of many pellets.

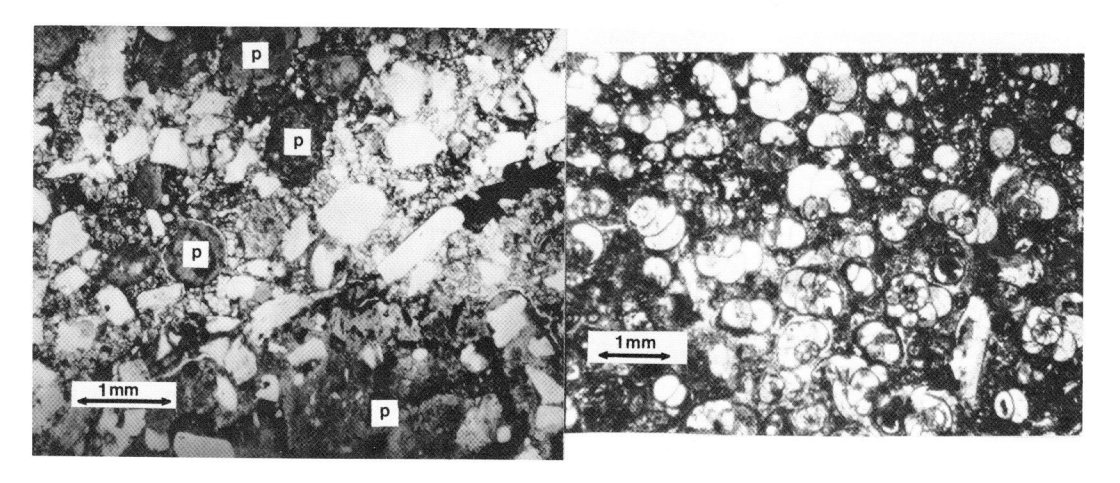

FIG. 6.—Left, photomicrograph (plane light) of glauco-conglomeratic phosphorite from a Miocene shelf-slope break sequence of the Santa Lucia Range, central California (see Fig. 1). Phosphorite clasts, mostly sand sized, appear darker (P), but a pebble of phosphorite enclosing detrital quartzofeldspathic silt stretches across the lower border of the photo. Right, photomicrograph (plane light) of a banktop foraminiferal coquina composed of upper bathyal species.

carbonate-evaporite-clastic facies tracts along ancient protoceanic or wrench margins display abrupt lateral facies changes.

Intrarc (backarc) extension is similar to protoceanic extension both in terms of structural style and environments. Thus, marginal basins involving continental crust (e.g., Japan Sea) are flanked by young rifted continental margins with characteristics similar to protoceanic settings (e.g., Red Sea). The major difference between the two settings is the presence of an active magmatic arc adjacent to one of the margins in the former setting. This results in the predominance of volcaniclastic detritus and the scarcity of sediment-starved shelves (Klein et al., 1979). Thus, shelf-slope breaks formed along the arc sides of marginal basins probably have characteristics intermediate between those of protoceanic margins and forearc settings in terms of quantity and type of detrital sediment.

SHELF-SLOPE BREAKS IN FOREARC SETTINGS

Subduction-controlled basins tend to consist of elongate troughs between structurally deformed ridges and/or magmatic arcs (Dickinson and Seely, 1979). Tectonic and magmatic events commonly are synchronous along the length of arc-trench systems (e.g., Ingersoll, 1982; Ingersoll and Dickinson, 1981), although variable geometry of subduction complexes, forearc basins and continental margins results in a variety of environments and deposits (Dickinson and Seely, 1979). Shelf-slope break deposits occur in forearc basins where tectonic disruption is minimal and in trench-slope basins where tectonic disruption is extreme (Fig. 3).

During the latest Cretaceous in California, the wide forearc basin began filling to sealevel along the east half of the Great Valley (Ingersoll, 1978, 1979, 1982). Excellent examples of prograding embankments topped by deltaic deposits occur in the "delta depocenter" of the southern Sacramento Valley (Figs. 1, 8). In this area of rapid sediment influx, the shelf-slope break is defined by the boundary between slope mudrocks and deltaic-shelf sandstones. The shelf-slope break and shoreline prograded westward as the deeper part of the basin to the west filled. A continental embankment similar to other types of continental margins characterized by rapid deltaic progradation (e.g., Gulf of Mexico) formed within this forearc basin. Farther south in the forearc basin, progradation of shelf deposits over turbidite and slope deposits occurred along the west side of the San Joaquin Valley during the Paleocene (Figs. 1, 9). A similar progradational margin characterized the Eocene forearc evolution of southwestern Oregon (Dott and Bird, 1979), where gravity-deposited sand-silt units are overlain by coal-bearing deltaic-shoreline deposits. In these settings, delta-front/delta-top transitions commonly correspond both in time and space with slope-to-shelf transitions.

Trench-slope breaks along actively accreting convergent continental margins are boundaries between highly deformed accreted material and slightly deformed forearc-basin deposits (Karig and Sharman, 1975; Moore and Karig, 1976). This boundary may have the character of a time-transgressive tectonic contact that migrates seaward and upward through time (Dickinson, 1975). In theory, as the subduction complex and forearc basin grow (Fig. 3), there may be a time during which the trench-

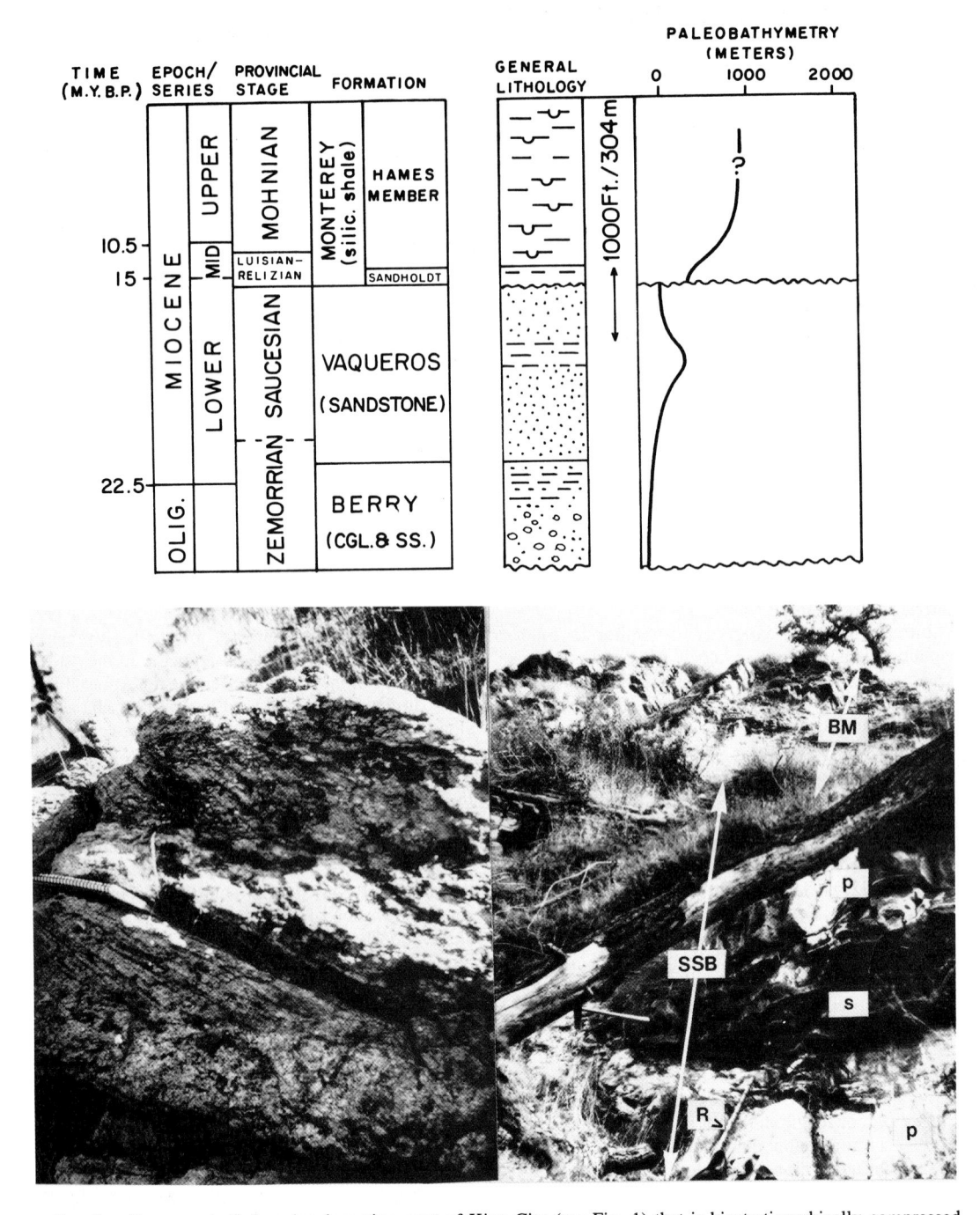

FIG. 7.—Neogene shelf-slope break section west of King City (see Fig. 1) that is biostratigraphically compressed (modified from Graham, 1976). Stratigraphic-paleobathymetric column shows transgression from Oligocene to late Miocene, with middle Miocene time compressed into a thin interval representing the clastic-starved shelf-slope break. Left photo shows underlying crossbedded lower Miocene shelf sandstones (notebook for scale). Right photo is taken looking directly up hill from the shelf sandstone at the middle Miocene shelf-slope break sequence (SSB) of shale (S) and pelletal phosphorite (P), overlain by basal part of upper Miocene basinal siliceous mudrock (BM). Measuring rod (R = 1.5 m) shown for scale.

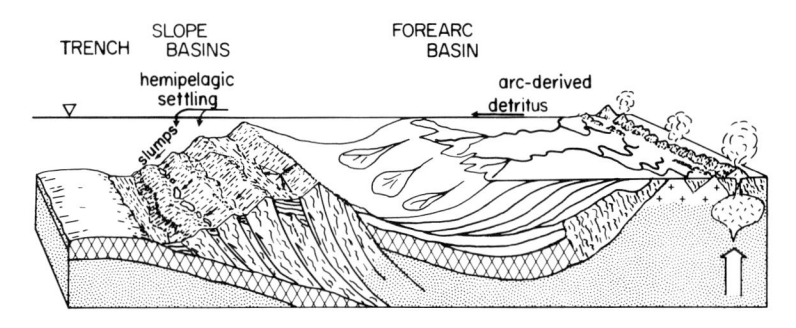

FIG. 8.—Schematic block diagram of the Great Valley forearc basin during the latest Cretaceous as prograding deltaic complexes filled the basin to sealevel along the east side (sediment derived from the Sierra Nevada magmatic arc). The Franciscan subduction complex formed a bathymetric barrier that trapped sediment within the forearc basin. The shelf-slope break is represented by coarse-grained shallow-marine/deltaic deposits overlying fine-grained slope deposits (also see Fig. 9). This stage in development of the forearc is transitional between stages 2 and 3 of Fig. 3. (After Moore et al., 1980).

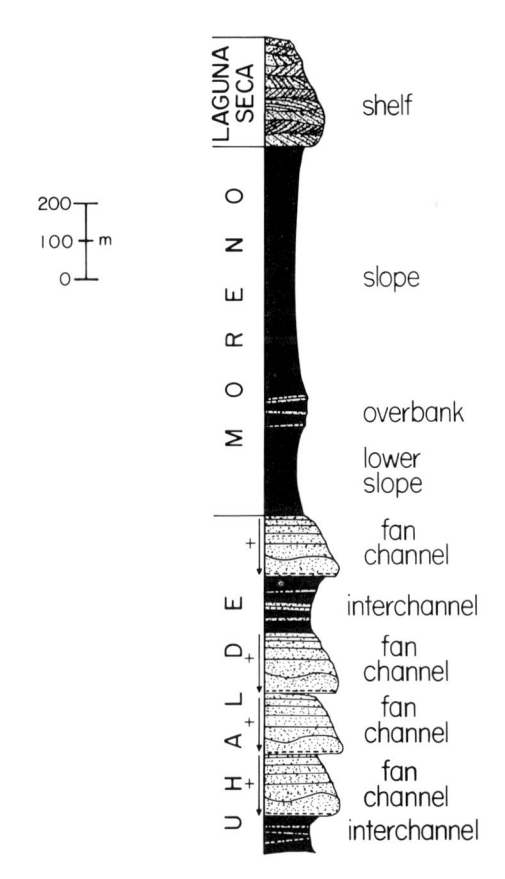

FIG. 9.—Schematic stratigraphic section from Laguna Seca area, west side of San Joaquin Valley (see Fig. 1). Deep-marine turbidite-fan sequences are overlain by slope mudrocks, which are covered by shallow-marine coarse-grained units. This progression accumulated as the forearc basin filled to shelf depths (see Fig. 8). The Cretaceous-Paleogene boundary occurs within the Moreno Formation (Dickinson et al., 1979).

slope break corresponds to the shelf-slope break (following filling of the forearc basin to shelf depths). At this time, undeformed, shallow-marine, arc-derived sediments would depositionally overlie highly deformed, deep-marine, arc-derived and/or accreted exotic sediments. Thus, ancient shelf-slope breaks could be identified by contrasts in structural style, paleoenvironments and provenance of sediments above and below the boundary. However, more commonly, both forearc and trench-slope basins are characterized by basinal and/or slope deposition (e.g., Bachman, 1982; Moore et al., 1980; Smith et al., 1979). This was the case during most of the evolution of the late Mesozoic and Paleocene forearc area of California (Ingersoll, 1978, 1982), so that the shelf-slope break was located along the east side of the forearc basin, as discussed above. By the Eocene, the forearc basin had filled to sealevel in most areas so that the shelf-slope break may have corresponded to the trench-slope break. However, since the Paleogene, the entire subduction complex has experienced considerable uplift and erosion so that this transition is poorly preserved (Dickinson et al., 1979). Erosional loss of ancient shelf-slope break deposits formed on top of subduction complexes probably is the rule. In addition, subduction complexes and forearc basins that later become involved in continental collisions (see below) usually are destroyed and/or so highly deformed that recognition of ancient shelf-slope breaks becomes difficult.

SHELF-SLOPE BREAKS IN SUTURE ZONES

Shelf-slope break deposits of continental margins involved in continental collisions become highly deformed, and are either partially subducted (thus removing them from view) or are uplifted and eroded. Therefore, it is not surprising that there are few ancient examples of such deposits. The classic association of preorogenic flysch (deep-marine deposits), and syn- and postorogenic molasse (shallow-marine and nonmarine deposits) in zones of continental collision (e.g., Alps, Himalayas and Ouachitas) is the natural result of sequential suturing of continental crust (Graham et al., 1975). The closing of a remnant ocean basin results in shallowing of depositional environments, so that a shelf-slope transition should occur during the final stages of suturing. However, this is the exact moment when structural disruption is at its peak, so that this transition is seldom preserved. Typically (e.g., Alps, Himalayas and Ouachitas), deep-marine deposits (including slope deposits) are thrust over shallow-marine and nonmarine deposits; in the process, shelf-slope break deposits either are partially subducted and obscured by imbricate thrust sheets, or are uplifted and eroded.

CONCLUSIONS

Deposits formed at the shelf-slope transition along ancient, tectonically active continental margins are preserved best in wrench-fault and forearc settings. Where the outer shelf was starved of sediment due to bypassing into deeper water, the shelf-slope break is represented by unconformities, nondetrital sediments and biostratigraphically compressed zones. Where sedimentation was rapid due to deltaic progradation, the shelf-slope break is expressed by slope mudrocks overlain by shallow-marine and nonmarine deltaic and related deposits (usually coarse-grained). Shelf-slope break deposits formed along protoceanic margins are destined to be buried under thick, rifted continental-margin sediments (miogeocline-eugeocline), whereas those formed on top of, or within, subduction complexes and those that are involved in continental suturing are not likely to be preserved.

ACKNOWLEDGMENTS

We thank G. T. Moore and D. J. Stanley for inviting our participation in this symposium. We also thank W. R. Dickinson and J. C. Ingle, Jr. for introducing us to many of the concepts and examples discussed in this paper. Reviews of the manuscript of S. B. Bachman and G. W. Moore contributed to improvements in its presentation.

REFERENCES

ATWATER, T., 1970, Implications of plate tectonics for the Cenozoic tectonic evolution of western North America: Geol. Soc. America Bull., v. 81, p. 3513–3535.

BACHMAN, S. B., 1982, The Coastal Belt of the Franciscan: youngest phase of northern California subduction, in Leggett, J. K., ed., Trench-Forearc Geology: Sedimentation and Tectonics on Modern and Ancient Active Plate Margins: Geol. Soc. London Spec. Pub. 10, p. 401–417.

BIRCH, G. F., 1979, Phosphatic rocks on the western margin of South Africa: Jour. Sed. Petrology, v. 49, p. 93–110.

CROWELL, J. C., 1974a, Sedimentation along the San Andreas fault, California: Soc. Econ. Paleontologists Mineralogists Spec. Pub. 19, p. 292–303.

———, 1974b, Origin of late Cenozoic basins in southern California: Soc. Econ. Paleontologists Mineralogists Spec. Pub. 22, p. 190–204.

DICKINSON, W. R., 1974, Plate tectonics and sedimentation: Soc. Econ. Paleontologists Mineralogists Spec. Pub. 22, p. 1–27.

———, 1975, Time-transgressive contacts bordering subduction complexes: Geol. Soc. America Abstracts with Programs, v. 7, p. 1052.

———, AND SEELY, D. R., 1979, Structure and stratigraphy of forearc regions: Am. Assoc. Petroleum Geologists Bull., v. 63, p. 2–31.

———, INGERSOLL, R. V., AND GRAHAM, S. A., 1979, Paleogene sediment dispersal and paleotectonics in northern California: Geol. Soc. America Bull., v. 90, Part I, p. 897–898, Part II, p. 1458–1528.

DOTT, R. H., JR., AND BIRD, K. J., 1979, Sand transport through channels across an Eocene shelf and slope in southwestern Oregon, USA: Soc. Econ. Paleontologists Mineralogists Spec. Pub. 27, p. 327–342.

EMERY, K. O., 1952, Continental shelf sediments of southern California: Geol. Soc. America Bull., v. 63, p. 1105–1107.

GORSLINE, D. S., AND EMERY, K. O., 1959, Turbidity-current deposits in San Pedro and Santa Monica basins off southern California: Geol. Soc. America Bull., v. 70, p. 279–290.

GRAHAM, S. A., 1976, Tertiary sedimentary tectonics of the central Salinian block of California [Ph.D. thesis]: Stanford, Stanford University, 510 p.

———, 1978, Role of Salinian block in evolution of San Andreas fault system, California: Am. Assoc. Petroleum Geologists Bull., v. 62, p. 2214–2231.

———, 1980, Notes on the Miocene section at Atascadero, California, in Blake, G. H., ed., Neogene Biostratigraphy of the Northern La Panza Range, San Luis Obispo County, California: Pacific Section, Soc. Econ. Paleontologists Mineralogists Field Trip Guidebook, p. 39–44.

———, DICKINSON, W. R., AND INGERSOLL, R. V., 1975, Himalayan-Bengal model for flysch dispersal in the Appalachian-Ouachita system: Geol. Soc. America Bull., v. 86, p. 273–286.

INGERSOLL, R. V., 1978, Paleogeography and paleotectonics of the late Mesozoic forearc basin of northern and central

California, *in* Howell, D. G., and McDougall, K., eds., Mesozoic Paleogeography of the Western United States: Soc. Econ. Paleontologists Mineralogists Pacific Section, Pacific Coast Paleogeography Symposium 2, p. 471–482.

———, 1979, Evolution of the Late Cretaceous forearc basin of northern and central California: Geol. Soc. America Bull., v. 90, Part I, p. 813–826.

———, 1982, Initiation and evolution of the Great Valley forearc basin of northern and central California, USA, *in* Leggett, J. K., ed., Trench-Forearc Geology: Sedimentation and Tectonics on Modern and Ancient Active Plate Margins: Geol. Soc. London Spec. Pub. 10, p. 459–467.

———, AND DICKINSON, W. R., 1981, Great Valley Group (sequence), Sacramento Valley, California, *in* Frizzell, V., ed., Upper Mesozoic Franciscan Rocks and Great Valley Sequence, Central Coast Ranges, California (Annual Meeting Pacific Section Soc. Econ. Paleontologists Mineralogists Field Trips 1 and 4): Soc. Econ. Paleontologists Mineralogists Pacific Section, p. 1–33.

KARIG, D. E., AND SHARMAN, G. F., III, 1975, Subduction and accretion in trenches: Geol. Soc. America Bull., v. 86, p. 377–389.

KLEIN, G. D., OKADA, H., AND MITSUI, K., 1979, Slope sediments in small basins associated with a Neogene active margin, western Hokkaido Island, Japan: Soc. Econ. Paleontologists Mineralogists Spec. Pub. 27, p. 359–374.

LOWELL, J. D., AND GENIK, G. J., 1972, Sea-floor spreading and structural evolution of southern Red Sea: Am. Assoc. Petroleum Geologists Bull., v. 56, p. 247–259.

MIDDLETON, G. V., 1973, Johannes Walther's law of the correlation of facies: Geol. Soc. America Bull., v. 84, p. 979–987.

MOORE, D. G., 1969, Reflection profiling studies of the California continental borderland: structure and Quaternary turbidite basins: Geol. Soc. America Spec. Paper 107, 142 p.

———, 1973, Plate-edge deformation and crustal growth, Gulf of California structural province: Geol. Soc. America Bull., v. 84, p. 1883–1905.

MOORE, G. F., AND KARIG, D. E., 1976, Development of sedimentary basins on the lower trench slope: Geology, v. 4, p. 693–697.

———, BILLMAN, H. G., HEHANUSSA, P. E., AND KARIG, D. E., 1980, Sedimentology and paleobathymetry of Neogene trench-slope deposits, Nias Island, Indonesia: Jour. Geology, v. 88, p. 161–180.

ODIN, G. S., AND MATTER, A., 1981, De glauconiarium origine: Sedimentology, v. 28, p. 611–641.

PARKER, R. J., 1975, The petrology and origin of some glauconitic and glauco-conglomeratic phosphorites from the South African continental margin: Jour. Sed. Petrology, v. 45, p. 230–242.

PIPER, D. J. W., NORMARK, W. R., AND INGLE, J. C., JR., 1976, The Rio Dell Formation: a Plio-Pleistocene basin slope deposit in northern California: Sedimentology, v. 23, p. 309–328.

SCHOLL, D. W., VON HUENE, R., VALLIER, T. L., AND HOWELL, D. G., 1980, Sedimentary masses and concepts about tectonic processes at underthrust ocean margins: Geology, v. 8, p. 564–568.

SEELY, D. R., 1979, The evolution of structural highs bordering major forearc basins: Am. Assoc. Petroleum Geologists Memoir 29, p. 245–260.

SMITH, G. W., HOWELL, D. G., AND INGERSOLL, R. V., 1979, Late Cretaceous trench-slope basins of central California: Geology, v. 7, p. 303–306.

UNDERWOOD, M. B., AND KARIG, D. E., 1980, Role of submarine canyons in trench and trench-slope sedimentation: Geology, v. 8, p. 432–436.

PART III
SEDIMENTATION AND DELTAIC INFLUENCE

DELTAIC INFLUENCES ON SHELFEDGE INSTABILITY PROCESSES

JAMES M. COLEMAN AND DAVID B. PRIOR
Coastal Studies Institute, Louisiana State University, Baton Rouge, Louisiana 70803

AND

JOHN F. LINDSAY
Exxon Production Research Company, Houston, Texas 77001

ABSTRACT

Large river systems deliver significant quantities of fine-grained sediment to continental shelf regions. In specific areas off deltas, deposition rates are rapid and the sediment may be involved in a variety of mass-movement processes on the subaqueous slopes (slumps and slides, debris flows, and mudflows), causing rapid sediment accumulation at shelfbreak depths and resulting in active progradation of the shelfedge. Seismically, the deposits appear as large-scale foresets and are commonly composed of *in situ* deep-water deposits alternating with shallow-water sediments transported by mass movement. On electric logs, sands within these units are sporadic and display sharp basal planes and blocky shapes. Progradation of the shelfedge deposits is generally accompanied by oversteepening and large-scale instability of the upper shelfbreak slopes. Deep-seated and shallow rotational slides move large volumes of sediments and deposit them on the adjacent slopes and upper rise. Extensive contemporaneous faults commonly form at the shelfedge. Continuous addition of sediment to the fault scarps, particularly by mass movement from nearby delta-front instability, causes large volumes of shallow-water sediment to accumulate on the downthrown sides of the faults, mostly forming large-scale rollover structures. Continued movement along the concave-upward shear planes commonly results in compressional folds and diapiric structures. Contemporaneous accumulation of shallow-water mass-movement deposits may occur in association with these structures.

Massive retrogressive, arcuate-shaped landslide scars and canyons or trenches can also form at the shelfedge owing to slumping and other mass-movement processes. Such canyons and trenches can attain widths of 10–20 kilometers, depths of 800 meters, and lengths of 80–100 kilometers. The Mississippi Canyon probably originated in this manner. The creation of such features by shelfedge instability results in the yielding of exceptionally large volumes of shallow-water sediment to the deep basins in the form of massive submarine fans. The infilling of depressions by deltaic progradation is rapid, forming large foresets near the canyon heads. The low strength of the rapidly infilled, underconsolidated sediments causes downslope creep or reactivation of failure mechanisms, resulting in multiple episodes of filling and evacuation.

INTRODUCTION

Shelfedges seaward of major deltas differ considerably from those off open coasts. Several factors are responsible for these differences:

1. Major fluvial systems deliver high sediment loads to the coast and adjacent shelves. In modern world deltas, this sediment yield varies from a relative low of 1.9×10^{10} kilograms off the Danube, to a high of 2.2×10^{12} kilograms off the Ganges-Brahmaputra. The modern Mississippi River delta transports some 860 billion kilograms annually to the coast, and during the Holocene alone (some 18,000–20,000 years) approximately 17×10^{16} kilograms have been delivered to the coast. This volume of sediment yield off deltas contrasts sharply to that of shelfedges off the east coast of the U.S. where, from a point source (Potomac River), sediment yield approximates 2.2×10^8 kilograms.

2. Most major fluvial systems carry a highly varied sediment load, but most carry at least 8–10 percent sand-sized particles. The high sediment yield results in excessive sedimentary loading, which causes high subsidence rates off the shelf. In the Mississippi River delta, average subsidence rates approach 1 m/century and in localized areas can be as high as 3–5 m/century. This rapid downbowing allows large volumes of sediment to be preserved over a rather short period of geologic time.

3. Rapid sediment introduction results in extremely high depositional rates, up to 2.5 m/year off the Mississippi Delta. Such accumulation rates do not allow pore fluids to escape, and the sediments are generally undercompacted, display excessive pore fluid pressures, and have extremely high water contents. These factors drastically decrease the sediment's shear strength and lead to a variety of subaqueous sediment instabilities.

4. Most fluvial systems carry a considerable volume of organic debris, and deltaic marine sediments often contain 0.5–1.5 percent organics. Biochemical degradation of these organics results in rapid production of methane gas. The presence of free gas in the pores leads to a further weakening of the sediments.

5. High sediment yields cause extremely rapid

Copyright © 1983, The Society of
Economic Paleontologists and Mineralogists

progradational rates. The Mississippi River delta progrades seaward at an average rate of 10 km/century. Thus eustatic sealevel changes do not drastically affect a prograding delta wedge. Even a rapid eustatic sealevel rise such as 0.5 m/century would have little effect on a delta building seaward at a rate of 10 km/century. As a result, transgressive units are often caused by changes in sites of the river's depocenter.

These various interacting factors result in a number of processes, many of them involving subaqueous sediment instabilities, that differ considerably from those operating on shelfedges where sediment yield is low. Through time, a distinctive set of environmental and seismic facies accumulates at the outer shelf-upper slope off deltas.

During the past 5 to 10 years, a concerted effort has been made to document these facies that front major deltas, especially off the Mississippi River delta. In the Mississippi Delta, research by Coleman (1981) indicates some of the characteristics of the sediment comprising the subaqueous delta front. Only within the last few years, however, has data been dense enough to document in detail some of the types and distribution of facies and the processes responsible for their formation.

Figure 1 shows the marine platform off the modern Mississippi River delta and the data base for the interpretation that follows. The boreholes are primarily soil foundation borings that have been utilized to determine the lithologic character and sediment and paleontological parameters of the various facies. In addition, samples have been used for radiometric dating to determine ages of various horizons. When this information is combined with seismic data, a time framework can be established for the marine deposits fronting the modern Mississippi River delta. The seismic grids are more dense in the shallower waters fronting the delta, but even in deeper waters lines are generally spaced closer than 700 m. The sensors normally utilized are 3.5–7 kHz subbottom profiler, lower frequency single-channel or multichannel sparker or minisleeve exploder systems, side-scan sonar systems, and precision depth recorder systems. Over a large percentage of the area, digitally acquired, scale-corrected side-scan sonar data were obtained. Some 15,000 line km of data have been gathered in this region since 1979.

SEISMIC FACIES

Utilizing the seismic and borehole data shown in Figure 1, several major features were found to characterize the shelfedge-upper slope off the Missis-

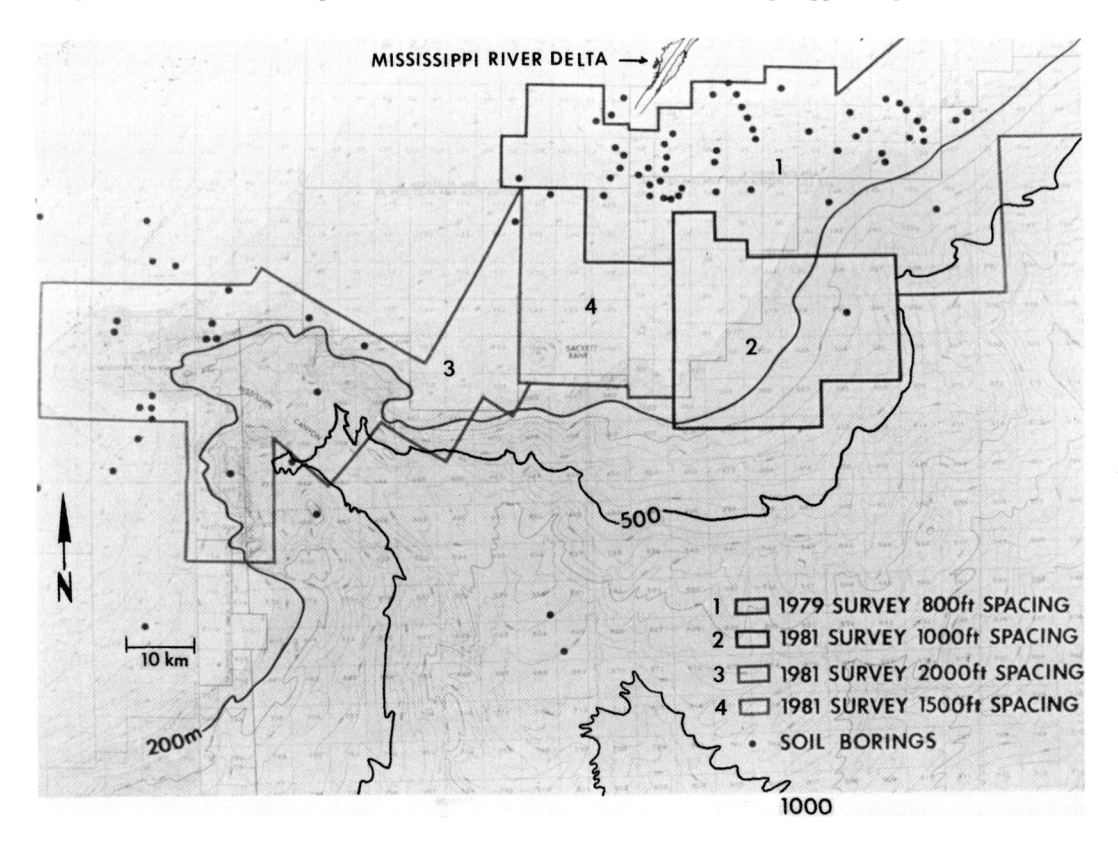

FIG. 1.—Shelf and slope off the modern Mississippi River delta showing data coverage. Depth in meters.

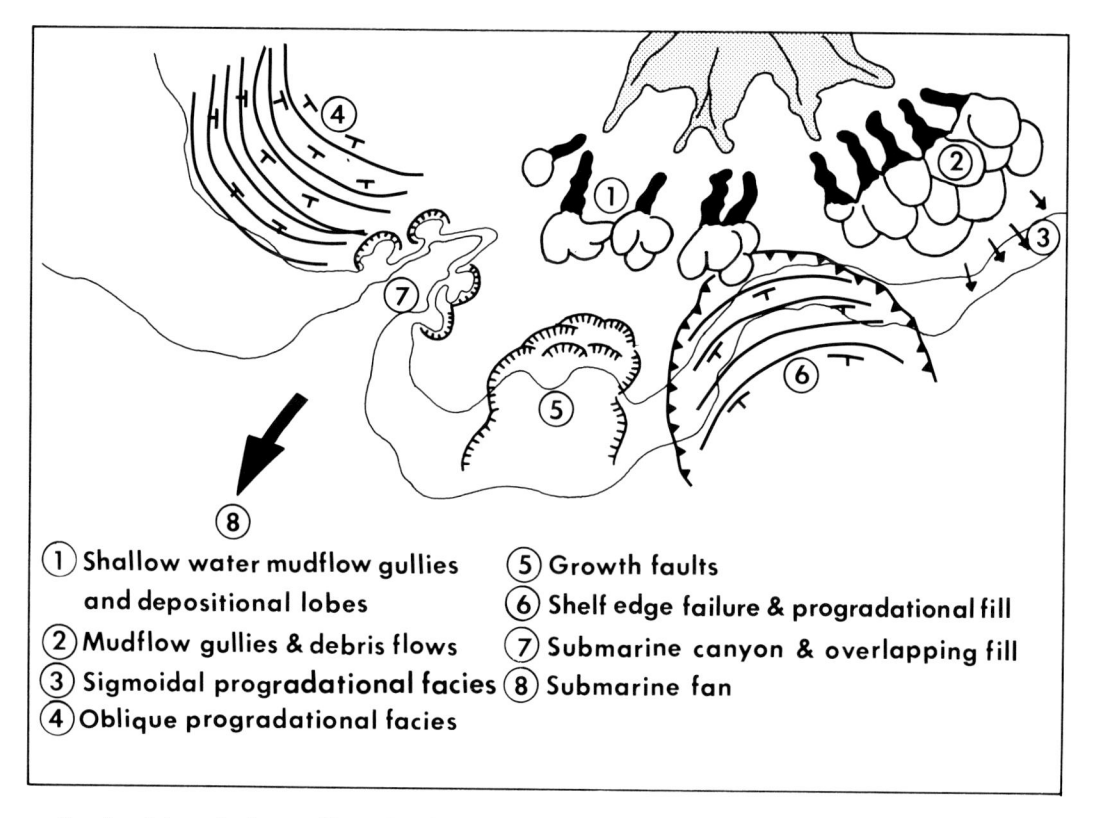

FIG. 2.—Schematic diagram illustrating the major sediment instability facies present off major river deltas.

① Shallow water mudflow gullies and depositional lobes
② Mudflow gullies & debris flows
③ Sigmoidal progradational facies
④ Oblique progradational facies
⑤ Growth faults
⑥ Shelf edge failure & progradational fill
⑦ Submarine canyon & overlapping fill
⑧ Submarine fan

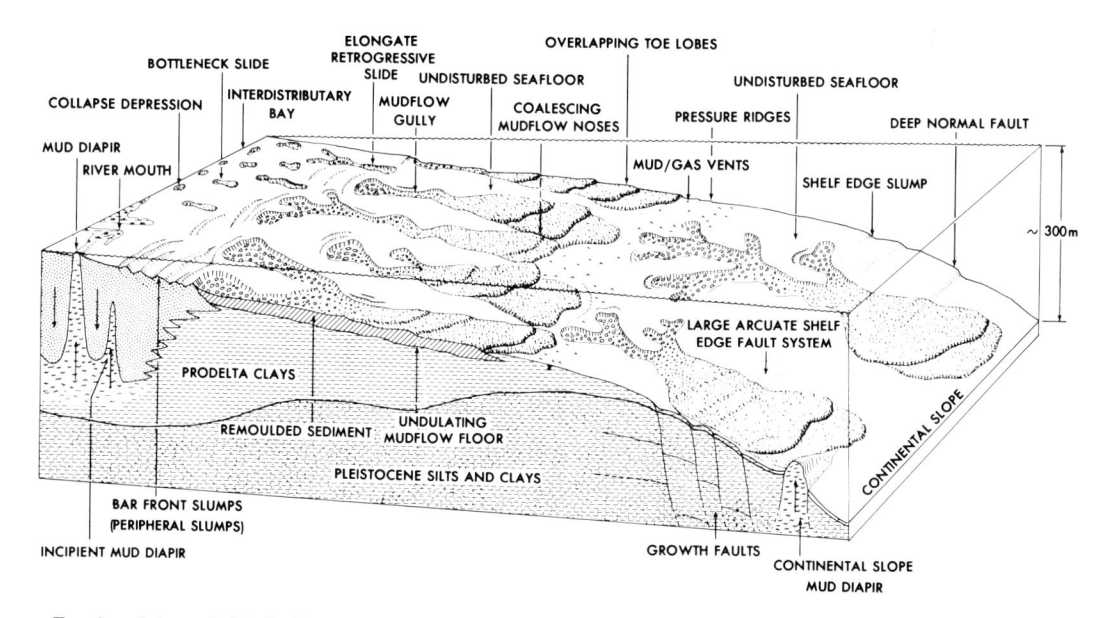

FIG. 3.—Schematic block-diagram showing the relationship of the various types of subaqueous sediment instabilities off the Mississippi River delta.

sippi Delta. These features are schematically illustrated in Figure 2 and include: (1) shallow-water rotational slumps and retrogressive landslide gullies; (2) retrogressive landslide gullies and large depositional debris lobes; (3) sigmoidal progradational units resulting from mass-movement processes; (4) outbuilding of shelfedge by shelf progradation (truncated oblique and sigmoidal progradational units); (5) contemporaneous or growth faults; (6) massive shelfedge instabilities and infilling oblique progradational units; and (7) large retrogressive failures forming major canyons and progradational infilling of the canyons. Each of these features will be described in the following discussions.

SHALLOW-WATER DELTA-FRONT INSTABILITIES

Figure 3 illustrates schematically the relationship of the variety of sediment instabilities found around the delta-front region of a single river distributary. All of these features have been described in detail by Coleman and Garrison (1977) and Prior and Coleman (1978, 1980). These major features, which affect the outbuilding of the shelfedge, include mudflow gullies and their depositional debris lobes. Extending radially seaward from each of the river mouth distributaries in water depths of 15–100 m are major elongate bathymetric gullies, first described by Shepard in 1955. Recent work has shown that the topographic channel features represent areas of seafloor instability (Coleman and Garrison, 1977). Each gully has a clearly recognizable area of rotational instability or shear slumps at its upslope margin and a long, sinuous, narrow chute or channel that links a depressed, hummocky source area upslope to composite overlapping depositional lobes on the seaward end. Narrow chutes alternate with larger bowl-shaped depressions containing hummocky, irregular, distinctive blocks or large clasts. Figure 4A is a side-scan sonar mosaic showing a few of these gullies off the Mississippi River delta. The widths of individual gullies range from 20–150 m at the narrow points to 350–1500 m at the widest points. In many cases adjacent gullies coalesce to form branching patterns. The sidewalls of the gullies are subject to instability and are responsible for variations in gully width and formation of large blocks that are later mass-moved downslope. The formation of elongate chutes of this type is very similar to the morphology associated with subaerial debris flows and some types of subaerial mudflow. The chutes generally emanate from upslope slump zones and constitute transport conduits for disturbed and remolded debris, together with displaced blocks of various sizes. Sediments are remolded as a direct consequence of disturbance of the sediment-water-gas system that accompanies slumping and represents fluidization/liquefaction mechanisms (Prior and Suhayda, 1979). The mechanisms of transport are probably charac-

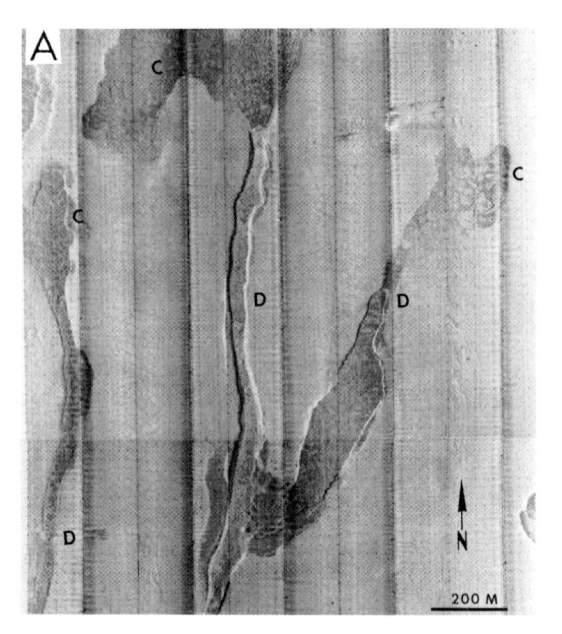

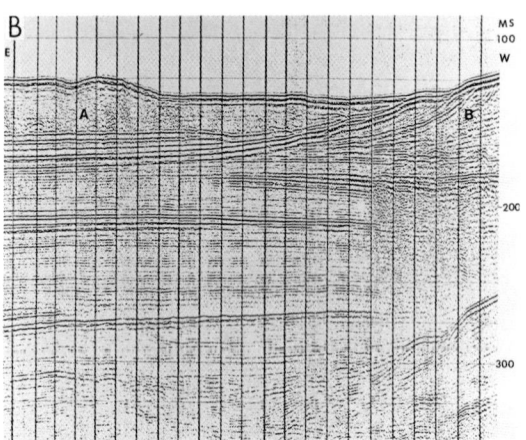

FIG. 4.—A, Side-scan sonar mosaic showing landslide gullies off the Mississippi River delta (C = source areas of failure, D = narrow chute). B, Sparker profile run across two mudflow lobes (A and B) off the Mississippi River delta. Navigation fix marks are 152 meters apart. Depth in milliseconds.

terized by a type of plug flow in which rigid plugs move over and within a zone of liquefied mud. The presence of partially disintegrated rafted blocks suggests laminar or plug flow rather than turbulent flow.

At the seaward or downslope ends of mudflow gullies, extensive areas of irregular bottom topography are composed of discharged blocky, disturbed debris. In plan view, this discharged debris is arranged into widespread overlapping lobes or fans. Because of the large number of gullies that front the present Mississippi River delta, the displaced debris from adjacent gullies coalesces, providing an almost continuous sinuous frontal scarp that forms the prograding shelfedge peripheral to the delta. Detailed mapping with side-scan sonar and subbottom profiling, however, shows that the depositional areas are composed of multiple overlapping lobes and are due to episodic discharge from individual gullies upslope. Each individual lobe, emplaced during a short period of time and composed of shallow-water displaced sediment, averages 10–15 m in thickness. Often individual blocks within the mudflows attain dimensions of 150–500 m in diameter and up to 15–20 m in thickness.

Episodic failures in the upslope gullies result in overlapping mudflow depositional lobes, and within a relatively short period of time (i.e., 200–300 years) these coalescing deposits form thicknesses up to 100 m. Since the mudflows are emplaced at the shelfedge episodically, the periods of time between the movement of displaced sediment are represented by normal vertical suspension settling. This settling results in capping of the acoustically amorphous mudflow deposits by acoustically stratified sediments. These alternating acoustic amorphous zones and stratified sediments are illustrated in Figure 4B, which is a seismic strike line run off the Mississippi River delta. In a dip section (Fig. 5A) it can be seen that the shelfedge has prograded some 15 km during the past 15,000 years (seismic horizon labeled A has been radiocarbon dated by borings within the area). The borings drilled through this wedge of sediments tend to show the alternating nature of the deposits. The acoustically amorphous zones represent disturbed, displaced sediment containing shallow-water fauna and other sedimentary inclusions indicative of shallow-water delta front deposition. The acoustically stratified units are generally composed of fine-grained clays containing normal shelfedge or upper-slope fauna. Within the past 15,000 years, a relatively short period of geologic time, the shelfedge has prograded seaward a distance of 15 km and laid down a wedge of oblique prograded seismic units in excess of 200 m thick. This wedge contains alternating zones of shallow-water and deep-water fauna. This situation contrasts sharply with continental margin progradation along sediment-starved coasts, where rates of 1 kilometer per million years are commonly reported (Boillot et al., 1981). Thus the mechanism of mass movement off a prograding delta is an extremely important process in outbuilding of the shelfedge.

OUTBUILDING OF SHELFEDGE BY DELTAIC PROGRADATION

Although most deltas can prograde seaward independently of sealevel changes, a falling sealevel promotes rapid progradation, especially of widespread shallow-water deltas. During the last glacial falling sealevel stage, several smaller deltas adjacent to the Mississippi River, as well as a few of the major Mississippi deltas, prograded seaward as a series of widespread shallow-water deltas. Each of the progradational units ranges in thickness from 30 to 55 m and progrades seaward for distances in excess of 100 km, with lateral dimensions of 50–70 km. Figure 5B is a minisparker seismic profile run east of the modern Mississippi River delta and seaward of the abandoned St. Bernard Delta. Core holes in the vicinity indicate that these progradational seismic units consist of sand and silt layers at the uppermost break in slope (top of the clinoforms) and silty clays and clays on the lower parts of the slope. The uppermost horizontal seismic reflections consist of thinly bedded silts and clays. At a point seaward, near the shelfedge (A in Fig. 5B), a series of sigmoidal units are commonly found that show few internal acoustic reflectors. A single cored boring in this series of units showed the sequence to be composed entirely of fine-grained clays containing a few scattered marine shells. Each sigmoidal package is approximately 60–70 m thick and 3–4 km long in a dip section. The slopes of progradational units range from 1.3 to 2.5 degrees. Radiocarbon dating of a shell horizon found approximately 15 m below the seafloor gave an age of 15,300 years B.P. This unit caps and often truncates some of the uppermost reflectors of the progradational units. Thus these prograding facies were formed during the last Late Wisconsin sealevel low stand (approximately 18,000–20,000 years B.P.) at an elevation of 60 to 90 m below present level.

GROWTH FAULTS

Another major type of sediment instability that has significant geologic importance is the arcuate rotational slumps and contemporaneous (growth) faults that commonly occur on the outer continental shelf in front of the prograding deltaic system. The plan view distribution of some of these features off the mouth of South Pass, Mississippi River delta, is shown in Figure 6. These large-scale features cut the modern sediment surface, often forming localized scarps on the seafloor with amplitudes up to 25–40 m. In most instances, the fault traces show numerous intersecting arcuate patterns that outline a much larger fault system. Lateral continuities of individual fault scars range from a few kilometers to as much as 8–10 km. Contemporaneous slump faults show recent movement along the shear planes, as the most recent surface horizons show offset. Movement along the fault is contemporaneous with sedimentation, and offset of individual beds on the subsurface increases with depth below the seafloor surface. Most of the faults shown in Figure 6 extend to depths of 700–800 m below the sea bottom before merging into a bedding plane fault. Offsets in the uppermost units are generally on the order

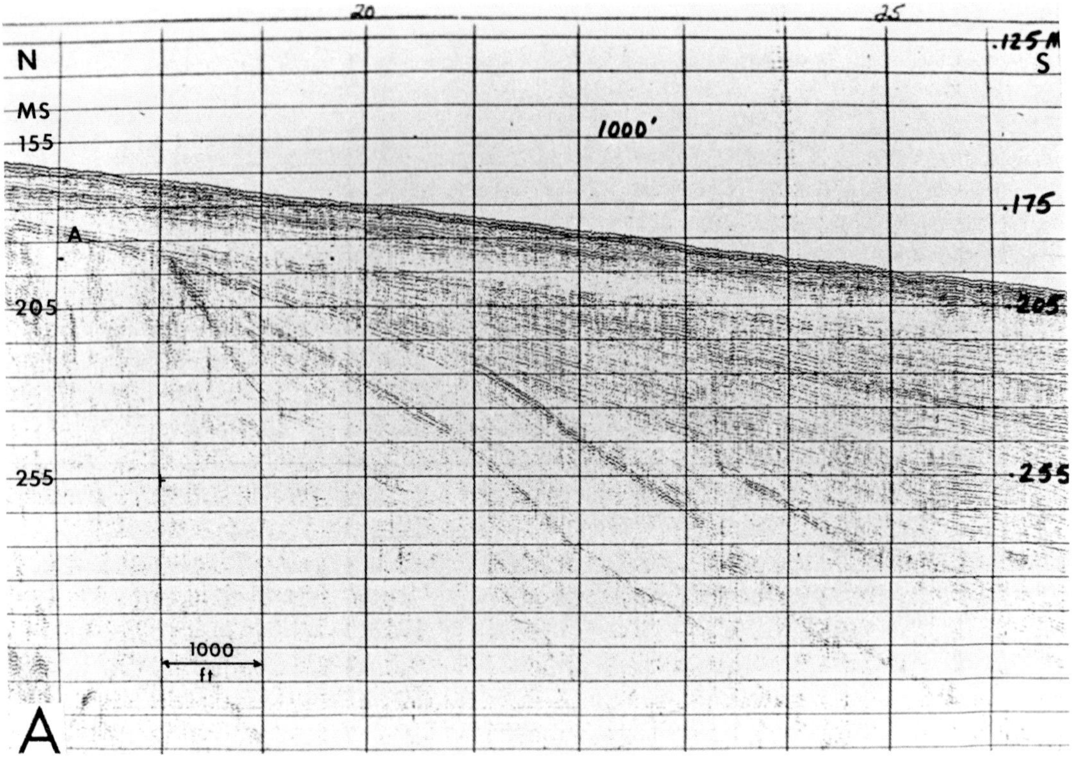
20
25
.125 M
N
S
MS
155
1000'
.175
A
205
.205
255
.255
1000 ft
A

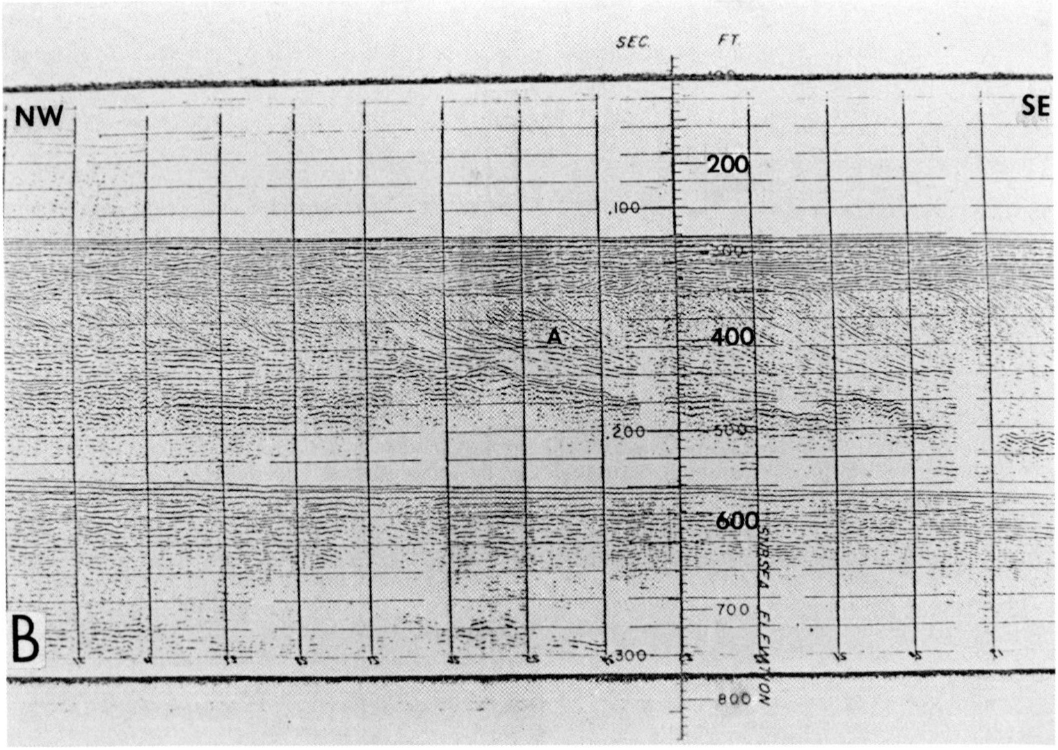
SEC.
FT.
NW
SE
200
.100
A
400
.200
500
600
SUBSEA ELEVATION
700
.300
800
B

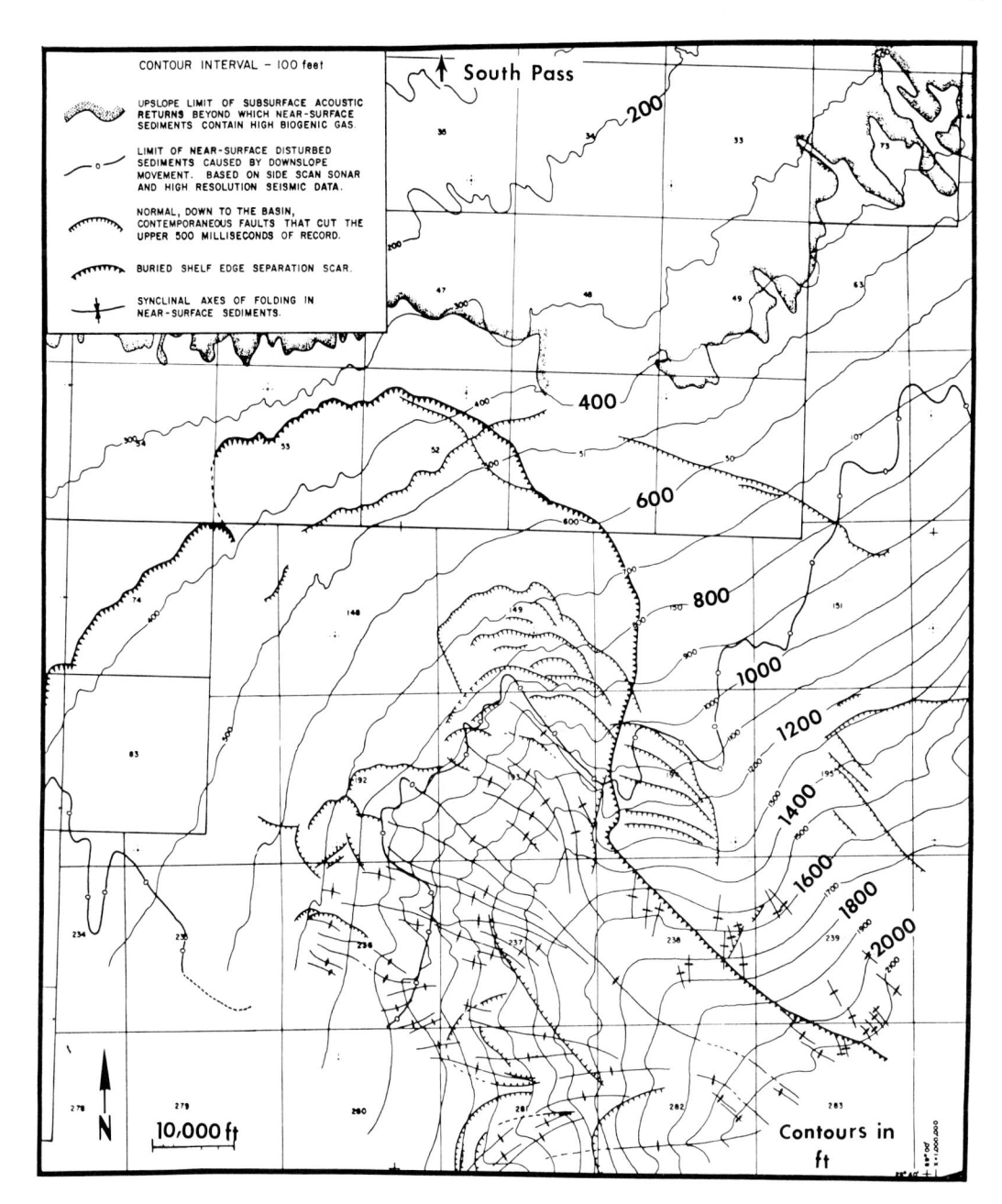

FIG. 6.—Map showing the distribution of growth faults, shelfedge failure, and fold axes off the modern Mississippi River delta.

FIG. 5.—*A*, Minisparker seismic profile run in a dip section showing a prograding shelfedge composed of alternating amorphous and acoustically stratified seismic units. Navigation fixes are 305 meters apart (depth in milliseconds). *B*, Shallow seismic section (dip line) showing shelfedge progradation in an area east of the modern Mississippi River delta (depth in feet).

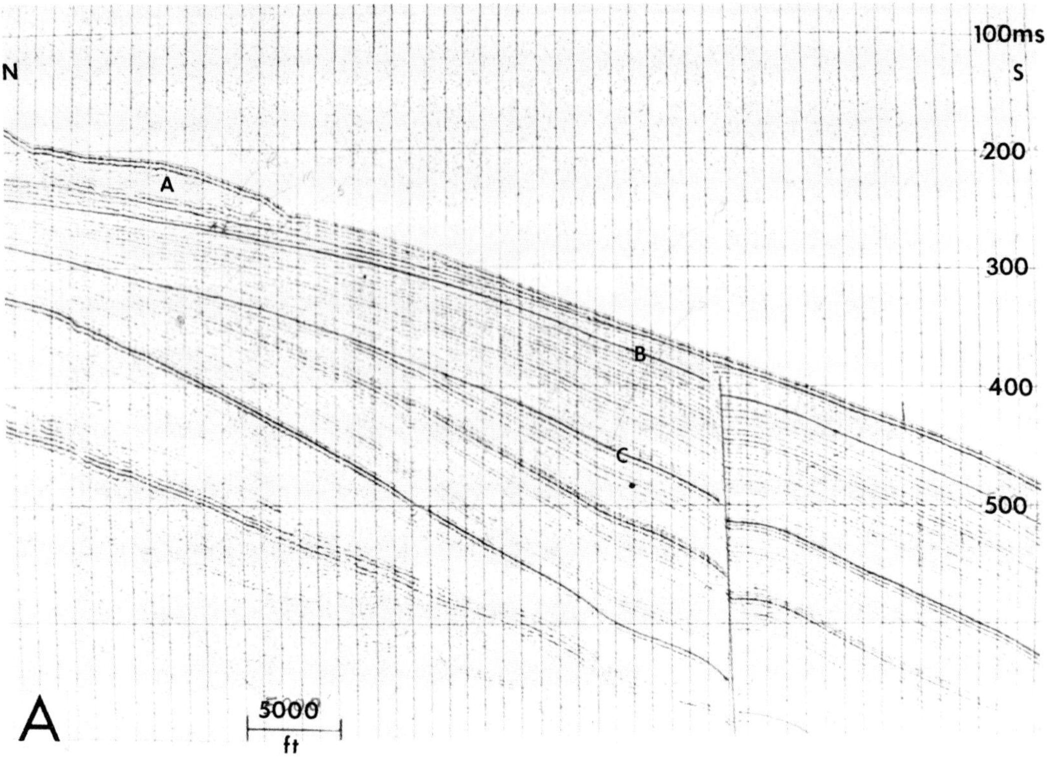
100ms
N
S
200
A
300
B
400
C
500
A
5000
ft

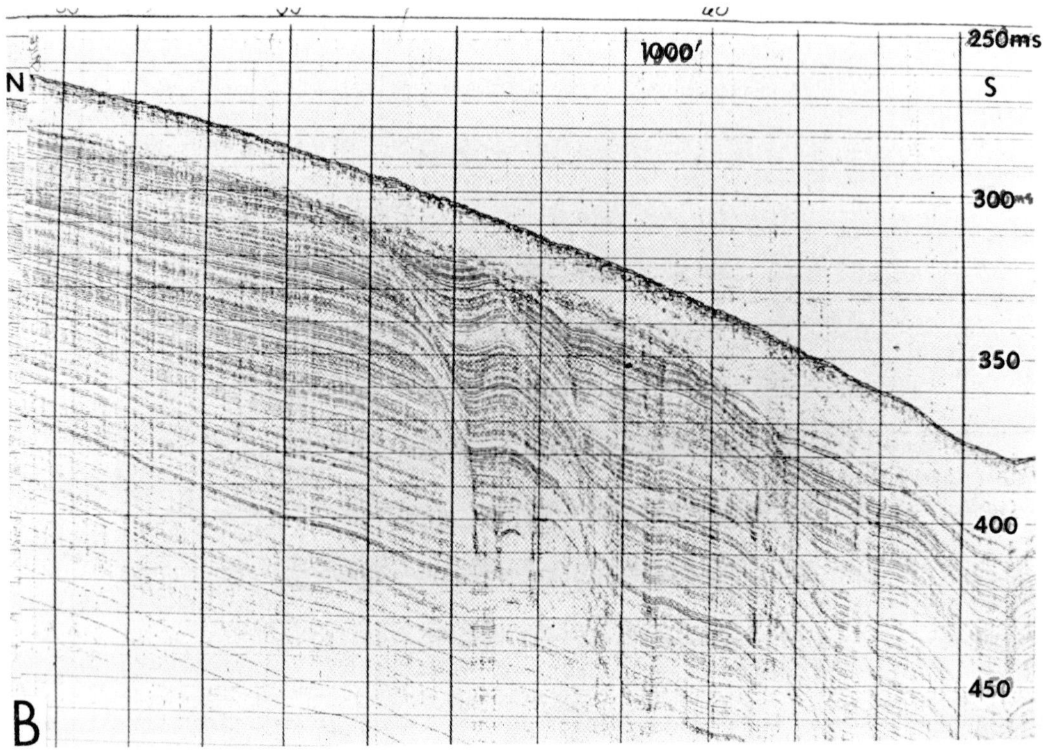
250ms
1000'
N
S
300
350
400
450
B

of 5–10 m, whereas at depth offsets of marker beds approach 70–80 m. The fault plane commonly displays a concave-upward geometry. Higher sediment accumulation rates are found on the downthrown side of the fault than on a similar bed on the upthrown side of the fault. Figure 7A illustrates this aspect and the increased offset of beds with depth. In water depths where these faults are found, normal vertical sedimentation rates are usually measured in centimeters per century. Cored borings in these fault systems, however, show extremely rapid accumulation rates, and many of the cores reveal displaced shallow-water sediment properties. Seismically these units often show amorphous returns capped by thinly parallel spaced acoustic reflections, which commonly indicate pelagic sedimentation. These relationships suggest that much of the material associated with the faults represents shallow-water mudflow deposits. As the surface mudflow crosses the scarps associated with these faults, larger volumes of sediment are deposited on the downthrown side of the fault (Fig. 7B). The blanketing of the fault zone by these mass-movement sediments eliminates surface scarps; however, continued movement on the fault will result in formation of new scarps and later blanketing by additional mudflow deposits. This type of interaction between surface mudflow movements and contemporaneous faults quite possibly plays a large role in maintaining continuing movement along fault planes.

A feature commonly associated with growth faults, and of extreme importance to petroleum accumulation, is rollover or reverse drag on the downthrown sides of the faults (Fig. 7A, B). The rollover tends to form soon after deposition of the sediment on the downthrown side. The mudflows have high water and gas contents and, as a result of early degassing and dewatering, volume changes occur in the sediment on the downthrown side. Pore waters and pore gases escape upward along the fault plane, thus decreasing the volume of sediment and allowing an early change in density to take place nearly contemporaneously with the development of the fault. Side-scan sonar data often show large mud volcanoes along the fault zone, attesting to early degassing and dewatering.

MASSIVE SHELFEDGE FAILURES

The shelfedge seaward of major deltas often shows massive failures in which the evacuated scar is later infilled with oblique progradational facies (Fig. 2). Figure 6 shows the surface trace of an extremely large failure that occurred in Late Pleistocene times and covers an area in excess of 18,700 square kilometers. Figure 8 shows a dip section seismic line (10 kilojoule sparker source) run across this feature. As can be seen, seismic reflections are abruptly truncated, indicating a failure surface and then an infilling of the scar by relatively steeply dipping progradational facies. These infilled sediments have been termed an "accretion unit." The steepest dips range from 6 degrees to 8 degrees and, as infilling proceeds, dip angles decrease to 3 degrees or less. This feature represents a massive failure at the shelfedge, where a shallow slab some 500 m thick failed. This failure moved 8600 km^3 of shelf and upperslope sediments downslope, where they accumulated on the rise at the base of the continental slope (upper part of the submarine fan). Radiocarbon dating of horizons from nearby borings document the time framework of this failure. Unit A (Fig. 8) has been radiocarbon dated in numerous borings and ranges in age from 29,000 to 35,000 years B.P. Thus the failure occurred at some time after formation of this reflection horizon. The infilling sequence then commenced and is represented by oblique progradational facies, as can be seen in Figure 8. A cored boring in the upper 200 meters of the infilled sequence indicates the presence of only fine-grained clays and silts and a total lack of any coarse detritus. Faunal content indicates deposition of beds in a marine environment alternating with displaced units containing shelf-depth fauna. Thus these clinoforms contain no coarse detritus, yet form a major part of the shelf progradation during Late Pleistocene times off the modern delta. Unit B is a thin shell horizon that is easily traced on seismic sections and has been radiocarbon dated on numerous borings. Dates from several samples indicate an average age of 15,500 years B.P. This unit forms an erosional unconformity across much of this area and is associated with the last low glacial sealevel stand and its subsequent rise during the Holocene. In some instances, sediment removal on the order of 20–50 m can be documented. This massive failure, then, formed post 35,000 years B.P., infilled by shelfedge progradation by 15,500 years B.P., and then experienced an erosional event associated with the last rise of sealevel. Based on the radiocarbon dating, the infilling occurred at an average rate of 1–6 m/year. Interestingly, the sediments that have experienced failure include a considerable amount of coarse deltaic sediments that were laid down in Early Wisconsin or Late Illionian times (estimate based on

---

FIG. 7.—A, Sparker profile (dip section) run across active growth fault in the Mississippi River delta area. A = upslope mudflow lobe, B = seismic reflection surface radiocarbon-dated at 15,500 years B.P., C = seismic reflection surface dated at 29,800 years B.P. B, Acoustic pulse seismic record run across growth fault showing amorphous capping by prograded mudflow. Depth in milliseconds.

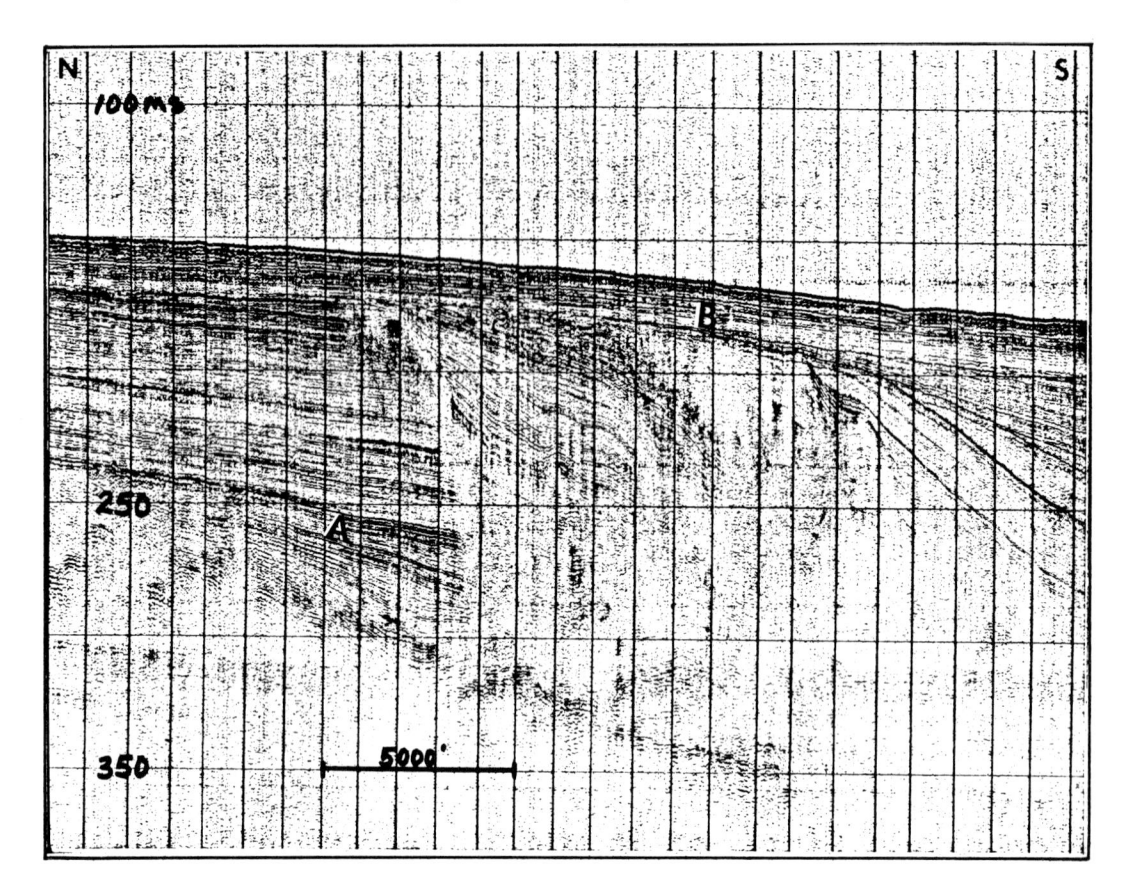

FIG. 8.—Sparker line (dip section) run across a massive shelf-edge failure in the Mississippi River delta. A = Age of horizon ranges from 29,000 to 35,000 years B.P. B = Age of horizon averages 15,500 years B.P.

borings outside the failure zone). Thus, these coarse deposits are now found in the submarine Mississippi fan, but lack of coring control in the fan prohibits speculation on their distribution or geometry. Walker and Massingill (1970), however, have mapped a large "slump mass" at the base of the slope that compares favorably with the volume of this failure.

SUBMARINE CANYONS

Submarine canyons or troughs are located on the continental shelves seaward of many modern large deltas (e.g., Mississippi, Congo, Amazon, Magdalena, Nile). The origins of these canyons and their fans and their association with the deltaic sequence have been discussed in the literature (Stuart and Caughey, 1976; Kenyon et al., 1978; Moore et al., 1978; Shepard et al., 1979; Damuth and Embley, 1981; Shepard, 1981). In an excellent review paper Shepard (1981) speculates, ". . . the discovery that wherever rivers have built deltas across the continental shelf, the foreset slope beyond is creased by valleys, some even of canyon dimension, makes one wonder if landslides may not be common in areas of forward-building sediment slopes." During the past several years, as the petroleum industry has leased areas in deeper waters off the shelf and upper slope of the Mississippi Canyon, detailed gridded seismic and side-scan sonar data and soil borings have been acquired, so that its mode of formation and age have been well documented.

The Mississippi Canyon or Trough was originally interpreted by Shepard (1937), and later by Gealy (1955), as resulting from massive submarine slumping in combination with diapir positioning. Bates (1953), Fisk and McFarlan (1955), and Phle-

FIG. 9.—A, Bathymetry of the region studied in the Mississippi Canyon. Seismic section A–A′ is shown in Fig. 10. B, Tracing of seismic section across the Mississippi Canyon at typical seismic vertical exaggeration of ×18 and a lower vertical exaggeration of ×2 of the same profile.

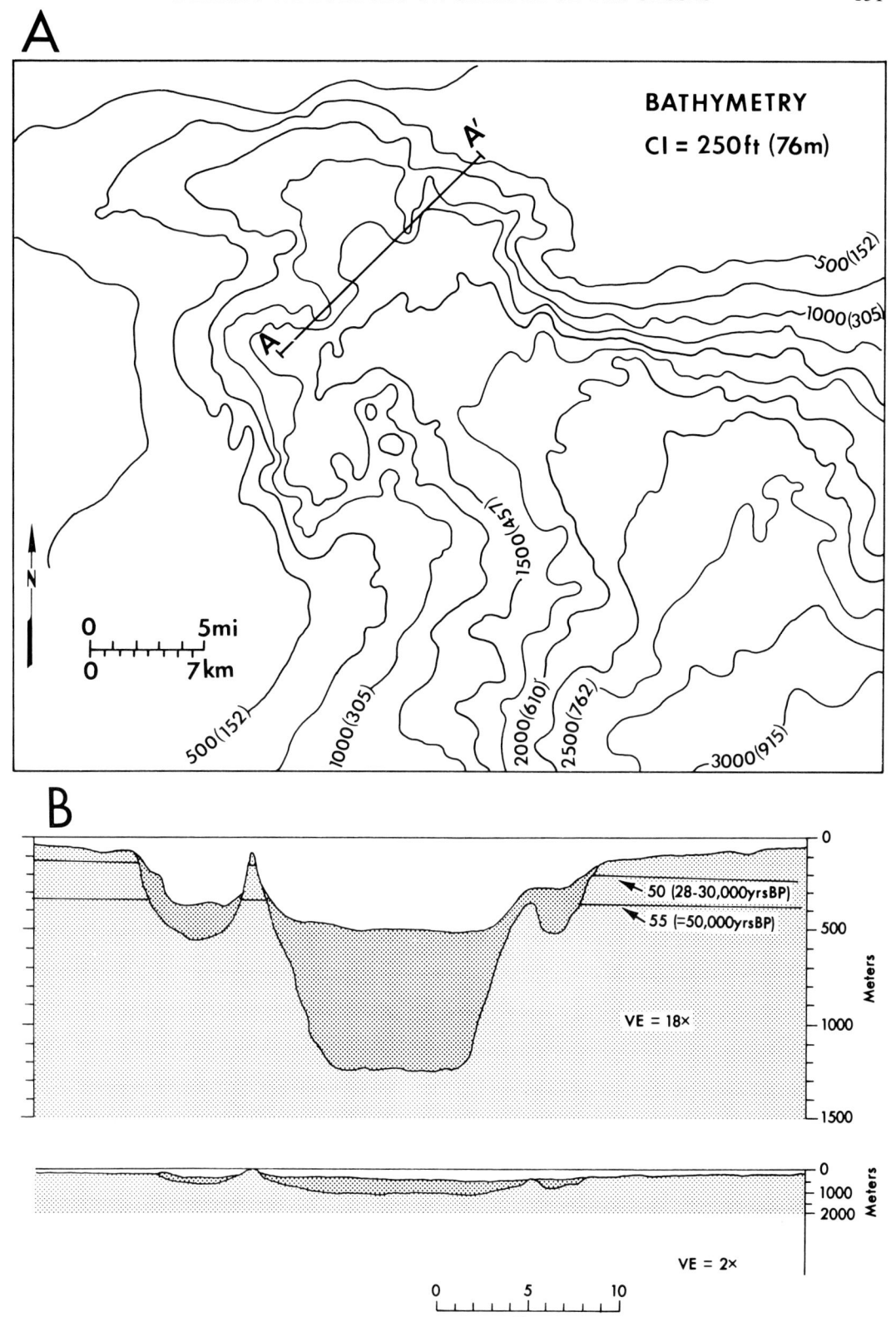

DELTAIC INFLUENCES ON SHELFEDGE PROCESSES 131

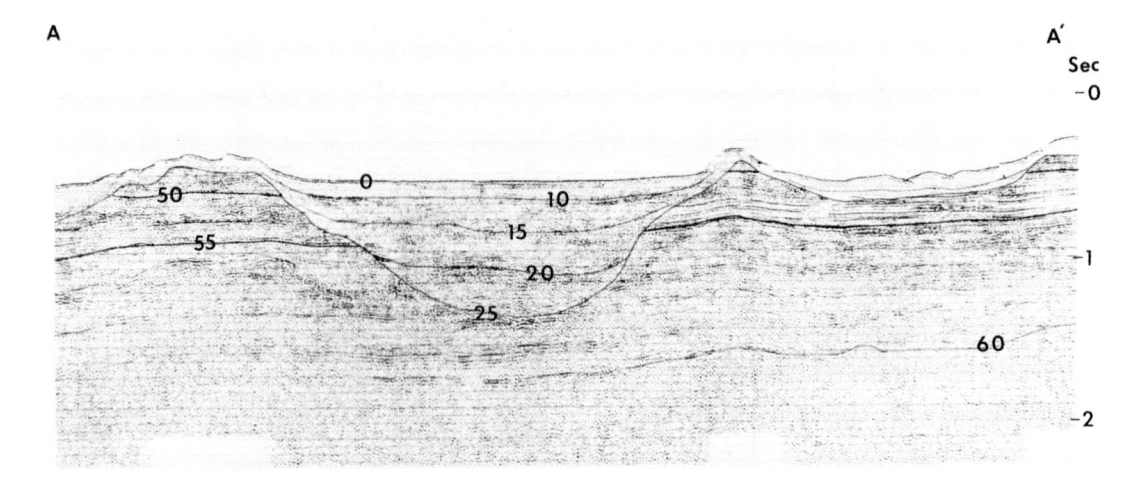

FIG. 10.—Multichannel (24 trace) seismic line run across the Mississippi Canyon. See text for identification of horizons. Location shown in Fig. 9A. Depth in seconds.

ger (1955) stress subaerial erosion and cutting of the canyon by sand-laden density flows. In 1979 Ferebee and Bryant (1979) summarized a large amount of seismic data and shallow cores obtained in the vicinity of the canyon and added considerably to knowledge of the geometry of the filled parts of the Mississippi Canyon.

The data collected in 1980 consist of approximately 2250 km of 12-channel sparker, minisparker, subbottom profiles and digitally acquired side-scan sonar data in the area shown in Figure 1 (area 3). In addition, numerous soil borings were obtained from which samples could be radiocarbon dated. These horizons were then traced laterally on the seismic traces. Figure 9A shows the bathymetry of the region studied in the Mississippi Canyon. In this area the canyon has a maximum relief of approximately 500 m. Figure 10 shows a 12-channel seismic line run across the canyon (see Fig. 9A for location). Note the depth to the base of the canyon, some 1080 m below sealevel, which presently has a sedimentary fill of over 700 m. The seismic horizons shown on Figure 10 have been traced across the entire region, corrected for velocity, and then depicted on a series of maps. In addition, several of these horizons have been radiocarbon dated, and older horizons have been paleontologically dated. These horizons are as follows:

Surface 60—Base of Illinoian Glacial Stage (350,000–400,000 B.P.)
Surface 55—Early Wisconsin (60,000–75,000 years B.P.)
Surface 50—Late-Mid Wisconsin (50,000 years B.P.)
Surface 40 (not shown on Figure 10)—Late Wisconsin (25,000–27,000 years B.P.)
Surface 25—Base of Canyon (22,000–25,000 years B.P.)
Surface 20—Slump Fill
Surface 15—Late Wisconsin (15,000 years B.P.)
Surface 10—Holocene (10,000–12,000 years B.P.)
Surface 0—Present seafloor

These surfaces display exceptionally good lateral continuity, and seismic ties can be made throughout the area. The seismic section, then, indicates that the canyon formed post horizon 40 (25,000–27,000 years B.P.), as it, as well as horizons 50 and 55, is cut by the base of the canyon. Infilling of the canyon commenced by 20,000 years B.P. and was virtually complete by 10,000 years B.P. Cores from the canyon fill indicate that since that time normal deep-water pelagic sediments have formed the uppermost fill. Thus, any mechanism proposed as producing the canyon must take into account the evidence that the formation of this massive feature occurred over an interval of approximately 5000 years or less. Figure 9B is a tracing of a seismic line across the canyon showing seismic horizons 40 and 50. Note that this section has a vertical exaggeration of ×18, whereas the lower diagram has only a ×2 exaggeration. The lower section shows that the canyon is indeed a very large, broad feature, with a width-depth ratio of 40:1.

Figure 11A represents structural contours on surface 55. Although the canyon truncates this surface, computer trend surface programs were used to reconstruct the surface as it existed prior to canyon

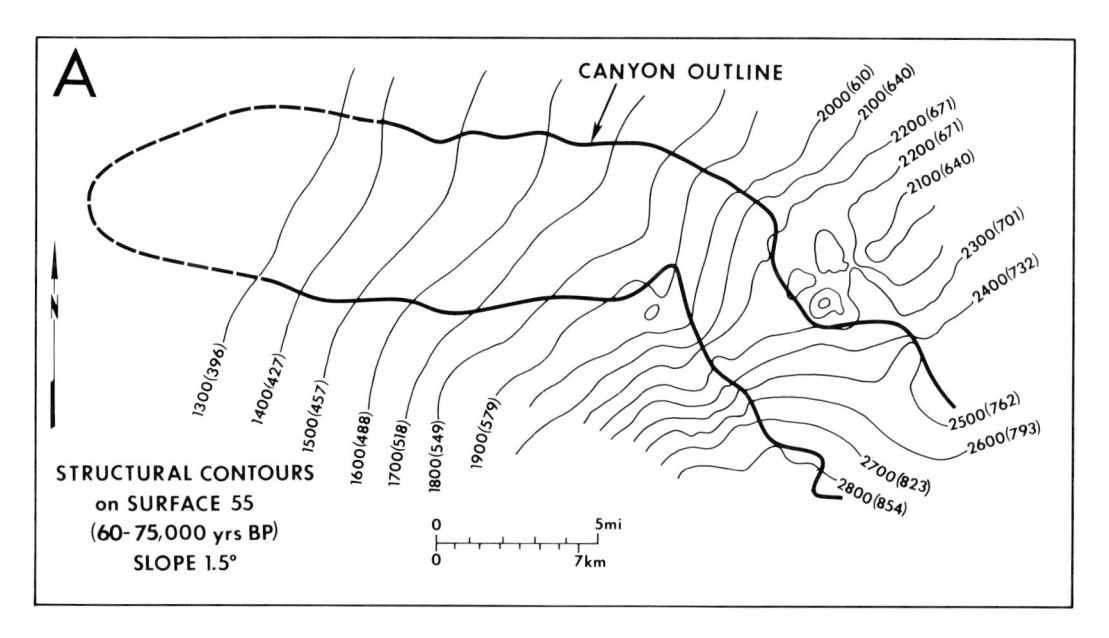

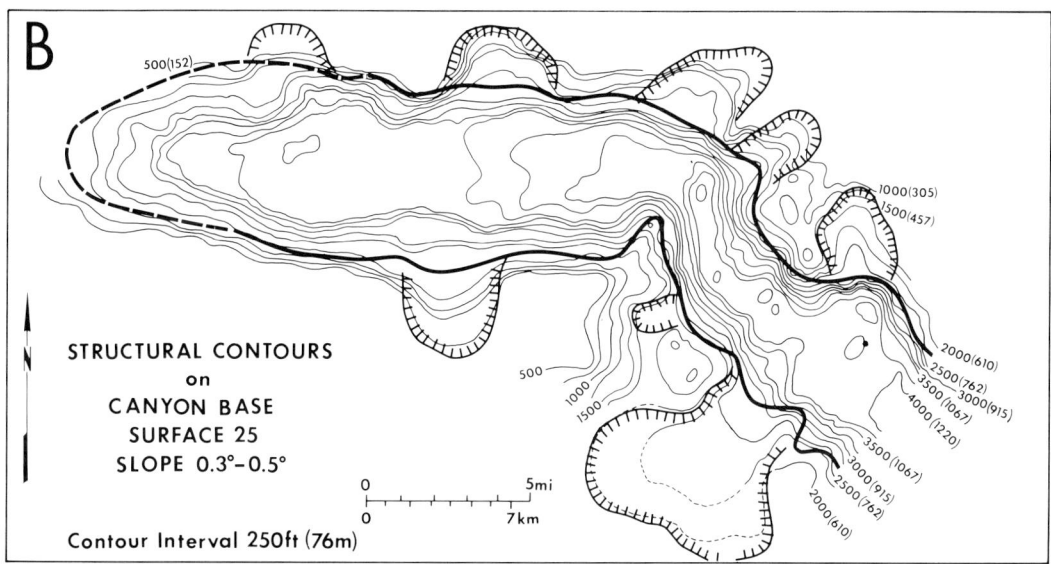

Fig. 11.—*A*, Structural contours on surface 55, Mississippi Canyon. *B*, Structural contours on surface 25, the base of the canyon.

formation. Note that the seaward slope of this surface is 1.5 degrees and that the canyon axis trends approximately at right angles to the slope of this surface. Figure 11B is a structural contour map of surface 25, or the base of the canyon. The head of the canyon is approximately 150 m below sealevel, and at the southeastern part of the canyon the base is some 1220 m below sealevel. The gradient along the canyon axis is 0.3 degree to 0.5 degree, much lower than that of the horizons through which the canyon is excavated. The total volume of material removed from the canyon is 1500 km^3. Note also that the canyon floor is rather flat and the side walls are fairly steep, slopes approaching 23 to 25 degrees. Such side-wall slopes were unstable, and large arcuate slumps formed along the wall as it attempted to regain stability. Slump scars are shown on Figure 11B. Thus infilling of the canyon takes place almost simultaneously with formation.

Figure 12A shows an isopach of the interval from

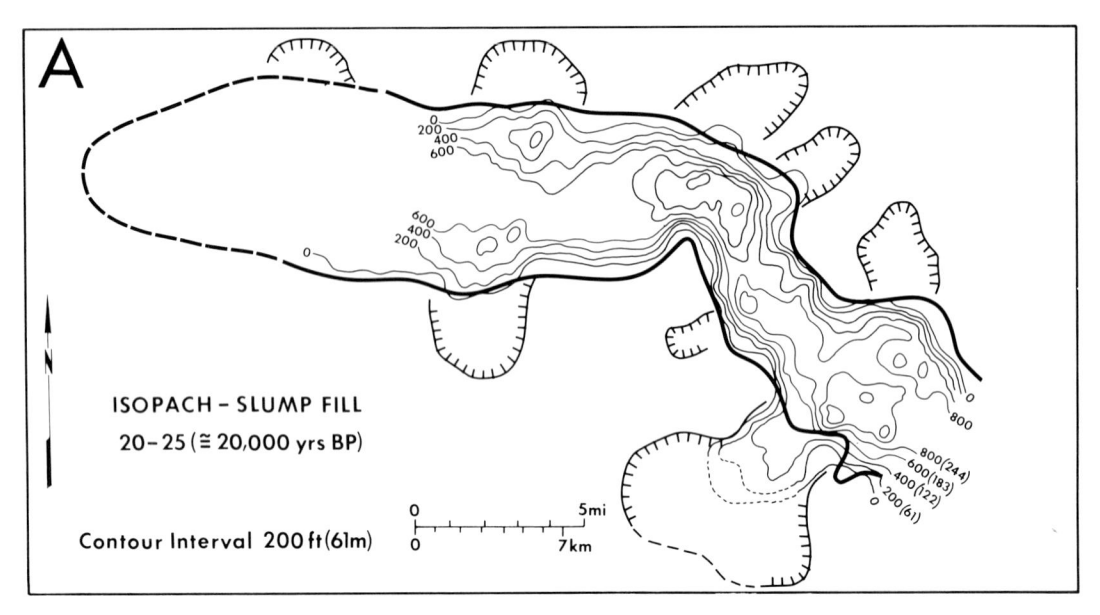

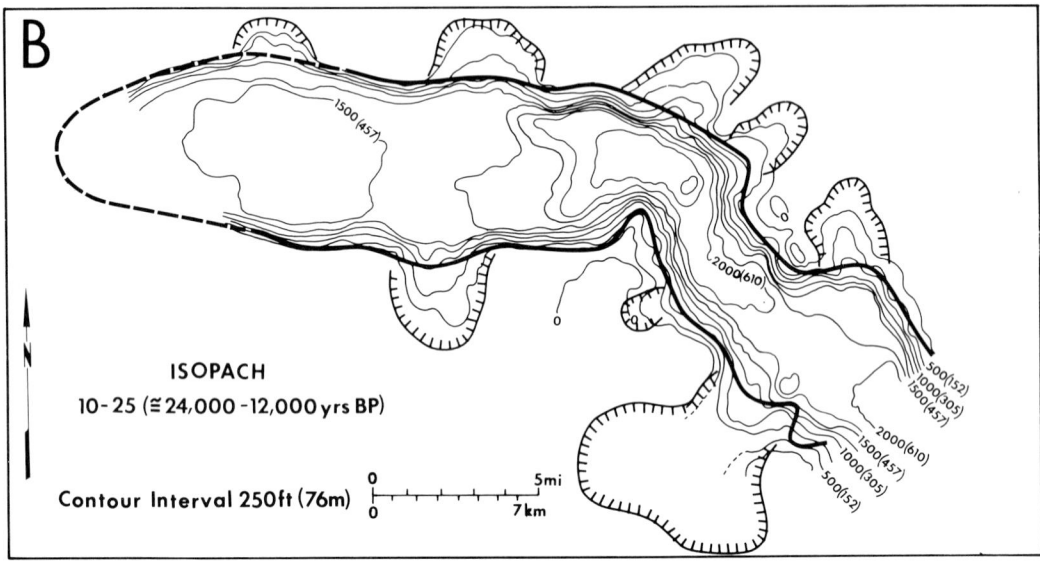

FIG. 12.—A, Isopach map of slump fill of the Mississippi Canyon (surfaces 20–25). B, Isopach map of fill in the Mississippi Canyon (surfaces 25–10).

surface 20 to surface 25, or the thickness of the slumped material. Note the thickened masses of sediment opposite the major slump scars on the canyon side walls. Borings through the canyon fill indicate that the lower unit displays steeply dipping, contorted bedding that could be attributed to sidewall instability. The total deltaic and slump fill is shown in Figure 12B, which is an isopach of the interval between surfaces 10 and 25. The sediments above surface 20 represent primarily finely laminated clays, with silt and fine sand stringers that carry shallow-water fauna. Radiocarbon dates and thicknesses of fill indicate an average sedimentation rate during the filling (post-slumping) ranging from 1.5 to 2.0 m/century. Considering that Late Pleistocene deltas were building in this vicinity, the sedimentation rate is not surprising. On seismic dip lines, large foresets can be distinguished.

Figure 13 is an isopach map of the latest fill, the pelagic sediments that have been accumulating during the last 10,000–12,000 years. Borings and drop cores taken within this unit indicate normal deep-

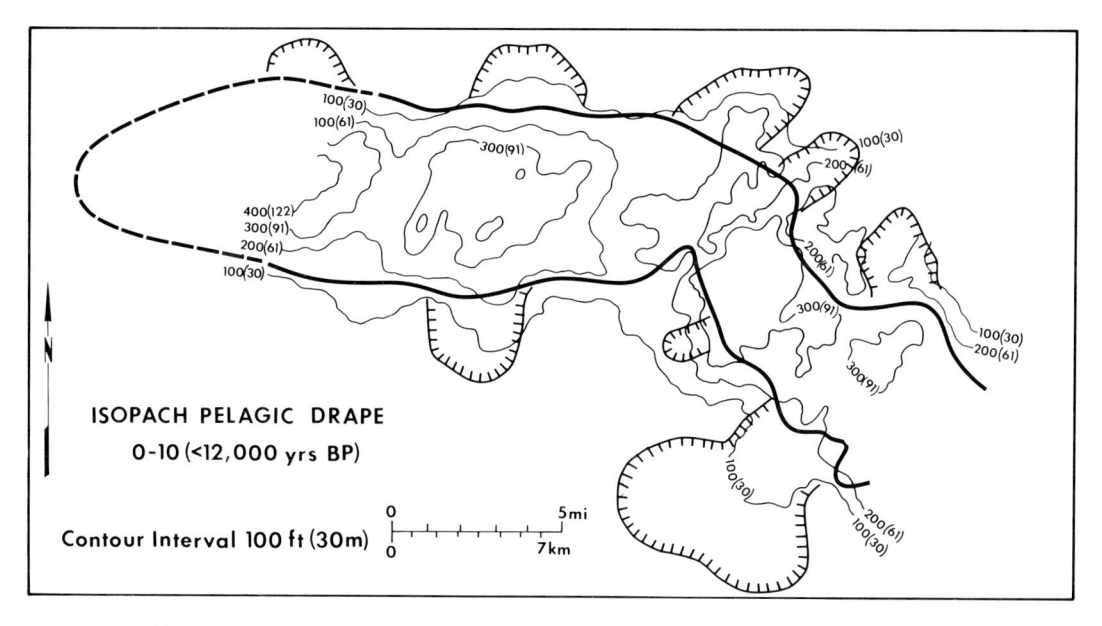

FIG. 13.—Isopach map of the pelagic drape in the Mississippi Canyon (surfaces 0–10).

water fauna consistent with present-day water depths. Occasionally a silty unit is present that contains shallow-water shelf faunal material. Note that even though the greatest thickness is present along the axis of the canyon, the material drapes over the adjacent walls.

Utilizing detailed data collected in the past few years, it is possible to reconstruct the events leading to formation of the canyon and its subsequent partial infilling. Surface 60 (Fig. 10), not cut by the canyon, was deposited during Illinoian-Early Sangamon times, approximately 350,000 to 400,000 years ago. This horizon was laid down in an upper deltaic environmental setting or a fluvial plain, as evidenced by the coarse nature of the sediments and the presence of numerous in situ peat stringers. Approximately 100,000 years ago, the major site of deposition shifted eastward of the canyon and the river constructed a major delta lobe in the vicinity of the present active Mississippi River delta. By approximately 80,000 years B.P., sealevel was similar to its present stage, and it is highly probable that surface 55 represents a time-transgressive unit formed during the subsequent drop in sealevel. During the period 65,000 years B.P. to approximately 50,000 years B.P. some 100 m of deltaic sediment was laid down, and surface 50 probably represents a time-transgressive unit formed across this delta mass during rising sealevel. By 25,000–27,000 years B.P., sealevel was lower than present, and surface 40 represents extremely shallow water deposition, with small coral and coralgal reefs present and associated with this surface. All of these units were deposited prior to formation of the canyon, and surface 40 can be seen to be truncated by the canyon wall throughout the area studied.

The single most difficult problem in reconciling the present data with earlier models is the timing of the formation of the canyon. The present data show definitely that the canyon formed post-surface 40 (25,000–27,000 years B.P.) and that the lower infilling of the canyon commenced by 20,000 years B.P. Thus given even the most generous interpretation, the canyon cannot have had more than 7000 years in which to form and remove 1500–2000 km^3 of material. Density currents are at best infrequent events, and are not the most probable erosional mechanism for cutting such a large feature in such a short time. Subaerial erosion by ancestral rivers is also not a likely mechanism, as the river would have had to erode to more than 1200 m below present sealevel in a short time during a period when sealevel could not have been lower than 100–150 m. In addition, the floor of the filled canyon (surface 25) has an axial slope of only 0.3–0.5 degree, and pre-canyon depositional surfaces have slopes of 1.5–2.0 degrees. The present surface of the canyon floor has a slope of 0.8 degree. Thus both the post-depositional surface and the present depositional surface are steeper than the base of the canyon floor. Any scouring action would be expected to cut a surface steeper than the depositional surfaces. The most likely explanation for the formation is large-scale slumping on an unstable continental margin. The most obvious indication of large-scale instability comes from the lowermost unit of canyon fill, where individual sedimentary masses can be related back to scars along the canyon wall. This first stage of canyon fill could thus be interpreted as the final phase of canyon formation, when the canyon walls themselves were achieving equilibrium. This indicates that such large-scale failures were possible. The overall morphology also suggests massive failures. The slope of the canyon base is low, 0.3–0.5 degree, which implies that failure continued until an equilibrium slope was reached, lower than the primary depositional slopes. In addition, the floor of the canyon is rather flat, and the canyon form is a shallow U-shape. Thus, it appears that the Mississippi Canyon formed as a series of successive failures, each one creating the upslope instability that triggered the next, and thus migrating rather rapidly upslope. Once initiated, the canyon then acted as its own conduit, directing the failed material basinward until equilibrium was reached. The primary instability was probably produced by rapid sedimentation close to the continental margin during an earlier phase of delta building. The shelfedge failure described earlier in this paper was probably a part of the same failure.

The irregular upper surface of the slump fill represents surface 20, and was completed by 20,000 years B.P. During the interval 20,000 years B.P. to 10,000–12,000 years B.P. a series of Late Wisconsin delta lobes deposited north of the present canyon were probably responsible for the major fill of the canyon. Since about 10,000 years ago, only pelagic sediments have been deposited in the area studied.

Although this is the only canyon associated with a delta that has such a large volume of detailed data and cored borings, the authors believe, even from limited data in similar situations, that mass-movement and large-scale instabilities play a major role in canyon formation. If this interpretation can be substantiated in other regions, it has major implications in the composition and distribution of sediments forming the sedimentary fans seaward of the canyons.

CONCLUSIONS

The data obtained on shelfedge regions seaward of major deltas indicate that mass-movement processes play a major role in shaping the shelfedge-upper slope morphology. The major conclusions are as follows:

1. The high sediment yield of major river deltas results in a wide variety of sediment instabilities at the shelfedge. These features and their infilling sequence display distinctive lithologies and seismic facies.

2. These shelfedge processes seaward of major deltas transport large volumes of shallow-water sediment to deeper water basins offshore. In late Pleistocene and Holocene times alone, some 12,000 km^3 of sediment has been moved off the shelf fronting the Mississippi Delta into deeper waters of the Gulf of Mexico. This material accounts for approximately 15 percent of the total fan volume. All of the material deposited has resulted from subaqueous sediment instabilities, and thus the lithologic nature of the material now residing in the upper part of the fan can be determined by borings on the adjacent shelf in the vicinity of the sediment instabilities.

3. Although the examples illustrated are off the Mississippi Delta, similar features, varying in scale and magnitude, have been documented off the Nile, Orinoco, Magdalena, Amazon, and Ebro. Unfortunately, borings for obtaining radiocarbon dates and lithologic data are unavailable and the mechanisms and timing of events are speculative.

4. Finally, the processes described in major delta areas are normal, everyday processes, occur rapidly, and are of large magnitude. They result in major sediment accumulations on the shelf, upper slope, and adjacent deep-water basins.

ACKNOWLEDGMENTS

Funds for this research were provided primarily by the Marine Geology Branch, U.S. Geological Survey, and the Coastal Sciences Program, Office of Naval Research. Data on specific offshore blocks were provided by numerous oil companies. Discussions with Drs. L. E. Garrison, A. H. Bouma, J. R. Hooper, and H. H. Roberts are greatly appreciated.

REFERENCES

BATES, C. C., 1953, Rational theory of delta formation: Am. Assoc. Petroleum Geologists Bull., v. 37, p. 2119–2162.

BOILLOT, G., MOUGENOT, D., AND REHAULT, J. P., 1981, Shelf break on modern passive margins: structure, sedimentation, and progradation (abstract): Am. Assoc. Petroleum Geologists Bull., v. 65, p. 905 (also their paper in this volume, p. 61–77).

COLEMAN, J. M., 1981, Deltas, processes of deposition and models for exploration: Minneapolis, Burgess Publishing Company, 2nd ed., 124 p.

———, AND GARRISON, L. E., 1977, Geological aspects of marine slope stability, northwestern Gulf of Mexico: Marine Geotechnique, v. 2, p. 9–44.

DAMUTH, J. E., AND EMBLEY, R. W., 1981, Mass transport processes on Amazon Cone: Western Equatorial Atlantic: Am. Assoc. Petroleum Geologists Bull., v. 65, p. 629–643.

FEREBEE, T. W., AND BRYANT, W. R., 1979, Sedimentation in the Mississippi Trough: Texas A & M Univ., College of Geosciences Tech. Rept. 79-4-T, 178 p.

FISK, H. N., AND MCFARLAN, E., JR., 1955, Late Quaternary deposition of the Mississippi River, in Poldervaart, A., ed., Crust of the Earth: Geol. Soc. America Spec. Paper 62, p. 279–302.

GEALY, B. L., 1955, Topography of the continental slope, northwest Gulf of Mexico: Geol. Soc. America Bull., v. 66, p. 203–227.

KENYON, N. H., BELDERSON, R. H., AND STRIDE, A. H., 1978, Channels, canyons, and slump folds on the continental slope between south-west Ireland and Spain: Oceanol. Acta, v. 1, p. 369–380.

MOORE, G. T., WOODBURY, H. O., WORZEL, J. L., WATKINS, J. S., AND STARKE, G. W., 1978, Investigation of Mississippi fan, Gulf of Mexico, in Bouma, A. H., Moore, G. T., and Coleman, J. M., eds., Framework, Facies, and Oil Trapping Characteristics of the Upper Continental Margin: Am. Assoc. Petroleum Geologists, Studies in Geology No. 7, p. 155–191.

PHLEGER, F. B., 1955, Foraminiferal faunas in cores offshore from the Mississippi Delta: Papers in Marine Biology and Oceanography, Deep-Sea Research, Supplement to v. 3, p. 45–57.

PRIOR, D. B., AND COLEMAN, J. M., 1978, Submarine landslides on the Mississippi delta-front slope: Louisiana State Univ. School of Geosciences, Geoscience and Man, v. 19, p. 41–53.

———, AND ———, 1980, Sonograph mosaics of submarine slope instability, Mississippi River delta: Mar. Geology, v. 36, p. 227–239.

———, AND SUHAYDA, J. N., 1979, Submarine mudslide morphology and development mechanisms: Proc. 11th Offshore Tech. Conf., Houston, Texas, Paper 3482, v. 2, p. 1055–1061.

SHEPARD, F. P., 1937, Salt domes related to Mississippi submarine trough: Geol. Soc. America Bull., v. 48, p. 1354–1361.

———, 1955, Delta front valleys bordering the Mississippi distributaries: Geol. Soc. America Bull., v. 66, p. 1489–1498.

———, 1981, Submarine canyons: multiple causes and long time persistence: Am. Assoc. Petroleum Geologists Bull., v. 65, p. 1062–1077.

———, MARSHALL, N. F., MCLOUGHLIN, P. A., AND SULLIVAN, G. G., 1979, Currents in submarine canyons and other seavalleys: Am. Assoc. Petroleum Geologists, Studies in Geology No. 8, 173 p.

STUART, C. J., AND CAUGHEY, C. A., 1976, Form and composition of the Mississippi fan: Trans. Gulf Coast Assoc. Geol. Soc., v. 26, p. 333–343.

WALKER, J. R., AND MASSINGILL, J. V., 1970, Slump features on the Mississippi fan, northeastern Gulf of Mexico. Geol. Soc. America Bull., v. 81, p. 3101–3108.

UNSTABLE PROGRADATIONAL CLASTIC SHELF MARGINS

CHARLES D. WINKER
Department of Geosciences, University of Arizona, Tucson, Arizona 85721

MARC B. EDWARDS
42 Eagle Court, Woodlands, Texas 77380

ABSTRACT

In some continental margin basins such as the northwestern Gulf of Mexico and the Niger Delta, large-scale slumping of the continental slope disturbs the topset-foreset geometry of the prograding shelf margin and thereby inhibits recognition of ancient shelfedges. As a result, concepts of shelf-margin dynamics have been underemphasized in explaining the structure and stratigraphy of such basins. Nonetheless, ancient unstable clastic shelf margins can be approximately located by criteria such as isopach maxima, timing of growth faulting, and the stratigraphic top of geopressure.

Gravity sliding of the continental slope creates a strongly extensional regime along the shelf margin, resulting in growth faulting and greatly enhanced subsidence rates. The corresponding compressional regime along the lower slope is important in initiating salt and shale structures; if the shelf margin progrades over these structures, diapiric activity can greatly complicate the style of growth faulting. High subsidence rates result in greatly expanded progradational cycles, which serve to distinguish shelf-margin deltaic sequences from deltas of the more stable shelf platform. Rapid fault movement along the shelf margin can hydraulically isolate shallow-water sandstones and juxtapose them against dewatering slope shales, thus allowing the development and maintenance of excess fluid pressure. These deep-water shales are probably a major source of both hydrocarbons and brines instrumental in diagenesis of geopressured deltaic sandstones.

INTRODUCTION

During the past two decades, a wealth of information has been acquired about the growth-faulted shelf margin of the northwestern Gulf of Mexico, primarily in the form of seismic reflection profiles. However, concepts of shelf-margin dynamics have not been applied widely to the interpretation of ancient analogs in the same basin, in contrast to the extensive application of models of deltaic sedimentation. For example, while numerous publications have mapped the distribution of various Tertiary and Cretaceous deltaic systems in the northwestern Gulf Basin, only a few generalized maps have been published of corresponding shelfedges (Hardin and Hardin, 1961; Woodbury et al., 1973; Martin, 1978). On the other hand, dynamics of shelf margins are closely related to principles of subsidence, growth faulting, diapirism, and excess fluid pressure (geopressure), subjects that have been discussed extensively in the literature. The main purpose of this paper is to review these subjects by using the unstable shelf margin as a unifying model and to speculate on possible implications of the model.

In the northwestern Gulf, the continental slope is a realm of gravity-driven sliding, slumping, and sediment transport on a wide range of scales, while the continental shelf is a realm of shallow-water deposition with a strong deltaic influence. The shelf margin represents the interface between these two realms, where the interaction between gravity-sliding and shallow-water sedimentation creates a unique association of structure and stratigraphy. Similar principles should apply to other continental margin clastic systems dominated by gravity tectonics, including the Niger Delta (Weber, 1971; Evamy et al., 1978; Girard, 1979), the MacKenzie Delta (Dailly, 1976; Lane and Jackson, 1980), the Baram Delta of eastern Malaysia (Scherer, 1980; Bol and van Hoorn, 1980), the Nile Delta (Ayout, 1980) and the adjacent continental slope of Sinai (Ben-Avraham and Mart, 1981) and Israel (Almagor and Wiseman, 1977; Garfunkel et al., 1979), the Amazon Cone (Huff, 1980), and continental margin basins of West Africa (Todd and Mitchum, 1977) and Brazil (Brown and Fisher, 1977).

CLASSIFICATION OF CLASTIC SHELF MARGINS

Sedimentary sequences on continental margins can be classified according to shelf margin migration, i.e., prograding, aggrading, or retreating (Fig. 1). This classification reflects the rate of sediment influx relative to subsidence, and the apportionment of sediment to deep-water (slope) and shallow-water (shelf) facies. In the case of progradational clastic shelf margins an additional subdivision is useful, based on gravitational stability. In the unstable case, large-scale slumping of the continental slope disrupts the original depositional geometry. This phenomenon can generate local subsidence rates along the shelf margin much greater than the regional rate and greater than would be possible for

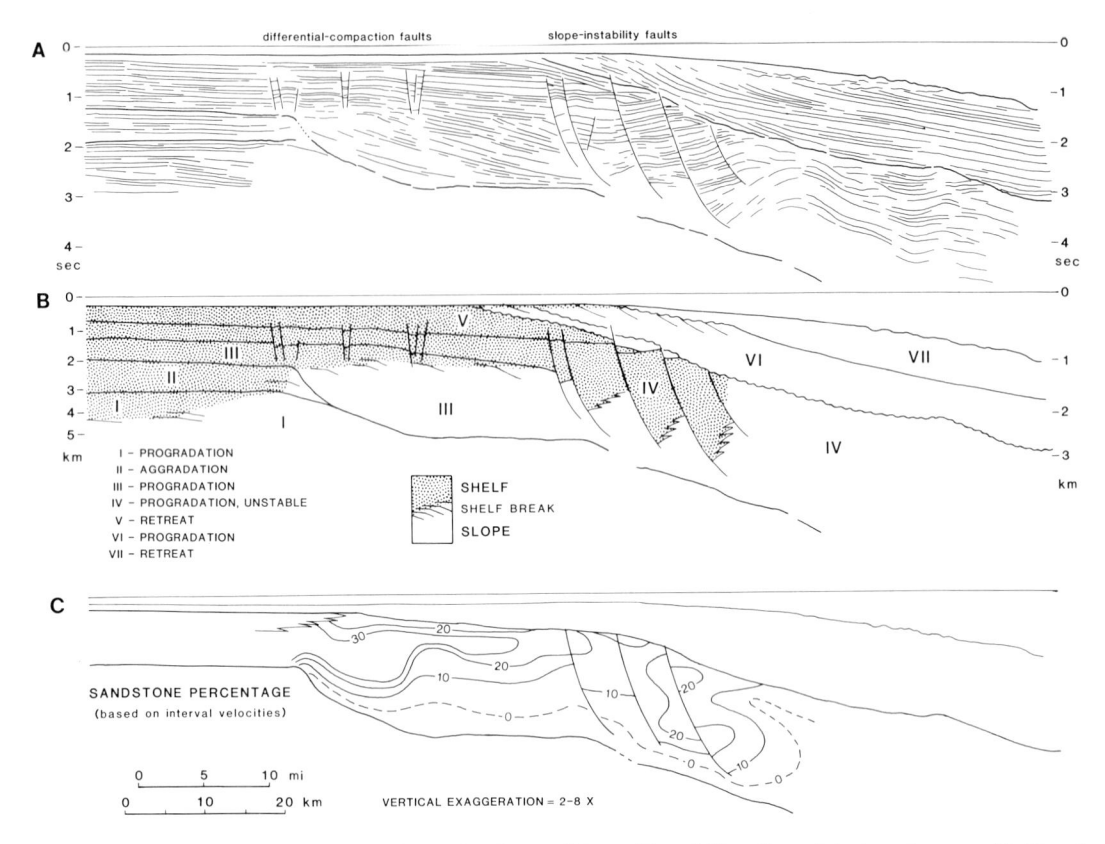

FIG. 1.—Interpretation of a seismic dip section from offshore West Africa (for original section, see Todd and Mitchum, 1977, their Fig. 9). A, Tracing of reflectors and interpreted faults. Faults generated by gravitational instability of slope are listric, and predominantly down-to-basin, and associated with contemporaneous compressional structures downdip. Faults generated by differential compaction of high-shale section (Sequence III) over carbonate bank and slope (Sequence II) are higher-angle, more symmetrically distributed, and not listric; displacement decreases to zero at depth. B, Interpretation of shelf and slope facies and classification of seismic sequences according to migration of the shelfedge. Exact position of shelfedge in growth-faulted sequence is speculative. Sequence IV demonstrates the initiation phase of unstable progradation, juxtaposing shallow-water sandstones against deep-water shale. Progradation did not proceed far enough for diapir-override phase (compare with Fig. 6 where diapir-override is well developed). Also compare with interpretation by Mitchum and Vail, (1977, their Fig. 3). C, Sandstone percentage (after Gralka et al., 1980), reflects high subsidence rate and rapid accumulation of shallow-water sediments deposited near the shelf margin. Net-sandstone map of Sequence IV would show major thickening across growth faults; percent-sandstone map should more accurately depict deltaic lobes (compare with Fig. 9).

a stable progradational margin. Consequently, in unstable progradational systems (unlike stable systems), sediments deposited near the shelf margin are a volumetrically substantial part of the total basin-fill and can be considered as a distinct megafacies. A seismic section from offshore West Africa (Fig. 1) illustrates the application of this classification scheme to seismic sequences and the contrasting geometry of stable and unstable prograding system. The potential for accumulation of great thicknesses of shallow water sediments in a short period of time is clearly evident.

RECOGNITION OF ANCIENT UNSTABLE SHELF MARGINS

Disturbance of the large-scale geometry of shelf topsets and slope foresets inhibits recognition of ancient unstable shelfedges by seismic stratigraphy or well-log correlation. Primarily for this reason, little mention has been made of them in the literature on such basins as the Gulf of Mexico and Niger Delta. However, it has been widely recognized that: (1) these basins have been filled by off-lapping depocenters migrating basinward (Wilhelm and Ewing, 1972; Antoine et al., 1974; Martin, 1978; Evamy et al., 1978; Weber, 1971); (2) the timing of maximum activity of regional contemporaneous faults becomes progressively later in a basinward direction (Thorsen, 1963; Dailly, 1976; Evamy et al., 1978); and (3) the stratigraphic top of geopressure also becomes progressively younger basinward (Dickinson, 1953; Harkins and Baugher, 1969; Stuart, 1970). These phenomena are related to progradation of the shelf margin and can provide alternate criteria for shelf margin recognition when more conventional criteria such as clinoform stratification, sedimentary structures, and faunal assemblages cannot be used due to inadequate or ambiguous data (Table 1).

Not surprisingly, there is a trade-off between accuracy of locating the shelfbreak and availability of the necessary data. Therefore, the less precise criteria tend to be more useful for regional mapping (Fig. 2), although they may be inappropriate for prospect-scale mapping. Along the modern shelf margin (Fig. 3), maximum displacements, expansion ratios of growth faults, and maximum thicknesses of progradational deltaic sequences typically occur within a few miles of the shelfbreak (Lehner, 1969). Harkins and Baugher (1969) demonstrated a close correspondence, with a similar degree of precision, between the shelfbreak and the updip limit of geopressure in the Miocene *Bigenerina* 'A' zone, offshore Louisiana. Weaver (1955) proposed that a flexure, defined as a change in regional dip and an increase in the rate of basinward thickening, represents the paleo-shelfbreak, although our experience is that the flexure usually occurs several miles updip of the corresponding shelfbreak.

Using these criteria, it is possible to map regional shelf margin trends of the northwestern Gulf of Mexico (Fig. 2) and the Niger Delta, primarily on the basis of published studies and data. Although

TABLE 1.—CRITERIA FOR LOCATING ANCIENT SHELF EDGES IN UNSTABLE PROGRADATIONAL SYSTEMS

	Criterion	Principle	Observed in: Quaternary	Observed in: Ancient	Figure and Literature References	Comments
(Most Precise)	Topset-foreset geometry	Stratification represents depositional relief of shelf and slope	X		Figs. 1, 3, C–F	Usually obscured by contemporaneous structural growth; below resolution of conventional seismic reflection data
	Lithofacies	Turbidites (sandstones) and disrupted, chaotic bedding (shales) characterize slope sediments	X	X	Berg, 1981	Ancient slope sediments seldom cored in this type of basin; "deep-water" sedimentary structures may occur in fairly shallow water
	Microfaunal assemblages	Neritic-bathyal transition marks shelfbreak	X	X	Woodbury et al., 1978; Stude, 1978; Biel & Buck, 1978	Faunas can mix and interfinger; neritic fauna may be reworked into slope sediments; species may change environmental range with time; disagreement among paleontologists
	Isopach maximum	Maximum sedimentation and subsidence rates occur at the shelfbreak	X	X	Fig. 3; Woodbury et al., 1973; Poag & Valentine, 1976	Usually insufficient downdip data to see thinning basinward of shelfbreak
	Maximum rate of growth faulting	Maximum extension rate and growth ratios occur at shelfbreak	X	X	Fig. 3; Lehner, 1969; Thorsen, 1963	Usually insufficient deep data to observe pre-maximum fault growth
(Least Precise)	Stratigraphic top of geopressure	Geopressure results from hydraulic isolation of shallow-water sandstones faulted against older slope shales		X	Fig. 5; Harkins & Baugher, 1969; Dickinson, 1953	Relationship fairly circumstantial and poorly documented, but can be used in areas of very sparse data (mud bights and paleo-tops)
	"Flexure"	Sharp increase in regional dip marks relict shelfedge		X	Weaver, 1955; Hardin & Hardin, 1961	In subsurface, "flexure" usually occurs several miles updip of shelf break; may be poorly defined

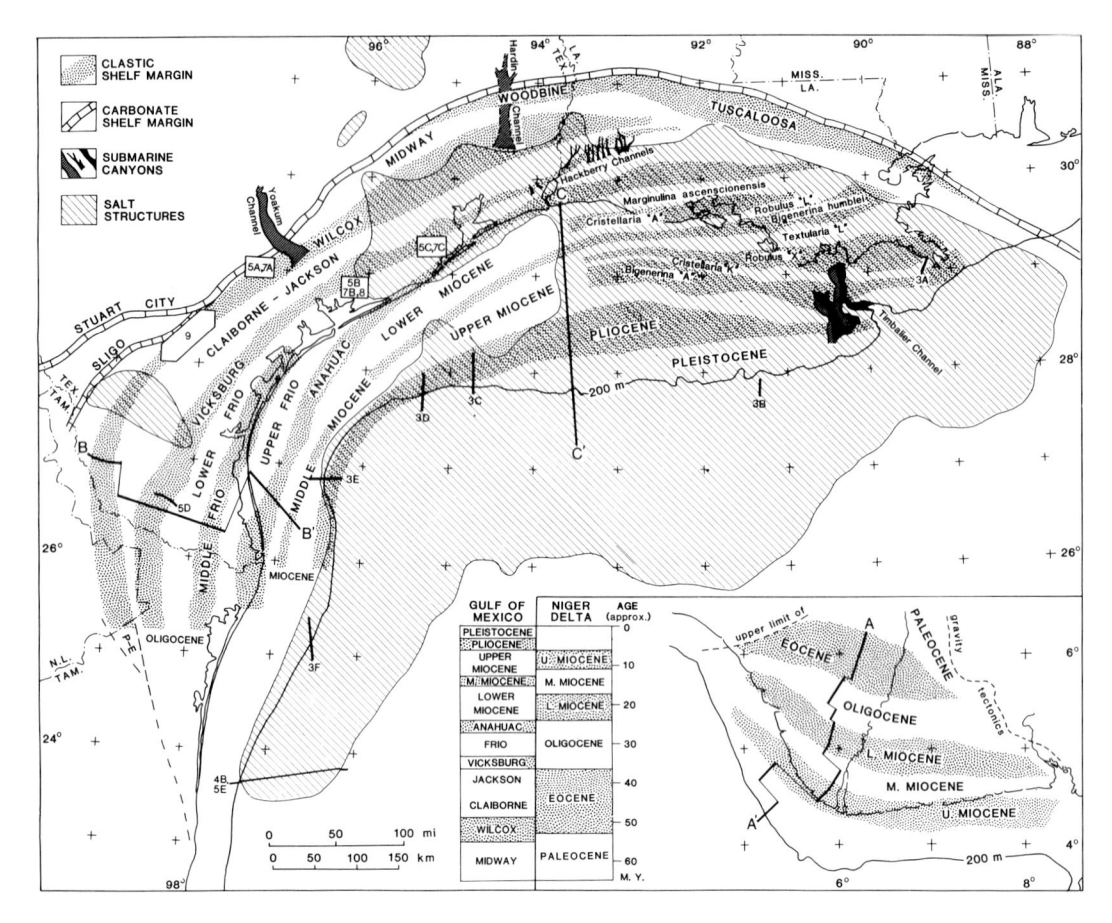

FIG. 2.—Regional shelf margin trends of the northwestern Gulf of Mexico and Niger Delta, based on isopach maxima, flexures (particularly in Niger Delta), timing of maximum growth-fault activity (particularly in Texas and northeast Mexico), and stratigraphic top of geopressure (particularly in Louisiana). Where necessary, shelf margin trends are extrapolated along regional growth-fault trends (particularly in Niger Delta). In Gulf Basin, Midway and Woodbine represent stable progradation; all other clastic sequences represent unstable progradation. Ages of submarine canyons are: Yoakum and Hardin, mid-Wilcox; Hackberry, mid-Frio; Timbalier, Pleistocene. Gulf Basin exhibits major shifts in progradation rates, whereas Niger Delta exhibits fairly steady progradation. Time scales and correlations between basins are approximate. (Based on Bebout et al., 1979; Busch, 1973, 1975; Christina and Martin, 1979; Dailly, 1976; Dickinson, 1953; Edwards, 1980, 1981; Evamy et al., 1978; Foss, 1979; Girard, 1979; Gregory, 1966; Hardin and Hardin, 1961; Harkins and Baugher, 1969; Hickey et al., 1972; Hoyt, 1959; Jones, 1975; Khan et al., 1975a, 1975b; Martin, 1978, 1980; Paine, 1968; Poag and Valentine, 1976; Thorsen, 1963; Woodbury et al., 1973, 1978; file data at University of Texas, Bureau of Economic Geology).

FIG. 3.—Dip sections of Quaternary shelf margin and vicinity, northwestern Gulf of Mexico. Locations of sections are shown in Fig. 2. A, Sand geometry typical of growth-faulted Tertiary deltas (compare section AA', Fig. 8) as observed in the modern Mississippi Delta. The modern Mississippi has not yet reached the shelfedge, but is closer to a true shelf margin delta than other Holocene models (after Friedman and Sanders, 1978, based on data from J. M. Coleman). B, Mixing and interfingering of shallow-water and deep-water fauna near present shelfedge may reflect downslope reworking of shallow-water fauna or oscillation of shelfbreak over several miles (after Woodbury and others, 1978). C, Tracing of a mini-sparker profile of late Pleistocene shelf margin delta (after Winker, 1980). D–F, Tracings of sparker profiles (KANE survey) of shelf margin deltas formed during late Pleistocene low-stands of sea-level. Note close association in most cases of shelfbreak with isopach maxima of time-stratigraphic units, maximum rate of fault movement, maximum thickness of progradational cycles, and maximum steepness of clinoforms. Section F is a rare example of a modern system cut by predominantly up-to-basin ("counterregional") faults (compare with Fig. 6). Depth conversion: 1 second = 2500–3000 ft.

A. SOUTHWEST PASS, MISSISSIPPI DELTA

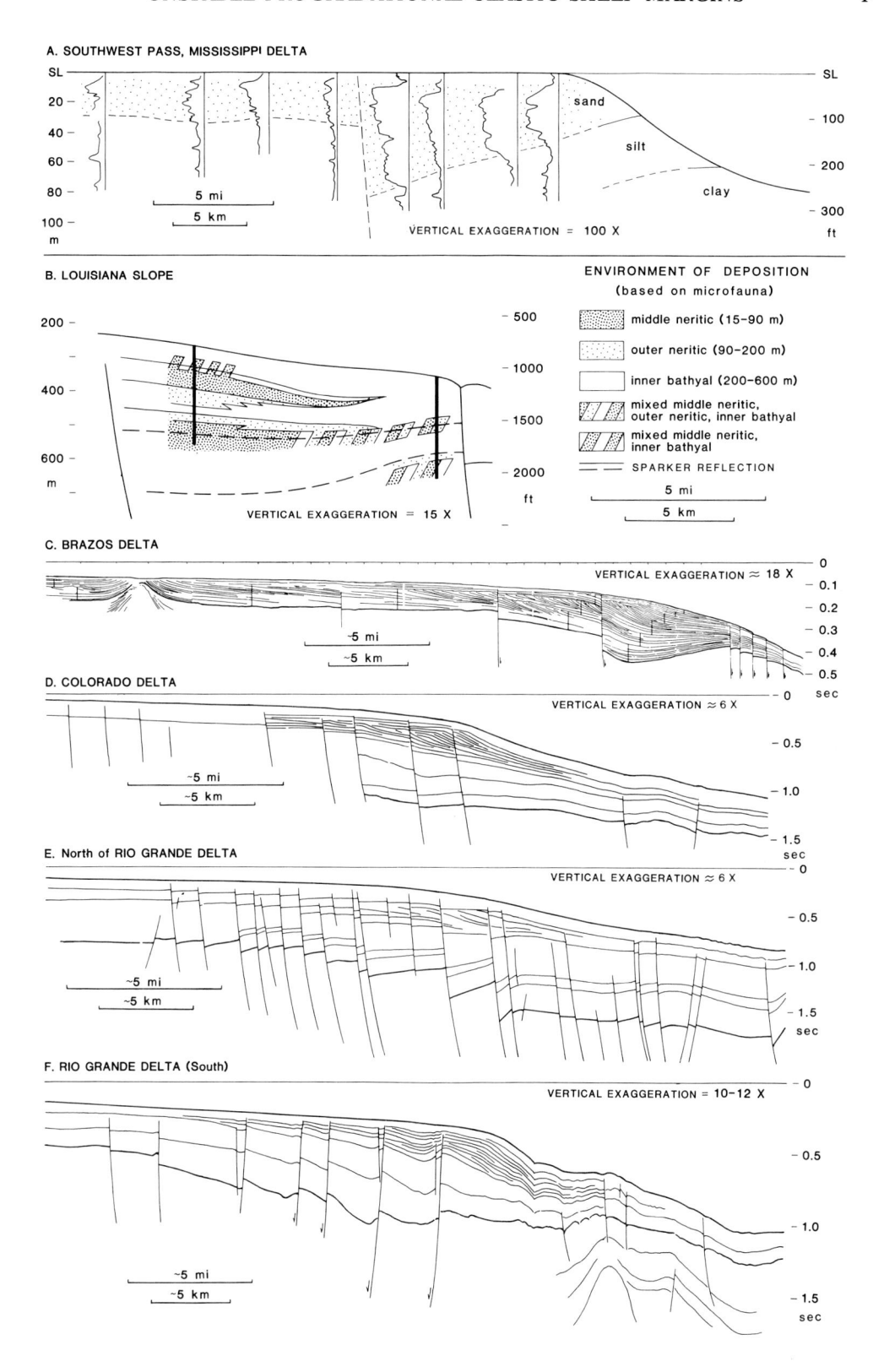

B. LOUISIANA SLOPE

ENVIRONMENT OF DEPOSITION
(based on microfauna)
- middle neritic (15-90 m)
- outer neritic (90-200 m)
- inner bathyal (200-600 m)
- mixed middle neritic, outer neritic, inner bathyal
- mixed middle neritic, inner bathyal
- — SPARKER REFLECTION

C. BRAZOS DELTA

D. COLORADO DELTA

E. North of RIO GRANDE DELTA

F. RIO GRANDE DELTA (South)

TABLE 2.—PROPOSED MECHANISMS FOR GULF COAST STYLE GROWTH FAULTING

Mechanism	References	Comments
Slope instability; gravitational creep	Rettger, 1935; Bornhauser, 1958; Cloos, 1968; Bruce, 1973; Crans et al., 1980	Difficult to document in ancient record because paleobathymetric relief and toe structures not well preserved
Sand loading on shale	Bruce, 1973	Does not explain growth faulting in all-shale sequences, or occurrence of toe structures far downdip of sand pinchout; may be important positive-feedback mechanism in some cases
Differential compaction	Carver, 1968; Bruce, 1973	Does not account for large extensional component or listric geometry; may cause faulting in some cases (Fig. 1)
Uplift of salt or shale ridges	Quarles, 1953; Ocamb, 1961	Does not account for asymmetry (predominantly down-to-basin) of most growth-fault systems; does not predict large extensional component of major faults; useful in explaining crestal and up-to-basin faults (Fig. 3F, 6)

these maps are too generalized for locating ancient shelfbreaks in prospect-scale work, they do reveal major changes in the rate of sediment influx. For example, they reveal major shifts in depocenters in the Gulf of Mexico in contrast to the regular progradation of a single depocenter in the Niger Delta.

STRUCTURAL MECHANISMS

In order to explain the circumstantial relationship of structural geometry to the shelf margin, it is useful to consider the mechanisms of growth faulting. Although a number of mechanisms have been proposed (Table 2), large-scale slope instability (also referred to as sliding, slumping, or creep) best explains the following features of regional growth-fault systems: (1) asymmetry (predominantly down-to-basin); (2) listric geometry, rollover, and large extensional component, all indicative of decollement; (3) contemporaneous faulting in all-shale sequences; and (4) localization of maximum extension rates along the contemporaneous shelf margin. Faults attributable to differential compaction (Fig. 1) or diapiric uplift have distinctly different geometries.

Slope instability involves three overlapping regimes: *translation* of the slope over a decollement surface or zone; *extension* along the top of the slope; and *compression* along the lower slope. Typical structural styles associated with these regimes are evident on seismic sections of the continental slope off northeastern Mexico (Fig. 4). Infinite-slope analysis, which deals with the translational regime, has been successfully applied to shallow-seated submarine slope failures (Prior and Suhayda, 1979; Almagor and Wiseman, 1977; Watkins and Kraft, 1979). A major implication of this analysis is the role of pore pressure in reducing effective stress, thus permitting failure of very low slopes (approximately 1°) typical of the continental slope of the northwestern Gulf. Similar analysis can be applied to larger-scale slope instability, but with qualifications:

1. The slope cannot be assumed to be infinitely long; buttressing effects at the toe are probably important.

2. Slumping does not occur as a single instantaneous failure, as predicted by infinite-slope analysis, but rather as a slow creep which is continuously active for millions of years. This suggests that the decollement is not a discrete surface of brittle failure but rather a zone of ductile or viscous deformation (Kehle, 1970).

3. Superimposed effects of "diapiric" structures driven by density inversion are not taken into account.

STRUCTURAL EVOLUTION

Regional Scale

Structural evolution can be considered on two scales: first, the regional evolution of shelf margin structural styles as the slope progrades into the basin, and second, the local evolution of structural styles as a specific area evolves from a slope environment to a shelf margin environment and finally a stable platform environment. On a regional scale, three phases can potentially develop if sufficient progradation takes place.

1. A *stable phase* is characterized by preservation of slope clinoforms (Figs. 1, 5A). In the Gulf Basin this phase is represented by the Woodbine (Upper Cretaceous) and Midway (Paleocene) trends (Fig. 2).

2. An *initiation phase* is characterized by closely and evenly spaced down-to-basin regional faults, usually without the complication of contemporaneous diapirs (Figs. 1, 5A, 6). In the Gulf Basin, this phase is represented by the Tuscaloosa (Upper

Cretaceous) and Wilcox (Paleocene-Eocene) trends (Fig. 2). Thrust faults and ridges or domes cored by salt or shale originate along the lower slope as compressional structures (Watkins and others, 1978; Buffler et al., 1979; Humphris, 1979).

3. A *diapir-override phase* begins when the shelf margin progrades over shale or salt domes formed during the initiation phase. This phase is represented by the Frio and younger trends in the Gulf of Mexico and post-Eocene trends in the Niger Delta (Fig. 6), but did not develop in the West Africa example because of insufficient progradation (Fig. 1).

The principal characteristic of the diapir-override phase is increasing structural complexity, caused by the interaction of regional horizontal extension and local uplifts. Previously formed diapirs tend to control the position of growth faults and thus cause more complex (strongly sinuous or arcuate) fault patterns in plan view than the evenly-spaced, subparallel patterns which typify the initiation phase. Crestal faults tend to develop over diapirs, and up-to-basin faults commonly develop on the landward side of shale ridges. Up-to-basin faults (Figs. 3F, 6) may be essentially symmetrical with down-to-basin faults ("back-to-back" faults of Evamy et al., 1978), or they may be the major structure-forming faults ("counter-regional" faults of Evamy et al., 1978; also see Spindler, 1977). As a general rule, the complexity and variety of shelf margin structural styles increases as progradation proceeds into the basin and increasingly mobile diapirs are overridden; a great variety of styles can be observed along modern unstable margins (Figs. 3, 4).

Local Scale

On a local scale, structural evolution can be reconstructed with a series of isopach maps if the appropriate paleobathymetric corrections are made. In the case of shallow water deposition where depositional relief is much less than the magnitude of contemporaneous structural relief, these corrections are not necessary for a first approximation (Fig. 7).

The first stage of local structural evolution, the compressional regime (Fig. 3), cannot usually be reconstructed because later deformation severely

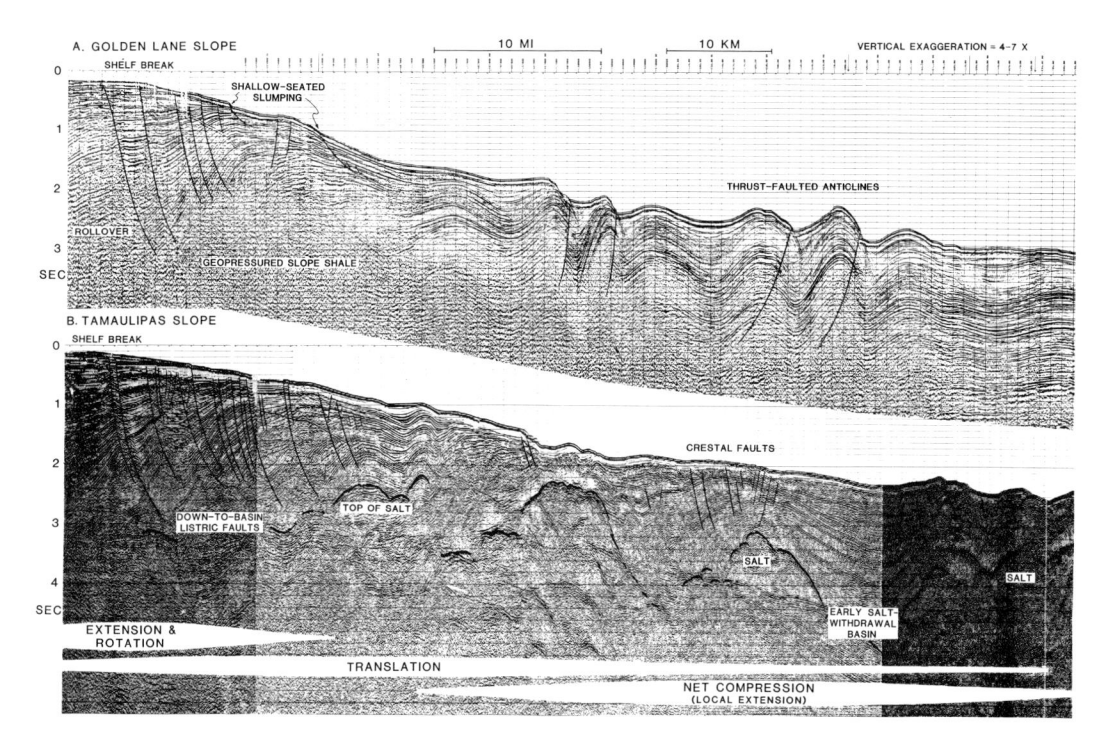

FIG. 4.—Seismic dip sections of the continental slope, western Gulf of Mexico (for original sections, see Watkins et al., 1976; Buffler et al., 1979). Sections illustrate origin of contemporaneous structures through deep-seated slumping of continental slope. Growth faults originate in tensional regime along top of slope; decollement is in undercompacted shale (A) or salt (B). Ridges cored by shale (A) or salt (B) originate in compressional regime along lower slope. These ridges may evolve into diapirs as shelf margin progrades over them. Most of the compression is probably accommodated by thrust faulting (which is rarely seen due to poor seismic penetration through salt and shale structures) rather than by folding. Location of section B is shown in Fig. 2. Approximate depth conversion: 1 sec = 2500 to 3500 ft.

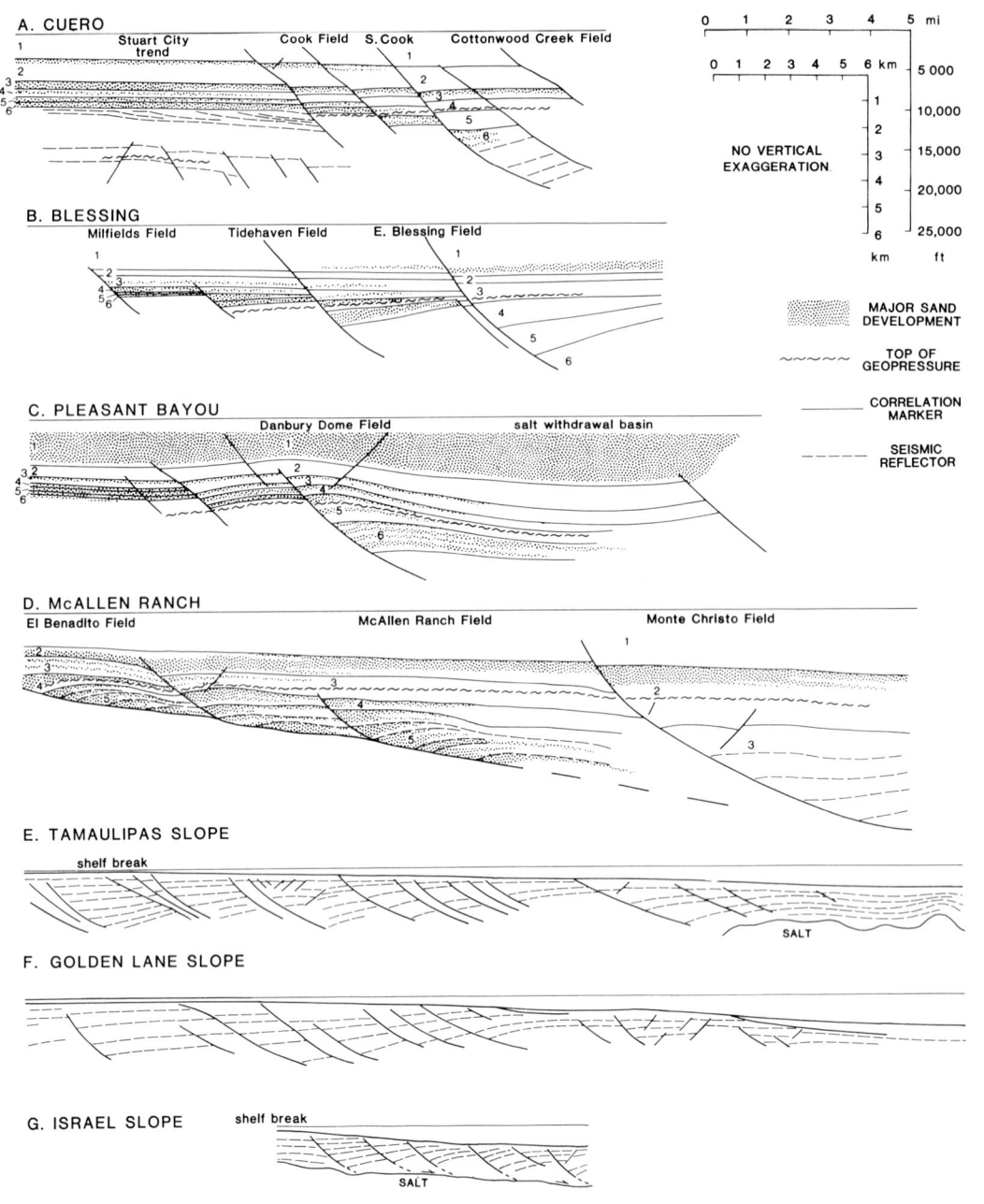

FIG. 5.—Dip sections of Tertiary growth-faulted shelf margin deltas (A–D) compared with Quaternary examples (E–G) at same scale, without vertical exaggeration. Locations of sections A–E are shown in Fig. 2. Areas A–D have been investigated as possible reservoirs for geopressured geothermal energy (Loucks, 1978; Bebout et al., 1978, 1979; Winker et al., 1981a, 1981b). Numbers in sections A–C correspond to isopach units in Fig. 7; highest numbered units are believed to represent deltaic sequences deposited closest to shelfedge; only these sequences are geopressured. In each case, geopressured sandstones are bounded landward by major growth faults, basinward by pinch outs, and above by transgressive shale wedges or sand-shale sequences. A, Transition from stable progradation, with clinoform stratification (Midway, immediately below unit 6) to initiation phase of unstable progradation (Lower Wilcox, unit 6). B, Fault spacing, rollover, and growth ratios for Frio trend are typically greater than for Wilcox trend. C, Diapir-override phase with late growth of Danbury salt dome. D, Shallow decollement is typical of Vicksburg trend in south Texas (Berg et al., 1979; Ashford, 1972). E–G, Analogous structures in modern upper-slope settings, interpreted from seismic sections. (E and F are based on section in Fig. 4; G is based on Garfunkel et al., 1979).

disturbs the early structures, because the lower-slope deposits are not usually penetrated by drilling, and because they are usually seismically transparent or absorbent. Consequently, the initiation of salt and shale structures is difficult to understand without studying modern examples (Lehner, 1969; Humphris, 1979; Buffler et al., 1979).

In contrast, the second stage, or extensional regime, is well known in the ancient record, and an extensive literature exists on the subject of growth faults (Table 2). The typical style is down-to-basin contemporaneous listric faults with "rollover" (reversal of dip from regional trend). In the absence of time-calibration of stratigraphy, relative rates of fault growth can be quantified by the *growth ratio* or *expansion index,* defined as the ratio of the downthrown thickness to upthrown thickness (Thorsen, 1963). With most growth faults, the largest growth ratios are observed in the deepest structures penetrated and decline steadily up-section (Fig. 7).

The final stage of local structural evolution is that of platform-style deposition, still within an extensional regime but a much weaker one. Most growth faults continue to move, but at a much-reduced rate (Fig. 7). This slower movement is still sufficient in many cases to create structural closure, however, even with simultaneous regional basinward tilting (Fig. 7A and B). Growth of salt structures (domes and withdrawal basins) is most apparent during this stage. In the case of salt tectonics, the structures formed during platform-style deposition may be entirely different from those formed near the shelf margin, as in the Frio trend of Brazoria County, Texas (Fig. 7C).

Shelf Margin Deltas

Attempts to develop models of deltaic sedimentation at the shelf margin are hampered by a lack of modern analogs, with the arguable exception of the modern lobe of the Mississippi Delta (Fig. 3A). Even this model does not explain many of the features observed in Tertiary deltaic sequences believed to be formed at the shelf margin. To a large extent we are dependent on studies of late Pleistocene deposition during low stands of sealevel (Leh-

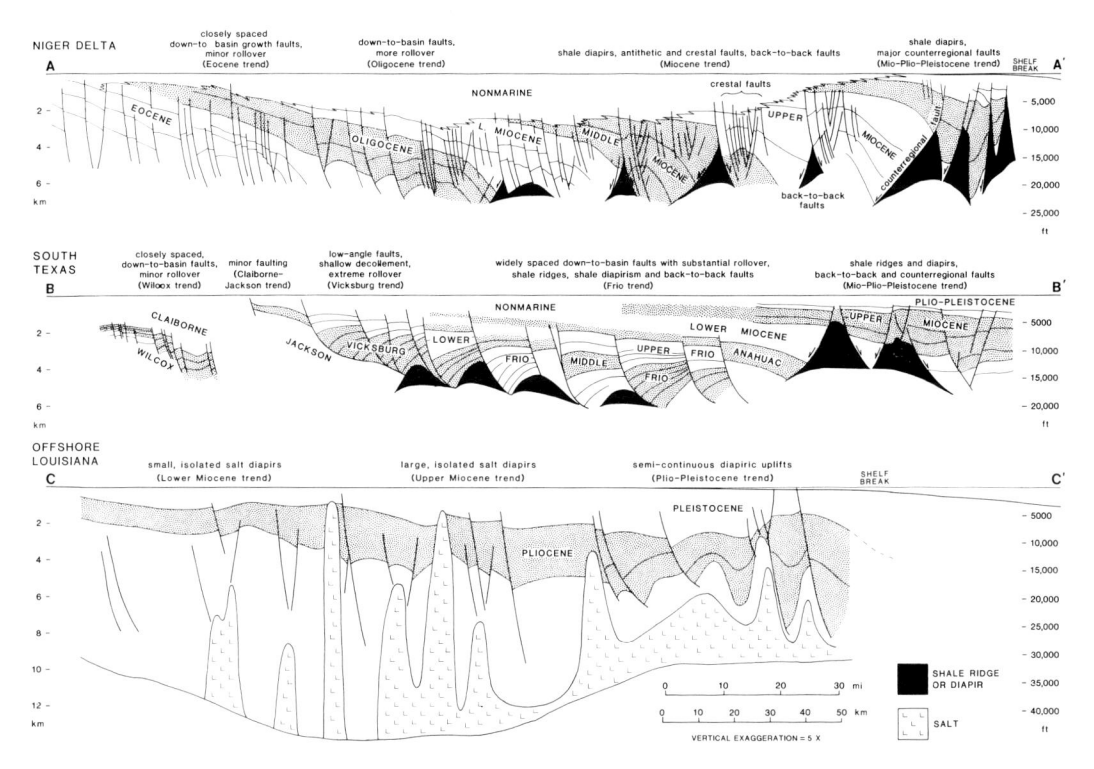

FIG. 6.—Generalized composite regional dip sections of northwestern Gulf of Mexico and Niger Delta, illustrating progressive changes in structural style associated with progradation of shelf margin (after Bebout et al., 1979; Evamy et al., 1978; Jones, 1975; Khan et al., 1975a; Woodbury et al., 1973). Locations of sections shown in Fig. 2. For each time-stratigraphic unit, shallow-water sediments form a basinward-thickening wedge, shelf margin position corresponds with maximum thickness, and slope sediments are mobilized into diapirs or "sheath" around salt domes. Closely spaced down-to-basin faults are typical of the initiation phase of growth faulting; contemporaneous shale ridges presumably formed approximately 50 mi downdip along the lower slope and were later overridden.

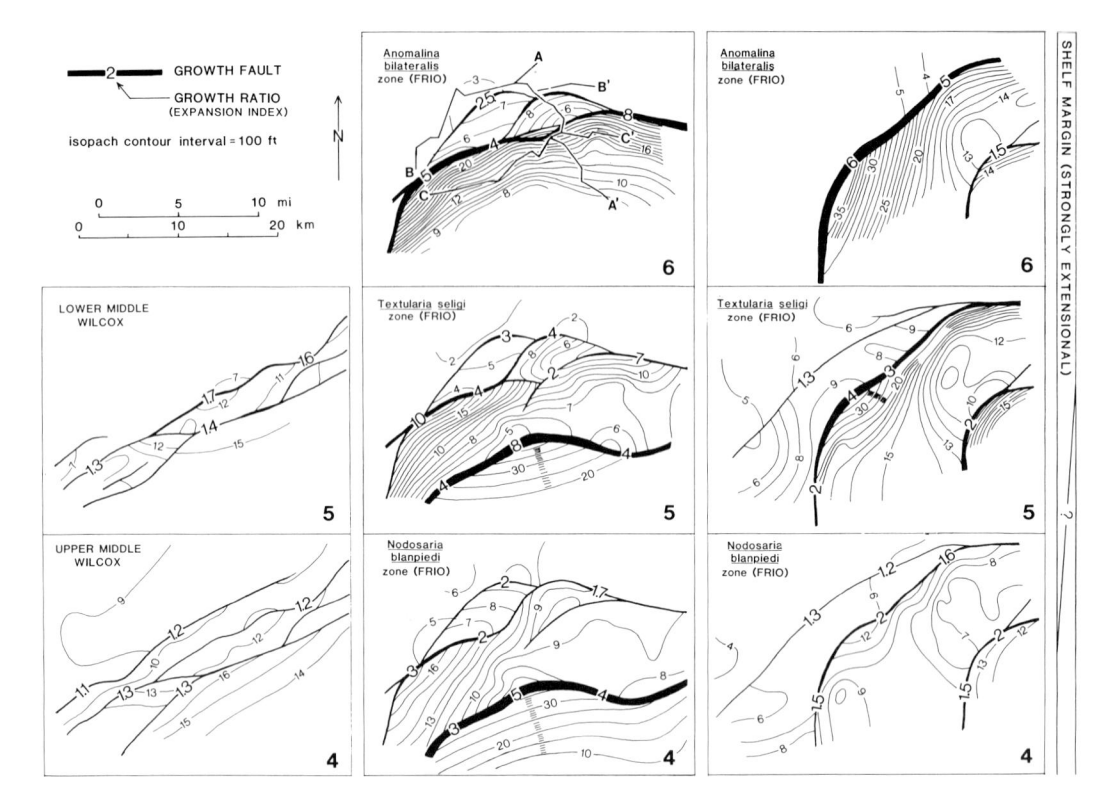

FIG. 7.—Structural evolution of three shelf margin deltaic sequences in Texas Gulf Coast Basin illustrated by successive isopach maps. Locations are shown in Fig. 2; dip sections of same areas are shown in Fig. 5; stratigraphic sections of Blessing area, unit 6, are shown in Fig. 8. Contour values × 100 ft. Shelf margin structures are characterized by maximum growth ratios and rollover; transition to shelf deposition is characterized by steady decline in growth ratios. Late stage is characterized by basinward tilting (A, B), or growth of salt domes and withdrawal basins (C). Note that in the case of salt tectonics, late contemporaneous structures may be entirely different from the early structures. Structural evolution is discussed in greater detail by Winker et al. (1981a, b).

ner, 1969; Berryhill, 1978, 1980; Sidner et al., 1978; Tatum, 1979; Berryhill and Trippet, 1980). Those studies have been based primarily on high-resolution seismic data, with sparse well and core control. As a result, they are useful for predicting the overall geometry of depositional cycles but less so for lithology and depositional environment.

The most distinctive feature common to deltas at unstable shelf margins is the rapid rate of subsidence, attributable to at least four effects: (1) flexural depression of the crust due to sedimentary loading on an elastic (or viscoelastic) lithosphere by the prograding depocenter (Walcott, 1972); maximum subsidence corresponds to the center of the applied load, which generally occurs at the upper slope or shelfbreak (Woodbury et al., 1973; Stuart and Caughey, 1977); (2) rapid extension at the shelf margin due to listric normal faulting, previously discussed, which causes a net regional thinning of the sedimentary section above the decollement zone; the resultant subsidence is concentrated on the downthrown side of faults; (3) salt withdrawal which may locally enhance subsidence; and (4) compaction of the thick section of recently deposited underlying sediments (an amplification mechanism).

Effects of rapid subsidence are most apparent in the geometry of progradational cycles. Along the modern shelf margin on the northwestern Gulf of Mexico, individual deltaic cycles tend to be thicker at the shelfbreak than at the inner shelf or mid-shelf (Fig. 3C). Clinoform stratification observed in high-resolution seismic lines is typically steeper for shelf margin progradation than for inner-shelf progradation (Winker, 1980).

In ancient sequences in the subsurface, clinoform stratification on this scale is not generally perceptible, either from log correction or on conventional CDP seismic sections. Progradation is, however, manifested as upward-coarsening cycles and funnel-shaped patterns on electric logs (Asquith, 1970); these cycles can be readily correlated within fault

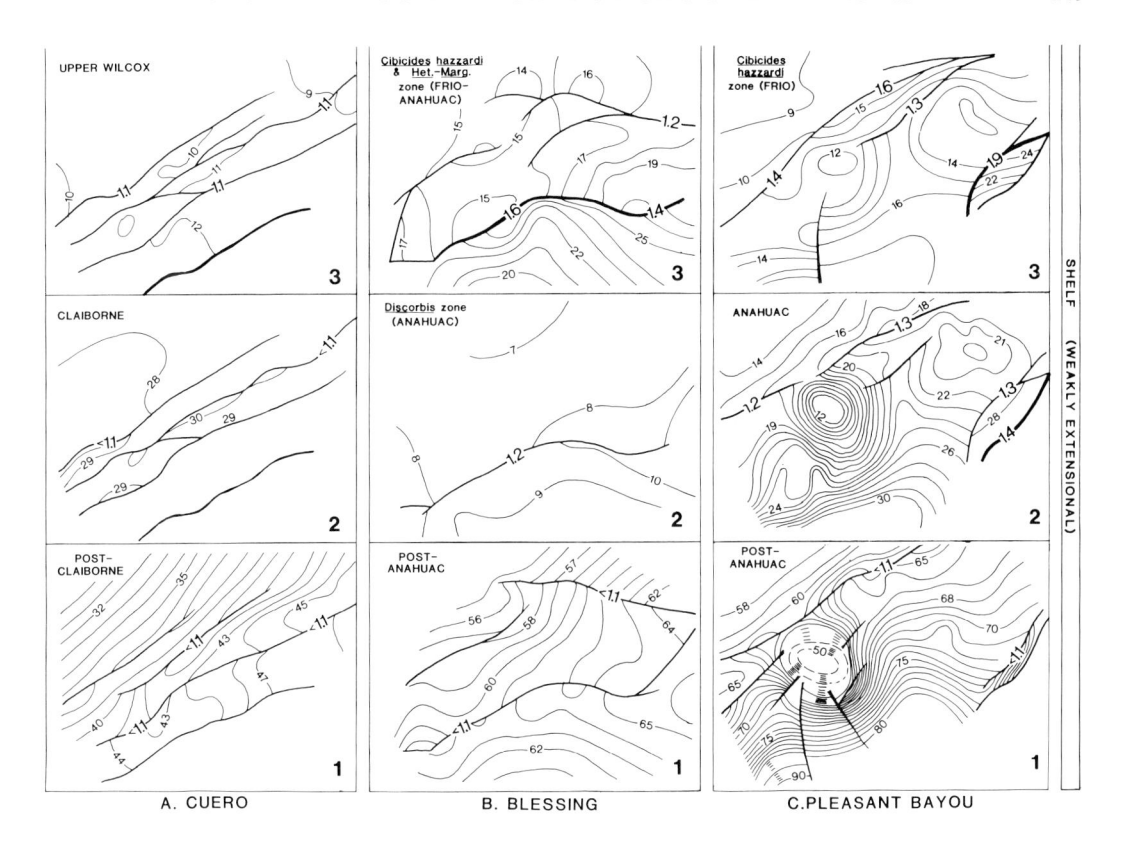

A. CUERO	B. BLESSING	C. PLEASANT BAYOU
UPPER WILCOX (3)	Cibicides hazzardi & Het.-Marg. zone (FRIO-ANAHUAC) (3)	Cibicides hazzardi zone (FRIO) (3)
CLAIBORNE (2)	Discorbis zone (ANAHUAC) (2)	ANAHUAC (2)
POST-CLAIBORNE (1)	POST-ANAHUAC (1)	POST-ANAHUAC (1)

SHELF (WEAKLY EXTENSIONAL)

blocks (Fig. 8). Such upward-coarsening shale-to-sandstone sequences, associated with growth faults and inferred to represent delta-front deposits (of shelf margin deltas by our interpretation) have been described from major stratigraphic units of the Gulf Coast Tertiary Basin: Paleocene Lower Wilcox Group (Fisher and McGowen, 1967), Eocene Upper Wilcox (Edwards, 1980, 1981), Oligocene Vicksburg Group (Han, 1981) and Frio Formation (Galloway and others, in press), Miocene (Curtis, 1970), and Pleistocene (Caughey, 1975a, b). In Tertiary sequences of the Gulf Coast Basin, substantial thickness variation of individual cycles is commonly observed, particularly in the downdip Frio trend (Fig. 8). Expansion of cycles across faults can be as great as 10:1, although more typically the growth ratio is less than 2:1.

Where a time-stratigraphic unit expands substantially across growth fault, its electric-log character typically changes markedly and may cause serious correlation problems. In the simplest style, the vertical sequence is essentially the same on both sides of the fault but is proportionally expanded on the downthrown side. In this case, the lateral continuity of individual sandstones is similar on both sides of the fault, but the thickest sandstones are on the downthrown side. A common complication of this pattern is the appearance of subcycles on the downthrown side and the breakup of individual thick sandstones into numerous thin sandstones at large amounts of expansion. In extreme cases, the electric-log character may be entirely different on the upthrown and downthrown sides (Fig. 8), with thin, laterally discontinuous sandstones on the upthrown side and thick, laterally continuous sandstones on the downthrown side. Our hypothesis for this last pattern is that at low subsidence rates, shifting channels (fluvial, distributary, and tidal) rework much of the section, reducing lateral continuity and even obscuring individual progradational cycles. At very high subsidence rates, channel reworking is much less important volumetrically.

Although the effects of subsidence rate on vertical sequence are readily apparent (Fig. 8), the extent to which contemporaneous structures directly control shallow water environments is less clear. Most Quaternary growth faults in the Gulf of Mexico do not have obvious bathymetric expression on the shelf (Figs. 3, 4) in contrast to the continental slope, which is characterized by bathymetric expression of contemporaneous structures (Martin and Bouma, 1978). As a general rule, sediment

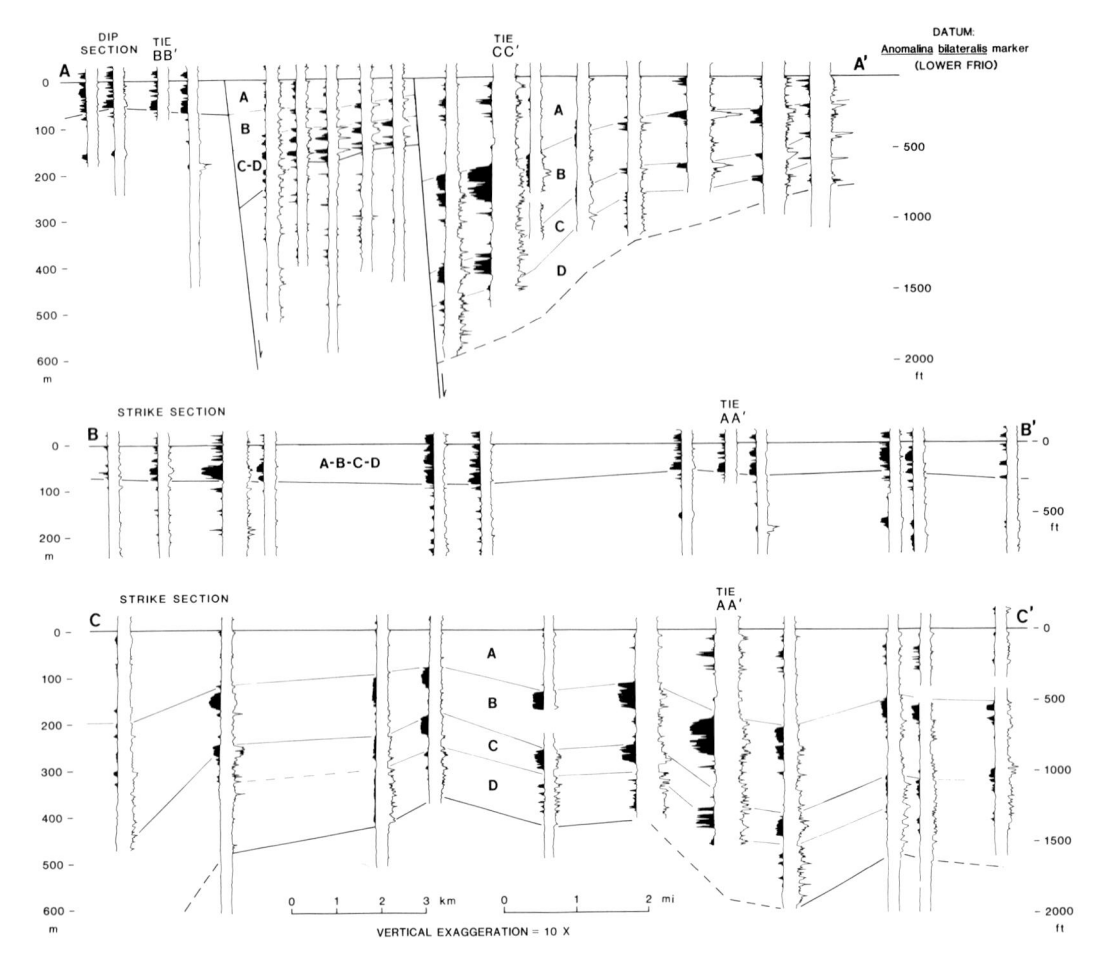

FIG. 8.—Stratigraphic sections of an Oligocene shelf margin delta, Blessing area, Matagorda County, Texas. Locations of sections are shown in Fig. 7B diagram 6; regional setting in Fig. 2. Differentiation of progradational cycles increases with subsidence rate, resulting in extreme changes of log character across growth faults. Where subsidence is slow, the sequence if probably dominated by channel reworking, resulting in lateral discontinuity of individual sandstones. Where subsidence is rapid, the sequence is dominated by alternating progradation and transgression, resulting in stacking of upward-coarsening cycles with better differentiation and greater continuity of individual sandstones. Compare with sand geometry in Fig. 3A, and progradational cycles in Fig. 3C.

supply to the shelf is sufficient to fill bathymetric irregularities as quickly as they are formed by structural growth. In ancient growth-faulted shelf-margin deltas, variations in subsidence rate appear to have influenced thickening of the progradational cycle and therefore caused lithologic variation related to preservation potential and reworking, but may not have been a major control on depositional environment.

On a regional scale the rapidly subsiding shelf margin acts as a major sediment trap permitting the accumulation of thousands of feet of shallow water deposits during a major regressive episode. With rapid slumping of the continental slope, translation of fault blocks is an important mechanism of mass transport into the deep basin, which could be expected to reduce relative importance of downslope sediment transport by other mechanisms such as

FIG. 9.—Sand distribution in plan view of an Eocene shelf margin delta, Texas coastal plain (after Edwards, 1980). Sandstone percentage (A) highlights deltaic lobes and provides prediction of down-dip pinch out. Net sandstone thickness (B) is strongly influenced by expansion of section across growth faults (C), resulting in patterns that would be difficult to interpret if faults were not recognized. Regional net-sandstone map shows an elongate strike-oriented high. Compare with Figs. 1C, 3A, and 8.

SLICK SAND
UPPER WILCOX

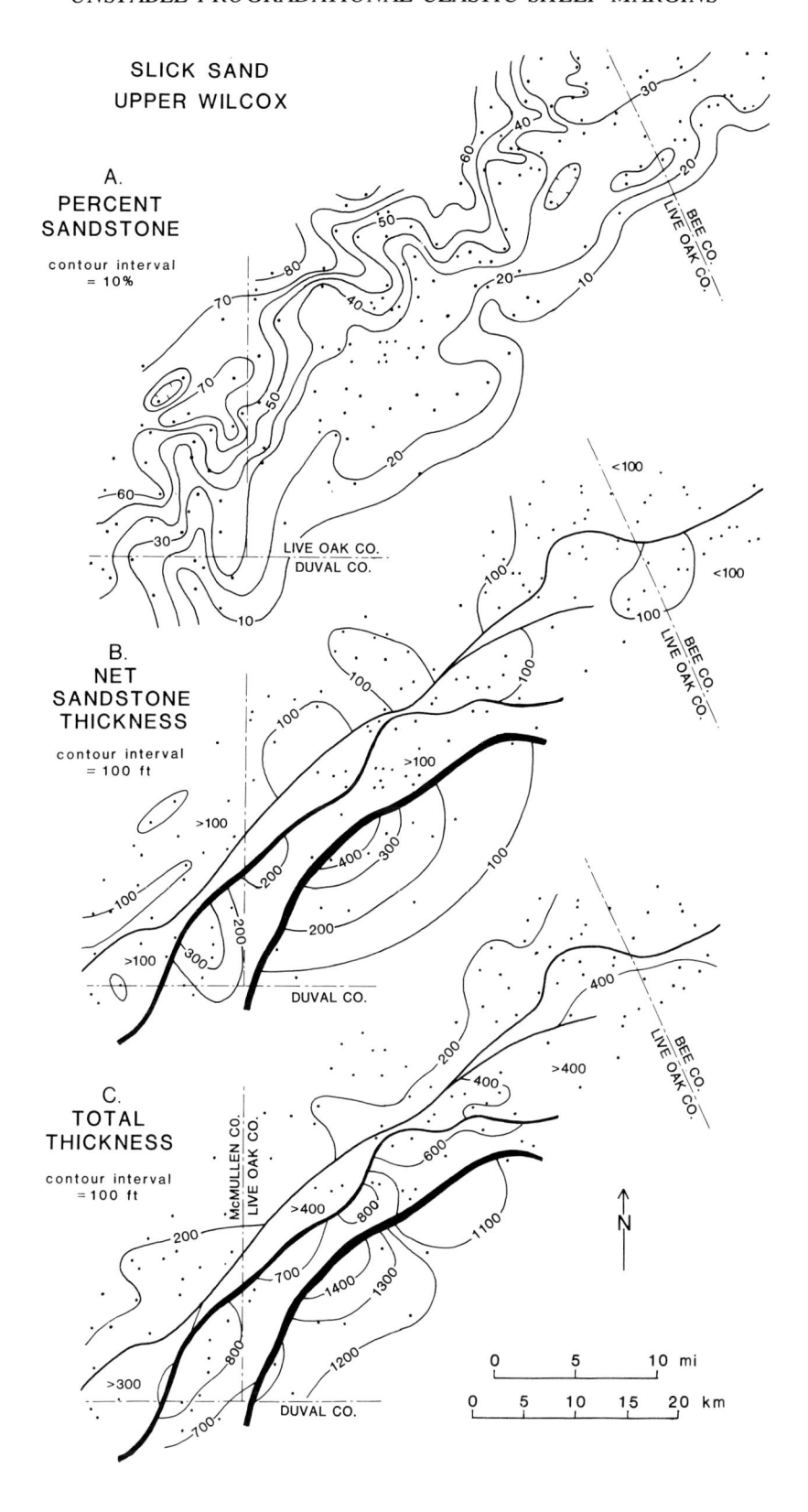

A. PERCENT SANDSTONE

contour interval = 10%

B. NET SANDSTONE THICKNESS

contour interval = 100 ft

C. TOTAL THICKNESS

contour interval = 100 ft

turbidity currents, mudflows, and shallow-seated slumps. Diapiric structures tend to create many small basins on the slope which trap sediment moving down the slope. As a result, the greatest sedimentation rates are generally found along the shelf margin, and regional isopach maps (Fig. 9C) for a particular time interval should depict the shelf margin as a strike-oriented depocenter or trough (Woodbury et al., 1973).

Maps of net sandstone thickness (Fig. 9B) for a particular time interval will show a similar pattern of regional strike-oriented maxima of sandstone thickness. To interpret depositional environments and to locate axes of fluvial sand input, maps of sand percentage (Fig. 9A) are more meaningful as they remove the element of differential subsidence and emphasize depositional control on lithofacies distribution.

GEOPRESSURE AND FLUID MIGRATION

In typical geopressured reservoirs in the Gulf Coast Basin (Fig. 5), sandstones are isolated on the landward side by large fault displacements (generated near the shelf margin) against older slope shales and on the basinward side by sandstone pinch-outs. The "cap" is typically a transgressive shale wedge or an interbedded sequence of sandstone and shale, and the "top" of geopressure is usually transitional over hundreds of feet (Harkins and Baugher, 1969; Fowler, 1970). This relationship with structural and stratigraphic geometry appears to hold true for a large number of Cenozoic fields regardless of age. Apparently, given the right stratigraphic and structural setting (characteristic of the shelf margin), geopressure can develop soon after deposition (Stuart, 1970), i.e., within a million years or so, and be maintained for at least 50 m.y.

Although generation of excess fluid pressure is a complex physical and chemical phenomenon, maintenance of excess pressure over extended periods of time is essentially a hydraulic problem and can be explained in terms of either a static model (perfect sealing) or a dynamic steady-state model (slow leakage with continuous replenishment of fluids). Recent studies on diagenesis of geopres-

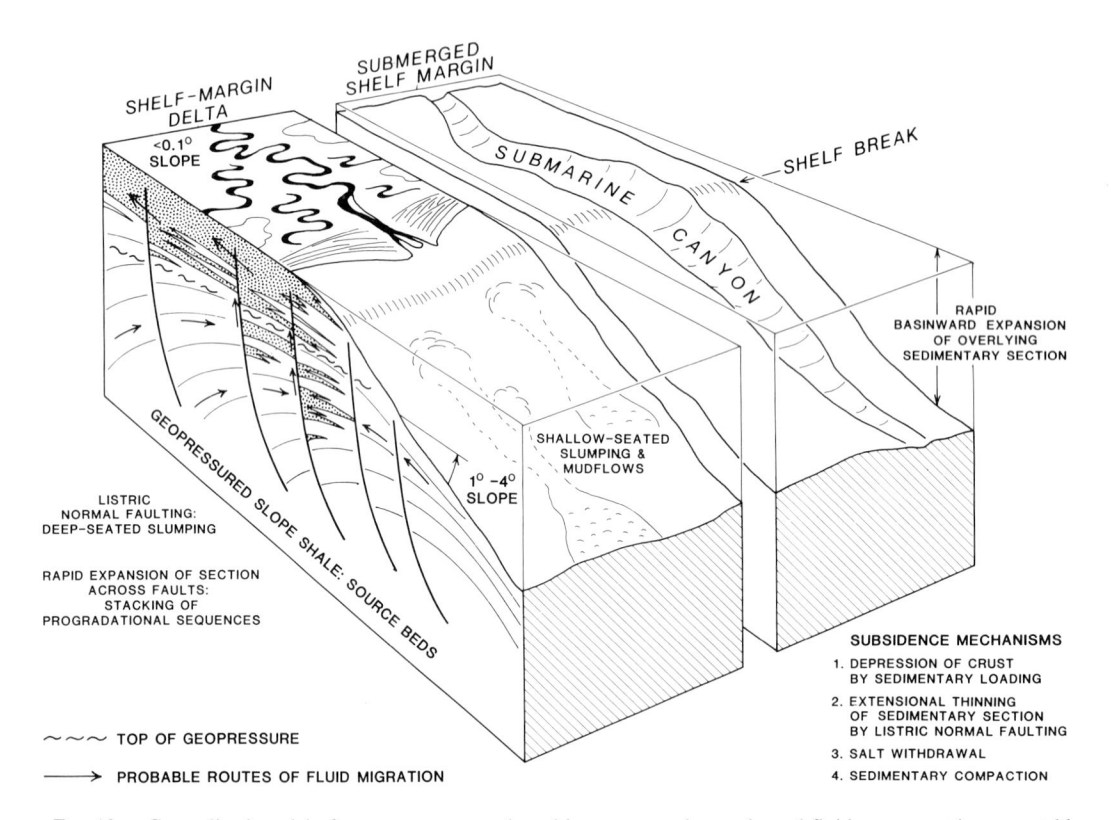

FIG. 10.—Generalized model of contemporaneous deposition, structural growth, and fluid movement in an unstable shelf margin system. Structural styles, depositional environments, and delta types are highly variable. Postulated fluid movement is based on a dynamic model of geopressure maintenance, with slow leakage up faults or across bedding planes; this model may be more useful than a static model for explaining hydrocarbon accumulation and diagenesis in deep geopressured sandstones.

sured shelf-margin sandstones in Texas (Loucks and others, 1977; Bebout et al., 1978; Milliken et al., 1981; Land and Milliken, 1981) suggest that large volumes of water must be flushed through these formations to account for the observed dissolution, replacement, and cementation (Boles, 1982; K. L. Milliken, personal commun., 1982). If we accept the hypothesis of early development of geopressure, the observed diagenesis would require a dynamic rather than a static model of geopressure maintenance. Distribution of hydrocarbons in the abnormally pressured Chocolate Bayou Field also appears to require a dynamic model (Fowler, 1970).

The most likely source for the required amount of water is the large volume of slope shale that underlies the shelf margin sandstones and is juxtaposed against them by faulting, but which is rarely penetrated by the drill and is therefore poorly described or understood. The probable role of slope shales in generating fluids is particularly significant in light of their high potential as source beds (Dow, 1978), in contrast to the generally poor potential of shallow water shale (Brown, 1979). In addition, slope shales should reach thermal maturity much sooner than would the younger shallow-water deposits in hydraulic continuity with them (Dow, 1978). From these considerations, we can expect that shelf margin systems are important not only as geopressured reservoirs, but also as part of the "plumbing" for hydrocarbons and diagenetic fluids (Fig. 10).

SUMMARY

1. Ancient shelf margins can be recognized and mapped on the basis of criteria such as timing of growth faulting, maximum rate of deposition, and top of geopressure, where data on sedimentary structures and microfaunal assemblages are not available, albeit with less precision.

2. Progradation of shelf margins can be used to explain the evolution of growth faults, diapirs, and related structures. As a general rule, the shelf margin represents a strongly extensional regime, and the lower slope a strongly compressional regime. The shelf margin structural style tends to increase in complexity as the shelf margin progrades into the basin, partly due to overriding of diapirs previously formed along the lower slope.

3. Unstable shelf margin deltas differ substantially in cross-sectional geometry from stable inner-shelf deltas, due primarily to differences in subsidence rate. Environments and facies tracts probably do not differ greatly between the two types of deltaic systems. Their stratigraphic expression, however, can be much different because high subsidence rates increase the differentiation of progradational cycles and reduce the relative importance of reworking by channels.

4. Rapid growth faulting along unstable shelf margins tends to hydraulically isolate shallow-water sandstones against dewatering slope shales, thus leading to the development of geopressured reservoirs.

ACKNOWLEDGMENTS

Work contributing to this paper was conducted as part of geopressured-geothermal research at the Bureau of Economic Geology under the direction of D. G. Bebout and R. A. Morton, with co-investigators M. M. Dodge, A. R. Gregory, R. G. Loucks, D. L. Richman, and B. R. Weise, with support from U.S. Department of Energy Grants DE-AS05-76ET 28461, DE-AC08-78ET 11397, DE-AC08-79ET 27111, and Gas Research Institute Grant 5011-321-0125. Principles of sedimentary tectonics and the significance of the shelf margin were introduced to the first author through lectures and unpublished work by R. O. Kehle. Further development of these concepts benefited greatly from discussions with M. M. Backus, M. E. Bentley, L. F. Brown, J. M. Coleman, W. E. Galloway, J. H. Han, K. L. Milliken, and L. Saugy. Research Associate L. A. Jirik and Research Assistants L. P. Chong, D. W. Downey, H. S. Hamlin, J. H. Han, J. L. Lawton, and R. A. Schatzinger contributed to the study of individual shelf margin systems. We also wish to thank R. T. Buffler of the Institute for Geophysics, University of Texas, and H. L. Berryhill, C. L. Holmes, and R. G. Martin, U.S. Geological Survey in Corpus Christi, for their assistance in making seismic data available to us, and the Institute for Geophysics for permission to publish two of their seismic sections. J. M. Coleman and R. T. Buffler provided valuable critical review of the manuscript.

REFERENCES

ALMAGOR, G., AND WISEMAN, G., 1977, The mechanism of submarine slumping in recent marine sediments on the continental slope off the southern coast of Israel: Marine Geotechnology, v. 2, p. 349–388.

ANTOINE, J. W., MARTIN, R. G., PYLE, T. G. AND BRYANT, W. R., 1974, Continental margins of the Gulf of Mexico, in Burk, C. A., and Drake, C. L., eds., The Geology of Continental Margins: New York, Springer-Verlag, p. 683–693.

ASHFORD, T., 1972, Geoseismic history and development of Rincon Field, South Texas: Geophys., v. 37, p. 797–812.

ASQUITH, D. O., 1970, Depositional topography and major marine environments, Late Cretaceous, Wyoming: Am. Assoc. Petroleum Geologists Bull., v. 54, p. 1184–1224.

AYOUT, M. K., 1980, Oil and gas prospects in Egypt, *in* Burollet, P. F., and Ziegler, V., eds., Energy Resources: Internat. Geol. Congr. 35, Colloq. C2, p. 231–241.
BEBOUT, D. G., LOUCKS, R. G., AND GREGORY, A. R., 1978, Frio sandstone reservoirs in the deep subsurface along the Texas Gulf Coast: their potential for production of geopressured geothermal energy: Austin, Univ. Texas, Bur. Econ. Geol. Rept. Invest. 91, 92 p.
———, WEISE, B. R., GREGORY, A. R. AND EDWARDS, M. B., 1979, Wilcox reservoirs in the deep subsurface along the Texas Gulf Coast: their potential for production of geopressured energy: Austin, Univ. Texas, Bur. Econ. Geol. Rept. to U.S. Dept. of Energy, 219 p.
BEN-AVRAHIM, Z., AND MART, Y., 1981, Late Tertiary structure and stratigraphy of North Sinai continental margin: Am. Assoc. Petroleum Geologists Bull., v. 65, p. 1135–1145.
BERG, R. R., 1981, Deep-water reservoir sandstones of the Texas Gulf Coast: Trans. Gulf Coast Assoc. Geol. Socs., v. 31, p. 31–40.
———, MARSHALL, W. D., AND SHOEMAKER, P. W., 1979, Structural and depositional history, McAllen Ranch Field, Hidalgo County, Texas: Trans. Gulf Coast Assoc. Geol. Socs., v. 29, p. 24–28.
BERRYHILL, H. L., 1978, South Texas continental shelf and continental slope: Late Pleistocene/Holocene evolution and sea-floor stability: U.S. Geol. Survey Open-File Rept. 78-514, 91 p.
———, 1980, Map showing paleogeography of the continental shelf during the low stand of sea level, Wisconsin glacial epoch, Port Isabel, 1° × 2° Quadrangle, Texas: U.S. Geol. Survey Map I-1254-E.
———, AND TRIPPET, A. R., 1980, Map showing structure of the continental terrace in the Port Isabel 1° × 2° Quadrangle, Texas: U.S. Geol. Survey Map I-1254-F.
BIEL, R., AND BUCK, G. H., 1978, Paleostructural mapping in bathyal environments: Trans. Gulf Coast Assoc. Geol. Socs., v. 28, p. 49–63.
BISHOP, R. S., 1977, Shale diapir emplacement in south Texas: Laward and Sheriff examples: Trans. Gulf Coast Assoc. Geol. Socs., v. 27, p. 20–31.
BOL, A. J., AND VAN HOORN, B., 1980, Structural styles in western Sabah offshore: Geol. Soc. Malaysia Bull., v. 12, p. 1–16.
BOLES, J. R., 1982, Active albitization of plagioclase, Gulf Coast Tertiary: Am. Jour. Sci., v. 282, p. 165–180.
BORNHAUSER, M., 1958, Gulf Coast tectonics: Am. Assoc. Petroleum Geologists Bull., v. 42, p. 339–370.
BROWN, S. W., 1979, Hydrocarbon source facies analysis, Department of Energy and General Crude Oil Company Pleasant Bayou No. 1 and No. 2 wells, Brazoria County, Texas, *in* Dorfman, M. H., and Fisher, W. L., eds., Proceedings, Fourth United States Gulf Coast Geopressured-Geothermal Energy Conference: Austin, Univ. Texas, Center for Energy Studies, p. 132–148.
BROWN, L. F., AND FISHER, W. L., 1977, Seismic-stratigraphic interpretation of depositional systems: examples from Brazilian rift and pull-apart basins, *in* Payton, C. E., ed., Seismic Stratigraphy—Applications to Hydrocarbon Exploration: Am. Assoc. Petroleum Geologists Memoir 26, p. 213–248.
BRUCE, C. H., 1973, Pressured shale and related sediment deformation: mechanism for development of regional contemporaneous faults: Am. Assoc. Petroleum Geologists Bull., v. 57, p. 878–886.
BUFFLER, R. T., SHAUB, F. J., WATKINS, J. S., AND WORZEL, J. L., 1979, Anatomy of the Mexican Ridges, southwestern Gulf of Mexico, *in* Watkins, J. S., Montadert, L., and Dickerson, P. W., eds., Geological and Geophysical Investigations of Continental Margins: Am. Assoc. Petroleum Geologists Memoir 29, p. 319–327.
BUSCH, D. A., 1973, Oligocene studies, northeast Mexico: Trans. Gulf Coast Assoc. Geol. Socs., v. 23, p. 136–145.
———, 1975, Influence of growth faulting on sedimentation and prospect evaluation: Amer. Assoc. Petroleum Geologists Bull., v. 59, p. 217–230.
CARVER, R. E., 1968, Differential compaction as a cause of regional contemporaneous faults: Am. Assoc. Petroleum Geologists Bull., v. 52, p. 414–419.
CAUGHEY, C. A., 1975a, Pleistocene depositional trends host valuable Gulf oil reserves. Part I: Oil & Gas Jour., Sept. 8, p. 90–94.
———, 1975b, Pleistocene depositional trends host valuable Gulf oil reserves. Part II: Oil & Gas Jour., Sept. 15, p. 240–242.
CHRISTINA, C. C., AND MARTIN, K. G., 1979, The Lower Tuscaloosa trend of south-central Louisiana: "You ain't seen nothing till you've seen the Tuscaloosa": Trans. Gulf Coast Assoc. Geol. Socs., v. 29, p. 37–41.
CLOOS, E., 1968, Experimental analysis of Gulf Coast fracture patterns: Am. Assoc. Petroleum Geologists Bull., v. 52, p. 420–444.
CRANS, W., MANDL, G. AND HAREMBOURE, J., 1980, On the theory of growth faulting: a geomechanical delta model based on gravity sliding: Jour. Petroleum Geology, v. 2, p. 256–307.
CURTIS, D. M., 1970, Miocene deltaic sedimentation, Louisiana Gulf Coast, *in* Morgan, J. P., ed., Deltaic Sedimentation, Modern and Ancient: Soc. Econ. Paleontologists Mineralogists Spec. Pub. 15, p. 293–308.
DAILLY, G. C., 1976, A possible mechanism relating progradation, growth faulting, clay diapirism and overthrusting in a regressive sequence of sediments: Canadian Petroleum Geol. Bull., v. 24, p. 92–116.
DICKINSON, G., 1953, Geological aspects of abnormal reservoir pressure in Gulf Coast Louisiana: Am. Assoc. Petroleum Geologists Bull., v. 37, p. 410–432.
DOW, W. G., 1978, Petroleum source beds in continental slopes and rises: Am. Assoc. Petroleum Geologists Bull., v. 62, p. 1584–1606.

EDWARDS, M. B., 1980, The Live-Oak Delta complex: an unstable shelf-edge delta in the deep Wilcox trend of south Texas: Trans. Gulf Coast Assoc. Geol. Socs. v. 30, p. 71–79.

———, 1981, Upper Wilcox Rosita Delta system of south Texas: growth-faulted shelf-edge deltas: Am. Assoc. Petroleum Geologists Bull., v. 65, p. 54–73.

EVAMY, B. D., HAREMBOURE, J., KAMERLING, P. KNAPP, W. A., MOLLOY, F. A., AND ROWLANDS, P. H., 1978, Hydrocarbon habitat of Tertiary Niger Delta: Am. Assoc. Petroleum Geologists Bull., v. 62, p. 1–39.

FISHER, W. L., AND MCGOWEN, J. H., 1967, Depositional systems in the Wilcox Group of Texas and their relationship to occurrence of oil and gas: Trans. Gulf Coast Assoc. Geol. Socs., v. 17, p. 105–125.

FOSS, D. C., 1979, Depositional environment of Woodbine sandstones, Polk County, Texas: Trans. Gulf Coast Assoc. Geol. Socs., v. 29, p. 83–94.

FOWLER, W. A., 1970, Pressures, hydrocarbon accumulation, and salinities—Chocolate Bayou Field, Brazoria County, Texas: Jour. Petroleum Tech., v. 22, p. 411–423.

FRIEDMAN, G. M., AND SANDERS, J. E., 1978, Principles of Sedimentology: New York, John Wiley & Sons, 792 p.

GALLOWAY, W. E., HOBDAY, D. K., AND MAGARA, K., 1983, Frio Formation of the Texas Gulf Coast Basin—depositional systems, structural framework, and hydrocarbon origin, migration, distribution, and exploration potential: Austin, Univ. Texas, Bur. Econ. Geol. Rept. Invest., in press.

GARFUNKEL, A., ARAD, A., AND ALMAGOR, G., 1979, The Palmahim Disturbance and its regional setting: Geol. Survey Israel Bull. 72, 56 p.

GIRARD, O. W., 1979, Petroleum geology of the Niger Delta, in Report on the Petroleum Resources of the Federal Republic of Nigeria: U.S. Dept. Energy, p. B1–B29.

GRALKA, D., YORSTON, H. J., WIDMIER, J. M., AND WEISSER, G. H., 1980, Prediction of reservoir sand in immature areas from seismic stratigraphic analysis of interval velocity, in The Sedimentation of the North Sea Reservoir Rocks, a Symposium in Geilo, Norway: Oslo, Norwegian Petroleum Soc., p. V 1–9.

GREGORY, J. L., 1966, A lower Oligocene delta in the subsurface of southeastern Texas: Trans. Gulf Coast Assoc. Geol. Socs., v. 16, p. 227–241.

HAN, J. H., 1981, Genetic stratigraphy and associated growth structures of the Vicksburg Formation, South Texas [Ph.D. Diss.]: Austin, Univ. Texas, 162 p.

HARDIN, F. R., AND HARDIN, G. C., 1961, Contemporaneous normal faults of Gulf Coast and their relation to flexures: Am. Assoc. Petroleum Geologists Bull., v. 45, p. 238–248.

HARKINS, K. L., AND BAUGHER, J. W., 1969, Geological significance of abnormal formation pressures: Jour. Petroleum Tech., v. 21, p. 961–966.

HICKEY, H. N., AND OTHERS, 1972, Tectonic map of the Gulf Coast Region, U.S.A.: Gulf Coast Assoc. Geol. Socs. and Am. Assoc. Petroleum Geologists.

HOYT, W. V., 1959, Erosional channel in the middle Wilcox near Yoakum, Lavaca County, Texas: Trans.. Gulf Coast Assoc. Geol. Socs., v. 9, p. 41–50.

HUFF, K. F., 1980, Frontiers of World Exploration, in Miall, A. D., ed., Facts and Principles of World Petroleum Occurrence: Canada Soc. Petroleum Geologists Memoir 6, p. 343–362.

HUMPHRIS, C. C., 1979, Salt movement on continental slope, northern Gulf of Mexico: Am. Assoc. Petroleum Geologists Bull., v. 63, p. 782–798.

JONES, P. H., 1975, Geothermal and hydrocarbon regime, northern Gulf of Mexico Basin, in Dorfman, M. H., and Deller, R. W., eds., First Geopressured-Geothermal Energy Conference, Proceedings: p. 15–89.

KEHLE, R. O., 1970, Analysis of gravity sliding and orogenic translation: Geol. Soc. America Bull., v. 81, p. 1641–1664.

KHAN, A. S., LATTA, L. A., OAKES, R. L., PERT, D. N., SWEET, W. E., SMITH, N., AND WALL, E. J., 1975a, Geological and operational summary, Continental Offshore Stratigraphic Test (COST) No. 1, South Padre Island East Addition, Offshore Texas: U.S. Geol. Survey Open-File Rept. 75–174.

———, ———, ———, ———, ———, AND ———, 1975b, Geological and operational summary, Continental Offshore Stratigraphic Test (COST) No. 2., Mustang Island, offshore south Texas: U.S. Geol. Survey Open-File Rept. 75-259, 32 p.

LAND, L. S., AND MILLIKEN, K. L., 1981, Feldspar diagenesis in the Frio Formation, Brazoria County, Texas Gulf Coast: Geology, v. 9, p. 314–318.

LANE, F. H., AND JACKSON, K. S., 1980, Controls on occurrence of oil and gas in the Beaufort-MacKenzie Basin, in Miall, A. D., ed., Facts and Principles of World Petroleum Occurrence: Canada Soc. Petroleum Geologists Memoir 6., p. 489–507.

LEHNER, P., 1969, Salt tectonics and Pleistocene stratigraphy on continental slope of northern Gulf of Mexico: Am. Assoc. Petroleum Geologists Bull., v. 53, p. 2431–2479.

LOUCKS, R. G., 1978, Sandstone distribution and potential for geopressured geothermal energy production in the Vicksburg Formation along the Texas Gulf Coast: Trans. Gulf Coast Assoc. Geol. Socs., v. 28, p. 239–271.

———, BEBOUT, D. G., AND GALLOWAY, W. E., 1977, Relationship of porosity formation and preservation to sandstone consolidation history—Gulf Coast Lower Tertiary Frio Formation: Austin, Univ. Texas, Bur. Econ. Geol. Circ. 77-5, 120 p.

MARTIN, R. G., 1978, Northern and eastern Gulf of Mexico continental margin: stratigraphic and structural framework, in Bouma, A. H., Moore, G. T., and Coleman, J. M., eds., Framework, Facies, and Oil-trapping Characteristics of the Upper Continental Margin: Am. Assoc. Petroleum Geologists, Studies in Geology No. 7, p. 21–42.

———, 1980, Distribution of salt structures in the Gulf of Mexico: map and descriptive text: U.S. Geol. Survey Map MF-1213.

———, AND BOUMA, A. H., 1978, Physiography of Gulf of Mexico, in Bouma, A. H., Moore, G. T., and Coleman, J. M., eds., Framework, Facies, and Oil-trapping Characteristics of the Upper Continental Margin: Am. Assoc. Petroleum Geologists, Studies in Geology No. 7, p. 3–19.

MILLIKEN, K. L., LAND, L. S., AND LOUCKS, R. G., 1981, History of burial diagenesis determined from isotopic geochemistry, Frio Formation, Brazoria County, Texas: Am. Assoc. Petroleum Geologists Bull., v. 65, p. 1397–1413.

MITCHUM, R. M., AND VAIL, P. R., 1977, Seismic stratigraphy and global changes of sea level, part 7: Seismic stratigraphic interpretation procedure, in Payton, C. E., ed., Seismic Stratigraphy—Applications to Hydrocarbon Exploration: Am. Assoc. Petroleum Geologists Memoir 26, p. 135–143.

OCAMB, R. D., 1961, Growth faults of south Louisiana: Trans. Gulf Coast Assoc. Geol. Socs., v. 11, p. 139–174.

PAINE, W. R., 1968, Stratigraphy and sedimentation of subsurface Hackberry wedge and associated beds of southwestern Louisiana: Am. Assoc. Petroleum Geologists Bull., v. 52, p. 322–342.

POAG, C. W., AND VALENTINE, P. C., 1976, Biostratigraphy and ecostratigraphy of the Pleistocene basin, Texas-Louisiana continental shelf: Trans. Gulf Coast Assoc. Geol. Socs., v. 26, p. 185–256.

PRIOR, D. B., AND SUHAYDA, J. N., 1979, Application of infinite slope analysis to subaqueous sediment instability, Mississippi Delta: Engineering Geology, v. 14, p. 1–10.

QUARLES, M., 1953, Salt-ridge hypothesis on origin of Texas Gulf coast type of faulting: Am. Assoc. Petroleum Geologists Bull., v. 37, p. 489–508.

RETTGER, R. E., 1935, Experiments in soft-rock deformation: Am. Assoc. Petroleum Geologists Bull., v. 19, p. 271–292.

SCHERER, F. C., 1980, Exploration in east Malaysia over the past decade, in Halbouty, M. T., ed., Giant Oil and Gas Fields of the Decade 1968–1978: Am. Assoc. Petroleum Geologists Memoir 30, p. 423–440.

SIDNER, B. R., GARTNER, S., AND BRYANT, W. R., 1978, Late Pleistocene geologic history of Texas outer shelf and upper continental slope, in Bouma, A. H., Moore, G. T., and Coleman, J. M., eds., Framework, Facies, and Oil-trapping Characteristics of the Upper Continental Margin: Am. Assoc. Petroleum Geologists, Studies in Geology No. 7, p. 243–266.

SPINDLER, W. M., 1977, Structure and stratigraphy of a small Plio-Pleistocene depocenter, Louisiana continental shelf: Trans. Gulf Coast Assoc. Geol. Socs., v. 27, p. 180–196.

STUART, C. A., 1970, Geopressures: New Orleans, Shell Oil Co., 121 p.

STUART, C. J., AND CAUGHEY, C. A., 1977, Seismic facies and sedimentology of terrigenous Pleistocene deposits in northwest and central Gulf of Mexico, in Payton, C. E., eds., Seismic Stratigraphy—Applications to Hydrocarbon Exploration: Am. Assoc. Petroleum Geologists Memoir 26, p. 249–275.

STUDE, G. R., 1978, Depositional environments of the Gulf of Mexico South Timbalier Block 54 salt dome and salt dome growth models: Trans. Gulf Coast Assoc. Geol. Socs., v. 28, p. 627–646.

TATUM, T. E., 1979, Shallow geologic features of the upper continental slope, northwestern Gulf of Mexico: Texas A & M Univ., Dept. Oceanogr. Tech. Rept. 79-2-T, 60 p.

THORSEN, C. E., 1963, Age of growth faulting in southwest Louisiana: Trans. Gulf Coast Assoc. Geol. Socs., v. 13, p. 103–110.

TODD, R. G., AND MITCHUM, R. M., 1977, Seismic stratigraphy and global changes of sea level, part 8; Identification of upper Triassic, Jurassic, and lower Cretaceous seismic sequences in Gulf of Mexico and offshore West Africa, in Payton, C. E., ed., Seismic Stratigraphy—Applications to Hydrocarbon Exploration: Am. Assoc. Petroleum Geologists Memoir 26, p. 145–163.

WALCOTT, R. I., 1972, Gravity, flexure, and the growth of sedimentary basins at a continental edge: Geol. Soc. America Bull., v. 83, p. 1845–1848.

WATKINS, D. J., AND KRAFT, L. M., 1978, Stability of continental shelf and slope off Louisiana and Texas: geotechnical aspects, in Bouma, A. H., Moore, G. T., and Coleman, J. M., eds., Framework, Facies, and Oil-trapping Characteristics of the Upper Continental Margin: Am. Assoc. Petroleum Geologists, Studies in Geology No. 7, p. 267–286.

WATKINS, J. S., LADD, J. W., BUFFLER, R. T., SHAUB, F. J., HOUSTON, M. H., AND WORZEL, J. L., 1978, Occurrence and evolution of salt in deep Gulf of Mexico, in Bouma, A. H., Moore, G. T., and Coleman, J. M., eds., Framework, Facies, and Oil-trapping Characteristics of the Upper Continental Margin: Am. Assoc. Petroleum Geologists, Studies in Geology No. 7, p. 43–65.

———, ———, SHAUB, F. J., BUFFLER, R. T., AND WORZEL, J. L., 1976, Seismic section WG-3, Tamaulipas Shelf to Campeche Scarp, Gulf of Mexico: Am. Assoc. Petroleum Geologists, Seismic Section No. 1, 1 p.

WEAVER, P., 1955, Gulf of Mexico, in Poldervaart, A., ed., Crust of the Earth: Geol. Soc. America Spec. Paper 62, p. 269–278.

WEBER, K. J., 1971, Sedimentological aspects of oil fields in the Niger Delta: Geologie en Mijnbouw, v. 50, p. 559–576.

WILHELM, O., AND EWING, M., 1972, Geology and history of the Gulf of Mexico: Geol. Soc. America Bull., v. 83, p. 575–600.

WINKER, C. D., 1980, Depositional phases in late Pleistocene cyclic sedimentation, Texas coastal plain and shelf, with some Eocene analogs, in Perkins, B. F., and Hobday, D. K., eds., Middle Eocene Coastal Plain and

Nearshore Deposits of East Texas: Soc. Econ. Paleontologists Mineralogists Gulf Coast Section, Field Trip Guidebook, p. 46–66.

———, MORTON, R. A., EWING, T. E., AND GARCIA, D. D., 1981, Depositional setting, structural style, and sandstone development in three geopressured-geothermal areas, Texas Gulf Coast: Austin, Univ. Texas, Bur. Econ. Geol. Rept. to U.S. Dept. Energy, 46 p.

———, ———, AND GARCIA, D. G., 1981, Structural evolution of three geopressured-geothermal areas in the Texas Gulf Coast, *in* Bebout, D. G., and Bachman, A. L., eds., Proceedings of the 5th Conference on Geopressured-geothermal Energy, U.S. Gulf Coast: Baton Rouge, Louisiana State Univ., p. 59–65.

WOODBURY, H. O., MURRAY, I. B., PICKFORD, P. J., AND AKERS, W. H., 1973, Pliocene and Pleistocene depocenters, outer continental shelf, Louisiana and Texas: Am. Assoc. Petroleum Geologists Bull., v. 57, p. 2428–2439.

———, SPOTTS, J. H., AND AKERS, W. H., 1978, Gulf of Mexico continental slope sediments and sedimentation, *in* Bouma, A. H., Moore, G. T., and Coleman, J. M., eds., Framework, Facies, and Oil-trapping Characteristics of the Upper Continental Margin: Am. Assoc. Petroleum Geologists, Studies in Geology No. 7, p. 117–137.

TOPOGRAPHY AND SEDIMENTARY PROCESSES IN AN EPICONTINENTAL SEA

DONALD L. WOODROW
Department of Geoscience, Hobart and William Smith Colleges, Geneva, New York 14456

ABSTRACT

The Devonian Catskill Sea of the Appalachian region was stratified much of the time as recorded by the deposition in it of black shale and evaporites. Submarine topography which developed along the southeastern perimeter of the sea consisted of a basin margin and a gently sloping clinoform constructed by sediment progradation. Intersection of the surface of stratification (pycnocline) with the seafloor marked the basin margin-clinoform junction and it caused a separation of sedimentary processes. Shoreward of the intersection, on the basin margin, bottom flow driven by various current-producing phenomena, transported, deposited and reworked sediment. A complex mosaic of facies resulted, including a typical suite of lenticular sandstone, bioturbated shales and mudstones, most of which are abundantly fossiliferous. Basinward of the intersection, on the clinoform, density currents originating at or shoreward of the intersection moved down the clinoform onto the basin floor. Relatively simple facies accumulated, including turbidites and other, pelagic sediments, with very few fossils. At the intersection, internal waves moved along the pycnocline, shoaled, broke and reworked sediment, leaving a record of relatively thick sandstones, some with hummocky cross-bedding. This separation of process is likely to have been a hallmark of sedimentation in epicontinental seas where stratification can be inferred.

INTRODUCTION

Most of the papers in this volume are concerned with sedimentary processes active at the shelfedge as that bathymetric feature is understood in the modern world ocean. The shelfedge is defined by a sharp increase in seafloor gradient even though the exact profile may vary greatly from place to place (Mitchum et al., 1977; Bouma, 1979). In epicontinental seas the configuration of the basin perimeter apparently was unlike that in the modern world ocean (Heckel, 1972). Thirty-two years ago, John Rich (1951) offered an interpretation of the bathymetric profile expected on the floor of an epicontinental sea (Fig. 1). Rich indicated that the break in seafloor gradient of the profile reflected wave-base just as he felt the shelfedge does in modern oceans. Rich's terminology gained limited acceptance (Asquith, 1970, 1974; Gutschick and others, 1976; Harrington and Hazlewood, 1962; Van Siclen, 1958), but recently Mitchum et al. (1977) have revived part of it without reference to process.

In this paper I examine the bottom topography of and sedimentary processes active along the perimeter of the epicontinental sea which was the locale of the Catskill Delta. The model derived here may have broader applications in like situations elsewhere.

CATSKILL SEA AND ITS BOTTOM TOPOGRAPHY

In middle and Late Devonian time various eastern and central parts of the North American continent were flooded to varying degrees by an extension of the large ocean referred to as the Catskill Sea (Woodrow and Isley, 1983). Reconstructions of the Devonian Old Red Continent (Fig. 2) illustrate this (Woodrow et al., 1973; Oliver, 1977; House, 1975; Dineley, 1975; Heckel and Witzke, 1979). The Catskill Sea, with its extension north into what is now Hudson Bay, was broad, shallow and lay near the equator. Connections to the world ocean to the west maintained sufficient exchange to ensure normal marine water, at least at the surface, over most of the basin as a normal circumstance. Fresh water influx was mainly from the south and east as indicated by the presence there of the Catskill Delta. Stream flow into the sea from the north apparently was minimal since in that area clastics are thin and carbonates predominate. The coast appears to have been microtidal (Walker and Harms, 1975; Woodrow and Isley, 1983), although Rahmanian (1979) presents evidence for a greater tide range at some locations. Bounding the sea on the east was a tectonic peninsula of the Old Red Continent. On the west were scattered islands and to the north were lowlands on the western part of the continent.

Over most of the Catskill Sea floor, topography was subdued, reflecting local tectonic features (Ettensohn and Barron, 1981) but configuration of the eastern and southeastern edge reflected the interplay of sediment influx, sedimentary processes and subsidence (Fig. 3). There, sand and mud (Walker, 1971; Walker and Harms, 1971) were introduced episodically (Woodrow et al., 1973) by relatively small streams whose locations apparently varied little with time (Sevon et al., 1978; Sevon, 1979).

Sediments were deposited as a variety of facies on the basin margin, a nearshore surface of low gradient (Fig. 1). Seaward of the basin margin was the clinoform, the gradient of which was slightly greater and on which turbidites were deposited.

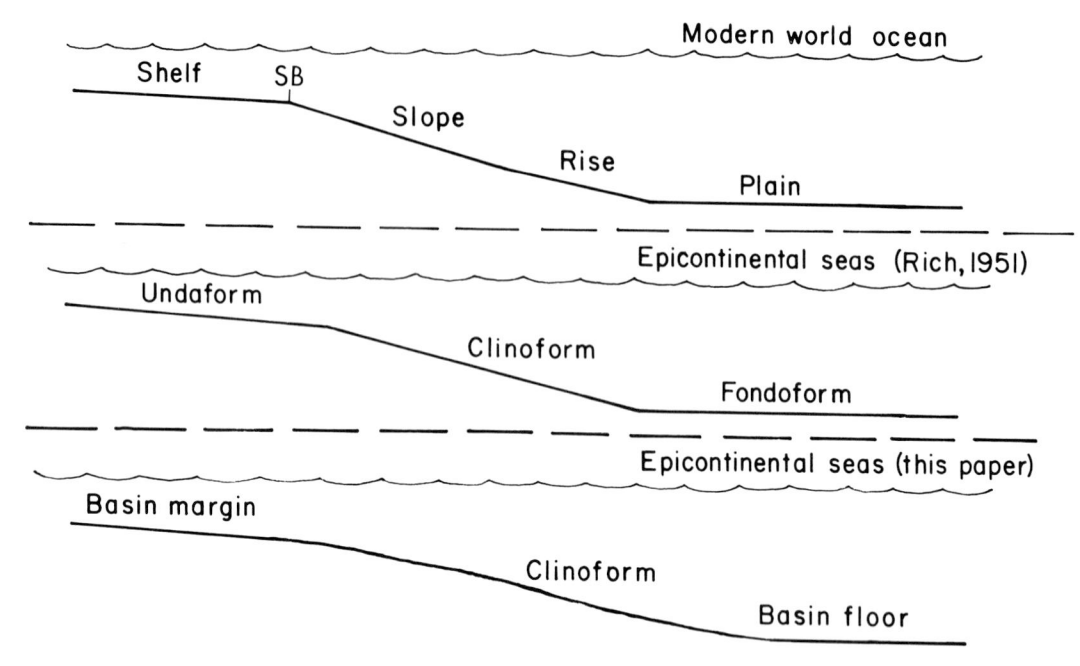

FIG. 1.—Terminology applicable to the topography of the ocean floor, modern and ancient; SB = shelfbreak.

Extending from the base of the clinoform was the basin floor, a nearly flat surface on which were deposited black shales. The basin floor does not concern us further in this paper.

The basin margin and the clinoform developed to varying degrees along the front of the delta but everywhere the clinoform lacks the features of steep-gradient oceanic slopes. There are no slump scars, deep channels or canyons, coarse-grained turbidites, channel-fill or levee facies, debris flows or strata formed as the result of wholesale slumping of sediment. Instead, the clinoform facies are made up of either thin-bedded, fine- to medium-grained turbidites with shale interbeds, or what Walker (1971) referred to as "slope shales."

STRATIFICATION IN THE CATSKILL SEA: IMPLICATIONS
FOR SEDIMENTARY PROCESSES

It is a premise of this paper that a well-defined surface of stratification (pycnocline) existed in the Catskill Sea throughout most of Late Devonian time (Fig. 3). Thick, black shales in western New York, Ohio and elsewhere in the eastern mid-continent attest to the existence of anoxic bottom waters in the sea (Byers, 1977; Ettensohn and Barron, 1981). This requires the establishment of a pycnocline, a surface developed in the water column between waters of differing density. In the modern ocean and in large lakes, stratification has pronounced effects both on the vertical distribution of life and on the transmission of mechanical energy introduced at or near the surface to the rest of the water column and to the sea or lake floor (Hutchinson, 1957, p. 426–430; Byers, 1977; Knauss, 1978, p. 65–67; Ettensohn and Barron, 1981). With stratification established, the bulk of the water column has a greatly reduced oxygen level, the extreme result of which is the establishment of anoxic bottom conditions. Anoxic waters impose an environmental limit to most benthonic forms forcing a segregation into distinct populations: those able to cope with the oxygen-depleted environment below the pycnocline and those needing the oxygen-rich environment above it (Bowen et al., 1974; Byers, 1977).

Existence of the pycnocline also provides a clear separation of sedimentary process. Above it the water is subject to wind stress and other current-generating phenomena like those active on the shelves in the modern world ocean (Swift, 1976). Below the pycnocline are waters of greater density which are little agitated and not subject to the deforming stresses typical of the shallows. Thus, the effects of bottom currents, generated by wind, for example, would be felt only in the surface waters above the pycnocline (for discussion of theory, see Knauss, 1978, p. 65–67). These and other currents generated in the shallows would be sufficient to move and rework sediments deposited shoreward of the intersection of the pycnocline with the seafloor. That intersection would be marked by a sandy sequence from which muds have been winnowed by the combined effects of shoaling internal waves and deep-reaching storm waves (Woodrow and Isley, 1983). Below the pycnocline only density currents would be effective. They would deposit turbidites derived from a linear source (Gorsline, 1978), that is, the basin margin and the pycnocline-seafloor intersection. Thus, two process zones are defined. One is above the pycnocline in the shallows where many current-generating processes are active. The second exists below the pycnocline where only density currents intrude. Between the two is the sandy facies marking the process boundary formed at the pycnocline-seafloor junction where the effect of internal waves is focused. The result is a complicated mosaic of shallow waters facies and a simple deeper water facies of turbidites and pelagic sediments with a distinctive set of sandstones between.

PROCESS-BOUNDARY IN ROCKS OF THE CATSKILL DELTA

The process-boundary is recorded in the sequence as a facies boundary. The basin margin facies are essentially those referred to as Chemung, Cattaraugus and Hamilton by Rickard (1975). The lithologic and biologic characteristics of these rocks have been documented elsewhere (for example: Sutton et al., 1970; Sutton and Ramsayer, 1975). Basin margin facies interfinger with the clinoform facies, referred to elsewhere as the Portage by Rickard (1975), characteristics of which are also well known (for example: DeWitt and Colton, 1978). Previous interpretation of these rocks has been that the Chemung and associated facies are those of a marine shelf and that the Portage facies are those of a marine slope. The exact nature of the slope is not often considered explicitly, although Glaeser (1979) provides such a discussion. The supposed shelf-slope junction has been thought comparable to a modern oceanic shelfedge.

Assuming that basin margin-clinoform seafloor configuration existed in the Catskill Sea and that the sea was stratified, a relationship between seafloor configuration and sedimentary process is suggested (Fig. 3). The depth range over which most bottom currents could work on sediments was controlled,

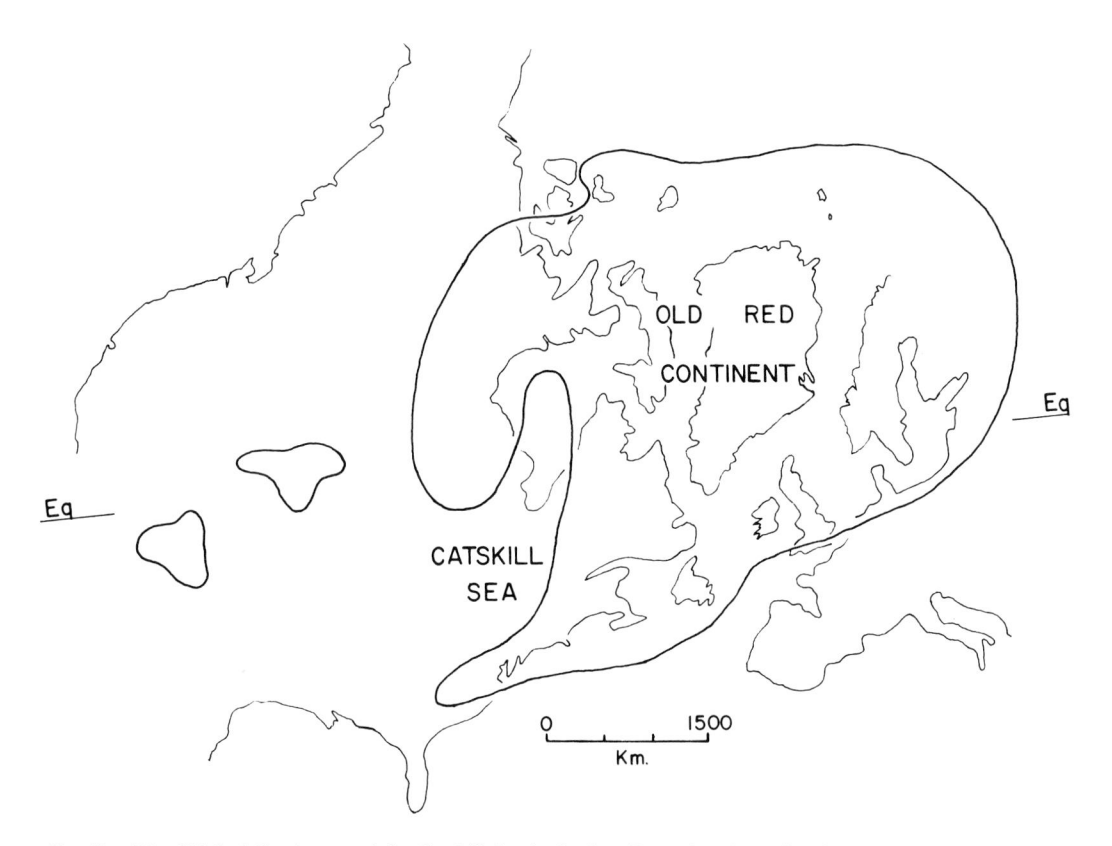

FIG. 2.—The Old Red Continent and the Catskill Sea in the Late Devonian. Modified from Woodrow et al. (1973). Position and outline of western island from Heckel and Witzke (1979). The Catskill Sea occupies part of Eastern Interior Sea of Gutschick and Sandberg (this volume, their Figs. 3, 5), and the "Black Shale Sea" of Ettensohn and Barron (1981). Eq = paleoequator.

in the main, by the position of the pycnocline. Its intersection with the floor marked the lower limit of meteorologic- and wave-generated bottom currents and most tidal effects. On the pycnocline itself moved internal waves which shoaled or broke on the floor, winnowed the sediments there, loosed masses of relatively coarse material down the clinoform and dispersed fine material in the water column. Below the pycnocline density currents swept the floor, carrying in relatively coarse clastics while fines were deposited both from the density currents and from the material dispersed along the pycnocline. Thus, the basin margin is made up of facies deposited as the result of many types of bottom currents operating in the well-mixed part of the water column. The clinoform is made up of facies deposited by density currents and from suspension in the unmixed waters below the pycnocline.

ROCK SEQUENCES REFLECTING PROCESS SEPARATION

Two examples of rock sequences deposited near the facies boundary and, therefore, at or near the inferred intersection of the pycnocline with the floor of the Catskill Sea are given below.

Late Devonian Genesee Group, Central New York

Main Features.—Rocks in the Genesee Group (Thayer, 1974; DeWitt and Colton, 1978) at or near the facies boundary between shallow-water and the deeper-water facies are seen in typical aspect near Watkins Glen, New York (Fig. 4). Found there are sequences of thin-bedded, fine-grained sandstones with b-c-d or c-d turbidite sequences and thin, dark-gray silty shales (Woodrow and Isley, 1983). Body fossils are rare and fragmented. Plant material occurs in quantity in a few beds but usually as fine fragments, thinly scattered across bedding surfaces and in ripple swales.

Figure 4 shows the major features of two of these sequences. In section *a* in Figure 4 both the bed thickness and sand-shale ratio increase upsection. The topmost sandstones are capped by interference ripples and contain coelenterate and brachiopod shell hash and fine plant material in far greater concentration than seen below. In section *b* of Figure 4, a thinner sequence is illustrated in which the sandstones are comparable to those seen at the top of the first section. Scattered exposures above both of these sections reveal that the next higher beds are thin, fine-grained sandstones, often bioturbated and sometimes lenticular, with greater numbers of body fossils. Channeling is not a common feature in either sequence. Individual beds extend across the entire width of an exposure with no change in thickness. No slump scars exist and no deformation exists to suggest bedding-place slip. No channel-conglomerates or levee deposits are found. The features illustrated and described here are typical of the rocks at or near the facies boundary throughout the Genesee Group.

Interpretation.—These strata have been interpreted as slope and base-of-slope deposits as reported by Rickard (1975). This interpretation is based on the lack of an abundant invertebrate fauna and the presence of turbidites. Although the facies boundary is clear and the contrast of rock type is obvious, there is no evidence for a change in gradient on the seafloor comparable to that found at modern ocean margins. Evidence for deep channeling, transport and deposition of conglomerates

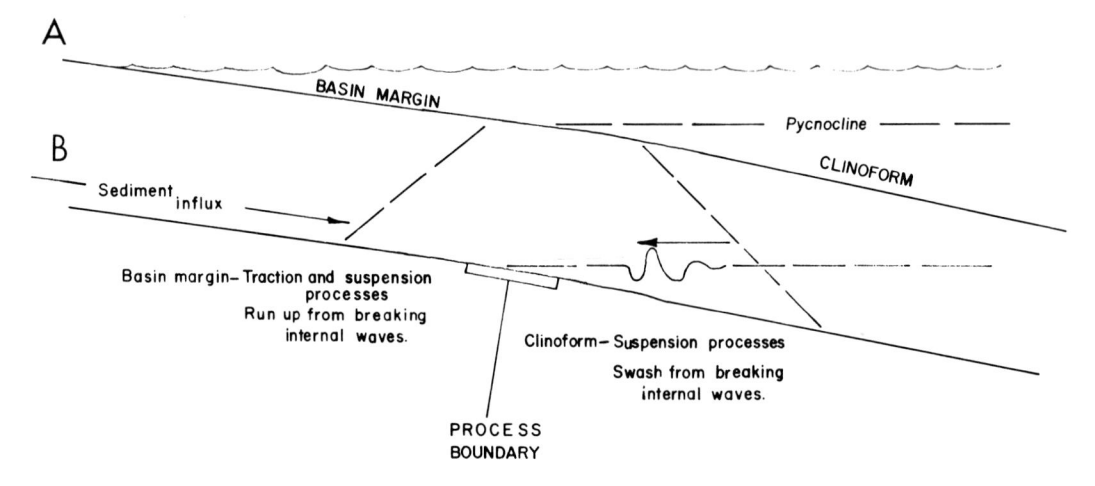

FIG. 3.—Bottom configuration and its relation to the pycnocline in the Catskill Sea. A, Pycnocline illustrated as sharply defined surface; may be tens of meters thick and marks lower boundary of shallow-water mixing. B, Expanded view of pycnocline-seafloor junction. Internal wave moving in from the right. Source of sediment to left. Intensification of sediment winnowing at process boundary. Density currents loosed down clinoform to the right with very fine-grained winnowing products probably trapped and dispersed along the pycnocline.

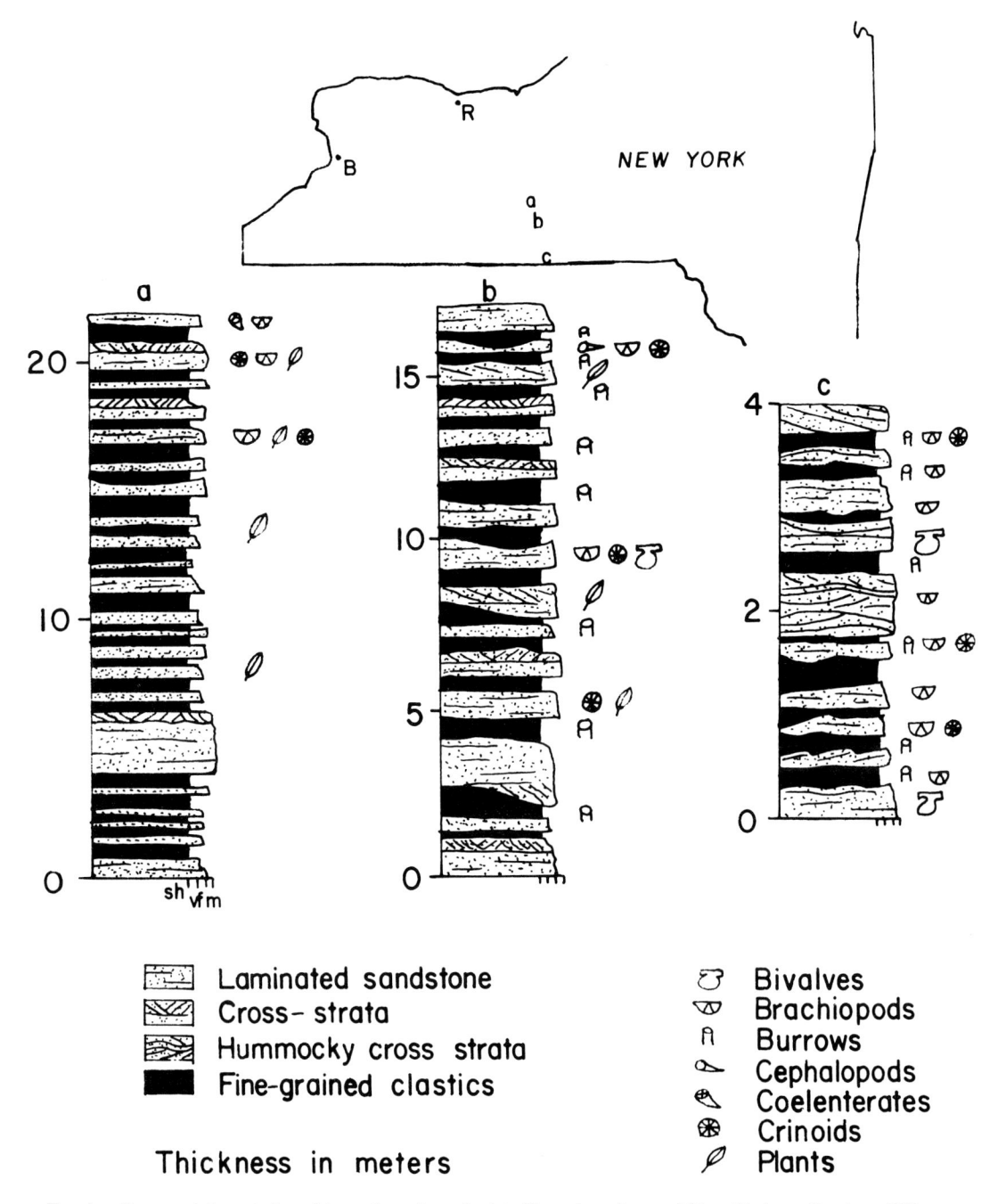

FIG. 4.—Representative stratigraphic sections from the late Devonian of central New York. *a*, Roadcut, NY route 414, 1.5 km northeast of Watkins Glen village. *b*, Roadcut, NY route 224, 1 km east of Montour Falls village. *c*, Roadcut, NY route 427, 1 km east of Wellsburg village. Sections simplified for clarity.

or slope failure is lacking. Instead one finds a sharp transition from evidence of traction processes to evidence for suspension processes. Shoreward of the process boundary there is evidence of sediment reworking: shallow channels, lenticular bedding, a wide variety of ripple types and ripple scales, shell-hash beds and intraformational conglomerates. Basinward of the boundary there are few channels, no conglomerates, fewer ripple types, no shell hash beds and sandstones persist, without thickness change, for at least hundreds of meters.

The facies boundary referred to above reflects a process boundary in the Catskill Sea, one which need not have been associated with a marked change in seafloor gradient.

Late Devonian West Falls Group, Central New York

Main Features.—The rocks of interest here are those at or near the shallow water/deep water facies boundary. One such sequence is exposed near Wellsburg, New York (Fig. 4, section c). Included there are fine-grained, laminated and cross-laminated sandstones, some of which contain coquinites. Found with the sandstones are gray silty shales. Some of the shales are fissile but most are bioturbated. An abundant fauna including brachiopods, pelecypods, bryozoans and corals typical of what Sutton and others (1970) have referred to as the outer platform is found in these rocks. Toward the base of the exposure is the thickest sandstone of the sequence (Fig. 4, section c). It is fine-grained, carries few shells and is cross-stratified. In it are preserved a variety of bedforms and rare, paper-thin, mud drapes. Shale clasts are strewn on a few internal bedding surfaces. These features resemble those of hummocky cross-bedding (Harms, 1975; Dott and Bourgeois, 1979; Hamblin et al., 1979; Hobday and Morton, 1980).

Interpretation.—These rocks have been interpreted as having developed near the seaward edge of the shelf (Rickard, 1975). Woodrow and Isley (1983) indicate that the sequences are the seaward edge of prograding, prodelta and delta-front sequences built onto the basin margin. The bases for these interpretations include the existence of the fauna and the characteristics of the rocks. The fauna suggests that an oxygenated bottom is required at least part of the time but the fissile dark-gray shales indicate that there were extended periods when the bottom was less well oxygenated and the bottom fauna restricted. The characteristics of the rocks require that mud deposition be the norm with intermittent incursions of sand and/or winnowing of mud to produce sandstones.

Sandstones at the base of the exposure are of particular interest. Their stratigraphic position indicates that they formed far from shore. Their characteristics record processes capable of producing: (1) a sandstone somewhat coarser than those above or below, (2) lenticular beds arrayed in discrete sedimentation units, (3) low, irregular bed forms (hummocky cross-beds) in a sequence otherwise notable for their absence, and (4) an essentially mud-free unit where sequences immediately above or below have a large mud component. These sandstones, and others like them found in similar facies contexts, record extraordinary events taking place far from shore. Two processes which might explain reworking of the bottom sediment and winnowing of the mud include: (1) currents developed during a single great storm or a closely spaced series of them, and (2) currents generated at the pycnocline including normal storm-related currents plus those generated by shoaling and breaking internal waves.

Application to Other Sequences

The conditions hypothesized for the Catskill may have application to any aqueous environment in which stratification of the water column can be demonstrated. The picture of process-separation by a pycnocline is likely to be a general feature, especially in the Paleozoic seas where black shales were deposited. Stratification has also been put forward as a significant feature of Mesozoic seas (Arthur and Schlanger, 1979) and the possibility of process separation as suggested in this paper cannot be ignored in that situation.

SUMMARY

1. The topography of the seafloor on the southeastern and eastern side of the Catskill Sea was more subdued than that found around modern oceans. Recognized there are the basin margin, a nearshore, low-gradient surface characterized by complicated facies. Seaward of the basin margin is the clinoform, a surface of somewhat greater gradient on which were deposited turbidites loosed from a linear source. Extending from the clinoform base was the basin floor, a nearly flat surface on which were deposited black shales.

2. The water column in the Catskill Sea usually was stratified, that is, a pycnocline developed there. The pycnocline provided effective separation of sedimentary processes. Various sediment-moving currents were active above the pycnocline while only a limited number were active below it.

3. Above the pycnocline on the basin margin were deposited a complicated set of facies. Below the pycnocline on the clinoform were deposited turbidites and pelagic sediments.

4. Stratification of the water column is an implied feature of many epicontinental seas. Where stratification can be reasonably inferred, it seems probable that separation of sedimentary processes existed as another feature of those seas.

ACKNOWLEDGMENTS

The comments of Ray Gutschick, the editors and an anonymous reviewer led to substantial improvements in this paper.

REFERENCES

ARTHUR, M. A., AND SCHLANGER, S. O., 1979, Cretaceous "oceanic anoxic events" as factors in development of reef-reservoired giant oil fields: Am. Assoc. Petroleum Geologists Bull., v. 63, p. 870–885.

ASQUITH, D. O., 1974, Sedimentary models, cycles and deltas: Upper Cretaceous: Am. Assoc. Petroleum Geologists Bull., v. 58, p. 2274–2283.

———, 1970, Depositional topography and major marine environments: Am. Assoc. Petroleum Geologists Bull., v. 54, p. 1184–1224.

BOUMA, A. H., 1979, Continental slopes, in Doyle, L. J., and Pilkey, O. H., Jr., eds., Geology of Continental Slopes: Soc. Econ. Paleontologists Mineralogists Spec. Pub. 27, p. 1–16.

BOWEN, Z. P., RHOADS, D. L., AND MCALESTER, A. L., 1974, Marine benthic communities in the Upper Devonian of New York: Lethaia, v. 7, p. 93–120.

BYERS, C. W., 1977, Biofacies patterns in euxinic basins: a general model, in Cook, H. E., and Enos, P., eds., Deep-Water Environments: Soc. Econ. Paleontologists Mineralogists Spec. Pub. 25, p. 5–17.

DEWITT, W., JR., AND COLTON, G. W., 1978, Physical stratigraphy of the Genesee Formation (Devonian) in western and central New York: U.S. Geol. Survey Professional Paper 1032-A, 22 p.

DINELEY, D. L., 1975, North Atlantic Old Red Sandstone—some implications for Devonian paleogeography, in Yorath, C. J., Parker, E. R., and Glass, D. J., eds., Canada's Continental Margins and Offshore Petroleum Exploration: Canadian Soc. Petroleum Geologists Mem. 4, p. 773–790.

DOTT, R. H., JR., AND BOURGEOIS, J., 1979, Hummocky cross-stratification—importance of variable bedding sequences analogous to the Bouma Sequence (abstract): Geol. Soc. America Abs. with Programs, v. 11, p. 414.

ETTENSOHN, F. R., AND BARRON, L. S., 1981, Depositional Model for the Devonian-Mississippian Black-Shale Sequence of North America: A Tectono-Climatic Approach: U.S. Dept. Energy, Document No. 12040-2, 80 p.

GLAESER, J. D., 1979, Catskill Delta slope sediments in central Appalachian basins; source and reservoir deposits, in Doyle, L. J., and Pikey, O. H., eds., Geology of Continental Slopes: Soc. Econ. Paleontologists Mineralogists Spec. Paper 27, p. 343–358.

GORSLINE, D. S., 1978, Anatomy of margin basins: Jour. Sed. Petrology, v. 48, p. 1055–1068.

GUTSCHICK, R. C., MCLANE, D. J., AND RODRIGUEZ, E. G., 1976, Summary of Late Devonian-Early Mississippian biostratigraphic framework in western Montana, in Tobacco Root Geological Society Guidebook: Montana Bur. Mines Geol. Spec. Pub. 73, p. 91–124.

HAMBLIN, A. P., DUKE, W. L., AND WALKER, R. G., 1979, Hummocky cross-stratification: indicator of storm-dominated shallow-marine environments [Abs.]: Am. Assoc. Petroleum Geologists Bull., v. 63, p. 460–461.

HARMS, J. C., 1975, Stratification and sequence in prograding shoreline deposits, in Harms, J. C., Southard, J. B., Spearing, D. P., and Walker, R. G., Lecture Notes: Soc. Econ. Paleontologists Mineralogists Short Course No. 2, p. 81–102.

HARRINGTON, J. W., AND HAZELWOOD, E. L., 1962, Comparison of Bahama land-forms with depositional topography of Neve Lucia dune reef-knoll, Nolan County, Texas: a study in uniformitarianism: Am. Assoc. Petroleum Geologists Bull., v. 46, p. 354–373.

HECKEL, P. E., 1972, Recognition of ancient shallow marine environments, in Rigby, J. K., and Hamblin, W. K., eds., Recognition of Ancient Sedimentary Environments: Soc. Econ. Paleontologists Mineralogists Spec. Pub. 16, p. 226–286.

———, AND WITZKE, B. J., 1979, Devonian world paleogeography determined from distribution of carbonates and related lithic paleoclimatic indicators, in House, M. R., Scrutton, C. T., and Bassett, M. G., eds., The Devonian System: Paleontological Soc., Spec. Papers in Paleontology 23, p. 99–123.

HOBDAY, D. K., AND MORTON, R. A., 1980, Lower Cretaceous shelf storm deposits: north Texas (abstract): Am. Assoc. Petroleum Geologists Bull., v. 64, p. 722.

HOUSE, M. R., 1975, Facies and time in Devonian tropical areas: Proc. Yorkshire Geol. Soc., v. 40, p. 233–288.

HUTCHINSON, G. E., 1957, A Treatise on Limnology, Volume I—Geography, Physics and Chemistry: New York, John Wiley and Sons, Inc., 1015 p.

KNAUSS, J. A., 1978, Introduction to Physical Oceanography: Englewood Cliffs, New Jersey, Prentice-Hall, Inc., 338 p.

MITCHUM, R. M., JR., VAIL, P. R., AND SANGREE, J. B., 1977, Seismic stratigraphy and global changes of sea level, part 6: stratigraphic interpretation of seismic reflection patterns in depositional sequences, in Payton, C. E., ed., Seismic Stratigraphy—Applications to Hydrocarbon Exploration: Am. Assoc. Petroleum Geologists Memoir 26, p. 117–133.

OLIVER, W. A., 1977, Biogeography of Late Silurian and Devonian rugose corals: Paleo. Paleo. Paleo., v. 22, p. 85–135.

RAHMANIAN, V., AND WILLIAMS, E., 1979, Tide-dominated deltaic-interdeltaic sedimentation, Upper Devonian, central Pennsylvania (abstract): Geol. Soc. America Abs. with Programs, v. 11, p. 50.

RICH, J. L., 1951, Three critical environments of deposition and criteria for recognition of rocks deposited in each of them: Geol. Soc. America Bull., v. 62, p. 1–20.

RICKARD, L. V., 1975, Correlation of the Devonian and Silurian rocks in New York State: N.Y. State Museum and Science Service Map and Chart Ser. No. 24 (map).

SEVON, W. D., 1979, Devonian sediment-dispersal systems in Pennsylvania: Geol. Soc. America Abs. with Programs, v. 11, p. 53.

———, ROSE, W. W., SMITH, R. C., II, AND HOFF, D. T., 1978, Uranium in Carbon, Lycoming, Sullivan and

Columbia Counties, Pennsylvania: Guidebook, 43rd Annual Field Conference of Pennsylvania Geologists: Pennsylvania Geol. Survey, 50 p.

SUTTON, R. G., BOWEN, Z. P., AND MCALESTER, A. L., 1970, Marine shelf environments of the Upper Devonian Sonyea Group of New York: Geol. Soc. America Bull., v. 81, p. 2975–2992.

———, AND RAMSAYER, G. R., 1975, Association of lithologies and sedimentary structures in marine deltaic paleoenvironments: Jour. Sed. Petrology, v. 45, p. 799–807.

SWIFT, D. J. P., 1976, Continental shelf sedimentation, in Stanley, D. J., and Swift, D. J. P., eds., Marine Sediment Transport and Environmental Management: New York, J. Wiley and Sons, p. 311–350.

THAYER, C. W., 1974, Marine paleoecology in the Upper Devonian of New York: Lethaia, v. 7, p. 121–155.

VAN SICLEN, D. C., 1958, Depositional topography—examples and theory: Am. Assoc. Petroleum Geologists Bull., v. 42, p. 1897–1913.

WALKER, R. G., 1971, Non-deltaic depositional environments in the Catskill Clastic Wedge (Upper Devonian) of central Pennsylvania: Geol. Soc. America Bull., v. 82, p. 1305–1326.

———, AND HARMS, J. C., 1971, The "Catskill Delta"—a prograding muddy shoreline in central Pennsylvania: Jour. Geology, v. 79, p. 381–399.

———, AND ———, 1975, Shorelines of weak tidal activity: Upper Devonian Catskill Formation—central Pennsylvania, in Ginsburg, R. N., ed., Tidal Deposits: New York, Springer-Verlag, p. 103–108.

WOODROW, D. L., FLETCHER, F. W., AND AHRNSBRAK, W. F., 1973, Paleogeography and climate at the deposition sites of the Devonian Catskill and Old Red Facies: Geol. Soc. America Bull., v. 84, p. 3051–3064.

———, AND ISLEY, A. M., 1983, Facies, topography and sedimentary processes in the Catskill Sea (Devonian), New York and Pennsylvania: Geol. Soc. America Bull. (in press).

PART IV
CARBONATE MARGINS

MODERN CARBONATE SHELF-SLOPE BREAKS

ALBERT C. HINE
Department of Marine Science, University of South Florida, St. Petersburg, FL 33701

HENRY T. MULLINS[1]
Moss Landing Marine Laboratories and San Jose State University, Moss Landing, CA 95039

ABSTRACT

Modern carbonate shelf-slope breaks are highly variable, complex features that are morphologically distinct from their siliciclastic counterparts because of the dominance of *in situ* organic sediment production and early diagenesis. Carbonate shelf-slope breaks are, in reality, carbonate margins commonly characterized by abrupt and rapid transitions of sedimentary facies, biological communities and physical energy. We recognize four major types of modern carbonate shelf-slope break margins: (1) reef-dominated rimmed margins; (2) atoll margins; (3) sand-shoal-dominated rimmed margins; and (4) non-rimmed margins.

Reef-dominated margins usually occur along windward, open-ocean settings, and their associated shelf-slope breaks tend to be abrupt and precipitous, with steep, seaward slopes. Such margins are characterized by distinct morphological and ecological zonation as well as spur and groove structure. Atoll shelf-slope break margins are circular to elliptical in map view and overlie oceanic, volcanic basement. These margins are the most precipitous and reef-dominated of all carbonate margins. Distinctly zoned reefs, with algal ridges at their seaward edges, generally separate deep, open-marine lagoons from steep (>50°), seaward slopes mantled with reef debris and talus. Sand shoal-dominated margins may also be abrupt and precipitous, but are essentially devoid of reefs at the surface. Such margins typically occur along leeward settings and are characterized by bank-parallel sand bodies composed mostly of non-skeletal and degraded skeletal grains which have buried earlier Holocene reefs. Similar shelf-slope breaks are also found in tide-dominated settings where strong tidal and storm currents flow on and off the shelfedge. These margins are characterized by bank-perpendicular sand bodies consisting mostly of oolitic grains. Shelf-slope breaks along non-rimmed carbonate margins are broad, subtle, non-reef features that occur in deeper waters (100–500 m). These carbonate shelfedges are characterized by a mix of non-skeletal and skeletal grains that grade up-dip into molluscan calcarenites and/or bioherms and down-dip into pelagic oozes.

Four primary processes control the location and gross geomorphology of carbonate shelf-slope breaks: (1) tectonism; (2) physical energy flux; (3) antecedent topography; and (4) sealevel history. Secondary processes such as biogenic barrier development, *in situ* sediment production, sediment transport and cementation serve to modify the gross structure of carbonate shelf-slope breaks. Analogous shelf-slope breaks should be recognizable in the rock record. Rimmed margins will be the easiest to identify, particularly on seismic reflection profiles, because of their abrupt changes in depth, sediment facies and biological communities. However, non-rimmed margins should also be recognizable on the basis of careful examination of lateral and vertical facies relationships.

INTRODUCTION

Shelf-slope breaks along modern carbonate depositional systems demonstrate a variability and complexity equal to, or perhaps even greater than, their terrigenous, siliciclastic counterparts. The main controlling factors are tectonic movement, sealevel fluctuations, antecedent topography and physical energy. Carbonate shelf-slope breaks are greatly influenced by autochthonous sediment production and early diagenesis resulting from biological and chemical processes.

The carbonate shelf-slope break is herein defined as a change in slope which marks the boundary between a relatively low gradient shelf and the upper extent of a more steeply seaward-descending slope. Modern carbonate shelf-slope breaks may be abrupt and precipitous or subtle, and may occur in water as shallow as a few tens of cm (Milliman, 1974) or as deep as 300 m (Logan et al., 1969).

The purpose of this paper is to examine and illustrate the various types of modern carbonate shelf-slope breaks. The dominant processes responsible for these products are identified and discussed and a number of criteria for recognition in the rock record are suggested.

Carbonate Margin Concept

When considering the geology of the edges of carbonate build-ups, geologists should be concerned with more than just a physiographic break in slope. In reality, the shallow-to-deep transition of sedimentary facies, biological communities and changes in physical processes combine to define a carbonate margin that separates deeper-water slopes and basins from relatively shallow-water lagoons.

Carbonate margins typically face open bodies of water and are thus the primary zones of physical

---

[1]Present address: Department of Geology, Syracuse University, Syracuse, New York 13210.

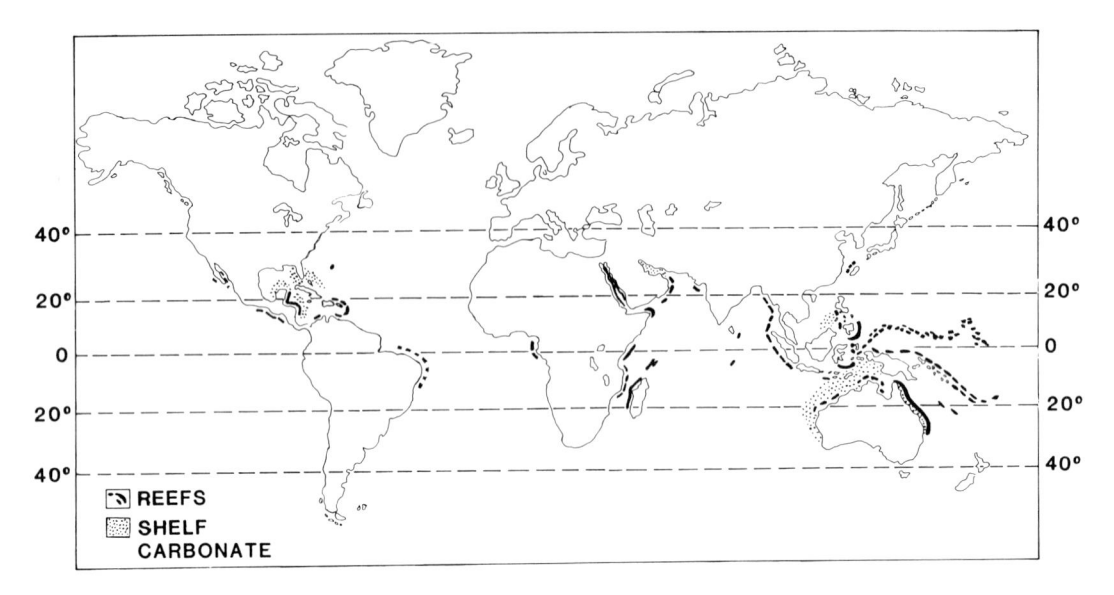

FIG. 1.—Distribution of major modern carbonate platforms and reefs (from Wilson, 1975, his Fig. 1-1). Note that most major carbonate-producing zones lie between 30°N and S latitudes.

energy absorption. Long-period, high energy waves as well as storm and tide generated currents affect the margin directly. As a result, well oxygenated, turbulent waters stimulate the growth of carbonate-producing organisms and act as a catalyst for submarine cementation.

When considering carbonate margins, ecologic reefs (Dunham, 1970) probably come to mind more than anything else. Certainly, for early investigators, spectacular fringing, barrier, and atoll reefs seen around the world attracted much attention and debate. However, recent work along modern carbonate margins has shown not only new complexities concerning ecologic reefs but also that some margins are reef-devoid and are dominated by lime muds, sands or bare rock surfaces (Hine and Neumann, 1977; Hine, et al., 1981a).

Distribution and Types of Modern Carbonate Build-Ups

The distribution of modern shallow-water carbonate build-ups is controlled by temperature, turbidity and salinity (Bathurst, 1975; Wilson, 1975). As a result, few examples are found outside of 30° north or south latitude, or in the vicinity of major rivers (Fig. 1). The primary carbonate sediments on

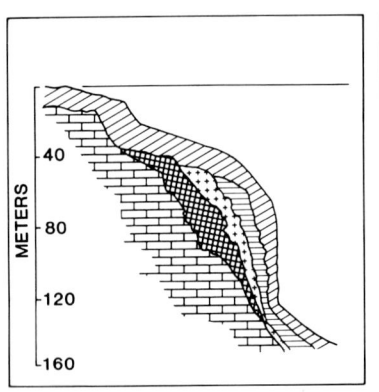

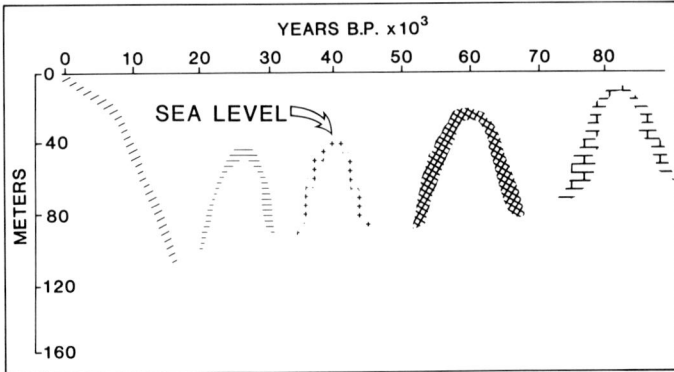

FIG. 2.—Sketch illustrating possible relationships between Late Quaternary sealevel fluctuations (right) and development of a reef-dominated carbonate shelf-slope break (left) on Belize margin. Note persistence and lateral migration (due to accretion) of the shelf-slope break. Different patterns correspond to specific Pleistocene sealevel events. From James and Ginsburg, 1979, their Fig. 7-6.

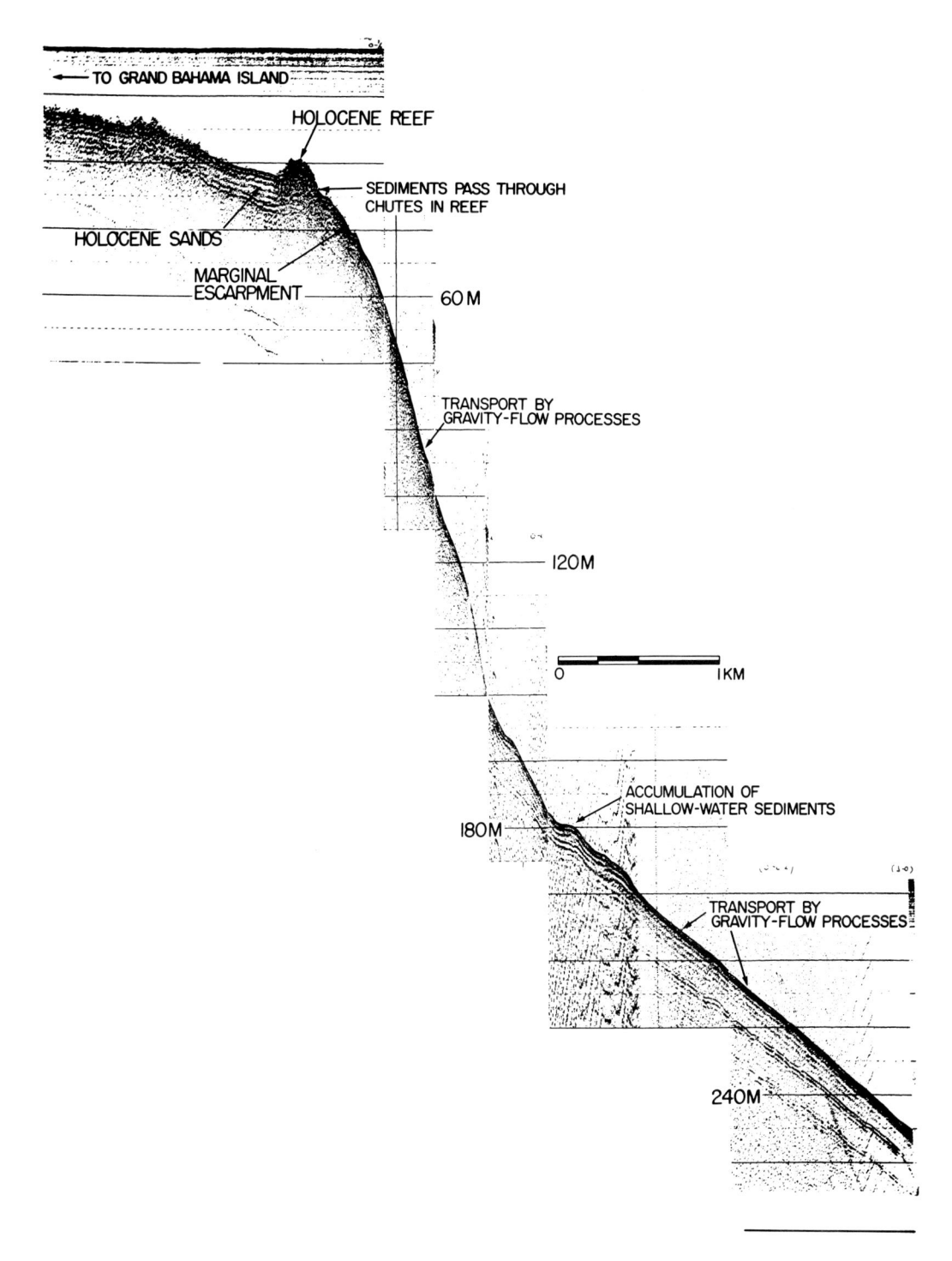

Fig. 3.—High-resolution seismic reflection profile (Uniboom) across shelf-slope break seaward of Grand Bahama Island. Note position of reef at the precise break in slope as well as the adjacent steep marginal escarpment (from Hine et al., 1981a, their Fig. 20). Vertical exaggeration ×25.

TABLE 1.—BELIZE SHELF MARGIN

	Reef Flat		Barrier Reef		
				Reef Front	
	Sand Apron	Pavement	Crest	Spur/Groove	Step
Sediments	c-vc sands, *Halimeda* plates, frags. of coral, coralline algae, encrusting forams, echinoids	v-c sands, gravels; coral frags., branches, gastropod shells, *Millepora; Homotrema,* rhodoliths; sand-*Halimeda,* coral, coralline algae frags.	c-vc sands; same as reef pavement	c-vc sands, gravels, coral rubble—mostly *A. palmata* 1.5 m thick, molluscs, rhodoliths, intraclasts of cemented grainstones	c-vc sand, rubble in chutes
Organisms	sparse *Thalassia testudinum* community	sparse encrusting corals, calc. algae	*Millepora,* dead coral, bare rock surfaces	upper-dead coral, zoanthid mat cover, *Millepora,* lower-*Agaricia agaricities, Acropora palmata, Millepora, Montastrea annularis, Porities asteroides,* coralline algae	overlapping plates of *M. annularis* and *Mycetophyllia* project out from cliff, coral cover decreases with depth, octocorals, coralline algae increase with depth
Rocks		rubble cemented by packstone/wackestone mortar, Mg calcite cements, extensive boring by endolithic sponges			
Depths/Geomorphology	<4 m, flat surface 1.5–10 km wide, large bedforms sometimes present	<4 m, 10–100 m wide, some small islands, smooth, undulating surface	awash, discontinuous rock ridge, meshes into shallow spur and grooves	awash to 15–21 m, grooves 1–5 m wide, spurs 6 m wide, grooves widen and deepen seaward	20–40 m, steep slope, coral hummocks and small winding chutes, cont. of shallow grooves

tropical and subtropical carbonate shelves are fragments of codiacian algae, hermatypic-scleractinian corals, molluscs, benthic foraminifera, coralline algae, and a prominent suite of non-skeletal grains such as ooids, peloids and aggregates (Milliman, 1974; Lees, 1975).

Because of their variability, carbonate margins are difficult to classify in an unequivocal fashion, which may be a function of our present lack of data and/or understanding. However, for simplicity and application, carbonate margins are classified here using cross-sectional morphology based, in part, on the suggestions of Ginsburg and James (1974), who grouped carbonate margins as either open or rimmed. Using this approach, we recognize four major types of carbonate margins: (1) reef-dominated rimmed margins; (2) atoll margins; (3) sand-shoal-dominated rimmed margins; and (4) non-rimmed margins.

Reef-dominated rimmed margins occur commonly along the windward edges of broad detached carbonate platforms such as the Bahama Banks, narrow insular shelves such as Jamaica, or along mainland attached platforms such as south Florida, Belize, or northeastern Australia. Atoll margins are also reef-dominated but differ from the first category in that a more or less continuous belt of reefs encircles a relatively deep, open-marine lagoon. Also, the basement beneath atolls consists of oceanic, volcanic basalts. Atolls are found in all the major

TABLE 1.—CONTINUED

Sand Slope	Fore-Reef		
	Brow	Wall	Deep Fore-Reef
sand-*Halimeda*, rubble—coral sticks, conch shells, rhodoliths	sand-*Halimeda*, meandering streams of mobile sands	sand-*Halimeda*, coralline algae, molluscs, forams; rubble-coral, perched on near horizontal surfaces, mud mixed in	rapid grain size decrease from proximal to distal (boulder scree to fine sand/mud) sands-*Halimeda*, coral, molluscs, corraline algae, benthic forams; distal seds-*Halimeda*, foram muds
clusters of *Penicillus*, *Halimeda*; sponges, octocorals on rubble	*M. annularis*, *Agaricia*-overlapping plates, corals cover 25% of surface; coralline algae, octocorals, sponges, *Halimeda* cover rest	biologically zoned; upper-shallow water reef community; lower-deeper water reef community of coralline algae, octocorals, ahermatypic corals, demo-, sclero-, endolithic sponges	encrusting epizoans, massive sponges, antipatharians, ahermatypic corals, crinoids, serpulid worms, forams, burrow mounds common
		brittle limestones, 50% coral, massive types more common than platey, lithified mudstone, packstone, grainstone matrix, *Halimeda* common to all three, mottled texture due to numerous generations of boring, infilling, and lithification, encrusting coralline algae	large talus-hard cavernous, similar to wall rocks; massive/platey corals, sed. infilled-*Halimeda* calcarenite Fe, Mn coatings
37–45 m, mostly featureless, gently sloping, small sand volcanoes	40–65 m, increasing steepness with depth (45–50°), irregular ledges, promontories, knobs, hummocks of corals, rollover to wall between 60–67 m	65–110 m, vertical, small angular ledges, caves, fissures, holes, pock-marks, smooth in places	110–220 m, *sloping:* (1) coarse boulder slope, (2) gentle slope of muddy sand, (3) flat slope of pelagic muds; *cliffed:* (1) upper talus slope, (2) ridge and furrow zone, (3) cliffs

oceans, but are by far most abundant in the Pacific; the best studied examples are Bikini and Eniwetok Atolls. Sand shoal-dominated margins are known mostly from the leeward edges of detached carbonate platforms such as the Bahamas where ecologic reefs are conspicuously absent at the surface. Non-rimmed margins lack an abrupt shelf-slope break and are essentially drowned platforms (Schlager, 1981); the best studied examples occur in the Gulf of Mexico along west Florida and Mexico's Yucatan Peninsula (Campeche Bank).

REEF-DOMINATED RIMMED SHELF MARGINS

Ecologic reefs are organically built and bound, rigid, wave-resistant structures (Dunham, 1970).

Such features are common in today's ocean at low latitudes (Fig. 1) along high-energy, open-ocean carbonate margins. In the Bahamas, coral reefs are most conspicuous along windward margins (Storr, 1964; Ginsburg and Shinn, 1964), but may also occur along leeward margins, where they are protected by islands or rock ridges from the offbank transport of carbonate sediment (Hine and Neumann, 1977; Hine et al., 1981a).

Shelf-slope breaks along modern reef-dominated carbonate margins typically are abrupt and precipitous, due to the vertical and lateral growth of reefs in response to Quaternary sealevel fluctuations (Fig. 2) as well as synsedimentary submarine cementation (Land and Goreau, 1970; Friedman et

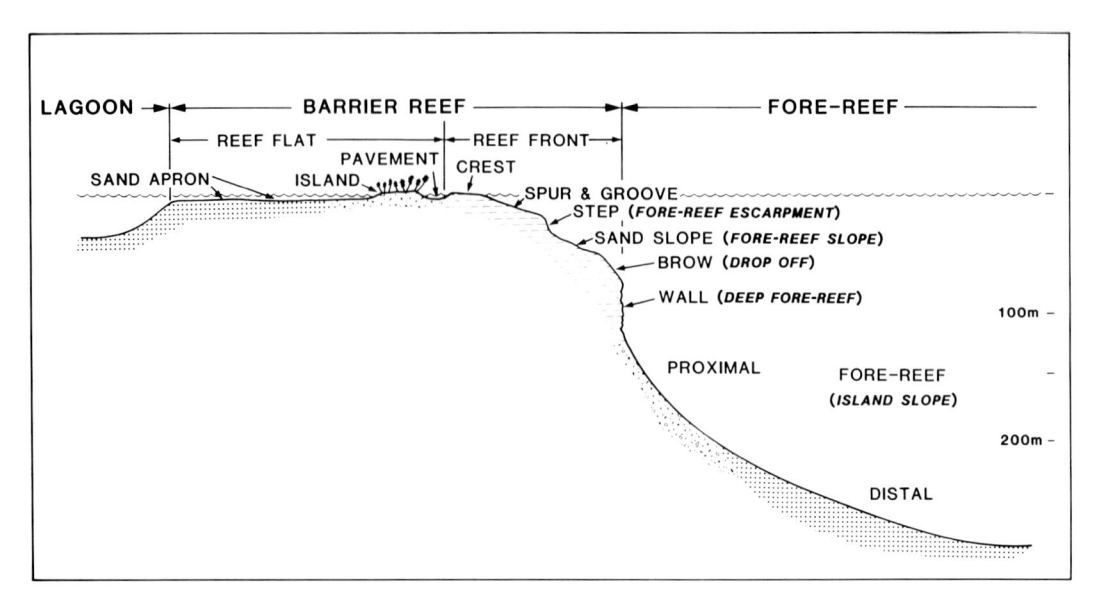

FIG. 4.—Generalized geomorphic cross-section across the Belize reef-dominated shelf-slope break (from James and Ginsburg, 1979, their Fig. 3-2).

al., 1974; James et al., 1976). Consequently, very steep (>45°) marginal escarpments are usually found immediately seaward of most modern reefs (Fig. 3). The Belize marginal escarpment, for example, consists of nearly vertical to overhanging slopes that extend from 65 m to at least 120 m of water (James and Ginsburg, 1979).

Most modern reefs appear to have developed on antecedent topographic highs. This is the case for the reefs of the northern Bahamas (Hine and Neumann, 1977), south Florida (Enos and Perkins, 1977), Belize (James and Ginsburg, 1979), Jamaica (Goreau and Land, 1974) and the Great Barrier Reef of Australia (Maxwell, 1968, 1973). When exposed during sealevel falls, antecedent highs form sites of new reef growth during the ensuing sealevel rise. As a result, the shelf-slope break not only persists through time but can accrete and migrate laterally (Fig. 2). Enos and Perkins (1977, p. 110), studying the south Florida margin, found that the pre-Holocene rock surface was so important in controlling modern sedimentation, they concluded that "the present shelf-break position differs only in detail from that of the latest Pleistocene shelf break."

Reef-dominated carbonate margins also exhibit distinct and ecological zonation. Back-reef, reef-flat, reef-front and fore-reef environments are usually readily recognizable (Table 1; Fig. 4). Along the south Florida margin, for example, the back-reef consists of skeletal sand/gravel accumulations as much as 7 m thick and 40 km long, dominated by coral fragments derived mostly from the adjacent outer reef (Enos and Perkins, 1977). Isolated patch reefs dominated by massive head corals such as *Montastrea* and *Diploria*, as well as the branching coral *Acropora cervicornis*, are also common in back-reef settings (Enos and Perkins, 1977).

Along the Belize barrier reef (Table 1; James and Ginsburg, 1979), the reef-flat is a 100 m wide, partially cemented, rubble pavement consisting of cobble to boulder-sized fragments of algal encrusted coral (Fig. 4), with occasional small islands or cays along its landward edge. Sands in this zone consist mostly of fragments of *Halimeda*, coral, coralline algae, encrusting foraminifera and echinoids.

The actual shelf-slope break occurs along the reef front, which consists of five sub-zones: (1) reef crest; (2) spur and groove; (3) "step"; (4) sand slope; and (5) "brow" (Figs. 4, 5, 6). The actual break in slope is commonly occupied by a ridge of coral and coralline algal growth that absorbs most of the physical energy from deep-water waves. In the Caribbean region, this zone is usually dominated by *Acropora palmata*, except locally where algal ridges are present (Adey and Burke, 1976;

FIG. 5.—*A*, Oblique aerial photograph of well-developed spur and groove structure along the reef seaward of Grand Bahama Island. *B*, Block diagram of shallow part of Belize reef-front illustrating spur and groove geomorphology (from James et al., 1976, their Fig. 4). Many of the features labelled in (*B*) can be seen in (*A*).

SHALLOW REEF MARGIN
BELIZE
(BRITISH HONDURAS)

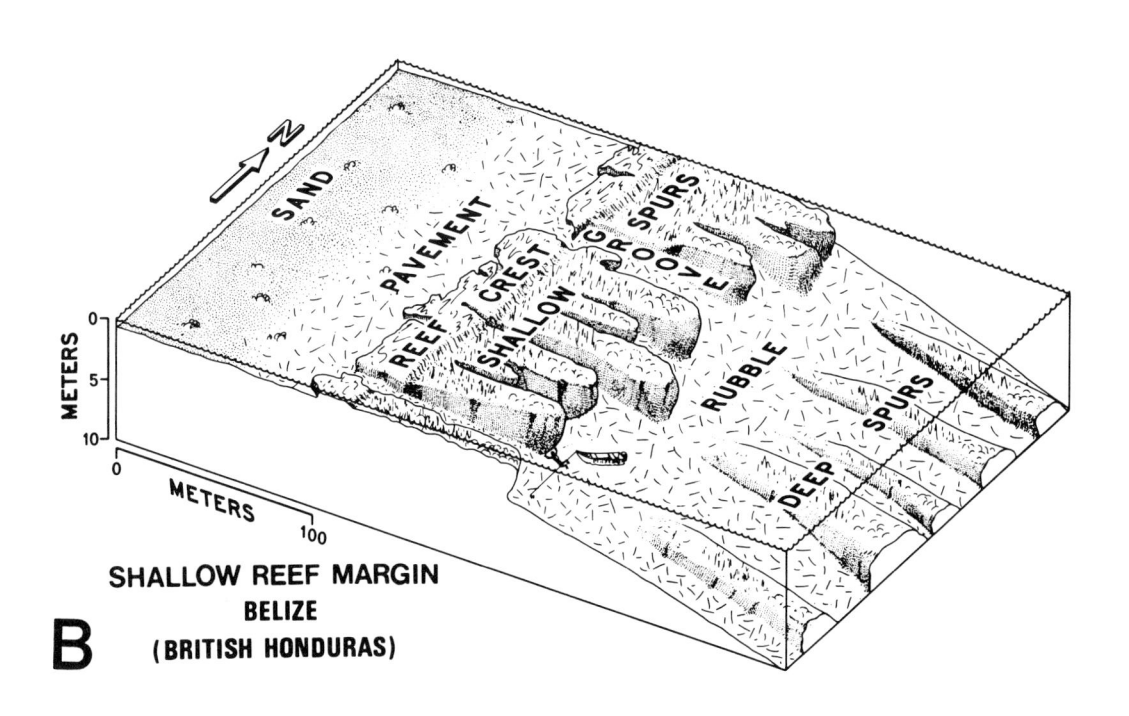

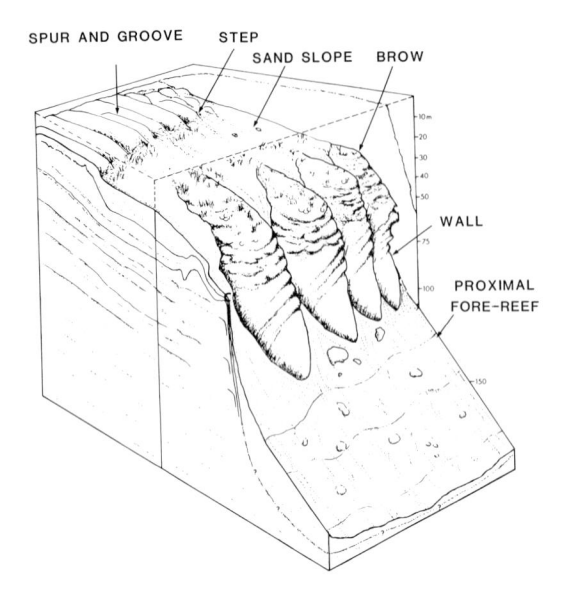

FIG. 6.—Block diagram of shallow spur and groove structure, step, sand slope, brow, wall, and proximal zone of deep fore-reef (periplatform sands). Terms correspond with those shown in Fig. 4. Drawing is of margin off Grand Bahama Island, Bahamas. Note large chutes and buttresses of the wall or marginal escarpment. This margin is actively transporting skeletal sand off the bank (from Hubbard et al., 1976, their Fig. 16).

Adey, 1978). Along the Great Barrier Reef of Australia, this zone consists mostly of coralline algae (Maxwell, 1968, 1973).

The fore-reef environment commonly consists of a nearly vertical wall as well as a submarine talus slope (Belize, for example). This slope is steep (35–40°) and has a proximal facies of extremely large limestone blocks (up to 10 m diameter) and coral plates within a coral-*Halimeda* sand matrix. In the Bahamas, Mullins and Neumann (1979) referred to this sediment slope as the peri-platform sand facies. The distal facies of the fore reef consist of finer bioturbated sands where the slope decreases to <10°. Further seaward, these coarser sediments grade into fine-grained "basinal" pelagic oozes (James and Ginsburg, 1979).

Along many fore-reef environments, such as those in the Bahamas (Hubbard et al., 1976) and Jamaica (Goreau and Land, 1974; Moore et al., 1976), large chutes facilitate the down-slope transport of sediment from the reef to deeper water environments (Fig. 6). Cross-sections of reef-dominated rimmed carbonate margins from the northern Bahamas are given in Figures 7A, B.

ATOLL MARGINS

Modern atolls form on top of oceanic volcanics (Ladd et al., 1970) and are typically circular to elliptical structures (1–130 km in diameter) rimmed by a more or less continuous belt of ecologic reefs that surround and separate a relatively deep, open-marine lagoon from the open ocean (Fig. 8A). Commonly, the ring of reefs is interrupted laterally by deeper water passes or channels which act as passageways for the free exchange of open-marine waters between the lagoon and the open ocean. Locally, these reefs are capped by islands, and this combination of reefs and islands acts as a natural buttress by absorbing most of the energy focused on the atoll by open-ocean waves.

Geomorphically, Pacific atolls can be divided into three major provinces: (1) lagoon; (2) reef; and (3) outer slope (Emery, 1948; Tracey et al., 1955; Wiens, 1962). The actual shelf-slope break, however, forms the boundary between the reef and the outer slope. The shelf-slope break of Bikini Atoll, as illustrated on unexaggerated profiles (Fig. 8A), is very abrupt and precipitous. The average slope between 0 and 366 m for Pacific atolls is 38° (Ladd, 1973); however, locally, slopes as steep as 68° have been measured (Emery et al., 1954).

Atoll reefs are distinct zones ecologically in belts that parallel the outline of the atoll itself (Fig. 8B). The seaward reefs along the windward margins of Bikini Atoll can be divided into six major zones extending from the open ocean toward the lagoon (Fig. 8B): (1) seaward slope; (2) algal ridge; (3) coral-algal zone; (4) outer "microatoll" zone; (5) main reef flat; and (6) inner "microatoll" zone (Wells, 1954). The main characteristics of each zone are summarized in Table 2.

Deposition and diagenesis along atoll margins appears to be cyclic. Drilling along the windward margins of Eniwetok, for example, has revealed six primary depositional sequences, all less than 600,000 years in age, that represent periods of reef growth during Quaternary high stands of sealevel. These sequences are separated by non-depositional unconformities capped by paleosols that developed during sealevel low stands associated with glacial episodes. Drilling has also indicated that shelfedge atoll reefs have undergone extensive early diagenesis in both marine and meteoric water environments (Goter, 1979).

SAND SHOAL-DOMINATED RIMMED SHELF MARGINS

Not all modern rimmed, carbonate shelf-slope breaks are reef dominated. Rather, some are characterized by extensive carbonate sand shoals that are oriented either parallel (Fig. 9) or perpendicular (Fig. 10) to the shelfedge (Ball, 1967; Halley et al., 1983). Such margins are typically open and allow tidal and storm-generated currents to flow unrestricted on and off the bank edge.

Most (but not all) of the bank-parallel sand bodies are located along the leeward sides of carbonate

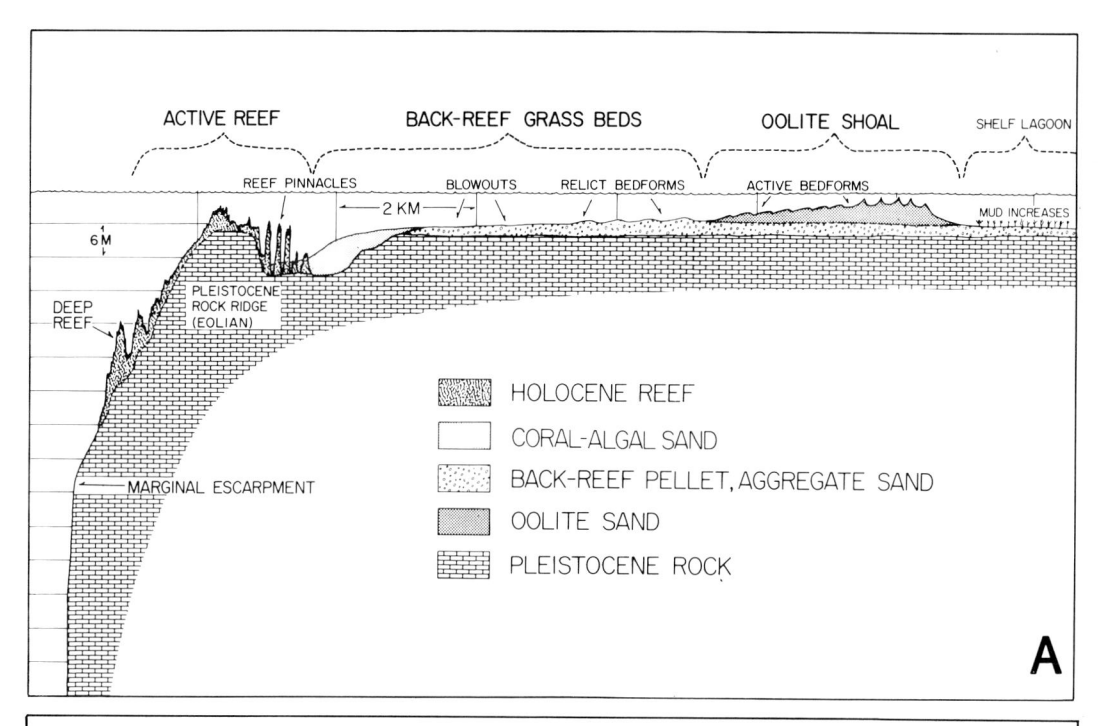

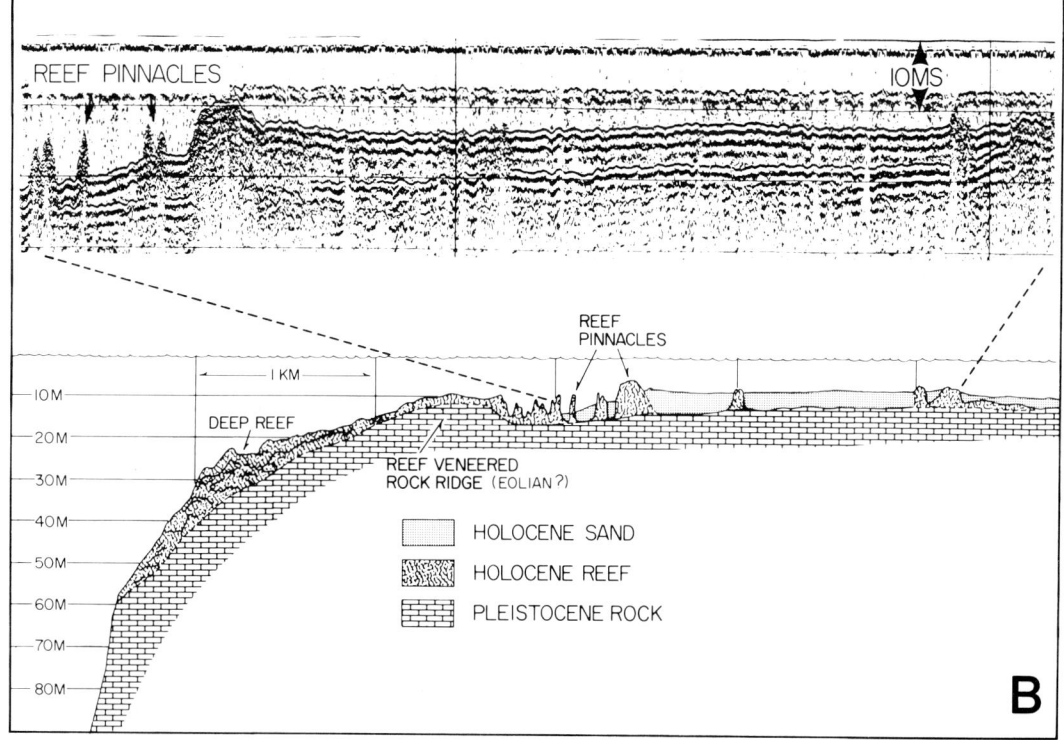

FIG. 7.—A, Generalized schematic cross-section across northwestern Little Bahama Bank. This is a windward, reef-rimmed margin (from Hine and Neumann, 1977, their Fig. 6). B, Diagram of interpretation of a high resolution seismic section across the margin of northwestern Little Bahama Bank. Also illustrated is a detail of the actual trace (from Hine and Neumann, 1977, their Fig. 7).

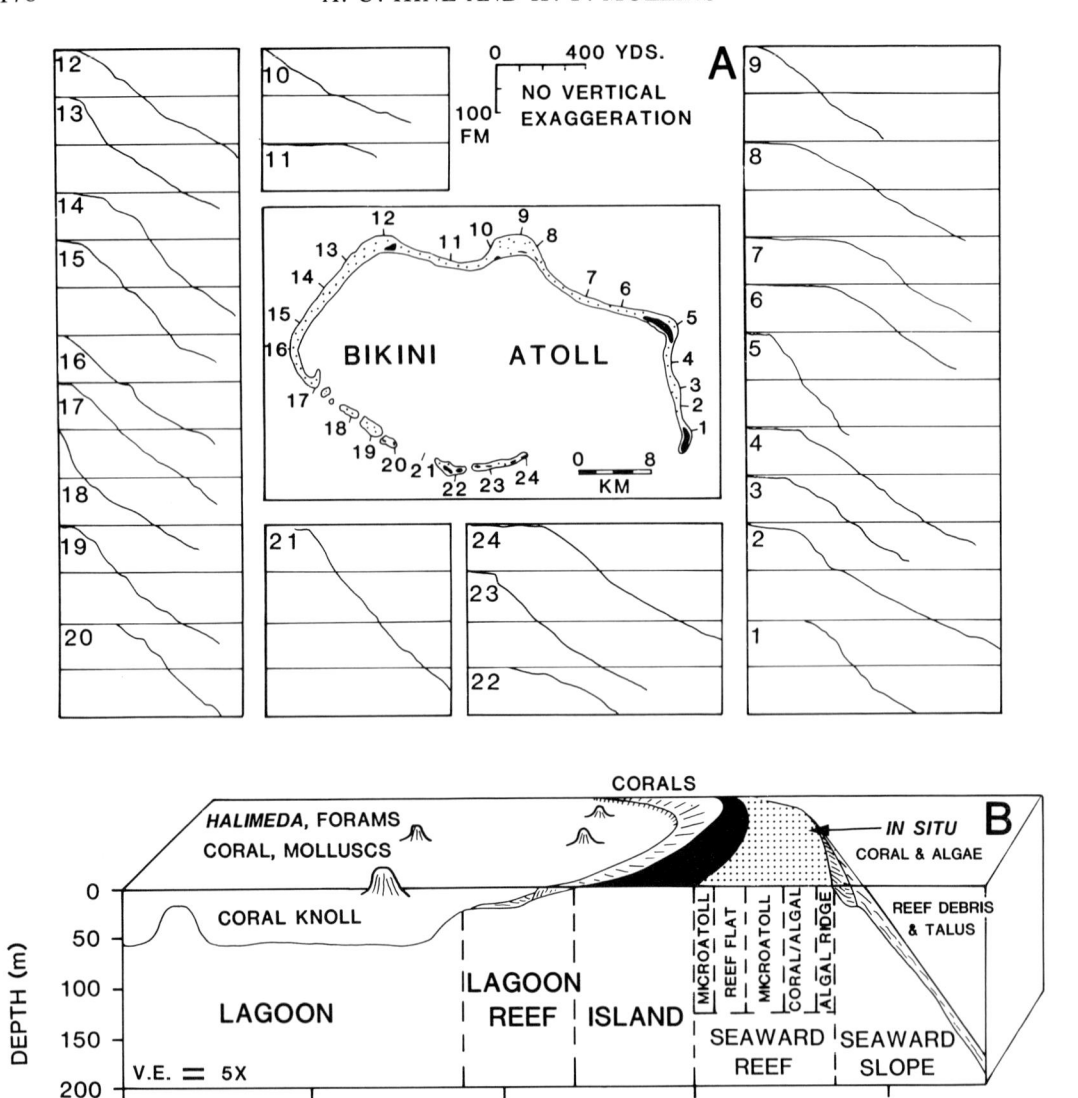

FIG. 8.—*A*, Topographic profiles (not exaggerated) across the shelf-slope breaks of Bikini Atoll (from Emery et al., 1954). *B*, Schematic block diagram model for Pacific atoll shelf-slope breaks based on a figure originally drawn by Wells (1954).

platforms. Here, the offshore directed physical energy flux[1] results in a net offbank transport of carbonate sediment (Neumann and Land, 1975; Hine et al., 1981b). Because of this offbank transport, early Holocene fringing reefs that developed during lower sealevel are inundated by sediment (mostly sands) (Fig. 11) once sealevel rises over the top of the adjacent platform (Hine and Neumann, 1977). Thus, a stratigraphic sequence of reefs overlain by carbonate sand bodies should characterize leeward carbonate bank margins, as illustrated in recent drill holes along the leeward edges of both Little and Great Bahama Banks (Beach and Ginsburg, 1980; Wilber, 1981).

High-resolution seismic reflection profiles across the western (leeward) edges of both Little and Great Bahama Banks typically show large sand bodies

---

[1] The physical energy flux refers to the flow of energy resulting from all water motion due to waves, tides, storms, and oceanic circulation.

overlying older rock ridges or reefs (Fig. 12). Such sand bodies tend to be broad (10 × 10 km), thick (up to 22 m), bankward-thinning blankets of sand (Hine and Neumann, 1977; Hine et al., 1981a). These sand bodies are also compositionally distinct, being dominated by a combination of grains such as peloids, ooids, altered skeletal grains and composite grains, which is in contrast with the fresh skeletal grains typical of windward margin sand bodies (Hine et al., 1981a).

Shallow rock drilling into these bank-parallel leeward sand bodies has demonstrated that they are characterized by sedimentary cycles correlative with Quaternary glacio-eustatic sealevel fluctuations (Wilber, 1981). Laterally discontinuous, unconformity-bound, depositional sequences are deposited during each relative rise of sealevel. An idealized depositional sequence from bottom to top consists of boundstone, packstone and grainstone (Wilber, 1981).

Shelfedge sand bodies trending obliquely or normal to the bank edge are found along large tidal-dominated embayments. In the Bahamas, good examples of these types of sand bodies are found at the ends of Tongue-of-the-Ocean (Fig. 10) and Exuma Sound, as well as portions of Little Bahama Bank (Fig. 13). These sand shoals are predominantly oolitic in composition and are mantled by large-scale bedforms. Individual shoals are up to 10 km long, 3 km wide and 10 m high (Ball, 1967).

NON-RIMMED MARGINS

Non-rimmed, open carbonate shelves (Ginsburg and James, 1974) or ramps (Ahr, 1973; Read, 1982) are gently seaward-dipping surfaces which lack a continuous elevated rim at the shelfedge to act as an energy barrier. As a consequence, open-ocean/deep-water waves are able to propagate across much of the shelf. Thus the break in slope at the seaward edge of modern non-rimmed margins is a subtle, relatively broad area (referred to as the shelfedge) which contrasts sharply with the abrupt, precipitous shelf-slope breaks of rimmed carbonate shelves.

According to Ginsburg and James (1974), non-rimmed shelves fall into two major groups: (1) those consisting of a mixture of quartz and carbonate sand which occur in facies belts parallel to the coastline, such as the west Florida shelf (Gould and Stewart, 1956; Doyle and Sparks, 1980; Doyle, 1981); and (2) those which have isolated organic banks along the outer shelf, such as Campeche Bank along Mexico's Yucatan Peninsula (Logan et al., 1969). Ginsburg and James (1974) have also pointed out a persistent facies pattern of not necessarily coeval sediments on non-rimmed shelves (Table 3): (1) *inner shelf*—molluscan calcarenite; (2) *shelfedge*—ooids, peloids and coralline algae; and (3) *continental slope*—planktonic foraminiferal oozes.

The continental shelf off west Florida is an example of a non-rimmed shelf that lacks major isolated organic build-ups along the shelf margin (Brooks, 1981) (Fig. 14). The continental shelf off west Florida is flat and smooth having a seaward dip of 0.4 m/km to a depth of 70 m that gradually increases to 1.6 m/km (Ginsburg and James, 1974). A bathymetric profile (Fig. 14) across the west

TABLE 2.—ATOLL SHELF/SLOPE BREAK CHARACTERISTICS (SEE FIGURE 8B)

Zone	Dominant Features
Seaward Slope	Spur and groove dominated—grooves infilled with reefal sands/gravels; spurs covered with living algae and coral; terrace at 22 m; hermatypic corals and algae decrease with depth; sediments are coral, red algae, *Halimeda,* forams, molluscs, boulder sized limestone blocks; poorly sorted coarse facies, fines to algal debris and ultimately foram sands at depth.
Algal Ridge	High energy environment; rich growth of crustose red calcareous algae—*Porolithon, Lithothamnion, Lithophyllum, Neogoniolithon;* corals subordinate to algae.
Coral-Algal Zone	Richest coral growth (*Acropora* dominant); 125 m wide; also nodular red algae common.
Outer Microatoll Zone	Subcircular reef masses up to 5 m in diameter; dominated by *Helipora* and *Acropora palifera;* sediments mostly benthic forams, coral and algal clasts.
Main Reef Flat	Horizontal rock surface barren of organic growth; undergoing erosion; covered by foram-algal mat.
Inner Microatoll Zone	Innermost reef zone; microatolls smaller; *Porites* and nodular red algae more common than in Outer Microatoll Zone; *Helipora* pavements form; islands may form; extensive beach-rock and boulder ramparts.
Sources	Ladd et al., 1950; Munk and Sargent, 1954; Wells, 1957; Emery, 1948; Schmalz, 1971; Emery et al., 1954.

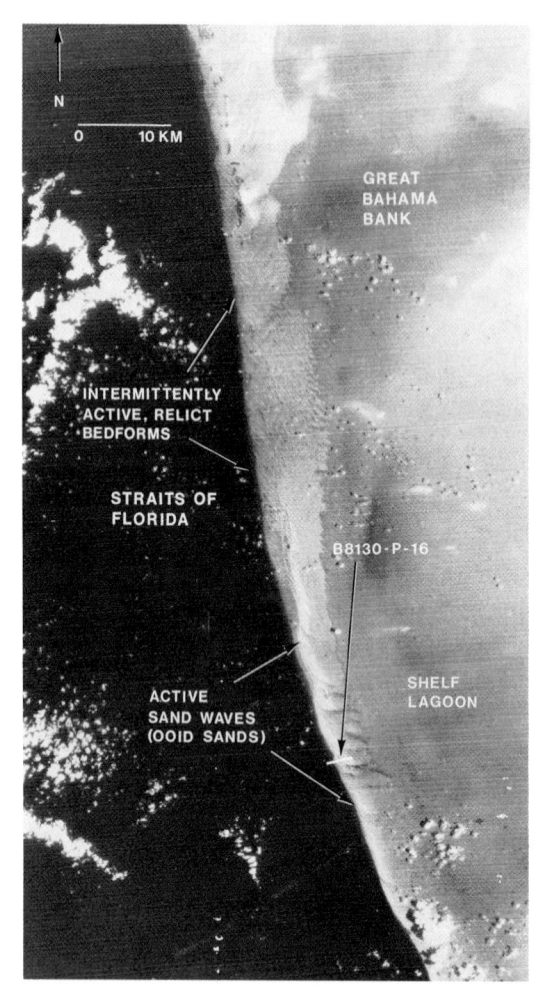

FIG. 9.—Landsat image (#261514563-4-01) of the leeward west-central edge of Great Bahama Bank. Dark area to left is deep water of the Florida Straits; gray area to right is shallow lagoon of Great Bahama Bank; white area between is a bank-parallel sand belt of mostly non-skeletal sand grains. Most reefs along this margin have been buried (see Figs. 11, 12). Note location of seismic profile B8130-P-16 shown in Figure 12.

Florida shelf (vertically exaggerated fifty times) illustrates the smooth, gently-sloping, ramp-like shelf and upper slope that clearly lacks a pronounced geomorphic shelf-slope break.

Campeche Bank, along Mexico's Yucatan Peninsula, is similar in many respects to the west Florida shelf (Fig. 14); the major difference being the presence of isolated, organic build-ups along the shelf-edge (Fig. 14; Logan et al., 1969; Logan, 1969). Exactly why these large isolated reefs are present along the shelfedge of Campeche Bank and not west Florida is poorly understood, although the Campeche Reefs appear to be localized on antecedent topography (Logan, 1969; Macintyre et al., 1977).

DISCUSSION

From a synthesis of the previous examples, the following geological processes appear to control the geomorphology, structure and distribution of organisms and sediments along carbonate shelf margins: (1) tectonism; (2) antecedent topography; (3) physical energy (waves, storms, tides, currents); and (4) sealevel history. Certainly, these are not independent processes, but are frequently interrelated. These four processes, when combined in varying degrees and allowed to operate over varying lengths of time, appear responsible for the range of cross-sections and topographic profiles previously described. Additional processes such as biogenic barrier development, sediment production, sediment transport and cementation are all highly significant, but are considered to be secondary and controlled by one or more of the four primary processes.

Tectonism

The tectonic setting and history of a carbonate shelf or bank controls the basic shape and orientation of the features. Areas such as the Bahamas (Mullins and Lynts, 1977), Belize (Dillon and Vedder, 1973), the Great Barrier Reef (Maxwell, 1968), West Florida/Campeche Bank (Antoine et al., 1974), and the Pacific Atolls (Ladd et al., 1970) have all had their own unique sequence of tectonic events which have shaped the pre-carbonate basement and have affected the ensuing carbonate accumulation by varying rates of subsidence or uplift. The locations of major banks, large open passes, interior seaways, major promontories, shelf relief and slope steepness all may have their origins tied to the underlying rock structure. However, the smaller, second-order features, such as individual reef tracts, islands, and sand shoals, respond to the higher frequency events such as sealevel fluctuations, storms, sediment transporting events and climatic changes. It is thus doubtful that tectonic activity, however rapid, could mask the effects of the higher frequency processes and control anything more than the general location of the shelf-slope boundary.

Physical Energy

Previously, a strong case was made for the importance of the duration, magnitude and direction of the physical energy flux on carbonate bank margins (Hine and Neumann, 1977; Hine et al., 1981a; Hine et al., 1981b). The physical energy flux: (1) provides nutrients so that the biological communities, including reef-building organisms, are stimulated; (2) transports sediments which can either aid the reef by removal of clastic and chemical wastes or can destroy the reef through inundation and burial; and (3) finally, enhances submarine cementation and diagenesis.

Along windward margins, the net energy flux is toward the shelfedge, which results in reef growth, island development, and when sufficient energy levels are attained, algal rim formation (Adey and Burke, 1976). These high-energy margins are characterized by well-developed spur and groove structure, distinct biological zonation, and an active offshelf flushing of sediments and debris, except in the surf zone, where material is transported landward to form wide, shallow flats.

When the net energy flux is off the shelfedge toward the sea, as along leeward margins, sediments formed within the lagoons can be carried to the margin where they bury older rock surfaces and reefs. Here, sand bodies accumulate and store sediment for transport to the adjacent deep-water environment during storms. If islands or rock ridges are present on these leeward margins, offbank sand movement is partially blocked and sand accumulates behind these barriers, which may allow reefs to develop seaward.

Antecedent Topography

A topographic surface can impart a significant feedback effect upon depositional processes (Ball, 1967). Along many carbonate shelf-slope breaks, the underlying Pleistocene topography controls the siting of new Holocene reefs; in essence, former reef rims beget new reef rims. Also, karst topography may play a role in new reef development (Purdy, 1974). Even topography due to subaerial fluvial drainage can control for formation of reef masses upon flooding (Maxwell, 1968). However, topographic highs are not mandatory for reef siting as some reefs have formed on top of seemingly flat surfaces (Garrett and Hine, 1979, in review).

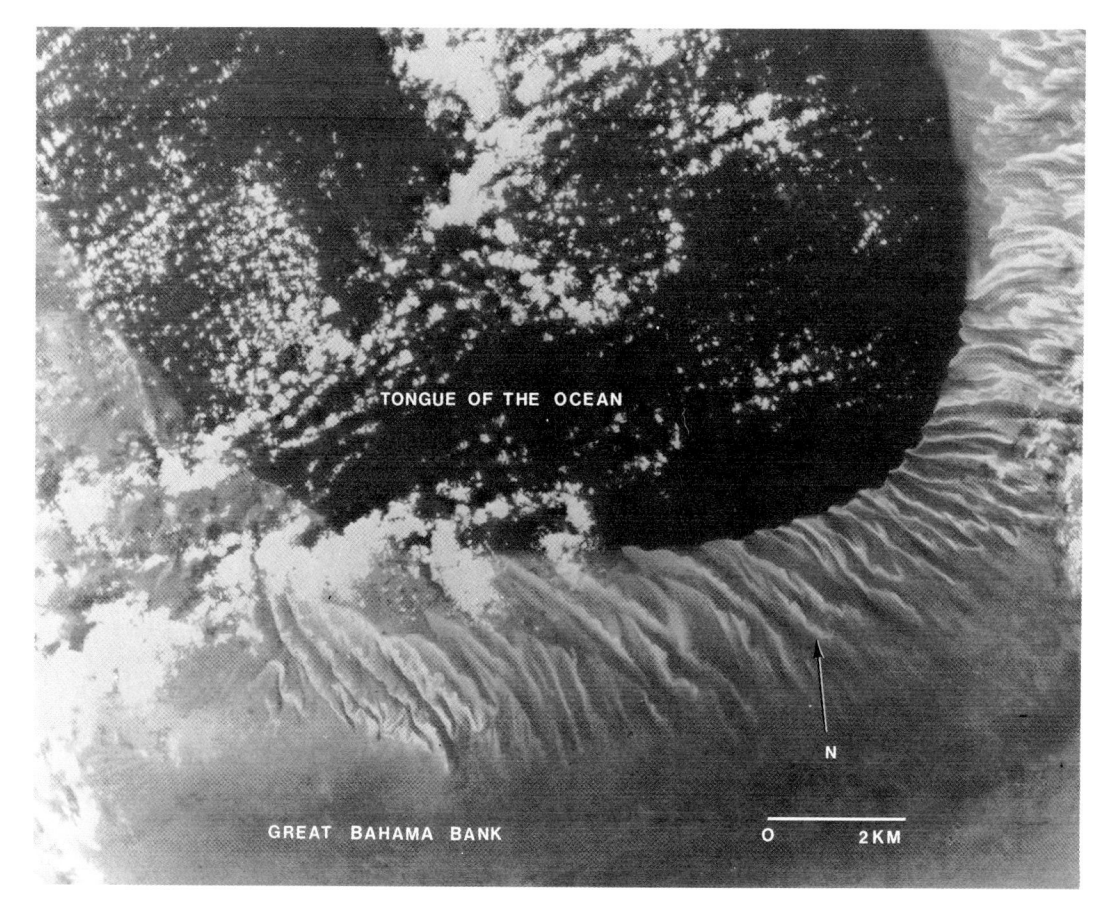

FIG. 10.—Landsat image (#530114295-4-01) of the southern end of Tongue-of-the-Ocean, Bahamas. Dark area at top is deep water; gray area at bottom is shallow lagoon of Great Bahama Bank; between is a series of carbonate sand shoals oriented perpendicular to the shelfedge. Strong tidal currents flow on and off the bank at the end of such embayments. No reefs are present along this modern carbonate bank margin. Sediments are predominantly ooids.

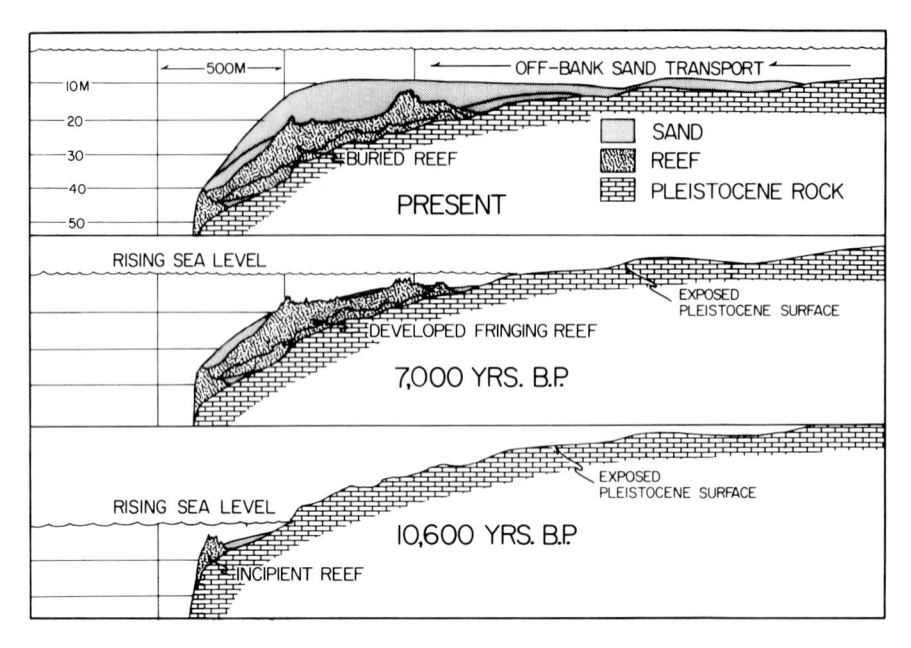

FIG. 11.—Growth history reconstruction of a leeward carbonate bank margin. Fringing reefs develop as sea-level rises, but are buried by off-bank transport once sealevel tops the bank (from Hine and Neumann, 1977, their Fig. 22).

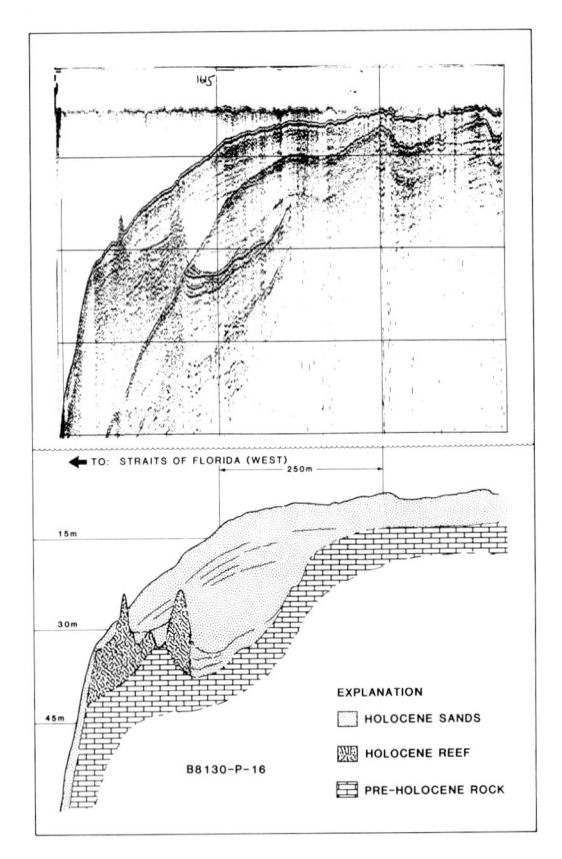

FIG. 12.—High resolution seismic trace (B8130-P-16) and accompanying interpretation from leeward, sand-dominated margin off western Great Bahama Bank, Bahamas (Fig. 9). Note thick sand body nearly burying earlier Holocene reefs. Surficial sands are oolitic. These sands are transported off the bank during storms.

FIG. 13.—Landsat image (#207415001-4-01) of the tide-dominated carbonate bank margin of Little Bahama Bank. Bankward (flood) oriented sand lobes can be seen to have migrated onto the adjacent lagoon.

TABLE 3.—Predominant Grain Types of Open Carbonate Shelves (Based on Ginsburg and James, 1974)

Area	Inner Shelf	Shelfedge	Continental Slope
Southeast U.S.	Molluscs, ooids, lithoclasts	Barnacles, molluscs, coralline algae	Planktonic foraminifera
West Florida	Molluscs	Coralline algae, ooids	Planktonic and benthonic foraminifera
West Africa	Bryozoans, molluscs	Molluscs, bryozoans, ooids, benthonic foraminifera	Planktonic and benthonic foraminifera
East India	Molluscs, bryozoa	Ooids	Planktonic foraminifera
Campeche Bank	Molluscs	Coralline algae, ooids, peloids, lithoclasts	Planktonic foraminifera

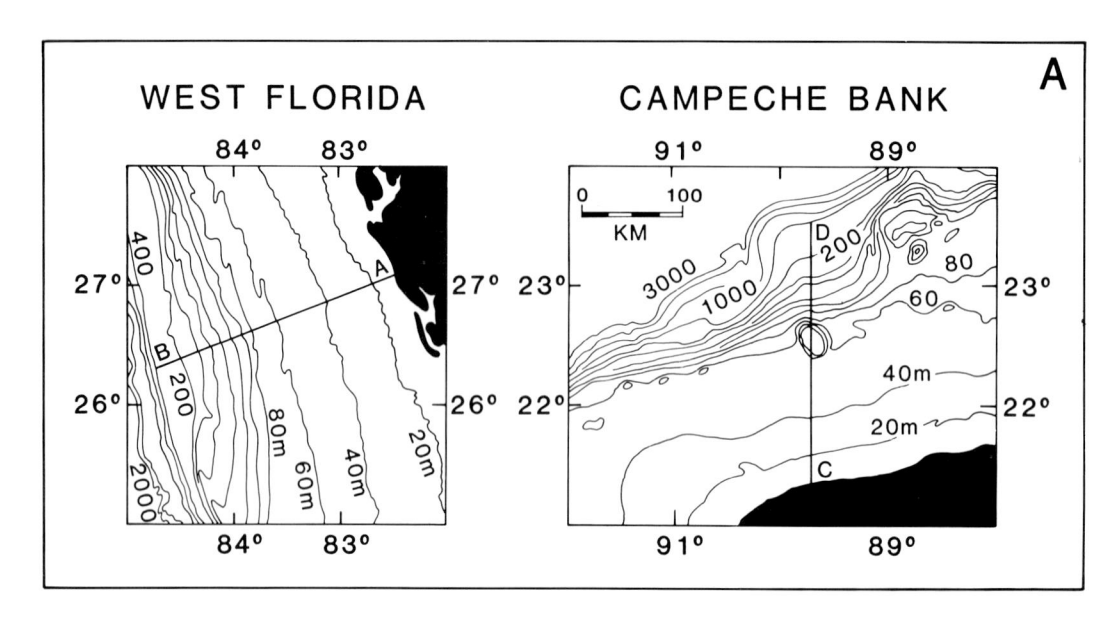

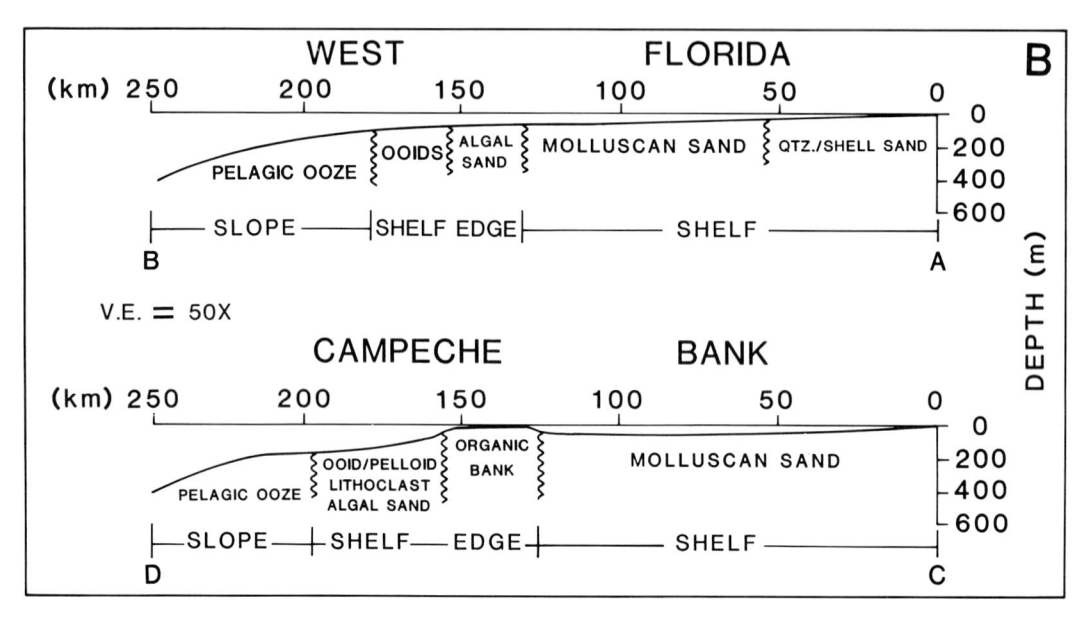

FIG. 14.—A, Generalized bathymetric maps for the non-rimmed carbonate margins of west Florida (left) and Campeche Bank (right) (based on Sorensen et al., 1977). B, Topographic cross-sections across the shelf-slope breaks of west Florida (top) and Campeche Bank (bottom). Note that even on these highly exaggerated (×50) profiles, the shelf-slope break is not pronounced. Generalized lateral facies patterns based on Gould and Stewart (1956) and Logan et al. (1969) are also shown.

Current patterns are also affected by pre-existing topographic highs. Rock ridges, for example, can accelerate flows which in turn stimulate ooid formation (Purdy, 1963). Even "soft" topographic features such as a field of sand waves are capable of blocking or redirecting currents (Hine, 1977).

Sealevel History

The fluctuation of sealevel, which forces physical, biological, and chemical processes to migrate both vertically and laterally, provides a complicating factor, not only in understanding the present, static shelf-slope profile, but also its internal cross-

TABLE 4.—CRITERIA FOR RECOGNITION OF DIFFERENT ANCIENT CARBONATE SHELF-SLOPE BREAKS

Type	Recognition Criteria
Rimmed, Reef-Dominated	Ecological zonation, coral growth form zonation (encrusting to massive to platey—shallow to deep); spur and groove structure; facies change—back reef rubble, sand bodies, patch reefs to *in situ* reef framework to fore-reef rubble, peri-platform sediments; extensive early diagenesis.
Atolls	Ecological zonation, facies changes similar to Rimmed Reef-Dominated Type; circular/elliptical reef trends; very abrupt/precipitous slope and facies changes; algal ridge zone common; volcanic basement.
Rimmed, Sand Shoal Dominated	Grainstones (high in non-skeletal constituents) covering reefs; cross-lamination structures in sands; submarine cemented zones; cyclic sedimentation consisting of unconformity bound reef-grainstone sequences.
Non-Rimmed	Absence of abrupt, precipitous changes in slope; gradual facies transition—inner shelf molluscan calcarenites to shelfedge algal, ooid, peloidal, lithoclast sands to slope pelagic oozes.

section (Fig. 2). The numerous terraces and the seemingly worldwide agreement of some of their depths found on carbonate margins may be products of both antecedent topography (Goreau and Land, 1974) and sealevel oscillations (Maxwell, 1968; James and Ginsburg, 1979). The level of the sea controls the flooding of back-reef environments which may, in turn, significantly impact the reef structure as cold and turbid waters form and flow seaward across the reefs (Lighty, 1977; Lighty et al., 1980).

The rate of sealevel movement and changes in these rates can control the formation of shelf-margin sand bodies (Hine and Neumann, 1977; Palmer, 1979; Hine et al., 1981a; Wilber, 1981). More importantly, these rates can control the type and internal structure of reefs whether they be starting up, catching up or keeping up with sea level (Schlager, 1981; Kendall and Schlager, 1981). The extent and thickness of sediment/rock packages formed along carbonate margins as a result of sealevel fluctuations are complex and have only been recently addressed through seismic data and the drill (Macintyre and Glynn, 1976; Goter, 1979; Beach and Ginsburg, 1980; Lighty et al., 1980; Wilber, 1981). The examination of these transgressive/regressive units and their subsequent diagenesis is one of the important areas of new research being conducted along carbonate shelf-slope breaks.

CRITERIA FOR RECOGNITION IN THE ROCK RECORD

Given good outcrop exposures or a number of closely-spaced core holes along a depositional dip section, carbonate shelf-slope breaks should be recognizable in the rock record (James and Mountjoy, this volume; Read, 1982). Ancient rimmed margins are probably more easily recognizable than open, non-rimmed shelf margins, particularly on seismic reflection profiles (Bubb and Hatlelid, 1977). Because of the rapid changes in paleodepths across ancient rimmed margins, the structures, fossils, grain sizes, sediment types, porosity, and permeability change abruptly as well. Overall, this should tell the investigator that the paleodepth has changed and a steep shelf-slope break has been located.

The more gently sloping, non-rimmed margins do not feature rapid changes in depth, and the paleoenvironments are not so nearly compressed laterally. Discovery and detection of the shelf-slope transition will have to come from the careful analysis of lateral facies relationships.

Some specific criteria that might be useful in recognizing ancient carbonate shelf-slope breaks are listed in Table 4. A more detailed discussion of ancient shelf-slope settings can be found in MacIlreath and James (1979), James (1979) and James and Mountjoy (this volume). However, caution should be exercised for, as Longman (1981) indicates, particularly for reef-rimmed margins, the Holocene may not be a suitable model from which to examine the ancient. This is primarily due to the modern dominance of scleractinian-hermatypic corals, the relative absence of modern intra-cratonic settings, the rapid rate of early-mid Holocene sealevel rise, and the relatively short period of time that modern reefs have existed.

CONCLUSIONS

The carbonate shelf-slope break, occurring primarily at lower latitudes, is defined as the change in slope marking the boundary between a relatively low gradient shelf and a more steeply seaward-descending slope. It is an important component of the geographically broader carbonate bank margin and it represents the primary zone of relatively rapidly changing depth and physical energy dissipation. Consequently, the shelf-slope break and the entire carbonate margin consist of a spectrum of facies, lithologies, organisms, and structures. The shelf-slope break is also an area of enhanced carbonate sediment production and early diagenesis due to elevated levels of turbulence and nutrients.

Four types of carbonate shelf-slope breaks are presented based upon setting and cross-sectional morphology. They are: (1) reef-dominated, rimmed; (2) atoll; (3) sand-shoal-dominated, rimmed; and (4) non-rimmed.

Rimmed margins dominated by reefs at the shelf-slope break are common in windward settings. They frequently have a back reef rubble zone, a reef-front, a deep fore-reef, a steep marginal escarpment, and a deep talus or debris slope. Generally there is also strong biological zonation across these features, and well developed spur and groove geomorphology.

Atoll shelf-slope breaks are the most abrupt and precipitous of any carbonate margins, being dominated by coral reefs and algal ridges. They are similar to reef-dominated rimmed margins but differ significantly in areal geometry (i.e., atolls are circular whereas reef-rimmed margins are generally more linear) and setting (atolls being in the open ocean on top of submerged volcanoes).

Rimmed margins having sand bodies and shoals are commonly leeward or tide-dominated. Oolite shoals are generally found in areas of strong tidal currents. Non-skeletal sands are in abundance and typically bury older rock surfaces, including reefs. Along leeward margins, these sands are also efficiently transported off the marginal escarpment to the deeper slopes.

No distinct geomorphic shelf-slope break occurs along open, non-rimmed margins. The boundary between shelf and slope is likely to be broad, gradational, subtle, and deep. On these modern ramps the shelfedge facies consists of various amounts of skeletal and non-skeletal sands that grade up-dip into molluscan calcarenites and/or bioherms, and down-dip into pelagic oozes with abundant planktonic foraminifera.

The four shelf margin types and the specific examples discussed indicate that four major processes control the nature of the carbonate shelf-slope break: (a) tectonism; (b) antecedent topography; (c) physical energy (waves, tides, storm currents) and (d) sealevel history. These primary processes control the level of other secondary factors such as reef development, sediment production and transport, and cementation.

ACKNOWLEDGMENTS

A review paper such as this would not have been possible without the assistance of many of our colleagues involved in modern carbonate research. We specifically would like to thank those individuals who allowed us to reproduce figures from their work. They are K. O. Emery, D. K. Hubbard, N. P. James, J. W. Wells, and J. L. Wilson. Special thanks are extended to D. J. Stanley for inviting us to participate in this symposium volume.

Financial support for writing this manuscript came from the Department of Marine Science at the University of South Florida and Moss Landing Marine Laboratories of the California State University System. Much of our work presented here was funded by the National Science Foundation through grants to the University of North Carolina, Chapel Hill (A. C. Neumann and A. C. Hine co-principal investigators). We also thank N. P. James and an anonymous reviewer for helpful suggestions.

REFERENCES

ADEY, W. H., 1978, Coral reef morphogenesis: a multidimensional model: Science, v. 202, p. 831–837.
———, AND BURKE, R., 1976, Holocene bioherms (algal ridges and bank-barrier reefs) of the eastern Caribbean: Geol. Soc. America Bull., v. 87, p. 95–109.
AHR, W. M., 1973, The carbonate ramp: An alternative to the shelf model: Trans. Gulf Coast Assoc. Geol. Socs., v. 23, p. 221–225.
ANTOINE, J. W., MARTIN, R. G., PYLE, T. G., AND BRYANT, W. R., 1974, Continental margins of the Gulf of Mexico, in Burk, C. A., and Drake, C. L., eds., Geology of Continental Margins: New York, Springer-Verlag, p. 683–693.
BALL, M. M., 1967, Carbonate sand bodies of Florida and the Bahamas: Jour. Sed. Petrology, v. 37, p. 556–591.
BATHURST, R. G. C., 1975, Carbonate Sediments and Their Diagenesis: New York, Elsevier, 658 p.
BEACH, D. K., AND GINSBURG, R. N., 1980, Facies succession, Plio-Pleistocene carbonates, northwestern Great Bahama Bank: Amer. Assoc. Petroleum Geologists Bull., v. 64, p. 1634–1642.
BROOKS, G. P., 1981, Recent carbonate sediments of the Florida Middle Ground reef system; northeastern Gulf of Mexico [unpub. M.S. thesis]: St. Petersburg, Univ. of South Florida, 137 p.
BUBB, J. N., AND HATLELID, W. G., 1977, Seismic recognition of carbonate buildups: Am. Assoc. Petroleum Geologists Memoir 26, p. 185–204.
DILLON, W. P., AND VEDDER, J. G., 1973, Structure and development of the continental margin of British Honduras: Geol. Soc. America Bull., v. 84, p. 2713–2732.
DOYLE, L. J., 1981, Depositional systems of the continental margin of the eastern Gulf of Mexico west of peninsular Florida: a possible modern analog to some depositional models for the Permian Delaware Basin: Trans. Gulf Coast Assoc. Geol. Socs., v. 31, p. 279–282.

———, AND SPARKS, T. H., 1980, Sediments of the Mississippi, Alabama, and Florida (MAFLA) continental shelf: Jour. Sed. Petrology, v. 50, p. 905–916.
DUNHAM, R. J., 1970, Stratigraphic reefs versus ecologic reefs: Am. Assoc. Petroleum Geologists Bull., v. 54, p. 1931–1932.
EMERY, K. O., 1948, Submarine geology of Bikini Atoll: Geol. Soc. America Bull., v. 59, p. 855–860.
———, TRACEY, J. I., JR., AND LADD, H. S., 1954, Geology of Bikini and nearby atolls: U.S. Geol. Survey Prof. Paper 260-A, 265 p.
ENOS, P., AND PERKINS, R. D., 1977, Quaternary sedimentation in south Florida: Geol. Soc. America Memoir 147, 198 p.
FRIEDMAN, G. M., AMIEL, A. J., AND SCHNEIDERMANN, N., 1974, Submarine cementation in reefs: example from the Red Sea: Jour. Sed. Petrology, v. 44, p. 816–825.
GARRETT, P., AND HINE, A. C., 1979, Probing Bermuda's lagoons and reefs: Am. Assoc. Petroleum Geologists Bull., v. 63, p. 455.
———, AND ———, 1983, Bermuda Atoll: Sedimentary evolution revealed by seismic stratigraphy: Geol. Soc. America Bull., in review.
GINSBURG, R. N., AND JAMES, N. P., 1974, Holocene carbonate sediments of continental shelves, in Burk, C. A., and Drake, C. L., eds., Geology of Continental Margins: New York, Springer-Verlag, p. 137–155.
———, AND SHINN, E. A., 1964, Distribution of the reef-building community in Florida and the Bahamas: Am. Assoc. Petroleum Geologists Bull., v. 48, p. 527.
GOREAU, T. F., AND LAND, L. S., 1974, Fore-reef morphology and depositional processes, north Jamaica: Soc. Econ. Paleontologists Mineralogists Spec. Pub. 18, p. 77–89.
GOTER, E. R., 1979, Depositional and diagenetic history of the windward reef of Enewitok Atoll during the mid to late Pleistocene [unpub. Ph.D. thesis]: Troy, N.Y., Rensselaer Polytechnic Inst., 241 p.
GOULD, H. R., AND STEWART, R. H., 1956, Continental terrace sediments in the northeastern Gulf of Mexico: Soc. Econ. Paleontologists Mineralogists Spec. Pub. 3, p. 2–19.
HALLEY, R., HARRIS, P. M., AND HINE, A. C., 1983, Bank margin sand accumulation, in Scholle, P. A., ed., Recognizing Carbonate Depositional Environments: Am. Assoc. Petroleum Geologists Memoir, in press.
HINE, A. C., 1977, Lily Bank, Bahamas: history of an active oolite sand shoal: Jour. Sed. Petrology, v. 47, p. 1554–1581.
———, AND NEUMANN, A. C., 1977, Shallow carbonate bank margin growth and structure, Little Bahama Bank: Am. Assoc. Petroleum Geologists Bull., v. 61, p. 376–406.
———, WILBER, R. J., AND NEUMANN, A. C., 1981a, Carbonate sand bodies along contrasting shallow bank margins facing open seaways in northern Bahamas: Am. Assoc. Petroleum Geologists Bull., v. 65, p. 261–290.
———, ———, BANE, J. M., NEUMANN, A. C., AND LORENSON, K. R., 1981b, Offbank transport of carbonate sands along open leeward bank margins, northern Bahamas: Mar. Geology, v. 42, p. 327–348.
HUBBARD, D. K., WARD, L. G., FITZGERALD, D. M., AND HINE, A. C., 1976, Bank margin morphology and sedimentation, Lucaya, Grand Bahama Island: Columbia, S.C., Univ. South Carolina Dept. Geol. Tech. Rept. 7-CRD, 36 p.
JAMES, N. P., 1979, Facies models 11: reefs, in Walker, R. G., ed., Facies Models: Geosci. Canada Reprint Ser. 1, p. 121–132.
———, GINSBURG, R. N., MARSZALEK, D. S., AND CHOQUETTE, P. W., 1976, Facies and fabric specificity of early subsea cements in shallow Belize (British Honduras) reefs: Jour. Sed. Petrology, v. 46, p. 523–544.
———, AND GINSBURG, R. N., 1979, The Seaward Margin of Belize Barrier and Atoll Reefs: Internat. Assoc. Sed. Spec. Pub. No. 3, 191 p.
———, AND MOUNTJOY, E. W., 1983, The shelf-slope break in fossil carbonate platforms: an overview, in Stanley, D. J., and Moore, G. T., eds., The Shelfbreak: Critical Interface on Continental Margins: Soc. Econ. Paleontologists Mineralogists Spec. Pub. 33, p. 189–206.
KENDALL, C. G. ST. G., AND SCHLAGER, W., 1981, Carbonates and relative changes in sea level: Mar. Geology, v. 44, p. 181–212.
LADD, H. S., 1973, Bikini and Eniwetok Atolls, Marshall Islands: in Jones, O. A., and Endean, R., eds., Biology and Geology of Coral Reefs, v. 1: New York, Academic Press, p. 93–111.
———, TRACEY, J. I., JR., WELLS, J. W., AND EMERY, K. O., 1950, Organic growth and sedimentation on an atoll: Jour. Geology, v. 58, p. 410–425.
———, ———, AND GROSS, M. G., 1970, Deep drilling on Midway Atoll: U.S. Geol. Survey Professional Paper 680-A, p. A1–A22.
LAND, L. S., AND GOREAU, T. F., 1970, Submarine lithification of Jamaican reefs: Jour. Sed. Petrology, v. 40, p. 457–462.
LEES, A., 1975, Possible influences of salinity and temperature on modern shelf carbonate sedimentation: Mar. Geology, v. 19, p. 159–198.
LIGHTY, R. G., 1977, Relict shelf-edge Holocene coral reef, southeast coast of Florida: Proc. Third Internat. Coral Reef Symposium, v. 2, p. 215–221.
———, MACINTYRE, I. G., AND NEUMANN, A. C., 1980, Demise of a Holocene barrier reef complex, northern Bahamas: Geol. Soc. America Abs. with Programs, v. 12, p. 471.
LOGAN, B. W., 1969, Coral reefs and banks, Yucatan Shelf, Mexico: Am. Assoc. Petroleum Geologists Memoir, No. 11, p. 129–198.

———, HARDING, J. L., AHR, W. M., WILLIAMS, J. D., AND SNEAD, R. G., 1969, Late Quaternary carbonate sediments of Yucatan Shelf, Mexico: Am. Assoc. Petroleum Geologists Memoir No. 11, p. 5–128.
LONGMAN, M. W., 1981, A process approach to recognizing facies of reef complexes: Soc. Econ. Paleontologists Mineralogists Spec. Pub. No. 30, p. 9–40.
MACINTYRE, I. G., AND GLYNN, P. W., 1976, Evolution of a modern Caribbean fringing reef, Galeta Point, Panama: Am. Assoc. Petroleum Geologists Bull., v. 60, p. 1054–1072.
———, BURKE, R. B., AND STUCKENRATH, R., 1977, Thickest recorded Holocene reef section, Isla Perez core hole, Alacran Reef, Mexico: Geology, v. 5, p. 749–754.
MAXWELL, W. G. H., 1968, Atlas of the Great Barrier Reef: Amsterdam, Elsevier, 258 p.
———, 1973, Sediments of the Great Barrier Reef province, in Jones, O. A., and Endean, R., eds., Biology and Geology of Coral Reefs, volume 1: New York, Academic Press, p. 299–345.
MCILREATH, I. A., AND JAMES, N. P., 1979, Facies models 12, carbonate slopes, in Walker, R. G., ed., Facies Models: Geosci. Canada Reprint Ser. 1, p. 133–143.
MILLIMAN, J. D., 1974, Marine Carbonates: New York, Springer-Verlag, 375 p.
MOORE, C. H., GRAHAM, E. A., AND LAND, L. S., 1976, Sediment transport and dispersal across the deep fore-reef and island slope (−55 m to −305 m), Discovery Bay, Jamaica: Jour. Sed. Petrology, v. 46, p. 174–187.
MULLINS, H. T., AND LYNTS, G. W., 1977, Origin of the northwestern Bahama Platform: review and reinterpretation: Geol. Soc. America Bull., v. 88, p. 1447–1461.
———, AND NEUMANN, A. C., 1979, Deep carbonate bank margin structure and sedimentation in the northern Bahamas: Soc. Econ. Paleontologists Mineralogists Spec. Pub. No. 27, p. 165–192.
MUNK, W. H., AND SARGENT, M. C., 1954, Adjustment of Bikini Atoll to ocean waves, Bikini and nearby atolls, Marshall Islands: U.S. Geol. Survey Professional Paper 260-C, p. 275–280.
NEUMANN, A. C., AND LAND, L. S., 1975, Lime mud deposition and calcareous algae in the Bight of Abaco, Bahamas: Jour. Sed. Petrology, v. 45, p. 763–768.
PALMER, M. S., 1979, Holocene facies geometry of the leeward bank margin of Tongue-of-the-Ocean, Bahamas [unpub. M.S. thesis]: Miami, Univ. Miami, 199 p.
PURDY, E. G., 1963, Recent calcium carbonate facies of the Great Bahama Bank—sedimentary facies: Jour. Geology, v. 71, p. 472–497.
———, 1974, Reef configurations: cause and effect: Soc. Econ. Paleontologists Mineralogists Spec. Pub. 18, p. 9–76.
READ, J. F., 1982, Carbonate platforms of passive (extensional) continental margins: types, characteristics, and evolution: Tectonophysics, v. 81, p. 195–212.
SCHLAGER, W., 1981, The paradox of drowned reefs and carbonate platforms: Geol. Soc. America Bull., v. 92, p. 197–211.
SCHMALZ, R. F., 1971, Formation of beachrock at Enewetak Atoll, in Bricker, O. P., ed., Carbonate Cements: Johns Hopkins Univ. Studies in Geology No. 19, p. 17–24.
SORENSON, F. H., SNODGRASS, L. W., REBMAN, J. H., MURCHISON, R. R., JONES, C. R., AND MARTIN, R. G., 1977, Preliminary bathymetric map of Gulf of Mexico region: U.S. Geol. Survey Open-File Map, Corpus Christi, Texas, 1 sheet.
STORR, J. F., 1964, Ecology and oceanography of the coral-reef tract, Abaco Island, Bahamas: Geol. Soc. America Spec. Paper 79, 98 p.
TRACEY, J. E., JR., CLOUD, P. E., JR, AND EMERY, K. O., 1955, Conspicuous features of organic reefs: Atoll Res. Bull. 46, 3 p.
WELLS, J. W., 1954, Recent corals of the Marshall Islands: U.S. Geol. Survey Professional Paper 260-I, p. 385–478.
———, 1957, Coral reefs: Geol. Soc. America Memoir 67, v. 1, p. 609–631.
WIENS, H. J., 1962, Atoll Environment and Ecology: New Haven, Yale Univ. Press, 532 p.
WILBER, R. J., 1981, Late Quaternary history of a leeward carbonate bank margin: A chronostratigraphic approach [unpub. Ph.D. thesis]: Chapel Hill, Univ. North Carolina, 290 p.
WILSON, J. L., 1975, Carbonate Facies in Geologic History: New York, Springer-Verlag, 471 p.

SHELF-SLOPE BREAK IN FOSSIL CARBONATE PLATFORMS: AN OVERVIEW[1]

NOEL P. JAMES
Department of Geology, Memorial University of Newfoundland, St. John's, Newfoundland A1B 3X5

ERIC W. MOUNTJOY
Department of Geological Science, McGill University, Montreal, Quebec H3A 2A7

ABSTRACT

The shelf-slope break is the zone which controls the evolution of fossil carbonate platforms and shelves because it is the locus of most rapid carbonate fixation, both organic and inorganic. The fossil record of carbonate platforms and shelf margins is biased. The best known and most studied shelf-slope breaks are of middle and late Paleozoic age, occuring in intracratonic basins. Those of early Paleozoic and Mesozoic through Cenozoic age, which occur mostly along continental margins, are poorly known. Shelf-slope breaks of Precambrian age are poorly known. Five recurring types of carbonate shelf-slope break are found in the fossil record: (1) stationary, (2) offlap, (3) onlap, (4) drowned and (5) exposed. The reefs and carbonate sand shoals at the break are the line source of most sediment deposited on the foreslope. In the case of drowned or exposed margins this sediment production is arrested, resulting in starved slope and basin margin sedimentation. Most examples in the rock record are a combination of these types. The nature of the break through time depends upon the types of organisms present and their paleoenvironment. When large skeletal metazoans were alive, barrier reefs formed at the margin and reef mounds grew on-shelf or downslope, but when only diminutive skeletal organisms occurred, the break is generally formed by sand shoals. Lithologies at the break are particularly prone to diagenetic alteration. Intensive early diagenesis tends to preserve texture but decrease porosity, whereas intensive late diagenesis generally destroys texture but creates good reservoir rock.

INTRODUCTION

The shelf-slope break in ancient carbonate complexes, whether the rim of an isolated platform or the margin of a continental shelf, was the crucial, yet often elusive, element of their anatomy. The zone was crucial because, unlike shelves of terrigenous clastic sediments, the facies developed at the carbonate shelf-slope break controls the way in which the platform evolves. It was here that the most diverse community of organisms grew, the most rapid accretion took place, the most intensive diagenesis probably occurred and the most rewarding hydrocarbon and mineral deposits may have accumulated.

This facies is elusive because it is relatively narrow and so chances of it outcropping or being intersected by drilling are low. Also because of the marked lithological differences between shelf carbonates and basin shales, it tends to be strongly deformed during orogenesis. As a result, the nature of the shelf-slope break is commonly interpreted rather than observed, and synthesized on the basis of information from surrounding facies.

The modern carbonate shelf-slope break, as illustrated in the preceeding article by Hine and Mullins (this volume), can be easily seen in present day shallow tropical seas and sometimes visited by submersible in deeper locations. This zone exhibits a wide range of deposits, the differences between which are largely the result of location in terms of latitude, configuration of the margin, closeness to terrigenous sedimentation, pre-existing topography, wave energy, and composition of the surrounding water mass. Such factors were equally important in the past and as these authors have demonstrated, similar styles of platform margins can be recognized in the fossil record.

In this chapter, we provide the added dimension of time to this synthesis. This perspective allows us to assess, from examples in the fossil record, the results of the interaction between several variables which in turn control the evolution of the shelf-slope break. The most important of these factors, in our view, are: (1) the tectonic setting in which the platform grew, (2) fluctuations in sealevel, (3) the dynamics of sedimentation at the margin proper, (4) the variation of reef-building organisms with time, and (5) diagenesis.

The interactions between these variables ultimately leads to a wider spectrum of types than in the modern ocean, but the definition of their character is less precise.

PREVIOUS WORK

The subject of facies development on fossil carbonate platforms, including the shelf-slope break, has been addressed by many authors in the last few years. The attributes of carbonate depositional environments in general are detailed by Wilson (1975),

---

[1]Dedicated to Wolfgang Krebs, tragically killed in November, 1981, who enriched so many with his warm human qualities and excellent science.

Copyright © 1983, The Society of Economic Paleontologists and Mineralogists

summarized by Sellwood (1978), modelled by James (1978, 1982) and profusely illustrated by many authors in Scholle et al. (1983). Particularly relevant to this overview of the shelf-slope break are recent papers in Toomey (1980) on European fossil reefs and in Cook and Enos (1977) on deep-water carbonate slope facies.

At the time of writing, two comprehensive articles documenting the different types of fossil carbonate platforms and the various stages in their development have recently appeared (Kendall and Schlager, 1981; Read, 1982). These latest contributions, read in conjunction with this overview, should provide a comprehensive guide to the nature of the carbonate shelf-slope break in the fossil record.

THE SETTING OF FOSSIL CARBONATE SHELVES AND PLATFORMS

The location of fossil carbonate shelves and platforms is tied to the distribution of crustal plates and the geometry of ocean basins through geologic time.

Precambrian.—Compared to the Phanerozoic, the nature of carbonate platforms in the Precambrian is poorly understood, except in local areas. Although known from shield areas in the USSR, Australia, Africa and North America, perhaps the best studied complexes are Proterozoic in age on and adjacent to the Slave Craton in northern Canada (Campbell, 1981). From this area, where carbonate platforms are developed within intracratonic basins (Campbell and Cecile, 1981), along the margins of evolving aulacogens (Hoffman, 1974) and rimming continental margins (Hoffman, 1973, 1980), it appears a wide spectrum of settings were present.

Early Paleozoic.—Cambro-Ordovician time is characterized by extensive, relatively flat, cratonic regions partly to completely covered by shallow, epeiric seas. The shelf-slope break was located along the edge of the continental shelf and faced relatively large ocean basins. Because they are continental margins, these examples have either been consumed during orogenesis or intensively deformed. The record is thus patchy and those margins that do remain are mostly found today in mountain belts. Aside from a few subsurface examples, most of these crop out in the Cordillera and Appalachian-Caledonian Crogenic belts, the Urals and in eastern Australia (Fig. 1A).

Middle and Late Paleozoic.—In the middle and late Paleozoic (Silurian to Permian) seaways were somewhat reduced in size but major downwarps developed on and between plates. As a result, almost all carbonate platforms of this age are developed as marginal shelves around, or isolated buildups within, these intracratonic basins (Fig. 1B). Even though most are buried in the subsurface, and so are of great economic interest, subsequent tectonics has exposed enough of these platforms that together they provide our most extensive inventory of fossil platform margins. The best studied examples come from the interior of North America, Australia and the Hercynian fold belt of western Europe.

Mesozoic.—With continental breakup in late Permian and early Triassic time, the major sites of carbonate sedimentation shifted to the northern and southern margins of the circumglobal Tethyan seaway and basins adjacent to it (Fig. 1C). Sedimen-

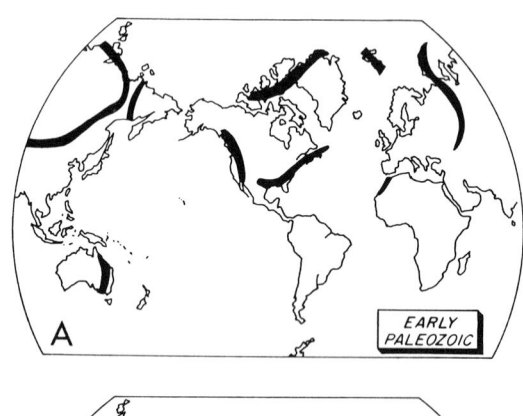

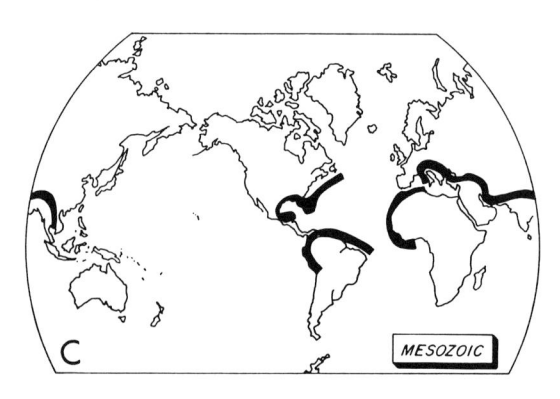

FIG. 1.—Generalized sketch maps illustrating the position of major carbonate platform margins during (A) Early Paleozoic (Cambro-Ordovician), (B) Mid-Late Paleozoic (Silurian-Permian) and (C) Mesozoic time.

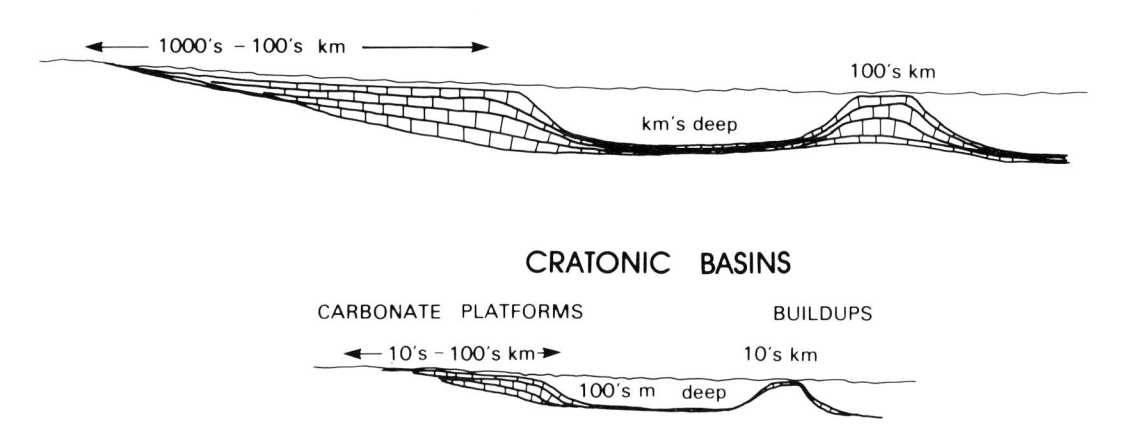

FIG. 2.—Diagram illustrating the difference in scale between continental margins or epicontinental platforms and isolated carbonate platforms in open ocean basins versus carbonate platforms and buildups developed in intracratonic basins.

tation was either in the form of continental shelves marginal to, or as isolated platforms within, the Tethys. These are now exposed and well known in central America, and the Alpine-Himalayan fold belt, and are significant structures in the subsurface of the Gulf of Mexico, the Bahamas, the eastern coast of North America, west coast of Africa and the Middle East.

Cenozoic.—This trend continues into the early Cenozoic, but with continental drift and accompanying plate collision in mid-Tertiary time, the Mediterranean was isolated as a separate sea and the Isthmus of Panama uplifted, thus destroying the Tethys as a circumglobal seaway. Consequently, post-Miocene carbonates are restricted largely to the Caribbean and Indo-Pacific. While some early Cenozoic platform margins are exposed in the Alpine-Himalayan chain, around the Gulf of Mexico and in Australia, and some Neogene examples are present on isolated islands in the modern tropics, most are buried in the subsurface.

Reduced to its basic elements, the carbonate shelf-slope break during any of these periods has the same makeup, whether it bounds a continental margin or small atoll reef. Obviously the scales are different and become important when comparing platforms developed in cratonic downwarps with those along plate boundaries. This is primarily because the life of an intracratonic basin is relatively short compared to that of a passive continental margin and sedimentation is commonly limited to one, possibly two, geologic periods. When compared directly (Fig. 2), continental margins are of a global scale and are many times larger than shelf-slope breaks in cratonic basins.

When considered with respect to tectonic setting, the record of the shelf-slope break in fossil carbonates is somewhat biased. The most intact margins, and those which are most intensively studied because of nearby hydrocarbon accumulations, occur in the cratonic interior, are relatively small and are of middle to late Paleozoic age. The large platforms of other ages are either buried or fragmented in orogenic belts, and so are less well known.

DYNAMICS OF PLATFORM MARGIN EVOLUTION

The slightly raised nature of facies at the margin of fossil platforms, together with their often extremely fossiliferous and 'reefy' nature, confirm that, like modern complexes, the rate of carbonate fixation was greatest at the shelf-slope break. Indeed, the shelf-slope break is the main production area in the 'carbonate factory'. Here nutrient supply is high and carbonate is readily available from the open ocean. The accumulations at the break are the result of the interaction between (1) high organic productivity, yielding skeletal sands and/or reefs, and (2) precipitation of $CaCO_3$ from seawater as ooid sands or as cements, both within reefs, and between particles. In this sense, the margin is like the rim of a bucket (Ladd, 1950) with the rim (shelf-slope break) consisting of rigid reefs or rapidly cemented and stacked sand shoals, while sediments produced on and around them are swept both leeward into the lagoon and seaward onto the adjacent slope. The reefs and sand shoals may grade into one another along strike, or may replace one another vertically, as the platform grows. Thus, the development of a platform is directly related to the growth potential of the rim.

The importance of this facies can be seen in numerous fossil examples, right from the inception of platform growth. Although there may be subtle differential relief on the seafloor that localizes the shelf-slope break, in many cases platforms rise directly from gently inclined, basin-sloping ramps (*in* Wilson, 1975). The lime sand/reef facies develop at the point on this slope where the correct balance between water depth, wave energy, nutrient supply and light, and conditions for carbonate fixation are optimum. Once formed, this zone in turn acts as a breakwater creating new environments; those shoreward are lower energy sites of relatively high carbonate accumulation because of their shallow water location; those seaward are areas of somewhat lower rates of carbonate deposition because of their deeper and darker location. Thus, the basic carbonate platform configuration is rapidly established but the key control remains the marginal facies.

Schlager (1981) has convincingly argued that carbonate production at the shelf-slope break of fossil carbonate platforms could more than keep pace with normal continental margin subsidence and eustatic changes in sealevel and that platform drowning and complete cessation of deposition only occurred at times of rapid sealevel rise, poor environmental conditions, or rapid subsidence induced by regional tectonics. The nature of the shelf-slope break is then, primarily, the product of the interplay between: (1) the rate of subsidence, and (2) the scale and timing of sealevel fluctuations. In the following discussion, we use the term "relative sealevel rise/fall" to refer to the combined effect of subsidence and sealevel movement. The interaction between carbonate production and relative sealevel movement has led to several different styles of shelf-slope break in the fossil record which we have termed (1) stationary, (2) offlap, (3) onlap, (4) drowned and (5) emergent. In all cases, there is little or no original depositional topography on the seafloor. Many examples are not clearly one type but a combination of several geomorphic forms.

STATIONARY MARGIN

If the rate of carbonate accretion is more or less matched by the rate of relative sealevel rise, then the shelf-slope break will remain more or less in the same geomorphic position (Fig. 3A). With time, as the platform increases in thickness and the relief between shelf-slope break and basin becomes progressively greater, the nature of the adjacent slope will change from a depositional to bypass mode (discussed in a later section).

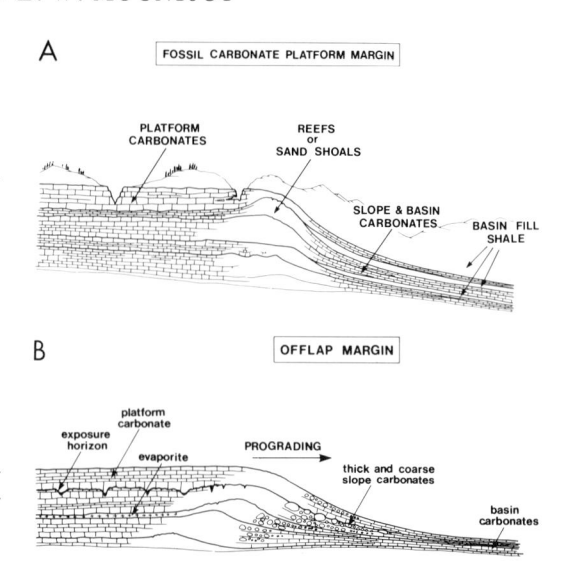

FIG. 3.—*A*, Sketch illustrating the main elements of a fossil carbonate platform margin. In this example the shelf-slope break is in a *stationary* mode, remaining more or less in the same position as the platform grew. *B*, Diagram of a carbonate platform in which the rate of accretion has exceeded the relative rate of sealevel rise and the shelf-slope is in the *offlap* mode, prograding over older slope deposits.

This is one of the most common types of carbonate platform margins. Excellent examples of stationary carbonate platform margins occur in many different parts of the world and are of different ages. Some of the more important examples include the Aphebian (Lower Proterozoic) of Great Slave Lake and Coronation geosyncline (Hoffman, 1974), Pennsylvanian of the Sverdrup Basin (Davies, 1977; Davies and Nassichuk, 1975), and those parts of Triassic platforms in the southern Alps and the Dolomites (Laubscher and Bernoulli, 1977; Wendt and Fursich, 1979; and Winterer and Bosellini, 1981).

When relative sealevel rise is slow, the platform is more or less flat with little relief between rim and shelf. If rise is rapid, and the shelf-slope break is reef-rimmed, then considerable relief may develop between the marginal rim and the deep water or shelf lagoon. If the rim is sand shoals, the particles may be cemented early but localized build-up is somewhat less than that of reefs. This is not because the fixation of carbonate is any less but, being sand, the sediments are swept into the lagoon and into deeper water more easily.

OFFLAP MARGIN

At times, when relative sealevel rise is outpaced by carbonate accretion, the shelf-slope break facies prograde out over older slope deposits (Fig. 3B).

FIG. 4.—Field examples of carbonate platform margins. *A*, Upper Permian Guadalupe Mountains, West Texas, view northeast from Guadalupe Peak across headwaters of Pine Springs Canyon. Illustrates offlap margin with thin-bedded, restricted lagoonal facies on left, grading into massive Capitan limestones (middle), and Capitan limestones prograding over dipping forereef strata on right (see Fig. 3B). *B*, Southeast onlap margin of the isolated Miette reef complex (Upper Devonian, Frasnian), exposed in steep dipping Rocky Mountain Front Ranges, east of Jasper, Alberta. View looking northwest illustrates onlap of basin formations Perdrix (P) and Mount Hawk (M) along a submarine unconformity over Cairn (C) stromatoporoid carbonates of the reef margin. *C*, Complexly deformed Middle Cambrian carbonates and shales of Main Ranges Mount Duchesnay immediately south of Field, B.C. Rocks are part of slope facies, lower (1c), middle (mc) and upper (uc) Chancellor Formation. The thicker carbonate units reflect tongues of shelf margin derived carbonate sediments equivalent to the Eldon and Pika Formations (see Fig. 9). Photo from Figure 24 of Cook (1975), courtesy Geological Survey of Canada.

Because carbonate sediments are in oversupply, the adjacent slope deposits are characterized by thick accumulations of many resedimented sands and conglomerates. This oversupply also leads to restricted circulation of open ocean waters on the platform behind the rim and, depending upon the amount of restriction, may lead to either evaporite formation or exposure with intensive diagenesis and numerous diastems.

One of the best examples of an offlap margin is the upper part (Capitan Limestone) of the Permian reef complex, (Fig. 4A; see Dunham, 1972; Hileman and Mazzullo, 1977). Other examples are the Late Mississippian to Middle Pennsylvanian Canyon Fiord Formation of Sverdrup Basin in the Canadian Arctic (Davies, 1977), and some Triassic build-ups in the Dolomites of northern Italy (Bosellini and Rossi, 1974).

In both of the preceeding cases (stationary and offlap), if sedimentation occurred under arid conditions, any restriction of circulation to the lagoon would result in elevated salinities and probably evaporite formation. Active growth of the margin facies alone may be enough to achieve this, as postulated by Bebout and Maiklem (1973) for the Middle Devonian platform evaporites of western Canada. Small-scale fluctuations in sealevel will, of course, enhance this tendency. The formation of evaporites in the lagoon, or on the platform, would result in saline waters which from time to time spill out over the rim and drastically affect the nature of the reef-building biota, and perhaps increase the formation of subsea cements and coated particles as in the Permian reef complex (Dunham, 1972; Hileman and Mazzullo, 1977).

In other instances, if there is considerable relief between the accreting margin and the lagoon or platform behind, the rim may serve as a barrier in the barred-basin model of evaporite formation (Kendall, 1978).

ONLAP MARGIN

A more complex situation seems to arise when carbonate production cannot keep pace with relative sealevel rise (Fig. 5). This most commonly occurs when subsidence increases dramatically due to tectonics (Schlager, 1981). The result is either onlap, or if the rise is rapid enough, complete inundation and drowning of the shelf-slope break, a circumstance treated in the next section.

Two situations occur during onlap, one in which there is gradual retreat of the margin shelfward over older platformal facies and another in which the margin moves shelfward in a series of steps. The reason for one or the other is unclear, but it may be due to the nature of the shelf-slope break, and/or the relative rates of sealevel rise.

If the break was mostly carbonate sand shoals, then the response to rising sealevel will be similar to that seen in siliciclastic sands, a gradual onlap of facies (Fig. 5, middle part).

If, on the other hand, this break is reefal, then the structures will build in place with the community not moving laterally so much as changing in place from a shallow-water to a deep-water assemblage as sealevel rises (Fig. 5). The growth rate of the reefs will progressively decline, the structure will gradually cease to be an effective barrier, and finally stop growing altogether. At some point in this development, conditions will become optimum for reef growth shelfward and a new community will start, only to suffer the same fate with time. In such instances, little sediment is transported sea-

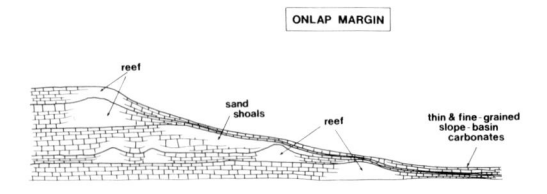

FIG. 5.—Sketch illustrating the response of the shelf-slope break to rapidly rising sealevel. In this onlap mode, two situations are possible: if reefs occupy the break then the onlap occurs in a series of steps; if sand shoals are at the shelf-slope break, then a classic gradual onlap occurs.

ward and so slope and basin deposits are thin and mostly carbonate muds unless a source of terrigenous sediment is available.

An excellent example of onlap occurs during the worldwide Middle and Upper Devonian (Frasnian part only) transgressions. The region in which this has been best documented is the subsurface Western Canadian Sedimentary Basin as illustrated in the north-south cross-section (Fig. 6). Carbonate build-ups transgressed or shifted southward further onto the craton from the Northwest Territories in the Middle Devonian in approximately three major stages: (1) Presqu'ile-Slave Point, (2) Swan Hills-Beaverhill Lake, and (3) Leduc-Woodbend (Fig. 6). The onlap of basin strata (Perdrix and Mount Hawk Formations) on the flank of the Leduc-Miette buildup exposed in the Alberta Rocky Mountains is illustrated in Figure 4B (Mountjoy, 1965).

DROWNED MARGIN

The most dramatic case of onlap occurs when rapid flooding occurs and the zone of high energy shifts far inward onto the shelf or platform (Fig. 7A). This type of flooding is termed drowning or inundation. Relatively deep water now covers the majority of the shelf or platform, placing it at or below the zone in which carbonate sediments form rapidly (below *ca.* 30 to 50 m), and the sediments are mainly muddy, often microskeletal. The shelf-slope break in such cases is a morphological feature only, inherited from earlier times and rarely illustrating a specific facies. The only record may be an abrupt appearance of sedimentary structures indicative of slope deposition such as synsedimentary folds, slumps, truncation surfaces and intraformation conglomerates. Slope and basin sedimentation is condensed or starved, especially in Paleozoic examples, because there is so little carbonate production on the deep shelf. In some instances, shore-derived terrigenous clastic sediments, trapped by the higher shorelines, may prograde over the carbonate platform (Stoakes, 1980) especially during still stands or regressive phases.

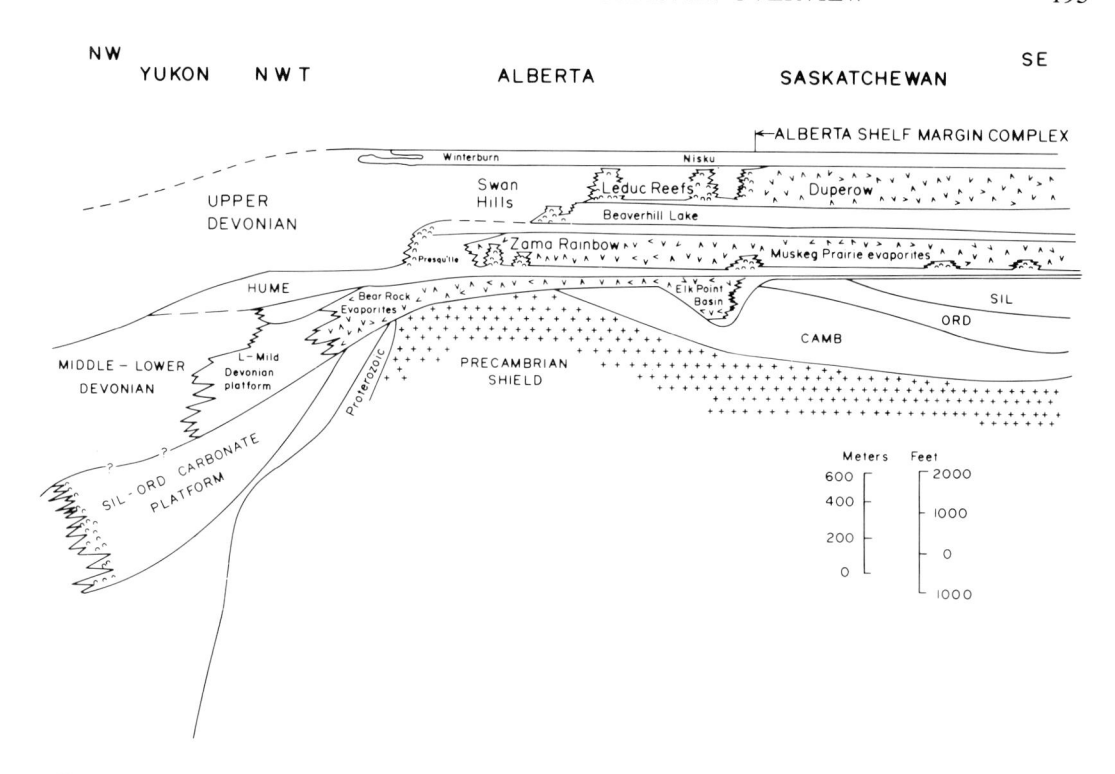

FIG. 6.—Subsurface diagrammatic cross-section from NW to SE through the western Canadian Sedimentary basin illustrating the stepped onlap mode during Late Devonian time. Modified from Bassett and Stout (1967).

This style of carbonate shelf is close to what Ahr (1973) has termed a 'Carbonate Ramp'. On such a platform, the high-energy zone is the shore line, with no obvious shelf-slope break and the carbonate facies pass progressively into deeper water lithologies (Fig. 15A). This type of shelf may develop during the initial stages of platform development (Wilson, 1975) or during drowning, as illustrated in Figure 7B. Read (1982) has termed this type of platform, when developed during inundation, a 'distally steepened ramp'. Brady and Rowell (1976) well illustrate the different facies to be expected from variations on this pattern of drowned shelves and ramps, using as their examples the Cambrian of the western United States.

One example of platform drowning can be seen in the Middle Ordovician Table Head Group of the Newfoundland Appalachians (Whittington and Kindle, 1963; Klappa et al. 1980). Here shallow-water platform strata are everywhere abruptly overlain by slope and basin sediments. The cause of this drowning is clearly tectonic subsidence of the continental margin related to the formation of a fore-deep during initial stages of the Taconic Orogeny.

An example in which the drowning appears to be due simply to rapid relative sealevel rise is found in the uppermost Jurassic and Lower Cretaceous of eastern Arabia (Glennie et al., 1974; Searle et al., 1983). In this instance, late Jurassic reef limestones along the shelfedge are abruptly overlain by deep-water radiolarian lime mudstones, marls, turbidites and debris flows. Although over 200 meters thick, these deep-water deposits grade upwards into Early Cretaceous shallow-water shelf carbonates, as carbonate sedimentation outpaced relative sealevel rise and the margin reverted to an offlap mode.

EMERGENT MARGIN

As is clear from the examples outlined, shallow carbonate platforms are particularly sensitive to minor fluctuations in sealevel. This is particularly true with a slight drop of sealevel (Fig. 7B) resulting in exposure of the entire platform and margin, shutting down the carbonate 'factory'. In addition, there can be minor erosion of the margin. The re-establishment of carbonate producers can, given sufficient time, produce a local narrow zone of shallow water carbonate sediments on top of the previous slope deposits. In addition, the circulation patterns and the supply of nutrients is drastically altered. All of these factors and events are reflected in the adjacent slope and basin environments by the lack of platform-derived carbonate sediments and often by starved sedimentation in the basin. Hence, both inundation and emergence lead to the same starved basin conditions primarily because the car-

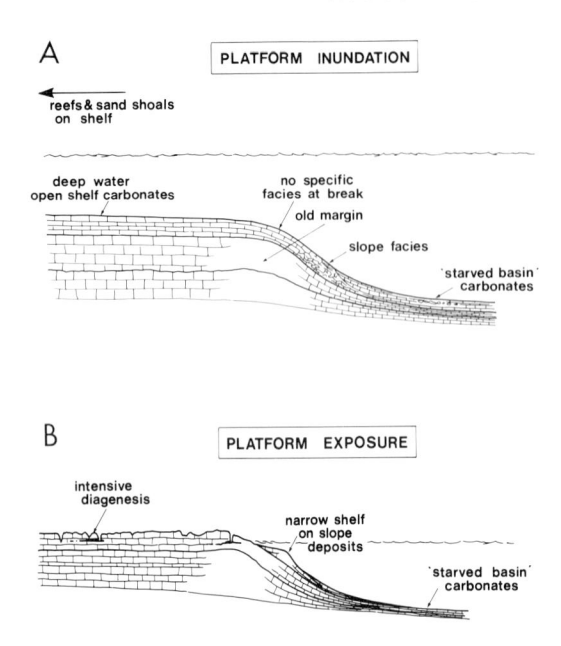

FIG. 7.—A, Sketch illustrating the style of deposition at the shelf-slope break during complete inundation of a carbonate platform. B, Diagram showing the effects of subaerial exposure of a carbonate platform.

bonate 'factory' has been largely shut down.

One of the best documented examples of emergence comes from the Pennsylvanian of the Sacramento shelf of southern New Mexico (Wilson, 1975). The platform was completely exposed resulting in exposure of platform margin algal mud mounds and the erosion of channels on the platform filled with terrigenous sand and conglomerate which forms a thin discontinuous unit across much of the platform and according to Wilson (1975, p. 216) was deposited as sealevel tended to rise.

COMBINATION PLATFORM MARGIN

While one of the above-cited margins may form for a relatively short period of time, the vagaries of sealevel fluctuation and subsidence, together with rates of carbonate production, does not allow them to persist for any length of time. Consequently, the shelf-slope break at the margin of most fossil carbonate platforms varies greatly in facies, configuration and location with time.

The Middle and Upper Devonian of Western Australia (Fig. 8; see Playford, 1980) provides a fine example whereby the margin changes with time. At least four types of platform margins occur in this summary schematic figure. Onlap margin in the Givetian, stationary (vertical) margins in the early and late Frasnian, inundation margin in the middle Frasnian, and offlap margin during much of the Famennian. Similar relationships are evident in the Devonian of western Canada and Europe (Meischner, 1971; Krebs and Mountjoy, 1972; Krebs, 1974, unpublished file data).

Another example is the Cambro-Ordovician shelf margin of western North America as exposed in thrust sheets of the Main Ranges of the Canadian Rocky Mountains (Fig. 2A). In this mountainous region, although many individual areas contain superb exposure, some of the stratigraphic relationships are obscured or complexly deformed (Fig. 4C) and so relationships must be inferred. The regional aspects of the succession together with the cyclicity (100 m+ large scale carbonate-shale cycles, termed grand cycles) of the Middle and Upper Cambrian are outlined by Aitken (1971, 1978). The western edge of the shelf appears to have been somewhat elevated and called the Kicking Horse Rim, a belt some 16 km wide and 130 km long. The shelf-slope break here varies through time from a spectacular escarpment (Fig. 9A; see McIlreath, 1977) to stationary and occasionally offlap modes (Figs. 7, 9B) yet remains more or less in the same geographic position for almost 50 my.

As spectacular as the Cathedral escarpment may appear in a mountain outcrop, it and the overlying offlap carbonate margins in the Cambrian are relatively insignificant features when placed in a regional perspective of continental dimensions (Fig. 10). The second layer down represents the entire Cambro-Ordovician succession with the Kicking Horse Rim occurring about the position of the arrow. The position of this carbonate margin is well inboard of the underlying edge of the cratonic crust, a characteristic of many Paleozoic carbonate margins around North America.

While eustatic changes in sealevel may be important, in the long term, tectonics appears to be

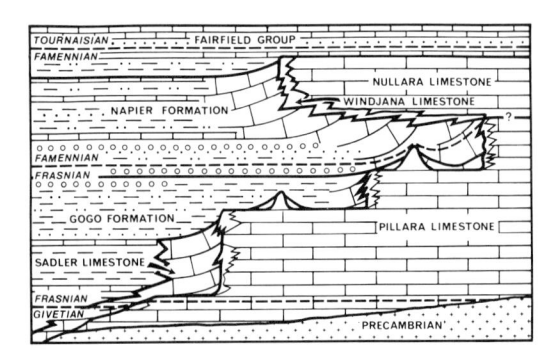

FIG. 8.—Diagrammatic cross-section of the Upper Devonian reef complex in the Canning Basin Western Australia illustrating the variation of the platform-margin, shelf-slope break facies through time. Specifically, onlap in Givetian, stationary to vertical margins in Early and Late Frasnian, inundation in middle Frasnian and offlap during much of the Famennian (from Playford, 1980).

the more overriding control of the style of the shelf-slope break. Tectonically controlled differential subsidence in terms of fault or flexure zones appears to have controlled, for example, the location of many margins (Aitken, 1971, 1978; Mountjoy, 1980).

The manner of subsidence of rifted continental margins is illustrated by Atlantic-type passive margins. For example, an Atlantic margin subsidence curve (Fig. 11) can be broken down into three components: (1) tectonic, due to cooling of the ocean crust, (2) sediment loading, and (3) sealevel rises, which also cause loading. Initially tectonic subsidence of new oceanic crust is uniform and related to the square root of time (Trehu, 1975, for the first 80 Ma in the case of the Nova Scotia shelf), and then changes abruptly to a slower rate of subsidence (Keen and Hyndman, 1979). Thus the tectonic subsidence of continental margins is faster immediately after continental breakup than later. This pattern of subsidence is apparently characteristic of passive continental margins and appears to apply to the Jurassic passive continental margin now exposed in the southern Alps in Italy (Winterer and Bosellini, 1981).

NATURE OF THE SHELF-SLOPE BREAK

Variations In Reef Builders With Time

In the foregoing discussion, it has been assumed that at any given time in the past, either carbonate sand shoals or reefs, or both, formed at the shelf-slope break. Implicit in this reasoning is the supposition that there was always a community of reef-building organisms that would, or could, grow at

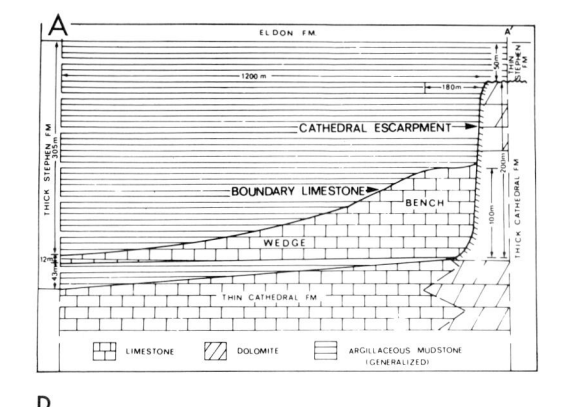

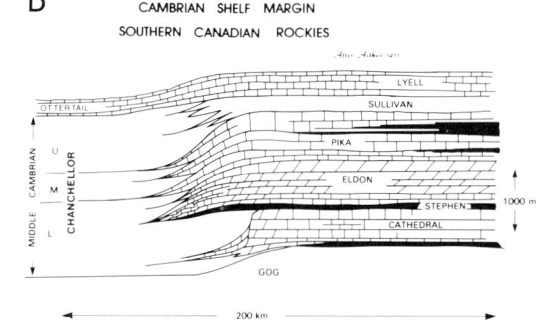

FIG. 9.—*A*, Diagram illustrating the vertical escarpment which comprised the shelf-slope break during Early Middle Cambrian time in the Southern Canadian Rockies (from McIlreath, 1977). *B*, Schematic cross-section of the entire Middle Cambrian shelf margin (Kicking Horse Rim) in the Southern Canadian Rockies (after Aitken, 1971).

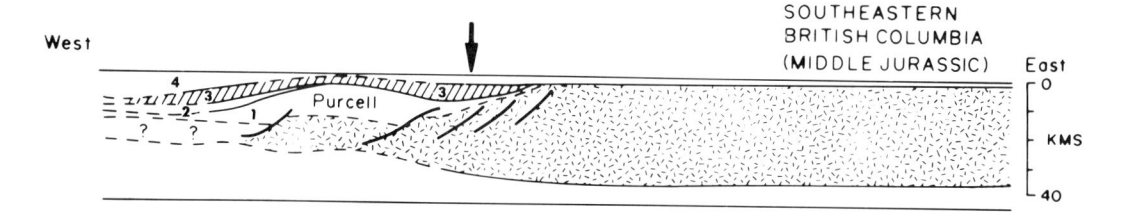

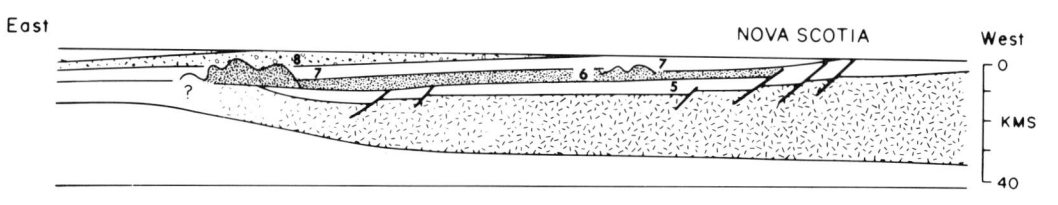

FIG. 10.—Comparison of reconstructed continental margin of western Canada at the end of Middle Jurassic time with that of the present Nova Scotia Atlantic margin off Canada, modified from Price (1981). Approximate positions of lower Paleozoic platform margins and Kicking Horse Rim are indicated by arrow.

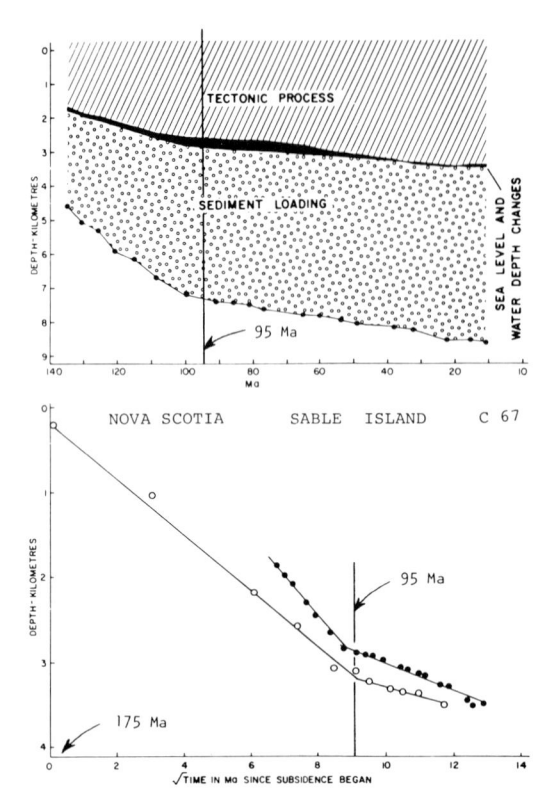

FIG. 11.—Upper diagram subsidence of basement on Nova Scotia shelf as a function of time (dots). Subsidence due to sediment loading and tectonics is shown; black area represents the effects of eustatic sealevel changes and paleo-water depth variations. The lower graph shows net tectonic subsidence of the same area plotted on $t^{1/2}$ scale (from Keen and Hyndman, 1979).

the shelf-slope break, which does not appear to be the case throughout much of geologic history.

In the Proterozoic, the shelf-slope break was often delineated by stromatolite bioherms and biostromes, composed of elongate mounds with deep channels between and oriented normal to the platform rim (Hoffman, 1974; Campbell and Cecile, 1981), very similar in morphology to later shelf-edge metazoan reefs.

The structure and composition of reefs throughout the Phanerozoic has been synthesized by James (1978, 1982b), building upon previous summaries by Heckel (1974) and Wilson (1975). A major conclusion appears to be that large, complex, zoned reefs like those in today's oceans coincide with times in which large, complex calcareous metazoans populated the shallow seafloor. At times when the larger organisms such as corals, stromatoporoids and rudists were absent, these reefs failed to develop (Fig. 12). In the large fossil reefs, there is commonly a vertical succession of communities, beginning with pioneer and colonization stages in which the organisms are mainly small, rooted pelmatozoans and calcareous algae together with an assemblage of little, delicate, commonly branching skeletal organisms (e.g., bryozoans, sponges, corals, stromatoporoids) and/or low-lying, encrusting foraminifers and algae. The bulk of the structures were made up of a more varied community comprising both members of the preceeding stages together with large numbers of bigger, more robust, metazoans in a wide variety of growth forms. It is at this stage that reefs grew into the zone of waves and swell and a pronounced windward-leeward zonation is commonly observed.

During much of geologic history, however, the large metazoans capable of growing in a wide variety of shapes and inhabiting most environments were absent. Instead, the calcareous benthos was dominated by smaller, more delicate organisms. Bioherms at these times were different and have been called mounds (Wilson, 1975) or reef mounds (James, 1978) or in those instances where there are few, if any, skeletal elements, carbonate mud mounds. Upon inspection, these mounds often exhibit the first two, pioneer and stabilization stages of growth but do not develop any further because of the lack of the bigger and more complex metazoans. Because the biota are so small and delicate these reef mounds developed in tranquil locations, on the platform behind a barrier at the shelf-slope break and/or downslope in deeper water, seaward of the break. This model also appears to be a satisfactory explanation of some Precambrian shelf margins as well (Aitken, 1981).

While this situation appears to hold true for most situations, some reef mounds appear to have developed at the shelf-slope break (McIlreath, 1977; Mountjoy and Jull, 1978; James, 1981), in agitated environments. The reasons for this apparent contradiction are not clear but may have to do with the precipitation of abundant synsedimentary cement, often in the form of micrite.

As a result of this variation in reef-building biota with time, the nature of the shelf-slope break differs dramatically from period to period. At times, when large stromatolites grew or a full spectrum of calcareous benthos lived, capable of producing large skeletons of all dimensions, extensive barrier reefs and/or skeletal sand shoals developed at the shelf-slope break. Reef mounds also formed, but on the platform and down-slope (Fig. 13A). During periods when only diminutive metazoans (or stromatolites) were present, the shelf-slope break was generally a series of carbonate sand shoals with the reef-mounds most often in an on-shelf or down-slope facies (Fig. 13B).

Intensity of Wave Action

As recognized by Wilson (1975), another key element in this equation is the intensity of wave

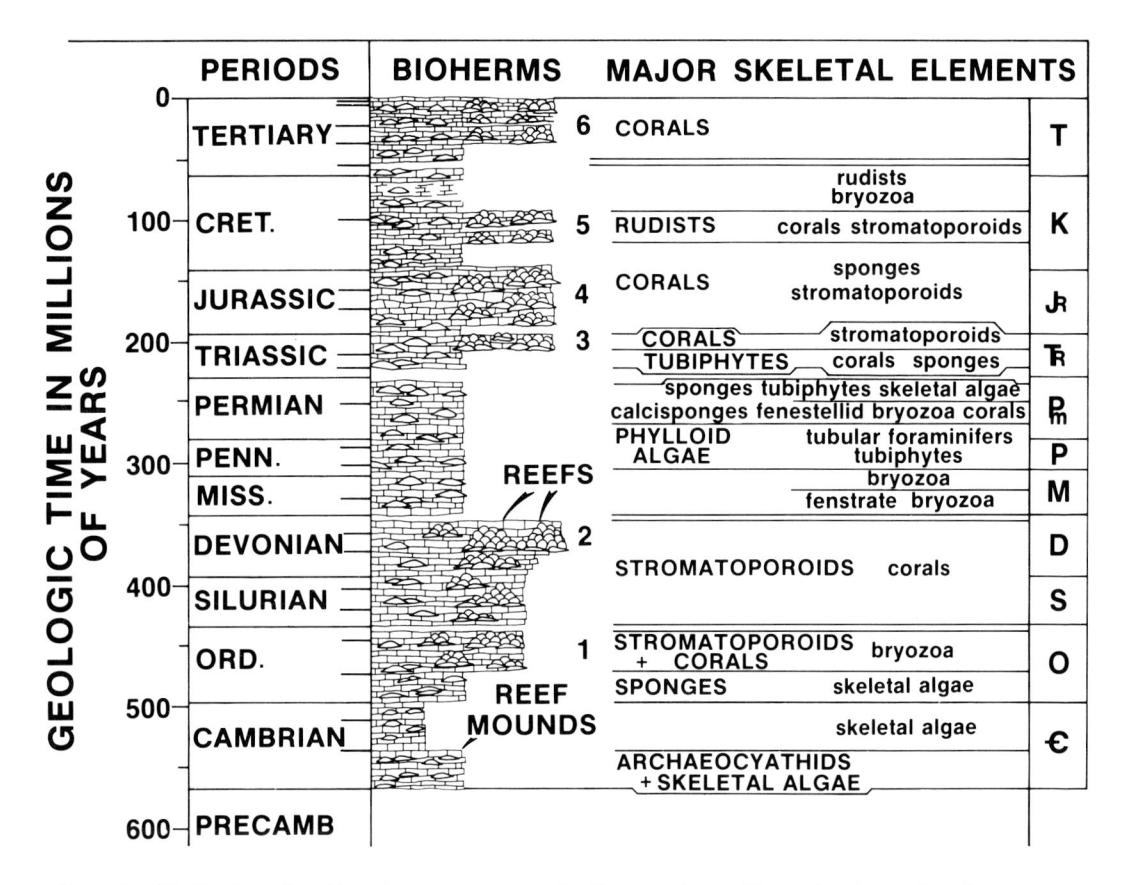

FIG. 12.—Idealized stratigraphic column representing the Phanerozoic and illustrating times when there appear to be no reefs or bioherms (gaps), times when there were only reef mounds, and times when there were both reefs and reef mounds and the organisms that built them (from James, 1983; permission Am. Assoc. Petroleum Geologists).

energy coming onto the shelfedge (Fig. 14). At times of high wave energy, barrier reefs like those of today, exhibiting a strong zonation, are found. On the other hand, when the shelf-slope break developed in quiet seas, a complex of carbonate sand shoals and low-relief, poorly-zoned, deeper water, equidimensional "knoll-reefs" are most common. These knoll-reefs also occur on the leeward sides of large, isolated platforms.

In total, there are three major types of fossil platform margins, as outlined by Wilson (1975). These types are: (1) sand shoals and downslope reef mounds, (2) sand shoals and reef-knolls, and (3) barrier-reefs (Fig. 14). When the types of reef-building biota, together with the paleotectonic locations of the shelfedge reefs, are considered in the context of geologic time, there is a clear temporal variation in style (Table 1). While this classification appears to hold for most cases in the geological record, at times when there are few large skeletal metazoans and relatively quiet seas prevailed there is in some intracratonic basins no clear shelf-slope break but instead a carbonate ramp without build-ups.

Rate of Shelf Margin Growth

The nature of the shelf-slope break and the rate at which upward growth takes place inevitably control the configuration of the slope itself. This is because the rim is the source of almost all sediment on the slope and the gradient of the slope is related to relief between shelf-slope break and basin plain.

Slope sediments are of two types: (1) resedimented, coarse-grained, shallow-water deposits, and (2) peri-platform ooze (McIlreath and James, 1978). The composition and frequency of the sediment gravity flows depend upon the relative rates of sealevel rise, as noted in the previous section, and the nature of the shelf-slope break as well as the tectonic setting. If carbonate sand shoals form the margin, then skeletal sand and oolite grain flows, or turbidites, together with debris flows containing cemented calcarenite clasts, are most common. Most of the clast conglomerates in these instances are slope-derived (Cook, 1979). In contrast, if the shelf-slope break is reef-rimmed, the debris flows, although rare are spectacular, often containing numerous and large blocks of early cemented reef-limestone and adjacent platform facies.

The fine-grained slope limestones have been called peri-platform ooze by Schlager and James (1978). This term emphasizes the fact that the particles come from two different sources: (1) episodic fallout of mud and silt stirred up on the platform during storms and swept seaward to settle out in deep water, and (2) the continuous rain of calcareous phytoplankton and zooplankton skeletons into deep water. While the first process can clearly operate throughout geologic time, the second is constrained by the record of planktonic calcareous organisms. Thus, during the Paleozoic, when there were few, if any, calcareous planktonic microfossils, most perennial fine-grained slope sediments were platform-derived. In the Mesozoic, however, with the appearance of coccolithophoroids in the Jurassic and planktonic foraminifers in the Cretaceous much of the periplatform ooze was, and remains, polygenetic.

The rate of growth of the shelf-slope break governs the declivity of the slope itself (McIlreath and James, 1978). When relief between the shelf-slope break and the basin floor is low, the slope is a depositional incline that gradually merges with the basin proper. On this depositional margin turbidites and debris sheets are deposited on the lower slope as well as in the basin (Fig. 15A), but because

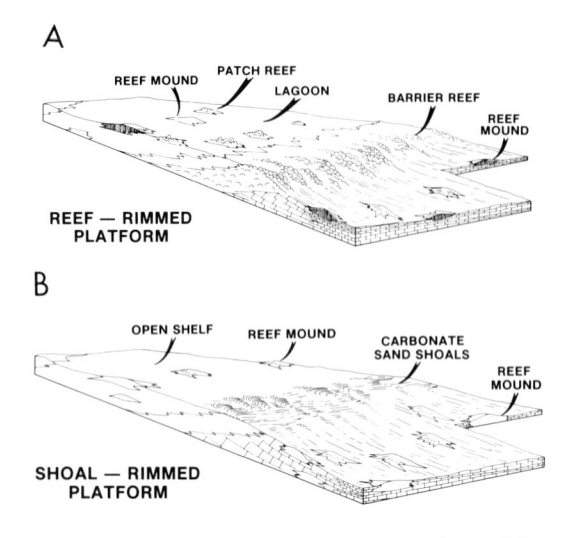

FIG. 13.—*A*, Sketch illustrating the disposition of facies on a carbonate platform at times when a complete spectrum of reef-building organisms was present (from James, 1983; permission Am. Assoc. Petroleum Geologists. *B*, Sketch illustrating the disposition of facies on a carbonate platform at times when only small delicate skeletal metazoans prevailed and most bioherms were reef mounds (from James, 1983; Am. Assoc. Petroleum Geologists.

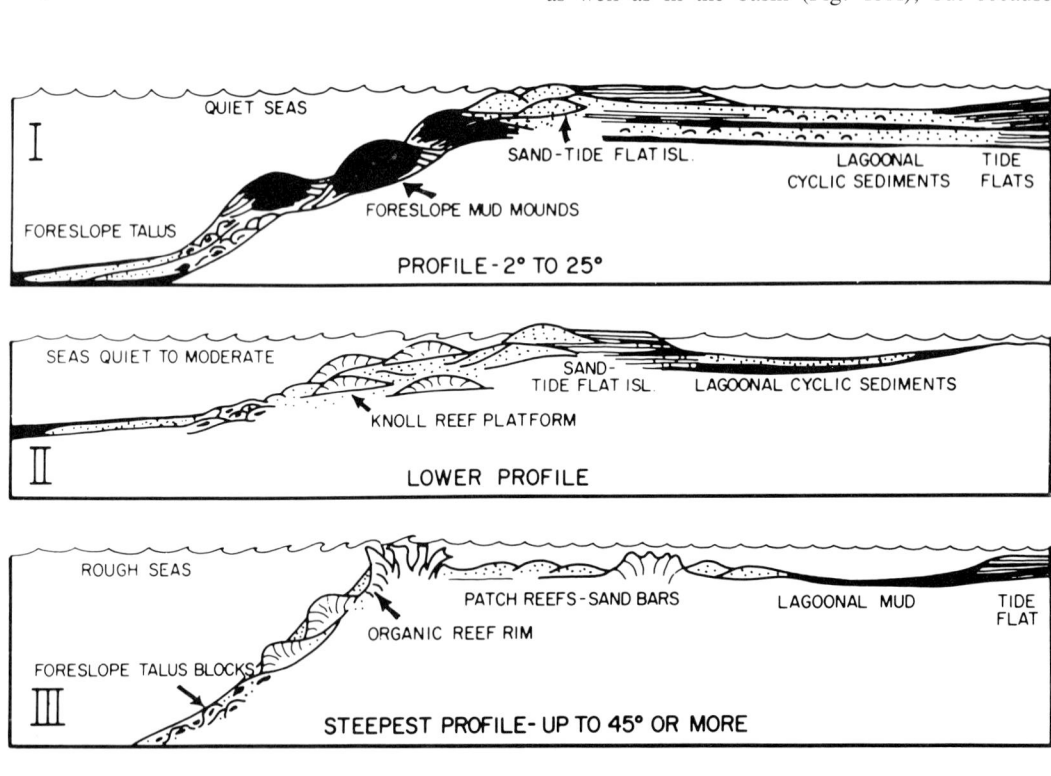

FIG. 14.—Three types of carbonate shelf margins: I, A shelf-slope break of carbonate sand shoals and down-slope lime-mud accumulation; II, a shelf-slope break of sand shoals and knoll reefs; and III, a shelf-slope break of a barrier reef (from Wilson, 1974; permission Am. Assoc. Petroleum Geologists).

TABLE 1.—EXAMPLES OF FOSSIL CARBONATE SHELF-SLOPE BREAKS

Time	Setting	Reef-Building Organisms	Shelf-Slope Break
Mesozoic-Cenozoic	Open ocean	Complete spectrum	III II
Late Paleozoic	Intracratonic	Small-delicate	I
Middle Paleozoic	Intracratonic	Complete spectrum	II
Early Paleozoic	Open ocean	Small-delicate	I
Precambrian	Open ocean	Stromatolites	II I

ramps have even lower gradients, slope sediments are normally much finer and consist of pelagic carbonates and shales. In other cases, where relief is considerable, the sands and debris flows, triggered at the shelf-slope break by oversteepening, collapse of the reef rim, or slumping of previously deposited sediments, sweep down across the relatively steep incline and accumulate out from the toe-of-slope on the adjacent basin plain. These by-pass-margins (Fig. 15B) are particularly common during the late stages of platform evolution. Schlager and Ginsburg (1981) have noted the change of platform margins from depositional to by-pass types in several fossil examples.

It is emphasized that because the provenance of most deepwater resedimented carbonates is the shelf-slope break, it is a line source (Schlager and Chermak, 1979) and not, as in equivalent terrigenous clastic deposits, a series of point sources emanating from submarine canyons. Another important contrast is that most of the sediment from the shelf-slope break is supplied at times of rising relative sealevel, when the carbonate production is greatest, whereas in siliciclastic deposits, resedimented deposits are most common when relative sealevel is falling and the sediments prograde to the shelf-slope break and spill over into deeper water. One result of this phenomenon is the classic situation of reciprocal sedimentation between clastics and carbonates (in the sense of Wilson, 1975), seen in numerous fossil complexes such as the Permian Reef Complex (van Siclen, 1958, 1964; Meissner, 1972; Wilson, 1975) and in the Mesozoic of eastern Arabia (Glennie et al., 1974; Searle et al., 1983).

DIAGENESIS

Facies that make up the shelf-slope break are more subject to diagenetic alteration than those in lagoonal and deep water settings. This diagenesis begins on the seafloor, contemporaneous with deposition, as subsea cement in the form of aragonite and Mg-calcite is precipitated in sediment and growth cavities of reefs and as intergranular cement in ooid sand shoals (Dravis, 1979; James and Ginsburg, 1979). Early cementation is commonly recorded in fossil platform margins and, by filling the void spaces, inhibits pervasive percolation of subsurface diagenetic fluids, thus aiding in the preservation of original textures and fabrics. On the other hand, this early cement, by occluding most porosity can make many of the "reefy" facies poor reservoir lithologies.

Because the shelf-slope break is slightly elevated, it is often emergent for short periods of time and so subject to meteoric diagenesis. The fact that the margin is at the boundary between lagoonal and open ocean waters as well as at the boundary between percolating fresh and salt water leads to mixing of different fluids in these sediments and often results in extensive alteration. At times, this may result in dissolution of aragonitic components, creating moldic to vuggy porosity. As many of the reef-building organisms are and were aragonite (stromatoporoids, scleractinian corals, rudist bivalves, gastropods, other bivalves and calcareous green algae) this phase is a critical one, with much of the record of the major biota being obliterated, but at the same time good potential reservoir rock being created. At other times, the margin facies may be dolomitized, possibly through the interaction of mixed meteoric and marine water, or percolating hypersaline waters in the shallow subsurface (see papers in Zenger et al., 1980). This partly obliterates the record of the facies, but sometimes greatly enhances porosity and permeability.

Later when buried in the subsurface, the shelf-slope break is at the interface between two different rock masses which will compact differently—the platform carbonates and the basin-fill shales or argillaceous limestones. Once again, the marginal facies appear to be subject to dolomitization in this environment. The probable mechanism involves migration of brines from slope and basin sediments as they are compacted into the relatively uncompacted and porous carbonates and reaction with the first limestones encountered—those at the shelf-slope break (Illing, 1959; Griffin, 1965; Mattes and Mountjoy, 1980).

The end-result of these processes is a wide variety of rock types from well-preserved, non-porous reef limestones to massive, porous, vuggy dolomites with barely a trace of their original composition remaining.

ECONOMIC ASPECTS

Those carbonate margins that contain suitable primary and/or secondary porosity form excellent

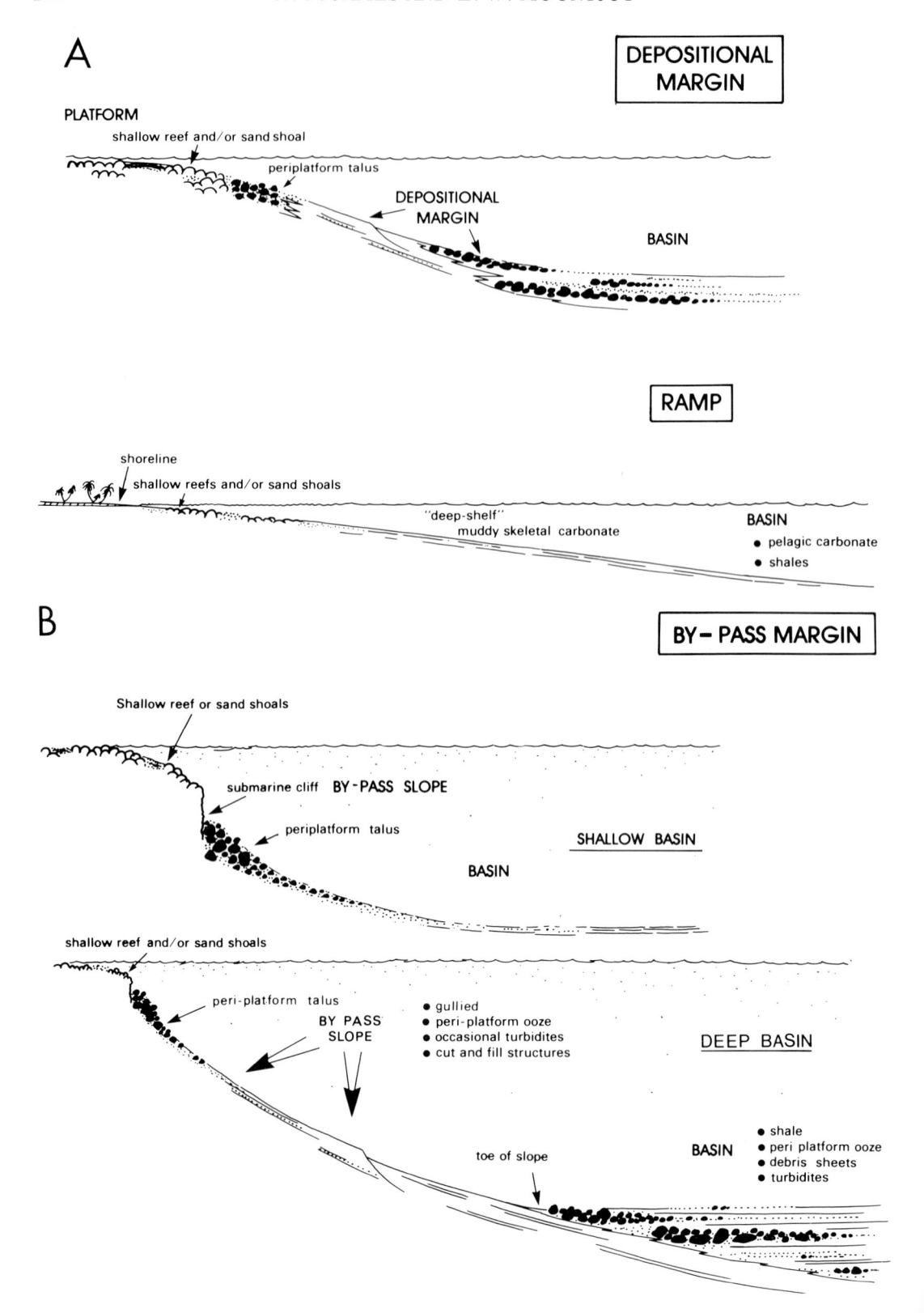

A

DEPOSITIONAL MARGIN

PLATFORM
shallow reef and/or sand shoal
periplatform talus
DEPOSITIONAL MARGIN
BASIN

RAMP

shoreline
shallow reefs and/or sand shoals
"deep-shelf" muddy skeletal carbonate
BASIN
- pelagic carbonate
- shales

B

BY-PASS MARGIN

Shallow reef or sand shoals
submarine cliff BY-PASS SLOPE
periplatform talus
SHALLOW BASIN
BASIN

shallow reef and/or sand shoals
peri-platform talus
BY PASS SLOPE
- gullied
- peri-platform ooze
- occasional turbidites
- cut and fill structures
DEEP BASIN
toe of slope
BASIN
- shale
- peri platform ooze
- debris sheets
- turbidites

reservoirs for petroleum and host rocks for base-metal (Pb-Zn) deposits. Often the adjacent deeper water shale facies, especially if deposited under anoxic conditions, are good to excellent source rocks. Some carbonate muds and turbidites are also good source rocks (Crevello et al., 1981). In the case of the onlap and inundation platform types (Figs. 5, 6, and 7A), these shales and carbonates may onlap the carbonate margins. If carbonate platforms formed on a recently rifted continental margin, the initial geothermal gradient would be high. The seaward side of the continental terrace wedge and slope would be the most deeply buried part and thus would be the first to become thermally mature. Unfortunately, most carbonate platforms dip seaward and continue to do so during burial. Hence compacted fluids and any hydrocarbons generated would move updip towards the craton until trapped against a permeability barrier.

Of equal economic significance are the many Pb-Zn sulphide deposits that are associated with carbonate platform margins, for example: (1) Middle Cambrian Kicking Horse Mines along the western margin of the Cathedral escarpment near Field, B.C. (Ney, 1954, 1957); (2) Middle Devonian Pine Point, N.W.T. located in a dolomitized Middle Devonian reef escarpment (Skall, 1975; Krebs, 1983); and (3) Triassic of the Alps (Maucher and Schneider, 1967). These carbonate margins have all undergone one or more extended periods of platform exposure, as the sulphide deposits occur in cave and karst systems which have been modified by later dolomitization.

Finally the shelf-slope break has a profound effect on the style of deformation in all mountain belts where these rocks normally crop out in thrust-fold belts. The shelf-slope break generally occurs where large thrust sheets with a distinctive stratigraphy change abruptly to complex and intricately folded, faulted and cleaved shaly slope rocks.

SUMMARY

1. The spectrum of carbonate sand shoals and reefs that occur at the shelf-slope break of fossil shelves and platforms formed because these rapidly accreting facies were able to keep pace with relative sealevel rise and so maintain the seafloor in shallow water.

2. The best exposed, and hence most studied, carbonate platform margins are of middle and late Paleozoic age and rim small shelves or buildups in intracratonic basins. The much larger early Paleozoic and Mesozoic-Cenozoic continental margin examples are relatively poorly known because they are either deformed in orogenic belts or are buried in the subsurface beneath modern shelves.

3. Active fixation of carbonate against a background of varying subsidence has produced different yet recurring facies patterns along the shelf-slope break. The most common carbonate shelf-slope breaks are designated as: (1) stationary, (2) offlap and (3) onlap.

4. Large changes in relative sealevel resulting in either drowning or exposure of the platform dramatically affect the nature of the shelf-slope break and result in starved basinal sedimentation. If drowned, the break is reduced to a declivity mantled by deeper water carbonates. If exposed, the break is subject to intensive diagenesis, and another small ineffective, narrow platform may be developed on top of older slope deposits.

5. The shelf-slope break was the site of most rapid carbonate accretion on the platform, in the form of reefs or lime sand shoals, and is the source of most of the sediment deposited on the foreslope.

6. The nature of the shelf-slope break through geologic time depends upon the paleobiology of reef-building organisms and their environmental setting: (a) Precambrian (Proterozoic)—carbonate sand shoals, stromatolite reefs along continental margins and within intracratonic basins, (b) Early Paleozoic—carbonate sand shoals and downslope reef mounds along continental margins, (c) Middle Paleozoic—knoll reefs and sand shoals and local large reefs forming barriers in intracratonic basins, (d) Late Paleozoic—carbonate sand shoals and downslope mounds in intracratonic basins, and (e) Mesozoic/Cenozoic—large barrier reefs, or sand shoals, or knoll reef complexes along continental margins.

7. The shelf-slope break lithologies are commonly affected more by diagenesis than surrounding facies and vary from superbly preserved limestones with abundant synsedimentary cement that are poor reservoir lithologies to completely obliterated, dolomitized carbonates that are excellent host rocks for economic hydrocarbon accumulations and base metal deposits.

ACKNOWLEDGMENTS

A review paper of this type is based on our joint experience and observations in different parts of the world often guided by colleagues too numerous to mention whose help, guidance, counsel, and friendship we cherish. They and others have helped and encouraged our ideas to evolve and crystallize. To all we extend our thanks and appreciation. A. C.

FIG. 15.—A, Sketch illustrating the morphology and sediments on a depositional margin and a carbonate ramp. B, Sketch illustrating the morphology of and sediments on by-pass margins fronting shallow and deep basins.

Hine and W. J. Myers kindly reviewed the original manuscript and offered many helpful suggestions towards its improvement. This research was supported mainly by the National Sciences and Engineering Research Council of Canada grants A2028 and A 9159.

REFERENCES

AHR, W. M., 1973, The carbonate ramp: an alternative to the shelf model: Trans. Gulf Coast Assoc. Geol. Socs., v. 23, p. 221–225.

AITKEN, J. D., 1971, Control of Lower Paleozoic sedimentary facies by the Kicking Horse Rim, Southern Rocky Mountains, Canada: Bull. Canadian Petroleum Geology, v. 19, p. 557–569.

———, 1978, Revised models for depositional grand cycles, Cambrian of the Southern Rocky Mountains, Canada: Bull. Canadian Petroleum Geology, v. 26, p. 515–542.

———, 1981, Stratigraphy and sedimentology of the Upper Proterozoic Little Dal Group, Mackenzie Mountains, Northwest Territories, *in* Campbell, F. H. A., ed., Proterozoic Basins of Canada: Geol. Survey Canada Paper 81-10, p. 47–71.

BASSETT, H. G., AND STOUT, J. G., 1967, Devonian of Western Canada, *in* Oswald, D. H., ed., Internat. Symposium on the Devonian System: Alberta Soc. Petroleum Geol., v. 1, p. 717–752.

BEBOUT, D. G., AND MAIKLEM, W. R., 1973, Ancient anhydrite facies and environments, Middle Devonian Elk Point Basin, Alberta: Bull. Canadian Petroleum Geology, v. 21, p. 287–343.

BOSELLINI, A., AND ROSSI, D., 1974, Triassic carbonate buildups of the Dolomites, northern Italy, *in* Laporte, L. F., ed., Reefs in Time and Space: Soc. Econ. Paleontologists Mineralogists Spec. Pub. 18, p. 209–233.

BRADY, M. J., AND ROWELL, A. J., 1976, An upper Cambrian subtidal blanket carbonate, Eastern Great Basin, *in* Robison, R. A., and Rowell, A. J., eds., Paleontology and Depositional Environments: Cambrian of Western North America: Brigham Young Univ. Geol. Studies, v. 23, p. 153–165.

CAMPBELL, F. H. A., 1981, Proterozoic Basins of Canada: Geol. Survey Canada Paper 81-10, 444 p.

———, AND CECILE, M. P., 1981, Evolution of the Early Proterozoic Kilohigok Basin, Bathurst Inlet—Victoria Island, Northwest Territories, *in* Campbell, F. H. A., ed., Proterozoic Basins of Canada: Geol. Survey Canada Paper 81-10, p. 103–131.

COOK, D. G., 1975, Structural style influenced by lithofacies, Rocky Mountain Main Ranges, Alberta, British Columbia: Geol. Survey Canada Bull. 223, p. 73.

COOK, H. E., AND ENOS, P., eds., 1977, Deep water carbonate environments—An introduction: Soc. Econ. Paleontologists Mineralogists Spec. Publ. 25, 336 p.

———, 1979, Ancient continental slope sequences and their value in understanding modern slope development, *in* Pilkey, O., and Doyle, R., eds., Geology of Continental Slopes: Soc. Econ. Paleontologists Mineralogists Spec. Pub. 27, p. 287–307.

CREVELLO, P. D., PATTON, J. W., AND GUENNEL, G. K., 1981, Source rock potential basinal carbonate muds, Bahamas [abs.]: Am. Assoc. Petroleum Geologists Bull., v. 65, p. 914–915.

DAVIES, G. R., AND NASSICHUK, W. W., 1975, Subaqueous evaporites of the Carboniferous Otto Fiord Formation, Canadian Arctic Archipelago: a summary: Geology, v. 3, p. 273–278.

———, 1977, Turbidites, debris sheets, and truncation structures in Upper Paleozoic deep-water carbonates of the Sverdrup Basin, Arctic Archipelago, *in* Cook, H. E., and Enos, P., eds., Deep Water Carbonate Environments: Soc. Econ. Paleontologists Mineralogists Spec. Pub. 25, p. 221–247.

DRAVIS, J., 1979, Rapid and widespread generation of recent oolitic hard-grounds on a high-energy Bahamian Platform, Eleuthera Bank, Bahamas: Jour. Sed. Petrology, v. 49, p. 195–209.

DUNHAM, R. J., 1972, Capitan Reef—New Mexico and Texas: Facts and questions to aid interpretation and group discussion: Soc. Econ. Paleontologists Mineralogists Permian Basin Sec., Spec. Pub. 72-14, 267 p.

GLENNIE, K. W., BOEUF, M. G. A., CLARKE, M. W. H., MOODY-STUART, M., PILAAR, W. F. H., AND REINHARDT, B. M., 1974, Geology of the Oman Mountains: Ned. Geol-Mijnbouwkd Genoot 31, 1–423.

GRIFFIN, D. L., 1965, The Devonian Slave Point—Beaverhill Lake and Muskwa Formations of northeastern British Columbia and adjacent areas: British Columbia Dept. Mines and Petroleum Resources Bull., v. 50, 90 p.

HECKEL, P. H., 1974, Carbonate buildups in the geologic record: a review, *in* Laporte, L. F., ed., Reefs in Time and Space: Soc. Econ. Paleontologists Mineralogists Spec. Pub. 18, p. 90–154.

HILEMAN, M. E., AND MAZZULLO, S. J., eds., 1977, Upper Guadalupian Facies Perian Reef Complex, Guadalupe Mountains, New Mexico and West Texas: Soc. Econ. Paleontologists Mineralogists Permian Basin Sec., Spec. Pub. 77-16, 508 p.

HINE, A. C., AND MULLINS, H. T., 1983, Modern carbonate shelf-slope breaks, *in* Stanley, D. J., and Moore, G. T., eds., The Shelfbreak: Critical Interface on Continental Margins: Soc. Econ. Paleontologists Mineralogists Spec. Pub. 33, p. 169–188.

HOFFMAN, P., 1973, Evolution of an early Proterozoic continental margin: The Coronation geosyncline and associated aulacogens of the northwestern Canadian Shield: Phil. Trans. Roy. Soc. London Ser. A., v. 273, p. 547–581.

———, 1974, Shallow and deep water stromatolites in lower Proterozoic platform-to-basin facies change, Great Slave Lake, Canada: Am. Assoc. Petroleum Geologists Bull., v. 58, p. 585–867.

———, 1980, Wopmay, orogen: A Wilson cycle of Early Proterozoic age in the northwest of the Canadian Shield, *in* Strangway, D. W., ed., The Continental Crust and its Mineral Deposits: Geol. Assoc. Canada Spec. Paper 20, p. 523–549.

ILLING, L. V., 1959, Deposition and diagenesis of some upper Paleozoic carbonate sediments in western Canada: New York, 5th World Petroleum Congr. Proc. Sec. 1, p. 23–52.

JAMES, N. P., 1978, Facies models 10: Reefs: Geosci. Canada, v. 5, p. 16–26.

———, 1981, Megablocks of calcified algae in the Cow Head Breccia, western Newfoundland: Vestiges of a Cambro-Ordovician platform margin: Geol. Soc. America Bull., v. 92, p. 799–811.

———, 1982, Depositional models for carbonate rocks, in Parker, A., and Selwood, B., eds., Sediment Diagenesis, a NATO Advanced Study Institute: Dordrecht, Holland, D. Reider Publishing Co., in press.

———, 1983, Reefs, in Scholle, P., Bebout, D., and Moore, C., eds., Carbonate Depositional Environments: Am. Assoc. Petroleum Geologists Memoir 33, in press.

———, AND GINSBURG, R. N., 1979, The deep sea ward margin of Belize barrier and atoll reefs: Internat. Assoc. Sedimentologists Spec. Pub. 3, 201 p.

KEEN, C. E., AND HYNDMAN, R. D., 1979, Geophysical review of the continental margins of eastern and western Canada: Canada Jour. Earth Sci., v. 16, p. 712–747.

KENDALL, A. C., 1978, Facies models, 14, subaqueous evaporites: Geosci. Canada, v. 5, p. 124–139.

KENDALL, C. G. ST. G., AND SCHLAGER, W., 1981, Carbonates and relative changes in sea level: Mar. Geology, v. 44, p. 181–212.

KLAPPA, C. F., OPALINSKI, P., AND JAMES, N. P., 1980, Middle Ordovician Table Head Group of western Newfoundland: a revised stratigraphy: Canada Jour. Earth Sci., v. 17, p. 1007–1019.

KREBS, W., 1974, Devonian carbonate complexes of central Europe, in Laporte, L. F., ed., Reefs in Time and Space: Soc. Econ. Paleontologists Mineralogists Spec. Pub. 18, p. 155–208.

———, 1983, Diagenetic sequence and burial history of Pine Point lead-zinc host rocks, Northwest Territories, Canada; Econ. Geology, in press.

———, AND MOUNTJOY, E. W., 1972, Comparison of central European and western Canadian Devonian reef complexes: 24th Internat. Geol. Congr., Sec. 6, p. 294–309.

LADD, H. S., 1950, Recent reefs: Am. Assoc. Petroleum Geologists Bull., v. 34, p. 203–214.

LAUBSCHER, H., AND BERNOULLI, D., 1977, Mediterranean and Tethys, in Nairn, A. E. M., Kanes, W. H., and Stehli, F. G., eds., The Ocean Basins and Margins, Volume 4A, The Eastern Mediterranean: New York, Plenum, p. 1–13.

MACQUEEN, R. W., 1976, Sediments, zinc and lead, Rocky Mountain Belt, Canadian Cordillera: Geosci. Canada, v. 3, p. 71–81.

———, 1979, Base metal deposits in sedimentary rocks: some approaches: Geosci. Canada, v. 6, p. 3–9.

MATTES, B. W., AND MOUNTJOY, E. W., 1980, Burial Dolomitization of the Upper Devonian Miette buildup, Jasper National Park, Alberta, in Zenger, D. H., Dunham, J. B., and Ethington, R. L., eds., Concepts and Models of Dolomitization: Soc. Econ. Paleontologists Mineralogists Spec. Pub. 28, p. 259–297.

MAUCHER, A., AND SCHREIDER, H. J., 1967, The alpine lead-zinc ores, in Brown, J. S., ed., Genesis of Stratiform Lead-Zinc-Barite Fluorite Deposits: Econ. Geology Monograph 3, p. 71–89.

MCILREATH, I. A., AND JAMES, N. P., 1978, Facies models—13. Carbonate slopes: Geosci. Canada, v. 5, p. 189–199.

———, 1977, Accumulation of a Middle Cambrian, deep water, basinal limestone adjacent to a vertical, submarine carbonate escarpment, Southern Rocky Mountains, Canada, in Cook, H. E., and Enos, P., eds., Deep-Water Carbonate Environments: Soc. Econ. Paleontologists Mineralogists Spec. Pub. 25, p. 113–124.

MEISSNER, F. F., 1972, Cyclic sedimentation in Middle Permian strata of the Permian Basin, West Texas and New Mexico, in Eram, J. C., and Chuber, S., eds., Cyclic Sedimentation in the Permian Basin, 2nd ed.: Midland, Texas, West Texas Geol. Soc., p. 203–232.

MEISCHNER, D., 1971, Clastic sedimentation in the Variscan geosyncline east of the river Rhine, in Müller, G., ed, Sedimentology of Parts of Central Europe: Frankfurt am Main, W. Kramer, p. 9–44.

MOUNTJOY, E. W., 1965, Stratigraphy of the Devonian Miette reef complex and associated strata, eastern Jasper National Park, Alberta: Geol. Survey Canada Bull. 110, 132 p.

———, 1975, Intertidal and supratidal deposits within isolated Upper Devonian buildups, Alberta, in Ginsburg, R., ed., Tidal Deposits: Berlin, Springer-Verlag, p. 387–395.

———, 1978, Upper Devonian reef trends and configuration of the western portion of the Alberta Basin, in McIlreath, I. A., and Jackson, P. C., eds., The Fairholme Carbonate Complex at Hummingbird and Cripple Creek: Canadian Soc. Petroleum Geology, p. 1–30.

———, 1980, Some questions about the development of Upper Devonian carbonate buildups (reefs), western Canada: Bull. Canadian Petroleum Geology, v. 28, no. 3, p. 315–344.

———, AND JULL, R. K., 1978, Fore-reef carbonate and bioherms and associated reef margin, Upper Devonian, ancient wall reef complex, Alberta: Canada Jour. Earth Sci., v. 15, p. 1304–1325.

NEY, C. S., 1954, Monarch and Kicking Horse Mines, Field, British Columbia, in Guidebook, Fourth Annual Field Conference, Banff-Golden-Radium, Alberta Soc. Petroleum Geologists, p. 119–1136.

———, 1957, Monarch and Kicking Horse Mine, in Structural Geology of Canadian Ore Deposits, Volume 2: 6th Commonwealth Mining and Metallurgy Congr., p. 143–152.

PLAYFORD, P. E., 1980, Devonian 'Great Barrier Reef' of Canning Basin, western Australia: Am. Assoc. Petroleum Geologists Bull., v. 64, p. 814–840.

PRICE, R. A., 1981, The Cordilleran foreland thrust and fold belt in the southern Canadian Rocky Mountains, in McClay, K. R., and Price, N. J., eds., Thrust and Nappe tectonics: Geol. Soc. London, Spec. Pub. 9, p. 427–449.

READ, J. F., 1982, Carbonate platforms of passive (extensional) continental margins: Types, characteristics and evolution: Tectonophysics, v. 81, p. 195–212.

SCHLAGER, W., 1981, The paradox of drowned reefs and carbonate platforms: Geol. Soc. America Bull., Part I, v. 92, p. 197–211.

———, AND JAMES, N. P., 1978, Low-magnesian calcite limestones forming at the deep-sea floor, Tongue of the Ocean, Bahamas: Sedimentology, v. 25, p. 675–702.

———, AND CHERMAK, A., 1979, Sediment facies of platform—basin transition, Tongue of the Ocean, Bahamas, in Doyle, L. J., and Pilkey, O. H., eds., Geology of Continental Slopes: Soc. Econ. Paleontologists Mineralogists Spec. Pub. 27, p. 193–208.

———, AND GINSBURG, R. N., 1981, Bahama carbonate platforms—the deep and the past: Mar. Geology, v. 44, p. 160–181.

SCHOLLE, P., BEBOUT, D., AND MOORE, C., 1983, Carbonate Depositional Environments: Am. Assoc. Petroleum Geologists, Memoir 33, in press.

SEARLE, M. P., JAMES, N. P., CALON, R. J., AND SMEWING, J. D., 1983, Sedimentology and structural evolution of the Arabian continental margin in the Musandam Mountains and Dibba Zone, United Arab Emirates: Geol. Bull. Soc. America, in press.

SELLWOOD, B. F., 1978, Shallow-water carbonate environments, in Reading, H. G., ed., Sedimentary Environments and Facies: New York, Elsevier, p. 259–303.

SKALL, H., 1975, The paleoenvironment of the Pine Point lead-zinc district: Econ. Geology, v. 70, p.22–47.

STOAKES, F. A., 1980, Nature and control of shale basin fill and its effect on reef growth and termination: Upper Devonian Duvernay and Ireton Formations of Alberta, Canada: Bull. Canadian Soc. Petroleum Geology, v. 28, p. 345–411.

TOOMEY, D. F., ed., 1980, European Fossil Reef Models: Soc. Econ. Paleontologists Mineralogists Spec. Pub. 30, 546 p.

TREHU, A. M., 1975, Depth versus (age)$^{1/2}$: a perspective on mid-ocean rises: Earth and Planetary Sci. Letters, v. 27, p. 287–304.

VAN SICLEN, D. C., 1958, Depositional topography—examples and theory, Am. Assoc. Petroleum Geol. Bull., v. 42, p. 1897–1913.

———, 1964, Depositional topography in relation to cyclic sedimentation, in Merriam, D. F., ed., Symposium on Cyclic Sedimentation: Geol. Survey Kansas Bull. 169, v. 2, p. 533–539.

———, 1972, A depositional model of late Paleozoic cycles on the eastern shelf, in Elam, J. C., and Chuber, S., eds., Cyclic Sedimentation in the Permian Basin: West Texas Geol. Soc. Pub. 69-56, p. 17–27.

WENDT, J., AND FURSICH, F. T., 1979, Facies analysis and palaeogeography of the Cassian Formation, Triassic, Southern Alps: Riv. Ital. Paleont., v. 85, p. 1003–1028.

WHITTINGTON, H. B., AND KINDLE, C. H., 1963, Middle Ordovician Table Head Formation, western Newfoundland: Geol. Soc. America Bull., v. 74, p. 745–758.

WILSON, J. L., 1974, Characteristics of carbonate-platform margins: Am. Assoc. Petroleum Geologists Bull., v. 58, p. 810–824.

———, 1975, Carbonate Facies in Geologic History: New York, Springer-Verlag, 471 p.

WINTERER, E. L., AND BOSELLINI, A., 1981, Subsidence and sedimentation on Jurassic passive continental margin, southern Alps, Italy: Am. Assoc. Petroleum Geologists Bull., v. 65, p. 394–421.

ZENGER, D. H., DUNHAM, J. B., AND ETHINGTON, R. L., eds., 1980, Concepts and Models of Dolomitization: Soc. Econ. Paleontologists Mineralogists Spec. Pub. 28, 320 p.

CARBONATE INTERNAL BRECCIAS: A SOURCE OF MASS FLOWS AT EARLY GEOSYNCLINAL PLATFORM MARGINS IN GREECE

HANS FÜCHTBAUER AND DETLEV K. RICHTER
Geologisches Institut der Ruhr-Universität Bochum, West-Germany

ABSTRACT

Internal breccias caused by dilation of slightly lithified limestones have been investigated in the Triassic and Jurassic of the island of Hydra (Greece). They occur between shelf platform and rift basin of the early Tethys and are in general composed of monomictic and closely fitted angular clasts of platform carbonates. The matrix consists frequently of reddish deep-water limestones with thin-shelled molluscs (filaments) and radiolarians; it was sucked-in from above as a result of dilation during brecciation. The vertical succession includes about 20–30 m beginning with (a) fractured and fissured shallow-marine limestones grading upward into (b) internal breccias and (c) mass flows (characterized by a low degree of fitting and considerable roundness). Internal breccias imply a very small lateral displacement; they provide an important source for mass flows near shelf-to-slope breaks. Repeated brecciation is typical. The breccias are most frequently composed of shallow-water limestones, but are overlain by basin sediments. This indicates that the brecciation was connected with tectonic downwarping. We suggest that these breccias were produced by large migrating flexures, and that such flexures are a tectonic alternative or substitute for faults.

In the Triassic and lower Jurassic limestones of the island of Hydra, five major breccia horizons are recognized. They correlate well with major tectonic phases in the early geosynclinal history of the northern and eastern Alps, in which internal breccias were found as well. This coincidence emphasizes the significance of such breccias in the evolution of geosynclines.

INTRODUCTION

Most mass flows occur in tectonically active areas. Many such flows containing carbonate components originate near shelf-to-slope breaks, and much information on this topic is presented in SEPM Special Publication 25 (Cook and Enos, 1977). The observations discussed in this paper offer one possibility of how carbonate mass flows can form. They are frequently underlain by, and gradually develop from, a specific type of breccia, the fragments of which fit against each other like a jigsaw puzzle. We made these observations on the island of Hydra (Greece) as well as in several places in the Northern Alps. Buggisch and Flügel (1980) reported similar observations from the Alps. Because we consider it important to distinguish these well-fitting breccias from mass flows, we assigned them a specific name ("internal breccias") and investigated their properties and occurrence. They are generally found in shallow-marine limestones but are overlain by deeper marine limestones. They provide a record of dynamic shelf-to-slope transition and, therefore, are relevant to the main topic of the present symposium volume. The detailed observations are documented in Richter and Füchtbauer (1981); in view of space limitations, this information is not repeated here.

PROPERTIES OF INTERNAL BRECCIAS

In many parts of the Calcareous Alps, but particularly in the Triassic and Jurassic of the small island of Hydra south of the Peloponnesus, Greece (Fig. 1), we have observed carbonate breccias. The fragments of these breccias fit so well together that there is little doubt about very small relative motion involved in their development (Fig. 2). Similar rocks have been described, for instance, from the Triassic and Jurassic of the Alps (e.g. Wiedenmayer, 1963, p. 621; Jurgan, 1969, p. 478; Schlager, 1969, pp. 301, 303; Rieche, 1971, p. 53), from the Cretaceous of Italy (Günther and Wachsmuth, 1970), and from the Cambrian of the U.S. (Keith and Friedman, 1977; Reinhardt, 1977; Cook, 1979). Fairbridge, in his Encyclopedia (1978, p. 84), termed them "autoclastic breccias." We prefer the term "internal breccia" which indicates their position and formation within units of carbonate rocks. In general, brecciation in these various publications is attributed to sliding. A series of observations, however, does not lend support to a slide origin for the breccias on Hydra. For example, we note that:

1. Many internal breccias consist of shallow-marine carbonate rocks that are overlain by red pelagic limestones.

2. The following sequence is typical, from base to top (Fig. 3): (a) shallow-water limestone, locally with calcite-cemented joints, (b) shallow-water limestone with fissures filled with a breccia of the same lithology, set in a red limestone matrix, (c) internal breccia of shallow-water limestone with a red limestone matrix filled-in from above (see e), (d) mass flow breccia consisting of the same shallow-marine limestone and the red matrix cited

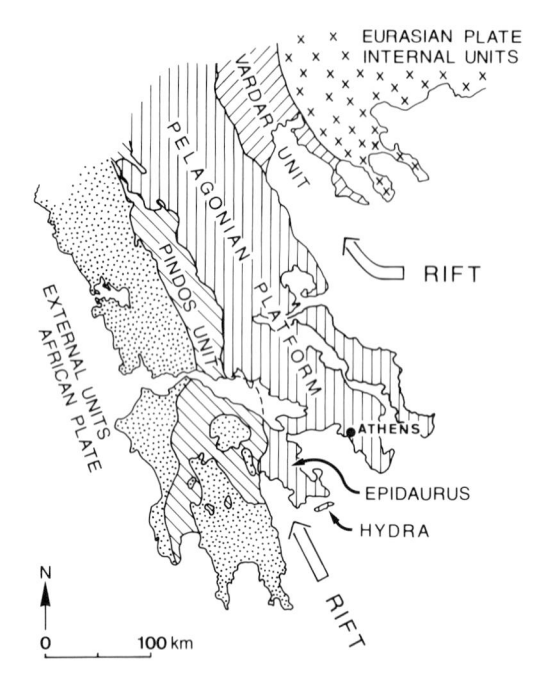

FIG. 1.—Simplified geotectonic map of Greece, showing the island of Hydra and two presumed Triassic rift systems.

above, and (e) red pelagic limestone with filaments (thin-shelled planktonic pelecypod fragments) and, occasionally, with radiolaria.

3. The breccia includes fragments of calcite veins as well as of older internal breccias.

Of these, observation (1) could perhaps suggest transport over a considerable distance from a platform downslope into a pelagic realm of prevailing red limestone deposition. Observation (2), however, is not compatible with such a long-distance transport, because during downslope sliding the entire shallow-marine limestone unit (and not only its surface layer) would probably have been broken, sheared, and brecciated. Observation (3) indicates a sequential formation over a long period of time; this would also be incompatible with brecciation of a slide mass which would imply sliding into a stable final position thus precluding repeated brecciation. We will reconsider these points in the last section.

A different scope is suitable for sedimentary conglomerates and for internal breccias. Whereas transport is the major focus in the case of conglomerates, origin by fracturing is envisaged in the case of breccias. The larger the pebbles of a conglomerate, the higher the transport energy; the smaller the fragments of an internal breccia, the higher the amount of energy required for fracturing. Another important difference is that conglomerates become better sorted during transport (i.e. the proportion of matrix decreases), whereas matrix content increases in breccias, particularly in debris flows (Hampton, 1972). This means that different textural properties are relevant for describing sedimentary conglomerates and breccias in general and especially internal breccias. The main properties to be used for quantification are: (a) in conglomerates = grain size and rock types; and (b) in breccias = matrix (or groundmass) content, grain-to-grain relation of the particles, roundness, and other properties considered below.

The matrix content can be estimated by point-counting. Because some breccias contain chemically precipitated cement instead of, or together with, sedimentary matrix, the term "groundmass" which includes cement and matrix is more appropriate.

The relation between the particles of internal breccias is governed by their fit against each other, i.e. the degree to which clast boundaries match those of adjacent clasts. "Fitting" is a term used to define the percentage of clasts that fit against other clasts; in measurements, each clast is counted only once.

The roundness of breccias as defined here differs markedly from the roundness of conglomerates; it is the percentage of clasts with at least one rounded edge. This means that many mass flow breccias have a roundness value of 100%, although clasts may display a considerable number of edges. It is possible to trace the transition of internal breccias into mass flow breccias in diagrams using fitting and groundmass as parameters (Fig. 4). In addition, roundness may be considered a useful indicator because it increases considerably from internal breccias to mass flows.

Moreover, it may be useful to record whether the groundmass is red or gray. Breccias with gray clasts and gray groundmass tend to be overlooked.

It is also important to determine whether a breccia is polymictic or monomictic. Because of the very small displacement of individual clasts, internal breccias should be termed "monomictic," even if rhythmic alternations of two or more rock types are brecciated.

In Figure 4, two additional breccia types are identified: fissure fill breccias which often have a high matrix percentage and small clasts, and the shear breccias which are mainly characterized by shear faces crossing the breccia and crushed clasts (the latter are sometimes difficult to distinguish from matrix). For these and other reasons the diagram in Figure 4 is not sufficient for the distinction of mass flow and shear breccias. The reader is directed to a more detailed discussion of breccia types provided elsewhere by Füchtbauer and Richter (1980) and Richter and Füchtbauer (1981).

Two breccia types are compared in Table 1. Similar reviews of the properties of different breccias

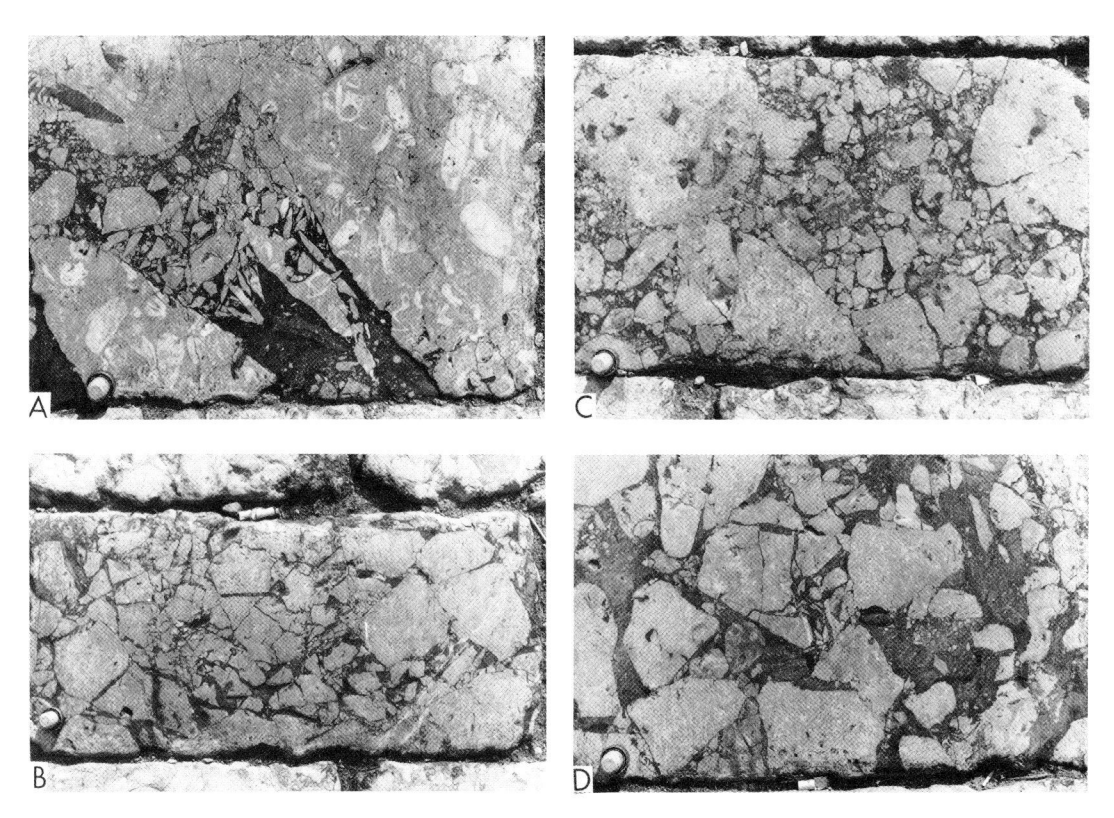

Fig. 2.—Internal and mass flow breccias. Examples are from Cenomanian rudistid limestones of the island of Dokos, now forming the pavement of the pier of Hydra Chora. Roundness and matrix content increase while "fitting" values decrease from A (internal) to D (mass flow). Scale = cork bottle stopper, 1.8 cm maximum diameter.

have been published by Blount and Moore (1969) and by Roehl (1981).

INTERNAL BRECCIAS IN THE HYDRA STRATIGRAPHIC SEQUENCE

Approximately 1500 m of mainly calcareous rocks of early Permian to late Jurassic age are exposed on the island of Hydra, Greece (Römermann, 1969). During this time span, the depositional environment shifted as follows: In the Permian—shallow-marine over the whole island; in the Triassic—shallow-marine in the northeast and pelagic in the southwest of Hydra; and in the Jurassic—pelagic over the whole island.

TABLE 1.—COMPARISON OF INTERNAL AND SHEAR BRECCIAS

Property	Internal Breccia	Shear Breccia
Clast shape	Irregular to isometric	Flat to lens-shaped
Clast lithology	Usually monomictic	Usually monomictic
Clast-to-clast relation	irregular	Closely packed and dissected by shear planes
Fitting	High	Low
Type of groundmass	Matrix sucked-in from above, or cement	Cataclastic clasts, or cement
Relative dislocation of clasts	Low	High
Fracturing of clasts	Several superimposed generations are distinctive	Different generations may be difficult to distinguish
Relation to stratification	Stratabound, but laterally discontinuous	Poorly stratified and discontinuous
Other terms used for the breccias	Dilation breccia (Roehl, 1981)	Tectonic Breccia (Blunt and Moore, 1969; Roehl, 1981)
	Autoclastic breccia (Fairbridge, 1978)	

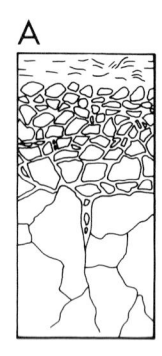

overlying red limestones of Scythian age were "internally" brecciated and then incorporated in a mass flow sandstone series.

Illyrian (=Upper Anisian; Middle Triassic).—A shallow-marine platform with massive oolitic and oncolitic limestones including reefoid rocks ("Eros" limestone) was deposited on top of the mass flow cited in A. Towards the south, a transition into a slope and a basin facies is inferred from the facies distribution and from observations in Adhami near Epidauros (Fig. 5). An internal breccia is developed in the upper 6 to 15 m of platform rocks, with fissures up to 10 m long beneath the breccia (Fig. 3). The breccia as well as the fissures are filled with a red micritic matrix. The degree of induration of the rock at the time of brecciation is recorded by the fact that the particulate limestones had only an even-style palisade cement when the red micritic limestone or marl matrix was filled in (Fig. 7). A blocky calcite cementation developed afterwards, separately from the red-matrix, infiltration phase.

About 2 km north of the reefoid rocks, the massive limestone grades into a thin-bedded lagoonal limestone. It is remarkable that this rock is slump-deformed and the folds are brecciated, especially within the deposit (Fig. 8). This breccia was not filled with red matrix, presumably because the permeability after brecciation was not sufficient to allow in-flow of mud from above. The fractures, however, were sealed by yellow dolomite; the fragments were, at that time, sufficiently lithified to resist dolomitization except marginal replacement. It is possible that the above-cited massive limestones were brecciated earlier than the thin-bedded lagoonal limestones.

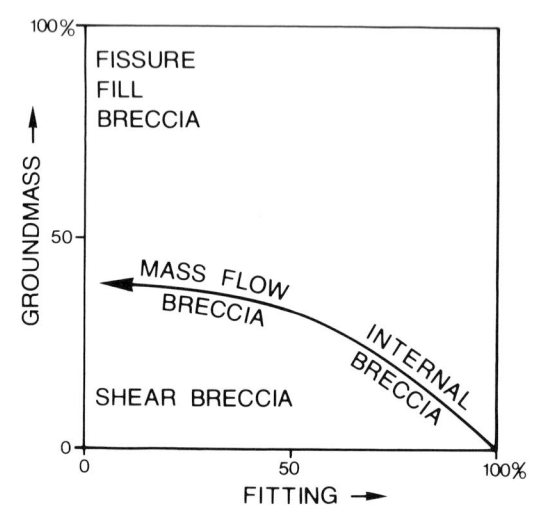

FIG. 3.—*A*, Schematic cross-section of an internal breccia recording only one brecciation event. *B*, Mass flow unit; *C*, internal breccia (both display a red matrix).

This progressive deepening, which is discussed below, was not a steady process. In general, a carbonate platform was re-established after each major subsidence. The major events, which are dated mainly by conodonts and occasionally by cephalopods, dasycladacean algae, and foraminifera, are summarized below and in Figures 5 and 6. For further details, see Richter and Füchtbauer (1981).

Scythian/Lower Anisian (Lower Triassic).—Upper Permian shallow-marine carbonates and the

FIG. 4.—Different breccia types positioned in a diagram emphasizing degree of fitting versus groundmass (groundmass = matrix + cement between fragments).

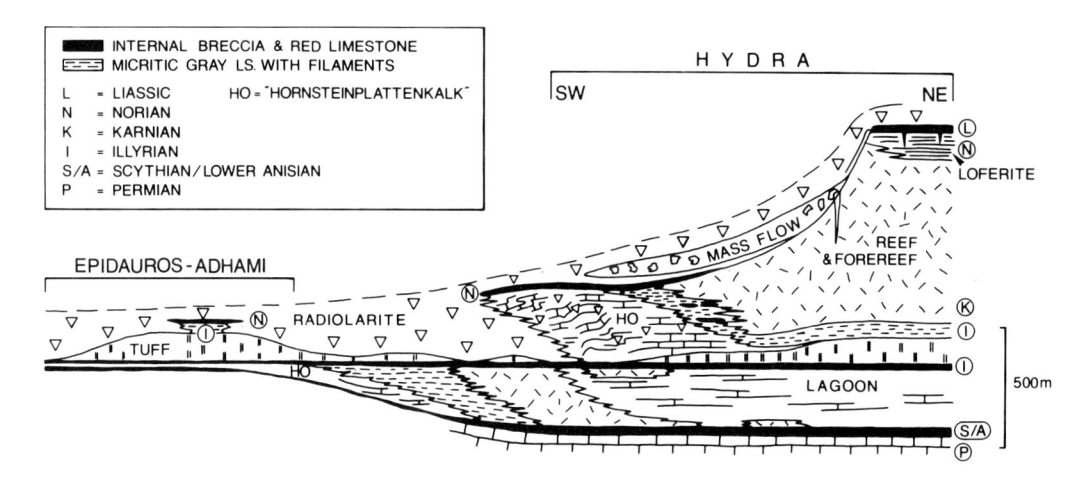

FIG. 5.—Schematic cross-section through the southwestern slope of the Pelagonian Platform showing the brecciation events (in black) and, from NE to SW, a lagoon-reef-slope (micritic limestone and "HO")—basin (radiolarite) sequence. Section is highly vertically exaggerated.

These monomictic internal breccias grade upwards into mass flows up to 10 m thick and containing polymictic and subrounded components. These include fragments of the red limestone filling the breccia (and overlying the mass flow, which in turn became brecciated, after consolidation and cementation). The overlying red limestone is also of Middle Triassic age and contains "filaments" (planktic thin lamellibranch shells), calcitized radiolaria and isolated cephalopods.

Lower Tuvalian (=lower Upper Karnian; Upper Triassic).—A distinct facies zonation was developed in late Triassic time from south to north, including radiolarites (inferred from the area of Epidauros, Argolis), bedded limestones with chert layers ("Hornsteinplattenkalk"), micritic limestones with turbidites or mass flows, and forereef to reef limestones (Fig. 5). Internal breccias have been observed only in the lower slope facies, in the micritic limestones, while slumps are developed in the "Hornsteinplattenkalk." In the reef, breccia-filled fissures presumably belonging to this event were formed.

Triassic-Jurassic Transition.—The reef limestones that comprise sponges, corals, bryozoans, echinoderms, benthic foraminifera, and red, green, and blue-green algae grade upward into a loferite facies with supratidal thin-bedded limestones and dolomites, partially reworked, and intercalations of lagoonal, megalodon limestones. The loferites and reef limestones are in places transected by fissures

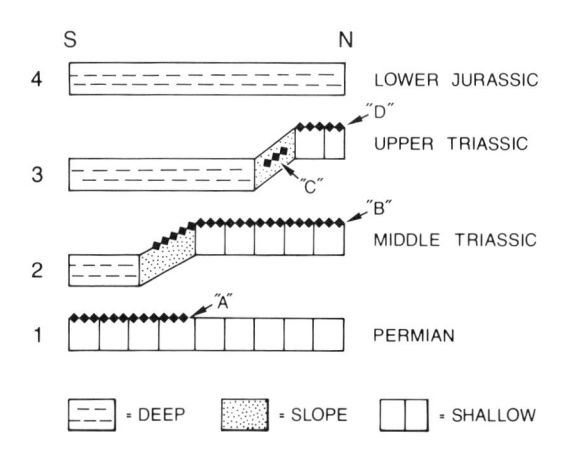

FIG. 6.—Scheme showing relation between internal brecciation and subsidence caused by a flexure which migrates towards the north. "A"–"D" = brecciation events described in text.

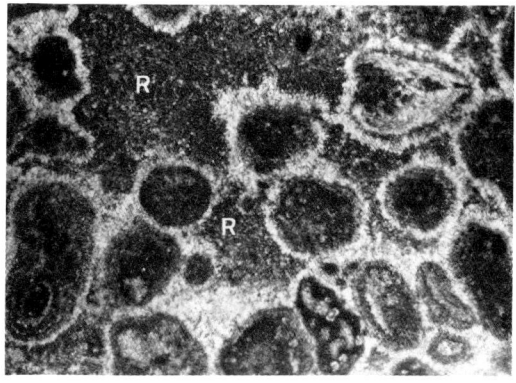

FIG. 7.—Peloidal "Eros" limestone clast of an internal breccia. Only fibrous calcite cement rims were present when the reddish lime mud (R) was filtrated into the clast. Length of photograph section is 2.8 mm.

FIG. 8.—Slump-deformed and internally brecciated lagoonal limestones of Upper Anisian age, eastern Hydra. Fractures are sealed by dolomite.

filled with cement or with gray and red micritic matrix. These fissures strike north–south and east–west and can be more than 50 meters long. The upper part of the loferites is deformed into an internal breccia with red micritic matrix. This brecciation occurred in early Jurassic time. Gradational change into a mass flow can be observed. At about the same time, mass flows also developed within the basinal bedded limestones.

Jurassic.—Red Liassic limestones above the reef were internally brecciated and again were overlain by mass flow breccias. During this phase the entire area subsided into deeper water, and radiolarites were widespread. A giant mass flow containing Jurassic limestones as well as fragments of Triassic reef rocks is located south of the town of Hydra and records the steep morphology of the former reef area (Fig. 5). Faults which also may have initiated this mass flow have not been found.

The events outlined above are schematically depicted in Figure 6. In each stage, the internal breccia is a precursor of a downwarping process which shifted towards the platform. This is shown in greater detail in Figure 5.

The stratigraphic-tectonic setting of the study area relative to Alpine orogenic trends is highlighted in Figure 1. The island of Hydra probably represents only a fragment of a Tertiary mélange, floating in a matrix of red radiolarian shale and ophiolitic material. The present strike of bedding on Hydra is now oriented approximately west-east; this presumably rotated from an original NNW–SSE strike. The island's position is on the southwestern edge of the "Pelagonian platform" (including Pelagonian, Subpelagonian, and Parnassos-Kiona zones) which is bordered on the SW by a rift belt comprising deep-water limestones and radiolarites of the Pindos unit. According to a study on intracontinental rifts, Burke (1980) states "in some rifts, it is hard to be sure whether a boundary is a fault or a steep monoclinal flexure" (l.c., p. 42).

Based on observations in the Mesozoic of the island of Hydra we consider the boundary between the rift zone and the Pelagonian platform as a flexure rather than a fault; this is discussed below. Flexures also play an important role in marked deformation of post-Miocene sequences in the Mediterranean as noted on continuous seismic profiles (Stanley, 1977, p. 109). The northeastern border of the Triassic rift and the subsequent drift system developed into a convergent plate margin, including subduction towards the northeast, with ophiolites and mélanges in early to middle Tertiary time (Altherr, 1981). This mélange zone can be followed as far as Yugoslavia (Dimitrijević and Dimitrijević, 1973), although the peninsula of Argolis is not included by these authors.

As shown in Figure 9, events A to E are not limited to the local geology of Hydra; these events also are significant in the eastern Alps. Major events in the northern Calcareous Alps (1–5) and the southeastern Alps (Drau) (I–IV), respectively, include:

1. The "Reichenhaller Wende" (i.e. turning point) of Schlager and Schöllnberger (1974) which is characterized by the beginning of widespread carbonate sedimentation, and is not directly comparable with our event A. Event (I), the first tectonic phase of Bechstädt et al. (1976), however, is correlatable with A, although a little younger.

2. The "Reiflinger Wende" indicating a widespread transition from shallow- to deep-marine sedimentation, and (II) which correlates remarkably well with B, which is one of the two most important events (B, D) recorded on Hydra.

3. The "Reingrabener Wende," and (III), the "Raibler" event, are small events in the Alps as well as on Hydra, where they occur at different times during the Karnian (C).

4. The "Adneter Wende" and (IV) correlate well with D. These events are similar in terms of their geological significance. In the eastern Alps and on Hydra they imply a marked foundering of the shallow-water carbonate platforms. According to Bernoulli and Jenkyns (1974), this subsidence applies to the whole Alpine-Mediterranean region and is presumably the most important event in the early geosynclinal history of the Alpine system. At the same time, radiolarites were deposited towards the northeast.

5. The "Ruhpoldinger Wende," may correlate with event E. This event was subsequent to the transformation of platforms into seamounts associated with an *in-situ* brecciation and the superposition of pelagic sediments in the Pliensbachian of northern Italy (Bernoulli et al., 1979, p. 201). At this time, the whole island of Hydra was buried by radiolarian mud.

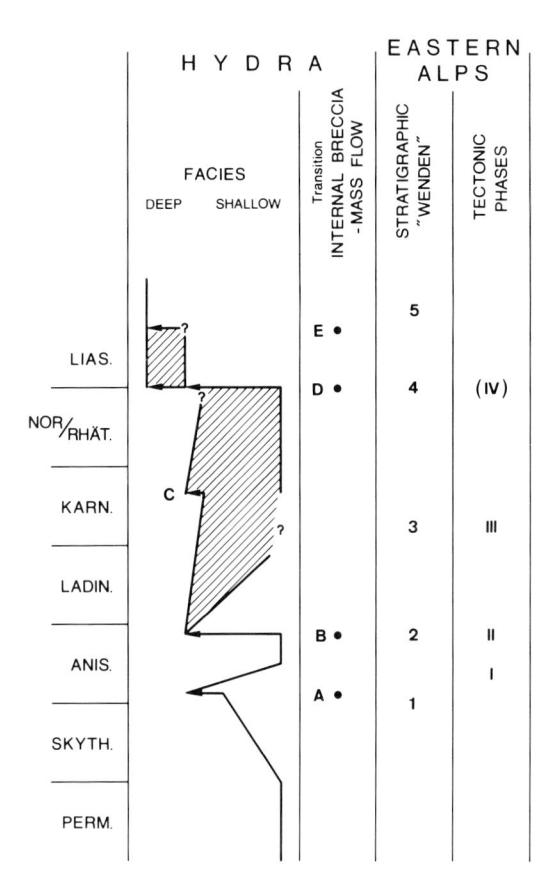

FIG. 9.—Comparison of tectonic events on the island of Hydra (A–E; arrows indicate time of flexure formation) with the changes in the northern Calcareous Alps (1–5, from Schlager and Schöllnberger, 1974) and the tectonic phases in the Drau area, eastern Alps, (I–IV, from Bechstädt et al., 1976). The left boundary of the shaded areas indicates the environment in the basin (Liassic) and on the lower slope (Triassic); the right boundary corresponds to the basin ridge (Liassic) and reef areas (Triassic).

ORIGIN OF HYDRA INTERNAL BRECCIAS

It is difficult to explain why sediments respond upon dilatation by brecciation rather than by faulting. A key to this problem, possibly, lies with the conditions of formation of internal breccias. They occur: (a) in limestones which were brecciated shortly after deposition, i.e. before thorough lithification, (b) in tectonically active zones, and (c) on slopes or on platforms which transformed into a slope position during brecciation. These three conditions may all be necessary for internal breccias to form; they are discussed below.

1. Three different responses during dilatation are depicted in Figure 3 from a reefoid platform. They may be due in part to increasing cementation, in part to decreasing dilatation both in downward direction. At the surface, the earliest cementation was only able to keep the grains in contact so that upon dilatation brittle rock fragments could form; these latter became rounded by friction and during mass flow transport. Below the surface, internal brecciation occurred in granular limestones (or rudites). Small angular chips suggest a certain degree of cementation, but Figure 7 indicates that only small cement rims were present at the time of brecciation. About twenty meters below the surface, only fissures or fractures developed which are now, for the most part, cemented by calcite. On the other hand, thin-bedded lagoonal limestones display slump folds with internal brecciation (Fig. 8) down to more than 100 m below the surface.

2. The relation between internal breccias and tectonically active zones suggests an influence of tectonic stress (or earthquakes) on brecciation. This idea has been recently promoted by Redwine (1981) and Roehl (1981) for the breccias in the Monterey Shale of California. They observed that portions of the Monterey units, embrittled by dolomitization, contain open extension fractures and even breccias. The development of such closely fitted "dilation breccias" is most abundant in areas of strike-slip faulting and is explained by an initial compression and subsequent elastic dilatation followed by continued stress and associated excess pore-fluid pressures. The difference between these breccias and the internal breccias described in the present paper is that the latter formed near the sediment surface and, therefore, were filled with the overlying soft sediment, whereas the Monterey breccias were filled with oil from adjacent shales or sealed with dolomite (as in the Illyrian lagoonal limestones of Hydra). In the case of both occurrences, repeated episodic dilatation is characteristic. That such natural hydraulic fracturing also was effective on Hydra is suggested by cemented partings in fine-bedded red and gray limestones (which are not brecciated). Breccias modified by overpressure previously were described by Masson (1972).

3. As shown in Figure 5, and schematically in Figure 6, internal breccias occur only in positions near the shelf-to-slope break, which, due to flexing, are gradually transformed into slopes or abyssal plains (Fig. 6, sequence 1–4). A possible alternative is faulting with rupture at depth but with near surface displacement by warping or folding rather than faulting. Such structures have been reported by Yeats et al. (1981, p. 194) from Quaternary sediments in Southern California, where they result from seismic shaking. Such shocks, together with the temporary compression and abnormal (high) pore pressure, may have produced internal breccias by hydraulic fracturing at a certain stage of lithification of the limestones.

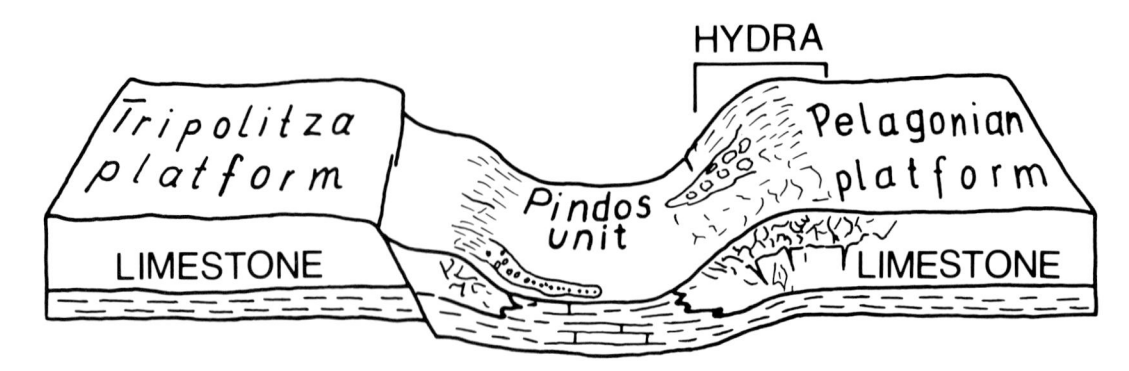

FIG. 10.—Interpretation of the Peloponnesus rift stage (Triassic/early Jurassic), emphasizing the flexure-generated breccias examined on the island of Hydra.

The flexure model is supported by early cemented fissures oriented predominantly in the tectonic bc-direction, i.e. parallel to the platform edge and to the strike of the inferred flexure, and perpendicular to the bedding. That tilting, rather than faulting, was the most important process is also suggested by the frequency of slump folds in sediments which were mechanically suitable to slumping, i.e. thin-bedded deep-water micrites with chert layers ("Hornsteinplattenkalk") and thin-bedded lagoonal limestones.

The flexure is part of the newly-forming rift basin which during Jurassic and Cretaceous time became an oceanic abyssal plain (Fig. 10). This can be described as the broadening of an early geosyncline (see Bernoulli et al., 1979) and is reminiscent of the onlap sequence formed by transgression out of the North Sea rift basin (Bjørlykke et al., 1979, their Fig. 15).

For the Cenomanian internal breccia of the island of Dokos (Fig. 2) and for breccias of event C, lateral limitations have been observed. Such restrictions are also reported by Redwine (1981).

The alpine tectonic activity which produced the mélange enclosing the geological fragment forming Hydra cannot be responsible for the internal breccias. We conclude this on the basis of the different structural inventory of the mélanges which are shear breccias characterized by pinch-and-swell, boudins, and lens-shaped clasts (see Table 1). Such rocks can be observed only at the margins of the island and are independent of the horizons of internal brecciation. Moreover, the contemporaneous red matrix is not compatible with an origin at the time of mélange formation. Moreover, short-term eustatic sealevel lowering cannot be responsible for the internal breccias, because the latter clearly differ from shrinkage cracks and contain deep-water matrix only.

SUMMARY

In conclusion, we suggest that:

1. Internal breccias are indicative of tectonically active shelf-to-slope breaks and—less frequently—of slope and basin ridge positions adjacent to shelf platforms.

2. Internal breccias occur mainly in flexure zones at the margins of developing geosynclines. The migration of flexures towards the inner shelf implies alternating dilatation and compression which, affected by earthquake-induced pore fluid overpressure, may explain the internal breccias.

3. Internal breccias, positioned close to the shelfedge, may serve as potential sources of mass flow debris.

ACKNOWLEDGMENTS

We thank D. J. Stanley, S. J. Culver and R. B. Halley for reviewing the manuscript and for providing many suggestions, and W. Schlager for valuable discussions.

REFERENCES

ALTHERR, R., 1981, Lower Miocene granitoids in the attic-cycladic crystalline complex (Greece)—Petrology and paleotectonic significance [abs.]: Internat. Symposium on the Hellenic Arc and Trench (H.E.A.T.), April 8–10, 1981, Athens, p. 3–4.

BECHSTÄDT, TH., BRANDNER, R., AND MOSTLER, H., 1976, Das Frühstadium der alpinen Geosynklinalentwicklung im westlichen Drauzug: Geol. Rundschau, v. 65, p. 616–648.

BERNOULLI, D., AND JENKYNS, H. C., 1974, Alpine, Mediterranean, and central Atlantic mesozoic facies in relation to the early evolution of the Tethys, in Dott, R. H., Jr., and Shaver, R. H., eds., Modern and Ancient Geosynclinal Sedimentation: Soc. Econ. Paleontologists Mineralogists Spec. Pub. 19, p. 129–160.

——, Kaelin, O., and Patacca, E., 1979, A sunken continental margin of the Mesozoic Tethys: the northern Apennines: Symposium Sédimentation Jurassique W. Européen, Assoc. Sédim. Franç. Publ. Spéc. No. 1, p. 197–210.

Bjørlykke, K., Elverhøi, A., and Malm, A. O., 1979, Diagenesis in Mesozoic sandstones from Spitsbergen and the North Sea—A comparison: Geol. Rundschau, v. 68, p. 1152–1170.

Blount, D. N., and Moore, C. H., Jr., 1969, Depositional carbonate breccias, Chiantla Quadrangle, Guatemala: Geol. Soc. America Bull., v. 80, p. 429–442.

Buggisch, W., and Flügel, E., 1980, Die Trogkofel-Schichten der Karnischen Alpen-Verbreitung, geologische Situation und Geländebefund.—Carinthia II, 36: Sonderheft (Hrsg.: E. Flügel), Klagenfurt, p. 13–50.

Burke, K., 1980, Intracontinental rifts and aulacogens, in Continental Tectonics, Studies in Geophysics: Washington D. C., National Acad. Sci., p. 42–49.

Cook, H. E., 1979, Ancient continental slope sequences and their value in understanding modern slope development, in Doyle, L. J., and Pilkey, O. H., eds., Geology of Continental Slopes: Soc. Econ. Paleontologists Mineralogists Spec. Pub. 27, p. 287–305.

——, and Enos, P., eds., 1977, Deep-water Carbonate Environments: Soc. Econ. Paleontologists Mineralogists Spec. Pub. 25, 336 p.

Dimitrijević, M. D., and Dimitrijević, M. N., 1973, Olistostrome melange in the Yugoslavian Dinarides and late Mesozoic plate tectonics: Jour. Geology, v. 81, p. 328–340.

Fairbridge, R. W., 1978, Breccias, sedimentary, in Fairbridge, R. W., and Bourgeois, J., eds., The Encyclopedia of Sedimentology: Stroudsburg, Dowden, Hutchinson & Ross, p. 84–86.

Füchtbauer, H., and Richter, D. K., 1980, Breccias: Criteria and observations: Internat. Assoc. Sedimentologists, 1st European Meeting, Bochum 1980, p. 52–55.

Günther, K., and Wachsmuth, W., 1970, Submarine Brekzien und Sedimentationslücken im Mesozoikum der nordwestlichen Toskaniden: Neues Jahrb. Geol. Paläont., Abh., v. 134, p. 57–100.

Hampton, M. A., 1972, The role of subaqueous debris flow in generating turbidity currents: Jour. Sed. Petrology, v. 42, p. 775–793.

Jurgan, H., 1969, Sedimentologie des Lias der Berchtesgadener Kalkalpen: Geol. Rundschau, v. 58, p. 461–501.

Keith, B. D., and Friedman, G. M., 1977, A slope-fan-basin-plain model, Taconic sequence, New York and Vermont: Jour. Sed. Petrology, v. 47, p. 1220–241.

Masson, H., 1972, Sur l'origine de la cornieule par fracturation hydraulique: Eclogae Geol. Helv., v. 65/1, p. 27–41.

Redwine, L., 1981, Hypothesis combining dilation, natural hydraulic fracturing, and dolomitization to explain petroleum reservoirs in Monterey Shale, Santa Maria area, California, in Garrison, R. E., and Douglas, R. G., eds., The Monterey Formation and Related Siliceous Rocks of California: Soc. Econ. Paleontologists Mineralogists Pacific Sec. Spec. Pub., Bakersfield, California, p. 221–248.

Reinhardt, J., 1977, Cambrian off-shelf sedimentation, central Appalachians, in Cook, H. E., and Enos, P., eds., Deepwater Carbonate Environments: Soc. Econ. Paleontologists Mineralogists Spec. Pub. 25, p. 83–112.

Richter, D. K., and Füchtbauer, H., 1981, Merkmale und Genese von Breccien und ihre Bedeutung im Mesozoikum von Hydra (Griechenland): Zeitschr. deutsch. geol. Ges., v. 132, p. 451–501.

Rieche, J., 1971, Die Hallstätter Kalke der Berchtesgadener Alpen [disser.] Berlin, T. U., D83, 173 p.

Roehl, P. O., 1981, Dilation brecciation—a proposed mechanism of fracturing, petroleum expulsion and dolomitization in the Monterey formation, California, in Garrison, R. E., and Douglas, R. G., eds., The Monterey Formation and Related Siliceous Rocks of California: Soc. Econ. Paleontologists Mineralogists Pacific Sec. Spec. Pub., Bakersfield, California, p. 285–315.

Römermann, H., 1969, Geologie der Insel Hydra (Griechenland): Geologica et Paleontologica, Marburg, v. 2, p. 163–171.

Schlager, W., 1969, Das Zusammenwirken von Sedimentation und Bruchtektonik in den triadischen Hallstätterkalken der Ostalpen: Geol. Rundschau, v. 59, p. 289–308.

——, and Schöllenberger, W., 1974, Das Prinzip stratigraphischer Wenden in der Schichtfolge der Nördlichen Kalkalpen: Mitt. Geol. Ges. Wien, v. 66/67, p. 165–193.

Stanley, D. J., 1977, Post-Miocene depositional patterns and structural displacement in the Mediterranean, in Nairn, A. E. M., Kanes, W. H., and Stehli, F. G., eds., The Ocean Basins and Margins, Vol. 4A: The Eastern Mediterranean: New York, Plenum Press, p. 77–150.

Wiedenmayer, F., 1963, Obere Trias bis mittlerer Lias zwischen Saltrio und Tremona (Lombardische Alpen). Die Wechselbeziehungen zwischen Stratigraphie, Sedimentologie und syngenetischer Tektonik: Eclogae geol. Helvet., v. 56, p. 529–640.

Yeats, R. S., Clark, M. N., Keller, E. A., and Rockwell, T. K., 1981, Active fault hazard in southern California: Ground rupture versus seismic shaking: Geol. Soc. America Bull., Pt. I, v. 92, p. 189–196.

PART V
PROCESSES AND SEDIMENT DYNAMICS

FACTORS THAT INFLUENCE SEDIMENT TRANSPORT AT THE SHELFBREAK

HERMAN A. KARL, PAUL R. CARLSON, AND DAVID A. CACCHIONE
U.S. Geological Survey, Menlo Park, California 94025

ABSTRACT

Because the shelfedge bridges shallow and deep ocean environments, sedimentary processes characteristic of each of these provinces interact at the shelfbreak to influence sediment transport in the benthic boundary layer. Processes at the shelfedge and mechanics of sediment transport have been inferred from data gathered in many regional shipboard sampling and surveying investigations. Grouping these processes into two major categories—geologic factors and oceanic factors—aids in conceptualizing the complex system of sediment dynamics at the shelfedge. Geologic factors include tectonism, sediment supply, sediment size, shelf width, depth of the break below sealevel, gradient of the upper slope, and bathymetric irregularities. Oceanic factors include fronts between water masses, boundary currents, meteorologically-induced currents, tides, internal waves, and surface waves. Although any of these factors may operate simultaneously on any continental margin, their relative importance varies with time and space; i.e., one, two, or several of these factors may dominate shelfedge sediment transport on a given continental margin or at any given time.

Few investigators have actually measured the water and sediment motions in the benthic boundary layer at the shelfedge. Regional sediment-transport data are of limited value as long as the various factors of the forcing mechanisms have not been properly studied and correlated with the flow field and sediment activity at the bottom. Sophisticated instruments deployed for long periods of time are necessary to acquire data adequate for an assessment of the forcing mechanisms that control sediment transport. The few existing measurements of this type support the concept that shelfedge processes differ with place and time among continental margins and on any given continental margin. It follows that caution should be exercised when one attempts to generalize about the shelfedge transport system.

INTRODUCTION

The outermost part of the continental shelf is a physiographic boundary between shallow water marine and deep ocean depositional environments. Diverse oceanographic phenomena that usually affect sediment transport only at either one or the other of these environments, but not both, interact at the shelfedge to produce complex processes and complex patterns of sediment transport. For example, boundary currents and frontal systems, such as the Gulf Stream, influence sediment movement at the upper slope and outer shelf, but not at the innermost shelf; on the other hand, surface waves exercise fundamental control over sediment transport at the shoreface, but (unless they are exceptionally large) barely affect or affect not at all bottom sediment movement on the continental slope. Yet, both these phenomena commonly influence sediment transport at the shelfedge. Viewed in this light, the shelfbreak is as important and as marked a boundary between environmental settings as is the shoreface, where continental and nearshore marine facies meld.

Numerous workers have studied beach, estuarine, and nearshore sedimentation: modern and ancient shoreface deposits have been described, sediment transport processes have been identified and measured, and the dynamics of the coastal region have been simulated and modeled numerically with computers (Inman and Bagnold, 1963; Rigby and Hamblin, 1972; Davis and Ethington, 1976; Komar, 1976). Consequently, a great deal is known about the shoreface as a transitional boundary that spans two dissimilar depositional provinces.

Investigations of the shelfbreak, especially of bottom and near-bottom sediment transport, are not nearly so numerous, diverse, specific, nor thorough as studies of the shoreface. In fact, a great deal of our knowledge of shelfbreak geology and sediment dynamics is derived incidentally from studies primarily of the continental shelf or slope which happen to include the shelfbreak (see chapters in parts V and VI of this volume). Some workers have purposefully investigated the shelfedge, but most of these studies have dealt with its morphology, structure, and lithofacies (e.g., Emery and Uchupi, 1963; Doyle et al., 1968; Pilkey et al., 1971; Swift et al., 1971; Wear et al., 1974; Watkins et al., 1979). It has been theorized and calculated that various oceanographic and atmospheric processes produce the strong currents and turbulent mixing which are said to characterize shelfedge dynamics (Fleming and Revelle, 1939; Galt, 1971).

Physical oceanographers have recognized the importance of the shelfedge as a zone of upwelling and the other water-mass interactions (e.g., von Arx, 1962; Csanady, 1973; Lee, 1975; Mooers et al., 1976). From observations of suspended particulate matter, bottom sediment distribution, and bedforms (Cartwright and Stride, 1958; Lyall et al., 1971; Stanley et al., 1972), some workers have inferred strong currents at the shelfedge. None of these investigations have actually measured the speed and direction of the currents in the benthic boundary layer at the shelfedge. Yet, to understand sediment dynamics at the shelfbreak, it is essential to measure currents directly, and to measure them for the durations of the major physical events. Since most sediment transport occurs near the seafloor, these measurements should be taken in the benthic boundary layer.

MEASUREMENT OF SHELFEDGE PROCESSES

Sedimentary deposits at the edge of the continental shelf represent the time-averaged product of all the depositional processes affecting the shelfbreak. Geologists can gain some insight into these processes by investigating those deposits. The equipment and the techniques necessary to produce a basic characterization of the seabed are well known, and they are readily available to marine geologists (Table 1). These investigations would be field sampling programs that use research vessels. Such regional sampling plans provide data on the variation among bottom and near-bottom sediments at various places and the properties of the water near the bottom at the continental margin, but they provide little information about the mechanics of sediment transport. Sediment transport data are, obviously, of limited value if the forcing mechanisms (tides, waves, wind, storms, etc.) have not been properly sampled and correlated with the flow field and sediment activity at the bottom. Thus, in addition to the regional sampling of various sediment and water properties, it is necessary to monitor the variation with time of near-bottom water motions.

Processes at the shelfbreak span time-scales that range from less than a few seconds (for turbulent bursts and eddies within the bottom boundary layer) to months (for seasonal climatic variations). The many intermediate durations of significance pertain to periodic or quasi-periodic water motions caused by surface waves, tides, internal waves, and other currents. Episodic events such as turbidity currents and tsunamis occur in an irregular time pattern and are more difficult to categorize according to a characteristic time-scale.

To measure adequately the gamut of shelfedge oceanographic processes and to monitor sediment activity requires instruments capable of acquiring data on the many variables that occur along a variety of time-scales. Over the past 15 years sophisticated instruments to study many of the important oceanographic and geologic parameters have been developed (Sternberg and Creager, 1965; Butman and Folger, 1979; Cacchione and Drake, 1979; McGrail, 1983). Figure 1 shows, as an example of these systems, one of the tripod systems (Geoprobes) employed by the U.S. Geological Survey. Geoprobes have been used to collect time-series data on the important physical and geologic param-

TABLE 1. STANDARD SHIPBOARD INSTRUMENTATION USED IN SEDIMENT DYNAMICS RESEARCH

Device	Data Obtained	Inferences Made Possible
A. Geologic		
Acoustic reflection profile	Profiles of subbottom reflectors	Structure, stratigraphy, "composition" of margin deposits
Echo-sounder	Profile of seafloor	Bathymetry; "hardness" of substrate
Side-scan sonar	Plan-view images of seafloor	Bed forms, structure, texture and "hardness" of sea bed
Underwater video and camera systems	Direct observation and images of sea bed	Bed forms, texture, and composition of substrate, organisms, water turbidity, current activity
Grab sampler and corer	Surface and subsurface sediment samples	Texture and composition, internal structures and stratigraphy of substrate
B. Oceanic (water)		
C-T-D	Vertical profiles of temperature, salinity	Vertical density field, identification of fine stratification and microstructure
Current meter	Time-series data of current speed and direction	Identification of processes (mean flow, surface waves, etc); temporal variations in current velocity
Nephelometer	Vertical profiles of light back-scattering	Turbidity of the water column; vertical distribution of suspended sediment; relative concentration and size of particles
Transmissometer	Vertical profiles of light transmission	Turbidity of the water column; vertical distribution and relative concentration of suspended sediment
Water sampler	Discrete sample of part of the water column	Absolute concentration, size, and composition of suspended sediment

FIG. 1.—Geoprobe instrument system: A, Four vertically-stacked electromagnetic current meters; B, rotor-vane current meter; C, 35 mm camera and strobe; D, transmissometer/nephelometer; E, pressure sensor; F, thermistors; G, pressure housings containing power and data recorder systems; H, recovery system.

eters of bottom-sediment dynamics on continental shelves. (For a detailed description see Cacchione and Drake, 1979).

Although instrumented-tripods have been used in several experiments on the shelf, on only very few occasions have tripods been placed at the shelfedge or close to it (Butman et al., 1979; Karl et al., 1980; and Drake et al., 1980a). Because of the variability in the time-scales of the processes affecting the continental shelf, it is necessary to deploy instruments at the shelfedge for periods of months, preferably for over a year. These experiments would represent major investments in time, equipment, man-power, and money. Until such observational programs are undertaken, we can only fall back upon conceptual models of sediment dynamics, standard marine geologic studies, and theoretical models of water motions in order to deduce the factors that influence shelfbreak sedimentation and to infer the benthic boundary layer processes that operate in this zone.

In the following sections we identify the geologic and oceanic factors that can influence the erosion, transportation, and deposition of sediment at the shelfedge. After giving examples of how the relative importance of different factors varies with geographic location, we conclude with a brief discussion of existing measurements of benthic boundary layer processes at the shelfbreak.

FACTORS INFLUENCING SHELFEDGE
SEDIMENT DYNAMICS

We have chosen to focus upon the North American continental margins in order to illustrate the effects of geologic and oceanic factors upon shelfedge processes; however, we do not limit our considerations of shelfedge processes to this continent. A more inclusive treatment of shelfbreak sedimentation and other shelfbreak processes can be found in the synthesis by Southard and Stanley (1976).

Geologic Factors

A decade ago, Inman and Nordstrom (1971) presented a new tectonic and morphologic classification of coasts based on the broad scale effects of plate tectonics. This classification emphasizes certain differences between the Atlantic and Pacific coasts of North America: the Pacific continental margin is a collision (or leading-edge) coast characterized by active tectonism; the Atlantic margin, on the other hand, being a tectonically stable region situated in the middle of the North America plate, is a non-collision (or trailing-edge) margin.

Although the Atlantic coast of the U.S. indeed is a trailing-edge margin and the Pacific coast is a leading-edge margin, each coast differs substantially in many ways from the generalization developed by Inman and Nordstrom (1971). The Atlantic shelf is generally wide (mean width of 126 km), but it narrows in several areas; for example it is only a few kilometers wide off southern Florida. On the other hand, the usually narrow Pacific shelf (mean width of 20 km from Mexican to Canadian border), widens to nearly 100 km in the northern Gulf of Alaska (Carlson et al., 1977; Blankenship, 1978). Flood control dams have limited greatly the amount of sediment supplied to the Atlantic shelf by rivers, whereas the Eel River alone discharges more sediment each year to the northern California shelf than is discharged annually to the Atlantic shelf by all the rivers debouching on it (see Field et al., 1983). Even though steep, mountainous terrain abuts the shore of the Gulf of Alaska, modern coarse sediment occurs only on the innermost part of the shelf and accumulations of mud characterize much of the outer and middle parts of the shelf (Carlson et al., 1977; Molnia and Carlson, 1980). Except at the Aleutian Islands, no active trenches occur along the coast of the western United States; there, the base of the slope is characterized by submarine fans and accumulations of pelagic sediment.

These basic characteristics of the U.S. east and west coasts suggest a minimum of seven fundamental regional geologic factors which influence sedimentation at the shelfbreak: tectonism, shelf width, sediment supply, sediment size, depth of the shelfbreak, upper slope gradient, and bathymetric irregularities. Many other factors become locally important—glaciation in northern latitudes, or reef-building in the tropics. However, for illustrative purposes we will discuss only the regional factors in subsequent sections.

Reef-building adds another class of factors—biologic—to the geologic and oceanic factors which we have cited as probably significant influences upon shelfedge sedimentation. Fluctuations in the productivity of plankton affect the amount of pelagic material contributed to the bottom sediment. Infaunal and epifaunal organisms rework the bottom sediment and not only modify its surface and near-surface textural and physical properties, but also introduce particles into the water column above. It is unquestionable that biological activity influences sediment transport, and investigators have begun to quantify the importance of such activity on the continental shelf (Rhoads, 1963; 1974; Jumars et al., 1981; Nowell et al., 1981; and chapters in parts IV and VII, this volume). The following discussion, however, will be limited to oceanic and geologic factors.

Oceanic Factors

Oceanic factors include tides, surface waves, internal waves, fronts between water masses, and meterologically induced currents that create a wide range of unsteady and quasi-steady water motions that, in turn, influence sediment entrainment and dispersal, not only in the benthic boundary layer but also throughout the entire water column. Like the geologic factors, many of these oceanic factors will vary in importance along a given continental margin.

For instance, just as the location of North America with respect to the North American plate accounts for fundamental geologic differences between the east and west coasts, so does the position of the continent with respect to the Atlantic and Pacific Ocean basins account for differences between certain large-scale oceanic processes at each coast. The east coast defines the western boundary of the Atlantic and the west coast forms the eastern boundary of the Pacific. This seemingly banal observation takes on meaning as we realize that the type, the occurrence, and the intensity of various currents off any continent vary systematically in relation to the eastern and western margins of ocean basins.

Western boundary currents tend to be narrow and intense; the Gulf Stream, for example, is locally as narrow as 50 km and is less commonly as wide as 200 km (L. J. Pietrafesa, 1981, pers. commun.) and near the surface attains velocities which may range from 100 cm/sec to nearly 300 cm/sec (Stommel, 1958; von Arx, 1962). Eastern boundary currents are relatively broad and weak; the California Current, for example, is as wide as 1000 km and (although surface speeds as high as 100 cm/sec have been observed) its surface flow is usually less than 25 cm/sec (Reid, 1966; Hickey, 1979).

Besides boundary currents, at least two oceanic factors—upwelling and surface waves—differ in intensity between the east and west coast of the United States. Upwelling, although it may occur anywhere, is most common along the western coasts of continents; it is especially pronounced off the western United States (La Fond, 1966). Surface waves, of course, affect both coasts, but they tend to be larger and stronger along the west coast (Emery and Uchupi, 1963; Komar et al., 1972). Surface wave energy along each of the coasts appears to correlate with latitude; larger waves occur usually in higher latitudes (Katz, 1978).

EFFECTS OF GEOLOGIC AND OCEANIC FACTORS

Geologic and oceanic factors interact to influence sediment transport processes at the shelfbreak. One or several of these factors, however, may dominate on a given continental margin, thereby not only changing the composition and geometry of sedimentary deposits at the shelfedge, but also affecting the way in which sediment is transported to and from the break. Because of the number of factors involved, whose effects vary not only with place, but in many instances over time, it is impossible to describe here all their many combinations and the consequent effects. Basic differences between the east and west continental margins of the United States illustrate the consequences of such interactions among various geologic and oceanic factors. One of the most striking contrasts here (other than tectonism) is that a wide shelf and strong boundary current characterize the east coast, whereas a weak boundary current and narrow shelf typify the west.

A wide shelf means that the shelfbreak is at a great distance from any terrigenous sediment source. Furthermore, large amounts of the sediment discharged by rivers on the east coast of the United States are trapped by estuaries. Thus, most of the sediment being transported at the shelfbreak probably is material that has been resuspended from the bottom, rather than derived directly from river discharge. Storms and surface waves probably account for most of this resuspension (Butman et al., 1979). Shoreward propagating internal waves may also resuspend fine-grained sediment (Cacchione and Southard, 1974; Butman et al., 1979). The mechanics of resuspension by these processes will be described below. The meandering of the Gulf Stream displaces water masses at the shelfedge and so not only disperses the resuspended matter laterally (Lyall et al., 1971) but also affects migration of bed load sediment (Rona, 1970; Hunt et al., 1977; Pietrafesa, 1983). In addition to displacing water by meandering, the Gulf Stream produces eddies which replace coastal water with offshore water, thus complicating sediment dispersal patterns even further (Lee, 1975).

Fine-grained components of sediment, whether derived directly from rivers or resuspended nearshore, may be transported across narrow west-coast shelves to the shelfbreak (Drake, 1972; Kulm et al., 1975; Karl, 1976). The percentage of sediment introduced at the shore face to arrive at the shelfedge depends upon many factors, among them the amount of sediment supplied, the mean grain-size of those particles, the strength of longshore currents relative to cross-shelf currents, and the width of the shelf. Even on shelves as narrow as 20 km, 70% of the silt and clay in suspension nearshore may be deposited between the inner shelf and the shelfedge and not be advected directly to the shelfbreak and deposited there (Drake et al., 1980a). One effect of this high percentage of deposition on near-shore parts of the shelf is that coarse, relict deposits at the shelfbreak are commonly not covered by modern muds. For example, in the Gulf of Alaska relict glacial deposits at the shelfbreak remain exposed even though the Copper River discharges large amounts of fine-grained sediment (Carlson et al. 1977; Molnia and Carlson, 1980). On the other hand, wherever large percentages of fine-grained sediment do reach the shelfbreak, these sediments (unless they by-pass the shelfedge) would accumulate and could cover any coarse, relict material present (Karl, 1976; Karl et al., 1980).

Steep slopes may prevent the build-up of sediment at the shelfbreak even in areas that receive a steady supply of fine-grained material. When sediment accumulations attain a critical angle or, perhaps, a critical mass on a given slope, or they are sufficiently agitated, then slumping may occur and the material will be transported by mass wasting (Hampton et al., 1978; Field et al., 1983). Many factors affect submarine slumping; one of the most important among these, certainly on the west coast, is seismicity.

The California Current, the boundary current which flows southward along the west coast of the United States, is much weaker than the Gulf Stream which bounds the east coast. Even so, the California Current system does affect the dispersal of suspended particulate matter along the continental margin (Bernstein et al., 1977; Thornton, 1981). A counter-current flows beneath each of these great boundary currents, but whereas the counter-current on the east coast is usually too deep to affect the shelfedge, the counter-current on the west coast is shallow enough in some areas to affect the shelfbreak (Hickey, 1979). Upwelling is very pronounced off the west coast, but it is insignificant off the east coast. Several investigators have measured the distribution of suspended sediment along the U.S. west coast during upwelling season, but no systematic study has been made of the effects of upwelling upon the transport of suspended matter (Pak et al., 1970; Drake, 1972; Harlett and Kulm, 1973; Karl, 1976; McCave, 1979). One way that upwelling may influence sediment transport at the shelfbreak is by changing the density structure of the water column. The permanent pycnocline rises as upwelling begins and descends as upwelling ceases (Pillsbury, 1972). Temperature anomalies during upwelling (Huyer and Smith, 1974; Mooers et al., 1976) may influence the concentration and distribution of suspended sediment by causing density changes and affecting nutrient distributions (Pak et al., 1970; Peterson, 1977; McCave, 1979). Very fine particulate material may concentrate in narrow layers that have a steep pycnocline. This layering in the concentrations of suspended matter could reflect the layering in density profiles.

Surface waves have been observed to resuspend sediment on both continental margins. As the depth of the shelfbreak increases, waves of increasingly longer period are required to generate currents that are strong enough to affect sediment at the shelfedge. Observations of oscillation ripple marks at depths as great as 200 m indicate that the large waves characteristic of the northwestern U.S. coast do transport sediment at the shelfedge (Komar et al., 1972). Long-term observations of bottom currents and of near-bottom sediment transport on the mid-Atlantic continental shelf reveal that wave-induced bottom currents resuspend sediment during periods of low mean flow (Butman et al., 1979).

Instruments were not located at the shelfbreak to document the wave-induced entrainment of shelfedge sediments. However, it is reasonable to assume that sufficiently large waves will produce currents that affect the shelfedge, and it has been estimated that bathymetric irregularities at the shelfedge can refract waves in such a way as to intensify the currents in local areas by about four-fold (Ewing, 1973).

Internal waves propagate along density interfaces and within pycnoclines, and they are common in the main and seasonal thermocline of the world's oceans (LaFond, 1962; Wunsch, 1971). Such waves may influence the transportation of sediment in the marine environment (Emery, 1956; LaFond, 1961, 1962). For example, if the seasonal or permanent thermocline intersects the shelfedge, internal waves propagated in the thermocline would impinge upon the shelfbreak and could affect sedimentation there. When the ratio of the bottom slope to the wave characteristic slope is less than unity, as internal waves move across a sloping bottom, they may break and generate internal surf analogous to that of surface waves (Wunsch, 1968; Cacchione and Wunsch, 1974). Emery and Gunnersen's (1973) interpretation of various thermal sections over the southern California coast supports the theory of shoaling and of breaking internal waves. Laboratory experiments have demonstrated that breaking internal waves could entrain and transport sediment (Southard and Cacchione, 1972), and Cacchione and Southard (1974) have argued that in theory waves with frequencies and amplitudes commonly found in the ocean should be capable of moving fine-grained sediment covering the continental shelf.

Internal waves propagating shoreward across the continental slope and shelf off southern California have been inferred from thermistor records, current meter records, and surface slicks (Lee, 1961; LaFond, 1962; Summers and Emery, 1963; Cairns and LaFond, 1966; Cairns, 1967; Winant, 1974; Cacchione et al., 1976; Karl et al., 1980). Shoreward propagating internal waves have also been identified off the east coast on the basis of surface slicks observed on satellite images (Apel et al., 1975) and from analysis of current-meter records (Butman et al., 1979). The few current meter measurements and the even fewer concurrent observations of bottom sediment movement and internal waves suggest that internal waves can have significant effects on the transport of fine-grained bottom sediment and suspended materials (Cacchione et al., 1976; Butman et al., 1979; Cacchione et al., 1980; McGrail, 1983). However, many more field observations are necessary in order to evaluate the importance at the shelfedge of this mode of transport.

Currents generated by the diurnal and semi-diurnal tides influence the transportation of sediment on continental shelves to greatly varying degrees. For example, in two areas where giant bedforms have been attributed to tidal flow—on the North Sea shelf adjacent to the British Isles (Kenyon, 1970) and in Cook Inlet, Alaska (Bouma et al., 1977)—tidal currents are probably the dominant process governing sediment transport. In other areas, tides affect only those particles already suspended in the water column by other processes. The effect of tidal currents not only varies among large geographic areas, but it also varies between places on any given shelf. It was thought that tidal-current velocity should be greatest at the shelfedge (Fleming and Revelle, 1939; Kuenen, 1939), but recent analyses have shown this not to be true (Clarke and Battisti, 1981). Cross-shelf differences in tidal energy have been observed off Southern California but not enough systematic measurements are available to assess confidently the importance of tidal currents at the shelfedge relative to other areas of the shelf (Karl et al., 1980).

Atmospheric variables exert strong influence on currents at the continental margin. Meteorological factors that force oceanic motions include the transfer of momentum through wind stress, wind stirring of the water column, pressure adjustments, heat transfer, and mass water transfer through evaporation and precipitation (Mooers, 1976). Of these factors, wind- and pressure-generated currents are the most important affecting the dynamics of sediment transport on the continental margins.

Storm-generated currents resuspend and transport sediment on the continental shelf (Smith and Hopkins, 1972; Butman et al., 1979; Drake et al., 1980b; Cacchione and Drake, in press); they can, in fact, move significantly greater amounts of sediment than do the fair-weather events which prevail a larger percentage of time (Smith and Hopkins, 1972; Drake et al., 1980b). No unequivocal field observations are available to determine whether storm-generated currents are stronger at the shelfbreak than elsewhere on the shelf. However, from observations made at three locations on the outer part of the mid-Atlantic continental shelf, Butman et al. (1979) report that during winter storms the speeds of currents near the seabed were slightly greater at the location nearest the shelfedge. Furthermore, Galt (1971) has shown that in theory pressure-induced surges generated by storms moving off the continental shelf produce currents that flow parallel to isobaths and these currents are concentrated in the region of the shelfedge. It follows that storm-driven currents—the intensification of oceanic processes caused by storms—exert significant influence upon shelfedge sediment transport.

Other very significant local effects on shelfedge sediment transport are produced by bathymetric irregularities at the shelfbreak—most notably, submarine canyons. For example, submarine canyons affect internal waves of tidal and higher frequencies and surface waves. Alternating upcanyon and downcanyon currents have been attributed to internal tides and waves propagating shoreward along the canyon axis (Shepard et al., 1979). Because the canyon seems to channel these internal waves, the waves apparently travel faster in the canyon than on the continental shelf (Lee, 1961; Shepard et al., 1974, 1979; Winant, 1974). Features on computer-enhanced Landsat images have been interpreted as being groups of high-frequency internal waves generated at the shelfbreak around the head of Hudson Canyon by the semi-diurnal and diurnal tides (Apel et al., 1975).

Submarine canyons also refract shoaling surface waves. Waves refracted by canyons that head close to shore (e.g., the classic "sand" canyons of the west coast of the U.S.) produce zones of wave divergence and convergence at the shoreface (Inman et al., 1976); those heading far from shore (for example, canyons on the east coast of the U.S.) are thought to focus and dissipate wave energy in a similar manner at the shelfbreak and on the adjacent shelf (Ewing, 1973).

In addition to interacting with internal and surface gravity waves, submarine canyons may preferentially trap edge waves, those periodic longshore motions which are manifested either as standing waves or as progressive waves that travel parallel to the coastline (Inman et al., 1976). Also, submarine canyons may affect the displacement of water masses and, possibly, favor the development of spin-off eddies from large-scale boundary currents that flow transverse to the canyon axes.

Ample field evidence exists to suggest that canyon-modified currents either increase or reduce sediment transport around the canyon heads. Differential wave set-up at the beach and trapped edge waves cause longshore currents to converge on canyon heads, forming seaward-flowing rip currents (Inman et al., 1976). Suspended sediment may be siphoned from the adjacent shelf (Beer and Gorsline, 1971; Lyall et al., 1971; Karl, 1980). Sediment waves such as those observed around the heads of several submarine canyons were not found in areas away from the canyons (Knebel and Folger, 1976; Karl and Carlson, 1981a, b). The large sediment waves associated with three enormous canyons in the Bering Sea suggest the presence of unusually intense currents at the canyon heads (Fig. 2). The complex interaction of canyon-modified currents produces a corridor for the preferential transport of sediment which varies in intensity both spatially and temporally (Karl, 1980).

BENTHIC BOUNDARY LAYER MEASUREMENTS AT THE SHELFEDGE

Up to this point, we have discussed briefly the geologic and oceanic factors thought, by inference

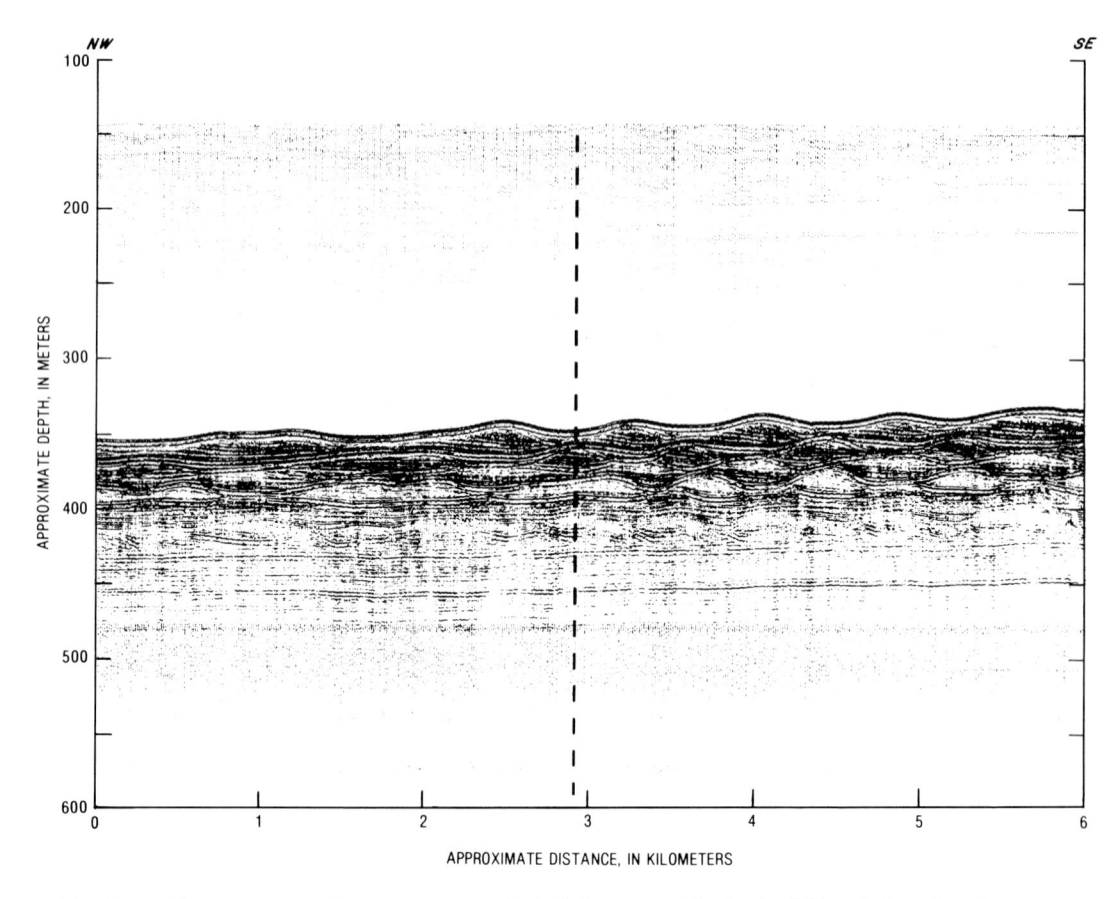

FIG. 2.—Sediment waves on the upper slope and shelfedge around the head of Navarinsky submarine canyon, northern Bering Sea.

from indirect evidence, to influence sediment transport at the shelfbreak. We have emphasized that these factors may vary in relative importance on different continental margins. It should be obvious that inference really sums up our present mode of understanding of the sediment transport processes at the shelfedge. In a paper published in 1976 (p. 365), Southard and Stanley opined that "until our observational knowledge of bottom currents and accompanying sediment transport [at the shelfbreak] is more extensive, . . . it does not seem to be an exaggeration to say that we are at the stage of knowing something, but not enough, about a lot of possible motions and having some, but too little, field data to guide our thinking." Since 1976 we have made slight progress beyond this stage, but we still need many more observations of the type in the following case history.

During the early spring of 1978, U.S. Geological Survey scientists investigated sediment transport on the San Pedro continental shelf off Southern California. The experimental design included regional geologic sampling, geophysical surveys, and oceanographic measurements to establish the spatial variability of the seafloor and the near-bottom water properties of the San Pedro shelf. In addition to shipboard regional surveys, bottom-moored instruments were deployed to determine temporal variability of several parameters such as pressure, temperature, light scattering and transmission, current speed and direction, and bed morphology. The instrument systems included one Geoprobe (G1B) deployed nearshore in 22 m of water and a second Geoprobe (G2) moored 67 m below sealevel at the shelfbreak; both operated for 40 days. Six other instrument systems were moored at strategic points on the shelf. Four of these systems had, in addition to a vector measuring current meter (VMCM), a transmissometer mounted in line below the VMCM (see Karl et al., 1980, for a detailed discussion of results).

The San Pedro shelf, approximately 30 km long ranges in width from about 20 km to about 3.5 km at the east and west boundaries. Flood-control dams

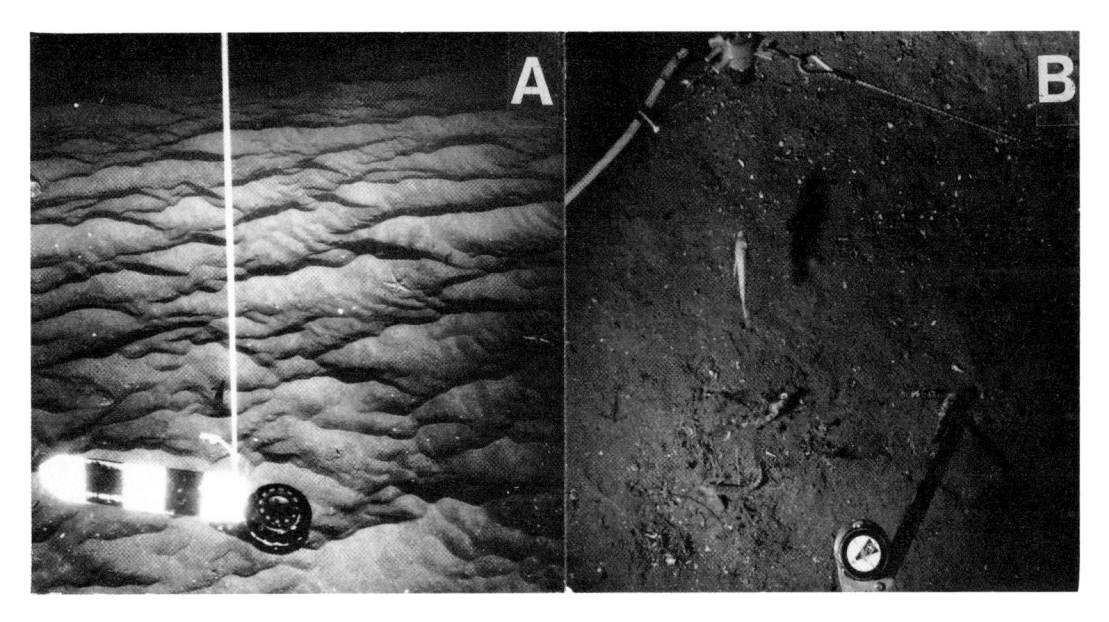

FIG. 3.—*A*, Substrate at the shallow (22 m) shelf site (G1B) in San Pedro Bay, California; *B*, Substrate at the shelfedge (67 m) site (G2) in San Pedro Bay, California. Compass is about 8 cm in diameter.

limit the amount of sediment discharged by the three rivers debouching into San Pedro Bay, and much sediment moving downcoast in the longshore drift is either intercepted by submarine canyons or blocked by the peninsula that lies just north of San Pedro Bay (Karl, 1976).

G1B was located on one of the patches of relict medium sand not covered by modern detrital sediment. Distinctive ripples characterized the substrate there (Fig. 3A). At G2, the substrate consisted of modern silty, fine sand typical of most of the sediment blanketing the shelf. The bed was biogenically reworked with no evidence of strong bedload motion (Fig. 3B).

Physical processes that affect sediment transport on the San Pedro shelf include surface waves, internal waves, tides, and a mean flow. Although these processes have affected the entire shelf, their intensity differed between the shallow (G1B) site and the shelfbreak (G2) site (Table 2). In general, all processes were more intense at the shallow site. Of the processes measured during the 40-day fair-weather period, surface waves had the greatest effect on bedload motion. Although no storms occurred during the experiment, mean flow near the bottom did respond almost instantaneously to a change in wind direction which appears to have initiated a local upwelling event. The dynamics of

TABLE 2.—Comparison of Sediment Transport Processes at the Benthic Boundary Layer on the Middle Shelf (G1B) and at the Shelfbreak (G2); from the San Pedro Shelf Experiment

Process	Shelf Site G1B	Shelfbreak Site G2
Surface waves (swell)	Typical periods 12–14 sec. and max. speeds 10–30 cm/sec.; significant bedload transport	Typical periods 12–14 sec. and typical max. speeds <10 cm/sec; insignificant bedload movement
Internal waves	Typical period 24 mins., not significant for initial motion of bedload; significant for suspended load transport	Typical period 6 hrs., not significant for bedload nor coarse suspended load transport
Tides	12-hr and 24-hr tides equally energetic; max. current speeds (~10–15 cm/sec) not significant for bedload motion	24 hr tide less energetic than 12 hr tide; max. speeds (~10–15cm/sec) not significant for bedload motion
Mean flow	Max. hourly avg. speed ~20 cm/sec, 40-day avg. speed ~1 cm/sec; southwesterly (off-shelf) transport; significant for suspended sediment dispersal	Max. hourly speed ~10 cm/sec, 40-day avg. speed ~3 cm/sec; southwesterly (off-shelf) transport; significant for suspended sediment dispersal

upwelling caused particles suspended either near the bottom or near the surface to be transported southwesterly off the shelf. Consequently, even though the mean flow was not capable of initiating bedload motion, it did control the net transport direction of material entrained in the water column; significantly, all through the 40-day experimental period that direction was persistently off shelf (Karl et al., in press).

Butman et al. (1979), from analysis of long-term records collected by tripods similar to Geoprobes, reported similar processes on the mid-Atlantic outer continental shelf. Storm-induced currents were the most significant factor in sediment transport during the winter; surface wave-generated currents resuspended sediment during periods of low mean flow. During stratified conditions in summer, packets of high-frequency internal waves generated bottom currents as strong as 20 cm/sec.

CONCLUSIONS

The shelfedge marks the boundary between two very dissimilar sedimentary environments—the continental shelf and the continental slope. Investigations of the shelfbreak and its environs have shown that, with rare exception, sediment deposited at the shelfbreak remains there only temporarily before it is entrained and transported to deeper ocean environments. A review of many outer-shelf and upper-slope studies suggests that the factors which control the transport of sediment at the shelfbreak are of two major kinds: geologic and oceanic. Geologic factors include (but are not limited to) shelf width, shelfbreak depth below sealevel, physiography, gradient of the upper slope at the shelfedge, sediment supply, and grain-size. Oceanic factors include (but are not limited to) fronts between water masses, tides, surface waves, internal gravity and body waves, and meteorologically-induced currents. Geologic and oceanic factors function interactively to influence sediment transport at the shelfedge. The relative importance of these factors varies not only spatially on different continental margins, but also temporally at a given location. An example of spatial variation would be the strong boundary currents, which may be important factors for entraining and dispersing fine-grained sediment along the western margins of ocean basins though not along the eastern margins. Factors which vary temporally include storm-induced currents and internal waves: the former are more important during the winter months; the latter during the summer when the water column is strongly stratified.

The processes which govern sediment transport at the shelfedge are poorly understood because few investigators have actually performed quantitative studies within the benthic boundary layer. Processes at the shelfbreak span time-scales ranging from a few seconds to months. Consequently, it is necessary to deploy very sophisticated instruments over very long periods of time to acquire adequate data about the factors that control sediment transport, so that we can correlate them with sediment activity and the bottom flow field. The few measurements now available indicate that the processes influencing shelfedge sediment dynamics vary significantly in both spatial and temporal specifics. Thus, caution should be exercised in any attempt to generalize about the processes affecting the transport of sediment at the shelfedge. The next phase of shelfedge studies must involve the long-term deployment of instrument systems in the benthic boundary layer, together with simultaneous measurements of seafloor changes.

ACKNOWLEDGMENTS

The San Pedro Shelf Experiment was funded under an interagency agreement between the U.S. Geological Survey and Bureau of Land Management.

We thank D. E. Drake, D. S. Gorsline, D. W. McGrail, and L. J. Pietrafesa for reviewing the manuscript.

REFERENCES

APEL, J. R., BYRNE, H. M., PRONE, J. R., AND CHARNELL, R. L., 1975, Observations of oceanic internal and surface waves from Earth Resources Technology satellite: Jour. Geophys. Res., v. 80, p. 865–881.
BEER, R. M., AND GORSLINE, D. S., 1971, Distribution, composition, and transport of suspended sediment in Redondo submarine canyon and vicinity (California): Mar. Geology, v. 10, p. 153–176.
BERNSTEIN, R. L., BREAKER, L., AND WHRITNER, R., 1977, California current eddy formation: ship, air, and satellite results: Science, v. 195, p. 353–359.
BLANKENSHIP, J. B., 1978, A comparison of shelf width and depth to shelfbreak: eastern, western and Gulf coasts, United States [unpub. rept.]: Los Angeles, Univ. Southern California, 20 p.
BOUMA, A. H., HAMPTON, M. A., WENNEKENS, M. P., AND DYGAS, J. A., 1977, Large dunes and other bedforms in lower Cook Inlet, Alaska: Proc. 9th Annual Offshore Technology Conf., p. 79–85.
BUTMAN, B., AND FOLGER, D. W., 1979, An instrument system for long-term sediment transport studies on the continental shelf: Jour. Geophys. Res., v. 84, p. 1215–1220.
———, NOBLE, M., AND FOLGER, D. W., 1979, Long-term observations of bottom current and bottom sediment movement on the mid-Atlantic continental shelf: Jour. Geophys. Res., v. 84, p. 1187–1205.
CACCHIONE, D. A., AND DRAKE, D. E., 1979, A new instrument system to investigate sediment dynamics on continental shelves: Mar. Geology, v. 30, p. 299–312.

———, AND ———, 1983, Measurements of storm-generated bottom stresses on the continental shelf: Jour. Geophys. Res., in press.

———, ———, DINGLER, J. D., WINANT, C. D., AND OLSON, J. R., 1976, Observations of sediment movement by surface and internal waves off San Diego, California [abs.]: Trans. Am. Geophys. Union, v. 58, p. 307.

———, KARL, H. A., AND DRAKE, D. E., 1980, Sediment transport by internal waves on the continental shelf off southern California [abs.]: Internat. Geol. Congr. Symposium, Sedimentary dynamics on continental shelves, Paris, 1980.

———, AND SOUTHARD, J. B., 1974, Incipient sediment movement by shoaling internal gravity waves: Jour. Geophys. Res., v. 70, p. 2237–2242.

———, AND WUNSCH, C., 1974, Experimental study of internal waves over a slope: Jour. Fluid Mech., v. 66, p. 223–239.

CAIRNS, J. L., 1967, Asymmetry of internal tidal waves in shallow coastal waters: Jour. Geophys. Res., v. 72, p. 3563–3565.

———, AND LAFOND, E. C., 1966, Periodic motions of the seasonal thermocline along the southern California coast: Jour. Geophys. Res., v. 71, p. 3903–3915.

CARLSON, P. R., MOLNIA, B. F., KITTELSON, S. C., AND HAMPSON, J. C., JR., 1977, Distribution of bottom sediments on the continental shelf, northern Gulf of Alaska: U.S. Geol. Survey Misc. Field Studies Map MF-876.

CARTWRIGHT, D. E., AND STRIDE, A. H., 1958, Large sand waves near the edge of the continental shelf: Nature, v. 181, p. 41.

CLARKE, A., AND BATTISTI, D., 1981, The effect of continental shelves on tides: Deep-Sea Res., v. 28, p. 665–682.

CSANADY, G. T., 1973, Wind-induced baroclinic motions at the edge of the continental shelf: Jour. Phys. Ocean., v. 3, p. 274–279.

DAVIS, R. A., JR., AND ETHINGTON, R. L., eds., 1976, Beach and Nearshore Sedimentation: Soc. Econ. Paleontologists Mineralogists Spec. Pub. 24, 187 p.

DOYLE, L. J., CLEARY, W. J., AND PILKEY, O. H., 1968, Mica: Its use in determining shelf-depositional regimes: Mar. Geology, v. 6, p. 381–389.

DRAKE, D. E., 1972, Distribution and transport of suspended matter, Santa Barbara Channel, California [unpub. Ph.D. diss.]: Los Angeles, Univ. Southern California, 358 p.

———, CACCHIONE, D. A., AND KARL, H. A., 1980a, Offshore transport of suspended matter in the bottom water on San Pedro shelf, California (abs.): Trans. Am. Geoph. Union, v. 61, p. 1000.

———, ———, MUENCH, R. D., AND NELSON, C. H., 1980b, Sediment transport in Norton Sound, Alaska: Mar. Geology, v. 36, p. 97–126.

EMERY, K. O., 1956, Deep standing internal waves in California basins: Limnol. Oceanogr., v. 1, p. 35–41.

———, AND GUNNERSON, C. G., 1973, Internal swash and surf: Proc. U.S. Natl. Acad. Sci., v. 70, p. 2379–2380.

———, AND UCHUPI, E., 1963, Western north Atlantic Ocean: Topography, rocks, structure, water, life and sediments: Am. Assoc. Petroleum Geologists Memoir 17, 532 p.

EWING, J. A., 1973, Wave-induced bottom currents on the outer shelf: Mar. Geology, v. 15, p. M31–M35.

FIELD, M. E., CARLSON, P. R., AND HALL, R. K., 1983, Seismic facies of shelfedge deposits, U.S. Pacific continental margin, in Stanley, D. J., and Moore, G. T., eds., The Shelfbreak: Critical Interface on Continental Margins: Soc. Econ. Paleontologists Mineralogists Spec. Pub. 33, p. 299–313.

FLEMING, R. H., AND REVELLE, R., 1939, Physical processes in the oceans, in Trask, P. D., ed., Recent Marine Sediments: Am. Assoc. Petroleum Geologists, p. 48–141.

GALT, J. A., 1971, A numerical investigation of pressure induced storm surges over the continental shelf: Jour. Phys. Oceanogr., v. 1, p. 82–91.

HAMPTON, M. A., BOUMA, A. H., CARLSON, P. R., MOLNIA, B. F., CLUKEY, E. C., AND SANGREY, D. A., 1978, Quantitative study of slope instability in the Gulf of Aslaska: Proc. 10th Offshore Technology Conf., Houston, Texas, v. 4, p. 2307–2318.

HARLETT, J. C., AND KULM, L. D., 1973, Suspended sediment transport on the northern Oregon continental shelf: Geol. Soc. America Bull., v. 84, p. 3815–3826.

HICKEY, B. M., 1979, The California Current system—hypothesis and facts: Progress in Oceanography, v. 8, p. 191–279.

HUNT, R. E., SWIFT, D. J. P., AND PALMER, H., 1977, Constructional shelf topography, Diamond Shoals, North Carolina: Geol. Soc. America Bull., v. 88, p. 299–311.

HUYER, A., AND SMITH, R. L., 1974, A subsurface ribbon of cool water over the continental shelf off Oregon: Jour. Phys. Ocean., v. 4, p. 381–391.

INMAN, D. L., AND BAGNOLD, R. A., 1963, Littoral processes in Hill, M. N., ed., The Sea: New York, John Wiley and Sons, p. 529–553.

———, AND NORDSTROM, C. E., 1971, On the tectonic and morphologic classification of coasts: Jour. Geology, v. 79, p. 1–21.

———, ———, AND FLICK, R. E., 1976, Currents in submarine canyons: an air-sea-land interaction: Ann. Rev. Fluid Mech., v. 8, p. 275–310.

JUMARS, P. A., NOWELL, A. R. M., AND SELF, R. F. L., 1981, A simple model of flow-sediment-organism interaction, in Nittrouer, A., ed., Sedimentary Dynamics of Continental Shelves, Developments in Sedimentology, v. 32: Amsterdam, Elsevier, p. 155–172.

KARL, H. A., 1976, Processes influencing transportation and deposition of sediment on the continental shelf, Southern

California Los Angeles [unpub. Ph.D. diss.]: Univ. of Southern California, 331 p.
———, 1980: Influence of San Gabriel submarine canyon on narrow-shelf sediment dynamics, southern California: Mar. Geology, v. 34, p. 61–78.
———, CACCHIONE, D. A., AND DRAKE, D. E., 1980, Erosion and transport of sediments in the benthic boundary layer on the San Pedro shelf, southern California: U.S. Geol. Survey Open-file Rept. 80-386, 54 p.
———, AND CARLSON, P. R., 1981a, Large sediment waves at the shelfedge, northern Bering Sea (abstract): Geol. Soc. America, Abst. with Programs, v. 13, p. 483.
———, AND ———, 1981b, Sediment waves in the head of Navarinsky, Pervenets, and Zhemchug submarine canyons, northwestern Bering Sea: U.S. Geol. Survey Circ. (Accomplishments in Alaska), 5 p.
———, DRAKE, D. E., AND CACCHIONE, D. A., 1983, Response of the suspended sediment transport system to continental shelf dynamics: Geo-Marine Letters, in press.
KATZ, S. D., 1978, A determination of the percentage of conterminous United States shore formed by deltas, beaches, beach barriers, estuaries, and sea cliffs [unpub. rept.]: Los Angeles, Univ. of Southern California, 126 p.
KENYON, N. H., 1970, Sand ribbons of European tidal seas: Mar. Geology, v. 9, p. 25–39.
KNEBEL, H. J., AND FOLGER, D. W., 1976, Large sand waves on the Atlantic outer continental shelf around Wilmington canyon, off-eastern United States: Mar. Geology, v. 22, p. M7–M15.
KOMAR, P. D., 1976, Beach Processes and Sedimentation; Englewood Cliffs, New Jersey: Prentice-Hall, Inc., 429 p.
———, NEUDECK, R. H., AND KULM, L. D., 1972, Observations and significance of deep-water oscillatory ripple marks on the Oregon continental shelf, in Swift, D. J. P., Duane, D. S., and Pilkey, O. H., eds., Shelf Sediment Transport: Process and Pattern, Stroudsburg, Pa.: Dowden, Hutchinson, and Ross, Inc., p. 601–619.
KUENEN, P. H., 1939, The cause of coarse deposits at the outer edge of the shelf: Geol. Mijnbouw, v. 1, p. 36–39.
KULM, L. D., ROUSCH, R. C., HARTLETT, J. C., NEUDECK, R. H., CHAMBERS, D. M., AND RUNGE, E. J., 1975, Oregon continental shelf sedimentation: interrelationships of facies distribution and sedimentary processes: Jour. Geology, v. 83, p. 145–175.
LAFOND, E. C., 1961, Internal wave motion and its geological significance, in Krishnan, M. S., ed., Mahadevan Volume; A Collection of Geological Papers in Commemoration of Professor C. Mahadevan: Hyderabad, India, Osmania University Press, p. 61–77.
———, 1962, Internal Waves, in Hill, M. N., ed., The Sea, Volume 1: New York, Wiley-Interscience, p. 731–755.
———, 1966, Upwelling, in Fairbridge, R. W., ed., The Encyclopedia of Oceanography: New York, Reinhold Publishing Corp., p. 957–959.
LEE, O. S., 1961, Observations on internal waves in shallow water: Limnol. Oceanogr., v. 6, p. 312–321.
LEE, T. N., 1975, Florida Current spin-off eddies: Deep-Sea Res., v. 22, p. 753–765.
LYALL, A. K., STANLEY, D. J., GILES, R. H., AND FISCHER, A., JR., 1971, Suspended sediment and transport at the shelf-break and on the slope, Wilmington canyon area, eastern U.S.A.: Jour. Mar. Technol. Soc., v. 5, p. 15–27.
MCCAVE, I. N., 1979, Suspended material over the central Oregon continental shelf in May 1974: I, concentrations of organic and inorganic components: Jour. Sed. Petrology, v. 49, p. 1181–1194.
MCGRAIL, D. W., AND CARNES, M., 1983, Shelfedge dynamics and the nepheloid layer, in Stanley, D. J., and Moore, G. T., eds., The Shelfbreak: Critical Interface on Continental Margins: Soc. Econ. Paleontologists Mineralogists Spec. Pub. 33, p. 251–264.
MOLNIA, B. F., AND CARLSON, P. R., 1980, Quaternary sedimentary facies on the continental shelf of the southeast Gulf of Alaska, in Field, M. E., and others, eds., Quaternary Depositional Environments of the Pacific Coast: Soc. Econ. Paleontologists Mineralogists Pacific Sec., Pacific Coast Paleogeogr. Symposium 4, p. 157–168.
MOOERS, C. N. K., COLLINS, C. A., AND SMITH, R. L., 1976, The dynamic structure of the frontal zone in the coastal upwelling region off Oregon: Jour. Phys. Ocean., v. 6, p. 3–21.
NOWELL, A. R. M., JUMARS, P. A., AND ECKMAN, J. E., 1981, Effects of biological activity on the entrainment of marine sediments, in Nittrouer, C. A., ed., Sedimentary Dynamics of Continental Shelves, Developments in Sedimentology, v. 32: Amsterdam, Elsevier, p. 133–153.
PAK, H., BEARDSLEY, G. B., AND SMITH, R. L., 1970, An optical and hydrographic study of a temperature inversion off Oregon during upwelling: Jour. Geophys. Res., v. 75, p. 629–638.
PETERSON, R. E., 1977, A study of suspended particulate matter: Arctic Ocean and northern Oregon continental shelf [unpub. Ph.D. diss.]: Corvallis, Oregon State Univ., 122 p.
PIETRAFESA, L. J., 1983, Shelfbreak circulation, fronts and physical oceanography: East and west coast perspectives, in Stanley, D. J., and Moore, G. T., eds., The Shelfbreak: Critical Interface on Continental Margins: Soc. Econ. Paleontologists Mineralogists Spec. Pub. 33, p. 233–250.
PILKEY, O. H., MACINTYRE, I. A., AND UCHUPI, E., 1971, Shallow structures: shelf edge of continental margin between Cape Hatteras and Cape Fear, North Carolina: Am. Assoc. Petroleum Geologists Bull., v. 55, p. 110–115.
PILLSBURY, R. D., 1972, A description of hydrography, winds, and currents during upwelling season near Newport, Oregon [unpub. Ph.D. dissert.]: Corvallis, Oregon State Univ., 163 p.
REID, J. L., JR, 1966, California current, in Fairbridge, R. W., ed., The Encyclopedia of Oceanography: New York, Reinhold Publishing Corp., p. 154–157.
RHOADS, D. C., 1963, Rates of sediment reworking by *Yoldia limatula* in Buzzards Bay, Massachusetts and Long Island Sound: Jour. Sed. Petrology, v. 33, p. 723–727.
———, 1974, Organism-sediment relations on a muddy sea floor: Oceanogr. Mar. Biol. Ann. Rev., v. 12, p. 263–300.

RIGBY, J. K., AND HAMBLIN, W. K., eds., 1972, Recognition of Ancient Sedimentary Environments: Soc. Econ. Paleontologists Mineralogists Spec. Pub. 16, 340 p.
RONA, P. A., 1970, Submarine canyon origin on upper continental slope off Cape Hatteras: Jour. Geology, v. 78, p. 141–152.
SHEPARD, F. P., MARSHALL, N. F., AND MCLOUGHLIN, P. A., 1974, "Internal waves" advancing along submarine canyons: Science, v. 183, p. 195–198.
———, ———, ———, AND SULLIVAN, G. G., 1979, Currents in submarine canyons and other sea valleys: Am. Assoc. Petroleum Geologists, Studies in Geology No. 8, 173 p.
SMITH, J. D., AND HOPKINS, T. S., 1972, Sediment transport on the continental shelf off of Washington and Oregon in light of recent current measurements, *in* Swift, D. J. P., Duane, D. B., and Pilkey, O. H., eds., Shelf Sediment Transport: Process and Pattern: Stroudsburg, Pa., Dowden, Hutchinson and Ross, Inc., p. 83–97.
SOUTHARD, J. B., AND CACCHIONE, D. A., 1972, Experiments on bottom sediment movement by breaking internal waves, *in* Swift, D. J. P., Duane, D. B., and Pilkey, O. H., eds., Shelf Sediment Transport: Process and Pattern: Stroudsburg, Pa., Dowden, Hutchinson and Ross, Inc., p. 83–97.
———, AND STANLEY, D. J., 1976, Shelf break processes and sedimentation, *in* Stanley, D. J., and Swift, D. J. P., eds., Marine Sediment Transport and Environmental Management: New York, John Wiley and Sons, p. 351–377.
STANLEY, D. J., FENNER, P., AND KELLING, G., 1972, Currents and sediment transport at the Wilmington Canyon shelfbreak, as observed by underwater television, *in* Swift, D. J. P., Duane, D. B., and Pilkey, O. H., eds., Shelf Sediment Transport: Process and Pattern: Stroudsburg, Pa., Dowden, Hutchinson and Ross, p. 621–644.
STERNBERG, R. W., AND CREAGER, J. S., 1965, An instrument system to measure boundary layer conditions at the sea floor: Mar. Geology, v. 3, p. 475–482.
STOMMEL, H., 1958, The Gulf Stream: Berkeley, Univ. of California Press, 202 p.
SUMMERS, H. J., AND EMERY, K. O., 1963, Internal wave of tidal period off southern California: Jour. Geophys. Res., v. 68, p. 827–839.
SWIFT, D. J. P., STANLEY, D. J., AND CURRAY, J. R., 1971, Relict sediment on continental shelves: a reconsideration: Jour. Geology, v. 79, p. 322–346.
THORNTON, S. E., 1981, Suspended sediment transport in surface waters of the California Current off southern California: 1977–1978 floods: Geo-Marine Letters, v. 1, p. 23–28.
VON ARX, W. S., 1962, An Introduction to Physical Oceanography: Reading, Mass., Radisen-Wesley Publishing Co., Inc., 422 p.
WATKINS, J. S., MONTADERT, L., AND DICKERSON, P. W., eds., 1979, Geological and Geophysical Investigations of Continental Margins: Am. Assoc. Petroleum Geologists Memoir 29, 467 p.
WEAR, C. M., STANLEY, D. J., AND BOULA, 1974, Shelfbreak physiography between Wilmington and Norfolk canyons: Jour. Mar. Technol. Soc., v. 8, p. 37–48.
WINANT, C. D., 1974, Internal surges in coastal waters: Jour. Geophys. Res., v. 79, p. 4523–4526.
WUNSCH, C., 1971, Internal waves: Trans. Am. Geoph. Union, v. 56, p. 233–235.
———, 1968, On the progression of internal waves upslope: Deep-Sea Res., v. 25, p. 251–258.

SHELFBREAK CIRCULATION, FRONTS AND PHYSICAL OCEANOGRAPHY: EAST AND WEST COAST PERSPECTIVES

LEONARD J. PIETRAFESA
Department of Marine, Earth and Atmospheric Sciences, North Carolina State University, Raleigh,
North Carolina 27650

ABSTRACT

A survey of fundamental physical oceanographic processes that may affect sediment distribution along shelfbreak regions is presented, emphasizing the Atlantic and Pacific coasts of the USA. The processes encompass the entire spectrum of known motions and are thus generic to all shelfbreak interfacial zones. These shelfbreak strips couple the bounded coastal oceans to the open seas, but there is no systematic pattern to this coupling. In the South Atlantic Bight, the Gulf Stream acts like a vibrating, permeable wall which can variously entrain shelf waters, flood the shelf with North Atlantic Central Water and violently mix shelf waters by towing whirling vortices through the outer shelf. Middle Atlantic Bight, New York Bight and Gulf of Maine shelfbreak processes contain many of the dynamic elements of their southeastern counterpart, but the relative importances of various random surface and offshore driving forces change.

Pacific coast shelfbreak processes tend to be less energetic than those on the Atlantic coast since the Pacific coast is missing a Western Boundary Current and because the shelf is narrow and deep. Subinertial frequency shelfbreak motions on the west coast are typically manifested across the entire shelf, while those on the east coast tend to be confined to a loosely defined band, which brackets the break. Principal Pacific coast circulation elements include forms of continental shelf waves and thermohaline driven and mechanically wind forced currents, as well as the California Current System. While high frequency edge waves and inertial currents are indigenous in similar fashion to all coasts, east and west coast tides are shown to be quite disparate, given tradeoffs between dominance of diurnal and semidiurnal constituents as a function of topographic constraint and strength of density stratification.

All of the shelfbreak zones are graced by thermohaline fronts. The fronts are progradational on the west and southeastern coasts and retrogradational on the northeastern shelf. These fronts are an integral ingredient of all aspects of physical processes at the shelfbreak strip. The interplay of bottom topography with the physics of the outer continental margin is significant. Bottom features such as shoals, bumps, ridges and canyons are shown to be regions of sediment erosion, deposition and draping. Moreover, these features are shown to be causally related to upwelling and downwelling phenomena and to the deflection and scattering of waves and currents. Both subtidal and supertidal frequency events are shown to be capable of initiating sediment motion and of suspending sediments, but lower frequency events are shown to be responsible for the bulk of sediment migration on the outer shelf and upper slope environs.

INTRODUCTION

Only within the last decade have both direct and passive measurements of water motions and buoyancy activity on the outer continental shelves of the United States been made. A data bank that will provide the scientific community a genetic view of the characterization of the physical oceanography of the continental shelf-slope interface from minutes to months to years in the temporal domain, and from meters to hundreds of kilometers in the spatial arena, is now emerging. Several important generalizations can be made from this testimony. These generalizations include shelfbreak physics important for circulatory dynamics which derive from: (1) tidal and inertial phenomena; (2) local eddy-like phenomena and long waves that impinge from the deep sea; (3) momenta and buoyancy fluxes and spatial variations thereof at the air-sea boundary on both synoptic and seasonal scales; (4) deep water influences that become important as a physical phenomenon persists; and (5) turbulent rectification of transient motions that may be a dominant ingredient of seasonal and continental length circulation patterns.

Physical oceanographic processes affecting the shelfbreak regions of both the west and east coasts of the continental United States are discussed in this chapter. Areas considered include the Middle Atlantic Bight, the New York Bight to New England, the South Atlantic Bight and the California to Washington margin (cf. Fig. 1). Summaries of relevant motions with periods longer than 10 minutes are depicted in Figure 2, and characterizations of frontal features indigenous to the regimes are shown in Figure 3. While the reader is referred to Karl, Carlson and Cacchione (1983) for a review of shelfbreak sedimentation, several implications for sediment dispersal along the shelfbreak sector are also made in the final section of this text.

The eastern U.S. continental margin between Cape Canaveral, Florida, and Cape Hatteras, North Carolina, is called the South Atlantic Bight (SAB). From Hatteras to the Canadian Maritimes one encounters the Middle Atlantic Bight (MAB), the

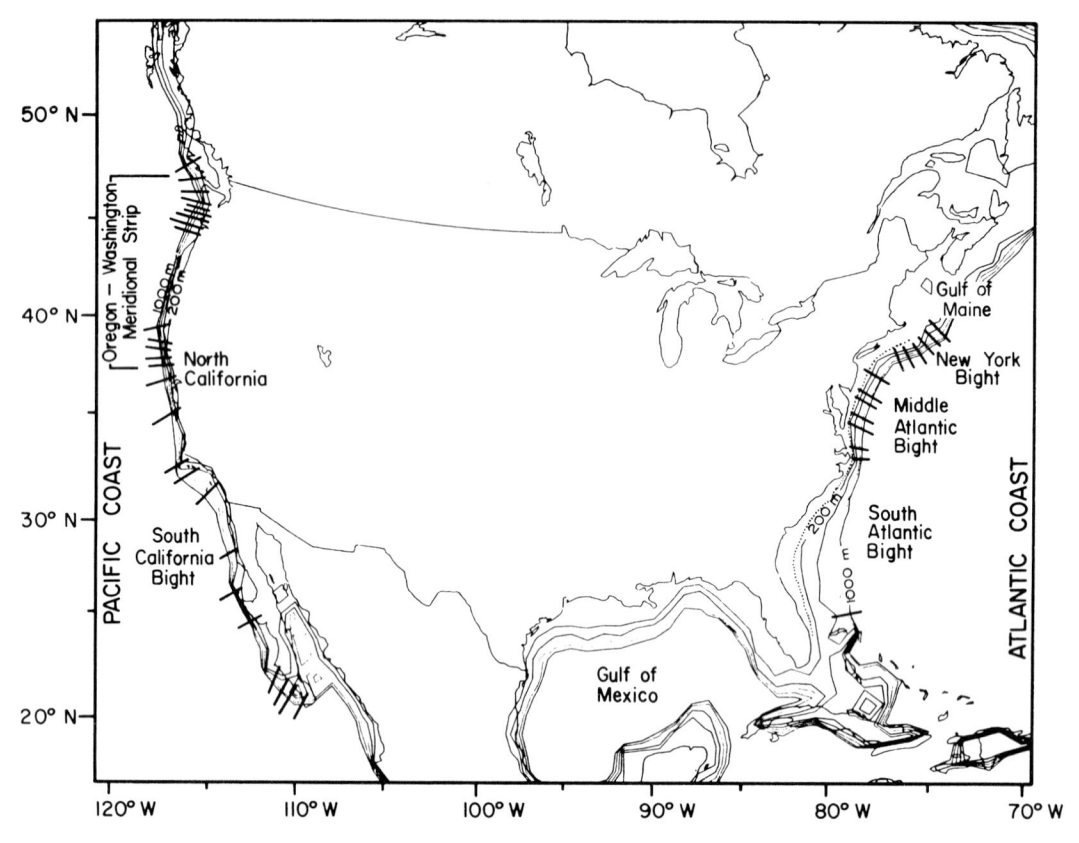

FIG. 1.—Mainland United States continental margins. Submarine canyons are depicted as straight lines across the continental slope.

New York Bight, the New England shelf, Georges Bank, the Gulf of Maine, and finally the Nova Scotian shelf (Fig. 1). The bathymetry of the shelfbreak in the SAB is rather uniform from Cape Canaveral, Florida, to Cape Fear, North Carolina. Extensive shoals extending from Capes Fear, Lookout and Hatteras, North Carolina, form a series of cuspate bays. The Gulf Stream flows to the northeast along the shelfbreak strip. Shelfbreak depths vary from 45 to 75 m in the SAB. The SAB shelf is up to 120 km wide at its midpoint and narrows to 25 km width at both Capes Canaveral and Hatteras. To the north of Cape Hatteras, the MAB, the shelf widens to 150 km and is generally twice as deep as that of the SAB. In addition, the MAB slope is cut by approximately twenty submarine canyons, features absent in the SAB. To the north of the MAB lies the New York Bight which is bisected by Long Island. The shelf to the north and off Cape Cod is several hundred kilometers wide. Farther north of Cape Cod lie Georges Bank, the Gulf of Maine and the Nova Scotian shelf.

The U.S. Pacific Coast (PC) is characterized by 100 to 200 m shelfbreak depths, has a cuspate shape from Point Conception to San Diego, California, and a relatively straight, meridional, uniform shelfbreak along Oregon and Washington. Off southernmost California the shelf is only 10 km wide, while farther north shelf widths reach 50 km. The West Coast continental margin is thus generally much narrower and also deeper than its East Coast counterpart. There are submarine canyons throughout the entire outer continental margin and an extension of the San Andreas fault, the Mendecino Escarpment, resides off San Francisco.

Shelfbreak fronts exist along both the Atlantic and Pacific coasts. These fronts are rather narrow hydrographic bands which affect a boundary between different but horizontally juxtaposed water masses. They are also regions of fluid convergence, strong vertical motions and a discontinuous thermohaline medium across which large oceanic scale motions lose their energy to smaller shelf scale motions via various cascading processes.

SHELFBREAK FRONTS

Although oceanic fronts are numerous in type, only three are discussed herein: (1) shelfbreak

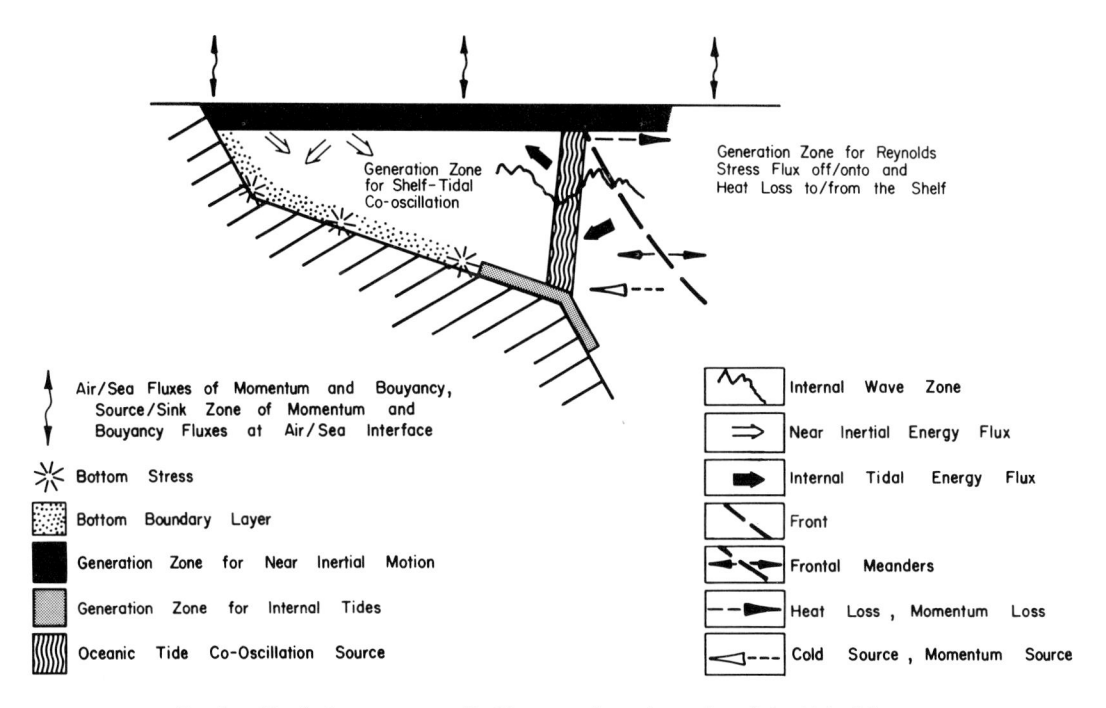

FIG. 2.—Physical processes applicable to continental margins of the United States.

fronts (characteristic of the MAB); (2) Western Boundary Current cyclonic fronts (characteristic of the SAB); and (3) upwelling fronts (characteristic of the MAB, SAB and, most prominently, the PC). Albeit, the Carolina Capes shoals also have some minor frontal features associated with them and there are estuarine plume frontal features in the regions of the Hudson (New York), Connecticut and Columbia (Washington, Oregon) rivers and both Chesapeake and Delaware bays. The three primary frontal types are characterized by cross-frontal widths on the order of 10 to 30 km and along-frontal length (parabathic) scales which range from tens to hundreds of kilometers. Persistent frontal motions have time scales in excess of several inertial periods. In the diabathic (transhelf) direction, there is appreciable geostrophic tendency, and the hydrostatic law rules the vertical. In the parabathic direction the momentum balance includes a mix of geostrophy, inertia and vertical turbulent eddy stress. Vertical and diabathic advection of mass and local time changes in density are balanced by the vertical diffusion of mass within the fronts. A prominent difference in the three principle types of fronts is their cross-shelf inclination or slope.

In the SAB and on the PC, the frontal isopleth slopes are aligned in the same sense as the diabathic topography (prograde fronts), whereas in the MAB the frontal isopleths tilt in opposition to the cross-shelf topography (retrograde front). Additionally, the MAB shelfbreak front (Fig. 3C) is a persistent year-round feature that straddles the shelfbreak, the PC front is a summer-time feature that crosses the shelfbreak (Fig. 3D) and the SAB front, the onshore wall of the Gulf Stream (Fig. 3E), intersects the surface above the shelfbreak. The following is a description of the MAB derived from discussions presented by Beardsley and Flagg (1975), Flagg (1977) and Mooers et al. (1978).

The MAB retrograde front separates shoreward surface waters of temperature (T) of 6° and salinity (S) of 35.5‰ from seaward surface waters of 8° and 36‰. Since T and S both increase seaward, the net density differential is reduced and the frontal discontinuity is not as sharp as it might be. Sigma-t is 26.3 in mid-front off Delaware and 26.5 off New England. The front rotates about 10° clockwise as viewed in a cross-sectional transect looking northeastward from Virginia and has a mean slope of 0.002; it affects a narrow sloping hydrographic band that separates vertically homogeneous coastal waters from somewhat denser, weakly stratified slope waters. The Gulf Stream cyclonic frontal zone typically lies from 20 to 150 km east of the shelfbreak from Cape Hatteras to New York. During spring and summer the near-bottom part of the front rises vertically and lies across the shelf at about mid-depth in the water column. Still, the surface temperature front can be observed seaward of the shelfbreak.

Along the shelfbreak zone of the PC, there are no wintertime fronts with the parabathic size or dia-

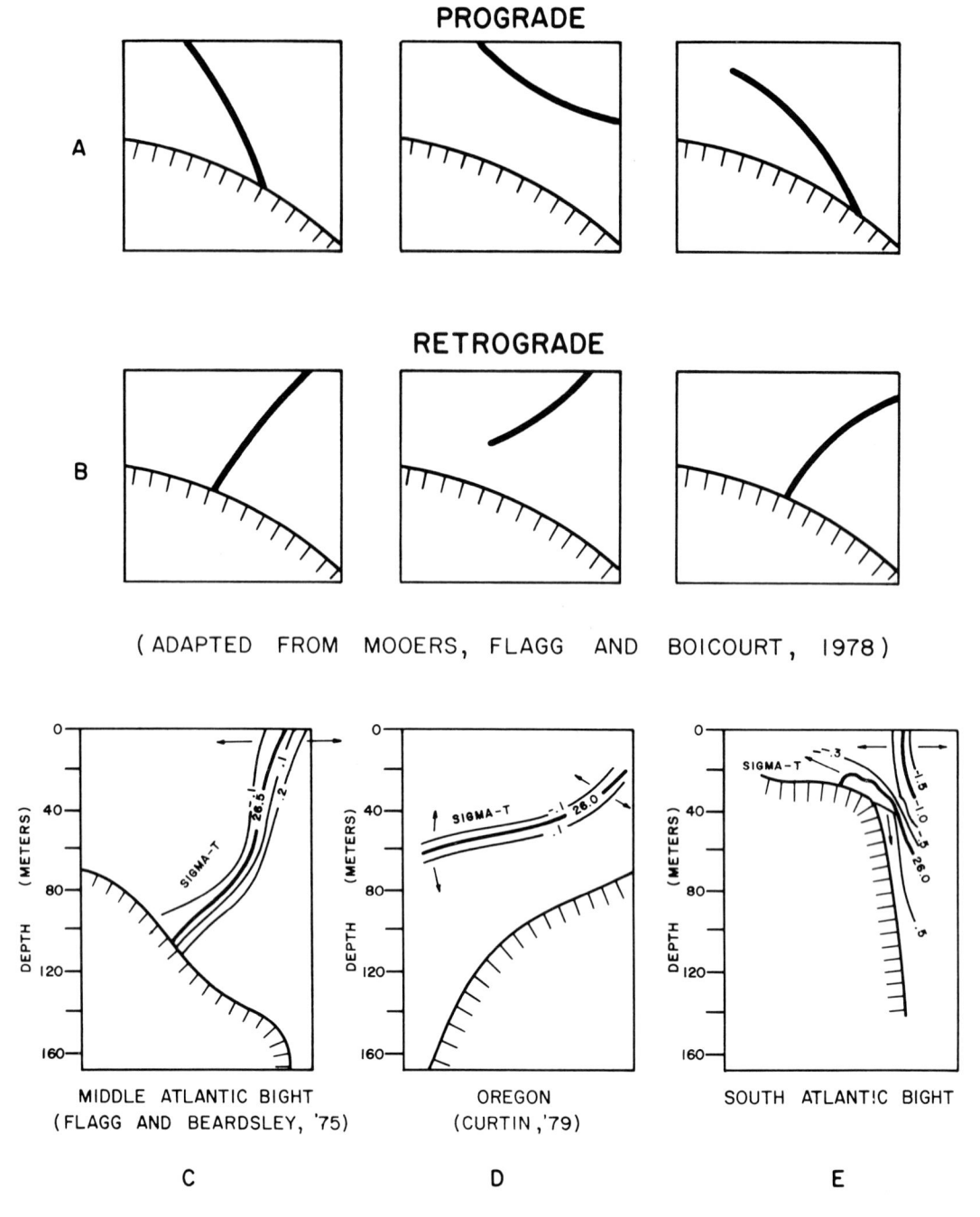

FIG. 3.—Shelfbreak fronts: (A) and (B) are nomenclature types, while (C), (D) and (E) are derived from actual observations. Arrows indicate directions of frontal movements.

bathic strength of those observed on the East Coast. According to Hickey (1979), the substantial summertime coastal upwelling front, which is observed on the Oregon and Washington shelves (Curtin, 1979), may not be as persistent off the central California coast. This contention may simply be based on the lack of data to confirm the existence of such a front. Albeit, there may be simply too few observations in the California regions to have detected the feature. It is clear from Curtin (1979) that the

Oregon-Washington summertime coastal upwelling front is progradational in nature and is caused by mechanically forced, wind-driven coastal Ekman divergence. Northerly winds drive surface waters offshore to the west, which requires a compensatory flow onshore to conserve volume and mass, and this water appears to originate at slope depths of 100 to 200 m. The upwelling process is hydrographically characterized by an upward-tilting pycnocline extending landward. This pycnocline generally cuts the surface at about 10 km from the coast, or about mid-shelf, to form a strongly horizontally stratified or hyperbaroclinic zone. Curtin (1979) has provided the most apt description of the upwelling cycle. His discussion of Oregon upwelling indicates that while the front is not, of itself, a shelfbreak process, it obviously transcends the break as it rises from a depth of about 150 m at 200 km offshore to about 80 m at the shelfedge (30 km offshore) to near (or intersecting) the sea surface at about 10 km from land. A seasonal summer temperature minimum is observed over the continental shelf, associated with coastal upwelling and a strong southward geostrophic current. A seasonal salinity maximum is influenced by runoff, via the Columbia River plume, as well as by coastal upwelling and horizontal advection. In the upper 25 m, neither temperature nor salinity is negligible in determining density; at depth, the salinity field generally tends to parallel the density field.

In the SAB, the shelfbreak front is progradational. This front is the shorewardmost or western boundary of the Gulf Stream, the Gulf Stream front or GSF. During the winter, shelf waters of 5 to 14°C and <35.9‰ are separated from axial Gulf Stream waters by this front. The summer presents a different picture as wind-driven upwelling occurs along the shelfbreak. Surface waters are driven offshore and the thermal contrast between shelf and Gulf Stream surface waters disappears with 27 to 29°C temperatures throughout. Surface salinities on the outer shelf are 35 to 35.9‰, while seaward of the front values exceed 36‰. Alternatively, nearbottom shelfbreak waters show marked contrast as shelf waters (22°C, 35–35.9‰) are separated from slope waters (12–19°, >36‰). The motion of the GSF dominates physical processes on the outer shelf and upper slope throughout most of the year. In the diabathic direction this front can bifurcate with one tongue jutting upward and intersecting the surface and with the other tongue extending across the shelf (Fig. 3E). As a result of recent moored current meter and satellite infrared (VHRR) observations made on both the U.S. East and West coasts, good kinematical and dynamical descriptors of the circulation are becoming available. Unfortunately, not all of the spatial scales of variability have been assessed in the regimes studied but geographical highlights can be presented. It is noted that the spectra from all of these shelfbreak regions are basically red, i.e., the energy of the currents increases with increasing period of motion (Fig. 4). As an example, it is known that motions with frequencies above 2 cycles per day (cpd), such as tidal harmonics, seiches, internal waves, edge waves, Kelvin waves and turbulence, occupy only about 3 and 17% of the respective near surface and near bottom total fractions of energy during the summer in the SAB (Pietrafesa, 1978). Wintertime counterparts are 6 and 8%, respectively. The highest frequency, shortest period motions which can be resolved on all of the shelves using the collective data is on the order of 144 cycles per day (cpd), i.e., 10 minutes. Surface wave induced motions are not included. At the lower periods (i.e., phenomena that occur within several tens of minutes to several hours), motions tend to be intermittent and are likely associated with internal wave bands at the

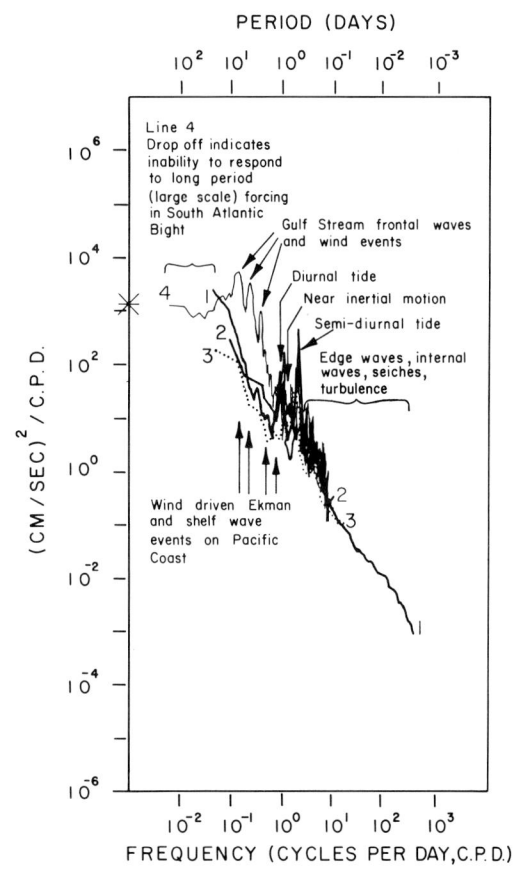

FIG. 4.—Kinetic energy spectra from currents at shelfbreaks of: Line 1, Del Mar, California (after C. D. Winant, 1980); Line 2, New England (courtesy of R. C. Beardsley); Line 3, Oregon shelf (courtesy of D. Halpern); Line 4, Onslow Bay, North Carolina.

lower period range or with edge waves, which have periods on the order of hours and which are impulsively generated by atmospheric events on all coasts.

The bulk of the internal wave activity occurs during summer on both the Pacific (Winant, 1980) and Atlantic (Butman et al., 1979) coasts when vertical stratification is at a maximum. Still, it is likely that the frontal zones of the SAB and MAB are active in the internal wave band year round since the fronts are omnipresent in these regions. These waves are probably nonlinear since the fronts can have close proximity to the bottom. One could assume that the internal waves are, in general, intermittent in occurrence, have lengths the order of a kilometer and appear as shoreward-moving progressive waves. Offshore current fluctuations, bottom topography and atmospheric events are all mechanisms which can generate internal waves in the zone of the maximum static stability, the front itself. We now move down the frequency spectrum toward motions with time-scales on the order of a half day to a day. This band covers the semi-diurnal and diurnal tides and near inertial motions.

TIDAL AND INERTIAL MOTIONS

Tidal motions are the only deterministic phenomena on the U.S. continental margins. In the MAB, the semi-diurnal tides account for the largest fraction of the total horizontal current variance within the front. In the SAB the principal constituents of the tide account for 6 to 10% and 12 to 20% of the respective total current variability in near-surface and near-bottom waters throughout the year. In both MAB and SAB regions the semi-diurnal tide is barotropic while the diurnal tide is baroclinic; both are clockwise polarized, with ellipse axes aligned preferentially onshore. On the East Coast, the observed dominance of the barotropic tide over the baroclinic tide is principally a function (assuming sufficient stratification) of the shelfbreak depth to shelf width ratio. On the East Coast this ratio is quite low since shelves are shallow and broad and thus the co-oscillation of the shelf at the semi-diurnal period dominates over that at a daily cycle (Clarke and Battisti, 1981). As these authors showed, semi-diurnal tides are amplified on wide, shallow shelves while diurnal tides are not amplified on wide, shallow shelves. Alternatively, on the deep, narrow California shelf, the barotropic and baroclinic components have comparable energy.

On the Oregon and Washington shelves, the daily tide is barotropic, whereas the M2 tide is baroclinic, satisfies the linear momentum balance for internal waves, and is a standing wave throughout the entire west coast (Rattray, 1957). Clearly, all of the baroclinic components depend heavily on the strength and persistence of shelfbreak frontal structure for their existence. Moreover, the shelfbreak region is the generation zone for internal tides. An interesting phenomenon that occurs during summer along the SAB shelfbreak is that increased vertical stratification suppresses the vertical extent of the bottom frictional layer. Consequently, near-bottom tidal ellipses from the summertime diurnal tide are more circularly polarized when the effect of friction is vertically constrained than during the winter when bottom currents become more rectilinear and friction can affect the tidal flow to greater height off the bottom. In general, bottom tidal currents lead upper level counterparts in phase by nearly an hour. The M2 tide in the SAB sector has diabathic current amplitudes of 10 to 30 cm/sec as a function of the width of the shelf. A composite of coastal barotropic tidal motions for both U.S. continental margins is presented in Figure 5. We now consider currents with periods close to a half pendulum day.

Inertial energy, which has been found to be present in all of the shelfbreak regions, has appreciable amplitude within local fronts, decreases downward, and appears to be impulsively induced by moving atmospheric storms. These phenomena are present on both the U.S. East and West coasts and can cause bottom stirring as well as contribute to vertical shear induced instabilities in shelfbreak fronts. This is particularly characteristic of the SAB and MAB fronts which are present during the winter when atmospheric storms are at a peak. Pietrafesa (1982) has found that inertial energy is not present in either current spectra or hodographs below 65 m water depth in the SAB.

SHELFBREAK VARIABILITY WITH PERIODS GREATER THAN TWO DAYS

Subinertial frequency current and density variability on all of the shelfbreak regions occurs primarily on time scales shorter than 2 weeks throughout the year due to atmospheric and offshore forcing. Response to seasonal forcing is weak in comparison and difficult to detect in most shelfbreak current meter data. Although there are significant seasonal cycles in atmospheric forcing, it is generally characterized by (a) the passage of cold fronts with 2 to 10 day time scales during the winter and spring and (b) by light and variable winds which frequently persist for periods of days to weeks during the summer. The intense fall and wintertime wind events tend to be downwelling favorable while the moderate spring and summertime winds are upwelling favorable. Thermohaline forcing varies with the annual run-off cycle, which is maximum in the spring and early summer and minimum in fall and winter, and with the seasonal change in atmospheric heating and cooling. This effect is prominent on the Oregon-Washington shelves and in the MAB but is not well pronounced in the Carolina Capes region due to the lack of river

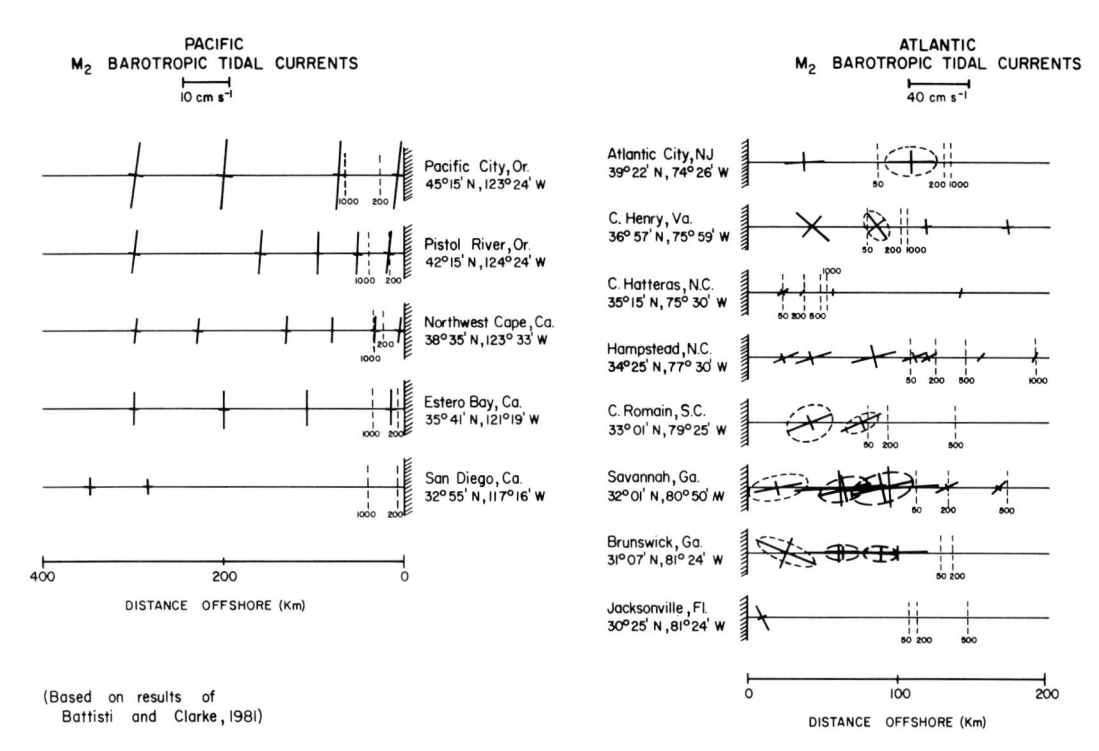

FIG. 5.—Semi-diurnal tidal ellipse axes on United States continental margins. Solid lines (—) are predictions of ellipse axes from Battisti and Clarke (1981), and broken line ellipses are based on actual observations.

sources. There is an occasional input of fresher MAB shelf water around Cape Hatteras into the Carolina Capes with the occurrence of "northeasters" in fall and winter.

The effect of these processes is to produce a seasonal cycle of shelf hydrographic properties on all outer shelves. During the winter, shelf waters become vertically homogeneous and horizontally stratified since atmospherically cooled surface waters are mixed downward by the wind and offshore waters are blocked from penetrating onto the shelf by the cold, dense water. The opposite effects occur during the summer and the water column is typified by strong vertical stratification. The transition from winter to summer hydrographic regimes occurs rapidly during the spring due to cross-shelf density-driven flows that may be triggered by atmospheric heating, increased precipitation and runoff and by strong upwelling events at the shelfbreak. Such shelfbreak events may be associated with the passage of Gulf Stream frontal filaments in the SAB or forced by strong wind-driven upwelling events off California, Oregon, Washington, the MAB and occasionally the SAB. The spring transition enhances the effects of upwelling at the shelfbreak by allowing offshore waters to penetrate farther onto the shelf via a bottom frictional layer.

On the outer shelf of the SAB, from Cape Canaveral to Savannah and from Cape Fear to Cape Hatteras, the northeasterly flowing Gulf Stream front (GSF) directly dominates the current and hydrographic signature year round through an accumulation of the effects of 2 day to 2 week period frontal events. The region between 32° and 33° latitude is a recirculating zone, referred to here as the "Charleston Gyre." Since the Gulf Stream butts against the outer continental margin of the SAB, there is no slope water south of Cape Hatteras and the shelfbreak has a direct coupling with deep North Atlantic Central Water. This is not the case north of Cape Hatteras or on the west coast where distinctive slope water is present along the outer continental margins.

The primary difference between shelfbreak processes in the SAB compared to those in the other U.S. regions is the presence of the Gulf Stream. Recent current meter and satellite observations, however, reveal a more variable current system and shelf response than previously envisioned. The principal features of the Gulf Stream western boundary are clearly revealed in satellite IR imagery (Fig. 6). Disturbances form in the cyclonic front and propagate to the north along the shelfbreak and slope region as stable and unstable waves. Stable waves are manifested as onshore/offshore displacements of the GSF. Occasionally,

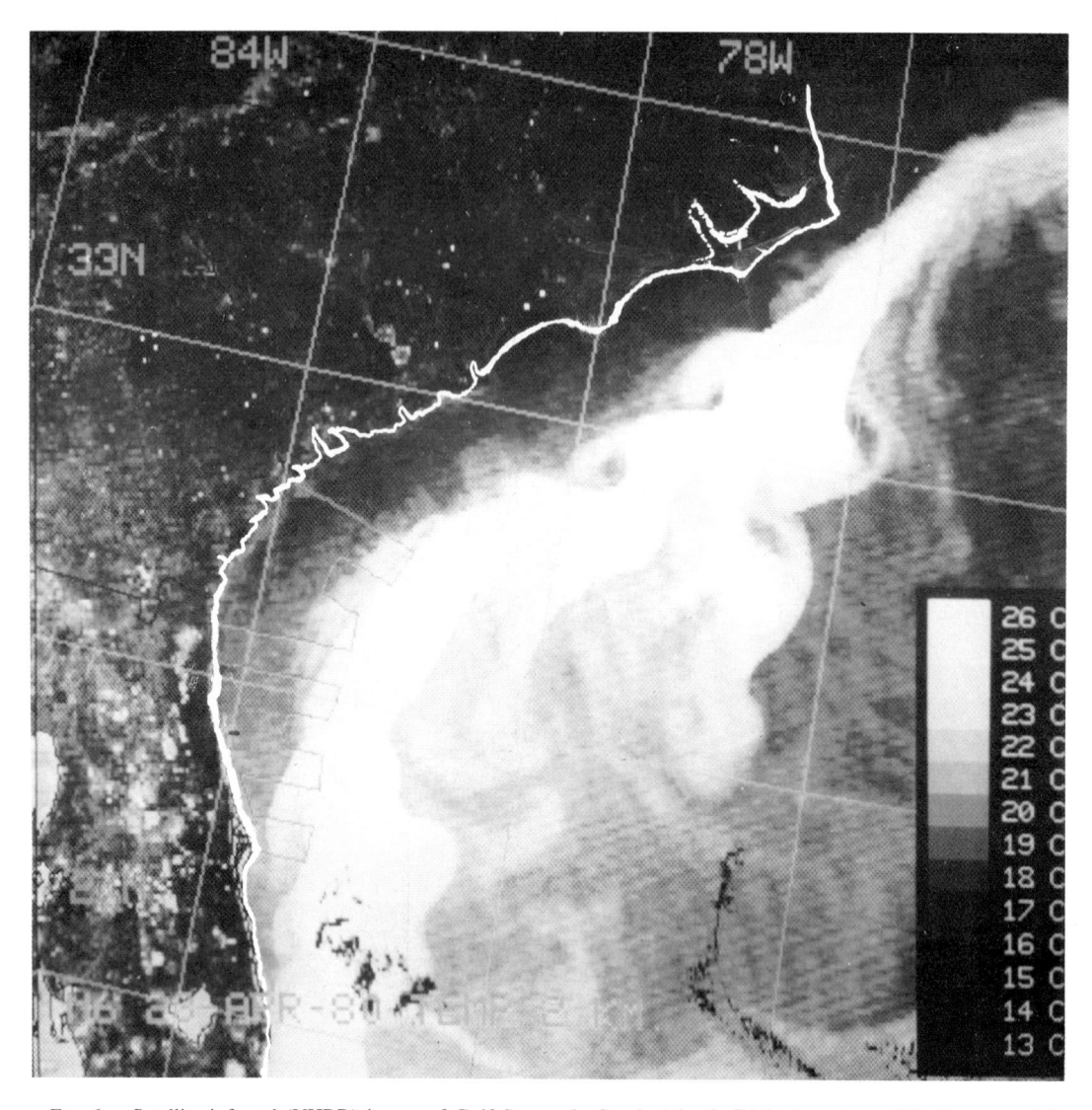

FIG. 6.—Satellite infrared (VHRR) image of Gulf Stream in South Atlantic Bight (courtesy of O. Brown and R. Evans). Note frontal events on western side of Gulf Stream. Well formed clockwise rotating filament-waves with associated counter-clockwise rotating troughs are clearly visible directly offshore of the Carolina Capes.

these waves grow in space as a function of time and break backward onto the shelf as the onshore portion of the waves fold into enlongated clockwise rotating warm core filaments, thereby transforming stable waves into folded back waves. South of 32°N the Gulf Stream follows the 100 m isobath, and east-west displacements of the front seldom exceed 25 km (Legeckis, 1979). North-south filament dimensions in this region range from 100 to 200 km. At 32°N a topographic feature known as the "Charleston Bump" (Pietrafesa et al., 1978) causes the Stream to deflect eastward downstream of the bump, forming a quasi-permanent meander (Rooney et al., 1978). Associated with this eastward meander is a region of active upwelling in the lee of the bump. Downstream of the bump, meanders grow in size with east-west displacements reaching 100 km and downstream dimensions of spin-off filaments can extend to over 300 km.

Subtidal frequency current information from the outer shelf-upper slope of the SAB (except for the Charleston Gyre region) indicate that the current variability, ranging from ±40 to ±80 cm/sec about a mean northward flow of 65 cm/sec in the winter and 55 cm/sec in the summer, is dominated by Gulf Stream spawned meander and filament events along the shelfbreak. Similar current fluctuations occur throughout the year, even though wind forcing undergoes a significant reduction in intensity during the summer. These energetic fluctuations are thus northeastward propagating waves with periods of 2.5 to 13 days, wavelengths from 100 to 300 kms and phase speeds of 20 to 75 cm/sec. The effect of Gulf Stream events at mid-shelf is found to be more pronounced in the summer than during the winter (Pietrafesa, 1981). This finding supports the previous notion that warm summertime shelf waters allow Gulf Stream events to penetrate farther onto the shelf than do cold and consequently more dense wintertime shelf waters.

There appears to be a little variation in intensity of Gulf Stream effects from year to year in the Carolina Capes region. Near-surface energy levels of currents on the outer shelf are about 1850 $(cm/sec)^2$, while on the upper slope they are about 1450 $(cm/sec)^2$. Thus in the shallower mid-outer shelf waters, wind effects as well as GSF effects are important. Near-bottom energy levels are of the order of 425 $(cm/sec)^2$ on the outer shelf and at the shelfbreak.

In the Charleston Gyre region a southerly mean flow of approximately 10 cm/sec exists year round throughout the water column at the shelfbreak. This flow is caused by a 3 to 4 cm drop in sealevel over a 50 km diabathic distance across the shelfbreak related to the offshore deflection of the GSF at the Charleston Bump. A similar countercurrent feature also appears along the shelfbreak of the MAB.

Along the shelfbreak strip of the MAB, and perhaps off of New England as well, a persistent 5 to 15 cm/sec southwestward flow persists year round through the entire water column. Beardsley and Winant (1979) and Bishop (1980), amongst others, define this current to be the westernmost boundary of the subpolar gyre, directly and ageostrophically driven by a surface slope which sets down to the southwest. Hopkins (1981), however, takes deference with these authors. He demonstrates that there is a drop of sealevel from the outer shelf toward the open ocean, which he contends is geostrophically driving a current to the southwest. A 5 cm drop in sealevel over a 50 km diabathic transect across the shelfbreak would easily account for the diabathic pressure gradient necessary to drive the current. In fact, Cheney and Marsh (1981) provide ample altimetric evidence in support of Hopkins's notion by showing that the surface of the ocean typically changes by the order of the few tens of centimeters over the 100 kilometer distance required to create the cross-shelf pressure gradient necessary to geostrophically drive the permanent southerly shelfbreak current observed in the MAB.

The MAB surface front is known to meander about its mean alongshore position on the order of 10 to 50 km onshore-offshore (Beardsley and Flagg, 1975), affecting propagating waves with periods from two days to two weeks and alongfront lengths of 150 km. Now, due to the fact that retrograde fronts are baroclinically stable, these meanders must be forced externally. The bottom portion of the front moves diabatically less than ± 8 km so that the surface front is undergoing the principal lateral excursions. Halliwell and Mooers (1979) have discussed the satellite descriptors of MAB and New England shelfbreak frontal statistics. They found that large warm core, anticyclonic Gulf Stream eddies propagate westward and impinge upon the MAB front. These large eddies are on the order of 100 km in diameter, extend 200 m downard and have rotational speeds of two knots so that the combination of their size and speed quite likely can cause an offshore motion of the front to occur via an intense entrainment, suction process.

Beardsley and Butman (1974) and Flagg (1977) have shown that the outer shelf of the MAB responds primarily to wind forcing during the winter and as such the MAB front also may respond to these energetic wind events. In fact, Flagg (1977) demonstrated that currents directly above the inclined front responded primarily to the alongshore component of the wind stress in a manner consistent with time-dependent coastal Ekman concepts. Flagg also showed that the slope of the front would decrease with southerly and increase with northerly wind stress pulses, respectively, and following the passage of atmospheric storms, the front would oscillate about some centroid located near the bottom for about 10 days, with a 3 to 4 day period of oscillation. Winds generally lead responding ocean currents by about 14 hours. Longshore currents over the outer shelf seem to be very coherent over distances on the order of 100 km and are highly correlated with both the longshore wind and the diabathic pressure gradient. These concepts are applicable to the New York Bight and New England outer shelves as well.

Buoyancy flux and thermohaline driven circulations can be influential on all of the U.S. continental margin regions. River runoff, detached boundary current filaments, warm core eddies, spring and fall air-sea exchanges and large horizontal density gradients all influence to various degrees outer shelf variability, especially under conditions of weak mechanical wind forcing. The work of Pietrafesa and Janowitz (1979) suggests that buoyancy forcing can be as powerful as moderate winds in affecting ocean processes at the shelfbreak on all coasts.

The outer continental shelf and upper slope of the U.S. Pacific seaboard appear to have some strikingly common features, despite a meridional extent of some 29° of latitude. The reader is referred to review articles with extensive bibliographies by Hickey (1979) and Winant (1980) for overviews of the dominant physical oceanographic processes throughout the PC. Hickey begins her review by discussing the oceanic climatology of the eastern Pacific Ocean. She indicates that the origins of the major longshore currents bordering the PC are all related to the West Wind Drift (WWD) current, the large eastward-flowing current which transits across the Pacific Ocean at about 40°N latitude.

A division of the WWD at the surface some 500 km from the Oregon-Washington coast results in the development of two current systems, the Alaska and California Currents. A third path, one slightly north toward Vancouver Island with an abrupt turn south to the California Current system, is also observed occasionally. The Alaskan Current is composed mainly of the northward flowing bifurcant of the WWD and Subarctic Current and spins counterclockwise about the Alaskan Gyre to become the Alaskan Stream. Within the Gulf of Alaska this current may extend to 2000 m. Since West North Pacific Water and Subartic Pacific Water both contribute to the WWD, and since the southerly flowing California Current is the principal recipient of the WWD, the water carried south tends to be cooler than adjacent water masses and is some 500 km wide. The California Current flows south to about 25°N latitude where it turns westward into the North Equatorial Current. In summer, the California Current System (CCS) originates farther to the north than during winter and migrates up onto the outer shelf. In winter the WWD breaks up farther south and a countercurrent develops shoreward of the California Current. The weather patterns of the eastern Pacific are the functional causes of seasonal variations in the ocean current systems. While the stationary North Pacific High pressure cell dominates the summer, the winter system is non-stationary and consists of a series of eastward moving lows. The High produces northerly winds which migrate from Baja California in April, to mid-California by May, to Oregon by July.

There are poleward countercurrents located inshore, i.e., east of the main equatorially-trending stream. Northward of approximately 35° the surface flow is called the Davidson Current. It develops along Washington and Oregon by September, appears to the south at Point Conception by October, is 80 km wide and travels at speeds of 20 to 45 cm/sec. By spring, the current has weakened and migrates north to 50° to become part of the Alaskan Current. South of 35°, surface poleward countercurrent flows can appear as far south as Baja California and occur during the late fall when winds are weakly directed to the south or are entirely absent.

The essence of Hickey's discussion is that shelfbreak currents are driven by either the longshore wind stress vector or by a quantity represented as the diabathic gradient of that vector, i.e., the curl of the longshore wind stress component, whichever is larger. During spring, when northerly (southward) winds are strong, the flow is southerly (southward) and in late summer, off the Baja, and fall, north of Point Conception, the flow becomes northerly (Davidson Current) with the dominance of a northward directed wind stress curl. In the California Bight this flow is also referred to as the inner part of the Southern California Eddy and may also be related to a flow separation caused when the southerly flowing California Current encounters Cape Mendocino.

The upwelling cycle on the Oregon-Washington shelf is the prominent summertime feature. This may be true off California as well. The along-shelf velocity field associated with large-scale summertime upwelling over the shelf has been inferred from direct current measurements (Huyer et al., 1975). In spring and summer, flow is typically southward at all depths but is strongest near the surface. In summer, though, when the stratification is sufficiently strong, a poleward undercurrent is often found. The southward surface flow forms a coastal jet centered at about mid-shelf and the poleward undercurrent magnitude increases with distance offshore over the shelf. Other than offshore surface Ekman transport, systematic large-scale structure in the cross-shelf velocity field has not yet been observed.

Long waves have long been known, or suspected, to be an integral part of the dynamics along the shelfbreak of the west coast. Mysak (1980a, b) provides rather exhaustive treatises on the descriptors of west coast waves which fit into the topographic or coastal trapped category and while these discourses will not be repeated here, some salient points will be made concerning the disposition of these events.

Two types of long waves of most interest for the west coast are Kelvin and Continental Shelf waves (CSW's). The classical Kelvin wave, which can exist at any frequency, travels alongshore with shallow water to its right and dies off exponentially in the offshore direction. It occurs along the slope-shelf boundary. Munk et al. (1970) estimated that on the West Coast, more than 70% of the semi-diurnal and 50% of the diurnal tidal amplitudes could be accounted for as barotropic Kelvin waves. Poincaré waves constituted the remainder. The Mendocino Escarpment, offshore of San Francisco, is viewed as a candidate for trapping a double Kelvin, an unusual oceanic wave feature in which the surface dies away exponentially in either direction from the break. Internal Kelvin waves are also assumed to be a part of shelfbreak phenomena under vertically stratified conditions.

Recently, Mysak (1980a) has investigated the appearance of large eddy-like structures which appear in satellite VHRR imagery along the Alaskan Stream. He sees these features as being caused by the atmospheric forcing of a baroclinic, longshore current which sheds baroclinic instabilities. Mysak also found large cold tongues of northwestward traveling water masses moving 10 to 15 km/day. These cold water masses are probably formed through a baroclinic process of the California Undercurrent off Vancouver Island. These events could be viewed as tornado-like features implanted in the Pacific coastal ocean. They have an appearance similar to the Gulf Stream frontal events shown in Figure 6.

Allen (1980) and others indicate that the alongshore time-dependent current behavior on the outer Oregon-California shelf is governed by a set of free and forced CSW's driven by the longshore wind stress and by the speed of moving weather systems. The shelfbreak region is peculiar in that it not only serves as a trapping zone but also as a zone of demarcation since the waves tend to be barotropic on the outer shelf and baroclinic over the upper slope. Therefore, the pressure gradients, both para- and diabathic, balancing the two modes of motion, are in opposition to one another. These waves may serve to bring turbulent energy from offshore to onshore regions along the west coast, much as the Gulf Stream meanders and filaments do in the SAB. A point to be made here is that CSW's can be detected in coastal sealevel data on the PC, while Gulf Stream frontal events cannot be detected in coastal tide gauges on the eastern seaboard. The reason is simply that the East Coast is wider than the diabathic width (or amplitudes) of the waves which are trapped at the shelfbreak in both the SAB and MAB (Wang, 1979; Chao and Pietrafesa, 1980).

TOPOGRAPHIC EFFECTS

The outer shelf and upper slope topography is thought to influence bottom intrusion events at the shelfbreak on all coasts. Arthur (1965) developed the notion of upwelling in the lee of capes, specifically to the south of Point Conception, California, to explain the cold bottom waters found there. Leming (1979) observed cold water to the north of Cape Canaveral, Florida, and Blanton et al. (1981) found cold bottom waters to be present not only surrounding Canaveral but to the north of the Carolina Cape Shoals as well. Leming and Blanton and others invoked Arthur's west coast notion of upwelling in the lee of capes to explain these cold pool features. Unfortunately, the concept of cape-induced vertical motions fails in general since there is no mechanism to provide for either upwelling or downwelling given an unsheared alongshore current on an f-plane over a flat shelf. An alternative and physically consistent notion is that, given a sheared and stratified current flowing over variable topography, upwelling and onshore flow will occur shoreward of the shelfbreak if the current maximum is located seaward of the shelfbreak and if the bottom isobaths separate, or diverge, in the downstream direction. If the bottom isobaths converge downstream of the current then downwelling and offshore flow will occur. In effect then, upwelling will occur in the cyclonic frontal zone, the inshore side, of a longshore current beneath which isobaths diverge atop the shelfbreak. This occurs on the northeast Florida shelf, from south of Cape Canaveral to Jacksonville, on the California coast to the south of Point Conception and on the northern side of the Carolina Cape Shoals. The important ingredient is the shape of the bottom as shown in Figure 7A.

The shape of the transition between the outer shelf and upper slope can also provide a topographic mechanism to cause shelfbreak upwelling. Simple, time-dependent coastal upwelling theory (Janowitz and Pietrafesa, 1980) indicates that density will be at a maximum directly above the shelfbreak when the quantity $h(h)_{xx}/2(h)^2_x$ is greater than unity; where h is the water depth, x is the offshore spatial coordinate, h_x is the diabathic bottom slope and h_{xx} is the rate of change of the slope. This inequality holds when the shelfbreak affects an abrupt, sharp transition from a relatively flat outer shelf (10^{-3} gradient) to a relatively steep upper slope (10^{-2} gradient). The resulting upwelled bump will be about 5 to 10 km wide in the diabathic direction. Figure 7B is a conceptual drawing of this phenomenon, which probably occurs locally off northern and far southern California, Oregon, Washington, North Carolina, the MAB and northeastern Florida.

SUBMARINE CANYONS AND SHELFBREAK CIRCULATION

The MAB and PC outer continental margins are intersected by submarine canyons (Fig. 1). These canyons, rather numerous in both regions, certainly affect the upper slope oceanography and also act as sediment depositories and by-ways. While physical oceanographic studies of these environments have been few in number (e.g., Stanley and Kelling, 1978), some recent results of the work of Hickey (1980) off Washington and of Hsueh (1980) are of note.

Hickey has found that the presence of a narrow submarine canyon, Astoria Canyon, does not affect the low frequency current structure about the canyon, so that the meridional upper slope-outer shelf flows remain fairly consistent over the canyon. Conversely, the parabathic isopycnals decrease in the canyon. Moreover, studies of Hickey in a broad canyon, the Quinault, show that meridional current fluctuations are coherent with those on the outer

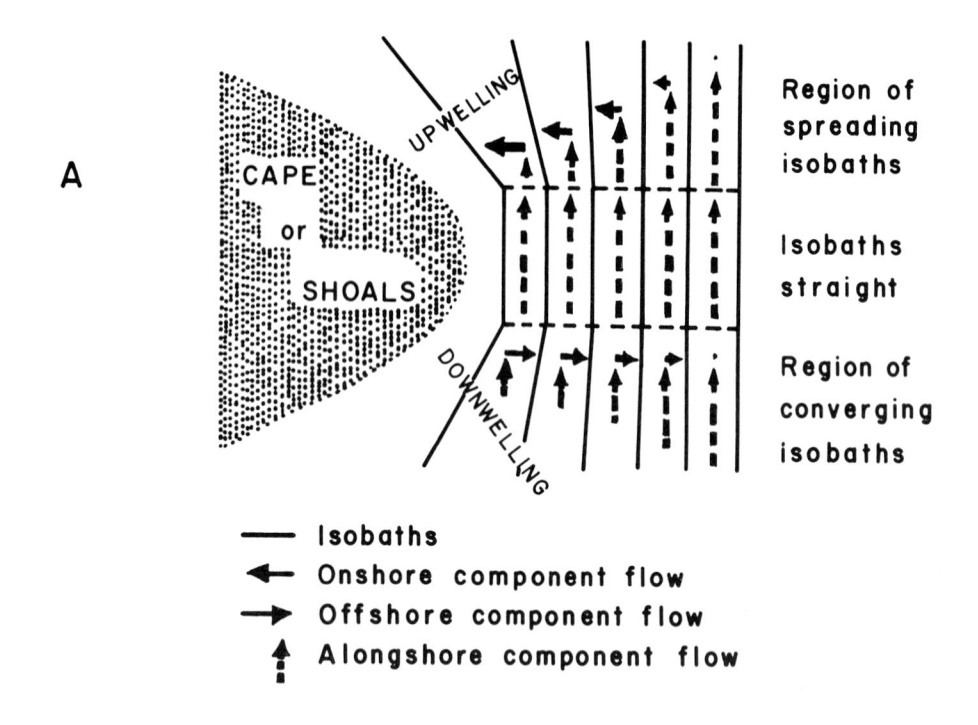

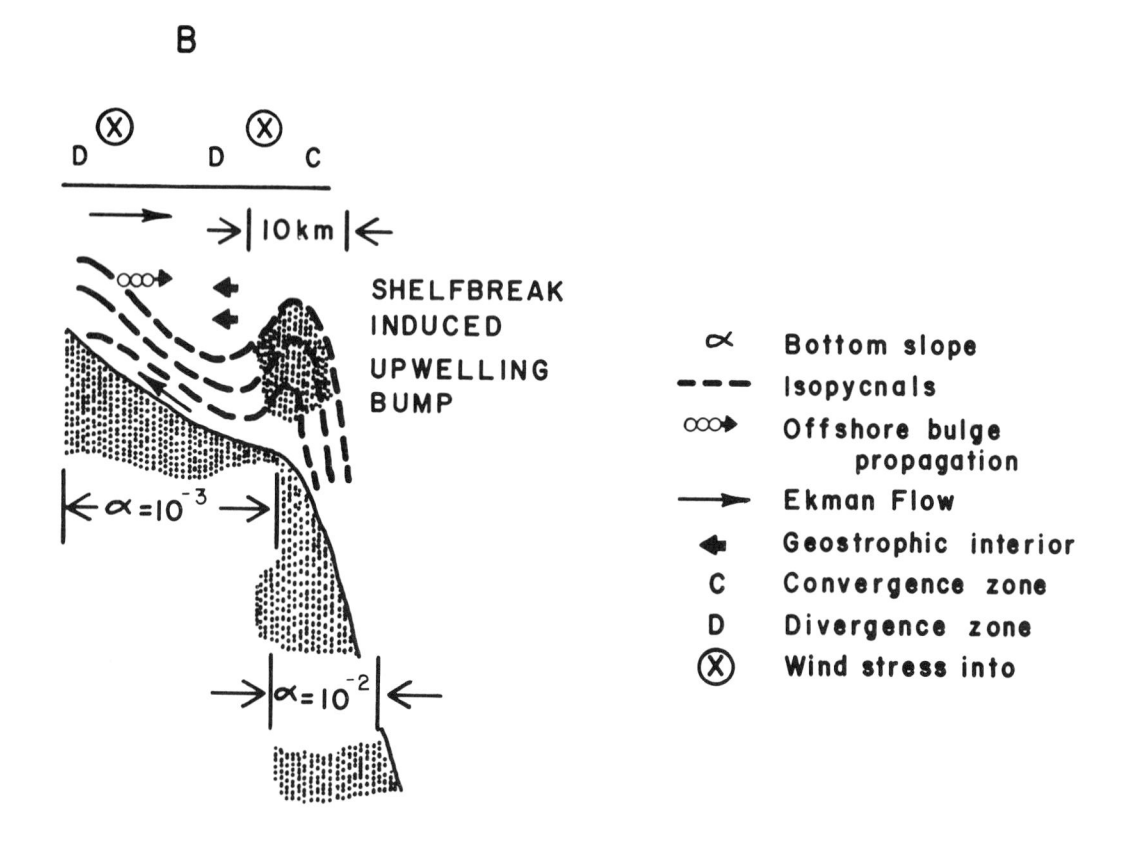

A

—	Isobaths
←	Onshore component flow
→	Offshore component flow
↑	Alongshore component flow

B

α	Bottom slope
---	Isopycnals
⚬⚬⚬→	Offshore bulge propagation
→	Ekman Flow
←	Geostrophic interior
C	Convergence zone
D	Divergence zone
⊗	Wind stress into

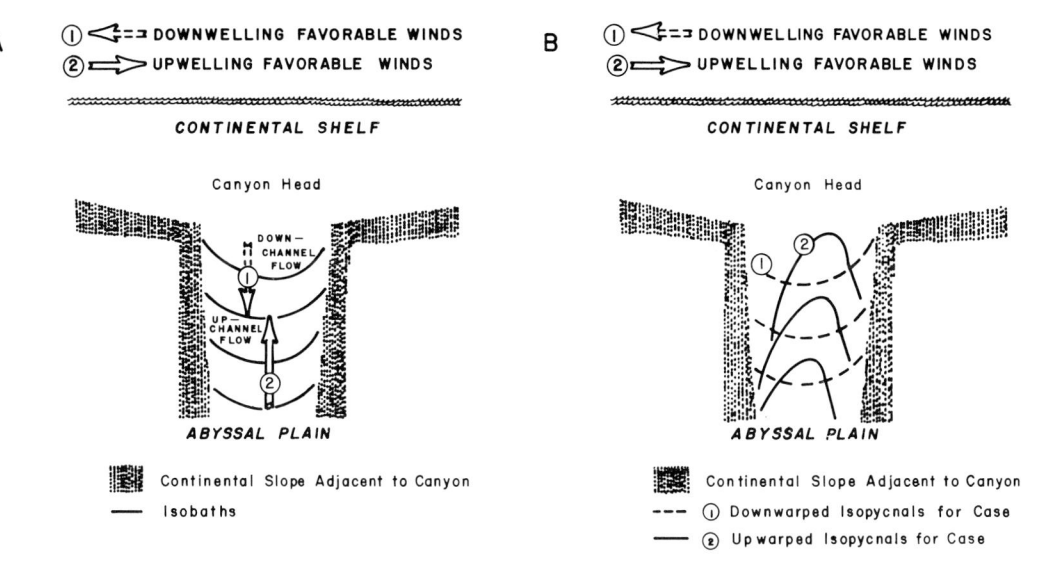

FIG. 8.—*A*, Up (down) channel flows caused by up (down) welling favorable alongshore winds. *B*, Isopycnal warping consistent with (A).

shelf. However, the amplitudes of the fluctuations are smaller by a factor of two over and within the canyon. Summertime data in both canyons suggest dramatic upwelling within the canyon with isopycnal downwarping above the axis and upwarping near the walls.

Hsueh (1980) reports that the average flow in the Hudson Valley (MAB) is shoreward at speeds of several centimeters per second. The valley flow at subtidal frequencies tends to be wind-driven and correlates well with local and regional winds; upchannel and downchannel flows are related to respectively westerly and easterly winds. The westerly winds induce a divergence of coastal waters compensated for by an upchannel flow. Clearly the downchannel flow helps flush the shelf of excess water accumulation accompanying easterly winds. Alongshore winds are also followed by upchannel or downchannel flow (Fig. 8). Hickey (1980) suggests that both southerly winds and poleward longshore currents can pull deep, cold water upchannel while simultaneously forcing warmer shelf water into the canyon. The opposite effect occurs for northerly winds or southerly flowing currents.

It is assumed that there are occasional cascading events which may appear as turbidity currents and flow down and out of the canyon valleys. These probably occur when outer shelf waters are cold and dense, and the longshore wind blows with the coast to its right, usually in the wintertime.

SHELFBREAK SEDIMENT DISPERSAL

In parts V and VI of this volume, companion articles such as those by Karl et al. (1983) present lucid discussions of sediment transport mechanics at the shelfbreak. It is of use here to discuss some of the implications of the preceding sections of this chapter on physical processes as they relate to sediment dispersal. In a classic study of sediment dispersal mechanisms on the Pacific Northwest shelf, Smith and Hopkins (1972) showed that significant transport of sediments only occur during substantial atmospheric storm conditions. Moreover, they showed that the suspended load mode of transport is extremely important relative to the bedload mode. Finally the authors determined that sediments fine enough to be suspended (3.5 phi and smaller) may be transported by low frequency currents while coarser sands are more simply trapped near the coast and tend to be continually reworked. This work is important since it shows that storms are the primary events of sedimentological significance and that low frequency currents provide the transport mechanism. Other authors, prior to this pioneering effort, had promoted the importance of waves and the bedload mode as the basis of a continuous movement of sediments while ignoring the overall importance of occasional, but nonetheless important, energetic wind-driven current events.

The U.S. West Coast can be categorized in a sedimentological way (Nittrouer, 1978) as one in

FIG. 7.—*A*, The interaction of bottom topography with an alongshore current. Note upwelling (downwelling) occurring in the region of diverging (converging) isobaths. *B*, Shelfbreak upwelling caused by a sharp shelfbreak interacting with an upwelling favorable wind.

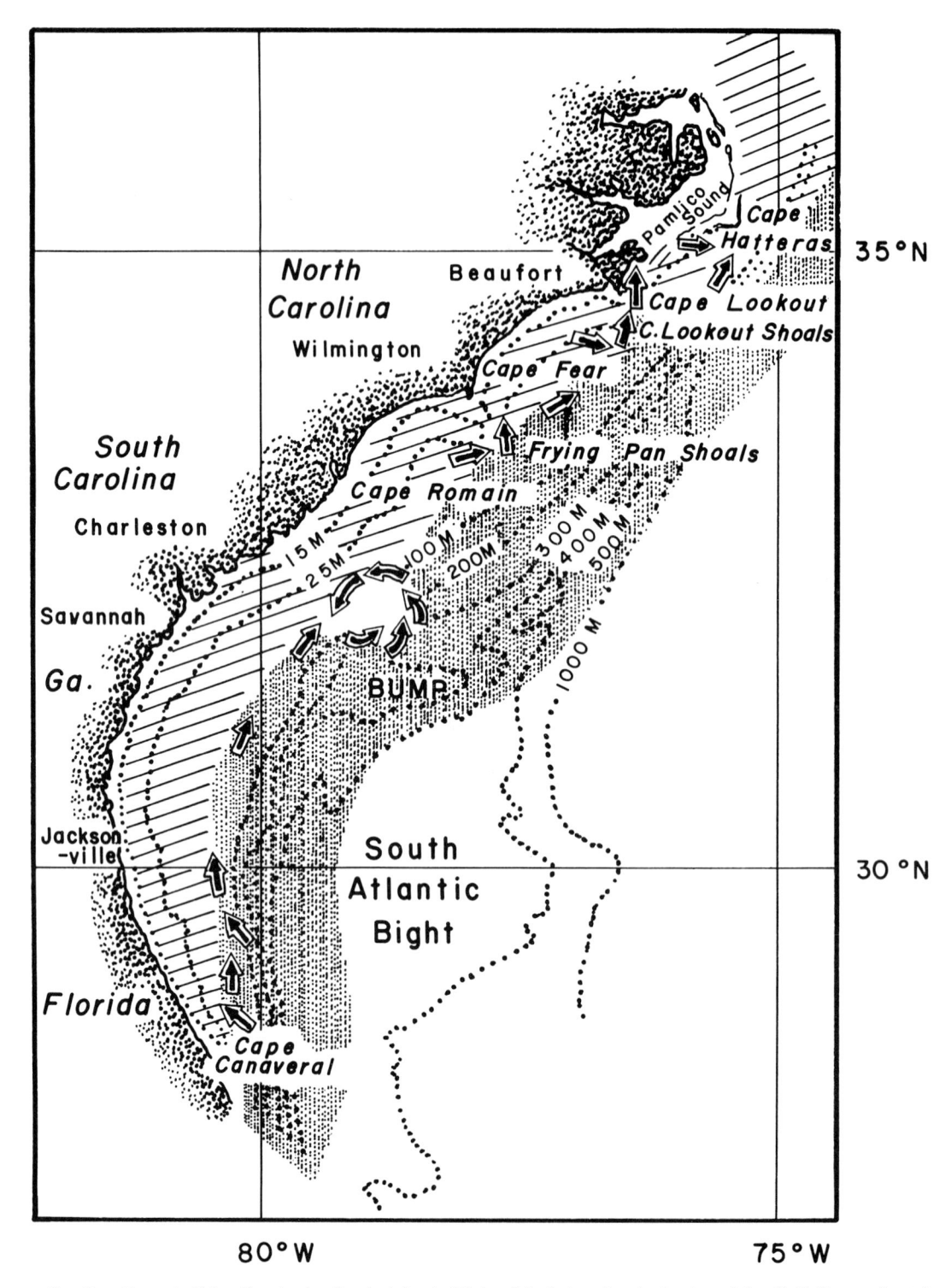

FIG. 9.—Cross-shelf banding in the South Atlantic Bight. Stippled region is dominated by Gulf Stream frontal events. Lined region is dominated by winds, tides, and gravity waves. Arrows indicate direction of sediment movement.

which fine silts and clays are present, with a number of riverine sources distributed throughout. These finer-grained materials can be suspended both by gravity waves and by other lower frequency currents as well, and moved offshore via an offshore flowing frictional bottom layer and subsequently deposited over the outer shelf and upper slope. Coarser sediments are simply reworked and essentially trapped on the shelf. Since most of the energetic activity occurs during the winter, it is during these severe storm periods that most of the sediment on the PC is probably reworked and transported. Nittrouer (1978) found this to be the case on the Washington-Oregon outer shelf.

In the MAB, Butman et al. (1979) reported sighting suspended silts and clays being transported 20 to 30 km to the south and 5 to 10 km offshore on the outer shelf during wind-forced downwelling events. The transport occurred most prominently during the winter when low frequency near-bottom currents in excess of 30 cm/sec provided both the means for suspension and mode of transport. Wave-induced bottom currents were found to resuspend sediments during periods of both high and low mean flow in the wintertime. During the summer, flocculant material covered the bottom and the minor activity observed was ascribed to mechanical stirring due to semi-diurnal tides and packets of internal waves and to bioturbation processes.

In the SAB, there are few, if any, reports of actual observation of sediment motion, but a wealth of data exists on the distribution of the sediments (Swift et al., 1972; Stanley and Swift, 1976). In the SAB, continental shelf sediments are largely relict sands with a large carbonate fraction (5–50%) and a small silt/clay fraction (<5%). The fine grained sediments are found on the upper continental slope but not on the shelf above the break. The SAB is, in general, silt-poor. Well-sorted carbonate sands are most prevalent. Alternatively, in the MAB, sediments are dominated by silts and clays which are carbonate poor. Doyle et al. (1968) and others contend that any fine-grained sands found nearshore in the SAB must have come from the outer shelf. Alternatively, Buss and Rodolfo (1972) suggest that the capes are regions where fine sands are being transported seaward. The sources of the fine-grained sands are considered by Buss and Rodolfo to be either the inlets or the newly deposited nearshore fines, but these contentions are subject to question as discussed below.

In a study of the movement of moderately sorted

FIG. 10.—Photomosaic of bottom, Cape Lookout shoals, North Carolina (courtesy of D. J. P. Swift). Sediment movement (→), shoals and capes due to interaction of bottom topography with alongshore current (⇒). Refer to Fig. 7A and text for details.

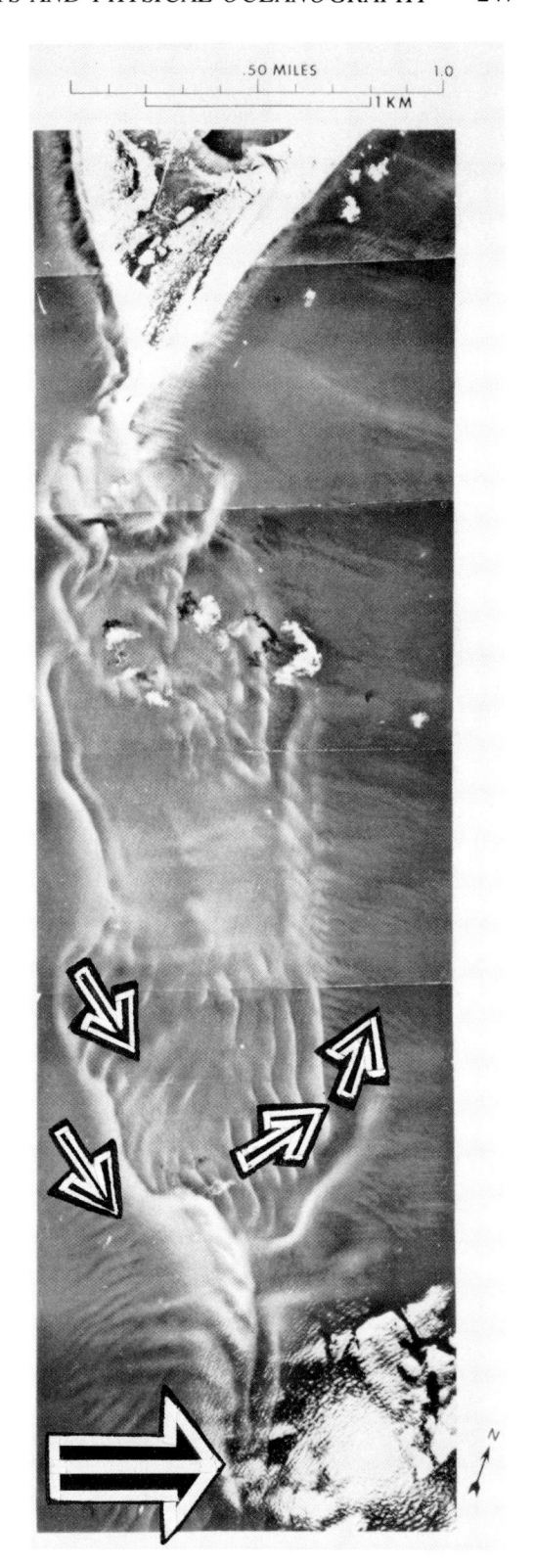

carbonate medium sand deposited under the Gulf Stream on the upper Florida continental slope, Wimbush and Lesht (1979) found that migration of these sediments occurred when current speeds exceeded 24 cm/sec about 3 m above the bottom. On the outer shelves of Florida, Georgia and the Carolina Capes, Pietrafesa (1981) has determined that 2 to 12 day period flows related to GSF events dominate the energetics of the system and that the current magnitudes of these events typically exceed 24 cm/sec. Moreover, the square root of the total variance 3 m above the shelfbreak is 22 cm/sec. Finally GSF events extend to the 33 m isobath in the Carolina Capes, as shown in Figure 9, and to the 40 m isobath off Florida and Georgia. Flows are directed to the northeast so that one can conclude that fine-grained material is being swept to the northeast, except on the northern sides of the Carolina Cape shoals and Cape Canaveral, where the isobaths on the outer shelf diverge. On the southern sides of the shoals, the migration of movable sediments should be offshore while on the northern sides the movement should be onshore. This hypothesis is borne out by the photomosaic of the bottom around Cape Lookout, North Carolina (Fig. 10).

During the winter, as atmospheric cold fronts pass, cold dense water is formed and driven seaward down toward and across the shelfbreak to ultimately cascade down the slope. During the summer dense bottom waters move up the slope and across the outer shelves. These phenomena occur on all of the U.S. continental margins.

Neither the PC nor the East Coast north of Cape Hatteras are fetch-limited so large amplitude sea and swell, capable of providing bottom shear stresses necessary for causing sediment migration, are observed. In the SAB, however, the shoals limit the extent of longshore fetch and the Gulf Stream acts as a low frequency filter of sea and swell. Bioturbation may be important, especially during the more quiescent summertime period, on all outer shelves.

CONCLUSIONS

Fine sediments with coastal sources are suspended and transported seaward over the shelfbreak on the West Coast and on the East Coast north of Cape Hatteras by a superposition of gravity and internal waves, semi-diurnal and diurnal tides, mainly semi-diurnal on the East Coast, and inertial and low frequency currents. These low frequency flows on the outer shelf are principally wind stress curl related on the West Coast and are superimposed on the CCS. Within and to the north of the MAB, circulation on the outer continental shelf can be forced directly by the wind and by a diabathic pressure gradient, i.e. a sea surface which slopes downward toward the ocean. In the SAB, Gulf Stream frontal waves continually sweep the shelfbreak clean of fine sediments while gravity and internal waves and both tidal and inertial currents are of lesser importance. Coarser sediments are reworked and bioturbative processes are present on all outer shelves. Capes and headlands should show sediment accumulations on the lee side of mean current flows and erosion on the upstream sides (Fig. 10). Wintertime is the period of major atmospheric event input. As a consequence, submarine canyons become repositories and byways for sediments (Fig. 8) since flow is into the canyons during winter when wave and current energies on the shelves are at yearly maxima and sediments are being actively transported.

ACKNOWLEDGMENTS

The author gratefully acknowledges the support of the U.S. Department of Energy under Contract DOE AS09-76-EY00902 for this work. Also, O. H. Pilkey, C. A. Nittrouer and D. J. DeMaster are thanked for their constructive reviews and helpful suggestions to better the manuscript. Finally, C. N. K. Mooers, C. N. Flagg, W. C. Boicourt, R. Beardsley, D. J. P. Swift, T. B. Curtin, D. Halpern, A. Clarke, D. Battisti, C. Winant, O. Brown and R. Evans are acknowledged for allowing their published figures to be used in this chapter.

REFERENCES

ALLEN, J. S., 1980, Models of wind-driven currents on the continental shelf: Ann. Rev. Fluid Mech., v. 12, p. 389–433.

ARTHUR, R. S., 1965, On the calculation of vertical motion in eastern boundary currents from determinations of horizontal motion: Jour. Geophys. Res., v. 70, no. 12, p. 2799–2803.

BATTISTI, D. S., AND CLARKE, A. J., 1981, A simple method for estimating barotropic tidal currents on continental margins with specific application to the M_2 tide off the Atlantic and Pacific coasts of the United States: unpublished manuscript.

BEARDSLEY, R. C., AND BUTMAN, B., 1974, Circulation on the New England continental shelf: response to strong wind storms: Geophys. Res. Letters, v. 1, no. 14, p. 181–184.

———, AND FLAGG, C. N., 1975, The water structure, mean currents, and shelf-water/slope-water front of the New England continental shelf: Seventh Liege Colloquium on Ocean Hydrodynamics, v. 10, p. 209–226.

———, AND WINANT, C. D., 1979, On the mean circulation in the Mid-Atlantic Bight: Jour. Phys. Oceanogr., v. 9, no. 3, p. 612–619.

BISHOP, J. M., 1980, A note on the seasonal transport on the Middle Atlantic shelf: Jour. Geophys. Res., v. 85, no. C9, p. 4933–4936.
BLANTON, J. O., ATKINSON, L. P., PIETRAFESA, L. J., AND LEE, T. N., 1981, The intrusion of Gulf Stream water across the continental shelf due to topographically induced upwelling: Deep-Sea Res., v. 28A, no. 4, p. 393–405.
BUSS, B. A., AND RODOLFO, K. S., 1972, Suspended sediments in continental shelf waters off Cape Hatteras, North Carolina, in Swift, D. J. P., Duane, D. B., and Pilkey, O. H., eds., Shelf Sediment Transport: Process and Pattern: Stroudsburg, Pennsylvania, Dowden, Hutchinson and Ross, Inc., p. 263–279.
BUTMAN, B., NOBLE, M., AND FOLGER, D. W., 1979, Long-term observations of bottom current and bottom sediment movement on the Mid-Atlantic continental shelf: Jour. Geophys. Res., v. 84, p. 1187–1205.
CHAO, S. Y., AND PIETRAFESA, L. J., 1980, The subtidal response of sea level to atmospheric forcing in the Carolina Capes: Jour. Phys. Oceanogr., v. 10, p. 1246–1255.
CHENEY, R. E., AND MARSH, J. G., 1981, Seasat altimeter observations of dynamic topography in the Gulf Stream region: Jour. Geophys. Res., v. 86, no. C1, p. 473–483.
CLARKE, A. J., AND BATTISTI, D. S., 1981, The effect of continental shelves on tides: Deep-Sea Res., v. 28, p. 665–682.
CURTIN, T. B., 1979, Physical dynamics of the coastal upwelling frontal zone off Oregon [Ph.D. diss.]: Miami, Florida, University of Miami, 317 p.
DOYLE, L. J., CLEARY, W. J., AND PILKEY, O. H., 1968, Use of mica in determining shelf depositional regimes: Mar. Geology, v. 6, p. 381–389.
FLAGG, C. N., 1977, The kinematics and dynamics of the New England continental shelf and shelf/slope front [Ph.D. thesis]: Cambridge, Massachusetts, Mass. Inst. Tech., Mass. Inst. Tech.—Woods Hole Oceanogr. Inst. Joint Program in Oceanogr., 207 p.
HALLIWELL, G. R., JR., AND MOOERS, C. N. K., 1979, The space-time structure and variability of the shelf water-slope water and Gulf Stream surface temperature fronts and associated warm-core eddies: Jour. Geophys. Res., v. 84, no. C12, p. 7707–7726.
HICKEY, B. M., 1979, The California Current system-hypothesis and facts: Prog. Oceanogr., v. 8, p. 191–279.
———, 1980, Pollutant transport and sediment dispersal in the Washington-Oregon coastal zone: Progress Rept. to the U.S. Dept. of Energy under Contract No. DE-AT06-76-EV-71025.
HOPKINS, T. S., 1981, Comments on the Mid-Atlantic Bight circulation: Rept. to the Dept. of Energy under Contract No. DE-AC02-76 CH00016.
HSUEH, Y., 1980, Scattering of continental shelf waves by longshore variations in bottom topography: Jour. Geophys. Res., v. 85, p. 1147–1150.
HUYER, A., HICKEY, B. M., SMITH, J. D., SMITH, R. L., AND PILLSBURY, R. D., 1975, Alongshore coherence at low frequencies in currents observed over the continental shelf off Oregon and Washington: Jour. Geophys. Res., v. 80, p. 3495–3505.
JANOWITZ, G. S., AND PIETRAFESA, L. J., 1980, A model and observations of time dependent upwelling over the mid-shelf and slope: Jour. Phys. Oceanogr., v. 10, p. 1574–1583.
KARL, H. A., CARLSON, P. R., AND CACCHIONE, D. A., 1983, Factors influencing sediment transport at the shelfbreak, in Stanley, D. J., and Moore, G. T., eds., The Shelfbreak: Critical Interface on Continental Margins: Soc. Econ. Paleontologists Mineralogists Spec. Pub. 33, p. 219–231.
KLINCK, J. M., PIETRAFESA, L. J., AND JANOWITZ, G. S., 1981, Continental Shelf circulation induced by a moving, localized wind stress: Jour. Phys. Oceangr., v. 11, p. 836–848.
LEGECKIS, R., 1979, Satellite observations of the influence of bottom topography on the seaward deflection of the Gulf Stream off Charleston, South Carolina: Jour. Phys. Oceanogr., v. 9, p. 483–497.
LEMING, T. D., 1979, Observations of temperature, current, and wind variations off the central eastern coast of Florida during 1970 and 1971: NOAA Technical Memorandum NMFS-SEFC-6, 197 p.
MOOERS, C. N. K., FLAGG, C. N., AND BOICOURT, W. C., 1978, Prograde and retrograde fronts, in Bowman, M. J., and Esaias, W. E., eds., Oceanic Fronts in Coastal Processes: Proc. Workshop Marine Sciences Research Center, May 25–27, 1977, p. 43–58.
MUNK, W., SNODGRASS, F., AND CARRIER, G., 1956, Edge waves on the continental shelf: Science, v. 123, p. 127–132.
———, ———, AND WIMBUSH, M., 1970, Tides offshore: transition from California coastal to deep-sea waters: Geophys. Fluid Dyn., v. 1, p. 161–235.
MYSAK, L. A., 1980a, Recent advances in shelf wave dynamics: Rev. Geophys. Space Phys., v. 18, p. 211–241.
———, 1980b, Topographically trapped waves: Ann. Rev. Fluid Mech., v. 12, p. 45–76.
NITTROUER, C. A., 1978, The process of detrital sediment accumulation in a continental shelf environment: an examination of the Washington shelf [thesis]: Seattle, Washington, Univ. of Washington, 243 p.
PIETRAFESA, L. J., 1978, Continental shelf processes affecting the oceanography of the South Atlantic Bight: Progress Report to the U.S. Dept. of Energy under Contract No. DOE-AS-0-76-EY00902, 228 p.
———, 1981, On the characterization of Gulf Stream frontal meanders and filaments in the Carolina Capes: submitted to Jour. Phys. Oceanogr.
———, AND JANOWITZ, G. S., 1979, On the effects of buoyancy flux on continental shelf circulation: Jour. Phys. Oceanogr., v. 9, p. 911–918.
———, BLANTON, J. O., AND ATKINSON, L. P., 1978, Evidence for deflection of the Gulf Stream at the Charleston Rise: Gulfstream, v. 4, p. 3, 6–7.

RATTRAY, M., JR., 1957, On the offshore distribution of tide and tidal current: Trans. Am. Geoph. Union, v. 38, p. 675–680.
ROONEY, D. M., JANOWITZ, G. S., AND PIETRAFESA, L. J., 1978, A simple model of deflection of the Gulf Stream by the Charleston Rise: Gulfstream, v. IV, p. 1–7.
SMITH, J. D., AND HOPKINS, T. S., 1972, Sediment transport on the continental shelf off of Washington and Oregon in light of recent current measurements, *in* Swift, D. J. P., Duane, D. B., and Pilkey, O. H., eds., Shelf Sediment Transport: Process and Pattern: Stroudsburg, Pennsylvania, Dowden Hutchinson and Ross, Inc., p. 143–179.
STANLEY, D. J., AND KELLING, G., eds., 1978, Sedimentation in Submarine Canyons, Fans, and Trenches: Stroudsburg, Pennsylvania, Dowden, Hutchinson & Ross, 395 p.
———, AND SWIFT, D. J. P., 1976, Marine Sediment Transport and Environmental Management: New York: John Wiley and Sons, Inc., 602 p.
SWIFT, D. J. P., DUANE, D. B., AND PILKEY, O. H., eds., 1972, Shelf Sediment Transport: Process and Pattern: Stroudsburg, Pennsylvania, Dowden Hutchinson and Ross, Inc.,.
WANG, D. P., 1979, Wind-driven circulation in the Cheasapeake Bay, Winter 1975: Jour. Phys. Oceanogr., v. 9, p. 564–572.
WIMBUSH, M., AND LESHT, B., 1979, Current-induced sediment movement in the deep Florida Straits: Critical parameters: Jour. Geophys. Res., v. 84, p. 2495–2502.
WINANT, C. D., 1980, Coastal circulation and wind-induced currents: Ann. Rev. Fluid Mech., v. 12, p. 271–302.

SHELFEDGE DYNAMICS AND THE NEPHELOID LAYER IN THE NORTHWESTERN GULF OF MEXICO

DAVID W. McGRAIL AND MICHAEL CARNES
Department of Oceanography, Texas A&M University, College Station, Texas 77843

ABSTRACT

An investigation of shelfedge sedimentary processes in the Gulf of Mexico has been underway for the past five years. It has consisted of *in situ* bottom boundary layer (BBL) experiments, time series observations using moored instruments, and hundreds of hydrographic stations. A ubiquitous nepheloid layer exists over the outer continental shelf in the BBL. It reaches a maximum thickness of 30 m when offshore flow near bottom stacks detached bottom boundary layers at the shelfedge. The shear stresses which maintain the sediment suspension are contributed by a superposition of many modes of motion. In the northwestern Gulf of Mexico, surface gravity waves, high frequency internal waves and tides do not appear to contribute significantly to the sedimentary processes at the shelfbreak. However, diurnal inertial oscillations do resuspend silt and clay at the shelfedge and transport that sediment to the offshore. Winter storms produce three types of phenomena that influence sediment transport: (1) direct, energetic, cross-shelf wind-driven flow; (2) production of dense, cool, saline bottom water that flows offshore under the influence of gravity; and (3) inertial oscillations which propagate to the bottom. The mean shelfedge flow was found to be west to east in the interior, with bottom waters oriented more southeasterly. The latter should contribute to a long term advection of sediment off the shelf. Flow on the bottom of the upper slope has been observed to be oriented to the northeast, suggesting a convergence in the BBL near the shelfbreak.

INTRODUCTION

Numerous banks occur along the shelfbreak in the northwestern Gulf of Mexico from near the Mississippi Delta westward to the point where the shelf axis swings to a southerly trend (Fig. 1). These banks are the surface expression of salt domes, many of which have potential for hydrocarbon traps in the subsurface. Some of the banks are capped by diverse hard-bank communities, and most are important in both sport and commercial fisheries.

Initial surveys of these banks to determine the nature of the biota on them and the setting in which they exist revealed the presence of a nepheloid layer at the seafloor over the whole region (Bright and Rezak, 1976, 1977). Investigations of the sediment and fluid dynamics attending the nepheloid layer phenomena on the outer continental shelf and upper slope off Texas and Louisiana have been in progress since 1977 (McGrail, 1978; McGrail and Horne, 1981).

These studies have involved two types of investigations: a reconnaissance survey of banks throughout the region, from just south of the Mississippi River west to the Texas-Mexico border, and an intensive longterm investigation at two of them, the East and West Flower Garden Banks (Fig. 2). This dual approach has provided a mechanism for obtaining a measure of both spatial and temporal variances in the processes of shelfedge sediment dynamics.

The purpose of this chapter is to provide a descriptive synopsis of water motions, boundary layer development, and sediment responses observed at the shelfbreak in the course of these investigations.

METHODS

Profiling Instruments

During the early stages of the field investigations, separate casts were taken by means of an STD, transmissometer, and profiling current meter at each hydrographic station. These casts provided nearly simultaneous measurements of salinity, temperature, horizontal velocity components, and transmissivity.

The profiling current meter readings were taken at 5 m incremental depths. At each 5 m increment the lowering was halted and a one minute vector average obtained. Navigational data were supplied by LORAC. Fixes were taken every 30 seconds so that the ship drift could be computed and subtracted from the measured velocity. The absolute accuracy of this early velocity profiling is difficult to assess except at those times when the profiles were obtained next to moored current meters and during *in situ* dye emission experiments. During approximately 12 hours of dye experiments in 1977, the speed measured approximately 2 m off the bottom by means of the profiling current meter agreed, to within ±5 cm/sec in speed and ±10° in direction, with that measured at 2 m above the bottom using flow visualization techniques.

In 1980 the researchers on this project designed and built an integrated profiling system called the Profiling Hardwired Instrumented Sensor for Hy-

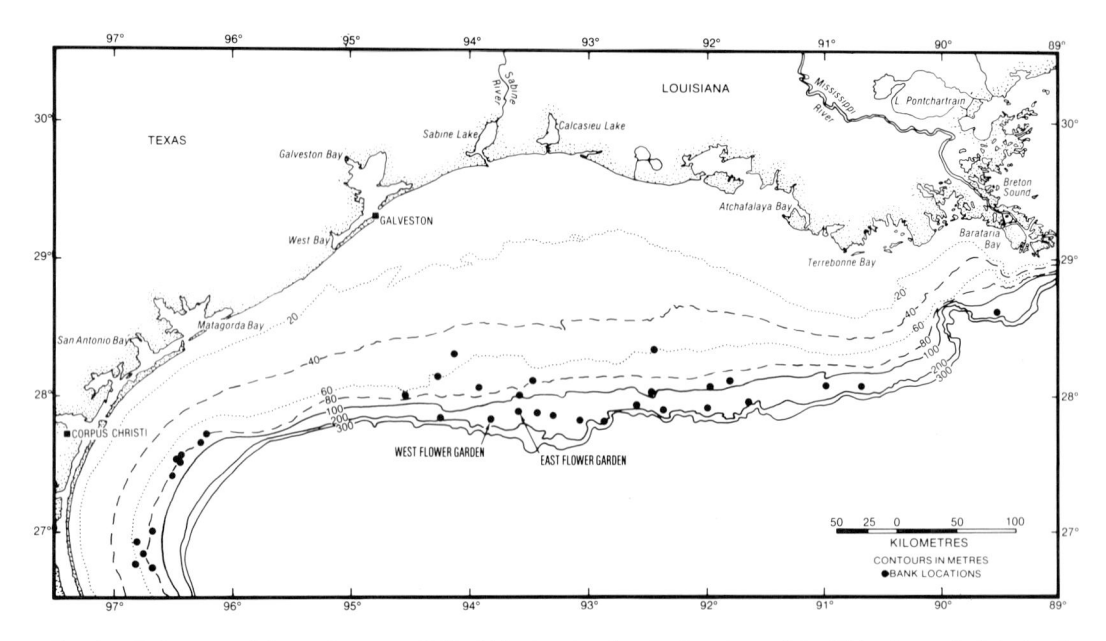

FIG. 1.—Index map of the northwestern Gulf of Mexico showing the locations of the 34 banks studied during the course of this investigation.

drography (PHISH). During lowering, the PHISH system provides simultaneous measurements of temperature, conductivity, transmissivity, current velocity, and depth. The system consists of: an underwater unit containing a CTD, an LED transmissometer, an electromagnetic current meter, two orthogonally mounted inclinometers, two bottom-tripped Niskin bottles, and a mechanically tripped magnetic switch as a bottom sensor, all mounted on a single stainless steel frame; a winch with 7-conductor armored cable; and a microcomputer-based data acquisition system in a portable electronics laboratory van.

The logging system in the lab van is comprised of an interface from the sea cable to a 64 K byte microcomputer with a 2.5 megabyte floppy disk system. The microprocessor is also interfaced to a precision LORAN C receiver with digital output so that continuous position data are recorded along with time and the data from the profiling package. The microprocessor is also interfaced to an incremental plotter to provide plots of all measured parameters in addition to the real time numerical data displayed on the CRT of the computer terminal. Power to the computer system is provided through an uninterruptable frequency and voltage stabilized source. Data acquisition is assured by carrying a redundant unit for every item except the plotter.

At the present data acquisition (1 Hz) and lowering (20 m/min) rates, approximately three sets of measurements can be taken for each parameter every meter. Orthogonal inclinometers provide continuous information on the orientation of the electromagnetic current meter so that contamination of the horizontal velocity components by vertical motions can be removed if necessary. Since the maximum inclination to date has been less than 10° from vertical, this correction has not been necessary. The compass in the electromagnetic current meter is monitored to assure that spinning of the unit does not induce spurious velocities. One bottom-tripped Niskin bottle provides water for salinity calibration, and the contents of the other are filtered to relate transmissivity to absolute concentrations of suspended sediment.

Four stations were occupied at each bank during the reconnaissance phase of the work. At each station the hydrographic casts described above were taken along with a grab sample of sediment for textural and compositional analyses.

Seasonal sampling has been carried out at the Flower Garden Banks. Until 1980, only 12 stations were occupied during each of the three cruises to the banks each year, and these were all clustered around the East Flower Garden Bank. Starting in September of 1980, seasonal sampling was done at both the East and West Flower Garden Banks, the cruises were increased to four per year, and the number of stations increased to as many as 40 per cruise.

Moored Instruments

Single point tautline current meter moorings were established in the vicinity of the East Flower Gar-

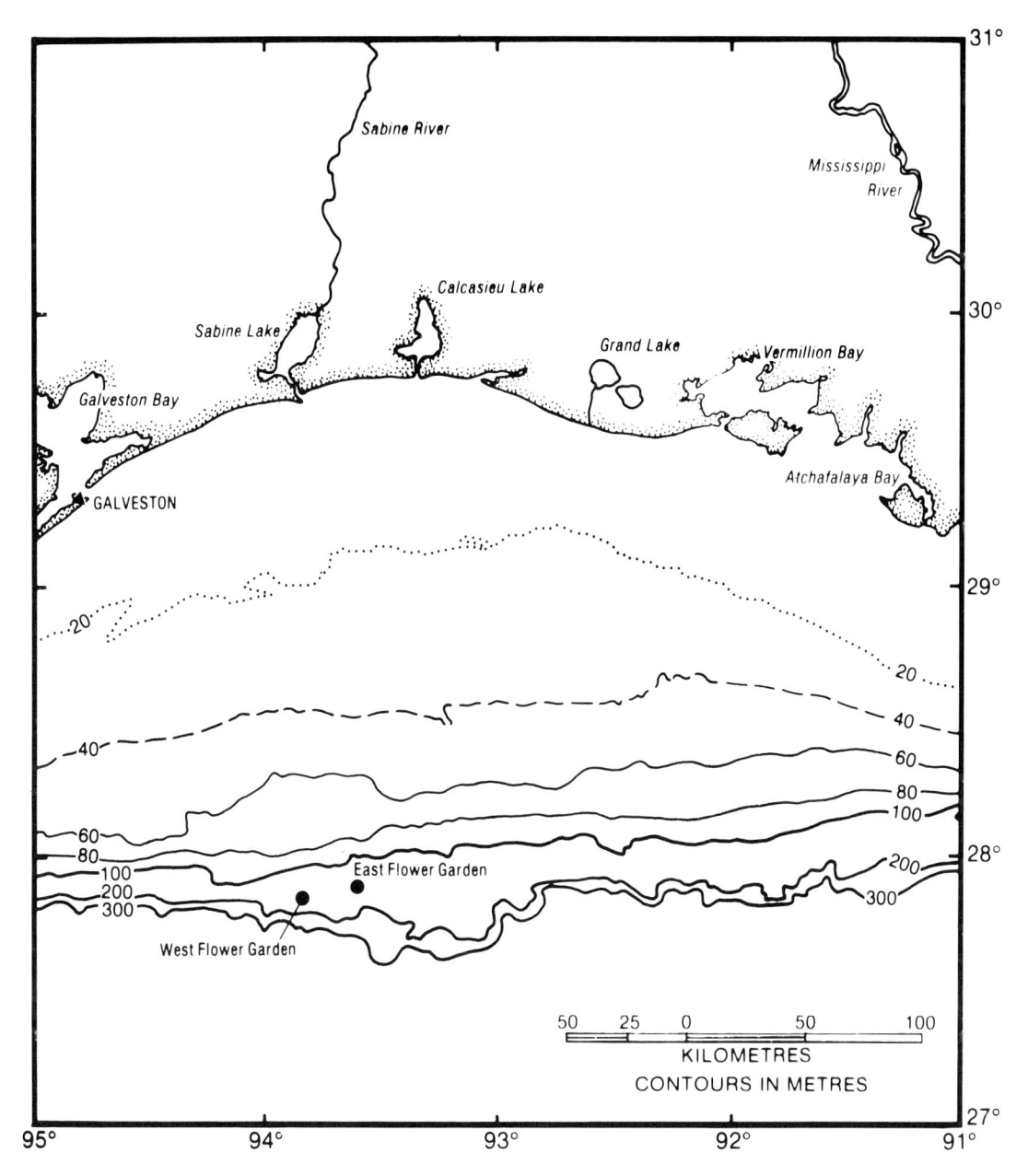

FIG. 2.—Location of the East and West Flower Garden Banks.

den Bank in January of 1979. Figure 3 provides statistics of deployments and Figure 4 depicts locations of the moorings. In addition, an electromagnetic current meter was mounted on a rigid pole placed near the top of the East Flower Garden Bank at about 30 m depth. Initially, each instrument on the tautline moorings sampled current velocity and temperature for one minute out of every six. The moorings were increased in number and instruments added so that as of March 1981 there were two moorings each at the East and West Flower Garden Banks. Each mooring had at least four current meters, and LED transmissometers were added to the lowest two meters. The sample rate was also altered so that one sample was taken every 20 minutes. The higher frequency original sampling rate was used to determine if velocity fluctuations due to short period internal waves would appear in the

A

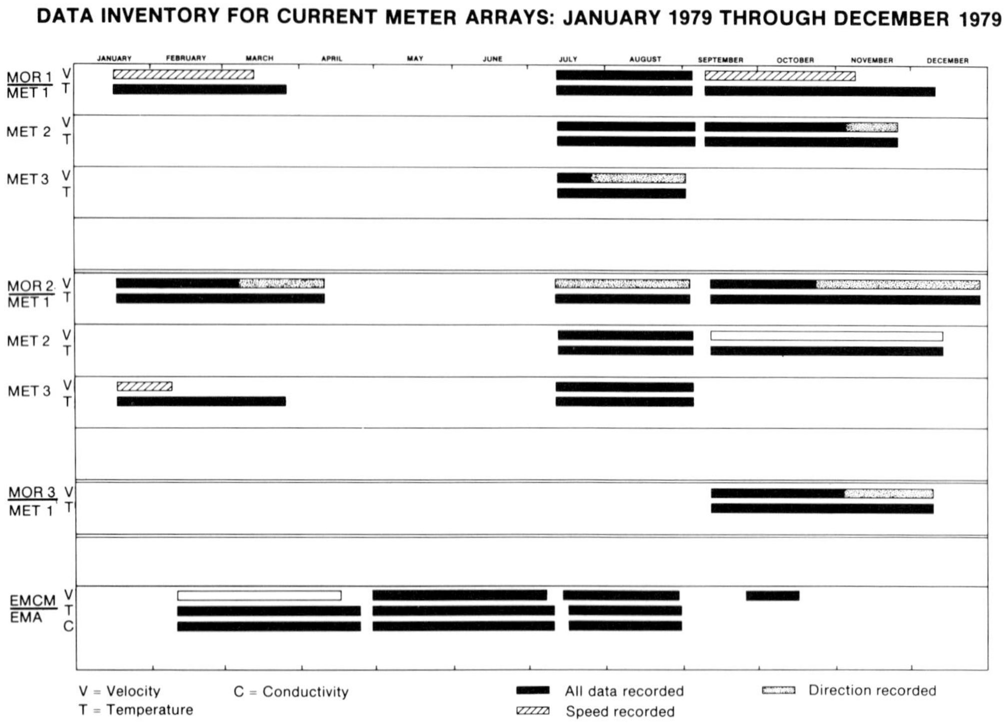
DATA INVENTORY FOR CURRENT METER ARRAYS: JANUARY 1979 THROUGH DECEMBER 1979
JANUARY
FEBRUARY
MARCH
APRIL
MAY
JUNE
JULY
AUGUST
SEPTEMBER
OCTOBER
NOVEMBER
DECEMBER
MOR 1 / MET 1 V T
MET 2 V T
MET 3 V T
MOR 2 / MET 1 V T
MET 2 V T
MET 3 V T
MOR 3 / MET 1 V T
EMCM / EMA V T C
V = Velocity
T = Temperature
C = Conductivity
All data recorded
Speed recorded
Direction recorded

B

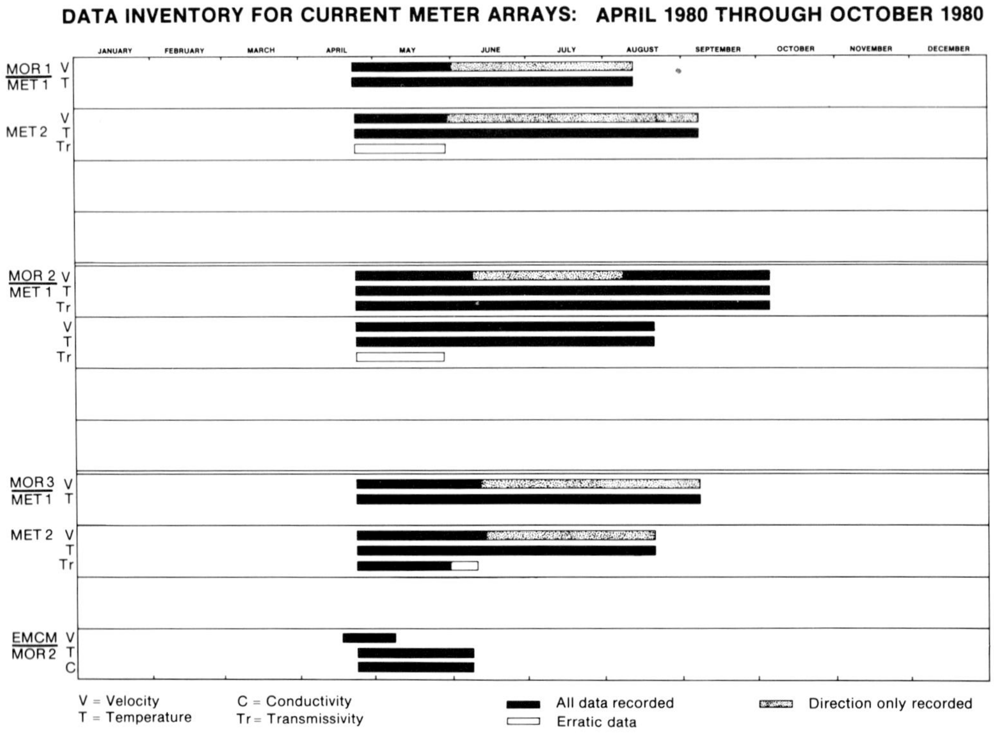
DATA INVENTORY FOR CURRENT METER ARRAYS: APRIL 1980 THROUGH OCTOBER 1980
JANUARY
FEBRUARY
MARCH
APRIL
MAY
JUNE
JULY
AUGUST
SEPTEMBER
OCTOBER
NOVEMBER
DECEMBER
MOR 1 / MET 1 V T
MET 2 V T Tr
MOR 2 / MET 1 V T Tr
V T Tr
MOR 3 / MET 1 V T
MET 2 V T Tr
EMCM / MOR 2 V T C
V = Velocity
T = Temperature
C = Conductivity
Tr = Transmissivity
All data recorded
Erratic data
Direction only recorded

SHELFEDGE DYNAMICS AND THE NEPHELOID LAYER

C
DATA INVENTORY FOR CURRENT METER ARRAYS: OCTOBER 1980 THROUGH JULY 1981

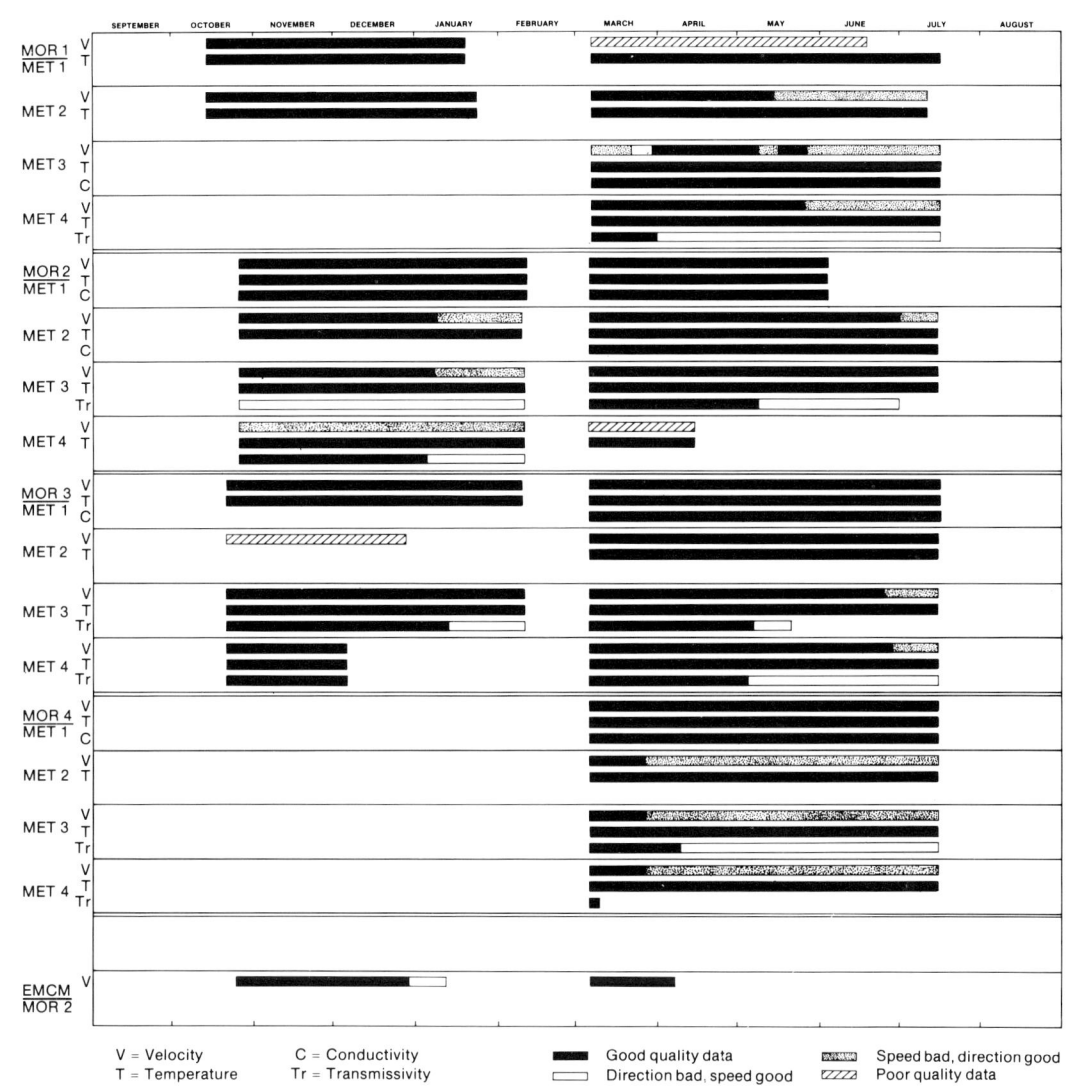

	V = Velocity	C = Conductivity	■ Good quality data	▨ Speed bad, direction good
	T = Temperature	Tr = Transmissivity	☐ Direction bad, speed good	▨ Poor quality data

FIG. 3.—*A–C*, Data inventory for current meter deployments during the period January 1979 through July 1981.

records. After it was determined that such very high frequency waves contribute negligibly to the fluid motions, the lower frequency sampling rate was used to extend the battery life in the instruments and reduce data reduction costs.

In Situ Boundary Layer Studies

In order to study the dynamics of the boundary layer in the vicinity of the East and West Flower Garden Banks, a dye emission system was developed. In its final form it consisted of a single point tautline mooring secured to its anchor by a quick-release snap hook. On the mooring were four small plexiglass canisters containing fluorescent dye, a timing circuit, and a battery pack. A small hole bored in the bottom of the canister served as the orifice for the dye. In the middle of the canister was a small diameter steel rod wound with copper wire. It ran the vertical length of the canister and acted as an electromagnet when the copper wire was charged by the timing circuit. Stretched over the open top of the canister was a rubber membrane

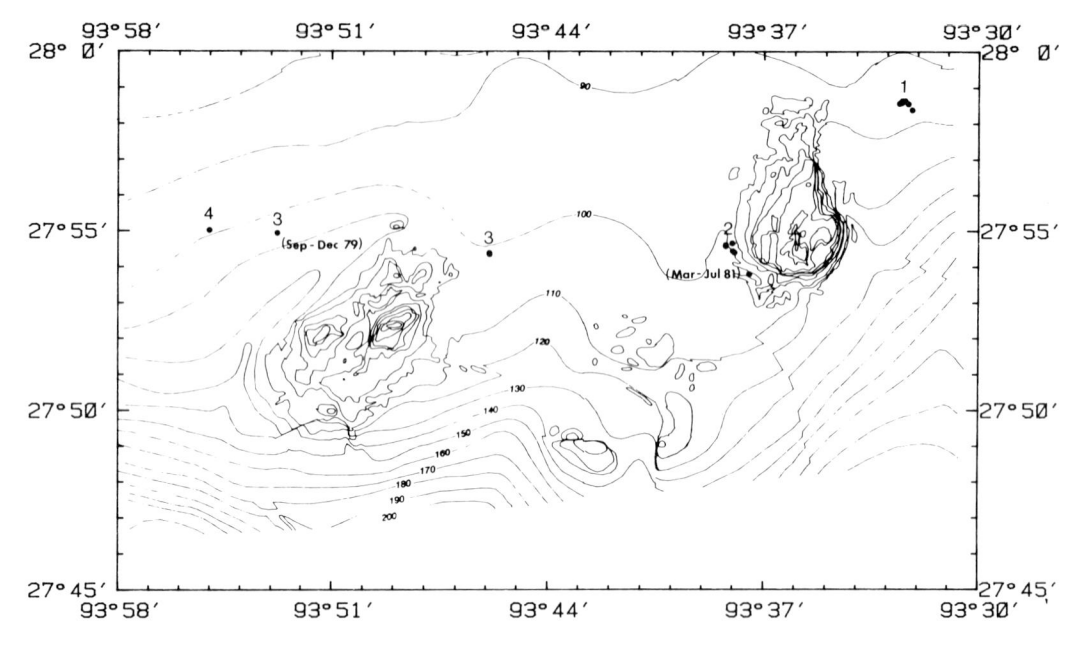

FIG. 4.—Locations of current meter moorings 1, 2, 3, and 4 for all deployments from January 1979 through July 1981.

with a small metal disk in the middle. Charging the electromagnet attracted the metal disk, causing the membrane to depress, thereby forcing out a pulse of dye. When the power to the magnet was turned off, the membrane relaxed, drawing in water for the next pulse.

Below each canister a thin rod with a paddle was mounted on the mooring wire so that it was free to rotate about the wire. The rod was marked with brightly colored stripes every 5 cm. These rods oriented themselves in the flow and served as a means of scaling the flow. The dye emittor arrays were lowered from the ship on bright yellow polypropylene line connected to the buoy by means of an acoustic release. Once the array was emplaced, the DRV DIAPHUS (a PC-14 two-man submersible) was launched. The yellow line was used to guide the submersible to the dye emittors on the bottom. When the observers were in place on the bottom in position to observe the dye, the acoustic release was triggered, separating the mooring from the surface and isolating it from motions induced on the surface buoy by surface gravity waves. The motions of the dye pulses were recorded from the submersible on both Super 8 mm color movies and video tape. The former are quantitatively analyzed by digitizing them on a back-projected digitizing table, and the latter by means of a video digitizer.

During the dye experiments, PHISH profiles were taken from the surface ship so that the motions of the nearbottom flow could be correlated with the conditions prevailing in the rest of the water column.

At the end of each experiment, the quick-release snap hook was opened by means of the mechanical arm on the submersible, and the dye emission package rose to the surface for recovery and reuse.

RESULTS

It is recalled that the Gulf of Mexico is a subtropical, microtidal (<0.5 m), small ocean basin with one of the world's largest rivers emptying into it. These attributes do not produce unique types of processes at the shelfbreak, but they do influence the relative importance of the various modes of motion present at the shelfedge. It is not possible, therefore, to extrapolate the results reported here to other regions without due consideration of the differences in geometry, latitude, and so forth.

Sedimentary processes are forced by fluid motions which have a variety of time and length scales, ranging from a few seconds (surface gravity waves) to seasonal variations. In the interest of simplicity and coherence, the results of this study are described here as a hierarchy of processes bearing ascending scales of length and time. It should be remembered that these processes coexist in time and space (cf. Southard and Stanley, 1976).

Surface Gravity Waves

At the high frequency end of the spectrum are surface gravity waves. In many areas, long period surface gravity waves can produce significant oscillatory velocities at the shelfbreak (see Komar et al., 1972, for example). However, in the Gulf of Mexico wave development is rather fetch limited and large-amplitude long period waves are rare. Bretschneider and Gaul (1956) using hindcast methods for the years 1950, 1952 and 1954 suggested that the most frequently occurring waves in the area of investigation were between 1 and 2 m height and 5 to 6 second periods. Waves of 5 to 6 second periods occurred an average of 23% of the time. Long period waves (>10 seconds) were reported to have existed only 22 hours per year, or 0.25% of the time. The largest wave they suggested for the study area was a 6 m wave with a 10 second period. Conditions suitable for formation of such a wave existed for only 4 hours per year.

During most of the year, the northwestern Gulf of Mexico is subject to rather gentle southeasterly winds which produce waves of only 1 m to 1.5 m height and 5 to 6 sec period as stated above. These, of course, produce no detectable motions at the depth of the shelfbreak. Starting in late September or early October, cold fronts begin to push out over the northern Gulf. With the approach of the cold front, the winds from the south increase in speed rapidly, frequently producing waves of 3 to 4 m height. As the front passes, the wind precipitously shifts to the north. Then, the wind returns to the gentle southerly flow. Some rare frontal passages, usually in mid-winter, are intense enough to raise waves of up to 7 m in height. It is only these super storms and occasional hurricanes which produce waves of sufficient amplitude and period to impinge upon the bottom at depths as great as 100 m. However, even a 10 m wave with a 12 sec period (a "monster" wave for the Gulf of Mexico) produces an oscillatory velocity of approximately 12 cm/sec at the shelfbreak depth of 135 m, according to Airy Wave theory (see Kinsman, 1965, for example).

It must be concluded from historical observations and theory that surface gravity waves do not contribute significantly, in a direct manner, to sedimentary processes in the shelfbreak region in the northern Gulf of Mexico.

Internal Gravity Waves, Tides, and Inertial Waves

From our earliest observations, it was thought that relatively high-frequency internal waves (those with periods of less than one hour) might contribute considerable energy to the bottom boundary layer on the outer shelf. Long term current meter records did not bear out this hypothesis. Only those oscillations with diurnal and longer periods appeared to produce sufficiently high velocities to induce significant sediment resuspension on the outer Texas shelf.

From late March through late November, the shelf waters in the northern Gulf of Mexico are sufficiently stratified to support rather energetic internal waves. These waves are analogous to wind driven waves except that they exist on density discontinuities within the water column rather than on the surface, hence the name internal waves. Their phase speed is much slower than that of a comparable surface wave because it depends on a gravity term adjusted by the density difference across the interface, the depth of water, and the relative depth of the discontinuity.

The highest frequency sustainable for the internal waves is the natural or Brunt-Vaisala frequency (N), which is a function of the vertical density gradient. The Brunt-Vaisala frequency is given by:

$$N = \left(-\frac{g}{\rho_o} \frac{\partial \rho}{\partial z} \right)^{1/2}$$

where ρ_o is the mean or reference density, z is the vertical coordinate (positive up), and g is the gravitational acceleration.

N may exceed 10 cph (cycles per hour) in the summer months and late fall over the outer continental shelf in the Gulf of Mexico. It decreases to 1 cph or less during the minimal stratification of the winter.

Internal waves, like their surface counterparts, are generated by tangential shear across a density discontinuity or by impulsive forces applied to the sea surface. The field observations were designed to use two methods for measuring the contribution from internal waves to the velocity field. First, the moored current meters recorded both temperature and velocity. Since internal waves deform isothermal surfaces during their passage, co-oscillation of the velocity and temperature with a phase shift between them could be attributed to the passage of an internal wave. Second, 12-hour anchor stations

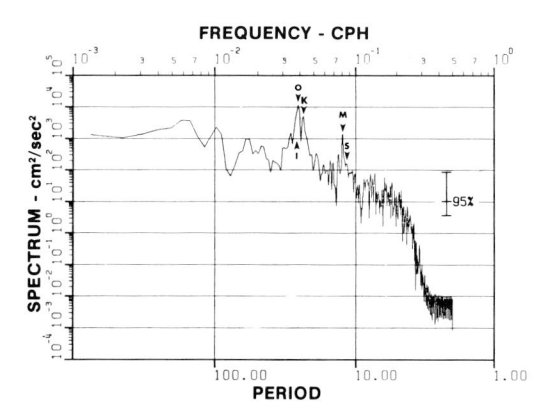

FIG. 5.—A typical autospectrum of the v (north-south) velocity component for winter. I marks the inertial period; O is the O_1, K is the K_1, M is the M_2 and S is the S_2 tidal constituents.

were occupied during submersible operations at which vertical profiles of temperature, salinity, and velocity were taken hourly during the dye emission studies.

Even with the sampling rate set at one every six minutes, no significant high frequency internal waves were observed in the moored current meter records. High frequency, in this case, is defined as greater than 0.125 cph.

More substantial internal waves were observed at the semi-diurnal and lower frequencies. Those at the diurnal frequency are attributable to both tides and inertial motions. The inertial period is given by

$$T_I = 2\pi f^{-1}$$

and

T_I is the inertial period
f is the Coriolis parameter $2\omega \sin\theta$
ω is the angular velocity of the earth in radians/sec (7.29×10^{-5} sec^{-1})
θ is the latitude.

Periods and frequencies for the major tidal constituents and inertial oscillations are shown in Table 1. Note that the O_1 constituent and the inertial frequency at the study site are separated by only 4×10^{-4} cph. It is therefore impossible to separate the inertial oscillations from the O_1 tidal oscillations, and only marginally possible with the K_1 tide.

Observations from the submersible during the passage of the crest of an internal wave with an apparent diurnal period revealed its effect on the fine sediment forming the substrate at the shelf-edge. This location was on the southwest side of a carbonate bank in about 80 m depth. A dye emission experiment was in progress at the time, and flow was toward the north. At the beginning of the experiment, the nearbottom velocity profile (as measured from the motion of the dye) fit the well known law of the wall with a logarithmic decrease in speed toward the bottom. Over the period of about an hour a nearbottom jet developed so that the flow within 1 m of the bottom was 5 to 10 cm/sec faster than that at 2 m above the bottom. This jet eroded fine sediment from the bottom and rapidly suspended it well above the bottom because of the sharply increased turbulence. The thickness of the jet grew and so did the turbulence. At last, the flow suddenly became vertical, rising at 10 cm/sec for approximately one minute. That would correspond to a vertical displacement of about 6 m. Naturally, the sediment suspended during the rapid nearbottom flow was carried upward during the vertical flow sequence. At the cessation of the vertical flow, the flow reversed with a turn to the south and development of a normal boundary layer velocity profile. The water remained turbid as the suspended sediment was carried back past the observation point.

Amplitude estimates of the u (east-west) and v (north-south) components of velocity for the K_1 (Luni-solar diurnal) and the M_2 (principal lunar) tidal currents were computed for the period 6 March to 16 July 1981 and are shown in Table 2. It is clear from this data that tidal currents at the shelfbreak off east Texas are negligible. Contributions from harmonics at 8- and 6-hour periods may be seen in the spectra (Figs. 5, 6) but amount to less than 1 cm/sec at the bottom.

Figure 7 shows a typical current meter record. The meter was at a depth of 58 m on a mooring to the northeast of the West Flower Garden Bank (see Fig. 4). The records were digitally filtered to reveal the various scales of motion, the tide, and particularly inertial oscillations. The 3-hour low-pass record is nearly raw data; only variations possessing times scales of 3 hours or less have been removed. The 28-hour low-pass filtered record shows the low frequency signal in the record. The band pass record is that portion containing oscillations with periods lying between 3 and 28 hours. The top panel

TABLE 1.—THEORETICAL FREQUENCY AND PERIOD OF MAJOR TIDAL CONSTITUENTS AND OF INERTIAL CURRENTS

Darwin name of Tidal Harmonic	Period (hrs)	Frequency (cycles/hour)
S_2	12.00	.0833
M_2	12.42	.0805
K_1	23.93	.04178
O_1	25.82	.03872
Inertial Period (27°55′N lat.)	25.56	.03912

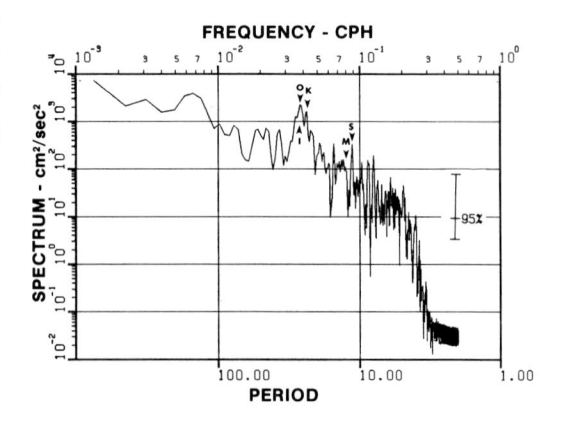

FIG. 6.—The autospectrum for the u (east-west) velocity component for the same current meter record as that shown in Fig. 5. Symbols are the same as those for Fig. 5.

TABLE 2.—TIDAL CURRENT AMPLITUDES IN CM/SEC AT MOORING 3 FOR 6 MARCH TO 16 JULY 1981

Meter #	Depth (m)	K_1 u	K_1 v	M_2 u	M_2 v
1	53	3.2	4.0	1.3	2.1
2	64	3.9	5.2	0.9	2.4
3	91	1.8	2.7	0.8	1.5
4	97	1.4	2.5	0.8	1.6

is the temperature record for the same period.

The presence and magnitude of inertial oscillations can be seen in the band passed record during the period from 10 to 22 May. The vectors rotate in a clockwise sense producing maximum cross-shelf speeds of 20 cm/sec to 40 cm/sec. These inertial currents may arise during spin down of some impulsively driven current (Neumann and Pierson, 1966). That is, when the wind ceases to drive a current, the balance of forces shifts to one between the Coriolis and centrifugal forces. The current turns to the right (northern hemisphere) under the Coriolis acceleration which becomes balanced by the centrifugal acceleration developed by the circular motion. Krauss (1973) points out that inertial oscillations are also produced during the spin-up process as the ocean currents come into equilibrium with the impulsively applied wind. In the latter case the currents oscillate about the equilibrium speed and direction at the inertial frequency. Under this condition the momentum penetrates the water column vertically in inertial oscillations much more efficiently than in increases of the mean speed.

Notice that the record of the meter only 4 m from the bottom (100 m depth) on the same mooring (Fig. 8) is dominated by inertial oscillation during the periods of strongest inertial oscillations in the 10 to 22 May period.

Both records possess low frequency modulations of the mean current that have a period of about 2 and 4 days. These may be quasigeostrophic shelf waves or a direct response to nearly periodic atmospheric forcing. The mean velocity for the meter at 58 m depth (Fig. 7) over this whole record was 18.4 cm/sec at 99.5°T. The mean velocity for the lower meter for the same period was 5.4 cm/sec at 149°T. Speeds at 4 m above the bottom exceeded 10 cm/sec slightly more than 16% of the time and reached a maximum of 30 cm/sec between late April and mid-June 1980.

Transmissivity was also recorded by the lower meter (Fig. 8) on this mooring. From this record it is apparent that the amount of sediment in the water at this height is poorly correlated with current speed. In fact, the lowest transmissivity appears to occur during periods of relatively low speed on-shore (northerly) flow. This would suggest that the greatest changes in suspended sediment concentrations are due to advective processes rather than local resuspension. On the other hand, it is obvious that the near-bottom velocities are sufficient to resuspend silt and clay most of the time, particularly during the inertial oscillations. It may well be that settling is sufficiently slow so that the water does not clear appreciably during the brief periods of low velocity flow. At any rate, it is certain that the strongest flow near the bottom is cross isobathyal and that during a substantial portion of the time sediment-laden water is running out across the shelfbreak.

Fall and winter conditions on the same mooring at a depth of 61 m are shown in Figure 9. The early portion of the record is dominated by very strong

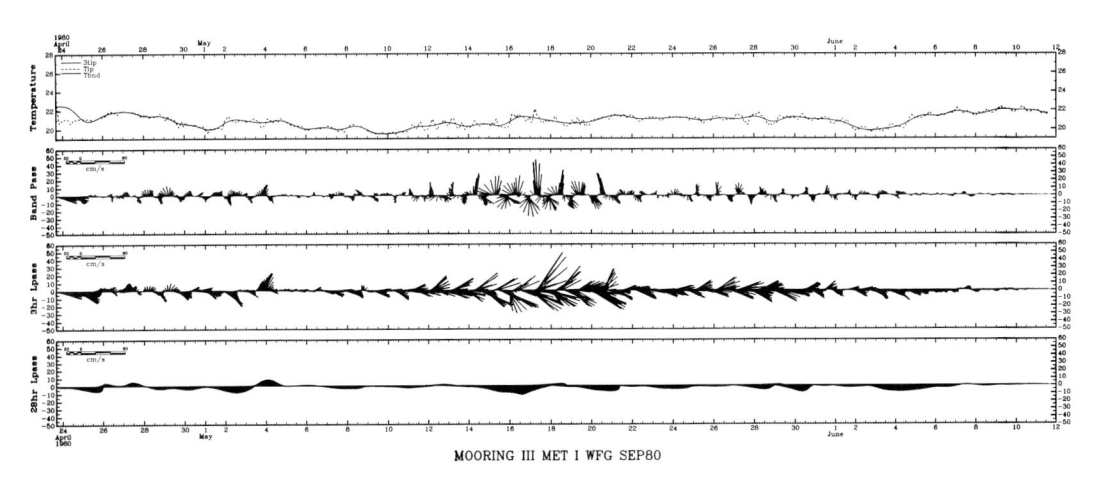

FIG. 7.—Velocity and temperature record for mooring 3, meter 1, which was at 58 m depth. Bandpass record contains only those signals having period of 3 to 28 hours. See Fig. 4 for location.

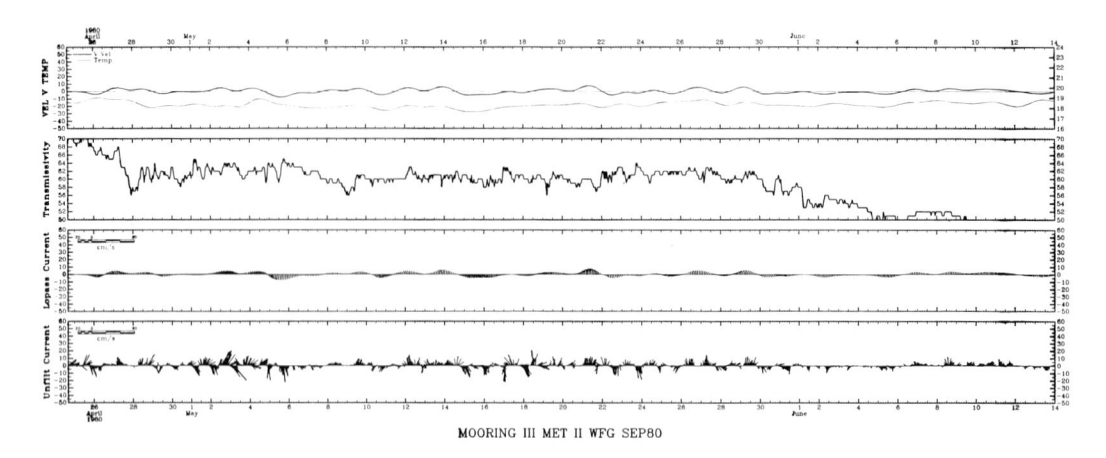

FIG. 8.—Velocity, temperature and transmissivity record for mooring 3, meter 2, which was a depth of 100 m or 4 m from the bottom. The lowpass records of velocity and temperature contain signals having periods of 28 hours or greater. See Fig. 4 for location.

inertial oscillations. This occurred during rather steady winds from the south. The first cold frontal passages took place during early October. This corresponds to the development of strong easterly flow modulated by cross-shelf oscillations that are well correlated with oscillations of the winds from southerly to northerly during the frontal passages.

It appears, therefore, that in the wake of storms or other major events, the inertial oscillations can induce significant near-bottom velocities capable of resuspending sediment and making it available to advective transport by the lower frequency currents.

Because the inertial currents are rotary, resuspended sediment can be swept over the shelfbreak. Differences in topography and substrate type in the alongshore direction lead to the development of turbid plumes which flow out along density interfaces during offshore flow; then the plumes become entrained in currents along the slope. This appears to be an important mechanism for sweeping fine sediment from the shelfedge and introducing it into the slope environment.

Storm Surge and Free Shelf Waves

Oscillations with periods of 2 to 4 days and 3 to 5 days were found to be very important in shelfedge processes (Figs. 7–9). These lower frequency current oscillations contribute to both resuspension and offshore transport of fine sediment at the shelfbreak.

As mentioned with reference to surface gravity waves, late fall and winter in the Gulf of Mexico are characterized by the quasi-periodic arrival of

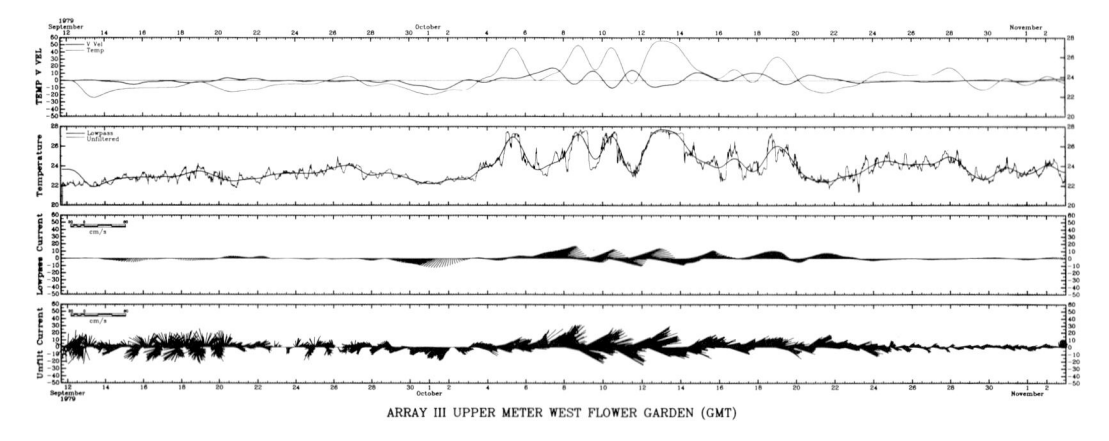

FIG. 9.—Velocity and temperature record for mooring 3, meter 1. The instrument was located at a depth of 61 m. See Fig. 4 for location. Lowpass record has 28 hour cutoff at the high frequency end of the filter.

cold fronts known as "blue northers" or just northers. These atmospheric disturbances have periods of from 3 to 5 days. The approach of the front is heralded by very strong southerly winds that drive the warm saline surface waters shoreward and homogenize them near the coast. As the front passes, the winds shift to the north, bringing cold dry air and clear skies over the water. This leads to a rapid drop in temperature of the nearshore surface waters through evaporation, convection, and radiant heat loss. The evaporation also causes an increase in nearshore salinity.

Nowlin and Parker (1974) have documented the response of the shelf waters to a cold air outbreak, and they suggest that it leads to the production of water with the T-S signature of intermediate depth water in the Gulf. The implication is that this water forms near the coast, then sinks and flows offshore because of its increased density. This happens during the time that waves and currents are at their maximum in the nearshore, so that quantities of fine sediment are entrained in the dense water. When this dense turbid water reaches the shelfbreak it remains attached to the bottom and spills down over the slope until it reaches a surface of equal density, then flows out as a detached plume from the bottom. This is not a continuous process of the winter regime. Rather, it occurs as pulses, the magnitude of each being proportional to the intensity of the generating cold front.

During March of 1981, PHISH stations were occupied along a transect from Galveston, Texas, to the upper slope south of the West Flower Garden Bank. The cross-section derived from these stations (Fig. 10) is illustrative of the various processes that occur in the aftermath of a frontal passage. Flow on the shelf is to the west in the interior with primarily offshore flow at the bottom. Bottom temperatures at station 28 and station 25 are less than 17°C, with salinities of approximately 36.4‰, suggesting that this is residual water formed on the inner shelf during cold air outbreaks just preceding the cruise. Because it is relatively dense, it flows obliquely off the shelf toward the southwest. The plume of turbid water extending out to station 20 implies that a pulse of 17°C shelf water ran off the shelf and became entrained in the slope flow somewhere to the west (upstream) of the transect. The data set for station 20 is shown in Figure 11. Notice that the near-bottom flow at this station is to the northeast, or onshore, which is in accord with the sharp up turn of the isotherms at the shelfbreak.

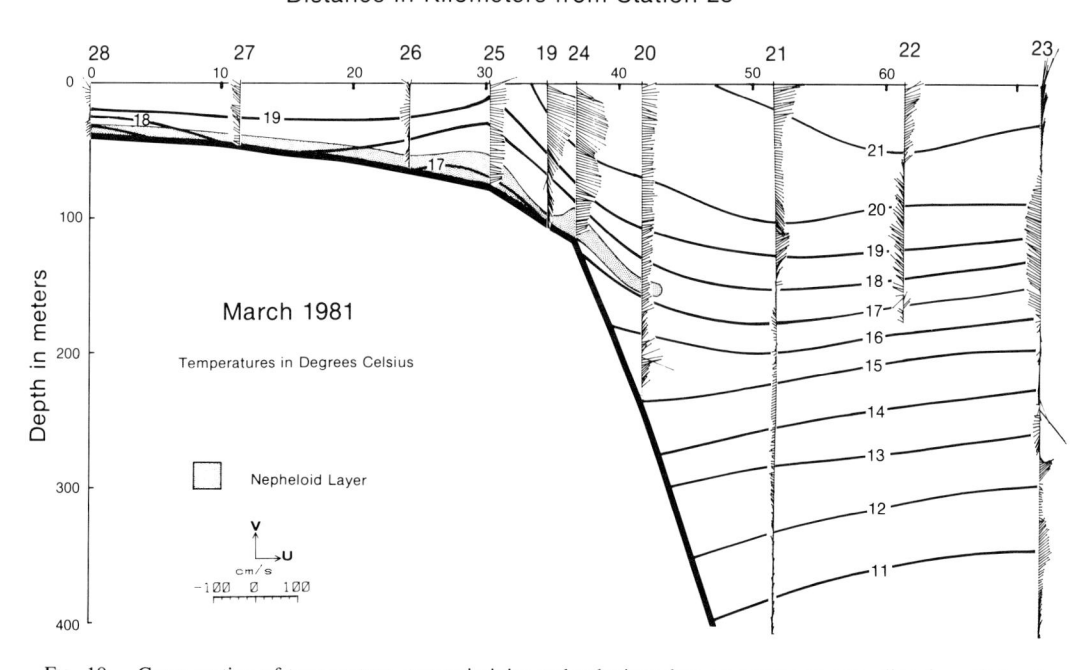

FIG. 10.—Cross-section of temperature, transmissivity and velocity taken on a transect extending from Galveston, Texas to the West Flower Garden Bank, then due south to the 1000 m isobath. See Fig. 2 for location of reference points. The line has been projected onto a north-south line which compresses the actual distance between stations 28, 27, 26, and 25. The velocity plots are made so that the origin of the vector is at the depth of observation, the length is proportional to the speed and it points in the direction of flow, assuming north is up and east is to the right. The nepheloid layer is defined here by transmissivities of 45%/m or less.

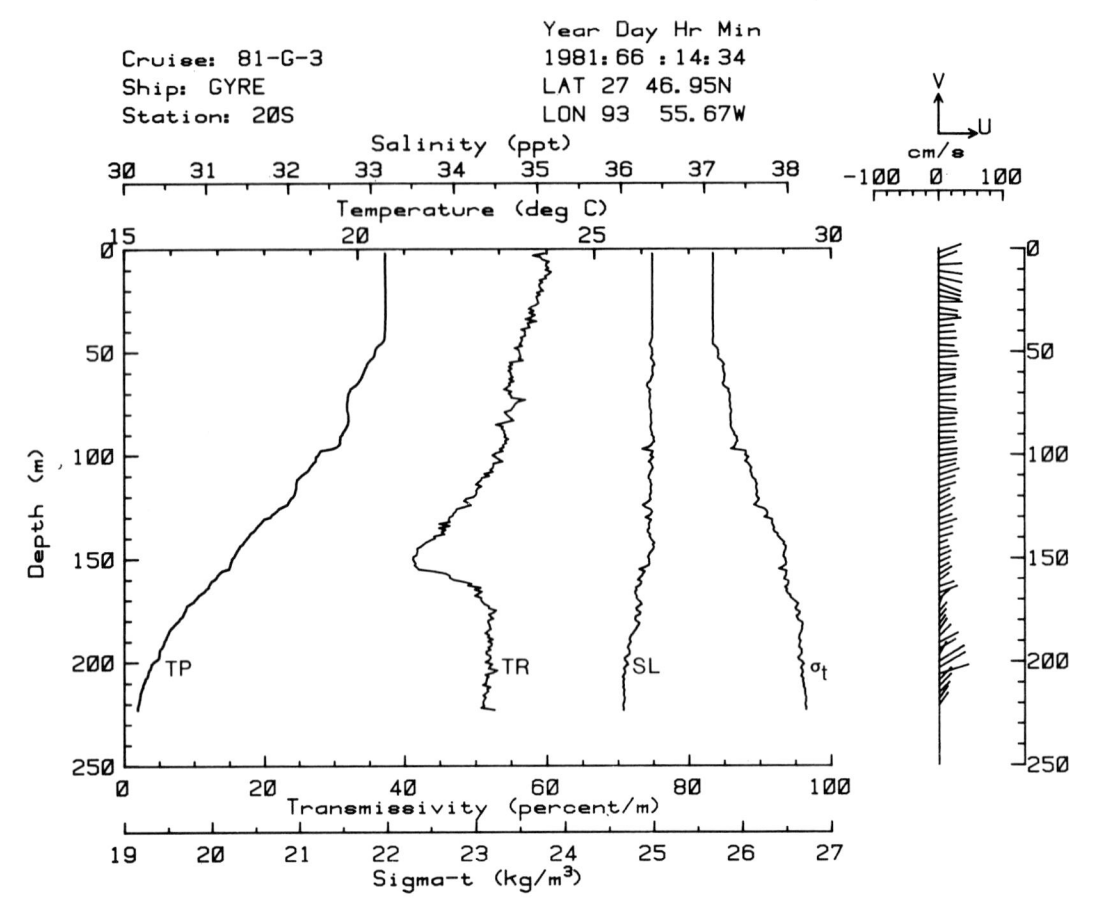

FIG. 11.—Station 20 from the March 1981 transect showing the type of information collected by means of PHISH system and the plume of turbid water detached from the shelf.

The plume appears at the decrease of transmissivity between 110 m and 150 m. From station 25 offshore to the upper portions of station 22 the flow is to the east. Velocities in the core of the current exceed 100 cm/sec.

The entire transect required approximately 14 hours to run so that it is only quasisynoptic. It must be remembered that each station is essentially a snapshot containing both the short-term mean flow and a variety of higher frequency modulations. The same cross-shelf transect as shown in Figure 10 was reoccupied on the 17th of March 1982 following a protracted period of southerly winds. The winds had subsided at the time of the crossing. Every station from 30 m depth to approximately 100 m possessed rather strong flow directed almost due south. Flow over the 100 m to about 400 m depth was to the east. Transmissivity in the entire water column below about 90 m was less than 40%/m, and a core of water possessing transmissivities of less than 30%/m extended off the shelfbreak for a distance of approximately 5 km.

In addition to the somewhat indirect effect of the cold air outbreaks, there is a direct forcing at the shelfedge. These disturbances induce strong cross-shelf oscillations in the predominantly alongshore interior flow that are coupled in frequency with the forcing. This energy propagates down to the bottom as strong inertial oscillations, with speeds of 15 to 25 cm/sec superimposed on weak mean flows directed obliquely on or offshore. Obviously, when the bottom flow is directed offshore it sweeps suspended sediment off the shelf and onto the slope and rise.

Following these strong flow events, bursts of inertial oscillations continue for 3 to 5 cycles. It is assumed that these latter represent the decay of the wind forced flow after the wind dies down.

The effects of hurricanes are indistinguishable in the current meter records from the direct forcing of the cold frontal passages.

Spectra from the current meter records and low-pass filtered plots of the records show the existence of the before-mentioned current oscillations with periods of 48 hours and 96 hours (Figs. 5, 6). At moorings far from the bank, the highest velocities in the upper portion of the water column are nearly parallel to the local isobaths. Even there, however, the nearbottom oscillations appear to be oriented in a more cross-shelf direction. These oscillations, therefore, also contribute to the offshore flux of fine sediment across the shelfbreak.

Aperiodic and Longterm Mean Flow

The existence of rings broken off from the Gulf Loop Current was brought to the author's attention by Dr. Takashi Ichiye when he requested that we take observations in the edge of a ring that lay in our trackline along the shelfedge. Because of the high particle velocities associated with these rings and their large size, they also appear to be capable of contributing to the offshore transport of silt and clay.

Elliott (1979) has done an extensive analysis of large-scale rings (radii approximately 250 km to 400 km) which break off the Loop Current after it enters the Gulf of Mexico through the Straits of Yucatan. These are analogous to Gulf Stream rings of the Atlantic Ocean. Elliott (1979) reported that between one and three rings form each year and migrate westward into the northwestern Gulf of Mexico. Some of these rings attach themselves to the slope just south of the Mississippi Delta, then propagate to the west at a rate of 2 to 4 km/day.

It has not been possible to separate or identify, with certainty, the passage of a ring in our data thus far. Since they are anticyclonic eddies, they should show up in the current meter records, first as a shift of the flow toward the northeast, then a rotation to the east, and finally a rotation to the southeast or obliquely offshore.

With persistent southeasterly winds over the northwestern Gulf of Mexico, we would expect that the surface waters should move onshore, setting up a pressure gradient that would drive mid-depth geostrophic flow toward the west. This, in turn, would create a bottom Ekman layer with flow toward the southwest or obliquely offshore. This appears to be the case inshore of the 70 m isobath (Fig. 10). At the shelfbreak, however, the mid-depth flow is persistently from west to east. That should set up an Ekman bottom boundary layer flow to the northeast or onshore. However, flow near the bottom at the shelfbreak varies from primarily southeasterly to northerly. The longterm mean flow is directed toward the southeast at the near bottom but with a speed of only 3 to 7 cm/sec. Therefore, the advection of suspended sediment along the shelfedge is oriented offshore and into the basin.

CONCLUSIONS

The waters of the shelfbreak are subject to motions with many time and length scales. These are superimposed on one another, producing energetic currents which vary in time and space. Many of these motions produce oscillations in the flow that are directed across the trend of the isobaths and sweep fine sediment off the shelf and into deeper waters. The flow is frequently of such magnitude that it could move sand if it were available at the shelfedge. These processes provide a very efficient mechanism for supplying quantities of sediment to the slope, rise, and deep sea. Similar processes should exist world-wide, but they will possess different periods and relative magnitudes.

ACKNOWLEDGMENTS

Acquisition and analysis of the data used in this report has been a group effort. James Stasny designed the electronics for PHISH and built it; in addition, he nursed and cursed our current meters into reliable operation and kept all of our electronic equipment working. Doyle Horne wrote all of the software for the PHISH system, designed the current meter arrays, and participated in all of the data acquisition. My students—David Huff, J. Stacy Jenkins, Fern Halper, Lauren Sahl, Jeffrey Hawkins, and Thomas Cecil—have been the backbone of the project. Rose Norman kindly edited the manucript and shepherded the graphics into existence, and Karen Dorman rendered the scribbled manuscript into typed form. This work was funded through the New Orleans Office of the Bureau of Land Management, Department of the Interior.

REFERENCES

BRETSCHNEIDER, C. L., AND GAUL, R. D., 1956, Wave Statistics for the Gulf of Mexico off Caplen, Texas: U.S. Army Corps of Engineers, Beach Erosion Board, Tech. Memo #86, 25 p.

BRIGHT, T. J., AND REZAK, R., 1976, A Biological and Geological Reconnaissance of Selected Topographical Features on the Texas Continental Shelf: Final Rept. to U.S. Dept. of Interior, Bur. Land Management, Contract #08550-CT5-4, 377 p.

———, AND ———, 1977, Reconnaissance of reefs and fishing banks of the Texas Continental Shelf, *in* Geyer, R. A., ed., Submersibles and Their Use in Oceanography: New York, Elsevier, p. 113–150.

———, AND ———, 1978, Current measurements and dye diffusion studies, *in* Bright, T. J., and Rezak, R., Northwestern Gulf of Mexico Topographic Features Study: Final Rept. to U.S. Dept. of Interior, Bur. Land Management, Contract #AA550-CT7-15, 629 p.

ELLIOTT, B. A., 1979, Anticyclonic rings and energetics of the circulation of the Gulf of Mexico [Ph.D. diss.]: College Station, Texas, Texas A&M Univ., 188 p.

KINSMAN, B., 1965, Wind Waves: Englewood Cliffs, N.J, Prentice-Hall, 676 p.

KOMAR, P. D., NEUDECK, R. H., AND KULM, L. D., 1972, Observations and significance of deep-water oscillatory ripple marks on the Oregon Continental Shelf, *in* Swift, D. J. P., Duane, D. B., and Pilkey, O. H., eds., Shelf Sediment Transport: Process and Pattern: Stroudsburg, Pa., Dowden, Hutchinson and Ross, p. 601–620.

KRAUSS, W., 1973, Methods and Results of Theoretical Oceanography: Berlin, Gebruder Borntraeger, 302 p.

MCGRAIL, D. W., 1978, Hydrography and suspended sediments, *in* Bright, T. J., and Rezak, R., eds., South Texas Topographic Features Study: Final Rept. to U.S. Dept. of Interior, Bur. Land Management, Contract #AA550-CT6-18, 772 p.

———, AND HORNE, D., 1981, Water and sediment dynamics [Flower Garden Banks], *in* Rezak, R., and Bright, T. J., eds., Northern Gulf of Mexico Topographic Features Study, Final Report to U.S. Bureau of Land Management, Contract #AA551-CT8-35: College Station, Texas, Texas A&M Tech. Rept. #81-2-T, v. 3, Part B, p. 9–45.

NEUMANN, G., AND PIERSON, W. J., JR., 1966, Principles of Physical Oceanography: Englewood Cliffs, New Jersey, Prentice-Hall, 506 p.

NOWLIN, W. D., JR., AND PARKER, C. A., 1974, Effects of a cold-air outbreak on shelf waters of the Gulf of Mexico: Jour. Phys. Oceanogr., v. 4, p. 467–486.

REZAK, R., AND BRIGHT, T. J., 1981, Northern Gulf of Mexico Topographic Features Study: Final Rept. to U.S. Dept. of Interior, Bur. Land Management, Contract #AA551-CT8-35, 901 p.

SOUTHARD, J. B., AND STANLEY, D. J., 1976, Shelf break processes and sedimentation, *in* Stanley, D. J., and Swift, D. J. P., eds., Marine Sediment Transport and Environmental Management: New York, John Wiley and Sons, p. 351–377.

MODERN SEDIMENT DYNAMICS AT THE SHELF-SLOPE BOUNDARY OFF NOVA SCOTIA

PHILIP R. HILL[1] AND ANTHONY J. BOWEN
Departments of Geology and Oceanography, Dalhousie University, Halifax,
Nova Scotia, B3H 3J5, Canada

ABSTRACT

Long term current-meter data from outer shelf, shelfbreak and slope sites off Nova Scotia have been compared with sediment textures in the same area to assess whether they are in equilibrium. Currents on the shelf and shelfbreak are strong with maximum velocities exceeding 50 cm s^{-1}. Sediment grain-size distributions were dissected into near-Gaussian medium sand subpopulations and non-Gaussian tails. These subpopulations were interpreted dynamically as representing bed-load (coarse tail), suspended load with "dynamic settling" (central subpopulation) and suspended load with "passive settling" (fine tail). Below 500 m water depth, only the fine-tail subpopulation is seen. The modal size of the central subpopulation corresponded well, in most cases, to u$_*$ estimates from Shields' criterion and to the observed maximum currents.

Sediment textures can be explained by modern dynamic conditions. Sand transport is dominantly in suspension and in an alongslope direction with a small downslope component. Medium sand is transported only during short periods of high flow, whereas fine sand transport is during more continuous weaker flow. Permanent deposition occurs at a point downslope where the currents rarely exceed the suspension criterion for the size of particle concerned. Slight differences between inferred u$_*$ gradients at two slope areas, separated by 150 km, are tentatively interpreted as reflecting the effects of topographic Rossby waves, formed by Gulf Stream eddies impinging on the slope.

INTRODUCTION

Sediment transfer across the shelf-slope boundary is generally poorly understood. Qualitative models exist for slopewards transport of both fine-grained sediments (McCave, 1972) and sand-sized material (Stanley et al., 1972), but there has been little success in developing quantitative models. Thus, it has been impossible to determine which of the many possible processes (e.g. waves, currents, tides, internal waves) are important for sediment transfer across the shelfbreak. Undoubtedly, the relative importance of each process changes with the setting, so that different shelfbreak zones must each be assessed independently.

In this paper, we have attempted an analysis of quantitative hydrographic and sediment textural data, to determine first order transport and depositional patterns and processes. The major aim was to decide whether it is necessary to invoke catastrophic processes, such as turbidity currents generated at the shelfbreak, or catastrophic events, such as very large storms with geologically significant recurrence intervals, to explain the observed textures, or whether the textures are consistent with transport by "normal" events as observed on relatively short-term current records.

SETTING

Off Nova Scotia, the continental shelf is approximately 200 km wide. The outer shelf comprises broad banks, of water depth generally less than 150 meters, separated by deeper saddle areas (Fig. 1). The shelfbreak occurs at varying depths from about 100 m off Sable Island Bank to 250 m off a prominent saddle area named (informally) the Scotian Gulf (Fig. 1). King (1970) recognized two shelf-edge surficial sand and gravel lithologies. On the banks, relatively well-sorted sands and gravels were found, while saddle areas were covered with relatively poorly-sorted sediments of similar mean size. King attributed these differences to shoreline winnowing of the shallower bank sediments during postglacial transgression. Upper-slope sands were grouped by King with the relatively poorly-sorted sand and gravel unit.

Water circulation on the shelf is complex, with contributions from Gulf of St. Lawrence and Slope Water sources. Important seasonal incursions of slope water onto the shelf have been noted by Houghton et al. (1978). The Slope Water is a mixture of Labrador Current Water, North Atlantic Central Water and Gulf Stream Water (Gatien, 1975). Local conditions within the Slope Water are therefore strongly influenced by fluctuation of the Labrador Current and Gulf Stream.

METHODS

Since December 1975, a program of continuous current monitoring has been in progress on the

---

[1] Present address: Atlantic Geoscience Centre, Bedford Institute of Oceanography, P.O. Box 1006, Dartmouth, Nova Scotia B2Y 4A2 Canada.

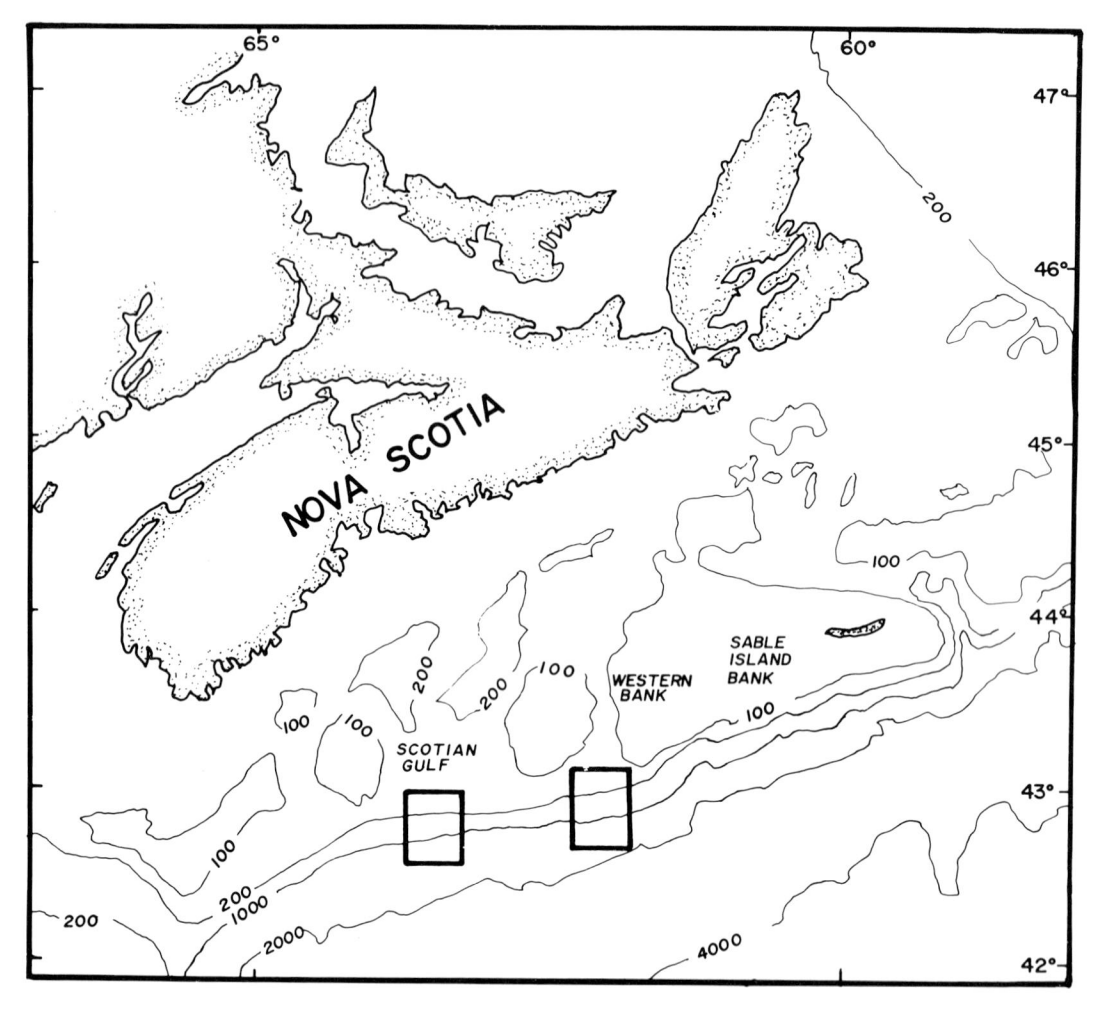

FIG. 1.—Chart of the Nova Scotian continental margin, showing two study areas detailed in Fig. 2. Bathymetry in meters.

Nova Scotian shelf and slope by B. Petrie and P. C. Smith of the Bedford Institute of Oceanography. We have access to data covering the two year period, December 1975 to January 1978. Meter arrays were deployed at shelf, shelfbreak and slope locations in the Scotian Gulf area (Fig. 2) with individual meters at near-surface, near-bottom and various intermediate positions (Lively, 1979). Contiguous temperature and conductivity (salinity) readings were taken and all data were averaged over one hour periods. Only data from the bottom current meters are presented here as they pertain to bottom sediment movement. Basic mooring data for the three sites are shown in Table 1. The current meters were deployed at approximately 20 m above the seafloor, which in all cases was within the bottom mixed-layer (Table 1), as determined from CTD profiles.

Bottom sediment samples were taken with Van Veen and Shipek grabs from a ship located by Loran C and satellite navigation in two areas of the Nova Scotian margin: (a) off the Scotian Gulf, and (b) off Western Bank (Figs. 1, 2). Below 500 m water depth in both areas, gravity-core top samples were used. Recovery from the grab samples was variable. Samples containing substantial amounts of gravel were often relatively small in volume, indicating only shallow penetration. Very small samples and other samples where washout was suspected were rejected on a qualitative basis, so that only samples which could be reasonably assumed representative of the bottom surficial sediment were used in the subsequent analysis. Even so, statistically significant amounts of the gravel fraction were not obtained. Grain size analyses were conducted on the sand fraction after wet sieving to remove silt

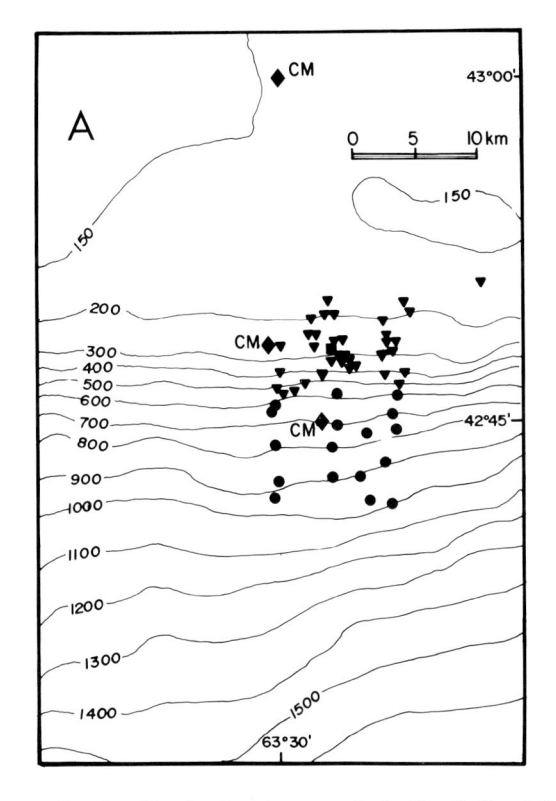

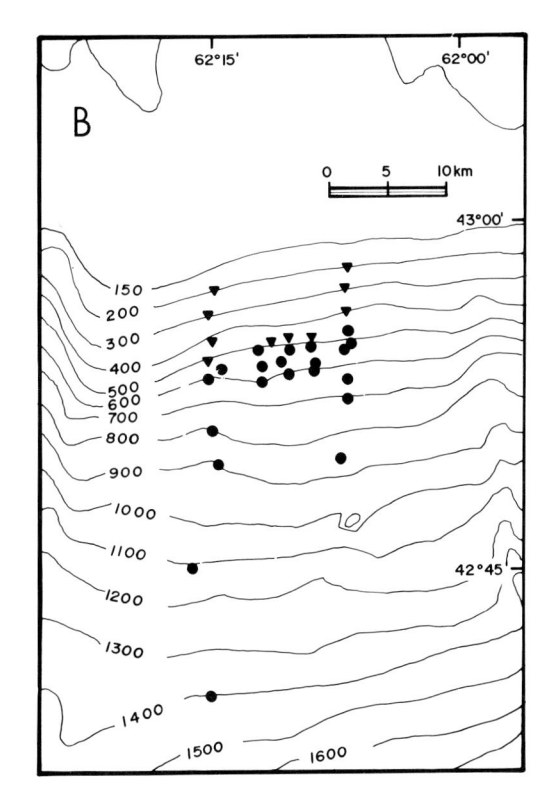

FIG. 2.—Details of study areas: *A*, Scotian Gulf and *B*, Western Bank. Triangles = grab-samples; dots = gravity and piston-cores; CM = current meter. Bathymetry in meters.

and clay particles. Standard, calibrated sieves at $^1/_4$ phi intervals were used. The fine fractions of muddy samples were analyzed by a pipette method.

CURRENT METER OBSERVATIONS

Shelf

The current meter located 20 m above the bottom in 170 m water (Fig. 2) shows a general eastward drift of bottom waters with a long period oscillation between northeastward and southerly daily motions. Petrie and Smith (1977) demonstrated that oscillations with a period of less than ten days correlate with wind stress variations measured on nearby Sable Island, and can be directly attributed to meteorological forcing. Longer period motions

do not correlate with wind stress events; their origin is thought to be related to the shoreward propagation of topographic Rossby waves (Petrie and Smith, 1977; Louis et al., 1983).

High current speeds are associated with wind forced events which occur predominantly in the winter months. Peak daily velocities reach values as high as 60 cm s^{-1} during the winter, but during summer, they rarely exceed 30 cm s^{-1} (Fig. 3). Some very large events appear on the record during the winters of 1976 and 1977 (Fig. 3) with current velocities exceeding 100 cm s^{-1} in a north to northeastward direction. The events are short-lived (generally 24 to 48 hours) but the high speed motions are maintained through at least 100 m of water column. These motions do correlate with slightly

TABLE 1.—MOORING DATA FOR THE THREE CURRENT-METER MOORINGS. THICKNESS OF BOTTOM MIXED-LAYER ESTIMATED FROM SALINITY AND TEMPERATURE PROFILES

Station	Latitude	Longitude	Water Depth	Instrument Depth	Thickness of Bottom Mixed-Layer
S6 Shelfbreak	42°48.0'N	63°30.0'W	250 m	230 m	60 m
S1 Shelf	42°48.0'N	63°30.0'W	250 m	230 m	60 m
S3 Shelf	42°45.0'N	63°30.0'W	710 m	690 m	~250 m

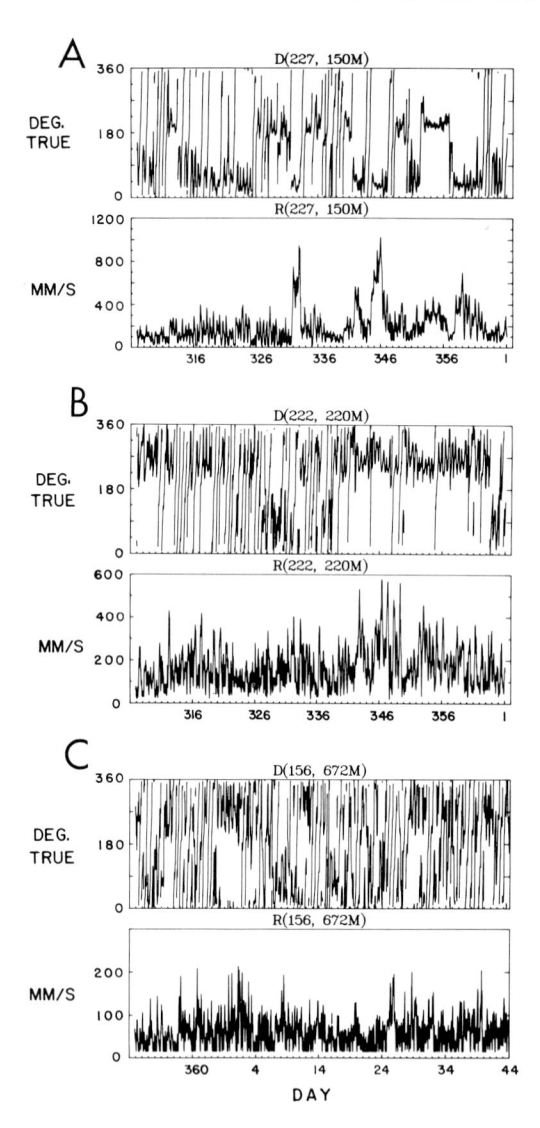

FIG. 3.—Current meter records for three stations shown in Fig. 2A. Readings were averaged at hourly intervals. A, Shelf; B, Shelfbreak; C, Slope. Courtesy of B. D. Petrie, Bedford Institute of Oceanography.

higher wind stresses; the water at lower depths moves onshore (northward) during westerly (eastwards) winds. They are probably related to the Scotian Gulf topography; a numerical model would be needed to resolve the problems since non-linear terms in the momentum balance appear to be significant (B. Petrie, pers. comm.).

Shelfbreak and Slope

Motions at the shelfbreak and on the slope are dominated by strong currents parallel to the contours (Fig. 3). Long period reversals of the current are typical, although net drift is to the west in the Scotian Gulf area as would be expected from the water mass distribution. Peak velocities of between 30 and 50 cm s^{-1} are characteristic of these strong motions at the shelfbreak and 15 to 30 cm s^{-1} on the slope at 700 m water depth (Fig. 3). Superimposed on the current drift is the effect of the semidiurnal (M2) tide. This causes most of the short period variability shown at all stations in Figure 3. During the long-period strong flows, the tide merely interferes to produce slight amplification and damping effects. When flow is minimal, the tide can dominate the water motions and current vectors follow an ellipsoidal pattern, similar to other tidally dominated seas (e.g. North Sea, see McCave, 1971). Current velocities during these periods are substantially lower, generally less than 20 cm s^{-1} at both locations.

The very strong wind-forced motions, observed on the shelf, do not correlate with higher bottom current velocities either at the shelfbreak or on the slope. However, strong near-surface currents flow to the south (offslope) at the shelfbreak during these events.

GRAIN-SIZE DISTRIBUTIONS

Textural Analysis

Various methods have been used to aid the interpretation of grain-size distributions (Blatt et al., 1980, p. 43 ff). Most make the assumption that a size distribution closely approximates a Gaussian distribution when plotted on a logarithmic size scale. Several moment measures around the distribution have been used to measure deviation from the Gaussian model, to outline trends and to distinguish environments (Folk, 1966). Other workers have used probability graph paper to distinguish several lognormal sub-populations within a single sample (Visher, 1969; Middleton, 1976). Still others have suggested that the distributions may be better described by combined logarithmic tails (Bagnold, 1937; Barndorff-Nielsen, 1977; Bagnold and Barndorff-Nielsen, 1980).

In the light of these differing opinions, great care is required when making quantitative comparisons between size distributions, as the parameters used may vary according to the assumptions made. The interpretations in this chapter are based on the following method of analysis. The size distributions were first plotted on Gaussian probability paper as cumulative frequency curves (Fig. 4). Most curves consist of three straight line segments, suggesting approximation to Gaussian subpopulations. However, in the tail regions, significant deviations from the lognormal distribution are often observed, which suggests that the Gaussian assumption may not be valid for at least part of the distribution. This is no surprise as some depositional models do not predict

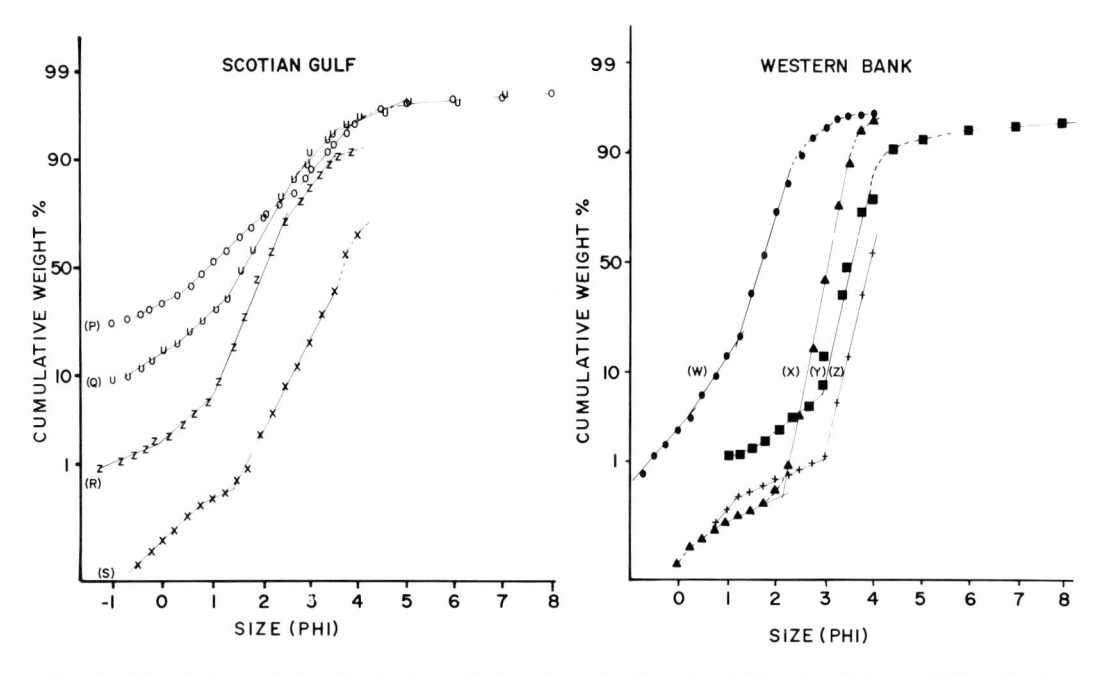

FIG. 4.—Cumulative grain-size distributions of selected samples from the shelfbreak and slope off Nova Scotia.

Gaussian behaviour (e.g. McCave and Swift, 1976). Generally, the central regions show the best adherence to the Gaussian model.

Weight proportions for each size class were recalculated from the cumulative curves where the sieve calibration indicated intervals were non-standard. These values were used in graphical dissections of the curves following the method described by Dalrymple (1977). Reasonable fits could be obtained in the central regions (Fig. 5), but problems were encountered with many of the tails as expected from the probability plots. A second dissection method was therefore used. As the central subpopulation is almost always the best sorted and the most lognormal, it was extracted first, rather than start at the tails as in Dalrymple's method. The tails were subsequently replotted by simple subtraction from the lognormal models of the central subpopulation (Fig. 6).

Beyond the initial interpretation of three main subpopulations, the only assumptions of the second procedure are the relative proportions of the subpopulations (the same proportions as the best-fit Gaussian dissections were used) and that the central subpopulation is near Gaussian. The straightline plot of the central subpopulation suggests that this second assumption is close to the truth, and when the subpopulation is very well sorted, the overlap

TABLE 2.—Estimates of Shear Velocity (u_*) from Observed Currents and Grain-Size Distributions at the Shelfbreak and Slope

Water Depth (Meters)	Current Estimates u_* (Max)	Current Estimates u_* (Av. Max)	Grain Size Estimates $u_* = w$ (Mode)	Shields u_* (a)
Scotian Gulf				
250	2.7	1.3	2.6	2.7
300	—	—	2.4	2.7
400	—	—	1.5	2.2
500	—	—	1.2	1.4
600	—	—	<1.0	1.2
700	1.6	0.7	<1.0	<1.0
Western Bank				
200	—	—	3.2	3.0
300	—	—	1.1	1.3

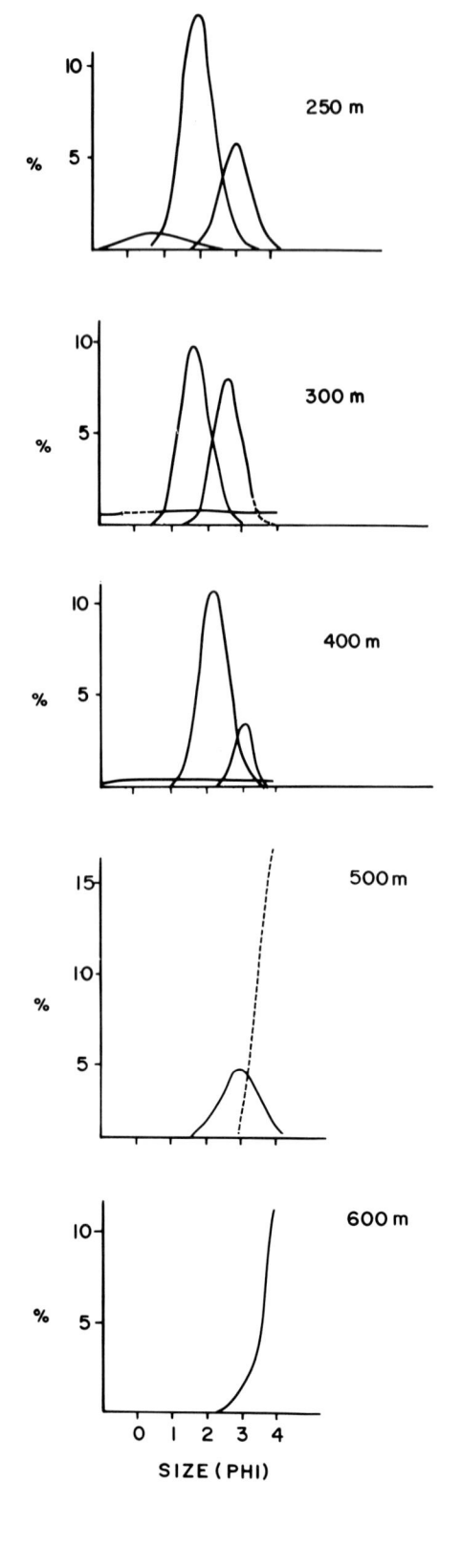

with other populations is minimal. An advantage of this method is that it does not force a Gaussian model on either tail when there does not appear to be any justification for it in the probability plots. However, even forcing the Gaussian model onto the central part of the curve is probably partially in error. For one thing, there is no direct information on the nature of subpopulation overlap. Some of the odd shapes in the fine tail subpopulations (Fig. 6) may be artifacts of the method and suggest that the central subpopulation is in fact truncated at its fine end. The method is useful for indicating the trimodality of the size distributions using minimum assumptions, but the results should not be considered unique solutions.

Interpretation of Size Distribution

Trimodality in sand samples from various environments has been recognized in several other studies (Visher, 1969; Moss, 1972; Middleton, 1976; Dalrymple, 1977) but agreement is not unanimous on the dynamic interpretation of the individual subpopulations. Perhaps the most widely accepted interpretation of trimodality is that of Middleton (1976), who felt that the subpopulations represented bedload (coarse tail), intermittent suspension (central subpopulation) and suspended load (fine tail). Essentially, our data support these contentions. The erosion and suspension criteria for sand size quartz particles have been established experimentally for plane bed conditions and monomodal sand (Fig. 7). At the Scotian Gulf shelfbreak (250 m) the central subpopulation has a distinct mode at close to 2 phi (Fig. 5). If a 2 phi particle were carried in suspension, the required shear velocity would be capable of eroding and transporting, in bedload, particles a little larger than 0 phi (from Shields criteria, Fig. 7) which is reasonably consistent with the size distribution of the coarse tail (although some coarser clasts are present and the gravel fraction is statistically unrepresentative. A shear velocity of 2.6 cm s^{-1} is close to the values obtainable directly from current meter velocities during maximum conditions, when a reasonable drag coefficient (C_D = 0.003) is assumed (Table 2). Using the intersection point of the coarse and central subpopulations to give a shear velocity estimate, as suggested by Middleton (1976), requires higher values in the order of 5.5 cm s^{-1}. Such a method, however, is prone to considerable error (Middleton, 1976; Dalrymple, 1977), being especially sensitive to the amount of subpopulation

FIG. 5.—Size distributions of samples from the Scotian Gulf area, dissected according to the method of Dalrymple (1977), assuming Gaussian subpopulations. Broken lines indicate significant non-fit to Gaussian model.

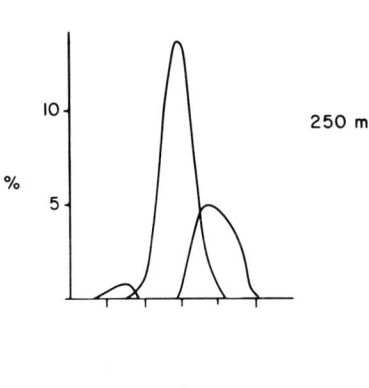

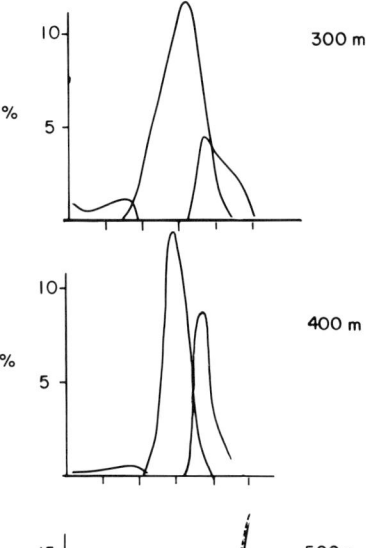

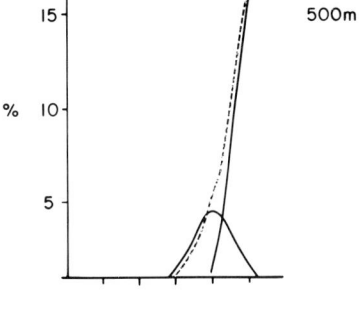

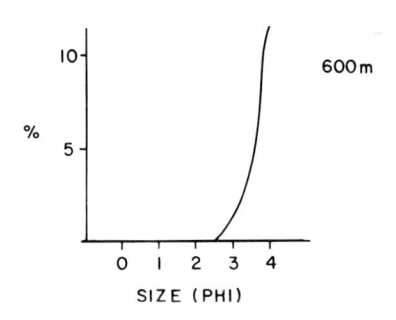

overlap and the assumed population proportions, not to mention the assumption of lognormal subpopulations. It seems more appropriate to use the mode of the central population rather than the tail as it is a relatively well determined point whichever method is used.

The fine subpopulations appear to be strongly asymmetric (Fig. 6) as predicted by the McCave and Swift (1976) model. The tails fine from an apparent mode of between 2.5 and 3.0 phi in every sample above 500 m water depth. This consistent value corresponds very closely to the size of particles that moves directly into suspension when eroded (Fig. 7) without a bedload stage. This suggests a depositional mechanism for the evolution of the separate subpopulations, from an original population moving as suspended load. Particles dropping out of suspension, with size greater than 2.75 phi, must pass through a phase of bedload transport before final deposition, whereas smaller sizes will settle almost passively, directly onto the bed. This can have two effects: (a) rates of transport decrease markedly in the bedload phase, and/or (b) winnowing may sort size classes in the bedload. One or both of these effects may cause the apparent evolution of two subpopulations during the depositional stage of sediment transport. The decreased transport rate means that over any transport distance, bedload material becomes more concentrated as the suspended fines are more rapidly transported away. Over a longer period, winnowing out of the fines deposited along with the bedload would improve the sorting of the coarser population, making it more distinct from the fine population.

This interpretation of the fine tail subpopulation is supported by the downslope trends shown by the central populations. The central mode fines in a downslope direction until about 500 m water depth (Fig. 6). Below this depth only the fine tail subpopulation is significant. The undissected curve of the 500 m sample still shows a distinctive kink at 3 φ, where, as in coarser samples, passive settling takes over from "dynamic" settling. By 700 m water depth, the mode is in the silt range, which suggests shear velocities of less than 1.0 cm s^{-1} (Miller et al., 1977). This is compatible with observed current velocities at that depth (Table 2), if a drag coefficient of 0.003 is again assumed.

Samples from the Western Bank slope show a similar pattern (Fig. 8) with two main differences: (a) the subpopulations appear to be better sorted, and (b) the central subpopulation fines to 2.75 phi at a shallower depth (200 to 300 m). The former may represent a difference in source material.

FIG. 6.—Size distributions of samples shown in Fig. 5, dissected by removal of central subpopulation as explained in text.

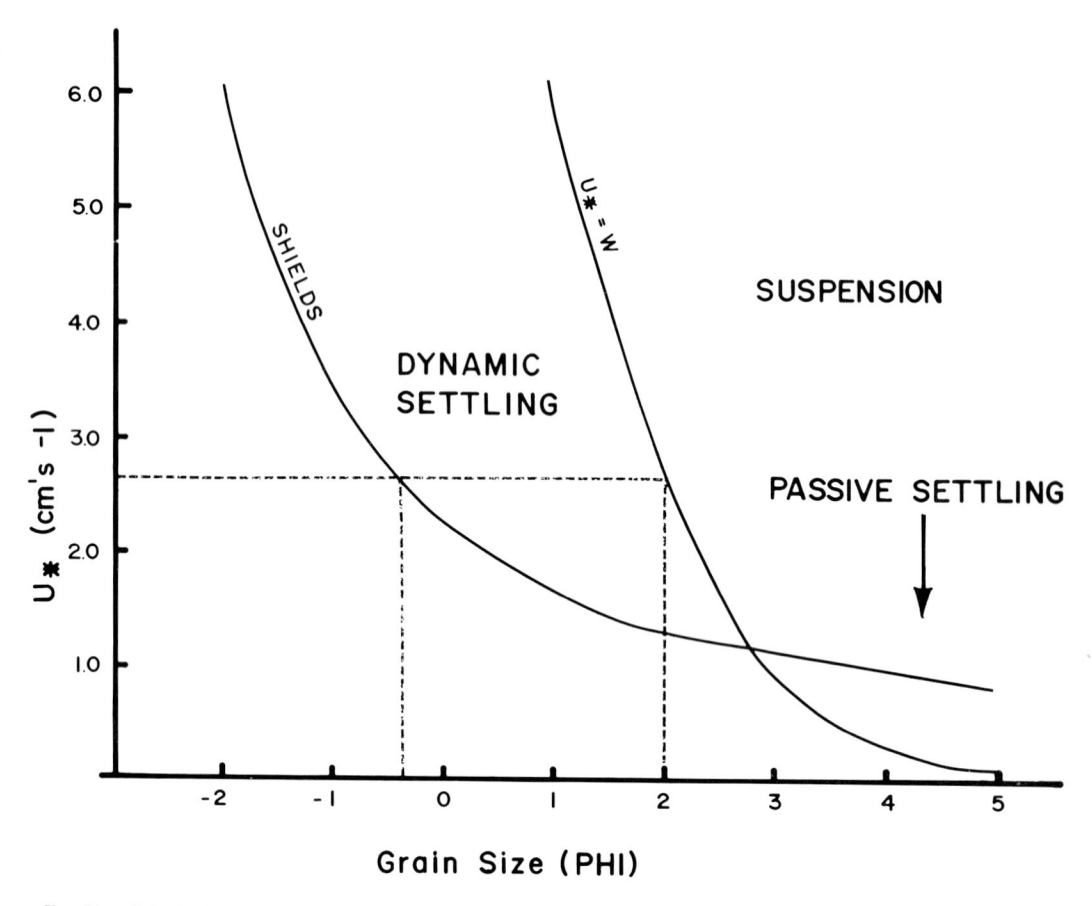

FIG. 7.—Criteria for initial movement and suspension of quartz grains in water at 20°C, plotted on phi size scale, after Blatt et al. (1980, p. 103).

King's (1970) map indicates that sand on Western Bank is better sorted than sand in the Scotian Gulf.

SEDIMENT TRANSPORT PATHS

Sources

Distinct downslope trends in the two coarsest populations (Figs. 6, 8) suggest that coarse sand and gravel is derived quite locally from the outer shelf. In the Scotian Gulf, sediment textures are variable, poorly sorted and contain large gravel fractions (Fig. 9). Some textures appear to be only slightly modified from a very poorly sorted source material. One sample taken in this area contained a few gravel clasts in the grab jaws with stiff sandy, gravelly mud plastered on the outside of the sampler. These textural properties and the hummocky nature of the sea bottom in the area suggest that glacial till is present in the Scotian Gulf. In places, it is very close to the sediment surface and probably covered with a thin gravel lag. The large bottom current velocities observed in the Scotian Gulf (up to 110 cm s^{-1}) would be capable of erosion of the till, even with the protective gravel lag. However, these high velocities are always directed onto the shelf and probably result in transport of sediment into adjacent Emerald Basin rather than onto the slope. The currents may, however, expose and erode large areas of till and partially sorted sand near the shelfedge, which may eventually supply the slope.

Conditions on Western Bank are not well known. Most of the bottom is sandy (King, 1970) and relatively well sorted. Parts of the bank may be relict (King, 1970, 1979), but sand waves and ripples have been documented on other areas close by (Stanley et al., 1972; King, 1970), indicating active sediment transport on some areas of the banks.

Compared to fine sediment, it is likely that the coarser sand is supplied to the slope from relatively local sources on the shelf (Fig. 10). Transport probably occurs across the shelfbreak only during the more severe conditions of strong flow (Fig. 3). Fine sand, on the other hand, is transported for a much

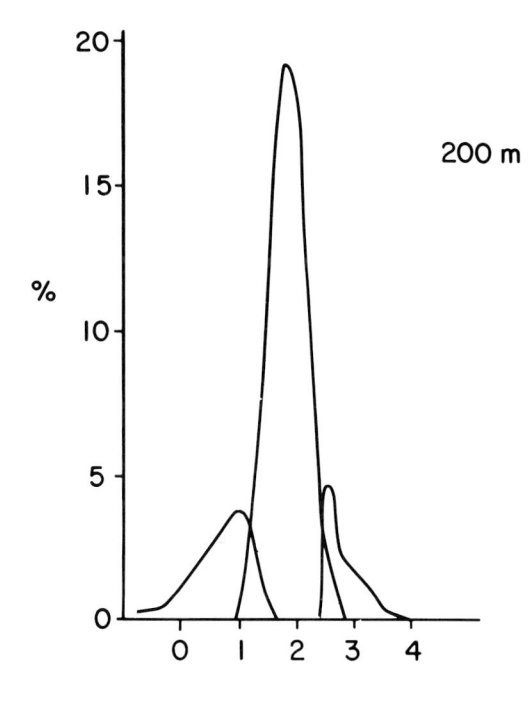

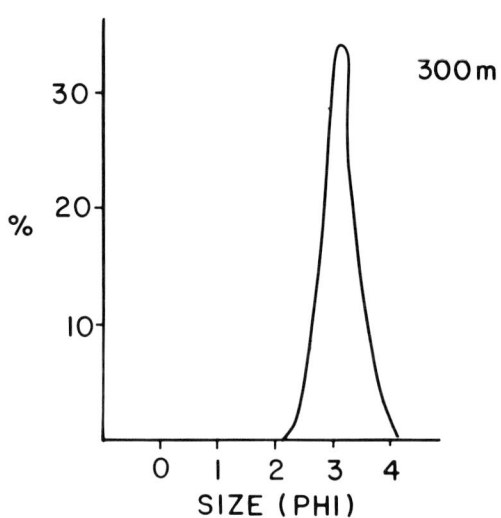

FIG. 8.—Size distributions of samples from the Western Bank area, dissected by removal of central subpopulation as explained in the text.

larger proportion of the time, requiring relatively low shear velocities (Fig. 7). Thus, sources of fine sand are potentially much more numerous. Heavy mineral abundances in fine sands suggest a shelf source (Hill, 1979), but can't be used more exactly due to the complex heavy mineral pattern on the Scotian shelf (James and Stanley, 1968). Fine sand and mud is probably supplied to the slope along the whole length of the Nova Scotian outer shelf (Fig. 10).

Transport and Deposition

The textural data suggest that sand transport is dominantly in suspension. However, the detailed interpretation of the separate subpopulations suggests that different rates of net transport apply to each subpopulation. From the current data (Fig. 3) and shear velocity calculations (Table 2), it can be seen that suspended load transport of medium sand-sized material can occur only during short periods of high flow. As a result, medium sand transport is directed essentially along isobaths with the strong flows. The current directions during these periods are presumably variable over the short-term and would result in some net transport of sediment obliquely downslope. Deposition of the sand would be temporary in upslope areas where the current maxima consistently exceed the suspension criterion. For any particular grainsize, deposition would be greatest at the point where $u_* = w$ (suspension criterion, Fig. 7) because the transport rate for that size is suddenly reduced, when less efficient bedload transport takes over from suspended transport as the primary mechanism. Thus, modal sizes should give a good indication of the local maximum shear velocity (Fig. 10).

The finer sediment, which essentially makes up the fine tail subpopulation, is maintained in suspension at lower velocities, and, consequently, transport is much more continuous and efficient. Fine sediment transport would follow the mean flow direction more closely. This is variable over long periods and sediment dispersal is essentially both alongslope and downslope. Long-term averages of hourly current vectors (Table 3) suggest that there is a distinct downslope component to the average drift at the shelfbreak. Diffusive and density processes may also contribute significantly to the downslope transport. Settling of fines may occur ubiquitously during quiet periods, but can only be permanent at a point downslope where the environment is sufficiently quiet to allow accumulation of sediment. Cores at 500 m depth indicate a net deposition of fine sand and mud at an average accumulation rate of approximately 5 cm/1000 yrs. We envisage a type of mechanism as suggested by McCave and Swift (1976) where time-dependent deposition from suspension allows introduction of sediment into the quieter downslope environment.

DISCUSSION

This chapter attempts to explain some of the gross textural changes across the shelfbreak and slope in terms of depositional mechanisms and transport paths. The correlation between texturally-derived and current-derived shear velocities is reasonable and suggests that the textural pattern is a response to the normal current regime in the Scotian Gulf area. The lack of current data makes it impossible to say whether exactly the same processes are important in the Western Bank area. In fact, there are interesting differences between the two areas. Figure 10 shows the position of shear velocity contours as derived from sediment textural data in the two areas. Contours of the same value are found at relatively shallower depth and the shear velocity gradient seems to be higher in the Western Bank area. Although the shelfbreak off Western Bank is also shallower, these are still unexpected results.

Recent data allow us to speculate on the problem. According to Louis et al. (1983), the Scotian Gulf regularly experiences the effects of Gulf Stream eddies. Most seem to form at around longitude 65° W and impinge locally on the slope. As a result, topographic Rossby waves are formed on the slope which have an average period of about 20 days and an amplitude in velocity terms of about 10 cm s^{-1}. The current meters at the shelfbreak and slope of the Scotian Gulf show the effects of these waves for approximately 65% of the total 1.5 years of record examined (B. Petrie, pers. comm.). Simple cal-

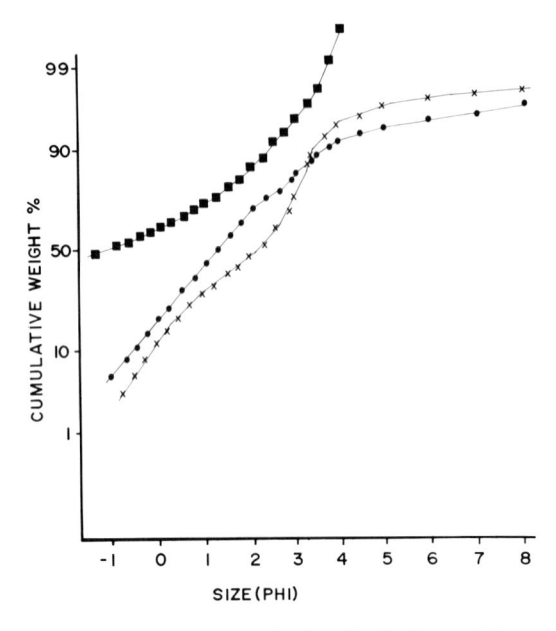

FIG. 9.—Cumulative grain-size distributions of three typical samples from the Scotian Gulf shelf. Note the variability and poorly sorted nature of the sediments.

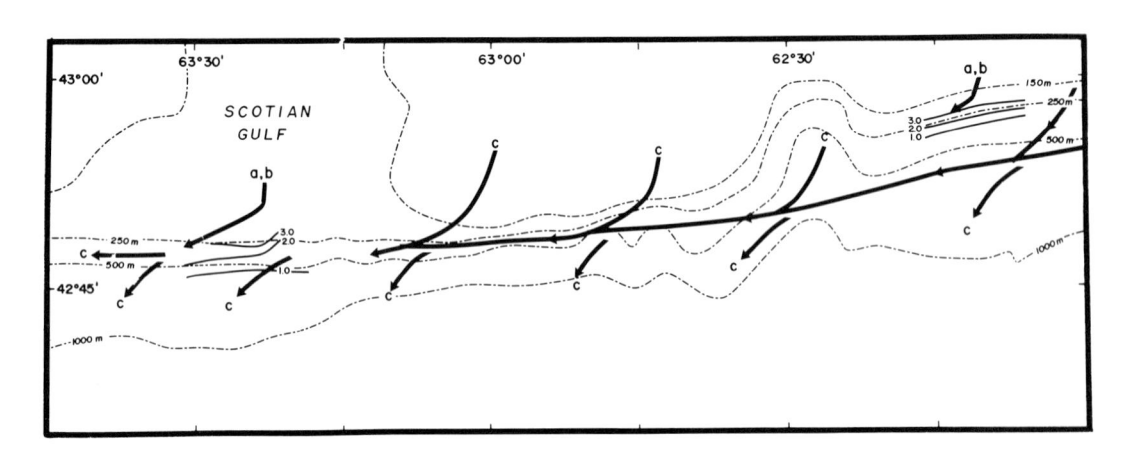

FIG. 10.—Map of part of the Nova Scotian margin showing u$_*$ contours derived from sediment texture in the two study areas, and schematic depiction of sediment transport routes. a = coarse tail subpopulation (bedload); b = central subpopulation ("intermittent" suspension); c = fine tail subpopulation ("continuous" suspension).

TABLE 3.—LONG-TERM AVERAGES OF HOURLY CURRENT VECTORS, INDICATING NET DRIFT DIRECTIONS AS AVERAGE VECTORS

Station	Depth (meters)	No. of Hourly Samples	U (+ve to E) cm s^{-1} Mean	Stand. Dev.	V (+ve to N) cm s^{-1} Mean	Stand. Dev.
S1	230 m	4479	−2.7	12.4	−1.2	10.8
	230 m	9232	−2.7	13.7	−1.7	10.9
S3	690 m	6283	−1.0	5.0	2.0	5.6

culations (subtraction of 10 cm s^{-1} from current speeds in the Scotian Gulf) suggest that the shear velocity pattern would be similar in the two areas if the topographic wave effect were removed. The Western Bank area is much farther away from the main area of eddy generation so that the topographic wave effect would be considerably less important here.

This demonstrates that our results are only locally specific, but the methods used would be usefully employed in other areas. More complex studies could be attempted if detailed information on bottom boundary layer dynamics and bedforms can be obtained. Sampling remains, however, a major problem, despite the fact that considerable efforts were made to use only representative samples. There is no control on the length of time represented by the samples, that is, whether the sample represents yesterday's event or an average of the last hundred years of events.

CONCLUSIONS

Sediment textures are in equilibrium with local flow dynamics on the Scotian shelfbreak and slope. Transport of medium sand across the shelfbreak occurs during periods of strong flow and is essentially directed alongslope, but downslope transport is likely due to local variability in the flow. Fine sediment transport is with the net flow and significant downslope transport occurs during periods of net offshore water movement or due to diffusive or density effects during quiet periods. Catastrophic gravity-driven processes and rare extreme events may contribute to net transport, but are probably less important.

ACKNOWLEDGMENTS

This study was made possible by the data and advice given to us by Brian Petrie of Bedford Institute of Oceanography. Sediment samples were collected through the cooperation of the captain and crew of CSS *Dawson*. Thanks to D. Piper, D. Huntley and R. Dalrymple for helpful discussions. Support for the study was provided by NSERC Canada grants to A. J. Bowen and D. J. W. Piper of Dalhousie University.

REFERENCES

BAGNOLD, R. A., 1937, The size grading of sand by wind: Proc. Royal Soc. London, Series A, v. 163, p. 250–264.

———, AND BARNDORFF-NIELSEN, O., 1980, The pattern of natural size distributions: Sedimentology, v. 27, p. 199–207.

BARNDORFF-NIELSEN, O., 1977, Exponentially decreasing distributions for the logarithm of particle size: Proc. Royal Soc. London, Series A, v. 353, p. 401–419.

BLATT, H., MIDDLETON, G., AND MURRAY, R., 1980, Origin of Sedimentary Rocks: Englewood Cliffs, New Jersey, Prentice-Hall, 782 p.

DALRYMPLE, R. W., 1977, Sediment dynamics of Macrotidal Sand Bars, Bay of Fundy [unpub. Ph.D. thesis]: Hamilton, Canada, McMaster Univ., 634 p.

FOLK, R. L., 1966, A review of grain size parameters: Sedimentology, v. 6, p. 73–93.

GATIEN, M. G., 1975, A Study of the Slope Water Region, South of Halifax [unpub. M.Sc. thesis]: Halifax, Canada, Dalhousie Univ., 135 p.

HILL, P. R., 1979, Late Quaternary Sediments on the Nova Scotian Continental Slope [unpub. Rept.]: Halifax, Canada, Dalhousie Univ., Dept. Geol., 109 p.

HOUGHTON, R. W., SMITH, P. C., AND FOURNIER, R. O., 1978, A simple model for cross-shelf mixing on the Scotian Shelf: Jour. Fish. Res. Bd. Canada, v. 35, p. 414–421.

JAMES, N. P., AND STANLEY, D. J., 1968, Sable island Bank off Nova Scotia: Sediment dispersal and recent history: Am. Assoc. Petroleum Geologists Bull., v. 52, p. 2208–2230.

KING, L. H., 1970, Surficial Geology of the Halifax-Sable Island Map Area: Canada Dept. Energy, Mines and Resources, Mar. Sci. Paper No. 1, 16 p.

———, 1979, Aspects of regional surficial geology related to site investigation requirements-eastern Canadian Shelf: Soc. Underwater Tech., Internat. Conf. Offshore Site Investigations, 34 p.

LIVELY, R. R., 1979, Current Meter and Meteorological Observations on the Scotian Shelf, December 1975 to January 1978: Bedford Inst. Oceanogr. Data Series BI-D-79-1, 2 volumes, 280 p., 368 p.

LOUIS, J. P., PETRIE, B. D., AND SMITH, P. C., 1982, Topographic Rossby waves on the Continental Slope off Nova Scotia: Jour. Phys. Oceanogr., v. 12, p. 47–55.

MCCAVE, I. N., 1971, Sand waves in the North Sea off the coast of Holland: Mar. Geology, v. 10, p. 199–225.

———, 1972, Transport and escape of fine-grained sediment from shelf areas, *in* Swift, D. J. P., Duane, D. B., and Pilkey, O. H., eds., Shelf Sediment Transport: Process and Pattern: Stroudsburg, Pa., Dowden, Hutchinson and Ross, p. 225–248.

———, AND SWIFT, S. A., 1976, A physical model for the rate of deposition of fine-grained sediments in the deep sea: Geol. Soc. America Bull., v. 87, p. 541–546.

MILLER, M. C., MCCAVE, I. N., AND KOMAR, P. D., 1977, Threshold of sediment motion under unidirectional currents: Sedimentology, v. 24, p. 507–527.

MIDDLETON, G. V., 1976, Hydraulic interpretation of sand size distributions: Jour. Geology, v. 84, p. 405–426.

MOSS, A. J., 1972, Bedload sediments: Sedimentology, v. 18, p. 159–219.

PETRIE, B., AND SMITH, P. C., 1977, Low frequency motions on the Scotian Shelf and slope: Atmosphere, v. 15, p. 117–139.
STANLEY, D. J., SWIFT, D. J. P., SILVERBERG, N., JAMES, N. P., AND SUTTON, R. G., 1972, Late Quaternary Progradation and Sand "Spillover" on the Outer Continental Margin off Nova Scotia, Southeast Canada: Smithsonian Contrib. Earth Sci. No. 8, 88 p.
VISHER, G. S., 1969, Grain size distributions and depositional processes: Jour. Sed. Petrology, v. 39, p. 1074–1106.

PART VI
SEDIMENTARY RESPONSES TO PROCESSES

THE MUDLINE: VARIABILITY OF ITS POSITION RELATIVE TO SHELFBREAK

DANIEL JEAN STANLEY
Division of Sedimentology, Smithsonian Institution, Washington, D.C. 20560

SUNIT K. ADDY
ARCO Exploration Co., Midcontinent Province, Denver, Colorado 80217

E. WILLIAM BEHRENS
Institute for Geophysics, The University of Texas at Austin, Galveston, Texas 77550

ABSTRACT

The mudline, the depth of substantially increased silt and clay content and the level below which deposition prevails on continental margins, often occurs near, but is only rarely coincident with, the shelf-to-slope transition. An evaluation of the mudline off the U.S. Mid-Atlantic States and northern Gulf of Mexico highlights marked differences between depth and position of this horizon and those of the shelfbreak, and is summarized in four relationships. *Type I* = Off Cape Hatteras and portions of the west Florida shelf, offshelf transport of sand-size material results in a mudline position well below the shelfbreak. Spillover at Cape Hatteras, where the mudline occurs at 800–1000 m, is a response to the powerful NE flow of the Gulf Stream that tangentially crosses the narrow, shallow shelf. *Type II* = The shallower depth of the mudline (200–400 m, or distinctly below the shelfbreak) off the Mid-Atlantic States between Norfolk and Wilmington canyons, and off Panama City, Florida, margin identifies the long-term signature of energy concentrated on the seafloor; erosion results from the interplay of several mechanisms, including fronts, tides, and internal waves. The mudline at these localities thus defines the position where, over time, shear-induced resuspension has largely exceeded the threshold required for sediment transport. *Type III* = The near-coincidence of the mudline (130–175 m) with the shelfbreak at the head of Hudson Canyon is a response to physical oceanographic parameters and to offshelf spillover; involved are the intersection of density fronts separating Shelf and Slope Water, and the channelizing effect of the canyon head cut deeply into the outer shelf. *Type IV* = Considerable shoaling of the mudline and a marked departure between this level and the shelfbreak occur on margins where large amounts of sediment are supplied. Broad asymmetric shoreward swings of the mudline on the Gulf of Mexico margin west of DeSoto Canyon record Mississippi and other river input and its extensive lateral dispersal by regionally important water mass flow. Along many continental margin segments, the mudline is an erosion-deposition boundary whose position relative to the shelfbreak on a margin is the long-term resultant of several factors including sediment supply, offshelf spillover by a plexus of fluid-driven processes and gravity flows, shelfbreak morphology, structural framework, sediment stability and eustacism.

INTRODUCTION

The concept of a texturally graded continental margin showing seaward-fining of sediments, with sand primarily on the shelf, silt on the slope and clay on the rise, developed in light of results from oceanographic expeditions in the nineteenth century. Early lead-line sounding surveys on continental margins in different oceans indicated that there is a tendency for grain-size to decrease in a direction seaward of the outer shelf. A natural outgrowth of this observation is the long-held tenet that the bulk of deep-sea sediments, derived from continents and shelves, are transported over the shelf-edge, onto the slope and beyond (Doyle et al., 1979). The name "mud-line" was used in a general fashion until the early part of this century to designate the depth, usually on the continental slope, beyond which silt and clay, rather than sand, are deposited (Stetson, 1939).

This term has recently been revived, but with more precision and added genetic significance, on the basis of dense bottom sampling surveys off the U.S. Mid-Atlantic States continental margin (Stanley and Wear, 1978). In the region between Wilmington and Norfolk canyons these authors recognized a seaward sequence consisting of three major lithofacies: (a) an outer shelf facies, extending to about 130 m near the shelfbreak and consisting of gravelly (largely shell and shell hash), coarse to medium grained sand with low (5%) mud content; (b) a transitional facies of gravelly muddy sand, muddy sand, and some sandy mud, on the uppermost slope where textural and compositional parameters change very rapidly between the break and about 250–300 m; and (c) a slope facies consisting of finer-grained foraminiferal-rich silty muds (to >75% silt and clay fractions) on the upper slope below 300 m (Fig. 1A). The mudline was defined as the lower boundary of the transitional facies at 250 to 300 m in this region by Stanley and Wear (1978, p. M27, and their figures 2 and 3), and it is at this level that the proportions of clayey silt, silty clay and clay no longer increase significantly with depth.

Copyright © 1983, The Society of
Economic Paleontologists and Mineralogists

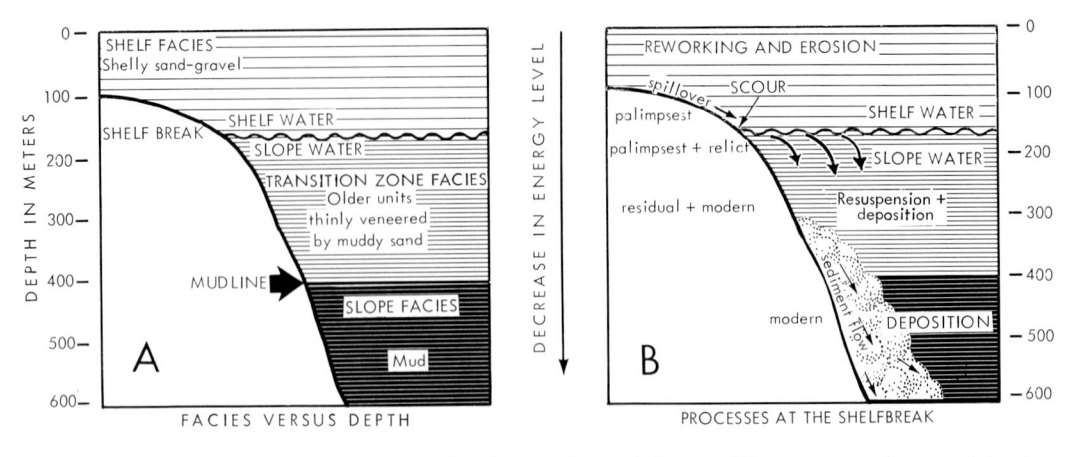

FIG. 1.—Diagrams showing (A) distribution of surficial sediment facies and (B) processes and zones of dominant erosion, alternating resuspension and deposition, and deposition along a shelf-to-slope transect. Note that the mudline, positioned in this hypothetical case below the shelfbreak, occurs at the boundary between the transition facies and slope facies.

An example of a similar textural distribution and comparable mudline position on the northern Gulf of Mexico margin is shown in Figure 2A. There, in the northeastern Gulf near DeSoto Canyon, mud percent begins to increase significantly at depths between 60 and 100 m. Average mud content in the upper few meters of sediment (representing both Holocene and Pleistocene) rises from less than 10% on the shelf to about 95% at a depth of about 300 m. It then continues to increase, but at a barely perceptible rate, to about 1,000 m depth.

It has been proposed that in some sectors of the U.S. East Coast margin the mudline records a long-term separation of energy levels. More specifically, Stanley and Freeland (1978) suggest that current speeds necessary to erode fine sediments decrease, sometimes significantly, in frequency of occurrence on the slope below the mudline. The shelfbreak and uppermost slope above the mudline, on the other hand, comprise a zone of almost continuous resuspension where there is only temporary or minor accumulation of fines. The probable relationships among mudline, sediment facies, transport processes and morphology are shown schematically in Figure 1B.

The definition of shelfbreak used here is "the point of the first major change in gradient at the outermost edge of the continental shelf; it is located on the landward point of the transition zone between the continental shelf and slope" (Wear et al., 1974, p. 40; see also Vanney and Stanley, 1983). As more field data are evaluated, it is recognized that the mudline often occurs near the shelf-to-slope transition, but is only occasionally coincident with the shelfbreak proper. This chapter, focusing on mappable attributes (primarily sediment texture, and to a lesser extent composition), considers the implications of differences between the depth and position of the mudline and those of the shelfedge. Using the relatively well-documented U.S. East Coast and northern Gulf of Mexico margins as examples we recognize several mudline types and attempt to identify the major factors responsible for their development.

The premise of this study is that the variability of mudline position along a margin is a probable response to, not one, but several factors. These include (a) sedimentation factors per se—sediment supply, offshelf spillover mechanisms, interplay of fluid-driven processes—and (b) the effects of regional setting—outer margin morphology, structural framework, seismic activity, slope instability, and eustatic evolution.

DATA BASE AND MUDLINE TYPES

Numerous surficial sediment sampling surveys involving bottom grabs and cores, in some cases supplemented by visual observation (submersible, bottom camera, television), provide a basis for defining the mudline on various U.S. continental margins. Useful seafloor studies off the East Coast showing surficial textural distributions on the outer shelf and slope include those by Trumbull (1972), Milliman et al. (1972), Hollister (1973) and Keller et al. (1979). Grain-size maps plotted from denser bottom sampling surveys more accurately define the mudline off the New York Bight near the head of Hudson Canyon, off the Mid-Atlantic States and off Cape Hatteras (charts by, respectively, Stanley and Freeland, 1978; Stanley and Wear, 1978; and Newton et al., 1971). The mudline position in the northern Gulf of Mexico may be estimated from several sources (among others, Gould and Stewart, 1955; Curray, 1960; Ludwick, 1964; Bouma, 1972;

Doyle and Sparks, 1980). Although sediment sampling sites are irregularly spaced, close control is available in a few sectors such as the northeastern Gulf-DeSoto Canyon area (Addy et al., 1979), off the Mississippi delta (Coleman, 1981) and the South Texas outer continental shelf (Shideler, 1977). Many of these references also provide information on fluid-driven processes, a term which in this paper refers to the transport of sediment particles by water motion caused by one or more mechanisms (e.g., tidal current, wave action, geostrophic current, internal waves, etc.).

The mudline on both continental margins is observed to vary in depth and distance from the coast and also relative to the shelfbreak position. Essentially four types are recognized on the basis of mapped regional textural facies distributions:

Type I.—The mudline occurs at considerable depth on the upper- and, in a few cases, mid-slope (300 to >1000 m).

Type II.—The mudline occurs below the shelfbreak on the uppermost slope (to approximately 200—400 m).

Type III.—The mudline is nearly coincident with the shelfbreak (the latter ranging from about 100 to 175 m in many instances).

Type IV.—The mudline occurs considerably shoreward of the shelfedge (<100 m).

Examples of these four types, selected from transects in the northern Gulf of Mexico, are depicted in Figure 3. A discussion of each type and consideration of possible origins is presented in the following four sections of this paper. The locations of the mudline examples selected on the two margins are shown in Figures 4 and 5A.

TYPE I EXAMPLES

On the East Coast margin, the mudline occurs at considerable depths below the shelfbreak only locally, such as along the axes of the more active submarine canyons, particularly those south of New England, and on the mid-slope off Cape Hatteras. The canyon sites record the downslope displacement of coarse material by gravitational sedimentary processes of vaious types, including turbidity current and debris flows, down submarine valley walls and axes. At Cape Hatteras (Fig. 4, Type I), the shelf is narrowest (about 20 km wide), and the

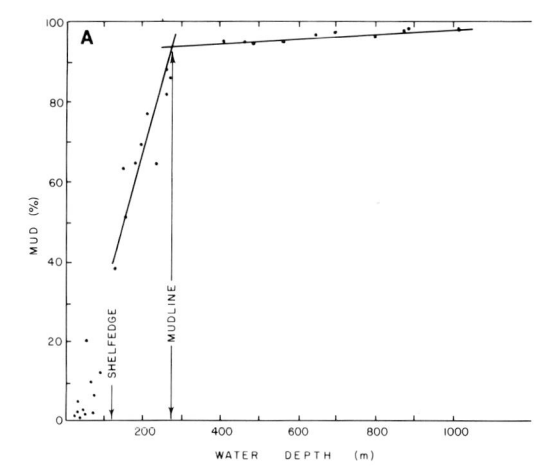

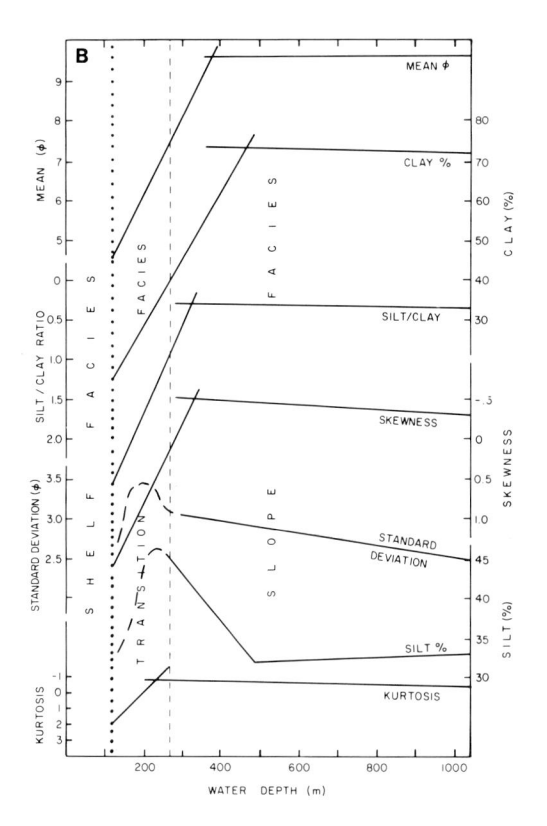

FIG. 2.—Sediment textures relative to depth and mudline definition. All data are from transects near Profile L in Figs. 3 and 5A. A, Percentage mud versus depth, showing mudline at about 260 m. Lines are calculated regressions. The correlation coefficient for the data between the shelfbreak and the mudline is .910, that for the deeper, upperslope data is .869. Each data point represents at least four analyses from a piston core. B, Depth distributions of seven additional textural parameters. Regression lines (solid) are based on 8 to 12 data points within each facies and on similar degrees of correlation as in A (data points are omitted for clarity). Mean, skewness, kurtosis, % clay, and silt/clay ratio behave similarly to % mud with a mudline definable by the intersection of two lines. Standard deviation and % silt peak in values near the mudline.

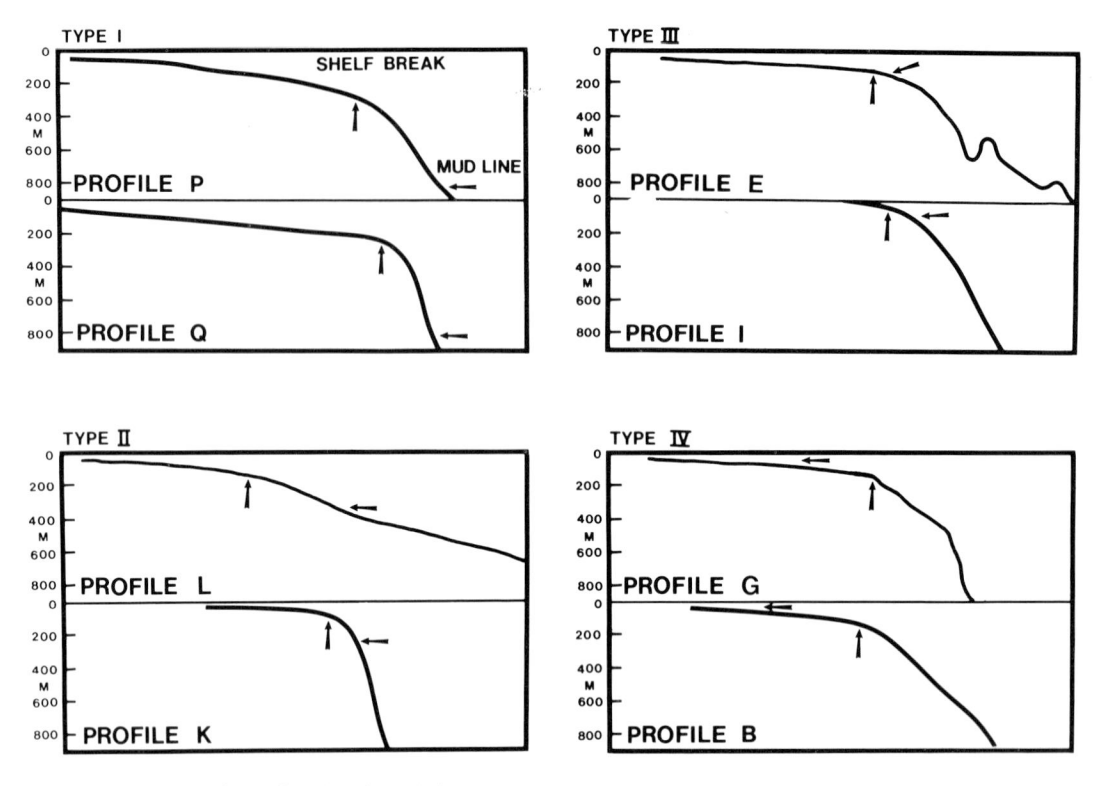

FIG. 3.—Bathymetric profiles showing relative positions of mudline (horizontal arrow) and shelfbreak (vertical arrow) for four mudline types in the Gulf of Mexico. Locations of the profiles are shown in Fig. 5A.

break is shallowest (~50 m) of any off the New England and Mid-Atlantic States. The upper slope here is also appreciably steeper, averaging 9° (Rona, 1969). Textural maps (Newton et al., 1971; Keller et al., 1979) based on the results of grab, coring and submersible dive projects show that the percentage of sand decreases sharply, while the percentages of clay and silt increase rapidly, on the mid-slope at about 1000 m (Fig. 6A). Deposition is a response largely to the effects of the northeast-flowing Gulf Stream that almost continuously erodes the seafloor and drives sediments tangentially off this narrow, shallow shelf. The mapping of sand wave orientation near the shelfedge (Hunt et al., 1977), observation of recently formed coarse sand ripple marks on the upper slope to a depth of about 400 m (Stanley et al., 1981, their fig. 7), and recovery of sand in cores from greater depths on the slope (Fig. 6A; also Keller et al., 1979), provide evidence of active offshelf spillover and continuous reworking along the upper slope by the Gulf Stream (Fig. 6D). Accumulation of finer-grained sediments prevails below 1000 m (Fig. 6B, C), that is, below the effect of major Gulf Stream erosion and above the slower moving Slope Countercurrent (Fig. 6E).

In the northern Gulf of Mexico, the Type I mudline is mapped on the west Florida margin where it is estimated to occur at depths of at least 700 to 1000 m (Fig. 5A, Type I). The mudline position as shown is an approximate one: its depth is at some point between that of mapped carbonate sand (50 to 80%) on the slope (about 4°) reported by Doyle and Sparks (1980), and the base of the west Florida Escarpment where sediment contains less than 10% sand (Bouma, 1972). The mudline relative to the shelfbreak is shown in profiles P and Q in Fig. 3. This occurrence is in an area distant from important terrigenous input by rivers; and the amounts of suspended particulate matter transported across the broad (220 km) shelf are small. Sediments on the west Florida platform comprise coarse quartz sand nearshore and algal, coral, shell and oolitic sand

FIG. 4.—Continental margin off northeastern United States showing distribution of sand on the outer shelf and slope (percentage mean sand data from Keller et al., 1979; modified with permission). Arrows depict examples of mudline types discussed in text: Type I (Cape Hatteras margin), Type II (most of the margin between Norfolk and Hudson canyons), Type III (shelfedge at head of Hudson Canyon). Depth in meters.

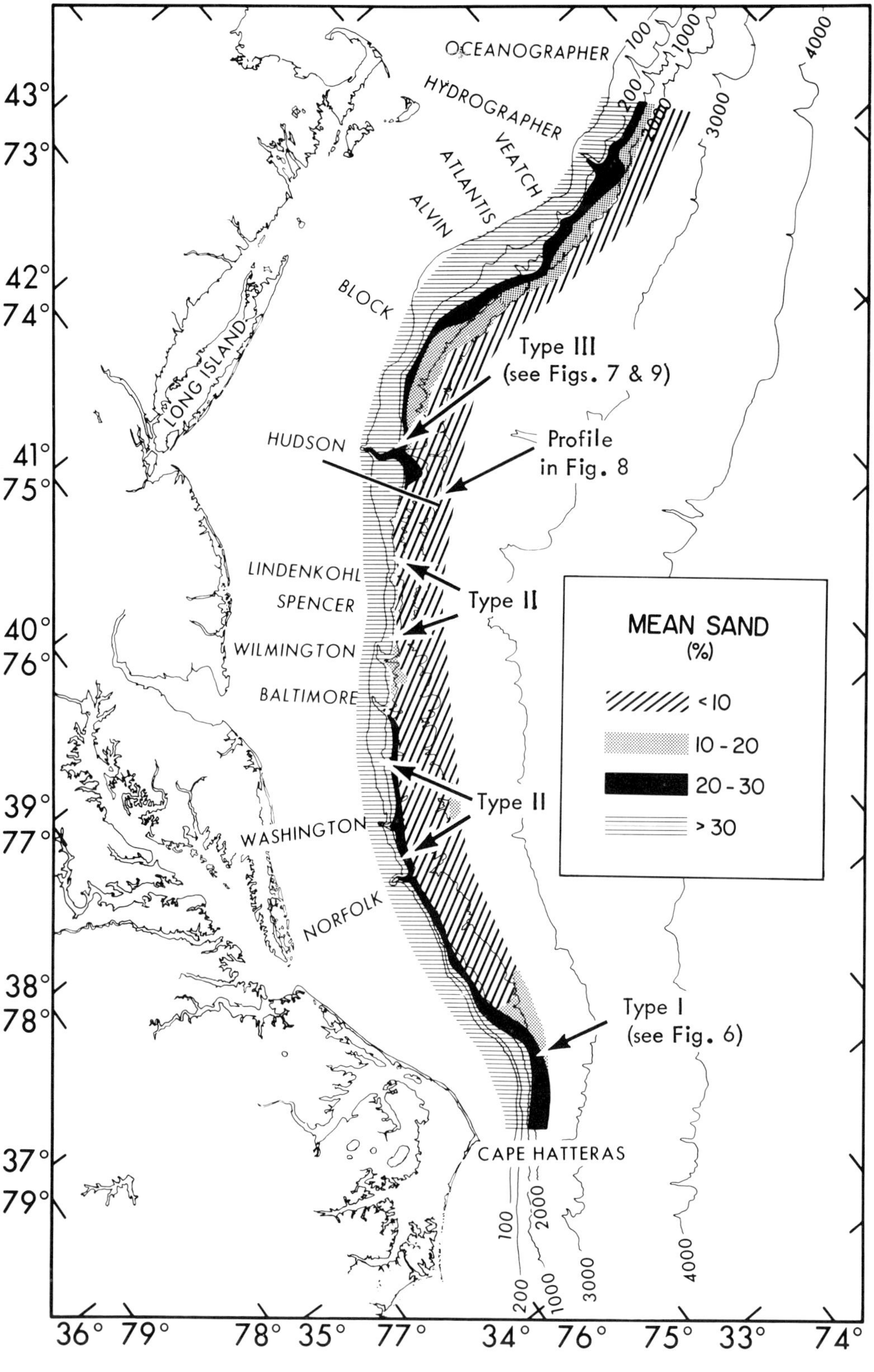
OCEANOGRAPHER
HYDROGRAPHER
VEATCH
ATLANTIS
ALVIN
BLOCK
LONG ISLAND
HUDSON
LINDENKOHL
SPENCER
WILMINGTON
BALTIMORE
WASHINGTON
NORFOLK
CAPE HATTERAS
Type III (see Figs. 7 & 9)
Profile in Fig. 8
Type II
Type II
Type I (see Fig. 6)
MEAN SAND (%)
<10
10 - 20
20 - 30
>30
100
200
1000
2000
3000
4000
43°
73°
42°
74°
41°
75°
40°
76°
39°
77°
38°
78°
37°
79°
36° 79° 78° 35° 77° 34° 76° 75° 33° 74°

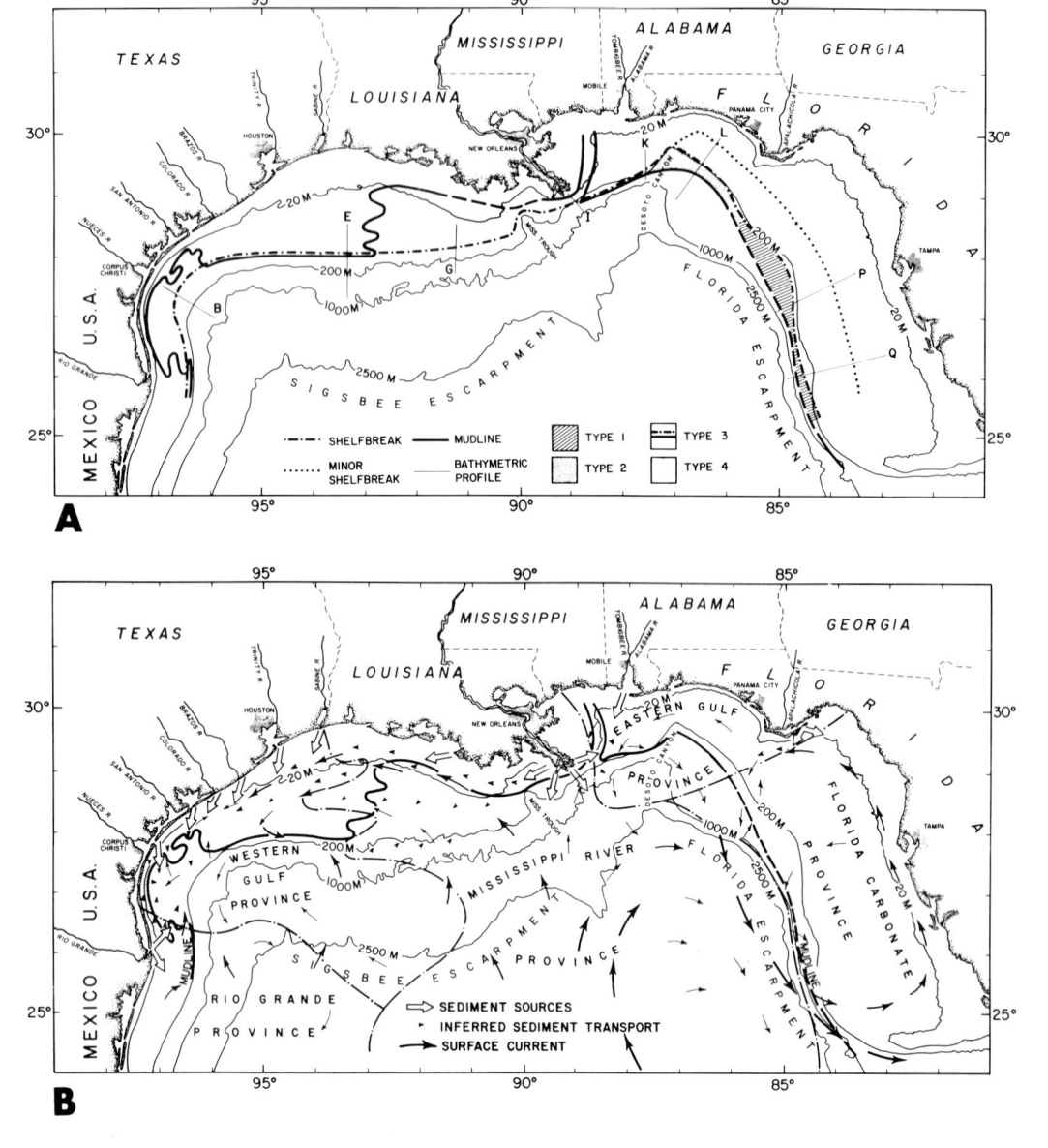

FIG. 5.—Mudline (depicted by heavy line) on northern Gulf of Mexico margin. A, Distribution of mudline types relative to shelfbreak. Bathymetric profiles of lettered lines are shown in Fig. 3. B, Relationship of mudline to sedimentary mineral provinces (e.g. Davies, 1972), sediment sources, surface currents, and inferred sediment transport. Currents are synthesized from Leipper (1954), Curray (1960), and Griffin (1962). Weights of current arrows record current speed and/or persistence. Lightest arrows depict seasonally varying or opposing flow directions.

further seaward. A limited amount of quartz sand is transported by longshore currents to its present location while the carbonate sands are primarily erosional products from the numerous algal and coral mounds and pinnacles on the platform (Uchupi and Emery, 1968). This region is affected by numerous, often intense, storms. It thus appears that the sand (comprising coarse biogenic components) forming the surficial drape on the slope is shed from the adjacent shelf by bottom currents (cf. Griffin, 1962). Spillover off the west Florida shelf is perhaps somewhat less active and more discontinuous than on the Cape Hatteras margin that is affected by the continuous and powerful Gulf Stream

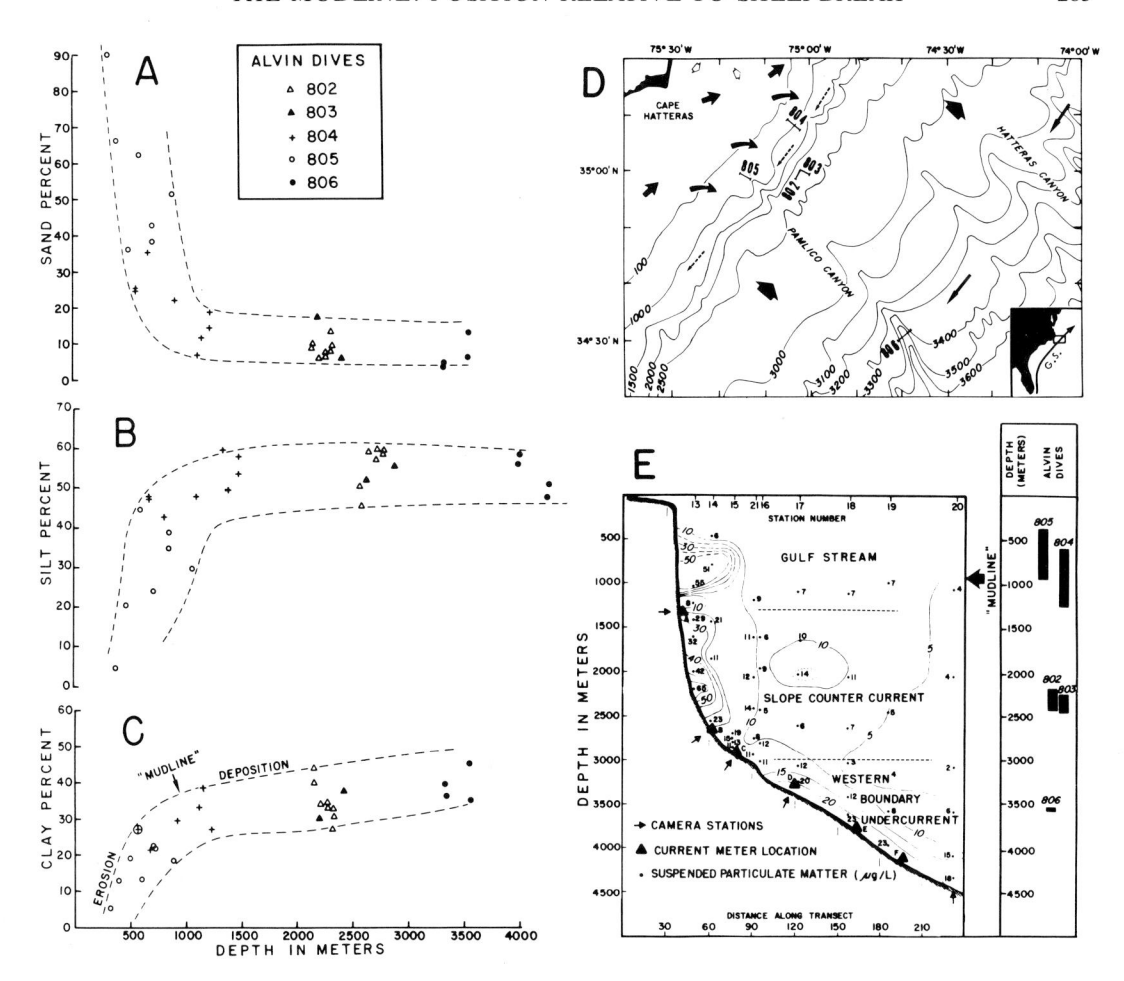

FIG. 6.—*A–C*, Graphic plots showing percentages of sand, silt and clay versus depth on the Cape Hatteras margin. Based on this data, mudline depth of about 1000 m is mapped on the slope. *D*, Heavy NE-trending arrows depict the axis of the gulf Stream (G.S.); thinner arrows depict the Gulf Stream trend on the shelf; curved arrows indicate offshelf spillover resulting in a Type I mudline. Open arrows show deflection of the S-trending Mid-Atlantic Shelf Current; thin dashed arrows depict a weak Slope Countercurrent; long arrows on rise, oriented SW, show the Western Boundary Undercurrent. Depth in meters. *E*, Profile across Cape Hatteras margin showing relation between mudline position (arrow) and the Gulf Stream and Slope Countercurrent contact on the upper slope (modified from Stanley et al., 1981).

flow. Doyle and Sparks (1980) indicated that periodic winnowing of bottom sediment takes place during hurricanes and the passage of storms in winter months. In the absence of an important proximal fluvial source of mud in the northeastern Gulf, the drape of sand spilled offshelf (resulting from current transport on the outer shelf) depresses the mudline well below the break. Major sources of finer sediment on the west Florida slope are fine carbonate nannoplankton and the clockwise west Florida or Loop Current (Cochrane, 1972) carrying distally-derived Mississippi River sediment (Fig. 5B). The mudline in this region roughly denotes the boundary between the Florida Carbonate Province and Mississippi Province (cf. Davies, 1972).

TYPE II EXAMPLES

The mudline off the Mid-Atlantic States between Cape Hatteras and the New York Bight most commonly occurs on the uppermost slope below the shelfbreak, particularly along the margin between canyons (Fig. 4, Type II). The shelf in this region is considerably wider (75–100 km) and deeper (about 100–130 m) than off Cape Hatteras, and the upper slope away from canyon and gullied areas is less steep (<6°). Extensive grab sampling along sixty transects shows a rapid transition from gravelly sand (largely granule-size shell hash and coarse sand-size terrigenous components) to muddy sand to sandy mud to clayey silt. The boundary between the last two textural types defines the mudline be-

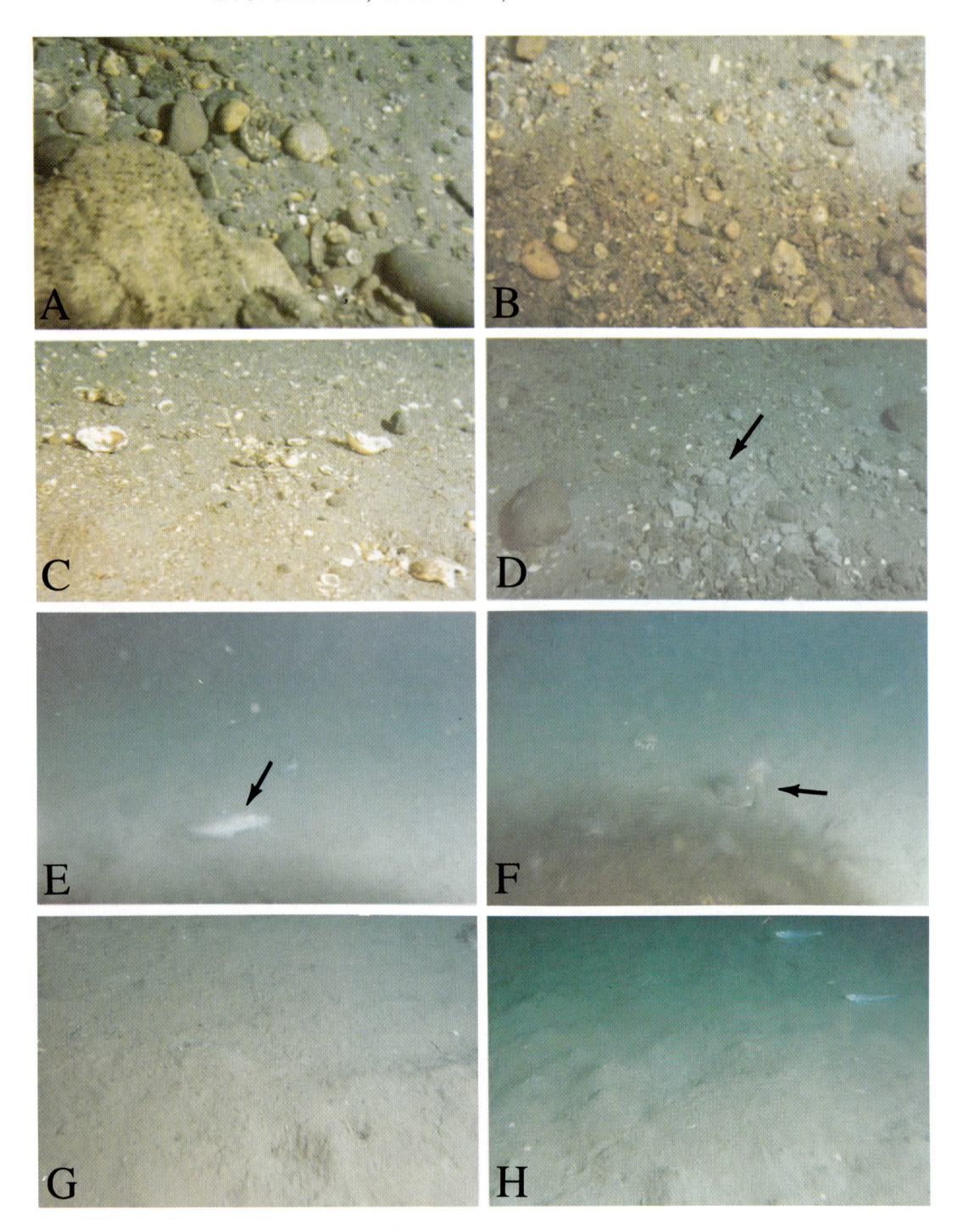

FIG. 7.—Photographs collected on a submersible survey made at the head of Hudson Canyon (*Nekton-Beta* dive locations are shown in Fig. 9). Outer shelf sediments include gravely sand (A,B), sometimes very coarse, derived from reworked pre-Holocene Hudson delta sediments; the dominant surficial facies is shelly sand (C). Locally, clasts of older (Pleistocene) blue-gray stiff clay (D, arrow) occur near outcrops on the outer shelf and upper slope. The transition facies zone (muddy sand) on the uppermost slope is characterized by poor visibility resulting from almost continuous resuspension; large amounts of suspended material markedly reduces clarity above the seafloor (arrows denote poorly-defined fish in E, and current-deflected anemone in F). G, H, Sandy silt and clayey silt prevail on the upper slope below the mudline.

tween 250 and 300 m (Fig. 1; see also Stanley and Wear, 1978, their figs. 2 and 3). At present, much of the sediment presently carried by rivers, including the fines, is trapped in estuaries (Meade et al., 1975). The coarse outer shelf facies (Fig. 7A–D) is identified as largely reworked relict (palimpsest) and includes mineral components of Appalachian origin, modern shell, and—locally—large clasts and grains of residual pre-Holocene lithologies [for example, blue stiff shales (Fig. 7D, arrow) and glauconitic iron-stained sandstone] eroded from underlying sedimentary deposits which form the outermost shelf and uppermost slope (to ~200 m). The much finer-grained surficial foraminifera-rich silty mud on the slope below 300 m is identified as a deposit of Holocene to modern origin (Fig. 7G–H).

The transitional facies on the uppermost slope, between the break and about 300 m, includes sediments intermediate in texture and composition between those of the outer shelf and those below the mudline. This depth range is interpreted as one where sediments likely come to rest temporarily, i.e., a zone of periodic resuspension (Fig. 1B). Evidence for this are the high amounts of measured suspended material in the water column (Fig. 7E–F; see also Lyall et al., 1971; Meade et al., 1975) and poor seafloor visibility as viewed in submersible, underwater camera and television surveys (Stanley and Fenner, 1973).

Resuspension is a response to bottom currents that periodically exceed threshold velocities needed to erode fine sediment. It is believed that this fluid-driven energy is induced by a plexus of mechanisms, including surface waves, tidal currents, atmospheric pressure-induced (wind stress) currents, internal waves (Southard and Stanley, 1976) and the effect of shear between major water masses and oceanic fronts that migrate across this outer margin (cf. Ruzecki and Welch, 1977; Karl et al., 1983; Pietrafesa, 1983). Seafloor erosion and resuspension of sediment is intensified in winter as a function of storm and in summer by internal wave activity, while temporary deposition probably prevails in summer when sediments concentrated along pycnoclines are released to the seafloor. The mudline forming the lower boundary of the transitional facies appears associated with the intersection of the slope with the maximum depression level of the density interface separating Shelf Water and Slope Water (Fig. 8). Thus, the position of the mudline in this region is believed to represent the level at which deposition prevails over the long-term as a result of the interplay of several physical oceanographic phenomena. It is suspected that its position shifts somewhat up- and downslope seasonally in response to fluctuations of bottom current erosion and to episodic changes of suspended sediment concentrations introduced above the outer margin (Fig. 1B).

The Type II mudline is also mapped along part of the northern Gulf of Mexico margin, specifically the eastern Gulf province off the northwestern Florida and Alabama shelf. Extensive sampling in the DeSoto Canyon area (Fig. 5A, Type II; see Fig. 3, profiles K and L) shows distinct correlations between mud (silt + clay) percent (Fig. 2A) and also several other grain size parameters (mean size, clay percent, skewness, silt-clay ratio; Fig. 2B) with depth. Shelf width in this region is approximately 110 km, and a very gradual shelfbreak occurs at roughly 130 m. Coarse sand comprises much of the outer shelf facies (to >90% sand), and this grades seaward to sandy mud and, at depths greater than 300 m, to clayey mud with less than 5% sand. Determination of mineralogical provenance (cf. Davies, 1972) shows predominant sediment input from rivers, including the Apalachicola, draining the southern Appalachians. These terrigenous materials are distributed and deposited mostly nearshore by seasonally alternating inner shelf, longshore cur-

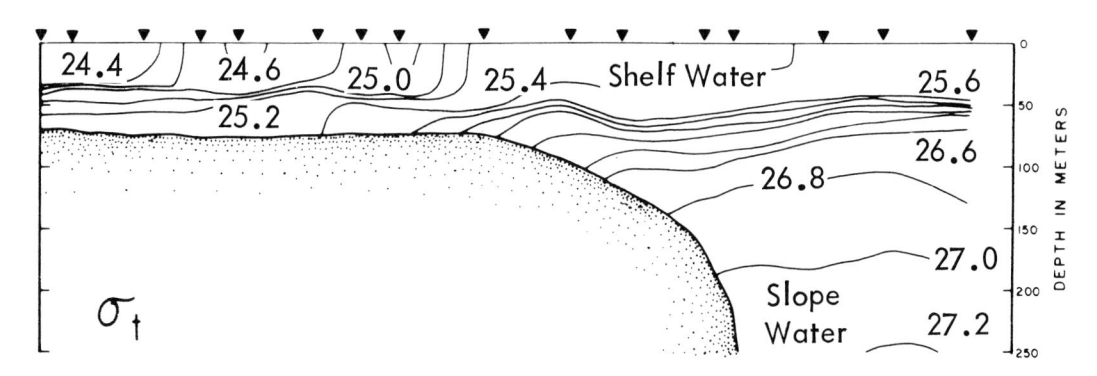

FIG. 8.—Profile showing stratification and a marked pycnocline (as highlighted by water density gradients) which separates Shelf Water and Slope Water near the shelfbreak immediately south of Hudson Canyon. The mudline in this region is positioned near the shelfedge (Type III), approximately at the intersection of the slope with the density interface separating the two water masses. Data used to compile the profile was collected in October, 1974 (modified from Gordon et al., 1976).

rents (Griffin, 1962). A similarly seasonally migrating convergence, influenced by winter storms and hurricanes, carries some of this material off-shelf where it combines with Mississippi River mud and is moved eastward and southward by the more persistent Loop Current which dominates offshore circulation in the eastern Gulf of Mexico (Fig. 5B; see also Leipper, 1954; Griffin, 1962). Algal mounds near the shelfbreak off Alabama (Addy et al., 1979) also provide small amounts of coarse (gravel and sand-size) biogenic material. However, the mudline on the Alabama slope is shallower than on the west Florida slope cited in the previous section. This probably reflects the transport to this area of a greater supply of fine-grained Mississippi river sediment.

TYPE III EXAMPLES

In several sectors of the two margins examined in this study, the mudline is coincident with, or lies immediately above or below the shelfbreak. One well-documented example off the Mid-Atlantic States is the head of the Hudson Canyon where the shelfedge lies about 160 km southeast of New York Harbor (Fig. 4, Type III). This large canyon head, about 5 km wide, is incised 31 km into the shelf. Bottom sediment sampling, coupled with direct observations made on submersible dives (Stanley and Freeland, 1978) show the rapid seaward transition between an outer shelf gravel-shell-sand facies and a deeper (>130–175 m) soft gray-green mud deposit consisting of sandy clayey silt (Fig. 9). Locally, the absence of a surficial veneer of fines reveals an underlying coarse sandy gravel facies (Fig. 7A, B) identified as a reworked lag of fluvial and beach deposits that were originally laid down during Quaternary low eustatic stands. During maximum low sealevel, the much enlarged ancestral Hudson River released enormous quantities of sediment at its mouth, thus forming a large delta bulge close to the shelfedge. In other canyon head areas where the veneer of modern surficial muds is absent, outcrops of stiff blue-grey clay (Fig. 7D), probable lagoonal deposits of late Pleistocene age (cf. Gibson et al., 1968), are exposed in many sectors between 100 and 175 m.

The presence of outcrops, the absence of fines at the shelfedge proper, and poor visibility of the bottom seaward of it (a function of important amounts of suspended sediment in the water mass) all have been noted on submersible dives, even those made during fair-weather conditions. Moreover, currents to speeds of 30 cm/sec have been measured above the bottom (Keller and Shepard, 1978). There is strong indication, therefore, that at the shelfedge relatively high bottom current energy results from various physical oceanographic phenomena such as storm- and internal wave-induced currents (Hotchkiss and Wunsch, 1982). The mudline, for instance, lies close to the depth at which there is intersection of the density front separating Shelf Water and Slope Water with the slope (Fig. 8), at about 150 m during the winter (Gordon et al., 1976; Bowman, 1977). Other fluid-driven processes likely to produce seafloor erosion near the shelfedge include tidal currents, flow in the Hudson Shelf Valley, and the up- and downslope movement of water masses due to the chanellizing effect of the canyon head cut deeply into the shelf. Fines eroded from the outer shelf and uppermost slope, a zone of almost continuous resuspension, include reworked clayey silt from relict fluvial deposits and small amounts of particulate matter transported across the shelf (Meade et al., 1975). Silt and clay accumulate more permanently below 200 m, presumably in a sector characterized by a lower energy regime. Thus, the mudline in this area, as in the case of Types I and II, identifies the depth below which the current velocities, which exceed threshold needed to erode fine sediment, diminish appreciably in frequency (Stanley and Freeland, 1978). Similar conditions probably occur at the heads of other canyons off the U.S. East Coast.

The presence of a Type III mudline is also noted along several portions of the northern Gulf of Mexico (Fig. 5A, Type III): (a) immediately west of the DeSoto Canyon margin, (b) immediately southeast of the Mississippi Birdfoot delta (Fig. 3, profile I), (c) head of the Mississippi Trough, (d) locally, off west Louisiana and east Texas (Fig. 3, profile E), and (e) at the shelfedge seaward of the Rio Grande (Shideler, 1977).

Area (a) occupies a transition zone extending from the clastic-poor Florida Carbonate Province toward the enormous mud supply of the Mississippi River. The large suspension load from the onshore source results in deposition of mud at shallower depths, including some pro-delta muds on the shelf, and thus a shoaling of the mudline to near the shelfedge. On the shelf, beyond the Mississippi Delta proper, the Mississippi-Alabama sand facies (Ludwick, 1964) indicates sufficient fluid-driven energy exists here to bypass modern fines across the shelf to the upper slope, resulting in a mudline near the shelfedge. Long-term net drift in the north-central Gulf is westward (Leipper, 1954), and the eastward sediment transport is probably by a combination of diffusion and intermittent currents.

In area (b), coarse sediments are entrapped in the delta proper, and little sand-size material bypasses the shelf. Much Mississippi River mud, however, is transported through the distributaries and is deposited seaward of the delta, in part by slumping, over large areas of the northern Gulf. The seaward growth of the delta results in the near-coincidence of the mudline and shelfedge. Here, therefore, sediment supply rather than marine fluid-driven processes is the controlling factor.

In area (c), the Type III mudline occurs at

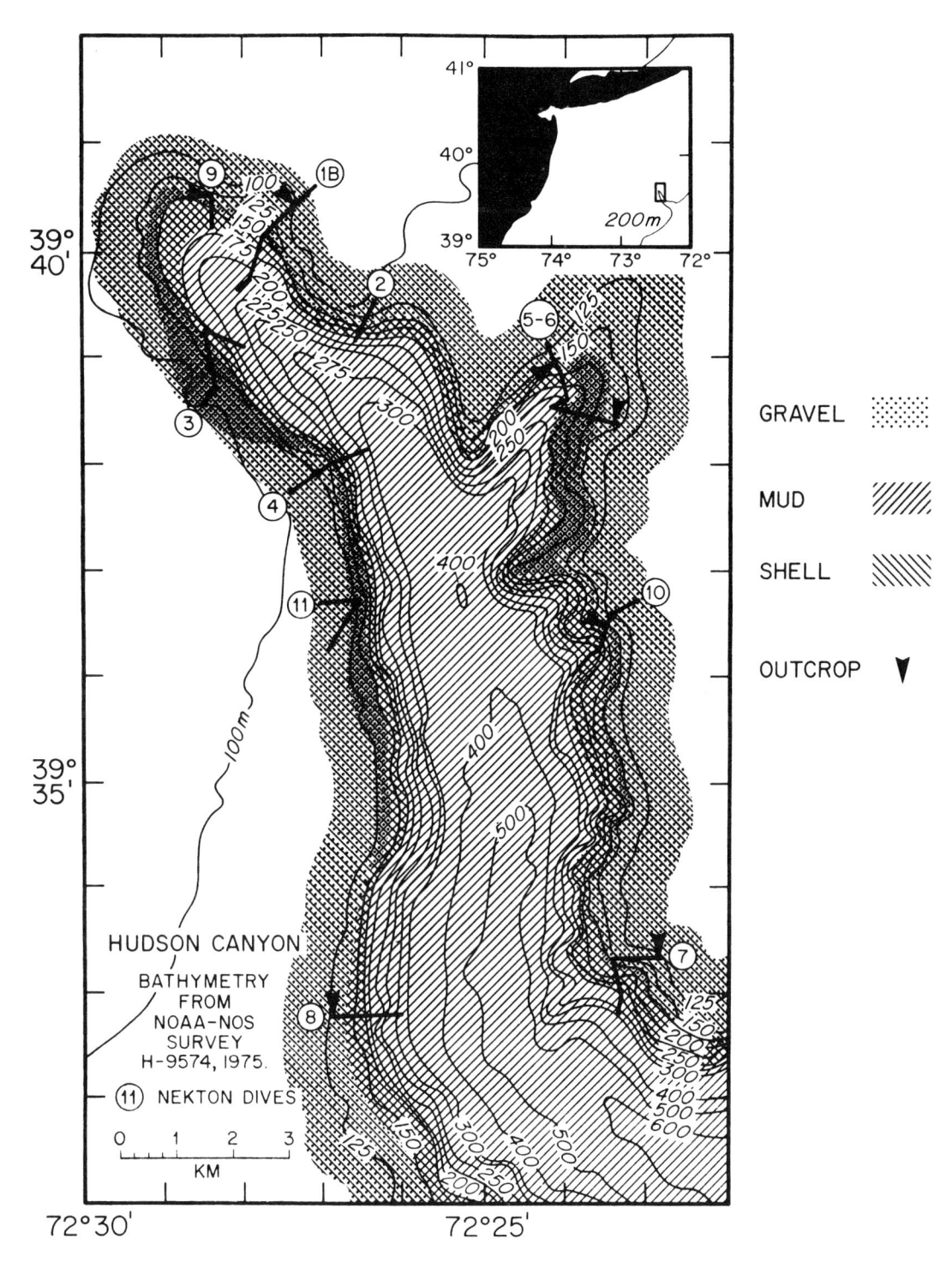

FIG. 9.—Surficial sediment distribution showing increased mud at, and just below, the shelfbreak at the head of the Hudson Canyon. This site, approximately 160 km southeast of New York Harbor, illustrates a Type III mudline. Depth in meters. From Stanley and Freeland, 1978.

the head of the Mississippi Trough, a large depression cut during lowered sealevel (perhaps at about 19,000–15,000 years B.P.) when a former mouth of the Mississippi extended to the head of this large submarine valley (Curray, 1960). This geomorphic feature (an erosional incision in the shelf, modified by retrogressive slumping) brings the shelfbreak close to the mudline. The mudline in this region occurs on the shelf in response to Mississippi River input (see discussion in Type IV section). Thus, in some cases, shelfedge configuration and proximity to supply prove to be controlling factors in producing a Type III mudline.

Off the east Texas and southwestern Louisiana coasts, Curray (1960) mapped a facies boundary between silty clay and sand-silt-clay at about 120 to 150 m while the shelfbreak occurs from about 70 to 110 m. Off the mouth of the Rio Grande the sand content drops from over 50% at a depth of 91 m to about 1% at 130 m (Behrens, unpublished file data). These areas, (d) and (e) respectively, are more distal from the large Mississippi mud source than the Louisiana Shelf but not so far removed as the Florida Shelf. Other rivers (Brazos, Colorado, Rio Grande) have contributed enough early Holocene sediment to this northwest Gulf shelf area for it to have distinctive mineralogic characteristics (the Western Gulf and Rio Grande Provinces, cf. Davies, 1972). These shelf deposits display a patchy sediment distribution pattern consisting of shelly sands and muds of early Holocene age (e.g., Nelson and Bray, 1970), some of which are reworked relict or palimpsest. New and/or reworked fines are carried in suspension westward on the inner shelf (Curray, 1960) and southward and offshelf at the shelfedge (D. W. McGrail, personal comm., 1982). Thus here, as in area (a), a small but important supply of fines, displaced by shelf winnowing and bypassing processes and transported by offshore currents, provides sufficient mud deposition on the uppermost slope to place the mudline near the shelfbreak. A continuous nepheloid layer has been recorded between shore and shelfedge (D. A. McGrail, personal comm., 1982). Algal and coral reefs that flourished during Pleistocene low sealevel stands near the shelfedge in this area (Uchupi and Emery, 1968) provide small amounts of carbonate sands to parts of the outer shelf and the shelfbreak. Offshelf transport of this or other coarse material and/or uppermost slope winnowing apparently depresses the mudline locally to as much as 50 m below the shelfbreak, but it is unlikely that it is depressed enough (100 m or more) to classify any of this area as Type II.

In sum, the Type III mudline in the northern Gulf is the result of a combination of several factors, the most important of which are an important supply of fluvial material, distance from these sources of input, variability in the shelfbreak configuration (depth, distance from shore), and the role of bottom currents which rework relict sediments and induce transport of fines along the margin and across the shelfbreak. In contrast, there are fewer of the above controlling factors for Type III on the East Coast margin, and the dominant condition is intense reworking of relict sediments by bottom currents near the outer edge of a broad shelf. In consequence, the Type III mudline is more prevalent in the Gulf than on the Atlantic margin.

TYPE IV EXAMPLES

Mudline positions above the shelfbreak and on the mid to outer shelf are locally important off southern New England (such as the mud patch south of Georges Bank) and the Mid-Atlantic States. However, as noted in the preceding section, the generally deep mudline position off large sectors of the U.S. East Coast is in part due to (a) the preferential deposition of fine sediment in estuaries and coastal to inner-shelf sectors and, consequently, low amounts of silt and clay transported across the broad shelf, and (b) the moderate to intense reworking and seaward winnowing of fine-grained material from the shelfedge landward or to the slope by the plexus of fluid-driven processes concentrated on outer shelf to upper slope environments.

In contrast, the Type IV mudline is mapped along extensive portions of the northern Gulf of Mexico margin (Fig. 5A) where input from sediment-laden rivers and bypassing the inner shelf clearly play a dominant role. Three important areas are recognized. (a) East of the Mississippi Delta the mudline extends onto the shelf and delineates the position of the relict prodeltaic muds of the Saint Bernard Delta (based on data from Ludwick, 1964); this latter depocenter was abandoned at the beginning of the formation of the presently active Mississippi Birdfoot delta. (b) From the Mississippi Birdfoot delta westward (to south of western Louisiana, 89°W to 92°W longitude), the mudline lies on the shelf. This shallow mudline (see profile G in Fig. 3) is directly related to both relict and modern delta complexes (Coleman, 1981), and in particular to the westwardly build-up of fine-grained sediment that extends from near the coastline to the shelfedge, and beyond onto the slope. Prevailing wind-generated surface currents and weak western current (Curray, 1960) carry enough suspended load from the Mississippi to deposit a muddy cover over extensive parts of the shelf, and to form a nearshore mudline at a depth of about 20 m (Fig. 5A). A dominant downwelling appears to affect the east–west oriented shelf west of the Mississippi River (D. W. McGrail, personal comm., 1982). The western limit of this Type IV area coincides with the boundary between the Mississippi and Western Gulf Provinces (cf. Davies, 1972; Fig. 5B).

The third area (c) occurs on the south Texas shelf between the Colorado-Brazos and Rio Grande delta systems (Fig. 3, profile B). Using sediment composition and textural data as well as hydrographic information, Curray (1960) concurred with earlier studies showing that this area is the site of convergence of both semipermanent surface currents and wind-driven longshore currents (Fig. 5B). These currents have distributed finely suspended particulate matter over much of the south Texas shelf. More recent beach sediment (Hayes, 1964; Watson, 1971) and drift-bottle studies (Watson and Behrens, 1970) support the existence of such a convergence of shelf currents. Alternatively, Shideler (1978) considered existing hydrographic data to be inadequate and proposed that the present sediment distribution is accounted for by net offshore and net southward, coastwise transport of sediment derived largely from the ancestral Brazos-Colorado Delta. These reworked relict materials have been deposited, along with modern river-derived and shelf-eroded mud, immediately south of this delta system.

Whether or not net sediment transport extends farther south than this Type IV area (Fig. 5A), continuing Holocene sediment dispersal processes have carried enough mud to and deposited it in this area to extend the mudline to the inner shelf (Fig. 10).

Mineralogical evidence (van Andel, 1960) suggests that local sediment sources (Brazos and Colorado Rivers, Cavallo and Aransas Passes), probably with some contribution from the Brazos Santiago Inlet and Rio Grande to the south, have provided much of the mud of this shelf deposit.

DISCUSSION

Mudline Position and Interplay of Dominant Variables

The texture of surficial sediment on margins, hypothetically, grades from sand (on the shelf) to silt (on the slope) to clay (on the rise), with a mudline positioned close to the shelfbreak. It is apparent from our review of two North American margins that this is not the usual case, and that the mudline position relative to the shelfbreak is a function of several variables. Some variables are more effective than others. On the margins studied, three of the consistently most important factors are shelf width, sediment supply (reworked coastal and shelf-derived as well as fluvial), and magnitude of fluid energy on the shelf, at the shelfbreak and on the upper slope (Fig. 1). Both margins examined are characterized by a generally broad, gently sloping shelf, fairly consistent shelfdepth (for the most part approximating 100 m), a slope with generally mod-

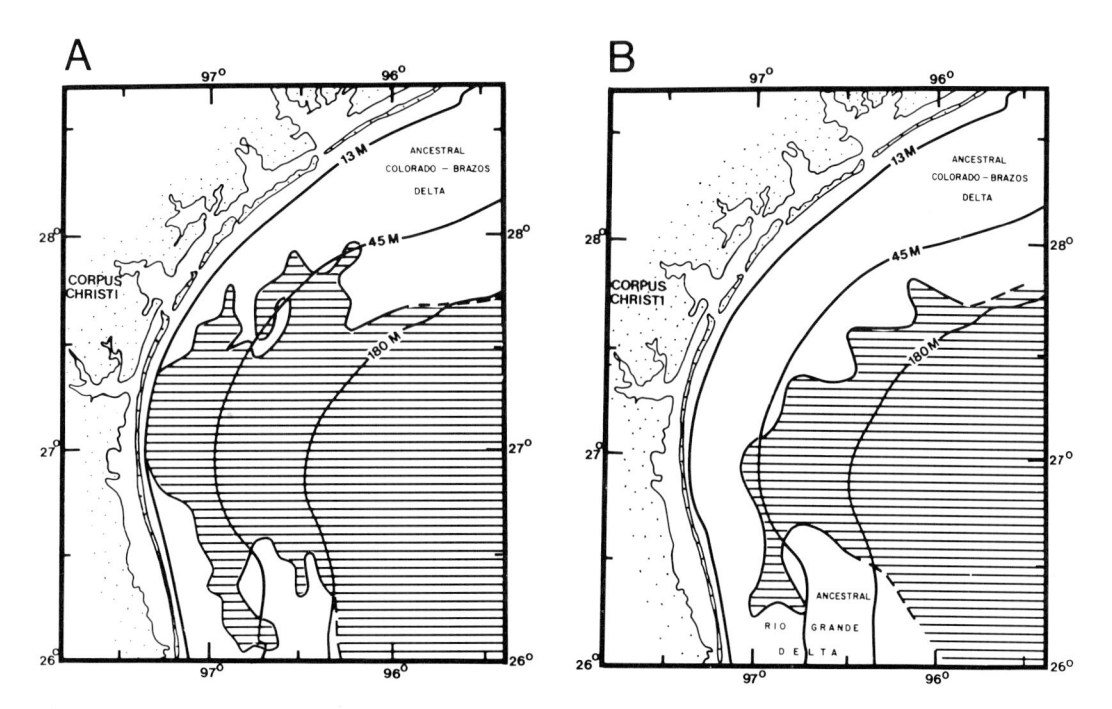

FIG. 10.—South Texas continental shelf, illustrating Type IV mudline. Shaded areas (bounded by mudline) have less than 11% sand. A, Modern surficial sediment distribution based on grab sample data (to 16 cm depth). B, Earlier Holocene sediment distribution based on upper 160 cm of gravity core section. Note seaward displacement of mudline during lower eustatic sealevel stand. Redrawn, with permission, from Shideler (1977, his Figs. 3 and 8).

erate gradient, and a relatively stable tectonic framework: they both would be considered passive margins. However, they differ in availability of sediment supplied to the shelf and fluid energy on the outer shelf and slope. Whereas the Atlantic outer shelf receives minimal fluvial-derived sediment at present, the northern Gulf has large areas receiving very high fluvial input as well as areas of minimal river supply. The margins also differ in that the Atlantic has higher tidal ranges and is exposed to a larger, higher latitude ocean and consequently to more frequent periods of high wave energy (chapters in Part V of this volume). This higher energy regime is capable of periodically eroding surficial sediments on the outer shelf and uppermost slope, sometimes extensively. The Gulf, on the other hand, is a semi-enclosed basin, with a more moderate, trade-wind dominated, subequatorial climate. Although it is a somewhat lower energy environment, strong flows are measured at the shelfedge (McGrail and Carnes, 1983).

The inter-related effects of the three variables on the mudline (sediment supply, fluid energy, shelf width) may be shown by a simplified set of models. The models may be depicted graphically by displaying fields of mudline types on graphs with axes representing low, moderate and high values of two of the three variables. Figure 11 shows fields of mudline types on graphs showing fluid energy versus sediment supply. The three shelf width conditions (narrow, moderate, wide) are shown on three separate graphs. These diagrams indicate that regardless of shelf width, mudline Type I prevails under high energy and low sediment supply conditions, and Type IV under low energy and high sediment supply. Mudline Type I results from high energy conditions which prevent deposition of mud on the outer shelf and erode and winnow it from the upper slope, as long as sediment supply is low. The offshelf spillover that depresses the mudline would be greater on a narrow shelf (e.g., Cape Hatteras) than on a wide one. Thus, this field is largest on the narrow width shelf graph (Fig. 11A, including much of the areas representing moderate levels of both energy and supply), and progressively diminishes on graphs representing moderate (Fig. 11B) and wide (Fig. 11C) shelves.

In contrast, mudline Type IV results from a large sediment supply, but insufficient energy to transport it offshelf (e.g., South Louisiana margin). Shelf width seems to affect the field of mudline Type IV less than it does Type I. Low energy is less likely to induce bypassing of mud across a wide shelf than a narrow shelf. It may be that part of this field—a wide shelf with extremely high sediment supply and low energy—can exist only ephemerally, because such conditions would lead to rapid deltaic progradation to the shelfedge (e.g., the Mississippi Birdfoot delta).

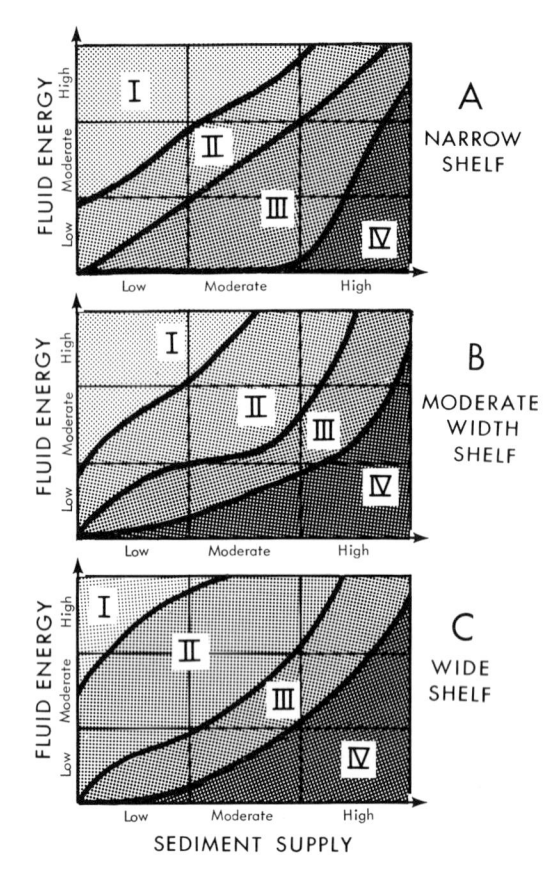

FIG. 11.—Simplified models showing inter-related effects of sediment supply, fluid energy and shelf width on the mudline. Fields of mudline Types (I, II, III and IV) are predicted as a function of the three variables. Further explanation is provided in text.

The field occupied by Type II mudline appears to prevail under conditions where energy levels control deposition more effectively than does sediment supply. That is, regardless of absolute levels, fluid energy is sufficient to bypass essentially all mud supplied to the shelf. Moreover, bypassing influences transport not only to the shelfedge, but further on the slope, depressing the mudline below the shelfbreak. As the shelf width increases and spillover to mid-slope depths (Type I mudline) decreases, upperslope Type II mudline becomes more prevalent. In sectors where the shelf is broad but lacks a substantial fluvial sediment supply, a moderate current system would cause offshelf spillover of sand-sized materials (e.g., off the Mid-Atlantic states).

The field occupied by Type III mudline occurs where more sediment is supplied than fluid processes can effectively erode, winnow and bypass beyond the shelfbreak (e.g., locally on the Alabama

Shelf west of DeSoto Canyon; head of Hudson Canyon). Shelf width *per se* seems to have less effect on this mudline type than on Type II.

This type of modeling, albeit overly simplified, helps explain why the mudline position is considerably more variable on the northern Gulf of Mexico margin where all four types are common than on the East Coast margin where Type II prevails. That is, there is greater variation in the three factors modeled—especially in sediment supply—on the Gulf margin. Although this conceptual classification was developed primarily for these two margins, it incorporates major factors which apply to all margins.

Influence of Tectonics and Associated Factors

Other factors in addition to those discussed above play a role, sometimes significant, in determining the mudline position over extensive portions of margins. For example, structural framework, which inevitably affects every margin to some extent, must be taken into account. Although both margins examined fall in the general passive margin classification, there are essential differences between the two. The Atlantic margin is a subsiding one and, locally, has undergone warping and rebound in response to the effect of glaciation and eustacism (Emery and Uchupi, 1972). The northern Gulf of Mexico margin is somewhat more stable, except in the region of the Mississippi delta where major subsidence has resulted from sediment loading (Martin and Case, 1974). Moreover, extensive areas of the Gulf are undergoing active salt tectonics (Martin, 1980) which locally alters the configuration of the margin, including the shelfbreak. This factor may locally displace the mudline by inducing failure of the upper slope reaches or by generating a local source of carbonate (reefal) sand that could spill down the adjacent, steep slope. Growth faults are common in the Gulf, some related to salt tectonics (Martin, 1978), others to sediment loading such as off the Mississippi delta. Active displacement of sediment along such faults positioned near the shelfbreak could modify, at least slightly, the position of the mudline.

It might be expected that the role of tectonics on mudline development would be even more important in seismically active (converging, subducted) margins. For example, in island arc settings characterized by high relief of the adjacent land mass, narrow shelves, and frequent earthquake tremors, the mudline would be depressed (Type I) as a result of offshelf spillover and failure of sediment onto steep inner trench slopes. Often associated with such margins is volcanic activity, both subaerial and submarine, which may be a significant source of coarse volcaniclastic sediments (Dickinson, 1974). The periodic addition, as overburden, of such material and their downslope displacement related to earthquake tremors would also result in mudlines positioned well below the shelfbreak.

The structural framework not only determines shelf width, earlier discussed as a primary factor of mudline control, but also shelf configuration and associated features. Significant in this respect are shelfbreak depth, the location of submarine canyon heads, particularly those heading on the shelf, and the shape of the shelf-slope break (Wear et al., 1974). Mudline position, for example, would be shallower in the case where canyons act as mud traps (for example, head of Hudson Canyon), but depressed (the more prevalent case) where the axes of valleys serve as active channels through which coarse shelf sediments are transported downslope (Fig. 4). Shallow, abrupt shelfbreaks also favor depression of the mudline by bringing outer shelf sediments into a higher fluid energy regime, promoting offshelf spillover (Cape Hatteras margin). Spillover across an abrupt shelf-slope break and onto a steep slope favors sediment transport, sometimes to great depths. Also of structural influence are depressions related to high amplitude folds on the slope parallel to the shelfedge (Buffler et al., 1978); such depressions can act as traps for sediment and thus arrest any further downward displacement of the mudline below the shelfedge.

Role of Deposition and Related Variables

Sediment supply, of primary influence in determining mudline position (Fig. 11), may be affected locally or regionally by either tectonically or geomorphologically caused changes, or combinations thereof. Any phenomenon affecting deposition clearly is a significant variable. Examples include marked alterations in drainage patterns induced by stream capture or a major shift in delta distributaries (e.g., shift of the outflow of the Mississippi to the Atchafalaya River).

Mudline position also may be influenced by factors affecting sediment stability. Instability resulting from changes in physical properties and reduction in shear strength leads to failure resulting, in turn, in offshelf spillover and downslope sediment movement. For example, excess pore water pressure associated with high sedimentation rates, coupled with pulses of deposition and sediment loading, commonly lead to slumps in delta fronts (Coleman et al., 1983, this volume). Increased slope instability and slumping may also result from *in situ* gas production or gas seepage through the sediment column from deeper sources. In the Gulf of Mexico, gas seepage associated with salt diapirs and growth faults is sometimes concentrated near the shelf margin (Addy and Worzel, 1979).

On a smaller scale, the role of bioturbation, on either tectonically active or passive margins, is to be considered. As they rework the seafloor, benthic organisms alter the primary fabric of surficial sediments which, on the slope, may decrease stability and lead to sediment failure (Stanley, 1971; Blake and Doyle, 1983). In the presence of offshelf water movement, both gas seepage and bioturbation may lead to mud winnowing by providing fine-grained sediment to the water column (i.e., resuspension by non-fluid-driven processes).

Sedimentation at the shelfedge also may be modified by a variety of human activities, such as dumping, and it is conceivable that these effects could induce some changes in the mudline position. For example, ocean dumping of disposable waste has increased steadily. In 1970 there were 246 known disposal sites off the coasts of the U.S., several of them near or beyond the shelfedge and involving the dumping of more than 48 million tons of waste in the ocean (Cargo and Mallory, 1974). The largest single source of waste reworked by man and dumped on the shelf is continuous dredging operations to keep harbors and rivers navigable. Continued operations of this type along some margins may affect the sediment supply factor and thus, in time, alter the mudline model depicted earlier. Moreover, more than 60 nations are known to be actively engaged in offshore exploration and production of hydrocarbons. On land, enthusiasm for building dams may significantly affect the sediment supply factor of mudline control (Fig. 11); the High Aswan Dam on the Nile River is a case in point. These and other activities, such as nuclear blasts, laying of underwater pipelines and cables, and discharging highly saline brine obtained from deep wells and from excavation of salt mines near the coast, also could influence mudline displacement, at least locally.

Mudline Position and Sealevel Changes

A final consideration is the effect on mudlines of sealevel changes through time. To a first approximation, we would expect mudlines to rise or fall proportionally with sealevel. However, effects on the geographic and relative mudline positions may vary considerably with mudline type. Type IV mudlines, for example, would have been virtually eliminated by the late Pleistocene sealevel lowering of 100 m or more to near the present shelfedge. In marked contrast, the transgression which responded to glacial melting would have been accompanied by the rising of mudlines to their present positions. A well-documented case occurs on the south Texas margin where grab sample data define the present mudline on the inner shelf while short core results place the earlier Holocene mudline farther offshore in the midshelf region (compare A and B in Fig. 10; see also Shideler, 1977). Although this example pertains to a Type IV mudline, types II and III would also migrate similarly in response to glacial eustatic oscillations.

Concurrent with Pleistocene sealevel lowering, the delivery to the shelfedge of most fluvially transported sediment would drastically increase sediment supply to the upper slope and thus would likely depress Type III and II mudlines distinctly more than the extent of sealevel lowering, at least near the point-sources of sediment. Each mudline type would then tend to change into a deeper type, with types II and I becoming prevalent.

Mudline positions distal from river sources may be less affected by sealevel changes and thus may record long-term averaging of the energy of sediment-transporting fluid processes. Such averaging would most likely be expressed by the moderately deep Type II mudlines. A well-documented example of this comes from the Desoto Canyon region of the Gulf of Mexico (between profiles K and L; Figs. 3 and 5A) where two core transects comprising 20 cores, each at least six meters long, were examined. Ages and sedimentation rates in two cores from the same area reported by Emiliani et al. (1975) indicate that these cores span a time period of at least 30,000 years. Thus the textural data shown in Figure 2 record a long-term term averaging (late Pleistocene to Holocene) of the energy of sediment-transporting fluid processes. The averaging effect of using several (at least four) samples from each core produces highly correlated texture—depth relationships for eight grain-size distribution parameters. These plotted relationships enable us to objectively map the mudline in this region. The mudline depth (about 260 m) indicated by the percentage of mud (Fig. 2A) does not coincide with similar sharp changes in the distribution pattern of seven other textural parameters (Fig. 2B). However, all parameters change rapidly with depth through a transition zone and then become almost constant as shown by the low slope of the regression lines below a depth of 200 m to 460 m. The eight parameters shown in Figure 2 record the significant decrease in competence of sediment transporting mechanisms on the upper slope below 120 m. Individual parameters reflect different aspects of the fluid energy regime as it changes with water depth. For example, the marked alteration in the rate of change of percentage of clay (<4 mμ) well below the mudline records the role of resuspension by weaker currents which extends the transition zone for this parameter from about 260 to 460 m.

Pre-existing Type I mudlines would be little affected because of their depth well below both glacial period and present sealevels. Type I mudlines, more extensive during low stands than at present, correlate with the narrow shelf model (Fig. 11A) discussed earlier.

Mudline: Its Significance

In sum, this general review emphasizes that the mudline position relative to the shelfbreak is controlled by the interplay of numerous variables. These variables are not equally effective in determining the mudline position, and each margin will reflect a different combination of controlling factors. It is of note that the mudline is in many cases an effective indicator of sediment transport competence and, in some cases, of the boundary between erosion and deposition. As such, the mudline serves as an energy-level marker, and identification of its position is of considerable value to physical oceanographers attempting to define conditions on outer continental margins. Moreover, modeling some of the mudline-controlling variables has potential value in interpreting ancient continental margin environments and, specifically, in attempting to identify the shelfedge in the rock record.

SUMMARY

1. The mudline is the depth on the outer continental margin where proportions of clayey silt, silty clay and clay no longer increase significantly with depth; this mappable horizon often occurs near, but is only occasionally coincident with, the shelfbreak.

2. A survey of surficial sediment distributions on North American margins, namely off the Mid-Atlantic States and in the northern Gulf of Mexico, highlights marked discrepancies between mudline depth and the shelfbreak, and reveals the presence of four mudline types, coded I to IV.

3. Type I mudline occurs at considerable depth on the upper- to mid-slope (300 to >1000 m); examples are mapped on the Cape Hatteras and west Florida margins where currents result in offshelf spillover of sand onto the slope.

4. Type II mudline occurs on the uppermost slope to approximately 200 to 400 m, and examples are mapped off large sectors between Norfolk and Wilmington canyons, and on the Panama City, Florida, margin; its position records where, over time, shear-induced resuspension has largely exceeded the threshold of sediment transport.

5. Type III mudline is nearly coincident with the shelfbreak, the latter commonly ranging from 100 to 175 m; examples include the head of Hudson Canyon lying at a depth near the intersection of density fronts separating Shelf and Slope waters, and sectors of the Mississippi delta margin in proximity to important fluvial supply and transport to the shelfedge of these sediments by fluid-driven energy.

6. Type IV mudline occurs considerably shoreward of the shelfedge (<100 m), and is mapped where large amounts of sediment are supplied, particularly in the northern Gulf, west of DeSoto Canyon where Mississippi and other river input is laterally dispersed by the regionally important flow of water masses.

7. We suggest that the position of the mudline on all margins is largely a response to the interplay of three controlling factors on the outer shelf, shelfbreak and upper slope: shelf width, sediment supply (reworked coastal and shelf-derived, and fluvial), and magnitude of fluid energy.

8. In modelling the mudline position, it becomes apparent that regardless of shelf width, Type I prevails under conditions of high energy and low sediment supply (prevalent on narrow, shallow shelves such as Cape Hatteras), and Type IV results from a large sediment supply but insufficient energy to transport it offshelf (e.g. off the south Louisiana margin).

9. Type II mudline appears to prevail under conditions where energy levels control deposition more effectively than does sediment supply (e.g. off the Mid-Atlantic States); Type III mudline occurs where more sediment is supplied than fluid processes can effectively erode, winnow and bypass seaward, beyond the shelfbreak (e.g. locally on the Alabama margin west of DeSoto Canyon).

10. To properly model the mudline position, other variables also must be integrated: outer margin configuration, structural framework, and factors likely to reduce sediment stability and induce failure (seismic activity, high sedimentation rates off deltas, salt tectonics and growth faults, bioturbation, and increased human activity near the shelfedge).

11. The mudline position has been considerably displaced through time, largely as a result of eustatic sealevel oscillations: during Pleistocene minimal lowstands, Type I mudlines became prevalent, Type III mudlines were displaced downslope becoming Type II, and Type IV did not develop.

12. Our survey indicates that the mudline, in many cases, serves as an energy-level marker that helps define the boundary between erosion and deposition; this concept is useful to physical oceanographers determining conditions on outer margins, and also can be used by geologists interpreting ancient continental margin environments.

ACKNOWLEDGMENTS

The authors thank A. H. Bouma, D. W. McGrail, and B. A. McGregor for reviewing an earlier version of this chapter and providing helpful suggestions. Appreciation is also expressed to R. Bennett, A. L. Gordon, G. H. Keller and G. L. Shideler for permission to modify and reproduce as figures several of their previously published illustrations. Funding used for preparation of the study was obtained from Smithsonian Institution Scholarly Studies grant no. 1233S201 awarded to D. J. S. This is University of Texas Institute for Geophysics Contribution no. 531.

REFERENCES

ADDY, S. K., BEHRENS, E. W., HAINES, T. R., SHIRLEY, D., AND WORZEL, J. L., 1979, Correlation of some lithologic and physical characteristics of sediments with high frequency subbottom reflection types: Offshore Tech. Trans., Paper No. 3569, p. 1869–1877.

———, AND WORZEL, J. L., 1979, Gas seeps and subsurface structure off Panama City, Florida: Am. Assoc. Petroleum Geologists Bull., v. 63, p. 668–675.

BLAKE, N. J., AND DOYLE, L. J., 1983, Infaunal-sediment relationships at the shelf-slope break, in Stanley, D. J., and Moore, G. T., eds., The Shelfbreak: Critical Interface on Continental Margins: Soc. Econ. Paleontologists Mineralogists Spec. Pub. 33, p. 381–389.

BOUMA, A. H., 1972, Distribution of sediments and sedimentary structures in the Gulf of Mexico, in Rezak, R., and Henry, F. J., eds., Contributions on the Geological and Geophysical Oceanography of the Gulf of Mexico: College Station, Texas, Texas A&M Univ. Oceanogr. Studies, v. 3, p. 35–65.

BOWMAN, M. J., 1977, Hydrographic Properties; MESA New York Bight Atlas Monograph, 1: Albany, N.Y., New York Sea Grant Inst. 78 p.

BUFFLER, R. T., SHAUB, F. J., WATKINS, J. S., AND WORZEL, J. L., 1978, Anatomy of Mexican Ridges, Southwestern Gulf of Mexico, in Watkins, J. S., Montadert, L., and Dickerson, P. W., eds., Geological and Geophysical Investigations of Continental Margins: Am. Assoc. Petroleum Geologists Memoir, 29, p. 319–327.

CARGO, D. N., AND MALLORY, B. F., 1974, Man and His Geologic Environment: Reading, Mass., Addison-Wesley Publishing Co., 548 p.

COCHRANE, J. D., 1972, Separation of an anticyclone and subsequent development in the Loop current (1969), in Capurro, L. R. A., and Reid, F. L., eds., Contributions on the Physical Oceanography of the Gulf of Mexico: College Station, Texas, Texas A&M Univ. Oceanogr. Studies, v. 2, p. 91–106.

COLEMAN, J. M., 1981, Deltas, Processes and Models of Deposition for Exploration (2nd edition): Minneapolis, Minnesota, Burgess Publishing Co., 124 p.

———, PRIOR, D. B., AND LINDSAY, J. F., 1983, Deltaic influences on shelfedge instability processes, in Stanley, D. J., and Moore, G. T., eds., The Shelfbreak: Critical Interface on Continental Margins: Soc. Econ. Paleontologists Mineralogists Spec. Pub. 33, p. 121–137.

CURRAY, J., 1960, Sediments and history of Holocene transgression, continental shelf, Northwest Gulf of Mexico, in Shepard, F. P., Phleger, F. B., and Van Andel, T. H., eds., Recent Sediments, Northwest Gulf of Mexico: Tulsa, Oklahoma, Am. Assoc. Petroleum Geologists, p. 221–265.

DAVIES, D. K., 1972, Deep sea sediments and their sedimentation, Gulf of Mexico: Am. Assoc. Petroleum Geologists Bull., v. 56, p. 2212–2239.

DICKINSON, W. R., ed., 1974, Tectonics and Sedimentation: Soc. Econ. Paleontologists Mineralogists Spec. Pub. 22, 204 p.

DOYLE, L. F., PILKEY, O. H., AND WOO, C. C., 1979, Sedimentation on the eastern United States continental slope, in Doyle, L. H., and Pilkey, O. H., eds., Geology of Continental Slopes: Soc. Econ. Paleontologists Mineralogists Spec. Pub. 27, p. 119–129.

———, AND SPARKS, T. N., 1980, Sediments of the Mississippi, Alabama and Florida (NAFLA) continental shelf: Jour. Sed. Petrology, v. 50, p. 905–916.

EMERY, K. O., AND UCHUPI, E., 1972, Western North Atlantic Ocean: Topography, Rocks, Structure, Water, Life and Sediments: Am. Assoc. Petroleum Geologists Memoir 17, 532 p.

EMILIANI, C., GARTNER, S., ELDRIDGE, K., AND OTHERS, 1975, Paleoclimatological analysis of late Quaternary cores from the northeastern Gulf of Mexico: Science, v. 189, p. 1083–1088.

GIBSON, T. G., HAZEL, J. E., AND MELLO, J. F., 1968, Fossiliferous rocks from submarine canyons off the Northeastern United States: U.S. Geol. Survey Professional Paper, v. 600-D, p. D222–D230.

GORDON, A. L., AMOS, A. F., AND GERARD, R. D., 1976, New York Bight water stratification—October 1974, in Gross, M. G., ed., Middle Atlantic Continental Shelf and the New York Bight: Lawrence, Kansas, American Soc. Limnol. Oceanogr., p. 45–57.

GOULD, H. R., AND STEWART, R. H., 1955, Continental terrace sediments in the northeastern Gulf of Mexico, in Hough, J. L., and Menard, H. W., eds. Finding Ancient Shorelines; A Symposium: Soc. Econ. Paleontologists Mineralogists Spec. Pub. 3, p. 2–20.

GRIFFIN, G. M., 1962, Regional clay-mineral facies—products of weathering intensity and current distribution in the northeastern Gulf of Mexico: Geol. Soc. America Bull., v. 73, p. 737–768.

HAYES, M. O., 1964, Grain size modes in Padre Island sands, in Scott, A. J., ed., Gulf Coast Assoc. Geol. Socs. Field Trip Guidebook: Gulf Coast Assoc. Geol. Socs., p. 121–126.

HOLLISTER, C. D., 1973, Atlantic continental shelf and slope of the United States—texture of surface sediments from New Jersey to southern Florida: U.S. Geol. Survey Professional Paper, v. 529-M, p. M1–M23.

HOTCHKISS, F. C., AND WUNSCH, C., 1982, Internal waves in Hudson Canyon with possible geological implications: Deep-Sea Res., v. 29, p. 415–442.

HUNT, R. E., SWIFT, D. J., AND PALMER, H., 1977, Constructional shelf topography, Diamond Shoals, North Carolina: Geol. Soc. America Bull., v. 88, p. 299–311.

KARL, H. A., CARLSON, P. R., AND CACCHIONE, D. A., 1983, Factors that influence sediment transport at the shelfbreak, in Stanley, D. J., and Moore, G. T., eds., The Shelfbreak: Critical Interface on Continental Margins: Soc. Econ. Paleontologists Mineralogists Spec. Pub. No. 33, p. 219–231.

KELLER, G. H., LAMBERT, D. N., AND BENNETT, R. H., 1979, Geotechnical properties of Continental Slope deposits—

Cape Hatteras to Hydrographer Canyon, *in* Doyle, L. J., and Pilkey, O. H., eds., Geology of Continental Slopes: Soc. Econ. Paleontologists Mineralogists Spec. Paper No. 27, p. 131–151.

———, AND SHEPARD, F. P., 1978, Currents and sedimentary processes in submarine canyons off the northeast United States, *in* Stanley, D. J., and Kelling, G., eds., Sedimentation in Submarine Canyons, Fans, and Trenches: Stroudsburg, Pa., Dowden, Hutchinson and Ross, p. 15–32.

LEIPPER, D. F., 1954, Physical oceanography of the Gulf of Mexico, *in* Galtsoff, P. S., coordinator, Gulf of Mexico, Its Origin, Waters and Marine Life: U.S. Dept. Interior, Fishery Bull. 89, p. 119–137.

LUDWICK, J. C., 1964, Sediments in northeastern Gulf of Mexico, *in* Miller, R. L., ed., Papers in Marine Geology: New York, McMillan Co., p. 204–238.

LYALL, A. D., STANLEY, D. J., GILES, H. N., AND FISHER, A., JR., 1971, Suspended sediment and transport at the shelfbreak and on the slope: Mar. Techn. Soc. Jour., v. 5, p. 15–27.

MARTIN, R. G., 1978, Northern and Eastern Gulf of Mexico continental margin; stratigraphic and structural framework, *in* Bouma, A. H., Moore, G. T., and Coleman, J. M., eds., Framework, Facies and Oil-trapping Characteristics of the Upper Continental Margin: Am. Assoc. Petroleum Geologists, Studies in Geology No. 7, p. 21–42.

———, 1980, Distribution of salt structures in the Gulf of Mexico: map and descriptive text: U.S. Geol. Survey Map MF-1213, 2 sheets, 8 p.

———, AND CASE, J. E., 1974, Geophysical studies in the Gulf of Mexico, *in* Stehli, F., and Nairn, A., eds., Ocean Basins and Margins, Volume. III, The Gulf of Mexico and Caribbean Sea: New York, Plenum, p. 65–106.

MCGRAIL, D. W., AND CARNES, M., 1983, Shelfedge dynamics and the nepheloid layer in the northwestern Gulf of Mexico, *in* Stanley, D. J., and Moore, G. T., eds., The Shelfbreak: Critical Interface on Continental Margins: Soc. Econ. Paleontologists Mineralogists Spec. Pub. No. 33, p. 251–264.

MEADE, R. H., SACHS, P. L., MANHEIM, F. T., HATHAWAY, J. C., AND SPENCER, D. W., 1975, Sources of suspended matter in waters of the Middle Atlantic Bight: Jour. Sed. Petrology, v. 45, p. 171–188.

MILLIMAN, J. D., PILKEY, O. H., AND ROSS, D. A., 1972, Sediments of the continental margin off the eastern United States: Geol. Soc. America Bull., v. 83, p. 1315–1334.

MOORE, G. T., 1973, Submarine current measurements—Northwest Gulf of Mexico: Trans. Gulf Coast Assoc. Geol. Socs, v. 23, p. 245–255.

NELSON, H. F., AND BRAY, E. E., 1970, Stratigraphy and history of the Holocene sediments in the Sabine-High Island area, Gulf of Mexico, *in* Morgan, J. P., ed., Deltaic Sedimentation: Soc. Econ. Paleontologists Mineralogists Spec. Pub. 15, p. 48–77.

NEWTON, J. G., PILKEY, O. H., AND BLANTON, J. O., 1971, An oceanographic atlas of the Carolina continental margin: Raleigh, North Carolina Dept. Conserv. Dev. Pub., 57 p.

NOWLIN, W. D., 1972, Winter circulation patterns and property distributions, *in* Capurro, L. R. A., and Reid, J. L., eds., Contributions on the Physical Oceanography of the Gulf of Mexico: College Station, Texas, Texas A&M Univ. Oceanogr. Studies, v. 2. p. 3–52.

PIETRAFESA, L. J., 1983, Shelfbreak circulation, fronts and physical oceanography: East and West coast perspectives, *in* Stanley, D. J., and Moore, G. T., eds., The Shelfbreak: Critical Interface on Continental Margins: Soc. Econ. Paleontologists Mineralogists Spec. Pub. No. 33, p. 233–250.

RONA, P. A., 1969, Middle Atlantic continental slope of United States deposition and erosion: Am. Assoc. Petroleum Geologists Bull., v. 53, p. 1453–1465.

RUZECKI, E. P., AND WELCH, C. S., 1977, Description of shelf water-slope water fronts in the Middle Atlantic Bight [abs.]: Trans. Am. Geoph. Union, v. 58, p. 888.

SHIDELER, , G. L., 1977, Late Holocene sedimentary provinces, South Texas outer continental shelf: Am. Assoc. Petroleum Geologists Bull., v. 61, p. 708–722.

———, 1978, A sediment dispersal model for the South Texas continental shelf, Northwest Gulf of Mexico: Mar. Geology, v. 26, p. 289–313.

SIDNER, B. R., GARTNER, S., AND BRYANT, W. R., 1978, Late Pleistocene geologic history of Texas outer continental shelf and upper continental slope: Am. Assoc. Petroleum Geologists, Studies in Geology No. 7, p. 243–266.

SOUTHARD, J. B., AND STANLEY, D. J., 1976, Shelf-break processes and sedimentation, *in* Stanley, D. J., and Swift, D. J. P., eds., Marine Sediment Transport and Environmental Management: New York, Wiley, Interscience, p. 351–377.

STANLEY, D. J., 1971, Bioturbation and sediment failure in some submarine canyons: Vie et Milieu, Supplement 22, p. 541–555.

———, AND FENNER, P., 1973, Underwater television survey of the Atlantic outer continental margin near Wilmington Canyon: Washington, D.C., Smithsonian Contrib. Earth Sciences 11, 54 p.

———, AND FREELAND, G. L., 1978, The erosion-deposition boundary in the head of Hudson submarine canyon defined on the basis of submarine observations: Mar. Geology, v. 26, p. M37–M46.

———, SHENG, H., LAMBERT, K. N., RONA, P. A., MCGRAIL, D. W., AND JENKYNS, J. S., 1981, Current-influenced depositional provinces, continental margin off Cape Hatteras, identified by petrologic method: Mar. Geology, v. 40, p. 215–235.

———, AND WEAR, C. M., 1978, The "mud-line:" an erosion-deposition boundary on the upper continental slope: Mar. Geology, v. 28, p. M19–M29.

STETSON, H. C., 1939, Summary of sedimentary conditions on the continental shelf off the East Coast of the United States, *in* Trask, P. D., ed., Recent Marine Sediments: Tulsa, Oklahoma, Am. Assoc. Petroleum Geologists, p. 230–244.

TRUMBULL, J. V. A., 1972, Atlantic continental shelf and slope of the United States-sand-sized fraction of bottom sediments, New Jersey to Nova Scotia: U.S. Geol. Survey Professional Paper, v. 529-K, 45 p.

UCHUPI, E., AND EMERY, K. O., 1968, Structure of continental margin off Gulf coast of the United States: Am. Assoc. Petroleum Geologists Bull., v. 52, p. 1162–1193.

VAN ANDEL, T. H., 1960, Sources and dispersion of Holocene sediments, Northern Gulf of Mexico, in Shepard, F. P., Phleger, F. B., and Van Andel, T. H., eds., Recent Sediments, Northwest Gulf of Mexico: Tulsa, Oklahoma, Am. Assoc. Petroleum Geologists, p. 34–55.

———, AND CURRAY, J. R., 1960, Regional aspects of modern sedimentation in northern Gulf of Mexico and similar basins, and paleogeographic significance, in Shepard, F. P., Phleger, F. B., and Van Andel, T. H., eds., Recent Sediments, Northwest Gulf of Mexico: Tulsa, Oklahoma, Am. Assoc. Petroleum Geologists, p. 345–364.

VANNEY, J. R., AND STANLEY, D. J., 1983, The shelfbreak: definition and classification, in Stanley, D. J., and Moore, G. T., eds., The Shelfbreak: Critical Interface on Continental Margins: Soc. Econ. Paleontologists Mineralogists Spec. Pub. No. 33, p. 1–24.

WATSON, R. L., 1971, Origin of shell beaches, Padre Island, Texas: Jour. Sed. Petrology, v. 41, p. 1105–1111.

———, AND BEHRENS, E. W., 1970, Nearshore surface currents, Southeastern Texas Gulf Coast: Contributions in Marine Sci., Univ. Texas, v. 15, p. 133–143.

WEAR, C. M., STANLEY, D. J., AND BOULA, J. E., 1974, Shelf-break physiography between Wilmington and Norfolk canyons: Mar. Technol. Soc. Jour. v. 8, p. 37–48.

SEISMIC FACIES OF SHELFEDGE DEPOSITS, U.S. PACIFIC CONTINENTAL MARGIN

MICHAEL E. FIELD, PAUL R. CARLSON, AND ROBERT K. HALL
U.S. Geological Survey, Menlo Park, California 94025

ABSTRACT

Pacific-style continental margins, such as that of western North America, are marked by large contrasts in the type of shelfedge sedimentary deposits and the processes that form them. The Pacific shelves of the United States are generally much narrower than the Atlantic shelves, and the source areas exhibit more relief. The greater relief of Pacific coast source terranes results in a relatively high rate of sedimentation in humid areas and fluctuating (areally and seasonally) sedimentation patterns and rates in semiarid areas. Sediment shed from the adjacent landmass is discharged, generally seasonally, onto the Pacific Continental Shelf at point sources. Many of the sediment sources of the northwestern United States and southern Alaska feed directly onto swell- and storm-dominated shelves. On such narrow unprotected shelves, sediment has a short residence time in submarine deltaic deposits before being remobilized and dispersed to outer-shelf and upper-slope environments.

Through study of high-resolution seismic-reflection profiles, we have identified four principal types of shelfedge deposits: (1) starved, (2) draped, (3) prograded, and (4) upbuilt and outbuilt. Each type of shelfedge deposit results from a characteristic balance between sedimentation rate and distributive energy (waves and currents) and is, therefore, characterized by distinctive seismic facies and bedding patterns. A special type, the cut-and-fill shelfedge, and a composite type consisting of two or more of the main depositional styles supplement the four principal types of shelfedge. Incorporated within each of these facies, especially on the upper slope, are chaotic deposits formed by slumps or slides, which are common along tectonically active margins.

INTRODUCTION

Many discussions on sedimentation and structure of the shelfedge treat it as a one-dimensional feature. It is, however, a *zone,* measuring hundreds to thousands of meters in width, that marks the transition from a continental-shelf to a continental-slope environment (Southard and Stanley, 1976). The types of sediment in these two environments may either be indistinguishable or differ markedly in texture and composition. In either case, the shelf-edge is the transition zone over which gradual or abrupt changes from shallow marine to deep marine occur. Recognition of deposits that are diagnostic of the shelfedge may facilitate recognition of adjacent facies and interpretation of local eustatic/tectonic history and of ancient sedimentary environments. The combination of factors that controls deposition at the shelfedge is fairly restrictive to that environment, and this relation suggests that the resulting deposits are diagnostic of their setting.

CONTROLLING FACTORS IN SHELFEDGE SEDIMENTATION

Deposition at the shelfedge is generally controlled by the same factors that control deposition elsewhere in the marine environment: the types and amounts of available sediment, and the types and amounts of energy available for transportation and redistribution. These factors are, in turn, controlled by a host of variables, ranging from the relief, petrology, and climate of the source region to the energy of seasonal storm waves over the depositional site. Most of these factors are too complex to evaluate except qualitatively. In the present context, the significant controls are the terrestrial and marine climate and morphology.

Terrestrial climate, as well as terrane, influences the amounts, rates, and types of sediment delivered to the shoreline. Thus, in humid areas, silt- and clay-size sediment is delivered in substantial amounts in a relatively continuous flow, in sharp contrast to arid to semiarid climates, where sand is a major product of weathering, and transport occurs sporadically or seasonally, so that large sediment loads are delivered to the coast and subsequently reworked.

The morphology of the continental margin can markedly influence paths of sediment transport (Vanney and Stanley, 1983). In a greatly simplified view, the inner margin can be divided into two main types of shelf morphology: simple or complex (Fig. 1). Morphologically simple shelves slope gently and are succeeded by a more pronounced declivity on the slope, whereas complex shelves are characterized by depressions or elevations of the seafloor. Irregularities on the shelf occur both as depressions (shelf valleys, canyons, small basins) and as elevated local bathymetric highs (uplifted anticlines, diapirs, reefs, fault blocks); both depressions and highs inhibit sedimentation on the shelfedge by acting as a sink or dam, respectively, for sediment transported across the shelf. The presence or absence of canyon systems, cut into the conti-

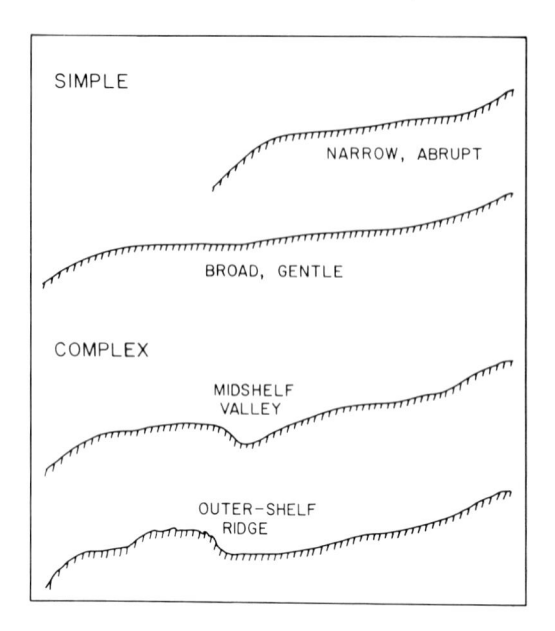

FIG. 1.—Typical profiles of U.S. Pacific Continental Shelf. Simple shelf profiles allow for direct sediment transport across shelf, whereas those exhibiting even minor relief, either positive or negative, tend to trap sediment emanating from coastal or inner-shelf region.

forms and ridgetops in the Southern California Borderland have narrow to wide shelves and minimal terrestrial sediment sources. The shelf off Oregon and Washington is generally wider (25–50 km) than off California, and the shelf off southern Alaska locally exceeds 100 km in width. Although most of the U.S. Pacific shelf has a simple morphology, some complex areas are present throughout the entire extent of the U.S. Pacific margin. Barriers to sediment transport occur as resistant bedrock outcrops off Oregon, Alaska, and central and southern California, as uplifted diapiric ridges off Oregon and northern California, and as shelf valleys in the Gulf of Alaska. Outcrop areas that exhibit very little relief (<5 m) are nonetheless efficient in trapping at least a portion of the sediment transported across the shelf. Most of the shelf is exposed to long-period waves and storm activity that is capable of transporting sediment grains at

nental margin, influences how currents affect the shelf-slope break. For example, canyons appear to have a focusing effect on such oceanographic phenomena as internal waves and storm waves, so that large bedforms are concentrated near some canyon heads (Karl et al., 1983).

The oceanographic climate of an area comprises the long-term patterns of all fluid flows (see chapters in Part V, this volume). These flows include both currents (tidal, density, geostrophic, and wind generated) and waves (surface-wind generated, internal, tsunamis). The magnitude of these processes dictates the directions of transport and the type and size of material transported to the shelfedge; these processes also control reworking of sediment at the shelfbreak, which may be a significant process (Karl et al. 1983).

U.S. PACIFIC SHELFEDGE: CONTRASTS IN SETTING

From Mexico to Alaska, the U.S. Pacific shelfedge consists of many variations in morphology and sediment type, which reflect corresponding variations in rock type, climate, and tectonic history. Maps of the shelfedge off California, Oregon, and Washington (Fig. 2) and off southern Alaska (Fig. 3) show marked differences in shelf width that are significant in regard to the amounts and types of material deposited on the shelfedge. Off southern and central California the shelf is generally very narrow, locally less than 2 km wide. Island plat-

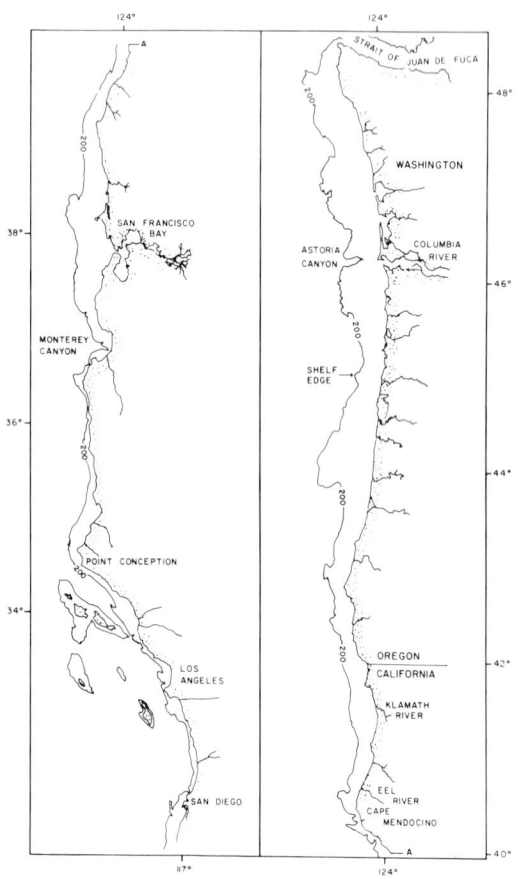

FIG. 2.—California, Oregon, and Washington continental margin, showing approximate position of shelfedge (200-m contour). Note that south of Cape Mendocino the shelf is generally narrow, and north of Cape Mendocino generally broad and gentle.

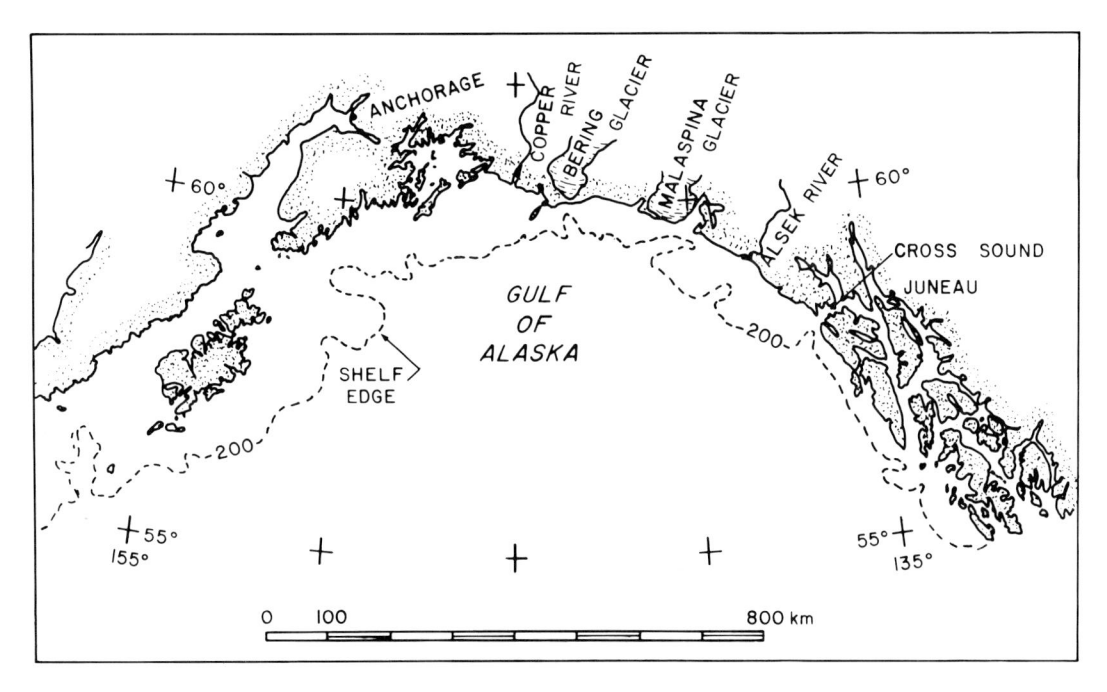

FIG. 3.—Gulf of Alaska, showing approximate position of shelfedge (200-m contour). Southern Alaska shelf is generally very wide and morphologically complex owing to presence of valleys, reentrants, ridges, and island banks.

shelfedge depths (Komar et al., 1972; Southard and Stanley, 1976). Superimposed on these predictable seasonal phenomena are sporadic tsunamis; Schatz (1965) reported that five notable tsunamis struck the shores of the Pacific Northwest during the period 1945–1965.

The climate along the Pacific coast varies from tropical-dry-continental to cold-temperate-marine (Table 1). These climatic variations influence weathering in the region and thus control to some degree the size and mineralogy of particles delivered to the coast, as well as the mode and the rates

TABLE 1.—CHARACTERISTICS OF THE U.S. PACIFIC SHELFEDGE

Setting	Shelf Width	Typically 20–50 km Range, <2 – >100 km
	Depth	Range 100–200 m
	Morphology	*Simple:* broad, gentle, to narrow, steep *Complex:* shelf valleys and ridges, submarine canyons
	Oceanic Conditions	Long-period swell Violent winter storms Tsunamis
Sediment Factors	Terrestrial Climate	Cold-temperate to marine (southern Alaska) Warm-temperate to marine (Pacific Northwest) Subtropical to wet-marine (northern California) Subtropical to tropical, to dry-continental (central to Southern California)
	Sediment Sources	Glacial melt water streams (southern Alaska) Large permanent streams (southern Alaska, Washington, Oregon, northern California) Small seasonal streams (central and Southern California)
History	Sealevel Changes	Eustatic transgressions and regressions Tectonic uplift and subsidence

at which they are delivered. South of Cape Mendocino, California, most streams are small point sources that carry significant amounts of sediment only during winter and spring rainfall periods. Rainfall north of Cape Mendocino is considerably higher, and streams are larger and carry material year round, although the summer discharge is greatly reduced relative to that in other seasons. In the Gulf of Alaska, summer melting of coastal glaciers contributes large amounts of fine material. The distribution of fluvial sediment reaching the coast is markedly asymmetric: of the 99 million tons of suspended sediment entering the Pacific between Mexico and Canada annually, 37% is delivered to the small section of northern California north of Cape Mendocino where the Eel, Mad, Klamath, and Smith Rivers discharge, and 16% is contributed by the Columbia River (Curtis et al., 1973; Griggs and Hein, 1979). The remaining 47% is discharged from numerous creeks and streams between Mexico and Canada, none of which contributes more than 5% of the total. In the Gulf of Alaska, the Copper River alone discharges more than 100 million tons of sediment to the ocean annually, more than all the streams of Washington, Oregon, and California combined (Curtis et al., 1973). Added to this volume is the contribution from numerous other streams, such as the Alsek River, to the Gulf of Alaska.

The entire Pacific shelf underwent multiple eustatic transgressions and regressions during the Quaternary. Superimposed on these changes are local tectonic influences, such as subsidence (Oxnard shelf) or uplift (Santa Monica and Crescent City, California; northern Gulf of Alaska), which alter depositional patterns on the shelfedge through changes in the energy regime or creation/destruction of sediment barriers.

DEPOSITIONAL PATTERNS ON THE U.S. PACIFIC SHELFEDGE

Rationale

Few data are presently available that enable the interpretation of precisely how, and at what rates, sediment grains are deposited on the Pacific shelfedge. The variables are numerous and complex, and without adequate long-term monitoring programs it is difficult to determine the source of particles, their means of transport to the shelfedge, and their history of resuspension and redeposition. A somewhat less definitive, but more practical, means of assessing the depositional history of shelfedge sedimentary units is to characterize the geometry of these units. Recognition of patterns in the distribution of different types of sedimentary unit permits, in turn, an inference of the controlling factors and aids delineation of facies relations.

We have examined several hundred high-resolution seismic-reflection profiles of the shelfedge off California, Oregon, Washington, and southern Alaska to establish empirically the dominant depositional patterns along the U.S. Pacific margin. This review resulted in the identification of four principal types of shelfedge: (1) starved, (2) draped, (3) prograded, and (4) upbuilt and outbuilt. Each type has a distinctive geometry and, by inference, a distinctive depositional history. In addition, there is a specialized shelfedge type (cut-and-fill) and a composite shelfedge type that reflects at least several sequential and distinct patterns of sedimentation. Each of these six types is discussed here in the framework of its configuration, the resulting sedimentary facies, and the factors that control its formation (sedimentation rate and available energy). Of necessity, these controls are discussed in a relative sense, without supporting quantitative data; such data are generally unavailable and, in most instances, not necessary for evaluation of the

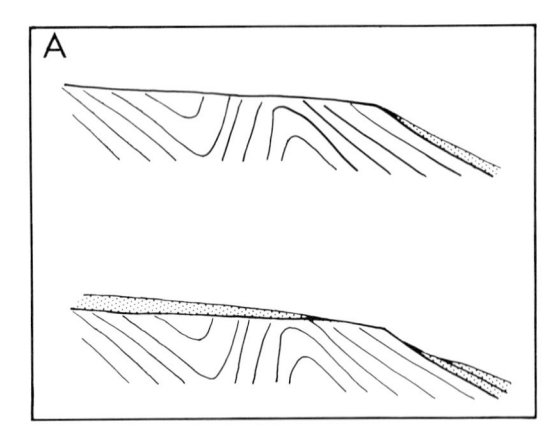

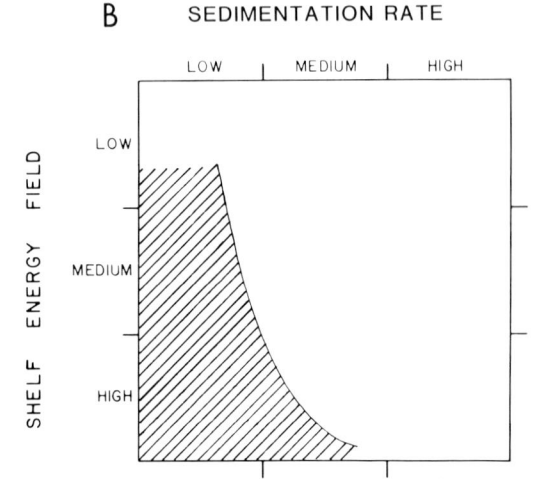

FIG. 4.—*A*, Depositional units on a starved shelfedge. *B*, Relation between sedimentation rate and shelf energy that results in a starved shelfedge-configuration.

models proposed here. For example, the sedimentation rates off major streams (e.g., the Columbia, Copper, Eel, and Santa Clara Rivers) are considered high, in contrast to rates in areas with limited or negligible sediment sources, such as the offshore ridges and island platforms in the Southern California Borderland and the Gulf of Alaska. Similarly, the term "shelf energy field" is a general term that includes all oceanic processes that result in sediment entrainment and movement on the outer shelf.

Type 1. Starved Shelfedge

Much of the present-day shelfedge of the U.S. Pacific margin is sediment starved. Starved shelfedges, which are common to shelves of both simple and complex morphology, generally result from a low rate of sedimentation, which in turn may result from a dearth of material reaching the shelf, blockage of the shelfedge by morphologic barriers, or a combination of oceanic processes that transport sediment away from or past the shelfbreak (Table 2). Some shelves are marked by an absence of sed-

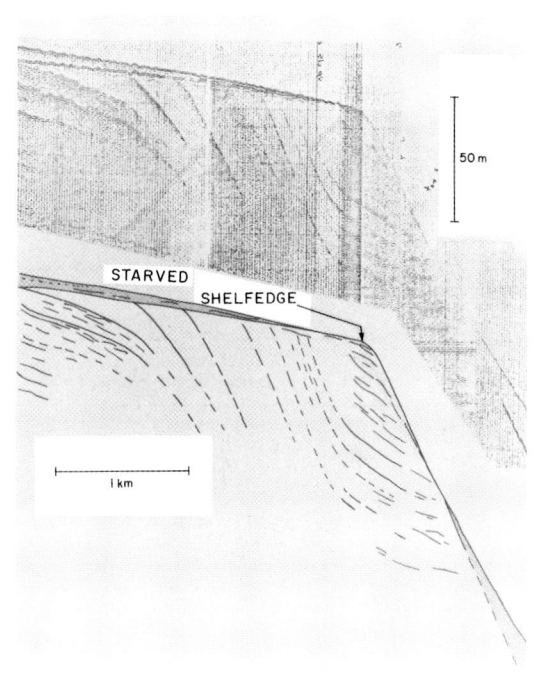

FIG. 6.—Typical starved shelfedge on Santa Rosa-Cortes Ridge in Southern California Borderland. Many shelf areas in Gulf of Alaska have profiles similar to those shown here and in Fig. 5.

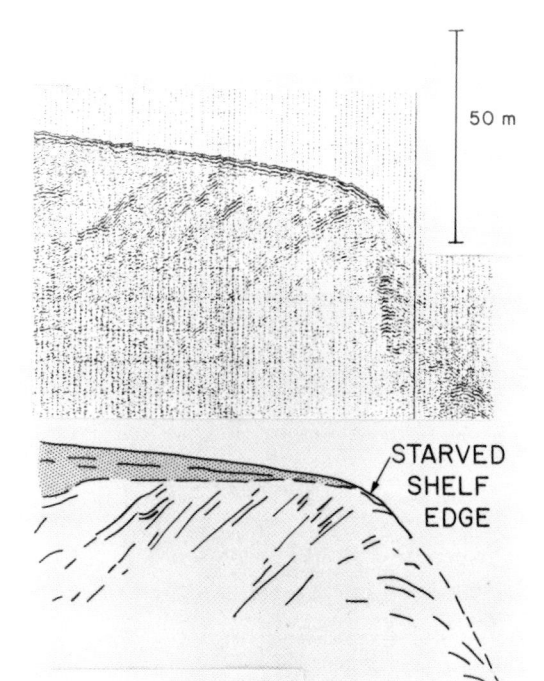

FIG. 5.—Typical starved shelfedge in Southern California Borderland south of Santa Rosa Island. Note absence of sediment on upper slope as well as at shelfedge.

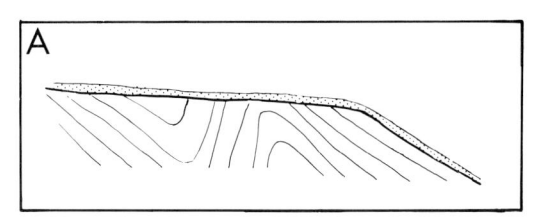

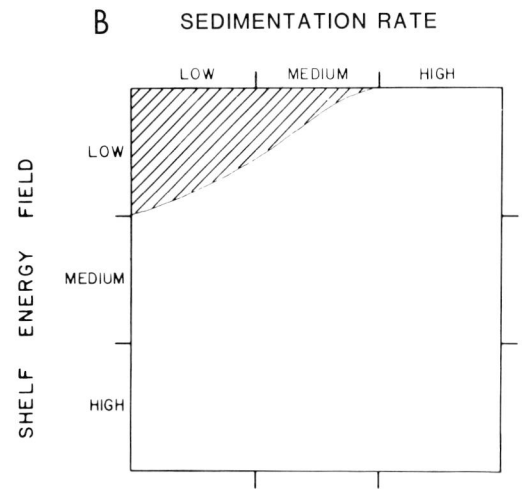

FIG. 7.—A, Depositional units on a draped shelfedge. B, Relation between sedimentation rate and shelf energy that results in a draped shelfedge configuration.

TABLE 2.—DEPOSITIONAL PATTERNS AND CONTROLLING FACTORS ON THE U.S. PACIFIC SHELFEDGE

Shelfedge Type	Controlling Factors	Depositional Patterns
1. Starved	a) Low sedimentation rate b) High-energy shelf environment c) Tectonic uplift d) Presence of outer-shelf barriers	a) Shelf sediment is absent or pinches out seaward b) Slope beds are subparallel to divergent. c) Shelfedge is barren d) No continuity between shelf and slope beds
2. Draped	a) Low sedimentation rate b) Low-energy shelf environment c) Subsidence d) Midshelf barriers	a) Thin seafloor-parallel beds from outer shelf to upper slope b) Continuous sedimentation across shelfbreak c) Very gradual facies transition from shelf to slope
3. Prograded	a) High sedimentation rate b) High-energy shelf environment c) Tectonic uplift d) Absence of shelf barriers	a) Shelf beds are thin to thick, parallel, and subhorizontal b) Slope beds are parallel to divergent c) Shelf-to-slope facies transition may be gradual or discontinuous (abrupt) across shelfbreak d) Slump blocks may be present
4. Upbuilt and Outbuilt	a) High sedimentation rate b) Low- to high-energy shelf environment c) Subsidence d) Absence of shelf barriers	a) Shelf-to-slope sedimentary sequence may consist of continuous subparallel beds (facies continuous across shelfbreak) b) Shelf-to-slope sedimentary sequence may be truncated with prograded shelf beds abutted by divergent slope beds (facies discontinuous at shelfbreak) c) Slump blocks likely present
5. Cut and Fill (special case)	a) Varying sedimentation rate b) High-energy shelf environment c) Eustatic sealevel oscillations d) Local uplift	a) Shelfedge erosional unconformity b) Prograding fill over unconformity
6. Composite	a) Change in sedimentation rate b) Change in shelf energy field c) Recent uplift and (or) subsidence	Stacked sedimentary units across shelf-to-slope zone that display all or some of the above patterns

iment across the width of the shelf, whereas others have a clearly defined wedge of modern sediment that thins seaward and pinches out landward of the shelfbreak (Fig. 4A). In either case, a marked discontinuity in sedimentary facies exists between the shelf and the adjacent slope, where onlapped sequences of parallel to divergent beds occur. Figures 5 and 6 show examples of starved shelfedges.

Maintenance of a starved configuration depends on a balance between sediment supplied and available energy for transport and removal (Fig. 4B). Although most starved shelfedges have low sedi-

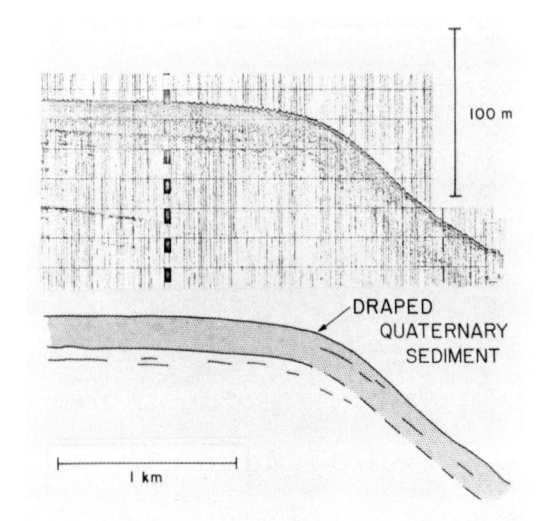

FIG. 8.—Typical draped shelfedge on Continental Shelf off northern Washington near Juan de Fuca Strait.

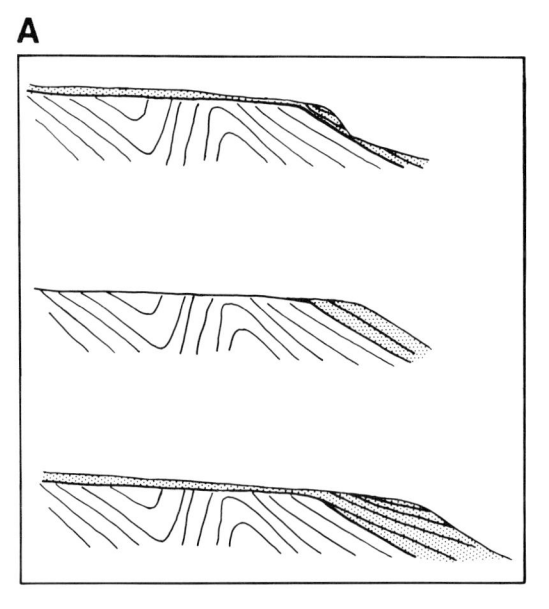

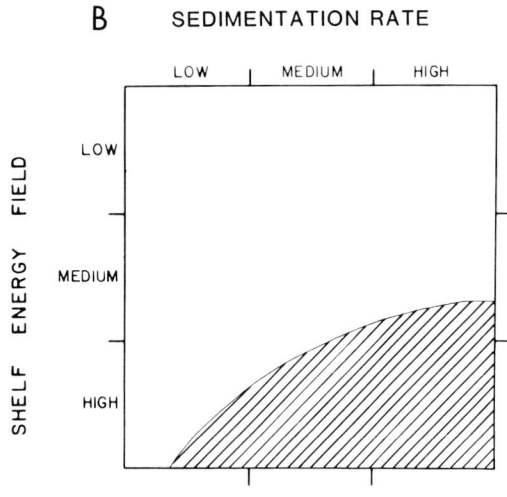

FIG. 9.—*A*, Depositional units on a prograded shelf-edge. *B*, Relation between sedimentation rate and shelf energy that results in a prograded-shelfedge configuration.

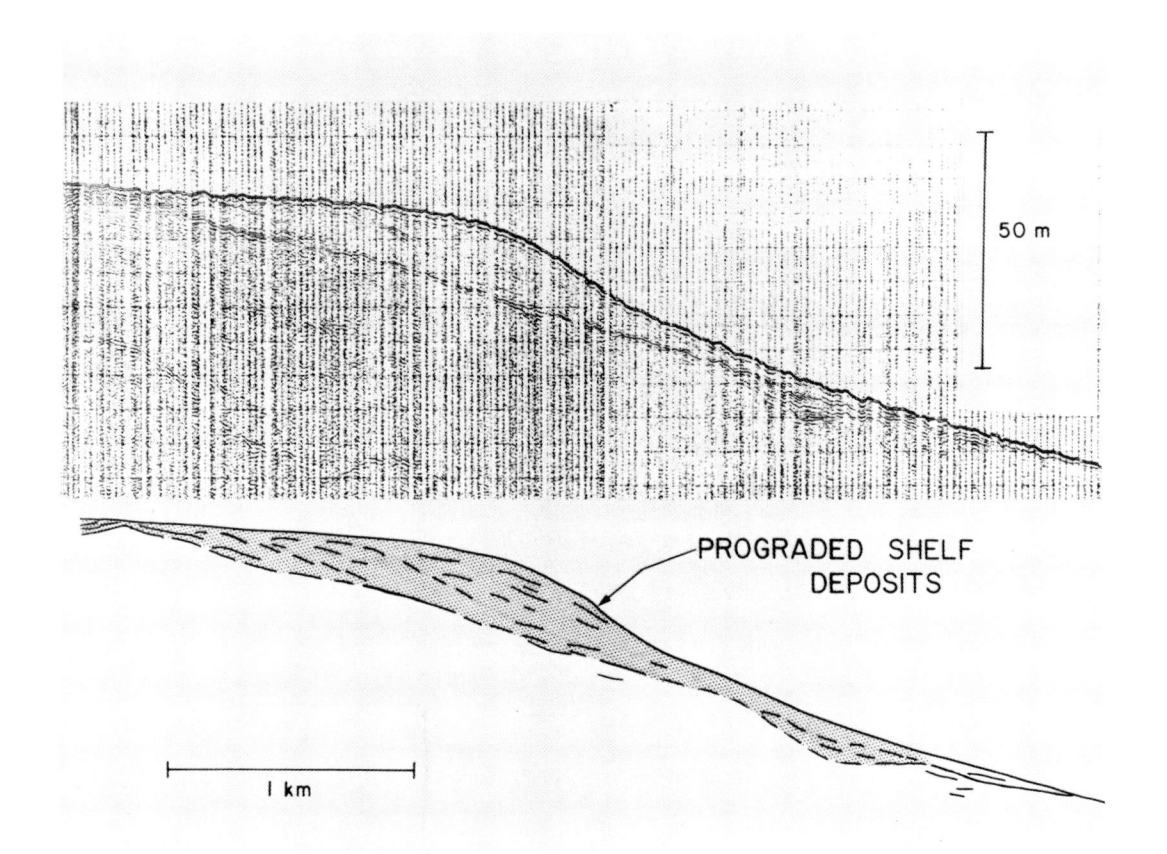

FIG. 10.—Typical prograded shelfedge similar to that shown in Fig. 9*A* (top), on Southern California mainland shelf off Santa Monica.

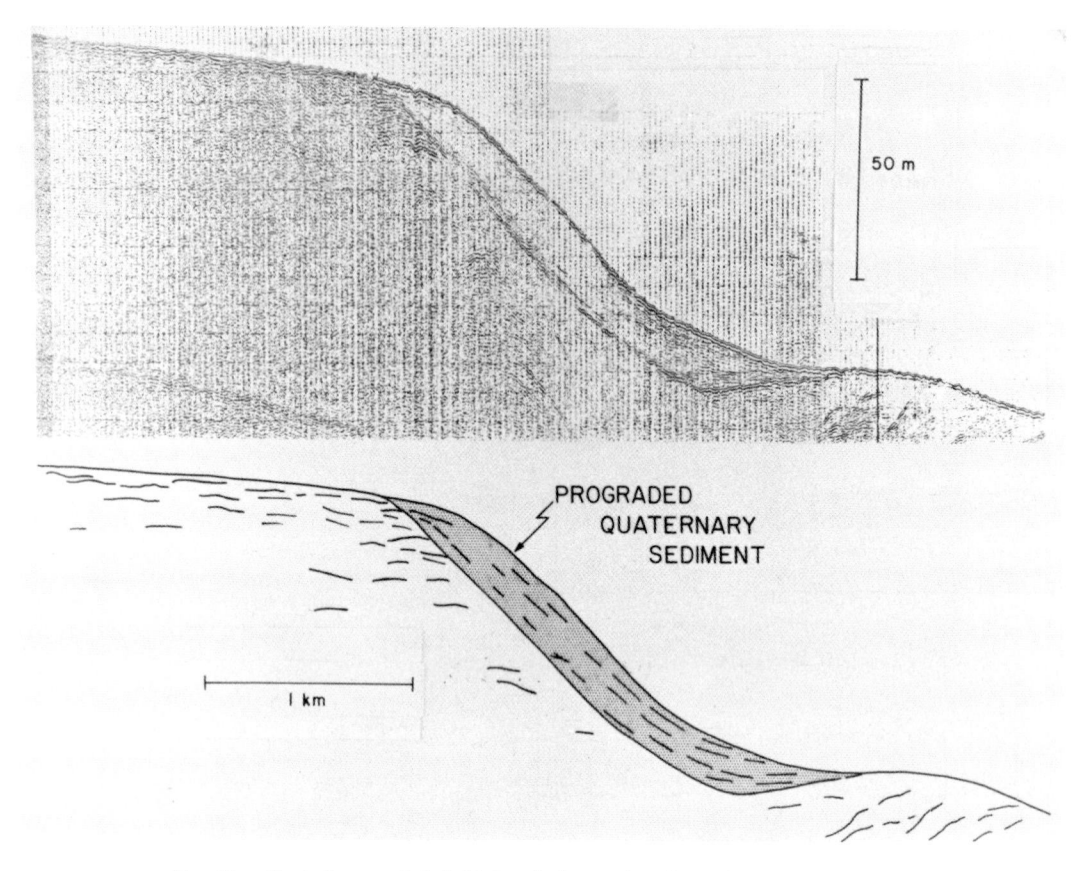

FIG. 11.—Typical prograded shelfedge similar to that shown in Fig. 9A (middle).

ment input, an increase in energy can accommodate a moderate increase in sediment supply and still maintain a barren shelfedge. Very low fields of energy, however, will lead to a draped-shelfedge configuration.

Type 2. Draped Shelfedge

Draped shelfedges are characterized by continuous uniform sedimentation from the midshelf areas, across the shelfbreak, to the upper slope (Fig. 7A). The resulting sedimentary sequence is a series of seafloor-parallel beds, commonly only several meters thick. If gradients in composition and texture occur, they are probably quite subtle and only detectable by careful analysis of sediment across the entire transect.

Low sedimentation rate combined with low distribution energy produces a draped shelfedge (Fig. 7B; Table 2). These factors may be enhanced by subsidence of the shelf, which decreases the energy level, or by the presence of midshelf barriers that entrap a large portion of the sediment emanating from the coast. The principal depositional mode on this type of shelfedge is probably by suspension fallout, which distributes a nearly uniform (in thickness and lithology) layer of material across the shelfbreak (Fig. 8); sand bodies and bedforms are rare. The energy-field/sedimentation rate relation is such that only mutually low values maintain a draped configuration (Fig. 7B). Draped shelfedges are relatively uncommon on the U.S. Pacific continental margin, owing in large part to the absence of areas along the west coast of North America that exhibit both low energy and a low rate of sedimentation.

Type 3. Prograded Shelfedge

Prograded shelfedges are a common and complex form of seismic facies that are characterized by lateral accretion or out-building. The form of the accretionary prism may vary considerably; the three most common types are illustrated in Figure 9A, and examples are shown in Figures 10 and 12. Sediment on the shelf is either a thin veneer or a uniform sequence composed of thin parallel subhorizontal beds (Fig. 9A; Table 2). In the zone from the shelfbreak onto the upper slope, beds range in attitude from parallel (middle, Fig. 9A), through

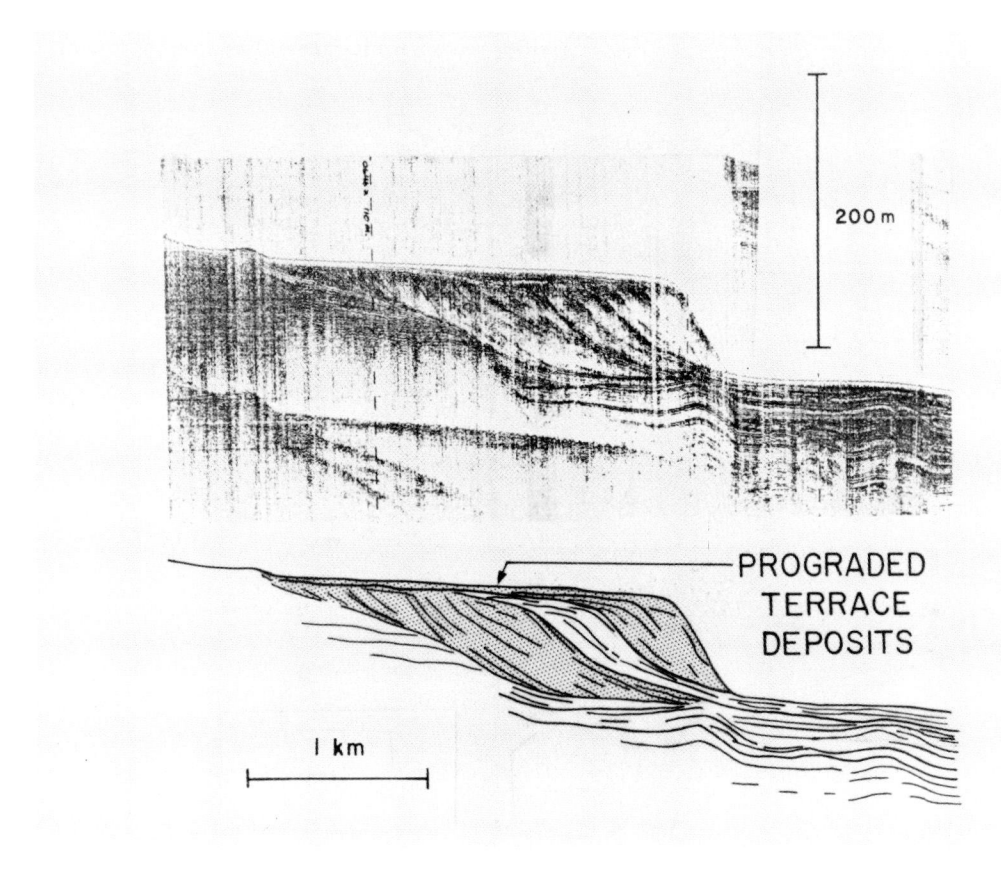

FIG. 12.—Configuration typical of a prograded shelfedge, or terrace deposit (similar to that shown in Fig. 9A, bottom); geophysical profile (from Greene and others, 1975) from Tanner Bank, Southern California Borderland.

subparallel (top, Fig. 9A), to divergent (bottom, Fig. 9A). Sediment thickness increases markedly at the shelfbreak (Field and Richmond, 1980). The transition in sediment composition and texture from the outer shelf to the upper slope may be either gradual or abrupt.

Prograded shelfedges are common in areas of relatively high sedimentation rates and high fields of shelf energy. Increasing sedimentation rates will cause progressive changes in the pattern of deposition, as shown in Figures 10 through 12. Uplift may be important locally in causing or amplifying these conditions. Shelf barriers, such as uplifted ridges or broad valleys, are generally absent or buried. Construction of a prograded shelfedge requires a balance between sediment input and the available hydrodynamic energy that transports and redistributes the sediment (Fig. 9B); any change in this balance may result in a change in the dominant seismic facies. For example, if a given prograded shelfedge has a high sedimentation rate, a decrease in energy level may yield an upbuilt and outbuilt shelfedge.

For a prograding shelfedge with a low sedimentation rate, however, a decrease in available energy will result in a draped shelfedge, and an increase in energy will induce scour and bypassing to form a starved shelfedge. The sediment composing prograded shelfedges is most likely clayey silt containing laminae and very thin beds of fine sand.

Deposition is probably episodic on most prograded shelfedges. Sediment is deposited on the shelf during major storms and subsequently reworked and transported across the shelf to deeper water. Rapid accumulation of unconsolidated sediment on a prograded shelfedge may lead to formation of medium-scale (100–1000 m) to large-scale (1–10 km) slump blocks on the shelfedge and upper slope (Field and Edwards, 1980; Field, 1981).

Type 4. Upbuilt and Outbuilt Shelfedge

Very high sedimentation rates result in a shelf-slope accretionary wedge that builds upward and outward from the shelfedge (Fig. 13). In some areas, the position of the shelfedge may remain

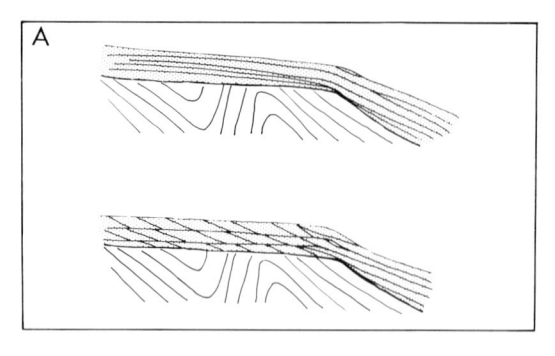

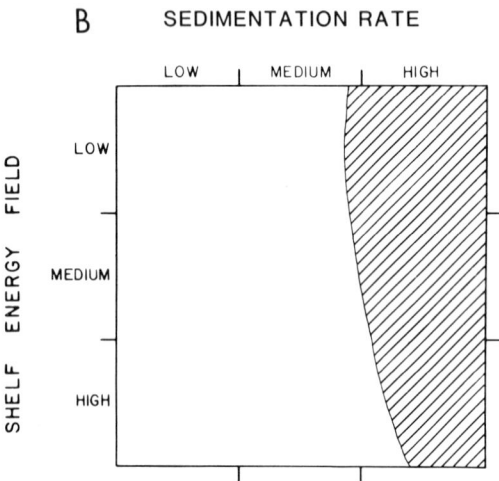

FIG. 13.—A, Depositional units on an upbuilt and outbuilt shelfedge. B, Relation between sedimentation rate and shelf energy that results in an up-built and outbuilt-shelfedge configuration.

constant or even migrate shoreward, but in all areas, significant deposition occurs on both the outer shelf and upper slope. The energy field may be low or high because the dominant factor is a large and nearly continuous supply of sediment (Fig. 13B). Two major types of upbuilt and outbuilt shelfedge are identified, each of which shows a distinctive depositional pattern. A continuous and abundant supply of sediment will result in a shelf-to-slope sedimentary sequence composed of continuous subparallel beds (top, Fig. 13A). In this case the energy field is low, and a gradual transition in sedimentary facies occurs across the shelfbreak. With a lesser amount of sediment input, the resulting configuration would be a draped shelfedge. Accretion may be enhanced by subsidence, either tectonic in origin or induced by sediment loading. High energy fields accompanying high sedimentation rates yield sediment bodies prograding outward from the shoreline across the shelf (bottom, Fig.

13A). As shown in Figures 13A and 14, repetition of the outbuilding process during marine regressions, followed by truncation of the sedimentary units during a marine transgression, results in a stacked sequence of prograding beds separated by local unconformities. Discontinuities related to sea level changes are less apparent in adjacent slope beds, and the shelfbreak marks a facies discontinuity between divergent slope beds and prograded shelf beds. The rapid and thick accumulation of unconsolidated sediment on an upbuilt and outbuilt shelfedge results in numerous medium- to large-scale sediment slides and sediment gravity flows, especially off major rivers, where fine-grained sediment rich in organic material accumulates in pulses of sedimentation (Molnia et al., 1977; Coleman and Prior, 1981, Field, 1981).

Type 5. Cut-and-Fill Shelfedge (Special Case)

The foregoing major types of shelfedge deposits are generally applicable to the entire Pacific shelf-edge, regardless of local morphology. The entire shelfedge was affected by eustatic changes in sea-level during the Quaternary, and so most depositional patterns on the shelfedge directly reflect these eustatic changes. Higher than average rates of erosion on some segments of the shelfedge, however, resulted in a conspicuous notching and subsequent infilling by sediment. This type of shelfedge, termed cut-and-fill, is characterized by a distinctive seismic facies (Figs. 15, 16). Cut-and-fill facies are formed by cutting of a concave-upwave erosional surface into the convex-upwave shelf-break morphology during a period(s) of low sealevel. Sediment infills the notched surface in a fashion dictated by sediment load and available energy; most commonly the fill is progradational in form, although it may also occur as a sediment drape. Depositional and erosional events that follow the infilling plane off the upper surface and bury it, so that, finally, a characteristic shelfedge sediment lens is preserved, bounded by unconformities.

Type 6. Composite Shelfedge

As local changes in sediment load or energy field occur in an outer shelf area, the pattern of sediment deposition at the shelfbreak likewise changes. Thus, in areas where a thick section of Quaternary sediment is present on the shelfedge, several episodes of deposition can be delineated. Each episode yields a starved, draped, prograded, or upbuilt and outbuilt shelfedge; the type of deposit formed during each depositional episode is controlled by the same factors as those listed in Table 2. Figure 17 shows a high-resolution seismic-reflection profile from the mainland shelfedge off Santa Monica in Southern California. We note that reconstruction of

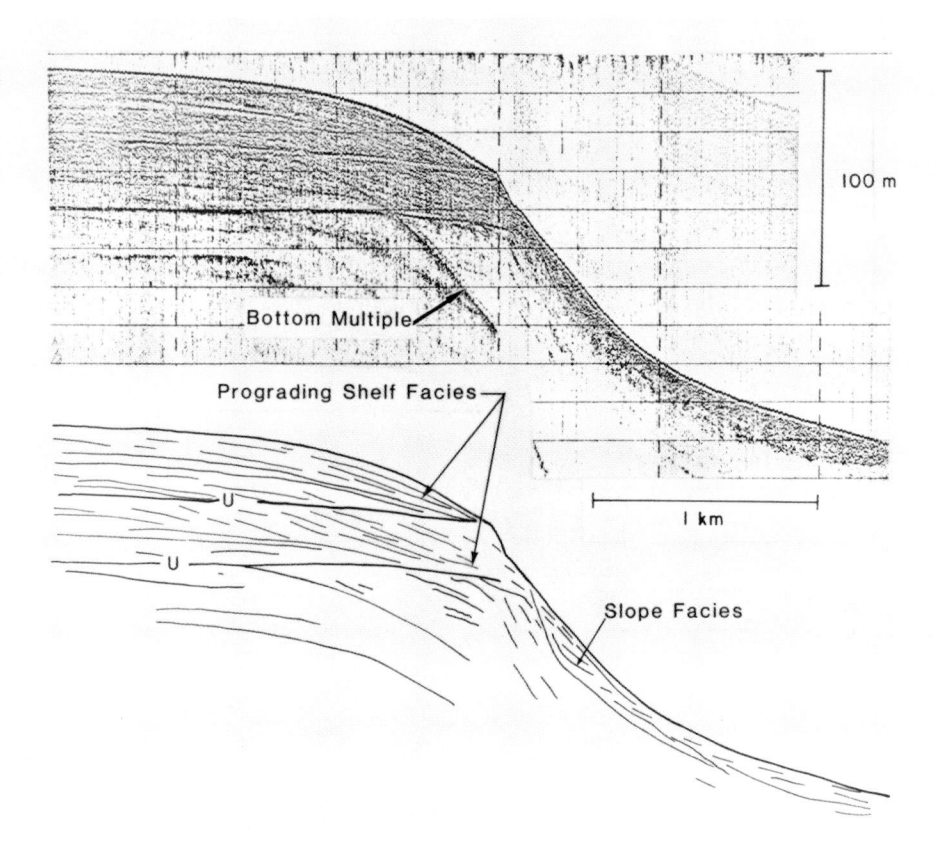

FIG. 14.—Typical upbuilt and outbuilt shelfedge similar to that shown in Fig. 13A (bottom). Profile was obtained from Southern California Continental Shelf off Oxnard, in vicinity of Santa Clara and Ventura Rivers. U, erosional unconformity.

individual units indicates at least five periods of shelfedge deposition, each of which can be characterized by one of the four main types of shelfedge seismic facies. Initial deposition after erosion and truncation by a Pleistocene transgression resulted in a seaward-migrating sedimentary unit and a starved shelfedge (unit 1, Fig. 17). A change in sediment supply or available energy, probably as a result of eustatic-controlled migration of the shoreline, led to development of a prograded shelfedge (unit 2, Fig. 17). Deposition of this unit was followed successively by a draped shelfedge, a starved shelfedge, and another draped shelfedge (units 3, 4, and 5, respectively, Fig. 17).

LIMITING FACTORS IN THE SEDIMENT-SUPPLY/ENERGY CONCEPT OF SEISMIC-FACIES DEVELOPMENT

The foregoing discussion was focused on the development of seismic facies at the shelfedge in response to sedimentation rate and available energy. This relation is based on empirical observations of the patterns of shelfedge deposition and an interpretation of the parameters that determine these patterns. Other factors, however, such as sediment size and shelf morphology, directly affect or are affected by sedimentation rate and shelf energy field. These other factors place some restrictions on the application of the sediment-supply/energy concept discussed in the previous section. For example, fine silt and clay, deposited at a certain rate, will respond differently to bottom currents than will coarse silt and sand deposited at the same rate, and thus will yield a different seismic facies. Similarly, the morphology of the shelf-slope break, its width, the presence or absence of sediment barriers, etc., will markedly influence the type of shelfedge deposits (see discussion in Stanley et al., 1983).

As a case in point, Figure 18 shows the sedimentary facies on the broad continental shelf of the Gulf of Alaska off Icy Bay. The facies distribution resembles a classic graded system, with sand nearshore, mud on the central shelf, and relict peb-

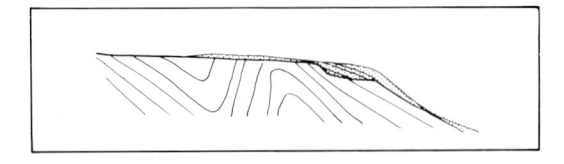

FIG. 15.—Depositional units on a cut-and-fill shelfedge.

bly mud on the outer shelf. The shelfedge is sediment starved, as is most of the shelfedge off southern Alaska. This particular configuration, however, is surprising and not what would be predicted on the basis of what is known about sediment input. The inner shelf undergoes sedimentation rates that range from 1 to 10 cm/yr; in addition, astonishingly high values of more than 100 cm/yr were calculated for cores collected within Icy Bay (Fig. 18). Despite these enormous amounts of sediment accumulation on or near the shelf, the combination of morphologic barriers (relict moraines) and distributive processes on the shelf result in no measurable sediment accumulation on the shelfedge. Thus, the recognition of a particular seismic facies on the shelfedge permits only an inference of relative sedimentation rates and processes at the shelfedge, not extrapolation to rates and processes over the entire shelf.

CONCLUSIONS

The form of Quaternary deposits on the Pacific shelfedge of the Western United States varies, and these variations are controlled chiefly, but not solely, by the morphology of the shelf, the pattern and rate of sediment accumulation, the energy of distributive processes on the shelf, and local tectonic and eustatic shelf history. We have identified four major types of seismic facies on the shelfedge by examination of high-resolution seismic-reflection profiles: (1) starved, (2) draped, (3) prograded, and (4) upbuilt and outbuilt. Each of these four

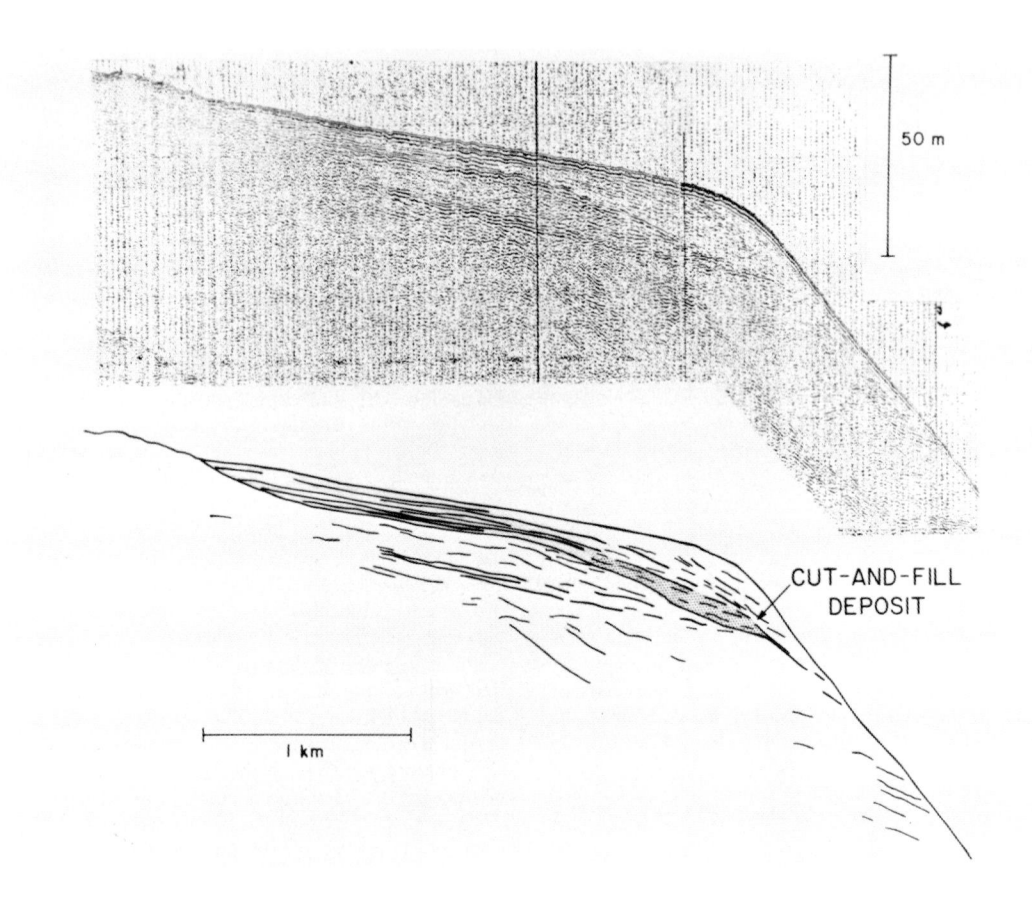

FIG. 16.—Typical cut-and-fill shelfedge from north of Anacapa Island in Santa Barbara Channel, Southern California. Erosion during a low stand of sealevel created notch in shelfedge that was subsequently infilled by prograded shelf deposits.

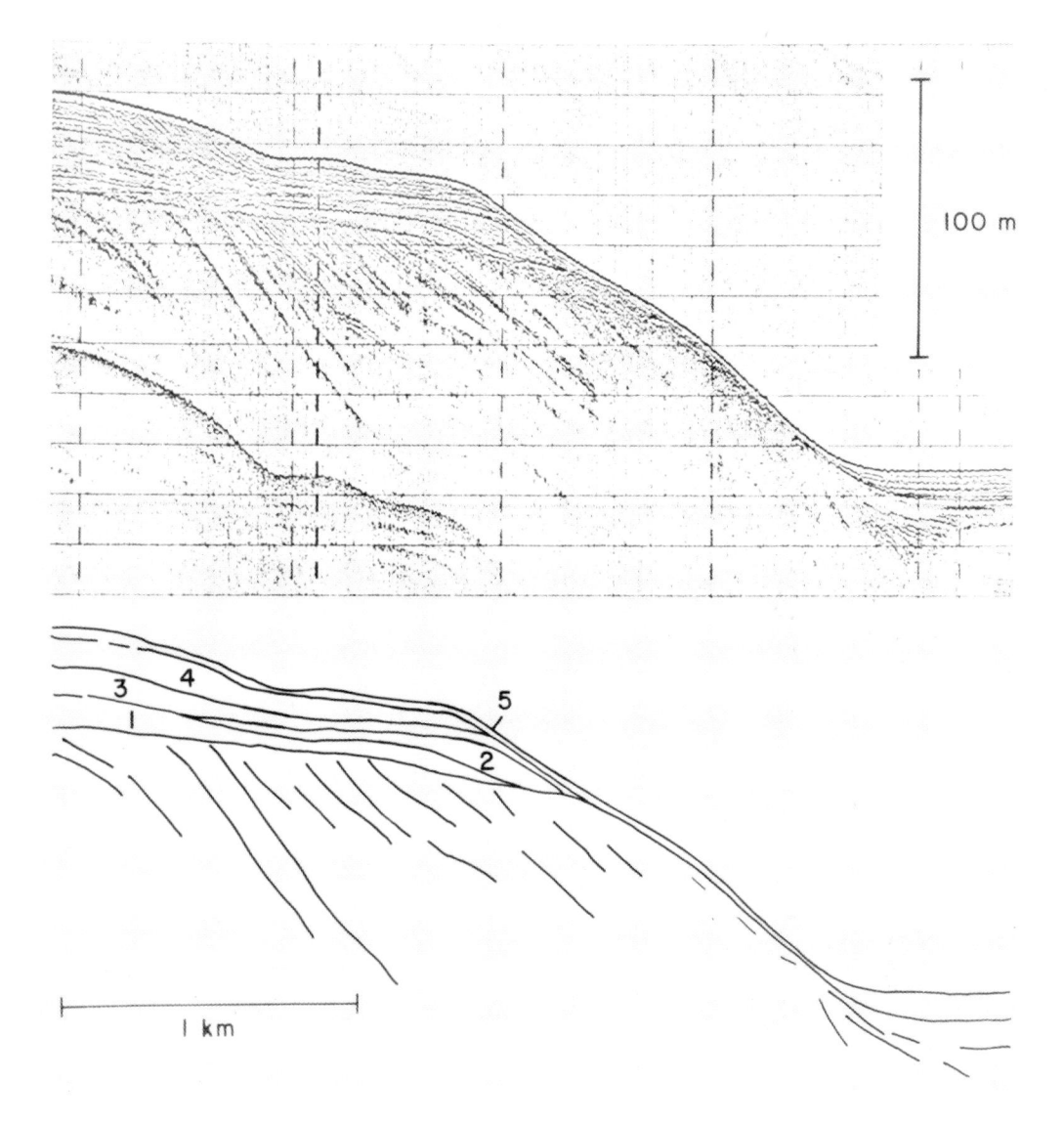

FIG. 17.—Typical composite shelfbreak from Santa Monica Bay, Southern California. Changes in sediment supply and available energy have created a series of stacked seismic facies that record each of those major changes: 1, starved; 2, prograded; 3, draped; 4, starved; and 5, draped.

seismic facies has a distinctive and diagnostic geometry and sedimentary facies that can be readily distinguished by geophysical techniques.

The characteristics of each deposit suggest that rate of sediment accumulation and available shelf energy (currents and waves) are the most important factors controlling the type of deposit; a change in either or both factors will result in a change in the seismic facies. In addition to these four main types of seismic facies, we identify two shelfedge facies indicative of other influences: (5) cut-and-fill and (6) composite. Cut-and-fill facies is restricted to areas of extensive shelfedge erosion during periods of lower sealevel stands, composite facies is a complex sedimentary unit formed by a series of depositional episodes.

Recognition of diagnostic shelfedge deposits has a twofold usefulness. It allows for interpretation of present and recent shelf history, especially with regard to identification of sedimentation styles; and it may aid in the interpretation of environments and conditions in marine sedimentary rocks.

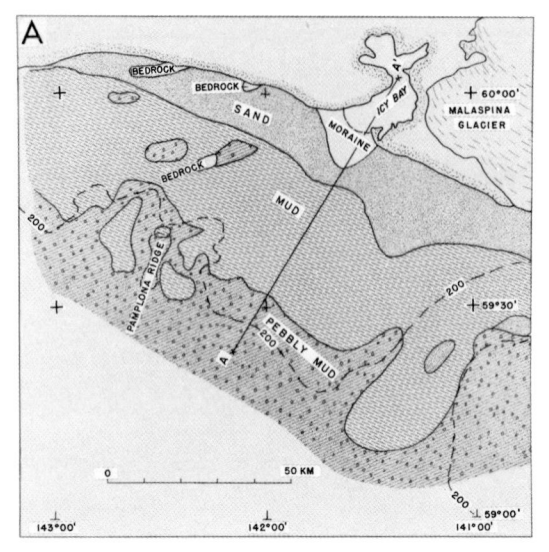

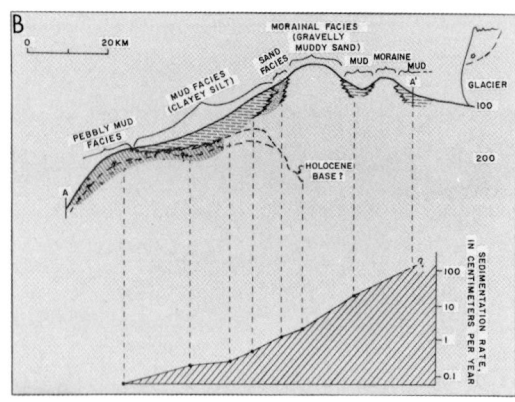

FIG. 18.—Sediment distribution map (A) and generalized profile (B), showing morphology, sedimentary facies, and sedimentation rates across shelf and upper slope off Ice Bay, northern Gulf of Alaska. Note that estimated sedimentation rate drops from more than 100 cm/yr to less than 0.1 cm/yr over a lateral distance of 80 km. Compiled from Carlson et al. (1977) and Molnia et al. (1980).

REFERENCES

CARLSON, P. R., MOLNIA, B. F., KITTELSON, S. C., AND HAMPSON, J. C., JR., 1977, Distribution of bottom sediments on the continental shelf, northern Gulf of Alaska: U.S. Geol. Survey Misc. Field Studies Map MF-876, 2 sheets, 13 p. pamphlet.

COLEMAN, J. M., AND PRIOR, D. B., 1981, Deltaic influence on shelfedge instability processes [abs.]: Am. Assoc. Petroleum Geologists Bull., v. 65, p. 912, (and this volume, p. 121–137).

CURTIS, W. F., CULBERTSON, J. K., AND CHASE, E. B., 1973, Fluvial-sediment discharge to the oceans from the conterminous United States: U.S. Geol. Survey Circ. 670, 17 p.

FIELD, M. E., 1981, Sediment mass transport in basins: controls and patterns, in Douglas R., Colburn, I., and Gorsline, D., eds., Depositional Systems of Active Continental Margin Basins: Soc. Econ. Paleontologists Mineralogists Pacific Sec. Short Course Notes, Los Angeles, p. 61–83.

―――, AND EDWARDS, B. D., 1980, Slopes of the southern California borderland: regime of mass transport, in Field, M. E., Bouma, A. H., Colburn, I. P., Douglas R. G., and Ingle, J. C., eds., Quaternary Depositional Environments of the Pacific Coast: Soc. Econ. Paleontologists Mineralogists Pacific Sec., Pacific Coast Paleogeography Symposium 4, Los Angeles, p. 169–184.

―――, AND RICHMOND, W. C., 1980, Sedimentary and structural patterns on the northern Santa Rosa-Cortes Ridge, southern California: Mar. Geology, v. 34, p. 79–98.

GREENE, H. G., CLARKE, S. H., JR., FIELD, M. E., LINKER, F. I., AND WAGNER, H. C., 1975, Preliminary report on the environmental geology of selected areas of the southern California continental borderland: U.S. Geol. Survey Open-File Rept. 75-596, 69 p.

GRIGGS, G. B., AND HEIN, J. R., 1979, Sources, dispersal, and clay mineral composition of fine-grained sediment off central and northern California: Jour. Geology, v. 88, p. 541–566.

KOMAR, P. D., NEUDECK, R. H., AND KULM, L. D., 1972, Observations and significance of deep-water oscillatory ripple marks on the Oregon continental shelf, in Swift, D. J. P., Duane, D. B., and Pilkey, O. H., eds., Shelf Sediment Transport: Process and Pattern: Stroudsburg, Pa., Dowden, Hutchinson, & Ross, p. 601–609.

KARL, H. A., CARLSON, P. R., AND CACCHIONE, D. A., 1983, Factors influencing sediment transport at the shelfbreak, in Stanley, D. J., and Moore, G. T., eds., The Shelfbreak: Critical Interface on Continental Margins: Soc. Econ. Paleontologists Mineralogists Spec. Pub. 33, p. 219–231.

MOLNIA, B. F., CARLSON, P. R., AND BRUNS, T. R., 1977, Large submarine slide in Kayak Trough, Gulf of Alaska, in Coates, D. R., ed., Landslides: Engineering Geology Reviews, Volume 3: Boulder, Colo., Geol. Soc. America, p. 137–148.

―――, LEVY, W. P., AND CARLSON, P. R., 1980, Map showing Holocene sedimentation rates in the northeastern Gulf of Alaska (scale 1:500,000): U.S. Geol. Survey Misc. Field Studies Map MF-1170.

SCHATZ, C. E., 1965, Source and characteristics of the tsunami observed along the coast of the Pacific Northwest on March 28, 1964 [Unpub. M.S. thesis]: Corvallis, Oregon State University, 39 p.

SOUTHARD, J. B., AND STANLEY, D. J., 1976, Shelf-break processes and sedimentation, in Stanley, D. J., and Swift, D. J. P., eds., Marine Sediment Transport and Environmental Management: New York, John Wiley & Sons, p. 351–377.

STANLEY, D. J., ADDY, S. K., AND BEHRENS, E. W., 1983, The mudline: variability of its position relative to shelfbreak, in Stanley, D. J., and Moore, G. T., eds., The Shelfbreak: Critical Interface on Continental Margins: Soc. Econ. Paleontologists Mineralogists Spec. Pub. 33, p. 279–298.

VANNEY, J. R., AND STANLEY, D. J., 1983, Shelfbreak physiography: an overview, in Stanley, D. J., and Moore, G. T., eds., The Shelfbreak: Critical Interface on Continental Margins: Soc. Econ. Paleontologists Mineralogists Spec. Pub. 33, p. 1–24.

ROLE OF SUBMARINE CANYONS ON SHELFBREAK EROSION AND SEDIMENTATION: MODERN AND ANCIENT EXAMPLES

JEFFREY A. MAY[1]
Department of Geology, Rice University, Box 1892, Houston, Texas 77251
JOHN E. WARME
Department of Geology, Colorado School of Mines, Golden, Colorado 80401
RICHARD A. SLATER
McClelland Engineers, 5450 Ralston Street, Ventura, California 93003

ABSTRACT

Heads of submarine canyons may occur anywhere on continental margins, from river mouths to continental slopes, producing a distinctive interface between shallow- and deep-marine environments. Inception of most canyons is subaerial, fluvially cut during lowered sealevel. Submarine mass flow also commences canyon formation. Submarine erosion shapes all canyons, and is especially effective in the headward region. Sliding and slumping are volumetrically most important as erosive agents, but sand spillover, bioerosion, sand flow, sand creep, and debris flow all play a part. Fluctuating channelized currents and low-velocity turbidity currents also erode and transport sediments.

Canyons alter shelfbreak circulation and sedimentation. They remove detritus from fluvial outflow, longshore transport, and cross-shelf drift, and may influence the position of rip currents. On narrow shelves, surface waves diverge over canyon heads, providing a transport corridor for the return of turbid water. Suspensates downwell along canyons as high-density nepheloid layers. Channelized currents winnow fines in upper canyon heads; focused internal tides and waves may actually break, producing more extensive erosion.

Although research on modern canyon systems has rapidly increased, detailed studies of ancient canyons remain sparse. An Eocene example from Southern California contains a tripartite fill representing progressive detachment from a nearshore source during a eustatic sealevel rise. Suspensate fallout, tractional flow, and mass-flow processes formed a basal amalgamated pebbly sandstone overlain by planar- to convolute-laminated sandstone, topped by variegated cut-and-fill mudstone channels. This tributary system fed the main canyon, filled with fining- and thinning-upward complexes.

The Pigeon Point and Carmelo Formations of coastal California and Tethyan submarine canyons of Czechoslovakia display similar fining-upward canyon fills. Contrasting fill sequences include coarse-grained units that dominate French Maritime Alps and New Zealand canyon complexes, and shales that plug canyons in the Gulf Coast, Sacramento Valley, and Israel. Shelf size and gradient, rates of eustacy, tectonism, and subsidence, and sedimentary-source input and migration interact to create this diversity of fills in ancient submarine canyons. Quantified analyses of canyon formation, maintenance, and fill, and application of sedimentary hydrodynamics to observed mass transport processes and their resultant ancient counterparts, are still needed.

INTRODUCTION

In comparison to the non-channeled shelf-slope break, submarine canyons are sites of a plethora of both subaerial and submarine erosional and depositional processes. Whereas submarine canyon heads actually may occur anywhere from river mouths to the lower continental slope, localized, often drastic depth changes at canyon heads produce a unique interface between shallow- and deep-marine hydrodynamic and sedimentologic conditions. It is well known that canyons are active conduits for transport of detrital material from nearshore and shelf environments to adjacent deep-marine basins, but the complex variables operative in the formation, maintenance, and fill of these sediment pathways through time and space are not fully understood.

Shepard and Dill (1966), Whitaker (1974, 1976), and Stanley and Kelling (1978) provide collections of ancient and modern case histories dealing with some of these processes. Scores of other authors have dealt with submarine canyons since Dana (1863) noted a sea valley off the Hudson River estuary. The purpose of this paper is to provide an overview of some of this work, plus that of the authors, specifically concentrating on the formation and role of submarine canyons near the shelf-slope break.

Following the usage of Shepard and Dill (1966), submarine canyons are herein considered as a specific type of sea valley, one usually with a V-shaped cross section, high and steep walls with rock outcrops, winding course, and numerous tributaries. An understanding of how these features evolve from the time of their inception to final filling, how they affect basin margin erosion and sedimentation, and how they interact with and influence marine circulatory patterns is economically and environmen-

---

[1]Present Address: Marathon Oil Company, Denver Research Center, P.O. Box 269, Littleton, Colorado 80160.

tally significant. Up-dip encasement of coarse-grained canyon-head sediments within finer-grained outer-shelf and upper-slope deposits may occur at the juxtaposition of continental, nearshore, and open-marine environments. This depositional setting is one of variable interstitial pore fluids, organic input, and tectonic processes. Hydrocarbon entrapment occurs not only within coarse-grained canyon fills, but also along flanks of systems plugged with fine-grained detritus (see, for example, Bornhauser, 1948; Hoyt, 1959; Galloway and Brown, 1973; Cohen, 1976; Almgren, 1978; Berg, 1979; Webb, 1979; Vormelker, 1980; Garcia, 1981). In developing models of basin evolution, exploration geologists must consider these variables to predict locations of favorable petroleum reservoirs, sources, and trap rocks or of stratiform mineralization. A knowledge of how canyons affect marine erosion, circulation, and sediment transport at the shelfedge and downcanyon may also help prevent ocean-wide pollution of the deep-marine environment by present-day offshore drilling and waste disposal.

SAMPLING ANCIENT VERSUS MODERN SYSTEMS

Although many modern submarine canyons have been studied, descriptions of ancient counterparts are relatively few. One reason is that these systems are sediment thruways rather than sites of active deposition. Even when plugged, their location along continental margins (settings of tectonic instability and multiple erosional episodes) apparently precludes numerous well-exposed, ancient examples. Buried canyons along margins of rapid subsidence cannot be directly studied and evaluated in detail. Fragmentary or poor surface exposures may lack critical criteria for distinguishing ancient submarine canyons from fan valleys (Whitaker, 1974; Stanley et al., 1978).

It also is often difficult to relate many modern canyon processes to their resultant deposits. Although ancient sequences provide insight into geologically significant tectonic and eustatic controls on canyon development and fill through time (Normark, 1974), modern canyons only display a transient depositional stage. Problems arise in relating ancient mass-flow units to the dynamics of modern canyon-related mass transport: (1) only the final flow process is preserved in a deposit's texture and structure, and the dominant lift or support mechanism(s) during transport may not be evident, (2) simultaneous processes of lift and/or support may operate in a single flow, (3) both traction and suspension may act as transport mechanisms within a single flow, and (4) buoyancy by a finer-grained matrix may not be recognized (Lowe, 1979).

A further complication is the observation gap between modern and ancient systems because of the difference in sampling scale (see Normark et al., 1979). Coring and dredging provide samples orders of magnitude smaller than those attainable at good outcrops. Visual observations using bottom photography or submersibles only begin to approach the detail easily viewed on outcrop; scuba is severely depth limited. Seismic methods (Fig. 1A) portray data on the same scale as outcrops, but lack comparable resolution.

Therefore, a unified interpretation of features and processes in modern submarine canyons and their ancient analogs is lacking. Although both systems have been described, quantification of the dynamics of basin-margin sedimentation as affected by canyons, and understanding of the factors controlling the formation and evolution of these conduits, are needed.

TIMING AND MEANS OF FORMATION

Submarine canyons dominate shelfbreak morphology in many areas of the world and, hence, patterns of shelfbreak circulation and sedimentation. The following is a review of ideas regarding how, why, and where canyons form. A variety of postulates explaining canyon formation were proposed in the late 1800's (Davidson, 1887; Dana, 1890; Le Conte, 1891; Lawson, 1893; Spencer, 1898, 1903; Smith, 1902). The debate was reactivated in the 1930's and continues to the present (see, for example, Davis, 1934; Daly, 1936; Kuenen, 1937; Johnson, 1938; Veatch and Smith, 1939; Bucher, 1940; Shepard and Emery, 1941). The main accepted hypotheses (among a multitude of others) include: (1) turbidity current erosion, (2) erosion by slow mass movement of sediments, (3) erosion by other bottom currents, and (4) drowning of subaerial valleys (Shepard and Dill, 1966). It is becoming

FIG. 1.—Stratigraphy and geologic history of East Coast United States submarine canyons. A, Line-drawing interpretation of seismic reflection (sparker) profile across the head of Carteret Submarine Canyon, New Jersey continental shelf. A probable buried channel indicates a multiple cut-and-fill history. Note the downflank slump. B, Sequential stages in the development of east coast submarine canyons: (1) deposition of relatively horizontal Cretaceous and Tertiary strata, (2) subaerial truncation during lowered sealevel stands, (3) rim upbuilding by Pleistocene and Holocene silts and clays draped over the continental margin during high sealevel stands, and (4) as represented in a close-up of the base of 3, widening of canyons by downflank slumping and sediment spillover, and scour of canyon floors and bases of canyon walls by downcanyon sediment movement (from Slater, 1981).

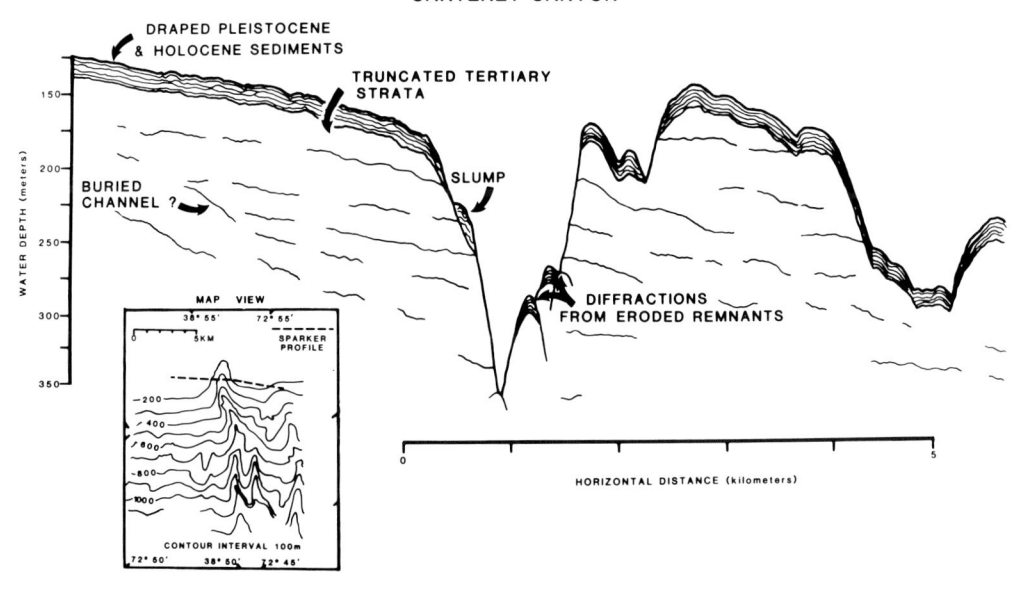

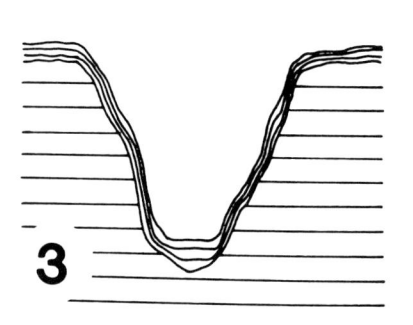

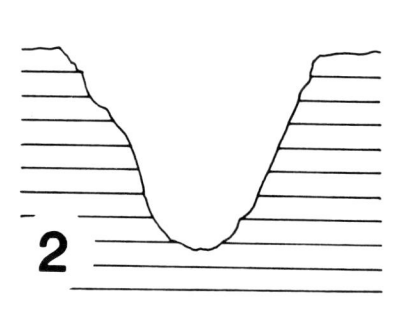

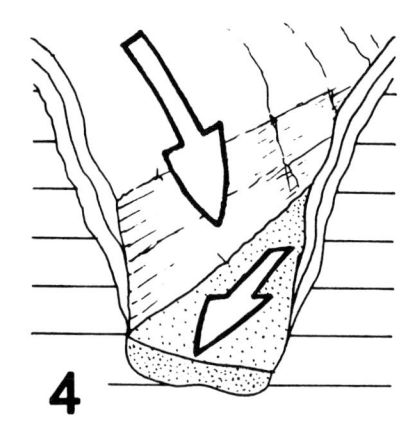

SUBMARINE CANYONS, EROSION AND SEDIMENTATION 317

evident that canyons are of composite origin, with these processes often acting in sequence or conjunction with one another (see Shepard, 1981).

Subaerial Inception

Fluvial erosion of emergent continental shelves and upper slopes during lowered sealevel stands apparently predominates in initiating submarine canyons. Most known canyon heads occur seaward of modern river valleys (Shephard and Dill, 1966). Buried shelf valleys, presumably of subaerial origin, extend from modern river and estuary mouths to canyon heads (see, for example, McClennan, 1973; Twichell et al., 1977; Knebel et al., 1979; Freeland et al., 1981; McGregor, 1981). Dendritic tributary patterns, cutting of single canyons equally well into crystalline and sedimentary units, and abundant canyon-head associations with shallow-marine and fluvial/deltaic formations all lend support to fluvial inception (Shepard and Dill, 1966; Starke and Howard, 1968; von der Borch, 1969; Swift et al., 1980). Some areas, such as western Corsica and Japan, contain submarine canyons almost directly connected to land canyons; the canyon head associated with the Congo River actually extends 20 miles (32 km) into the estuary (Buchanan, 1887; Shepard and Dill, 1966).

Seismic profiles from the low-gradient, eastern United States shelf indicate a variety of buried valleys of late Tertiary and Pleistocene subaerial origin. One example is a large buried channel that splits from the present-day topographic Hudson Shelf Valley, filled with fluvial and estuarine deposits and capped by shallow-marine units of a subsequent transgression (Knebel et al., 1979; Freeland et al., 1981). This abandoned valley, formed during subaerial exposure of the continental shelf, may have been responsible for funneling sediments southeastward and forming the distinctive northward hooks in the heads of Wilmington and Baltimore Canyons (Kelling and Stanley, 1970). Older, buried westward extensions of these two canyons formed during an earlier lowered sealevel stage (Kelling and Stanley, 1970; McClennan, 1973; Twichell et al., 1977; McGregor, 1981). Estuary-mouth scour during shoreface retreat, as well as fluvial processes, originated these and similar filled valleys on the east coast; their seaward extensions now form canyon heads more than 100 km from shore (Stanley, 1974; Swift et al., 1980; Slater et al., 1981).

Along the western United States and other high-gradient shelves, submarine canyon heads project much closer to shore; South Branch of Scripps Canyon is within 60 meters of the sea cliffs and many southern Baja California canyons extend into the surf zone (Shepard and Dill, 1966). Straight heads of Scripps and La Jolla Canyons may be traced inland into fault zones and incised fluvial valleys, indicating structural control on the subaerially eroded pathways (Dill, 1964; Shepard and Dill, 1966). Though probably not of subaerial origin, structural features likewise influenced the locations, trends, and shapes of submarine canyons along the Bering Sea margin (Scholl et al., 1970).

Ancient canyon sequences also yield evidence of submarine canyon inception by subaerial processes. The ancestral Sacramento River eroded the now-buried Princeton Submarine Valley; its course was structurally controlled by downwarping of the Great Valley Synclinorium (Redwine, 1978). Rather than eustatic sealevel changes, Almgren (1978) called upon sequential tectonic uplift to explain the initiation of three separate subaerially eroded systems in the Sacramento Valley: the Martinez, Meganos (Krueger and North, 1983), and Markley submarine canyons. Erosion of a well-exposed Eocene canyon north of La Jolla, California, has been correlated with a late Early Eocene eustatic sealevel drop (Lohmar and Warme, 1978, 1979; Lohmar et al., 1979). Similarly, cutting of Tethyan canyons in Czechoslovakia occurred during a Late Eocene to Early Oligocene lowstand (Picha, 1979). An extreme situation occurred during the Messinian desiccation of the Mediterranean, when canyons were initiated by the drastic incision of fluvial systems along the basin margin (Hsü, 1972; Bellaiche et al., 1979). Analogous desiccation of the South Atlantic basin may have been responsible for the formation of the Congo Canyon (Shepard, 1981).

Submarine Inception

Subaerial erosion cannot account for the initiation of all submarine canyons. Many occur offshore of coasts lacking fluvial valleys. Some of these may be older features of subaerial origin, decapitated from their adjacent river systems by lateral faulting, such as for the Monterey Canyon (Starke and Howard, 1968). But most canyons lacking alignment with onland rivers do possess a common feature—they are formed where a substantial sediment supply is brought to the canyon heads, usually on narrow, steep shelves. A common placement is upcurrent of headlands, where there is a decrease of longshore-current transport, accumulation of sediment, and subsequent downslope movement and erosion normal to contours (Davis, 1934; Crowell, 1952; Dietz et al., 1968). Another type of genesis arises from rapid submarine erosion into poorly consolidated shelf sediments directly seaward of fluvial input. Reimnitz and Gutiérrez-Estrada (1970) show that shifting sediment point-sources of the Rio Balsas delta have activated preexisting, partially filled canyons and lead to erosion of new ones, even those with dendritic tributaries, since the last sealevel rise. As another example, they cite the Riony River which, emptying into the Black Sea, has formed a new canyon only two decades after diversion of the river mouth. Felix and Gorsline (1970) showed similar lateral shifts in position of the Newport Submarine Canyon in response to sealevel changes and position of the Santa Ana River.

Sediment instability, mass flow, and submarine erosion may be caused by local tectonic activity as well as by rapid sediment input and buildup. In one example from New Zealand, a submarine shelf syncline acted as a depocenter. Syndepositional deepening of the trough and steepening of the sides may have triggered down-axis mass movement of coarse-grained, longshore-derived sediments (Herzer and Lewis, 1979). Retrogressive slumping and debris flows extended this depression shoreward and, combined with axial downcutting, produced a submarine canyon (see Farre et al., this volume).

Thus, it is possible that submarine failure of rapidly deposited, unstable accumulations of coarse-grained detritus may commence canyon cutting and continue its development headward. Submarine erosion at the shelfedge is probably accomplished by a variety of sediment-gravity flows and bottom currents (Shepard and Dill, 1966). Landslides, slumps, debris flows, and turbidity currents all may play a part in originating canyons at the shelfbreak (Shepard, 1981). Special geographic and sedimentologic conditions are necessary, though, and submarine processes alone probably did not account for the inception of most canyons. Instead, subaerial erosion appears to have initiated the upper reaches of most canyon systems during worldwide or local sealevel lowstands. Upon subsequent transgressions, canyons are shaped by submarine processes of erosion and deposition as outlined below.

MODERN SHELFBREAK CANYONS

Erosional Processes

Submarine erosion is of major importance in affecting canyon morphology. Slumping is volumetrically the most important activity shaping canyon heads, resulting in cirque-like, wide and broadly rounded scalloped heads with gentle side slopes (Figs. 1 and 2) (Stanley, 1974; Dill et al., 1975; Marshall, 1978). Indirect evidence of this process includes exceptionally large and rapid depth changes noted by sequential surveys of canyons (Shepard, 1951). A new canyon tributary appeared in Scripps Canyon following an earthquake and period of storm waves in 1950 (Shepard and Dill, 1966), and was repeated when a slump of 10^5 cubic meters of sediment opened a buried tributary to Scripps Canyon 400 meters long and 16 meters deep (Marshall, 1978). Sudden deepenings within unconsolidated detritus of canyon axes have also been noted. Canyon deepening simultaneous with opening of a new head took place in Los Frailes Canyon, Baja California; in Scripps Canyon, the sudden removal of up to 7 meters of the upper sediment surface has been recorded (Shepard and Dill, 1966).

Direct observations within canyon heads reveal evidence of sliding and slumping: rock slabs and slump units accumulate near the bases of canyon walls (Shepard, 1973; Stanley, 1974; Valentine et al., 1980), slip planes are evident along some walls (Malahoff et al., 1977; Malahoff and Fornari, 1979), and multiple slumps may produce stepped canyon walls (Slater et al., 1981). Conditions in submarine canyon heads are prime for slumping and sliding—rapid sediment buildup past the critical slope, earthquake shocks or cyclic stress by focused long-period or internal waves on unstable sediment masses, gas generation within and decay of organic materials binding detrital accumulations, bedrock joints, undercut ledges, networks of boring and burrowing organisms in rock walls, and upslope slumping may all cause the shear stress to exceed the internal or shear strength of a depositional unit (Chamberlain, 1964; Shepard and Dill, 1966; Stanley, 1971; Marshall, 1978).

Headward slumping may keep the canyon head in the nearshore zone during sealevel transgressions, especially on high-gradient shelves underlain by poorly consolidated strata or where direct sediment input from fluvial or longshore sources can be maintained (Stanley, 1974; Shcherbakov, 1978). The striking nickpoint between submarine canyons and adjacent stream valleys, due to the great difference between axial gradients of the two systems, and the concave-up longitudinal profile of submarine canyons, with the steepest portion in shallow water near the canyon heads, indicate that canyons reach a grade adjusted to sealevel (Shepard and Dill, 1966; von der Borch, 1969). This process is geologically very rapid. For example, La Jolla Canyon has retreated about 0.7 meters per year since early submarine surveying began (Shepard, 1973), and the axial valley within Tongue of the Ocean, Bahamas, may have eroded headward over 225 kilometers since the Early Cretaceous (Hooke and Schlager, 1980). Other systems eroding headward may slice across and capture drainage from adjacent canyons, analogous to subaerial stream piracy (Holcombe and Einwich, 1976; Herzer and Lewis, 1979).

Erosional processes other than slumping and sliding are also operative in submarine canyon heads. Undercut ledges and removal of rockfalls, occurring between sequential scuba or submersible observations, indicate active downcanyon erosional processes (Chamberlain, 1964). Abrasion along lower canyon walls produces polished surfaces and hourglass-shaped cross sections (Dill, 1964). Truncated holes of boring organisms and small, incised vertical furrows along upper canyon walls reflect lateral sediment input and erosion by sand spillover from the adjacent shelf (Dill, 1964). In many canyons, especially in the headward portion, the upper walls are localities of extensive bioerosion. Burrowing and boring organisms are dominant, often

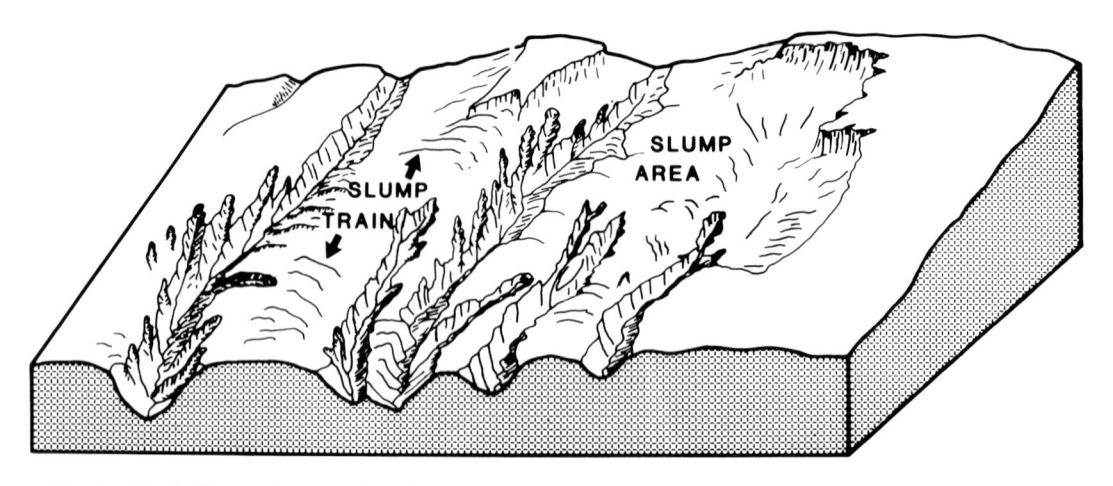

FIG. 2.—Block-diagram interpretation of a side-scan sonar record of a submarine canyon head, western Celtic Sea. Relief in areas of active canyon heads resembles badland topography, where small lateral gulleys converge on the main tributaries at small angles. Intercanyon slumping is predominant, occurring as large areas of hummocky topography being actively dissected and as slump trains moving downslope (modified from Belderson and Stride, 1969).

producing intricate "pueblo villages" (Chamberlain, 1964; Warme et al., 1978; Valentine et al., 1980).

Sediment Deposition and Transport

Coring and dredging of submarine canyon heads yield a variety of sediment types and bedding styles. Scuba and submersible observations permit more detailed mapping and analyses of these materials and direct viewing of bottom sediment transport (Fig. 3).

Along the California coast, littoral cells funnel sediment into canyon heads (Emery, 1960). Heterogenous mixtures of fine-grained sand and silt, often with abundant organic debris, are introduced from fluvial, nearshore, and offshore sources. These sediments spill over from the shelf along small sand chutes and move downcanyon by glacier-like creep and sand flow, abrading and plucking out adjacent material (Dill, 1964; Chamberlain, 1964). Sediment movement to 0.2 knots has been noted carrying rocks to six inches (15 cm) in diameter (Shepard and Dill, 1966). Stepped surfaces along the fill indicate episodes of catastrophic slumping.

Many canyon heads on the East Coast of the United States have much finer-grained axial fill and lack obvious continuous creeping and sliding of bottom sediments (Shepard and Dill, 1966). Direct coarse-grained sediment input is greatly diminished because these canyons head much farther from shore. However, coarse-grained relict shelf sediment is observed in upper canyon heads. In Hudson Submarine Canyon a distinct boundary is observed between erosion and deposition of this relict material (Stanley and Freeland, 1978). Impingement of currents at the shelfbreak suspends fine-grained components of the relict sediments; downcanyon currents then transport the winnowed material beyond a distinct "mudline" (see Stanley et al., this volume). This mudline divides an upper gravel-sand-shell facies and a lower mud facies, and represents separation of distinct energy levels necessary to erode fine-grained sediments (Fig. 3A). Similar sharp breaks in bottom sediment type and size are present in the heads of other East Coast canyons (Valentine et al., 1980).

Even though U.S. East Coast canyons head far from shore, they do funnel coarser-grained clastic detritus downslope to varying degrees, though not as actively as in those canyons off California. In Wilmington Canyon, relict shelly sand and silt is reworked and swept into the canyon during periods of storm and high tidal surges, mixing with unconsolidated and consolidated sediment along the upper canyon walls (Stanley, 1974). Ripples have been observed in these fine-grained sands and silts along the walls and floor in the shallow portions of the canyon. Below the mudline, accumulations of slumped sediment along the base of the canyon wall give way axially to a muddy fill containing pebbles, subangular cobbles, and large rock fragments (Stanley, 1974). Within the axis proper, this fill becomes a true pebbly mudstone (Fig. 3B) (Stanley, 1974). Thus, the dominant transport processes in proximal canyon settings that are detached from a nearshore source involve: (1) shelfedge scour and spillover of sand and silt, (2) reworking and transport of fines by channelized currents, (3) slumping down canyon walls, partially in response to sediment spillover, and (4) downcanyon transformation of laterally infilled sediment to debris flows (Got and Stanley, 1974; Stanley, 1974).

Submarine canyons also serve as preferential traps and conduits for suspended sediments. Relative to

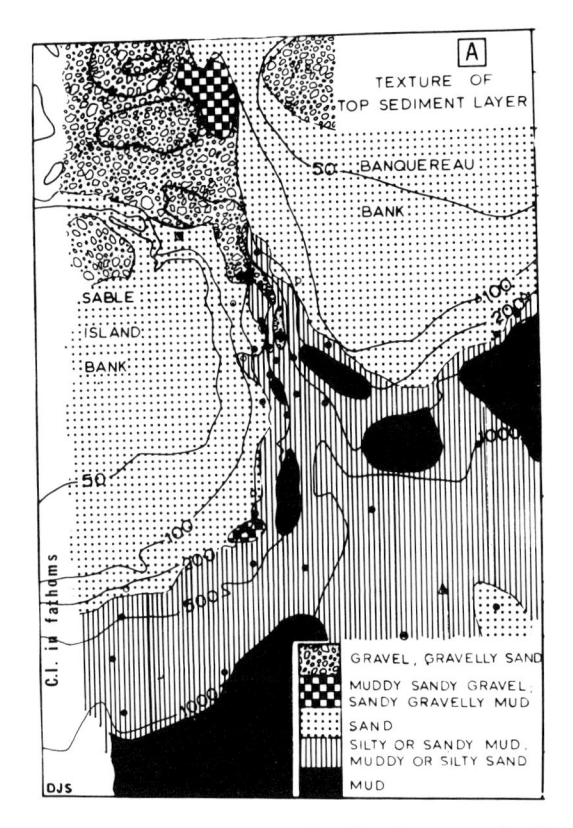

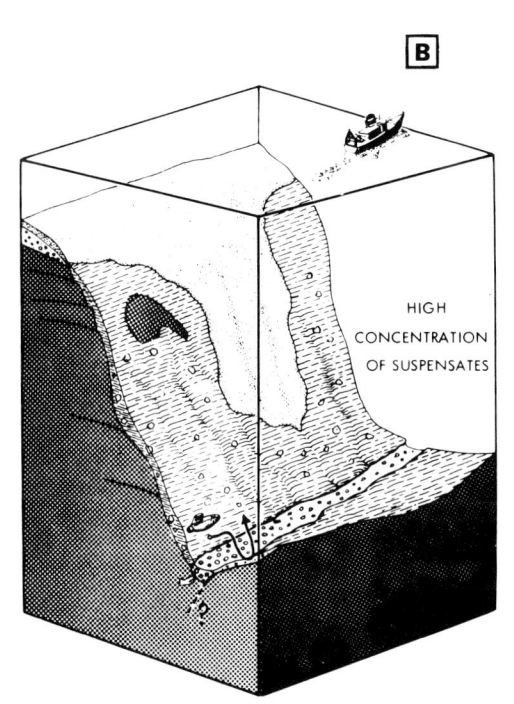

FIG. 3.—Sediment transport in submarine canyon heads. *A*, Textures of surficial sediments in The Gully Canyon, Sable Island Bank, Nova Scotia continental shelf. Relict, coarse-grained shelf sediments are winnowed by channelized currents in the upper canyon head and spill over into the head during storm events (from Stanley and Silverberg, 1969). *B*, Lateral transformation of sediment movement within Wilmington Canyon head, East Coast United States. Shelly, silty sand spills over the shelfedge and is reworked with canyon wall material by downflank slumping. These units, in turn, are transformed to axial debris flows which provide for canyon head flushing (from Stanley, 1974).

adjacent shelf and slope concentrations, Hydrographer, Hudson, and Wilmington canyons contain lenses of particle-rich, near-bottom water (Drake et al., 1976). Baker (1976) measured particulate concentrations within the water column of Willapa Submarine Canyon, Washington, that were up to three times those found on the adjacent open shelf and slope. These relatively dense nepheloid layers, also present in a number of California canyons (Drake and Gorsline, 1973), may contribute to downcanyon bottom flow (see following section).

Yet another depositional process associated with canyons is rim upbuilding combined with axial erosion. Many canyons display unconsolidated Pleistocene and Holocene materials mantling intercanyon areas and canyon margins, especially where shelf currents are relatively weak (Fig. 1). This process may considerably aid in increasing canyon relief. The Waria Canyon of New Guinea may have undergone at least 100 m of rim upgrowth at the shelfbreak during the last 25,000 years, and possibly has had its complete axial depth of 1500 m produced by the same process continuously since the Pliocene (von der Borch, 1969). As the Blake-Bahama platform subsided, sediment upbuilding along the rim of Tongue of the Ocean kept pace, forming exceptionally high walls (Hooke and Schlager, 1980). Shepard (1981) emphasized alternating periods of marginal buildup and canyon-floor erosion, dating back to the Cretaceous in New England examples, in forming exceptionally large canyon complexes.

Channelized Currents

Transport at the shelfbreak within canyons is not only by gravitational mass flow, but also by a variety of channelized currents. Currents in canyons were first noted in the late 1930's (Stetson, 1936; Shepard et al., 1939). Shepard and his associates have carried out systematic worldwide studies in this field (see, for example, Shepard, 1979; Shepard et al., 1979a). The presence of both graded sands and rippled sands and silts mantling canyon walls and forming the uppermost sediment layer in upper canyon heads attest to at least periodic tractional flow (Bouma, 1965; Shepard and Dill, 1966; Got and Stanley, 1974; Birdsall and Scott, 1976; Valentine et al., 1980). Current-meter measurements and observations from submersibles indicate that alternating upcanyon and downcanyon flow rarely ceases. These normal currents usually do not exceed 30 cm/sec (Shepard et al., 1979a), but are strong enough to move fine-grained sand.

Periodicity of upcanyon and downcanyon flow is related to the tidal range and water depth. The highest frequency flow reversals occur in shallow canyon heads, whereas reversals in deeper water approach the semidiurnal tidal cycles (Shepard et al., 1979a). Internal waves affect the tidally-derived flow components, advancing both upcanyon and downcanyon, possibly in response to adjacent fluvial outflow or tidal waves flowing along density-stratified water masses (Shepard et al., 1979a). In California canyons, the downcanyon component tends to be stronger, producing net downslope transfer of sediment; in contrast, many U.S. East Coast canyons have more equal upcanyon and downcanyon flow (Shepard et al., 1979a). Cross-canyon flow occurs as a result of wind influence (Shepard, 1979; Shepard et al., 1979a).

High-frequency flow reversals in shallow canyon heads are caused by the addition of other factors. Current pulses related to surf beat and rip currents are common, and periods of heavy swell and strong onshore winds likewise affect these shelfbreak environments (Reimnitz and Gutiérrez-Estrada, 1970; Shepard and Marshall, 1976). Current drift along the shelf introduces another component interacting with canyon head flow (Valentine et al., 1980).

Occasional strong downcanyon surges also occur in canyons, with speeds usually under 100 cm/sec. These are interpreted as turbidity currents related to periods of high discharge of adjacent rivers, storm surges, or exceptionally high tides (Shepard et al. 1979a). Significantly, these are common events, especially where large quantities of sediment are fluvially introduced to canyon heads (Reimnitz, 1971; Shepard, 1979; Shepard et al., 1979a). Where beaches or shelves are the primary sediment sources, these low-velocity turbidity currents are much rarer. However, even in these places, storm events may produce similar downcanyon flow (to 110 cm/sec in Hudson Submarine Canyon; Hotchkiss and Wensch, 1979), and high perigean tides may initiate equivalent upcanyon pulses (to 80 cm/sec at depths of 48 meters in La Jolla Canyon; Shepard et al., 1979b). One of us (R.A.S.) was caught in an apparent turbidity current while diving in Oceanographer Submarine Canyon. The submersible was enveloped in a cloud of sediment and pushed both downslope and cross-canyon. This density flow occurred during some of the highest tides of the year and during peak surface currents.

Effects on Shelf Circulation

Canyons are not only extremely important as sediment conduits and as channels for tidal, storm, and other currents, but their shallow portions also affect shelfbreak to nearshore sedimentation and marine circulation. Modern canyons passively trap sediment as longshore and boundary currents flow over the open shelf; canyons were probably even more active as sediment traps during lowered sealevel (Emery, 1960; Cleary and Conolly, 1974; Stanley et al., 1981). Not only is fine-grained sediment suspended and winnowed by channelized flow impinging at canyon heads, but internal tides and waves may actually break on the shelfedge, having been funneled upward along canyon axes and focused in the headward region. These can combine with longshore currents or eddies shed from boundary currents to cause extensive erosion of the shelf adjacent to canyon heads.

One example of erosion localized by a canyon head is from the shelf surrounding Hudson Canyon, which displays truncation of the upper Pleistocene silty sands. This area contains depressions with variable spacings (<100 m–2 km), depths (1–10 m), outlines, and bottom configurations (Knebel, 1979). Neither ponding of fine-grained sands nor the existence of a well-sorted sand sheet as found over much of the remaining shelf is present. During lowered sealevel, southwesterly currents and upcanyon channelized flow combined to scour a 740 km^2 region south of the canyon head; the eroded material was funneled downcanyon. To the north of the submarine canyon, the eroded area is much less extensive (75 km^2), reflecting a decrease in erosive strength of the longshore current after passing over the canyon head (Knebel, 1979). A combination of southwesterly drift along the outer shelf and convergence of internal waves and tides at the canyon head have maintained the relict erosional surface.

Further evidence of erosional processes by shelf currents associated with submarine canyons is the preferential development of canyon tributaries on the upcurrent sides of canyon heads. This is probably related to increased spillover of, and resultant erosion by, coarse-grained detritus transported by boundary or longshore currents (Shepard and Dill, 1966).

Canyons extending close to shore may provide locations for seaward-flowing rip-currents, especially when large amounts of wave-driven water is discharged onto adjacent beaches. These rip-currents can then transport fine-grained sand and silt directly to the canyon heads (Chamberlain, 1964). A similar transfer of sediment from the nearshore zone to canyon heads occurs across narrow shelves (Karl, 1980). Surface waves are refracted as they approach a canyon head, forming a zone of divergence along the shoreline. Local circulation drives nearshore currents back toward the lower energy divergent zone over the canyon head; consequently, turbid water moves seaward and downcanyon. This circulation system produces a corridor for preferential sediment transport on the continental shelf, where higher concentrations of total suspended particulate matter and coarse-grained surficial sediments indicate a zone of cross-shelf bedload and suspended load movement (Karl, 1980, and Karl et al., this volume).

ANCIENT SHELFBREAK CANYONS

Eocene of San Diego: A Case History

North of San Diego, California, Early Eocene lagoonal (Delmar Formation) and offshore barrier (Torrey Sandstone) deposits form part of a foundered shelf that probably was subaerially cut during a late Early Eocene eustatic sealevel drop (Lohmar and Warme, 1978, 1979; Lohmar et al., 1979). An erosional unconformity, with a dip of 5°, represents the floor of a canyon tributary that cut across the older, horizontally bedded, shallow-marine units (Fig. 4A). The eroded surface is stepped and plucked, and exhibits features of injection, pry-ups, erosional remnants projecting upward into the overlying fill (Fig. 4B), and undercut ledges (Fig. 5A).

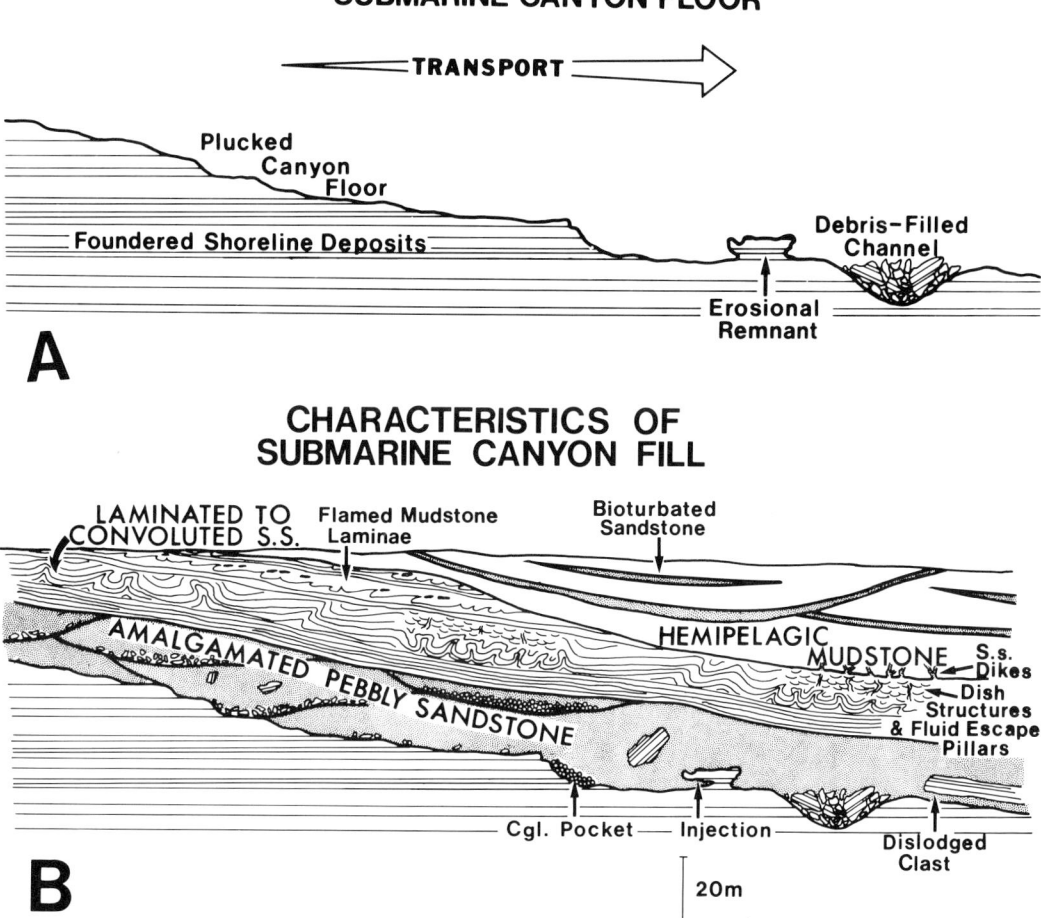

FIG. 4.—Characteristics of the Eocene Torrey Submarine Canyon tributary north of San Diego, California. A, Nearshore (lagoonal and barrier bar) deposits were subaerially cut during a late Early Eocene eustatic sealevel drop. The resultant canyon was further shaped during the subsequent sealevel rise, producing an irregular canyon floor. This erosive surface is plucked and stepped, and displays erosional remnants protruding upward, injection features, pry-ups, and intraclast-filled pockets. (See B for scale). B, The submarine canyon fill is tripartite, representing progressive detachment from the coarse-grained, nearshore source. The basal pebbly sandstone has cross-cutting, amalgamated units with intraclast-rich bases and laminated fill. The overlying planar-laminated sandstone becomes convoluted upward, and displays a concomitant increase in dish structures, fluid-escape pillars, flamed mudstone laminae, and clastic dikes (sedimentary structures are diagrammatic, not to scale). The uppermost cut-and-fill mudstone channels contain siltstone and sandstone layers that are variably graded, rippled, cross-bedded, bioturbated, or richly fossiliferous.

FIG. 5.—Features of the Eocene Torrey Submarine Canyon tributary north of San Diego, California. A, Undercut canyon floor (arrows) composed of lagoonal units eroded and redeposited as mudstone intraclasts in the overlying canyon fill. B, Eroded remnant of cross-bedded Torrey Sandstone projecting upward into the submarine canyon fill (unconformity marked by arrows and rock hammer). The overlying crosscutting units of pebbly sandstone have intraclast-rich bases and become laminated upward. C, Planar plant- and mica-rich laminae develop box-like and flamed folds (downcanyon transport to right). D, Resistant siltstone and sandstone layers enhance the cut-and-fill nature of channels dominated by hemipelagic mudstone. One channel is draped with mudstone and plugged with laminated sandstone (arrow). (Cliff face is approximately 70 meters high).

The overlying canyon fill proper is tripartite, consisting of a basal amalgamated pebbly sandstone, an intermediate planar- to convolute-laminated sandstone, and capping mudstone sequence. This succession probably represents progressive detachment of the canyon tributary from a coarse-grained, nearshore source during the subsequent Middle Eocene eustatic sealevel rise (Fig. 4B). A sealevel high stand and subsequent drop provided the impetus for final shelf progradation and consequent burial of the tributary system. We have named the complete Eocene canyon depositional package the Torrey Submarine Canyon.

The basal unit of the canyon fill is an amalgamated sandstone, displaying crosscutting and very poorly sorted, pebbly, granule to medium-grained, structureless to faintly laminated sandstone. Basal portions of many cut-and-fill structures contain matrix-supported mudstone ripup clasts that have irregular margins. Each separate cut-and-fill becomes laminated upward above its clast-rich portion (Fig. 5B). Outsize mudstone clasts to 5 meters in length "float" within the sandstone at all levels. The matrix is 95% sand-sized or larger grains, with a size distribution that is negatively skewed, polymodal, and truncated at 3 phi.

Generation of very coarse-grained, proximal deep-marine sediments is poorly understood. Such deposits may form from a variety of mass-transport processes (see Lowe, 1979; Nardin et al., 1979). In the basal Torrey Canyon unit, deposition by sandy debris flows, with a transition to turbidity currents or grain flows, is one possible mechanism, similar to that for two-layer mass-flow deposits described by Krause and Oldershaw (1979). Removal of the fine-grained (smaller than 3φ) material could have been accomplished by winnowing during backward streaming at the nose of such flows; mudstone clasts and pebbles may have been supported by a sand + mud + water matrix (see Lowe, 1979). High-density turbidity currents or liquefied flows are other means postulated as able to provide buoyancy of large clasts by a dense sand-water mixture during deposition (Lowe, 1979). Processes of rockfall and sliding are represented by large blocks (up to 10 meters across) of well-lithified mudstone and burrowed siltstone that rest singly upon the basal erosional surface or fill in scoured lows of the Eocene example. This material was apparently derived from the canyon walls.

The second canyon-fill unit, overlying the basal amalgamated pebbly sandstone facies, consists of planar- to convolute-laminated sandstone. This silty, medium- to fine-grained sandstone has parallel plant- and mica-rich laminae that become convoluted upwards into box-like and "flamed" folds (that are oversteepened in a down-canyon direction) (Fig. 5C). The coarsest size fraction begins at 3φ; the sandstone is moderately well-sorted and strongly positively skewed. Laminae become muddier and dewatering features increase upward. Apparently this material was overcharged with pore water, whose expulsion produced convolutions, dewatering pipes, dish-and-pillar structures and, in the uppermost portion, clastic dikes. Two separate mechanisms, grain flow and fluidized flow (Nardin et al., 1979), have been called upon in the literature to describe analogous units, and both or either may have been operating, possibly in conjunction with sand creep or turbidity currents.

Sand flows observed in modern canyons may represent either grain flows or fluidized flows. In the former, grain-dispersive pressure is the support mechanism; this may be important only on very steep slopes or as traction carpets beneath turbidity currents (Sanders, 1965; Middleton and Southard, 1977). Artificially produced grain flows in the head of Carmel Canyon move downslope tens of meters from the point of sediment input and form thin, laminated, and inversely graded deposits (Dingler and Anima, 1981). In contrast, fluidized flows may travel for much longer distances down relatively gentle slopes and be efficient transfer agents for thin units of fine-grained materials (Middleton, 1969; Nardin et al., 1979). The variety of dewatering structures in the Torrey Submarine Canyon indicates that fluidized flows, with the conversion to liquefied flows just prior to deposition (Nardin et al., 1979), may have been the dominant transport mechanism. Any of these flow types may have formed as transitions from turbidity currents (see Stanley et al., 1978; Walker, 1978).

The final facies consists predominantly of mudstone-filled channels on scales of tens of meters to hundreds of meters wide, and meters to tens of meters thick (Fig. 5D). The mudstone may represent passive hemipelagic deposition after the Torrey Submarine Canyon tributary head became removed from coarse sediment input during a eustatic high stand of sealevel. Meandering currents successively evacuated crosscutting channels (Fig. 6A). Thin-bedded sandstone and siltstone layers floor many of these mudstone-filled channels, and are graded, rippled, cross-bedded, or burrowed. Richly fossiliferous layers are also common, especially in the upper portion of this facies. The rippled and cross-bedded layers indicate deposition from both up-canyon and down-canyon currents, and graded sandstone may represent deposition from tails of turbidity currents which bypassed the head and flowed down-canyon. The extremely fossiliferous beds resemble those of the overlying shelf sequences and are probably spillover deposits resedimented from the adjacent shelf.

Cross-cutting channels with variegated fills thus characterize this Eocene proximal submarine canyon (Fig. 6B). Suspensate fallout of hemipelagic material is the predominant channel fill, with some intervals indicating channelized-current deposition. Graded turbidite sandstones and/or tidally influenced rippled and cross-bedded layers are locally prevalent. Some channels display a basal hemipelagic mudstone drape, then a fill of laminated sandstone. Still others contain a variety of fining- and thinning-upward sequences produced by meandering thalweg deposition.

Therefore, a whole suite of erosional and depositional processes are represented by these rocks. Whereas the specific hydrodynamics responsible for each type of unit may not be known, affinities with various modern flows may be evaluated. Many features found associated with modern canyons near the shelfbreak have analogs in this sequence described from the Eocene of San Diego.

Comparison with Other Ancient Examples

Sequences and deposits similar to those of the Eocene Torrey Submarine Canyon have been described from other ancient canyon systems. Perhaps the best studied is the Grès d'Annot channel fill in the French Martime Alps (Stanley, 1967, 1975; Stanley et al., 1978). The French canyon is dominated by massive, amalgamated sandstones similar to the basal unit of the San Diego example. These sandstones, containing little matrix and abundant pebbles and rip-up clasts, were designated "fluxo-turbidites" and later reinterpreted to result from a variety of processes similar to those observed in modern canyons: disorganized pebbly components represent debris flow deposits and massive sand-

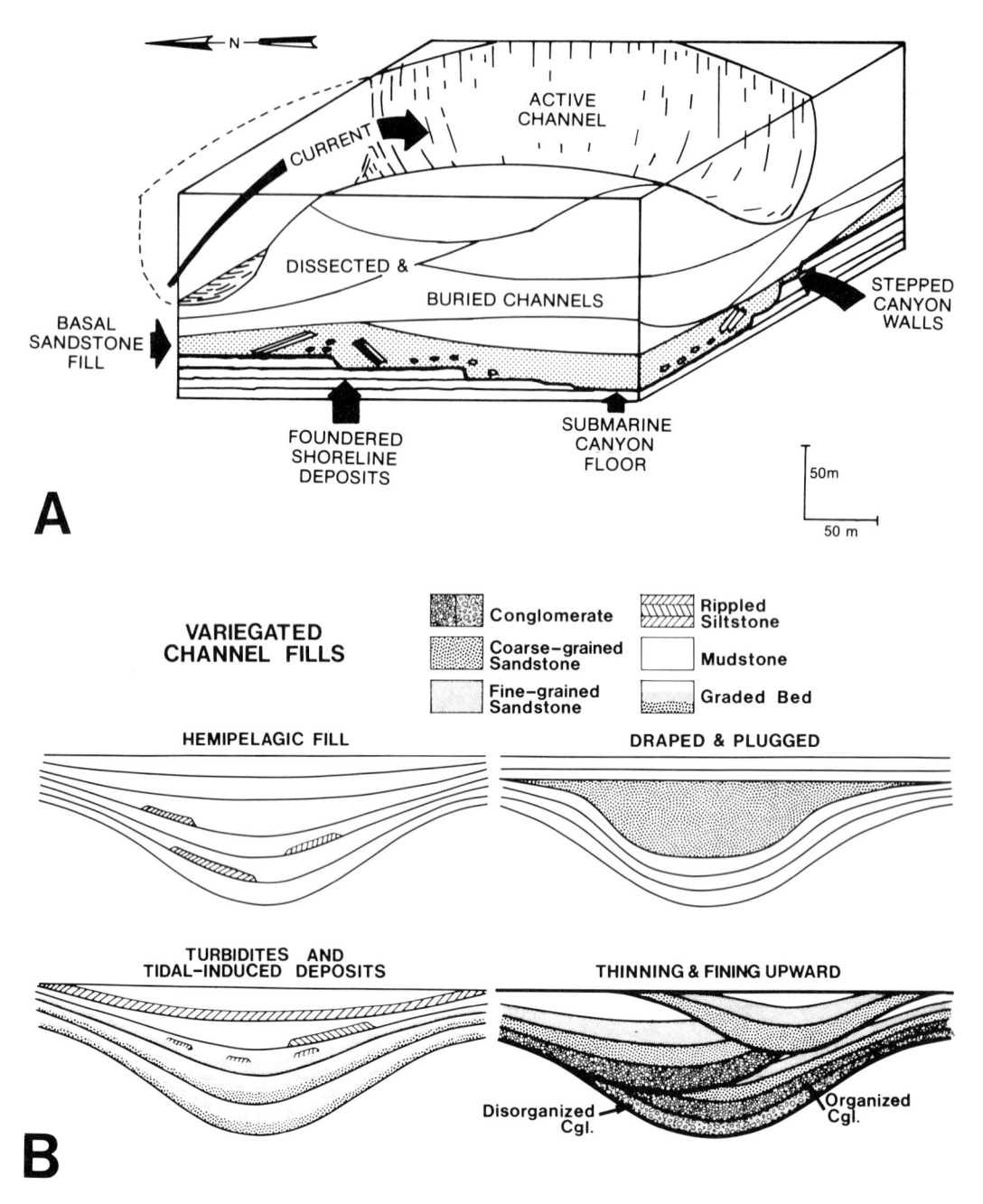

FIG. 6.—Lithologically varied channels characterize the Eocene Torrey Submarine Canyon head north of San Diego, California. A, Meandering currents periodically evacuated channels and moved sediment downslope. These channels in turn were plugged by a variety of materials. The scale and nature of such channels depend on canyon configuration and scale, shelf morphology, eustatic change, basin-margin subsidence and uplift, and source fluctuation. B, Variegated channel fills include those: (1) dominated by fine-grained suspensate fallout, (2) containing graded (turbidite) sandstone or rippled (tidally-influenced) siltstone, (3) draped with hemipelagic mudstone and plugged by sandstone, and (4) multiply cross-cut with fining- and thinning-upward sequences.

stones indicate deposition from creep and sand flow (grain flow and/or fluidized flow) (see Stanley and Unrug, 1972; Stanley, 1975; Stanley et al., 1978). Neither the interbedded sandstone-siltstone layers nor the poorly graded turbidites in the Annot units are directly comparable to deposits of the San Diego example, and irregular, discontinuous slump bodies are more common in the French canyon fill. Convoluted layers and units with dewatering features are rarer in the Annot system, and a thick hemipelagic sequence comparable to that of the Torrey Submarine Canyon is missing in outcrop.

The Balleny Group of southern New Zealand contains an Oligocene submarine canyon complex very similar to the Annot sequence (Carter and Lindqvist, 1977). Coarse- to very coarse-grained sandstone predominates in the Balleny Group. Basal coarse-grained breccia beds that were deposited by rockfall and debris flows are overlain by thickly bedded and cross-cutting structureless sandstones that contain scattered rip-up clasts and rare bioturbated sandy mudstone beds. Emplacement of these sandstones was interpreted as dominated by "inertia-flow" processes (debris flow and grain flow), slump-creep, and "fluxoturbidites" (grain flow carpet associated with turbidity currents) (Carter and Lindqvist, 1977).

Whereas the Annot and Balleny Group submarine canyon exposures are much sandier than the San Diego Eocene sequence, other ancient canyon sequences are much finer-grained. For example, fill of the buried Yoakum Canyon in Texas is almost uniformly silty shale (Hoyt, 1959), as is the Early Cretaceous Gevaram Canyon of Israel (Cohen, 1976). The buried Tertiary canyons of southern Sacramento Valley, California—Martinez, Meganos, and Markley canyons—are also predominantly filled with fine-grained units, with some interbedded sandstone, siltstone, and shale in the shallowest portions (Dickas and Payne, 1967; Almgren, 1978).

Closer approximations to the complete fill sequence observed in the Torrey Canyon tributary are provided by the Upper Cretaceous Pigeon Point and Paleocene Carmelo Formations of coastal California (Lowe, 1972; Clifton, 1981) and Late Eocene to Early Oligocene Tethyan submarine canyons in Czechoslovakia (Picha, 1979). All of these systems are upward-fining and display many analogous deposits. The basal Pigeon Point canyon fill consists of bedded sand-matrix conglomerates that were interpreted as grain flow and traction current deposits (Lowe, 1972). The next unit is a heterogeneous mixture of muddy sand- and mud-matrix conglomerate, sandstone, mudstone, and pebbly mudstone, indicative of mixed deposition from grain flows, debris flows, slumps, hemipelagic fallout, and possible turbidity currents (Lowe, 1972). The uppermost unit is dark gray mudstone with some pebbly mudstone and slumped mudstone in the lower portion, and with sparse siltstone and sandstone laminae in the upper half, indicating turbidity-current bypassing dominated by hemipelagic fallout (Lowe, 1972). Sandstones, conglomerates, pebbly mudstones, and contorted units in the lower Tethyan canyon fill indicate deposition by "proximal" and "distal" turbidity currents, grain flows, debris flows, and slumps (Picha, 1979). The upper fill is silty mudstone representing hemipelagic fallout (Picha, 1979). The Carmelo Formation canyon head likewise contains an upward-fining sequence. Basal conglomerates and cobbly mudstones pass upward through amalgamated pebbly sandstones with conglomeratic layers, into thick-bedded sandstones, and finally thin-bedded mudstones and sandstones (Clifton, 1981).

DISCUSSION

Knowledge of the processes and factors involved in submarine canyon formation, maintenance, and fill is far from complete. Better and new instrumentation, such as side-scan sonar and improved current meters and coring devices, will continue to increase understanding of modern canyon processes. But experimentation has lagged distantly behind the theorizing about mass-transport mechanisms, and is necessary before the dynamics of observed sediment-gravity flows, for example, and their correlative ancient deposits can be fully discerned.

The diversity in submarine canyon fill types and sequences results not only from hydrodynamic variations, but also from the interaction of such variables as shelf width and gradient, rates of eustacy, subsidence, and tectonics, and input and migration of sediment sources. Canyons that head near pulsating sediment sources and are dominated by erosion are designated as "proximal" by Dill (1981); rapid accumulations of poorly sorted, metastable detrital masses create slumps and sediment gravity flows. In contrast, Dill's (1981) "distal" canyons undergo quiescent fill with cohesive, fine-grained material. Episodic or progressive variations in sediment supply in turn may be controlled by factors such as climatic fluctuation, sealevel change, or river mouth migration. Local sealevel variation is affected by worldwide eustatic events and basin-margin uplift and subsidence. These eustatic and tectonic effects may be in or out of phase, and are further affected by continental shelf morphology. For example, submarine canyon heads along the broad, gentle shelves of the northern Gulf of Mexico and western Atlantic are more responsive to sealevel change; they were alternately active and inactive in response to Pleistocene glacial-eustatic cycles (Jacka et al., 1968). In contrast, canyons of the narrow, steep shelf of western North America have maintained their proximal, nearshore positions even during high Pleistocene sealevel stands (Jacka et al., 1968). Thus, a variety of factors must be considered to explain the fill sequence of any one ancient canyon complex. One goal is quantification of the interactive temporal and spatial variables that control submarine canyon evolution, in order to model the environmental significance and economic potential of these systems.

CONCLUSIONS

Submarine canyons at the shelfbreak trap coarse- and fine-grained bedload materials as well as suspended sediments, funneling them into adjacent marine basins. These systems also produce an abrupt interface between shallow- and deep-marine hydrodynamic and sedimentologic conditions. A complex mixture of sediment types is created by diverse subaerial and submarine processes in this environment.

Marine circulation affected by submarine canyons may overprint normal tidal and wind-driven currents along continental margins, thereby influencing shelf sediment concentrations and dispersal pathways. Surface waves diverge over canyon heads, establishing a differential energy gradient along narrow shelves and causing littoral sediment to flow back toward canyons along a preferential transport corridor. Canyons funnel turbidity currents and fluctuating up-canyon, down-canyon, and cross-canyon currents, causing localized winnowing and both upwelling and downwelling in the headward region. High-frequency internal waves are focused along canyon axes and may break adjacent to canyon heads. Boundary currents shed eddies while passing over canyon axes and may combine with focused up-canyon flow or longshore drift to cause erosion on the shelf.

Along the shelfbreak, older consolidated and unconsolidated sediments may be fluvially incised during subaerial exposure, or carved by submarine mass flows derived from rapidly accumulated sediment along narrow, steep shelves. In "distal" canyons heading far from active sediment sources, relict coarse-grained material may spill over from the shelf during storms and drape canyon walls. This detritus is reworked by channelized currents and becomes finer-grained and less gravelly down-canyon. Down-flank slumps incorporate this sediment and older wall material, widening and eroding canyons headward. Slumped sediment accumulates along bases of canyon walls and collects products of rockfalls and bioerosion. This amassed debris may be transformed to pebbly mudstones, which often move as debris flows away from canyon margins and down-canyon axes.

"Proximal" canyons heading near sources of unsorted, coarse-grained sediment display similar lateral infill, then more catastrophic downslope movement, as slumps and sand spillover are transformed to various sediment-gravity flows. Sandy, often organic-rich buildups of sediment prograde down the canyon head until the internal shear strength is overcome; the sediment mass may then give way to creep, sand flows, debris flows, and turbidity currents. Channelized currents rework surficial sediments, and exceptionally large slumps may flush canyon axes. Clean sand is rare; irregular sediment distributions characterize modern canyon heads, although grain size does tend to decrease further downcanyon.

Deposits in the heads of ancient submarine canyons thus also lack lithologic uniformity and display abundant cut-and-fill channels. Products of mass flow, tractional deposition, and hemipelagic fallout are all represented. Disorganized and organized conglomerates, pebbly sandstones, and massive sandstones ("fluxoturbidites" or "proximal" turbidites) are common axial deposits. These deposits are typical of high-density sediment-gravity flows on steep slopes, and have been interpreted as representing debris flows, high-concentration turbidity currents, and sand flows (grain flows and/or fluidized flows). Convolute-laminated sandstones may be products of fluidized flows and/or creep. Marginal to canyon walls, chaotic and brecciated slump units, pebbly mudstones (debris flows), and rockfalls are conspicuous. Turbidite sequences, traction deposits, and hemipelagic mudstones characterize deposits in more deactivated canyon heads.

Ancient canyons display a record of these processes complicated in time and space by: (1) the geologic setting, such as morphology and stratigraphy of the shelf and slope or tectonic history of the basin margin, (2) rates and degrees of eustatic sealevel rise and fall, (3) rates of and changes in basin-margin subsidence and uplift, and (4) type of and fluctuations in sediment supply. Truly active canyon development is characterized by axial downcutting, canyon widening, and rim upbuilding. Diverse fill types and sequences indicate fundamental changes in one or more of the above parameters and varying amounts of deactivation.

ACKNOWLEDGMENTS

Work on the Eocene Torrey Submarine Canyon was part of J. A. May's dissertation research, funded by grants from Marathon Oil Company, Amoco Production Company, Shell Development Company, Union Oil Company, the American Association of Petroleum Geologists, and the Geological Society of America. Jack Welch, Department of Parks and Recreation, was instrumental in permitting access for field work in Torrey Pines State Reserve. Philip R. Hill, Paul R. Carlson, and Donald G. McCubbin critically reviewed the manuscript and provided many helpful comments, and Daniel J. Stanley encouraged the manuscript's submittal. Special thanks to Karen A. Crossen for field, darkroom, and editorial assistance.

REFERENCES

ALMGREN, A. A., 1978, Timing of Tertiary submarine canyons and marine cycles of deposition in the southern Sacramento Valley, California, in Stanley, D. J., and Kelling, G., eds., Sedimentation in Submarine Canyons, Fans, and Trenches: Stroudsburg, Pa., Dowden, Hutchinson, & Ross, Inc., p. 276–291.

BAKER, E. T., 1976, Distribution, composition, and transport of suspended particulate matter in the vicinity of Willapa Submarine Canyon, Washington: Geol. Soc. America Bull., v. 87, p. 625–632.

BELDERSON, R. H., AND STRIDE, A. H., 1969, The shape of submarine canyon heads revealed by Asdic: Deep-Sea Research, v. 16, p.103–104.

BELLAICHE, G., COUMES, F., IRR, F., ROURE, F., AND VANNEY, J.-R., 1979, Structure of the French Riviera submarine canyons: Evidence of a polygenetic history from a submersible study ("Cyaligure" Campaign): Mar. Geology, v. 31, p. M5-M12.

BERG, R. R., 1979, Turbidite channel reservoirs in Canyon Sandstone: Roundtop area, Fisher County, Texas: Am. Assoc. Petroleum Geologists, Southwest Sec. Meeting Abs., v. 63, p. 1425.

BIRDSALL, B. C., AND SCOTT, R. M., 1976, Physical and biogenic characteristics of sediments from upper Hueneme Submarine Canyon, California coast [abs.]: Am. Assoc. Petroleum Geologists Bull., v. 60, p. 650.

BORNHAUSER, M., 1948, Possible ancient canyon in southwestern Louisiana: Am. Assoc. Petroleum Geologists Bull., v. 32, p. 2287–2290.

BOUMA, A. H., 1965, Sedimentary characteristics of samples collected from some submarine canyons: Mar. Geology, v. 3, p. 291–320.

BUCHANAN, J. Y., 1887, On the land slopes separating continents and ocean basins, especially those on the West Coast of Africa: Scot. Geograph. Mag., v. III, p. 217–238.

BUCHER, W. H., 1940, Submarine valleys and related geologic problems of the North Atlantic: Geol. Soc. America Bull., v. 51, p. 489–512.

CARTER, R. M., AND LINDQVIST, J. K., 1977, Balleny Group, Chalky Island, southern New Zealand: An inferred Oligocence submarine canyon and fan complex: Pacific Geology, v. 12, p. 1–46.

CHAMBERLAIN, T. K., 1964, Mass transport of sediment in the heads of Scripps Submarine Canyon, California, in Miller, R. L., ed., Papers in Marine Geology, Shepard Commemorative Volume: New York, The MacMillan Co., p. 42–64.

CLEARY, W. J., AND CONOLLY, J. R., 1974, Petrology and origin of deep-sea sediments: Hatteras Abyssal Plain: Mar. Geology, v. 17, p. 263–279.

CLIFTON, H. E., 1981, Submarine canyon deposits, Point Lobos, California, in Frizzell, V., ed., Upper Cretaceous and Paleocene Turbidites, Central California Coast: Soc. Econ. Paleontologists Mineralogists, Pacific Sec., Field Trip 6, p. 79–92.

COHEN, Z., 1976, Early Cretaceous buried canyon: Influence on accumulation of hydrocarbons in Helez Oil Field, Israel: Am. Assoc. Petroleum Geologists Bull., v. 60, p. 108–114.

CROWELL, J. C., 1952, Submarine canyons bordering central and southern California: Jour. Geology, v. 60, p. 58–83.

DALY, R. A., 1936, Origin of submarine canyons: Am. Jour. Sci., v. 31, p. 401–420.

DANA, J. D., 1863, A Manual of Geology: Philadelphia, T. Bliss & Co., 798 p.

———, 1890, Long Island Sound in the Quaternary Era, with observations on the submarine Hudson River channel: Am. Jour. Sci., Ser. 3, p. 425–437.

DAVIDSON, G., 1887, Submarine valleys on the Pacific Coast of the United States: Calif. Acad. Sci. Bull., v. 2, p. 265–268.

DAVIS, W. M., 1934, Submarine rock valleys: Geol. Rev., v. 24, p. 297–308.

DICKAS, A. B., AND PAYNE, J. L., 1967, Upper Paleocene buried channel in Sacramento Valley, California: Am. Assoc. Petroleum Geologists Bull., v. 51, p. 873–882.

DIETZ, R. S., KNEBEL, H. J. AND SOMERS, L. H., 1968, Cayar Submarine Canyon: Geol. Soc. America Bull., v. 79, p. 1821–1838.

DILL, R. F., 1964, Sedimentation and erosion in Scripps Submarine Canyon head, in Miller, R. L., ed., Papers in Marine Geology, Shepard Commemorative Volume: New York, The MacMillan Co., p. 23–41.

———, 1981, Role of multiple-headed submarine canyons, river mouth migration, and episodic activity in generation of basin-filling turbidity currents [abs.]: Am. Assoc. Petroleum Geologists Bull., v. 65, p. 918.

———, MARSHALL, N. F., AND REIMNITZ, E., 1975, In situ submersible observations of sediment transport and erosive features in Rio Balsas Submarine Canyon, Mexico: Geol. Soc. America, Abs. with Programs, v. 7, p. 1052–1053.

DINGLER, J. R., AND ANIMA, R. J., 1981, Field study of subaqueous avalanching [abs.]: Am. Assoc. Petroleum Geologists Bull., v. 65, p. 918.

DRAKE, D. E., AND GORSLINE, D. S., 1973, Distribution and transport of suspended particulate matter in Hueneme, Redondo, Newport, and La Jolla Submarine Canyons, California: Geol. Soc. America Bull., v. 84, p. 3949–3968.

———, HATCHER, P. G., AND KELLER, G. H., 1978, Suspended particulate matter and mud deposition in upper Hudson Submarine Canyon, in Stanley, D. J., and Kelling, G., eds., Sedimentation in Submarine Canyons, Fans, and Trenches: Stroudsburg, Pa., Dowden, Hutchinson, & Ross, Inc., p. 33–41.

———, ———, Pak, H., and Keller, G. H., 1976, Contrasts in concentrations of suspended matter in Hydrographer, Hudson, and Wilmington Submarine Canyons [abs.]: Am. Assoc. Petroleum Geologists Bull., v. 60, p. 667.

Emery, K. O., 1960, The Sea off Southern California: New York, John Wiley & Sons, 366 p.

Farre, J. A., McGregor, B. A., Ryan, W. B. F., and Robb, J. M., 1983, Breaching the shelfbreak: passage from youthful to mature phase in submarine canyon evolution, in Stanley, D. J., and Moore, G. T., eds., The Shelfbreak: Critical Interface on Continental Margins: Soc. Econ. Paleontologists Mineralogists Spec. Pub. 33, p. 25–39.

Felix, D. W., and Gorsline, D. W., 1971, Newport Submarine Canyon, California: an example of the effects of shifting loci of sand supply upon canyon position: Mar. Geology, v. 10, p. 177–198.

Freeland, G. L., Stanley, D. J., Swift, D. J. P., and Lambert, D. N., 1981, The Hudson Shelf Valley: Its role in shelf sediment transport: Mar. Geology, v. 42, p. 399–427.

Galloway, W. E., and Brown, L. F., Jr., 1973, Depositional systems and shelf-slope relations on cratonic basin margin, uppermost Pennsylvanian of north-central Texas: Am. Assoc. Petroleum Geologists Bull., v. 57, p. 1185–1218.

Garcia, R., 1981, Depositional systems and their relation to gas accumulation in Sacramento Valley, California: Am. Assoc. Petroleum Geologists Bull., v. 65, p. 653–673.

Got, H., and Stanley, D. J., 1974, Sedimentation in two Catalonian canyons, northwestern Mediterranean: Mar. Geology, v. 16, p. M91–M100.

Herzer, R. M., and Lewis, D. W., 1979, Growth and burial of a submarine canyon off Motunau, north Canterbury, New Zealand: Sedimentary Geology, v. 24, p. 69–83.

Holcombe, T. L., and Einwich, A. M., 1976, Geomorphology west of St. Croix, U.S. Virgin Islands—a case history of "submarine stream capture": Geol. Soc. America, Abs. with Programs, v. 8, p. 921.

Hooke, R. LeB., and Schlager, W., 1980, Geomorphic evolution of the Tongue of the Ocean and the Providence Channels, Bahamas: Mar. Geology, v. 35, p. 343–366.

Hotchkiss, F. S., and Wunsch, C., 1979, Dynamic ingredients of Hudson Submarine Canyon [abs.]: EOS Trans. Am. Geoph. Union, v. 60, p. 89–90.

Hoyt, W. V., 1959, Erosional channel in the Middle Wilcox near Yoakum, Lavaca County, Texas: Trans. Gulf Coast Assoc. Geol. Socs., v. 9, p. 41–50.

Hsü, K. J., 1972, When the Mediterranean dried up: Scientific American, v. 227, p. 26–36.

Jacka, A. D., Beck, R. H., St. Germain, L. C., and Harrison S. C., 1968, Permian deep-sea fans of the Delaware Mountain Group (Guadalupian), Delaware Basin, in Silver, B. A., ed., Guadalupian Facies, Apache Mountains Area, West Texas: Soc. Econ. Paleontologists Mineralogists Permian Basin Sec., Symposium and Guidebook, Pub. 6811, p. 49–90.

Johnson, D. W., 1938, The Origin of Submarine Canyons: New York, Columbia Univ. Press, 126 p.

Karl, H. A., 1980, Influence of San Gabriel Submarine Canyon on narrow-shelf sediment dynamics, Southern California: Mar. Geology, v. 34, p. 61–78.

Kelling, G., and Stanley, D. J., 1970, Morphology and structure of Wilmington and Baltimore Submarine Canyons, eastern United States: Jour. Geology, v. 78, p. 637–660.

Knebel, H. J., 1979, Anomalous topography on the continental shelf around Hudson Canyon: Mar. Geology, v. 33, p. M67–M75.

———, Wood, S. A., and Spiker, E. C., 1979, Hudson River: Evidence for extensive migration on the exposed continental shelf during Pleistocene time: Geology, v. 7, p. 254–258.

Krause, F. F., and Oldershaw, A. E., 1979, Submarine carbonate breccia beds—a depositional model for two-layer, sediment gravity flows from the Sekwi Formation (Lower Cambrian), Mackenzie Mountains, Northwest Territories, Canada: Canadian Jour. Earth Sci., v. 16, p. 189–199.

Krueger, W. C., and North, F. K., 1983, Occurrences of oil and gas in association with the paleo-shelfbreak, in Stanley, D. J., and Moore, G. T., eds., The Shelfbreak: Critical Interface on Continental Margins: Soc. Econ. Paleontologists Mineralogists Spec. Pub. 33, p. 409–427.

Kuenen, P. H., 1937, Experiments in connection with Daly's hypothesis on the formation of submarine canyons: Leidsche Geologische Medeelingen, v. VIII, p. 316–351.

Lawson, A., 1893, The geology of Carmelo Bay: Univ. California Dept. Geol. Bull., v. 1, p. 1–59.

Le Conte, J., 1891, Tertiary and post Tertiary changes of the Atlantic and Pacific coasts: Geol. Soc. America Bull., v. 2, p. 323–328.

Lohmar, J. M., May, J. A., Boyer, J. E., and Warme, J. E., 1979, Shelf edge deposits of the San Diego Embayment, in Abbott, P. L., ed., Eocene Depositional Systems, San Diego: Soc. Econ. Paleontologists Mineralogists Pacific Sec., Field Trip Guide, p. 15–33.

———, and Warme, J. E., 1978, Anatomy of an Eocene submarine canyon-fan system, Southern California Borderland: Offshore Tech. Conf. Preprints, p. 571–580.

———, and ———, 1979, An Eocene shelf margin: San Diego County, California, in Armentrout, J. M., Cole, M. R., and Terbest, H., Jr., eds., Cenozoic Paleogeography of the Western United States: Pacific Coast Paleogeography Symposium No. 3, p. 165–175.

Lowe, D. R., 1972, Submarine canyon and slope channel sedimentation model as inferred from Upper Cretaceous deposits, western California: Internat. Geol. Congr., Montreal, Proc. Sec. 6, p. 75–81.

———, 1979, Sediment gravity flows: their classification and some problems of application to natural flows and

deposits, *in* Doyle, L. J., and Pilkey, O. H., eds., Geology of Continental Slopes: Soc. Econ. Paleontologists Mineralogists Spec. Pub. 27, p. 75–82.
MALAHOFF, A., EMBLEY, R. E., PERRY, R. B., AND FEFE, C., 1977, Sedimentation processes on the continental shelf, slope, and upper rise near Baltimore Canyon [abs.]: EOS Trans. Am. Geoph. Union, v. 58, p. 1160.
———, AND FORNARI, D. J., 1979, Geologic observations from *Alvin* of the continental margin from Baltimore Canyon to Norfolk Canyon [abs.]: EOS Trans. Am. Geoph. Union, v. 60, p. 287.
MARSHALL, N. F., 1978, A large storm-induced sediment slump reopens an unknown Scripps Submarine Canyon tributary, *in* Stanley, D. J., and Kelling, G., eds., Sedimentation in Submarine Canyons, Fans, and Trenches: Stroudsburg, Pa., Dowden, Hutchinson, & Ross, Inc., p. 73–84.
MCCLENNAN, C. E., 1973, Great Erg buried channel on the New Jersey continental shelf: A possible continuation of the Pleistocene Schuylkill River to Wilmington Canyon: Geol. Soc. America, Abs. with Programs, v. 5, p. 194–195.
MCGREGOR, B. A., 1981, Ancestral head of Wilmington Canyon: Geology, v. 9, p. 254–257.
MIDDLETON, G. V., 1969, Grain flows and other mass movements down slopes, *in* Stanley, D. J., ed., The *New* Concepts of Continental Margin Sedimentation: American Geol. Inst. Short Course Lecture Notes, p. GM-B-1 to GM-B-14.
———, AND SOUTHARD, J. G., 1977, Mechanics of Sediment Movement: Soc. Econ. Paleontologists Mineralogists Short Course No. 3, 250 p.
NARDIN, T. R., HEIN, F. J., GORSLINE, D. S., AND EDWARDS, B. D., 1979, A review of mass movement processes, sediment and acoustic characteristics, and contrasts in slope and base-of-slope systems versus canyon-fan-basin floor systems, *in* Doyle, L. J., and Pilkey, O. H., eds., Geology of Continental Slopes: Soc. Econ. Paleontologists Mineralogists Spec. Pub. No. 27, p. 61–73.
NORMARK, W. R., 1974, Submarine canyons and fan valleys: Factors affecting growth patterns of deep-sea fans, *in* Dott, R. H., Jr., and Shaver, R. H., eds., Modern and Ancient Geosynclinal Sedimentation: Soc. Econ. Paleontologists Mineralogists Spec. Pub. 19, p. 56–68.
———, PIPER, D. J. W., AND HESS, G. R., 1979, Distributary channels, sand lobes, and mesotopography of Navy Submarine Fan, California Borderland, with applications to ancient fan sediments: Sedimentology, v. 26, p. 749–774.
PICHA, F., 1979, Ancient submarine canyons of Tethyan continental margins, Czechoslovakia: Am. Assoc. Petroleum Geologists Bull., v. 63, p. 67–86.
REDWINE, L. E., 1978, Tertiary Princeton Submarine Valley system beneath Sacramento Valley, California [abs.]: Am. Assoc. Petroleum Geologists Bull., v. 62, p. 2360.
REIMNITZ, E., 1971, Surf-beat origin for pulsating bottom currents in the Rio Balsas Submarine Canyon, Mexico: Geol. Soc. America Bull., v. 82, p. 81–90.
———, AND GUTIÉRREZ-ESTRADA, M., 1970, Rapid changes in the head of the Rio Balsas Submarine Canyon system, Mexico: Mar. Geology, v. 8, p. 245–258.
SANDERS, J. E., 1965, Primary sedimentary structures formed by turbidity currents and related resedimentation mechanisms, *in* Middleton, G. V., ed., Primary Sedimentary Structures and Their Hydrodynamic Interpretation: Soc. Econ. Paleontologists Mineralogists Spec. Pub. 12, p. 192–219.
SCHOLL, D. W., BUFFINGTON, E. C., HOPKINS, D. M., AND ALPHA, T. R., 1970, The structure and origin of the large submarine canyons of the Bering Sea: Mar. Geology, v. 8, p. 187–210.
SHCHERBAKOV, F. A., 1978, Some characteristics of sediment orogenesis on the continental margin of the Black Sea: Oceanology, v. 18, p. 575–578.
SHEPARD, F. P., 1951, Mass movements in submarine canyon heads: EOS Trans. Am. Geoph. Union, v. 32, p. 405–418.
———, 1973, Submarine Geology: New York, Harper and Row, 517 p.
———, 1979, Currents in submarine canyons and other types of sea-valleys, *in* Doyle, L. J., and Pilkey, O. H., eds. Geology of Continental Slopes: Soc. Econ. Paleontologists Mineralogists Spec. Pub. 27, p. 85–94.
———, 1981, Submarine canyons: Multiple causes and long-term persistence: Am. Assoc. Petroleum Geologists Bull., v. 65, p. 1062–1077.
———, AND DILL, R. F., 1966, Submarine Canyons and Other Sea Valleys: Chicago, Rand McNalley & Co., 381 p.
———, AND EMERY, K. O., 1941, Submarine topography off the California Coast: Canyons and tectonic interpretation: Geol. Soc. America Spec. Paper 31, 171 p.
———, AND MARSHALL, N. F., 1976, Currents in submarine canyons and their relation to deep-sea fans [abs.]: Am. Assoc. Petroleum Geologists Bull., v. 60, p. 721–722.
———, ———, MCLOUGHLIN, P. A., AND SULLIVAN, G. G., 1979a, Currents in submarine canyons and other sea-valleys: Am. Assoc. Petroleum Geologists, Studies in Geology No. 8, 193 p.
———, REVELLE, R. R., AND DIETZ, R. S., 1939, Ocean-bottom currents off the California coast: Science, v. 89, p. 488–489.
———, WOOD, F. J., AND SULLIVAN, G. G., 1979b, Perigean spring tides and unusual currents in La Jolla Submarine Canyon: Geol. Soc. America, Abs. with Programs, v. 11, p. 515.
SLATER, R. A., 1981, Submarine observations of the sea floor near the proposed Georges Bank lease sites along the North Atlantic outer continental shelf and upper slope: U.S. Geol. Survey Open File Rept. 81-742, 65 p.
———, TWICHELL, D. C., AND ROBB, J. M., 1981, Submersible observations of potential geologic hazards along the

mid-Atlantic outer continental shelf and uppermost slope: U.S. Geol. Survey Open File Rept. 81-968, 50 p.
SMITH, W. S. R., 1902, The submarine valleys of the California coast: Science, v. 15, p. 670–672.
SPENCER, J. W., 1898, On the continental elevation of the glacial epoch: Geol. Mag., v. 4, p. 32–38.
———, 1903, Submarine valleys off the American coast and in the North Atlantic: Geol. Soc. America Bull., v. 14, p. 207–226.
STANLEY, D. J., 1967, Comparing patterns of sedimentation in some modern and ancient submarine canyons: Earth and Planetary Sci. Letters, v. 3, p. 371–380.
———, 1971, Bioturbation and sediment failure in some submarine canyons: Vie et Milieu, Suppl., v. 22, p. 541–555.
———, 1974, Pebbly mud transport in the head of Wilmington Canyon: Mar. Geology, v. 16, p. M1–M8.
———, 1975, Submarine Canyon and Slope Sedimentation (Grès d'Annot) in the French Maritime Alps: Proc. IX Internat. Congr. Sediment., Nice, 129 p.
———, AND FREELAND, G. L., 1978, The erosion-deposition boundary in the head of Hudson Submarine Canyon defined on the basis of submarine observations: Mar. Geology v. 16, p. M37–M46.
———, AND KELLING, G., 1978, Sedimentation in Submarine Canyons, Fans, and Trenches: Stroudsburg, Pa., Dowden, Hutchinson, & Ross, Inc., 395 p.
———, PALMER, H. D., AND DILL, R. F., 1978, Coarse sediment transport by mass flow and turbidity current processes and downslope transformations in Annot sandstone canyon-fan valley systems, *in* Stanley, D. J., and Kelling, G., eds., Sedimentation in Submarine Canyons, Fans, and Trenches: Stroudsburg, Pa., Dowden, Hutchinson, & Ross, Inc., p. 85–115.
———, SHENG, H., LAMBERT, D. H., RONA, P. A., MCGRAIL, D. W., AND JENKYNS, J. S., 1981, Current-influenced depositional provinces, continental margin off Cape Hatteras, identified by petrologic method: Mar. Geology, v. 40, p. 215–235.
———, AND SILVERBERG, N., 1969, Recent slumping on the continental slope of Sable Island Bank, southeast Canada: Earth and Planetary Sci. Letters, v. 6, p. 123–133.
———, AND UNRUG, R., 1972, Submarine channel deposits, fluxoturbidites, and other indicators of slope and base-of-slope environments in modern and ancient margin basins, *in* Rigby, J. K., and Hamblin, W. K., eds., Recognition of Ancient and Modern Sedimentary Environments: Soc. Econ. Paleontologists Mineralogists Spec. Pub. 16, p. 287–40.
STARKE, G. W., AND HOWARD, A. D., 1968, Polygenetic origin of Monterey Submarine Canyon: Geol. Soc. America Bull., v. 79, p. 813–826.
STETSON, H. C., 1936, Geology and paleontology of the Georges Bank canyons, I. Geology: Geol. Soc. America Bull., v. 47, p. 339–366.
SWIFT, D. J. P., MOIR, R., AND FREELAND, G. L., 1980, Quaternary rivers on the New Jersey shelf: relation of seafloor to buried valleys: Geology, v. 8, p. 276–280.
TWICHELL, D. C., KNEBEL, H. J., AND FOLGER, D. W., 1977, Delaware River: evidence for its former extension to Wilmington Submarine Canyon: Science, v. 195, p. 483–485.
VALENTINE, P. C., UZMANN, J. R., AND COOPER, R. A., 1980, Geology and biology of Oceanographer Submarine Canyon: Mar. Geology, v. 3, p. 283–312.
VEATCH, A. C., AND SMITH, P. A., 1939, Atlantic submarine valleys off the United States and the Congo Submarine Valley: Geol. Soc. America Spec. paper No. 7, 101 p.
VON DER BORCH, C. C., 1969, Southern Australian submarine canyons: their distributions and ages: Mar. Geology, v. 6, p. 267–279.
VORMELKER, R. S., 1980, Texas Middle Wilcox channel: Oil and Gas Jour., v. 78, p. 136–154.
WALKER, R. G., 1978, Deep-water sandstone facies and ancient submarine fans: models for exploration for stratigraphic traps: Am. Assoc. Petroleum Geologists Bull., v. 62, p. 932–966.
WARME, J. E., SLATER, R. A., AND COOPER, R. A., 1978, Bioerosion in submarine canyons, *in* Stanley, D. J., and Kelling, G., eds., Sedimentation in Submarine Canyons, Fans, and Trenches: Stroudsburg, Pa., Dowden, Hutchinson, & Ross, Inc., p. 65–70.
WEBB, G. W., 1979, Entrapment factors in California turbidite and canyon-related pools [abs.]: Am. Assoc. Petroleum Geologists Bull., v. 63, p. 548–549.
WHITAKER, J. H., McD., 1974, Ancient submarine canyons and fan valleys, *in* Dott, R. H., Jr., and Shaver, R. H., eds., Modern and Ancient Geosynclinal Sedimentation: Soc. Econ. Paleontologists Mineralogists Spec. Pub. No. 19, p. 106–125.
———, 1976, Submarine Canyons and Deep-sea Fans: Modern and Ancient, Benchmark Papers in Geology No. 24: Stroudsburg, Pa., Dowden, Hutchinson, & Ross, 460 p.

SEAFLOOR CHARACTERISTICS AND DYNAMICS AFFECTING GEOTECHNICAL PROPERTIES AT SHELFBREAKS

RICHARD H. BENNETT
Naval Ocean Research and Development Activity (NORDA), Seafloor Division Code 360,
NSTL Station, Mississippi 39529

TERRY A. NELSEN
National Oceanic and Atmospheric Administration, Atlantic Oceanographic
and Meteorological Laboratories, 4301 Rickenbacker Causeway, Miami, Florida 33149

ABSTRACT

Variable geotechnical and sedimentological properties of sediments on both active and passive continental margins is the rule rather than the exception. Significant variability is observed in the physical, mechanical, and textural properties of sediments from the individual core to the regional level. Sediment properties and soil "state" are determined by the primary depositional properties and post-depositional processes active on and within shelfbreak deposits. Post-depositional processes play an important role in determining the ultimate nature and time-dependent changes in the mass physical and mechanical properties of submarine deposits. Fundamental geotechnical properties such as shear strength and compressibility from several shelfbreak areas are compared and related to eight shelfbreak sediment textural models. Sediment behavior such as resistance to erosion, slumping, consolidation, and liquefaction depend upon: (1) the basic sediment types present at shelfbreaks (textural model), (2) the fundamental geotechnical properties of the particular deposit, and (3) temporally and spacially variable physical and biological processes active in the shelfbreak zone.

Shelfbreak characteristics important to offshore engineering activities and future scientific studies on continental margins include: varied morphology with steep local slopes; rapid changes in physical oceanographic processes; variable erosion and sedimentation rates; variable offshelf sediment transport; onset of significant pelagic sedimentation; rapid textural changes; rapidly changing benthic community; high variability in geotechnical properties; and the onset of creep and mass wasting processes. Shelfbreak sediment types may be predominantly terrigenous, carbonate, or glacio-marine in origin. Although some of these characteristics can be found in other submarine environments, they can be identified as important elements of many passive and active continental margins.

INTRODUCTION

Geotechnical study of submarine sediment, an electrolyte-gas-solid system, includes the scientific and engineering investigation of the physical, mechanical, chemical, biological, microstructural, and acoustical properties of seafloor sedimentary deposits. Relatively little is known of the interrelationship of these properties which determine the ultimate nature and dynamic behavior of submarine sediment (soils in the engineering sense). Geologists and engineers commonly investigate these various properties and the sediment's response to natural or man-induced static and dynamic forces. Many of the fundamental geotechnical properties are used for soils classification, estimations of stress within the sediment, soil state ("liquid," plastic, solid), and for parameters in the analysis of seafloor stability (large and small-scale mass movement, liquefaction potential, and degree of consolidation). Geotechnical data obtained from *in situ* and laboratory tests and experiments are used as parameters to evaluate soil behavior and foundation characteristics for the emplacement of structures on the seafloor. Predictive capabilities and reliable engineering judgement are critical elements in the design and construction of offshore structures.

Thus, an understanding of geotechnical properties and their variability is critical to the cost-effective design, construction, and safety of offshore structures and to potential environmental impacts.

Geotechnical data are used by the geologist to study the fundamental nature and variability of submarine sediments to gain an understanding of geological processes and sedimentary patterns and changes in various depositional enviroments for different sediment types. A few studies of surficial submarine sediment and the volumes of data collected with the advent of the Deep Sea Drilling Project have demonstrated clearly the high variability of the mass physical and mechanical properties not only among different sediments but also within specific sediment types (Keller and Bennett, 1968, 1970, 1973; Bennett et al., 1970, 1980; Bryant et al., 1981). The mass physical, chemical, and mechanical properties are used to delineate diagenetic changes that sediments undergo during burial and through geologic time. These changes and the fundamental properties differ among sediment types (Bryant et al., 1981).

The sediment microstructure (fabric and physicochemical forces) and specifically the fabric (spatial arrangement and particle-to-particle associations) determine the fundamental properties of sediment during deposition and during post-depositional changes (Bennett et al., 1977; Bennett and Hulbert, 1982). The microstructure plays an important role in determining fundamental properties such as water content, porosity, permeability, wet unit weight, and the strength of sedimentary material under given states of stress. The microstructure is considered an important factor in determining the isotropic and anisotropic mechanical and acoustical properties of a sediment and its behavior under environmental forces and stresses. Examples of such forces can be seismic shock or surface wave activity (Almagor and Wiseman, 1977; Suhayda et al., 1976; Bennett and Faris, 1979; Yamamoto, 1980).

The more commonly used geotechnical properties that can be expressed in quantitative terms are presented in Tables 1 and 2. Although all of these properties are not referred to in the text, they are included in these tables to provide the reader with a broad example of geotechnical measurements, their units and application. Metric system symbols and units expressed in SI (Le Système International d'Unités) terms are now the international standard (Richards, 1974). Factors such as biological influence on the sediment, chemical effects, and microstructure are still in the qualitative and semi-quantitative stages of study and are not at present useful as parameters for numerical modeling (Bennett and Hulbert, in press; Richardson and Young, 1980). Thus, it is clear that geotechnical studies are of practical as well as scientific interest to geologists and engineers concerned with seafloor deposits.

Geotechnical investigations of continental margin shelfbreak deposits offer challenging scientific and engineering opportunities because of the virtual frontier of possible studies and because of the rather unique environmental setting of the shelfbreak as a major zone commonly characterized by textural changes, variability, and ocean dynamics. The few studies available have shown the highly variable nature of shelfbreak deposits in terms of composition, texture, morphologic settings, seafloor gradients, physical and mechanical properties, and dynamic processes. The geological setting, depositional environment, morphological character, and dynamic processes strongly affect the nature of the geotechnical properties and ultimately the stability of seafloor deposits. These variables establish the uniqueness of shelfbreak deposits which contrast with shelf, slope, and deep-sea basin sediment properties and processes. For the geologist, the above aspects are some of the scientific interests in the sedimentological and geotechnical properties of these deposits. The shelfbreak deposits, however, are equally important in the engineering sense because of the virtually constant seaward march of offshore activities.

Recognizing the limitations in the availability of shelfbreak data, this paper discusses and contrasts the geotechnical and sedimentological properties of surficial deposits of a few areas including sediments from both active and passive margins. Areas of interest include selected portions of continental margins: off the U.S. East and West coasts, Mississippi Delta area, the Gulf of Alaska, Norway, and the eastern Mediterranean. Relevant processes are discussed in reference to their importance in affecting and determining the fundamental nature and variability of the deposits. During the past decade, numerous papers were published concerning submarine slides, slumps, and other gravity-induced mass movements, sediment flows, and their resultant deposits. These papers have dealt essentially with the geophysical and geological (in limited cases, the geotechnical) interpretations of the deposits, and are too numerous to mention here and are beyond the scope of this study. Only those studies specifically addressing the geotechnical and related sedimentological aspects of submarine sediments at shelfbreaks will be discussed.

The shelfbreak is the juncture between the nearly horizontal ($<1°$) continental shelf and the more steeply dipping ($>4°$) continental slope. Shepard (1973) notes that on a global average this transition takes place 75 km from shore in a water depth of 130 m but that the variability about these values is considerable. This variability includes not only distance and depth but also the abruptness of the break as pointed out by Southard and Stanley (1976). They summarized break types into sharp and gentle. The former is subdivided into (a) abrupt, commonly found at canyon heads, and (b) gradual, which is typical of most shelf-slope junctures. The latter, gentle type, is represented by a poorly-defined break such as those off major deltas and tectonically tilted margins. In reviewing the criteria for determining the shelfbreak position, Wear et al. (1974) and Vanney and Stanley (this volume) considered the point at which the slope departs from the horizontal extension of the shelf as the most consistent method for defining its location. In this paper, the shelfbreak is considered not as a point or line, but rather a zone having a significant change in gradient seaward and acts as a bridge through which sediment is transported from the shelf to the slope. As such, shelfbreak processes are thus outershelf and upperslope processes. In the discussion to follow the variability of processes and products from the outershelf to the upperslope will be considered within the shelfbreak domain.

TEXTURAL VARIATIONS AND SHELFBREAK MODELS

Shelfbreak sediment type models can be defined in terms of major textural characteristics of deposits that occur on the outer shelf and upper slope of continental margins. Such textural models can be independent of mineralogical content and nongenetic in connotation. On the basis of general texture, sand, mud, and diamicton (Flint, et al., 1960) or mixtum (Schermerhorn, 1966; Reading, 1978) are possible inputs to models and include: sand-mud; sand-sand; mud-mud; mud-sand; diamicton-mud; diamicton-diamicton; diamicton-sand; and sand-diamicton combinations (Fig. 1). Most of these textural models have been found; however, diamicton-sand and sand-diamicton combinations are less certain, although margins off Norway and Alaska seem to possess these textural characteristics (Rokoengen, 1979; Carlson, 1978; Carlson et al., 1978). The models depict the textural changes as occurring at the shelfbreak, but these changes or gradational changes, may occur above or below the point where a major change in slope occurs. In addition, textural changes can occur laterally, such as slope parallel or slope normal, as well as vertically with depth below the seafloor.

The shelfbreak models are intended to reveal major textural types and variations that are found at continental margins. Recognition of these textural variations and combinations are important because of the vastly different geotechnical properties that are characteristic of the major sediment types. Different sediment types have characteristic mass physical and mechanical properties and possess particular "soil states." For example, "clean" sands have very low compressibilities, low water contents and porosities, and virtually zero cohesion but are relatively high in strength, the latter being directly proportional to the effective stress. Sands having low water contents are characteristically high in wet unit weight. Quartz sands usually possess specific gravities (grain densities) of about 2.65 (Lambe and Whitman, 1969, pg. 30), whereas carbonate sands may have grain densities of 2.70 – 2.80 (Keller and Bennett, 1970; Bennett et al., 1970; Demars et al., 1976.

In contrast, fine-grained terrigenous submarine sediments (muds) are generally highly compressible, have high water contents and porosities, and are characterized by low wet unit weights. Muds generally have low shear strength but possess cohesion and thus behave significantly different than sands during static and dynamic loading. Grain densities of terrigenous mud are highly variable depending upon the type and mixtures of minerals and pelagic components present having values as low as 2.27 to as high as 3.26 (Keller and Bennett, 1970). U.S. Atlantic continental slope grain densities average 2.71 off New England and average 2.72 off the mid-Atlantic slope (Keller et al., 1979) (this difference being insignificant), but locally the grain densities may be quite variable: 2.62–2.83 (slope south of

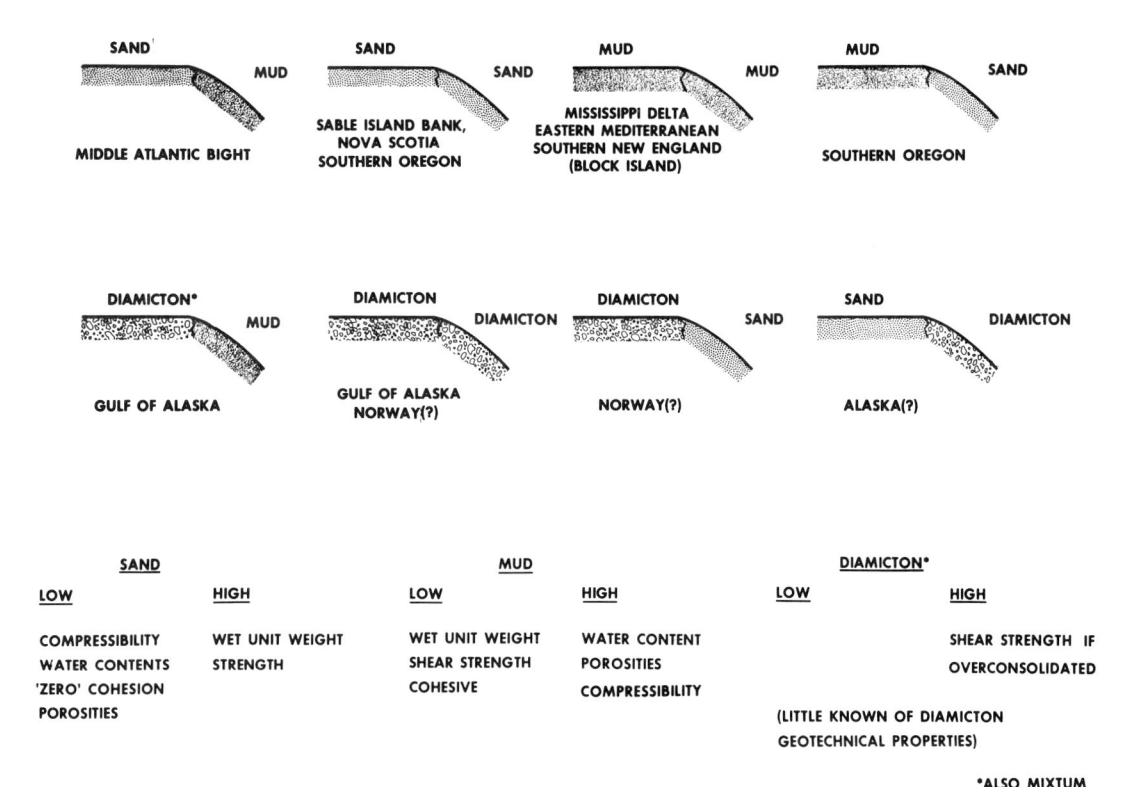

FIG. 1.—Shelfbreak models based on sediment textures, associating geotechnical properties to sand, mud, and diamicton deposits.

TABLE 1.—GEOTECHNICAL PROPERTIES

Mass Physical (Volume & Mass Relationships)	Symbol	Equation	Units	Application or Use	Key References
Water Content	w	$w = \dfrac{W_w}{W_s} \times 100$	$\dfrac{g}{g}\%$	Soil state, consolidation studies.	Terzaghi & Peck, 1967
Porosity	n	$n = \dfrac{V_v}{V} \times 100$	$\dfrac{cm^3}{cm^3}\%$	Soil state, consolidation studies.	Richards, 1962
Void Ratio	e	$e = \dfrac{V_s}{V_v}$	$\dfrac{cm^3}{cm^3}$ pure number	Consolidation studies; input parameter for engineering calculations.	Skempton, 1970
Degree of Saturation	S	$S = \dfrac{V_w}{V_v} \times 100$	$\dfrac{cm^3}{cm^3}\%$	Engineering studies (indicates amount of gas in soil mass).	Lambe & Whitman, 1969
Grain Specific Gravity*	G_s	$G_s = \dfrac{(\gamma_s)}{\gamma_o}$	$\dfrac{g/cm^3}{g/cm^3}$ pure number	Input parameter for calculations. Indicates significant mineralogical changes from sample to sample.	Lambe & Whitman, 1969
(grain density)	(γ_s)	$(\gamma_s) = \dfrac{W_s}{V_s}$	$\dfrac{Mg}{m^3}$		
Wet Unit Weight (wet bulk density)	γ_t	$\gamma_t = \dfrac{W}{V}$	$\dfrac{Mg}{m^3}$	Input parameter for calculations, i.e., total stress and stability studies.	Lambe & Whitman, 1969
Mechanical Shear Strength	τ_f	$\tau_f = \bar{c} + \bar{\sigma} \tan \phi$	kPa (psi)	Engineering studies and applications, stability studies, foundation assessment.	Calding & Odenstad, 1950; Evans & Sherratt, 1948; Lambe & Whitman, 1969
Cohesion	$\bar{c}$	$\bar{c} = \tau_f$ when $\bar{\sigma} = $ zero	kPa (psi)	Engineering studies and applications, stability studies, foundation assessment.	Calding & Odenstad, 1950; Evans & Sherratt, 1948; Lambe & Whitman, 1969
Sensitivity	S_t	$S_t = \dfrac{\tau_f}{\tau_{f_R}} = \dfrac{\text{undisturbed}}{\text{remolded}}$	pure number	Indicates strength loss potential when disturbed or remolded. Indication of liquefaction potential.	Kerr, 1963
Soil State (Indices) Atterberg Limits: Liquid limit	W_L	$W_L = \dfrac{W_w}{W_s} \times 100$	$\dfrac{g}{g}\%$	Indicates soil state; engineering applications; soil classification.	Atterberg, 1911; Casagrande, 1932, 1948
Plastic limit	W_p	$W_p = \dfrac{W_w}{W_s} \times 100$	$\dfrac{g}{g}\%$	Indicates soil state; engineering applications; soil classification.	Atterberg, 1911; Casagrande, 1932, 1948

*γ_o density of pure water of 4°C.

TABLE 1.—CONTINUED

Soil State (Indices)	Symbol	Equation	Units	Application or Use	Key References
Shrinkage limit†	SL	$SL = \dfrac{W_w{}^*}{W_s} \times 100$	$\dfrac{g}{g}\%$	Indicates soil state; engineering applications; soil classification.	Atterberg, 1911; Casagrande, 1932, 1948
Liquidity Index	I_L	$I_L = \dfrac{w - W_p}{W_L - W_p}$	pure number	Input parameter for engineering applications. Indicates natural soil state with reference to liquid limit.	Terzaghi & Peck, 1967
Plasticity Index	I_p	$I_p = W_L - W_p$	$\dfrac{g}{g}\%$	Classification purposes, engineering applications. Indicates range of water content in which soil mass will be plastic.	Terzaghi & Peck, 1967
Activity	a_c	$a_c = \dfrac{I_p}{\text{clay fraction} <2\mu m\ (\%)}$	pure number	Indicates amount of clay and general clay type present in soil.	Skempton, 1953

†Not commonly used in marine geotechnics.
*Water content at point that shrinkage ceases.

Baltimore Canyon, McGregor et al., 1979.)

Diamicton, a nongenetic term for an unconsolidated poorly sorted terrigenous sediment, has textures ranging from clay sized particles to pebbles, cobbles and/or boulders. Commonly such deposits are the results of glacial processes. Little seems to be known of the geotechnical properties of diamicton type deposits occurring in the marine environment. Relatively high shear strengths and high wet unit weights have been reported for overconsolidated tills off the Norway margin (Rokoengen et al., 1979). The occurrence of diamicton type sediments has been reported for the Icy Bay-Malaspina Glacier area of the slopebreak and slope (Carlson, 1978).

The soil "state" can be expressed in terms of the Atterberg limits (liquid and plastic limits) and the natural water content. The liquid and plastic limits are empirical "boundaries" based on the water content of fine-grained sediment for a specific type of test (Cassagrande, 1932, 1948). A soil (sediment) is said to behave plastically at water contents between the liquid limit and the plastic limit ($I_p = W_L - W_p$) and it is considered to be in a potentially "fluid state" at water contents above the liquid limit. A soil is considered to be semi-solid to solid at water contents below the plastic limit (Table 1 and Fig. 2). It is well known in marine geotechnics that many of the fundamental properties can be related to the soil "state." Specifically, knowledge of the natural water content, in reference to the Atterberg limits, gives a strong indication of: (1) how the deposits will behave under static or dynamic loads; (2) if the deposit is highly compressible; and (3) if the deposit is potentially susceptible to "liquefaction" type behavior if shocked or disturbed by natural or man-induced forces (seismic shock, wave loading, offshore construction, etc.). In a general sense, a soil mass having a water content approaching the liquid limit and beyond will generally display a relative decrease in the magnitude of its shear strength and wet unit weight, while on the other hand, as the natural water content decreases, the strength and wet unit weight will increase and may become less susceptible to disturbance and liquefaction (Fig. 2). Thus, the soil "state" and the other geotechnical properties are critical in determining the potential stability of the seafloor and its response to natural and man-induced loads (forces).

Mass transport processes (rock-falls, slides and sediment gravity flows) have been related to the Atterberg limits of sediments, specifically the liquid and plastic limits (Fig. 2; see Nardin et al., 1979). Unfortunately, in some cases, models of transport processes relate directly to coarse-grained sediments and, in the geotechnical sense, the liquid and plastic limits of a soil are defined only for fine-grained material. Geotechnically, sands do not have a liquid or a plastic limit and, in practice, if sediments contain appreciable amounts of sand size par-

TABLE 2.—GEOTECHNICAL PROPERTIES

Acoustical	Symbol	Equation	Units	Application or Use	Key References
Compressional Velocity	V_p	$V_p = \left(\dfrac{k + 4/3\,\mu}{\rho}\right)^{1/2}$	$\dfrac{m}{s}$	Geological & geophysical studies; sound propagation studies; practical & theoretical.	Hamilton, 1971
Bulk Modulus (incompressibility)	$k = \left(\dfrac{1}{\beta}\right)$	$\beta_{sw} = n\beta_w + (1 - n)\beta_s$ $k_{sw} = \dfrac{k_w k_s}{[n(k_s - k_w) + k_w]}$	$\dfrac{\text{dynes}}{\text{cm}^2}$	Input parameter to geoacoustical modeling.	Hamilton, 1971
Shear Modulus (rigidity)	μ	$3/4(\rho V_p^2 - k)$	$\dfrac{\text{dynes}}{\text{cm}^2}$	Input parameter to geoacoustical modeling.	Hamilton, 1971
Density (wet unit weight)	$\rho = \gamma_t$	$n\rho_w + (1 - n)\rho_s$	$\dfrac{g}{\text{cm}^3}\left(\dfrac{Mg}{m^3}\right)$	Input parameter to geoacoustical modeling.	Hamilton, 1971
Shear-Wave Velocity	V_s	$\left(\dfrac{\mu}{\rho}\right)^{1/2}$	$\dfrac{m}{s}$	Geological & geophysical studies; sound propagation studies; practical & theoretical	Hamilton, 1971

β = compressibility s = mineral solids
n = porosity w = water
ρ = density

	Size Classification				
Textures	Engineers		Geologists	Phi (ϕ)*	Key References
Boulder	>0.305 m	Boulder	$>25.6 \times 10^{-2}$ m	>-8	Krumbein & Pettijohn, 1938 Lambe & Whitman, 1969
Cobble	0.152 – 0.305 m	Cobble	$6.4 \times 10^{-2} - 25.6 \times 10^{-2}$ m	(−6) to (−8)	
		Pebble	$4 \times 10^{-3} - 6.4 \times 10^{-2}$ m	(−2) to (−6)	
Gravel	0.002 – 0.152 m	Granule	$2 \times 10^{-3} - 4 \times 10^{-3}$ m	(−1) to (−2)	
Sand	$6 \times 10^{-5} - 2 \times 10^{-3}$ m	Sand	$62.5 \times 10^{-6} - 2 \times 10^{-3}$ m	4 to (−1)	
Silt	$2 \times 10^{-6} - 6 \times 10^{-5}$ m	Silt	$3.9 \times 10^{-6} - 62.5 \times 10^{-6}$ m	4 to 8	
Clay	$<2 \times 10^{-6}$ m	Clay	3.9×10^{-6} m	>8	

*Phi = $-\log_2$ (diameter in mm).

ticles, they are removed from the sample prior to testing for the Atterberg limits. Perhaps a better index to relate soil "state" of sands to mass transport processes would be the use of critical void ratio (Taylor, 1948; Lambe and Whitman, 1969). If *in situ* void ratio is greater than the critical void ratio, the deposit is highly susceptible to flow. Determination of *in situ* geotechnical properties of sands and other coarse-grained clastics is difficult which makes this particular property hard to obtain. Nevertheless, the models of Nardin et al. (1979) are useful for gaining an insight into the fundamental nature of the original deposits that resulted in sediment gravity flows. Clearly, the ultimate "state" of a deposit and its related geotechnical properties are critical in determining the stability of a particular deposit in terms of potential mass movements, or susceptibility to erosional processes.

PROCESSES AT SHELFBREAKS AFFECTING SEDIMENT GEOTECHNICAL PROPERTIES

The ultimate soil "state" and related geotechnical properties of a sedimentary deposit, and its variability at any given time, depends upon a long se-

MASS TRANSPORT PROCESSES

ROCK FALL		
SLIDE		GLIDE
		SLUMP
SEDIMENT GRAVITY FLOW	MASS FLOW	DEBRIS FLOW-MUD FLOW
	FLUIDAL FLOW	GRAIN FLOWS
		LIQUIFIED FLOW
		FLUIDIZED FLOW
		TURBIDITY FLOW

SOIL STATE FOR COHESIVE SEDIMENTS

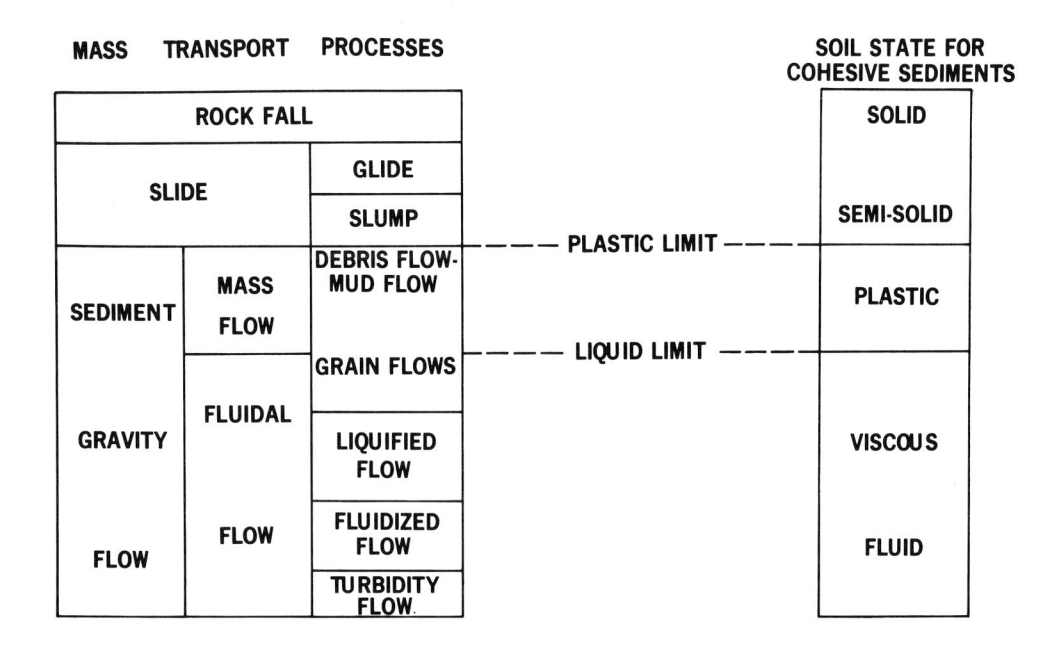

SOLID
SEMI-SOLID
PLASTIC
VISCOUS
FLUID

- - - PLASTIC LIMIT - - -
- - - LIQUID LIMIT - - -

(MODIFIED FROM NARDIN et al., 1979)

VISCOUS FLUID	PLASTIC	← SEMI-SOLID TO SOLID →
	LIQUID LIMIT	PLASTIC LIMIT

SOIL STATE

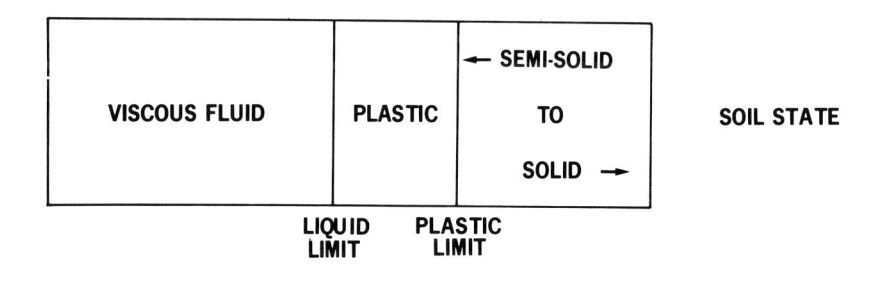

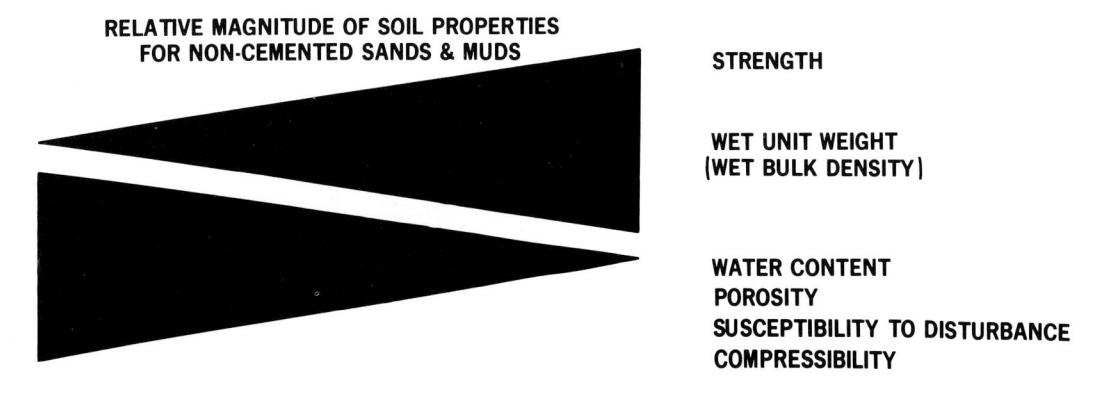

RELATIVE MAGNITUDE OF SOIL PROPERTIES FOR NON-CEMENTED SANDS & MUDS

STRENGTH

WET UNIT WEIGHT (WET BULK DENSITY)

WATER CONTENT
POROSITY
SUSCEPTIBILITY TO DISTURBANCE
COMPRESSIBILITY

FIG. 2.—Selected geotechnical properties related to soil "state" which are associated with sediment transport models proposed by Nardin et al. (1979).

ries of geological and oceanographic processes and events and upon the types of source material available and the environmental setting. These processes and factors produce primary and post-depositional sediment properties. The primary depositional textures, geotechnical properties, and soil "states" of submarine sedimentary deposits are determined by (1) particle size, (2) mineralogy, (3) particle size distribution and (4) microstructure (fabric and physico-chemistry). Sediment particles are derived from detrital and/or biogenic sources. While in the water column and at the sediment-seawater interface, these particles interact in response to the mechanical and electrical forces of the moderately high ionic strength seawater. Deposited particles form the primary properties of the sediment on the seafloor. Once on the seafloor the post-depositional processes come into play. These processes and effects can be grouped into three categories: (1) bio-geochemical; (2) physical/mechanical; and (3) mechanical. These processes and their effects modify the primary depositional properties and characteristics of the sediments and produce variability not only regionally but also locally (Fig. 3).

The more important bio-geochemical processes include biological activity (bioturbation), gas production (bubble and dissolved phases), chemical activity (authigenic mineralization), and intrastratal changes (salinity gradients). These bio-geochemical processes are generally considered to be long-term (slow) changes. On the other hand, physical/mechanical processes include wave effects (surface and internal) bottom currents, seismic activity, and in some cases in northern latitudes, ice gouging.

These processes are generally short-term ranging from seconds to minutes (surface wave activity producing pressure changes on the bottom) and seismic shock lasting seconds to minutes, to longer-term processes such as ice plowing lasting perhaps days, months, and years in certain geological environments. Mechanical processes include rapid or slow sedimentation and loading effects of the sediment, mass loading which probably occurs generally quite rapidly (slumping with removal of overburden at a specific location and mass loading of sediments downslope from the failure), consolidation and dewatering due to normal increases in sedimentation or rapid buildup of pore pressures due to mass loading; and creep processes which result in slow downslope movement of sediments. These post-depositional processes obviously can produce profound changes in the geotechnical properties, textures, and soil "state" and thus create both lateral and vertical variability in a sedimentary deposit regionally as well as locally (Fig. 4). All of these processes obviously do not act at the same degree in all sedimentary deposits nor do all the processes occur at every type of shelfbreak. Shelfbreak areas are characterized by a few predominant processes depending upon the particular sedimentary and oceanographic environment.

Quantification of the interaction of processes on sediments and their geotechnical properties is very limited and essentially non-existent at the shelfbreak. The following discussion will examine the best documented post-depositional processes (bioturbation, waves, currents and other physical oceanographic processes, and seismic activity) and

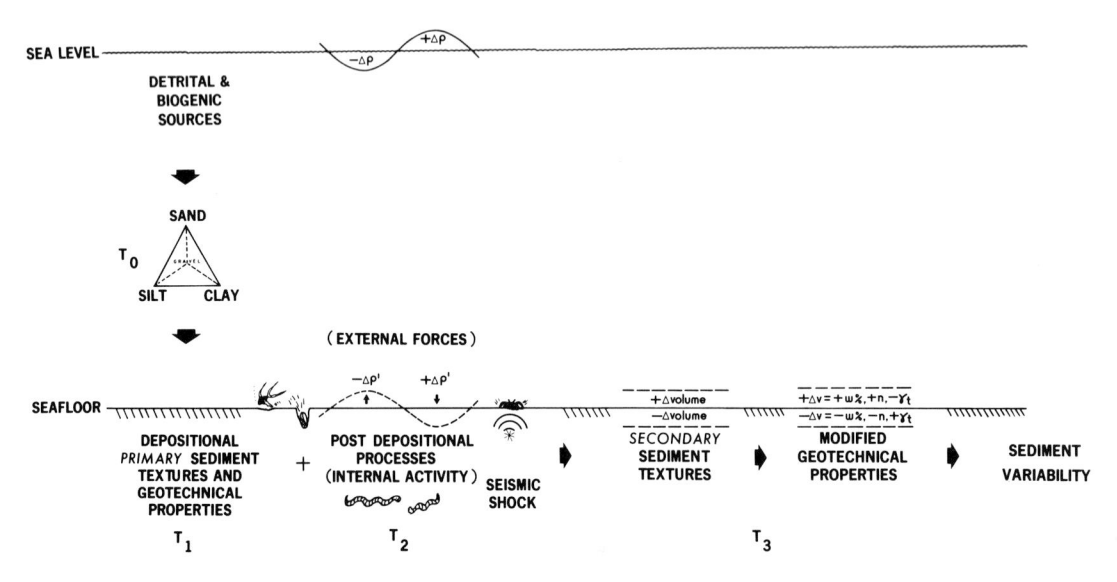

FIG. 3.—Diagram depicting the time-dependent processes from: availability of source material (T_0) to: primary depositional sedimentary textures and properties (sediment) [T_1] and post-depositional effects (T_2) which produce secondary properties and textures and modified geotechnical properties (T_3).

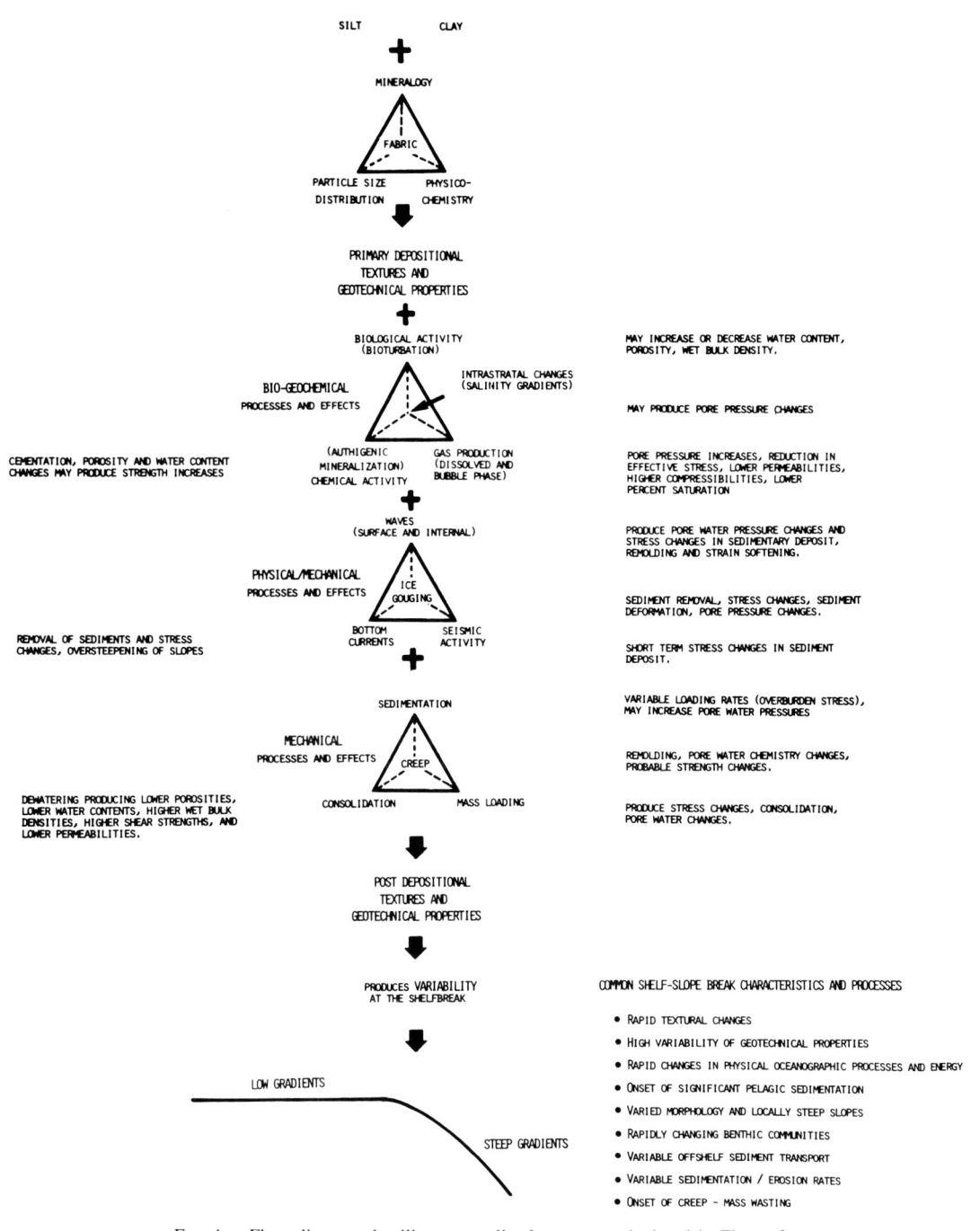

FIG. 4.—Flow diagram detailing generalized processes depicted in Figure 3.

relate them to potential effects on geotechnical properties at the shelfbreak.

Bioturbation

Bioturbation is the process of mixing sediment by the burrowing activity of benthic biota and is usually only absent or minimal in areas of extremely high sedimentation rates or anoxic bottom water (Rowe, 1974). Burrowing activities exist from the beach to the abyssal environments and is normally limited to the uppermost sediments. Rhoads (1970), in a study of subtidal muds, showed that the greatest density of active burrowers is in the upper 10 cm of sediment while at abyssal depths a similar conclusion was reached by Berger et al. (1979) who observed that active burrowing was limited to about 13 cm. Bioturbation also exhibits depth-related zonation. Rowe and Haedrich (1979) state that the greatest change in types of benthic biota occurs at the shelfbreak where a major biological boundary exists. This boundary, they note, marks the nominal depth (≈ 200 m) of the permanent thermocline and separates the highly variable light and temperature shelf environment, and its eurythermal fauna, from the more steady state slope environment with its stenothermal fauna. The latter, Rowe and Haedrich (1979) state, are highly depth dependent in that small changes across isobaths result in radical changes in species but, at a given depth, populations normally extend for great distances along isobaths. They note that the causes for such zonation are unknown and that it appears to be independent of sediment particle size, organic matter content, or sediment physical properties. Zonation occurs on the U.S. East Coast. This pattern, however, can be modified in areas of coastal upwelling and high productivity where sinking detrital organic matter creates oxygen minimum zones in the bottom waters at and near shelfbreaks. Rowe and Haedrich (1979) note that off Peru, southwest Africa and India such conditions exist at the shelfbreak where bottom biota are limited to microbes, nematodes and benthic foraminifera, thus limiting the degree of bioturbation.

The result of sediment bioturbation is an increase in the variability of geotechnical properties (Rowe, 1974; Richards, 1965). These changes in geotechnical properties include sediment fabric, grain size, sorting, porosity, water content, compressibility, shear strength, acoustical properties, and bottom stability (Rhoads and Boyer, in press; Richardson and Young, 1980). Rhoads (1974) states that mobile deposit feeders create a heterogeneous open-sediment fabric through ingestion/defecation/pelletization whereas extensive burrowing and tube building creates a tight homogeneous sediment fabric with random particle orientation and "random size distribution." Pelletization by mobile deposit feeders acts to modify grain size and hence sorting by producing large fecal pellets from constituent silts and clays which are then redeposited in the parent mud. In addition, pelletization can increase porosity through production of interpellet void space (Rhoads and Boyer, in press) while lowering shear strength (Rhoads and Boyer, in press) and making the sediment more susceptible to current erosion (Rowe, 1974). For the burrowing and tube building biota the results, however, can be quite different. In this group a cohesive and tight sediment fabric may result from extensive burrowing and tube building (Rhoads, 1974), but in muds, increased water contents also can result (Rhoads, 1970). Rhoads and Boyer (in press) point out:

"To facilitate burrowing and feeding, some metazoa, especially bivalves, also liquify the sediment by injecting water anteriorly into the bottom. This causes an instantaneous local increase in pore water pressure and the liquid limit of the sediment is temporarily exceeded. At this instant the organism moves forward into the liquified zone. In sands, this fluidized sediment represents a transient state as overburden pressure soon causes the sediment to collapse on itself. In cohesive silts and clays the sediment may remain dilated long after the burrowing organism has passed through, or otherwise processed the sediment."

Rhoads (1970) notes that in Buzzards Bay, Massachusetts, a water content (% dry weight) of 50% appears to separate firm from soft muds. He states that high clay content muds with greater than 50% water display false-bodied thixotropism and are potentially unstable in the presence of relatively weak bottom currents. In a study of bottom erosion Southard et al. (1971) concluded that as sediment water content increased, erosion resistance decreased. Under very high water contents Postma (1967) showed nearly an order of magnitude reduction in critical erosion velocity, in the less than 2 micron sediments, between water contents of 70% and 90%, respectively. On the other hand some burrowing species apparently increase sediment shear strength (Rhoads and Boyer, in press) and stability of naturally thixotropic sediments through the construction of membranous tubes (Rowe, 1974) or through the production of mucus for the binding of sedimentary particles. Rhoads and Boyer (in press) point out that pore water irrigation through the burrow tubes of spionid polychaetes is known to be associated with enhanced production of viscous and elastic binding mucus generated by sediment bacteria. This mucus fills the intergranular pore space increasing both sediment shear strength and erosion resistance through mucus binding of the particles. This increased resistance to erosion was demonstrated in an *in situ* mud erosion study conducted by Young and Southard (1978). Although not measuring mucus content *per se,* they found that as the carbon content of the sediment increased the critical erosion velocity was also elevated.

From the above discussions it can be seen that the interaction of bottom dwelling biota with sediment can enhance or reduce sediment erosion resistance and significantly alter the geotechnical properties. The discussion to follow will consider physical processes active at the shelfbreak and how these processes can alter sediments which may or may not have been previously modified by biological processes.

Waves

When landward propagating gravity waves reach a point where the depth of water over which they are moving is equal to one half their wavelength, they commence a process known as the shoaling transformation. This water depth, regarded by geologists as "wave base," is the onset of the interaction of wave-instigated water motion and the seafloor. Landward of wave base, the shoaling transformation reaches water depths where wave energy action on the seafloor can exceed the threshold of movement for the indigenous sediment. This energy is expressed as oscillatory bottom flow, capable of resuspending sediment and producing oscillatory ripple marks. Friedman and Sanders (1978) note that at global mean shelfbreak depths, waves having periods greater than 8 seconds begin to interact with the bottom. Komar (1976) shows that waves (period = 15 s) with heights of 3.5 m can move medium sand at water depths of 90 m and fine sand at depths up to 120 m, while waves with 6 m heights can move medium sand as deep as 125 m and fine sand at depths greater than 140 m. Hadley's (1964) calculations show that, for the Celtic Sea, storm-generated wind waves occurring one to two weeks per year should be capable of moving sediment to a depth of 300 m, but this probably represents an extreme case. Komar et al. (1972) used stereo bottom photography to study ripple symmetry, height and wavelength. They observed symmetrical ripples to a depth of 204 m on the outer shelf and concluded that surface storm waves could account not only for seasonal presence or absence of ripples but also for their time dependent pattern changes as well. The above sediment transport processes at and near the shelfbreak were the result of surface wave activity. However, in a stratified fluid, energy can also propagate along an internal density interface in the form of a wave generally referred to as an internal wave. In the ocean the interface necessary for the wave's existence is the pycnocline resulting primarily from thermal stratification. Internal waves can originate anywhere from deep-ocean regions to nearshore, but considerable evidence exists (Apel et al., 1975; Neuman et al., 1977; Butman et al., 1979) for their tidally-induced generation at or near the shelfbreak. In wavetank (Southard and Cacchione, 1972) and modelling studies (Cacchione and Southard, 1974) it has been demonstrated that shoaling internal gravity waves can transport sediment under realistic oceanic conditions. In a study designed to test this prediction, Neuman et al. (1977) observed the outer shelf waters south of Block Island for currents, water structure and acoustic signature. They observed the shoreward passage of high frequency internal waves under which a bottom nepheloid layer (acoustically detected) vertically expanded and contracted in sympathy with the passage of internal wave crests and troughs, respectively. The well-mixed bottom nepheloid layer systematically expanded and contracted in thickness from 8 to 2 m and contained up to two orders of magnitude more suspended sediment than the water directly above. It was the conclusion of Neuman et al. (1977) that sediment resuspension by internal wave action had taken place and under conditions similar to those predicted by Cacchione and Southard (1974). Additional observation of internal waves near the outer shelf adjacent to Wilmington Canyon (Butman et al., 1979) has demonstrated that the high-speed component of the waves' oscillatory flows was oriented offshore thus enhancing the potential for offshelf sediment transport through the shelfbreak zone. Mayer et al. (1983) points out that during the winter, when shelf water is vertically homogenous, the region adjacent to, and seaward of, the shelfbreak remains stratified because of the slope water. They conclude that winter storms have the potential of generating considerable baroclinic response near the shelfbreak. This raises the possiblity of internal wave-induced interaction with the shelfbreak and upper slope sediments during the winter.

These studies have shown that waves can and do interact with the substrate at and near the shelfbreak. These waves can be important in affecting the geotechnical properties at the shelfbreak for two reasons. First, as pointed out above, wave energy can be an important component of sediment transport thus influencing the nature of the sediment cover at or near the shelfbreak and hence the overall geotechnical properties as modeled in Figure 1. Another major geotechnical response of the bottom to wave action, through extraction of wave energy, is the visco-elastic response of fine-grained sediments and the subsurface variation in pore pressure. In the first case, Suhayda (1977) has shown in the Mississippi Delta area that vertical bottom oscillations of fine-grained sediments were correlated with the passage of surface gravity waves in a manner analogous to the motion of the near-bottom nepheloid layer cited above in the Neuman et al. (1977) study near the outer New England Shelf. Suhayda (1977) further noted that the energy extracted from the surface waves for the bottom response exceeded that which was extracted by bottom friction. In a survey paper, Lee and Focht (1976a) conclude that cyclic loading, such as that induced by both oceanic and seismic waves, can have significant effects on clay soils. They point out that all clays develop significant cyclic strains under pulsating stresses through the development of large excess pore pressures that in turn yield softer and weaker sediments. In a subsequent paper they (Lee and Focht,, 1976b) consider actual surface wave data and apply it to: (1) wave loading theory; and (2) cyclic storm loading design of offshore structures. In the second case the passage of surface gravity waves were shown to have a direct effect on *in situ* pore water pressures to a depth of at least 12 m below the sediment-water interface in the fine-grained sediments of the Mississippi Delta (Bennett et al., 1982). As seen in Figure 5, the "hydrostatic"pressure at 6.5 m, which records the direct water column pressure variation in response to wave passage, is essentially synchronous with the change in pore water pressures at the same subbottom depths. This pressure is attenuated with depth below the seafloor. The pressure response due to tidal activity was also demonstrated during the investigation (Bennett and Faris, 1979; Bennett et al., 1982). Although not measured at the shelfbreak the above two examples, in conjunction with the earlier cited outer shelf resuspension study (Neuman et al., 1977), clearly indicated a direct dynamic response of the sediment to the passage of both surface and internal gravity waves. Studies of this nature have added critical *in situ* verification of the response of the sediment to dynamic physical processes and have aided in the development and verification of stability analysis models of the seafloor (Yamamoto, 1980).

Currents

Although waves are probably the best quantified dynamic physical oceanographic process which directly affect sediment geotechnical properties, unidirectional ocean currents also play a role. At and near the shelfbreak a major oceanographic discontinuity commonly exists which affects the magnitude of the bottom currents. As an example, in the western North Atlantic from Cape Hatteras northward almost to the Norwegian Sea, a sharp, permanent, seaward-inclined ocean water front exists at or near the shelfbreak (Beardsley and Winant, 1979). Although the exact depth at which this front intercepts the seafloor may vary within the Middle Atlantic Bight from 80 m (Mooers et al., 1979) to as deep as 200 m (Voorhis et al., 1976) the front forms a semipermeable barrier between Slope Water and the relatively fresher and cooler Shelf Water (Beardsley et al., 1976; Voorhis et al., 1976; Beardsley and Winant, 1979; Mooers et al., 1979). In a study of the front on the New England continental shelf during the late spring, Voorhis et al. (1976) found that the Shelf Water was moving westward at 8.7–14 cm/sec whereas Slope Water appeared stationary with no detectable westward mean drift. This illustrates that considerable shear can exist at a front, and although not addressed by Voorhis et al. (1976) this shear, near the bottom, may be expressed as turbulence capable of influencing sediment resuspension and transport.

This oceanographic discontinuity plays a major role in determining where net erosion and/or sedimentation takes place and the resulting transition from coarse to fine-grained sediments. This transition reflects current intensity and has been formally recognized as the "mudline" (see Stanley et al., this volume). The "mudline" represents the time-integrated boundary between net erosion and deposition of fine-grained sediments. The position of this boundary, relative to the shelfbreak, then dictates which of the shelf to slope textural models (sand-mud models), shown in Fig. 1, dominates an area and hence the overall geotechnical character of the shelfbreak sediments.

In addition, slope currents can cause net erosion and undercutting of strata leading to oversteepening

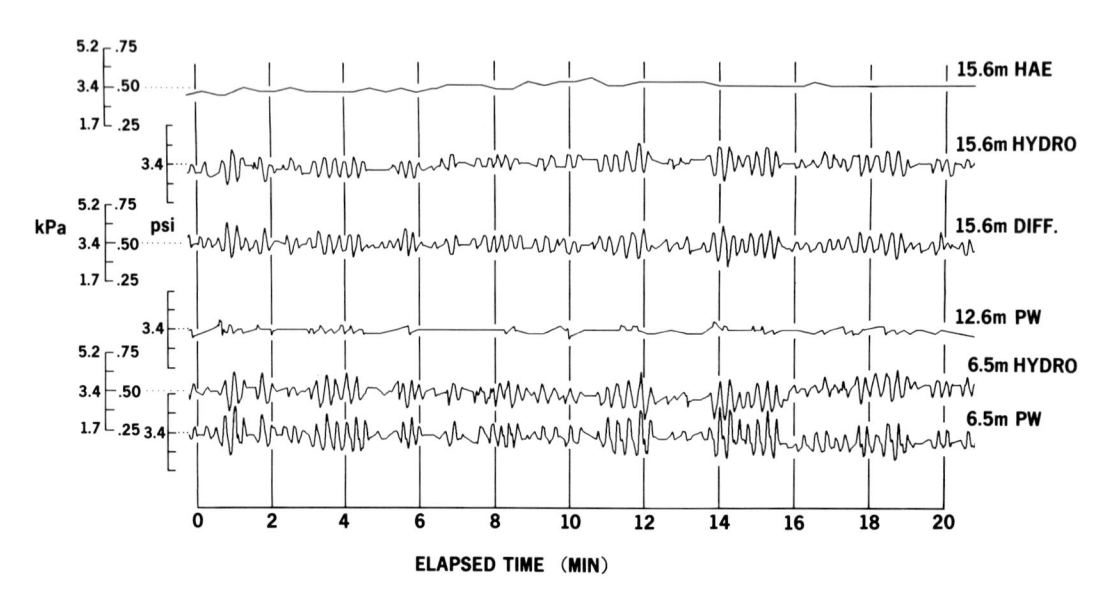

FIG. 5.—Time series data depicting the response of pore water pressure and "hydrostatic" pressure to short period wave activity; Mississippi Delta (after Bennett et al., 1983).

of deposits and sediment failure. Sediment failure of this nature has been suggested for strata ranging from several tens of meters (Lewis, 1970) to entire stratigraphic sequences (Rona, 1969).

Other Processes

Finally, of a more speculative nature, there are other physical oceanographic processes such as major current spinoff eddies which may affect the shelfbreak areas. For example, flow instabilities in the Gulf Stream east of the U.S. Middle Atlantic Bight manifest themselves as meanders which may become detached from the main flow as either cyclonic (Cheney and Richardson, 1976) or anticyclonic (Saunders, 1971) eddies. It has been estimated that an average of three eddies per year enter the Middle Atlantic Bight with a mean residence time of three months each according to Morgan and Bishop (1977). From salinity data, it was shown that rings of this nature impinging on the slope can extend to a depth of at least 400 m. Saunders (1971) notes that anticyclonic eddies can extend to depths of 1000 m and have surface currents up to 75 cm/s. Although the above studies of eddies were concerned with physical oceanographic problems and not sediment geotechnical properties, the implications for energy sources of this magnitude and frequency of occurrence is clear. In dealing with the energy loss from a major eddy, Schmitz and Vastano (1977) observed a cyclonic ring that shoals and dissipates on the Blake Plateau. They noted that as the ring shoaled, it underwent a strong interaction with the Plateau which was expressed by marked asymmetry in the ring structure. They calculated that after reaching the Plateau, the ring lost about half its kinetic energy. In a similar study of a shoaling ring, Cheney and Richardson (1976) suggest that the bottom of the ring formed a bottom frictional layer which accelerated its decay rate by a factor of four upon shoaling. Although to date unquantified, flows of this nature should be considered among shelfbreak processes and are certainly potential sources of sediment transport energy which may yield erosion, oversteepening of slopes and eventually sediment failure.

Seismic

In a final example of physical processes acting to modify geotechnical properties, the concepts of cyclic loading by seismic waves will be considered. In the terrestrial environment the effects of seismic activity on soil structure is long recognized. As an example, it was shown by Seed et al. (1975) that seismic shaking during the San Fernando earthquake of 1971 induced very high pore-water pressures at the base of an earthen dam. Subsequent liquefaction and accompanying lowered shear strengths of the dam's foundation led to soil failure in the form of a major slide. In the marine environment this cause and effect relationship is more difficult to show by nature of the general paucity of observational data and theoretical understanding of marine sediments and seismic activity. From a seismological point of view Spudich and Orcutt (1982) point out: (1) Almost no seafloor ground-motion data exists for earthquake magnitudes greater than three, (2) theoretical understanding of seismic wave propagation in the marine environment is incomplete, and (3) it has not yet clearly been demonstrated how to extrapolate to a seafloor environment ground motions observed or predicted on land.

Even though deficiencies of this nature exist, modelling of shelfbreak and slope failures resulting from earthquakes has proceeded. A modelling study of this nature, based on slope angle, shear strength of the sediments and horizontal components of earthquake-induced acceleration (expressed as a percentage of gravity), was carried out by Morgenstern (1967). Almagor and Wiseman (1977) have recently showed reasonable agreement between theory and geotechnical and geophysical observations of the continental slope off Israel. However, the existence of extensive mass creep and small rotational slumping phenomena at the shelfbreak off Israel (Neev et al., 1976; Almagor and Garfunkel, 1979) cannot be explained using this model. Analysis of the static force equilibrium that exists within the shelfbreak and upper slope sediments indicates that the sediments are sufficiently strong to sustain the slopes at the shelfbreak and upper slope, even if subjected to the effects of earthquake accelerations greater than those detrimental to the steepest parts of the continental slope. It is suggested that these creep phenomena reflect long-term deterioration in shear strength of the sediments due to repeated earthquake loading effects (Almagor, 1980).

GENERAL SHELFBREAK CHARACTERISTICS

With the above processes in mind, and considering the various types of shelfbreak (textural) models, a few seemingly common shelfbreak characteristics can be summarized as follows: (1) onset of significant pelagic sedimentation, (2) variable sedimentation/erosion rates, (3) high variability of geotechnical properties, (4) rapidly changing benthic communities, (5) variable offshelf sediment transport, (6) varied morphology and locally steep slopes (gradients), (7) rapid textural changes, (8) rapid changes in physical oceanographic processes and related energy environment, and (9) onset of creep and mass wasting processes.

The above summary is intended to point out processes and characteristics that can be identified with shelfbreak areas. Clearly some of these characteristics may, and in some cases do, occur in other submarine environments; however, these nine characteristics and processes, discussed at length in many chapters of this volume, appear to be common elements to many shelfbreaks of both passive and active margins. The above list calls attention to the highly variable nature of shelfbreaks and the uniqueness of the shelfbreak environment in contrast to shelf, slope, and deep sea basin environments that have been more thoroughly studied.

GEOTECHNICAL PROPERTIES OF SHELFBREAK AREAS

Geotechnical data were abstracted from available publications describing geological and engineering properties of deposits of a few selected continental margin areas including: U.S. East and West coasts, Gulf of Mexico (Mississippi Delta area), Alaska (Icy Bay area), eastern Mediterranean, and Norway (Figs. 6, 7). Despite the limited data available for shelfbreak areas, significant comparisons can be made. The various studies from which the data were abstracted were not concerned specifically with shelfbreak processes and therefore sampling stations were not generally consistent with this aspect of continental margin sediment properties and processes. Most of the geotechnical data from the U.S. Atlantic margin were collected from upper slope cores whereas outer shelf and upper slope core data were available for a limited area of the Mississippi Delta and the eastern Mediterranean. Data from Norway include shelf and slope sampling stations. Data from the U.S. west coast off California were taken from continental slope cores and the information gathered for the Alaska area was abstracted primarily from slope cores. Fortunately, textural data were available for the outer shelf and upper slope for most of the areas investigated.

Several shelfbreak textural models (Fig. 1) were observed for the areas studied and geotechnical data were available for various sediment types (Figs. 6, 7) including: clays, silts, "sands," in various admixtures, and glaciomarine clay and tills (diamicton). Lateral variability in texture is greatest at shelfbreaks off the U.S. Atlantic, Norway, and Alaska, whereas the Mississippi Delta areas and the eastern Mediterranean display little variability in textural type at the outer shelf and upper slope where mud–mud sediments occur (Fig. 1). Although gross sediment type may be consistent within an area,

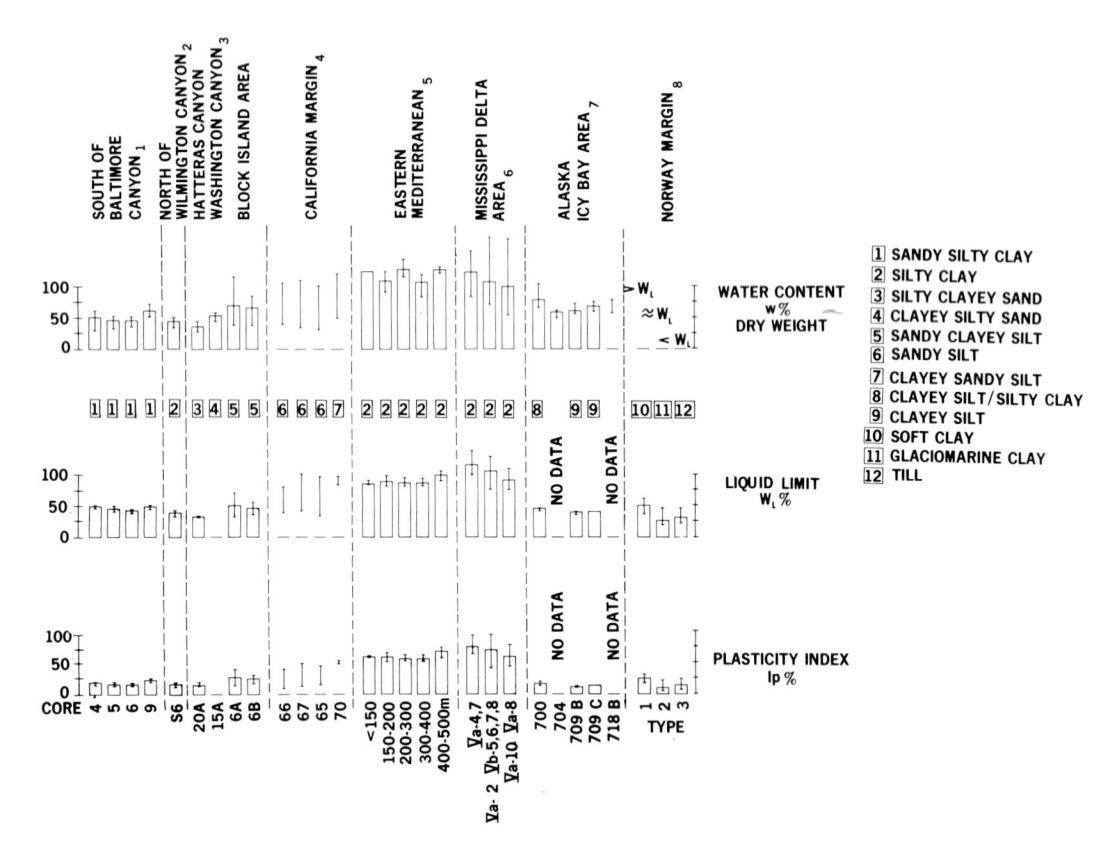

FIG. 6.—Average values and ranges observed in water content, liquid limit, plasticity index, and sediment texture for selected shelfbreaks from: 1. McGregor et al., 1979; 2. Bennett et al., 1977; 3. Keller et al., 1979; 4. Edwards et al., 1980; 5. Almagor, 1978; 6. Booth and Garrison, 1978; 7. Carlson et al., 1978; 8. Rokoengen et al., 1979.

variability in actual percentages of sand-silt-clay is commonly quite high laterally and with depth below the seafloor. As mentioned earlier, high variability in textures can occur due to various types of activity on and within the sediments (Fig. 3). Textural characteristics are fundamental in determining the nature of other mass physical and mechanical properties of submarine sediments, and sand-silt-clay percentages can be correlated often with other geotechnical properties such as water content, porosity, and shear strength.

Water Content

As expected, silty clays display the highest average water contents (percent dry weight) with values often greater than 100% (eastern Mediterranean and Mississippi Delta; see Fig. 6). Considerable variability is observed in the water contents for the Delta area compared with the other areas. Although average values for water content are not available for the California study area, the range of values is noticeably high in these sandy silts. Considerably lower average water contents, generally 65% and less, are observed for the U.S. Atlantic margin sediments (exception: Washington Canyon area and Block Island area) as compared to the other areas although water content reported for the Alaska area averages between 65% and 75% (Fig. 6).

Liquid Limit

Of significance is the almost ubiquitous occurrence of high water contents that are greater than the liquid limits (Fig. 6, see Bennett et al., 1980). Again, the Mississippi Delta and eastern Mediterranean areas show the greatest liquid limits which are generally less than the natural water contents of the sediments. The Norway margin soft clays display water contents higher than the liquid limits; water contents are approximately equal to the liquid limits in the glaciomarine sediments but less than the liquid limits in the overconsolidated tills.

Liquidity/Plasticity Indices

The plasticity index ($I_p = W_L - W_p$) indicates the range in the water content for which the sediment will behave plastically. The liquidity index (a function of the natural water content and plasticity index, Table 1) is often quite high and commonly greater than unity for surficial submarine sediment. High values are observed for the eastern Mediterranean and Mississippi Delta areas. Relatively low values for plasticity index are observed for the Icy Bay area and the U.S. Atlantic margin as compared to the eastern Mediterranean (Fig. 6). These major differences in water content, liquid limit, and plasticity index can be related to the textural character and mineralogy of the sediments. The eastern Med-

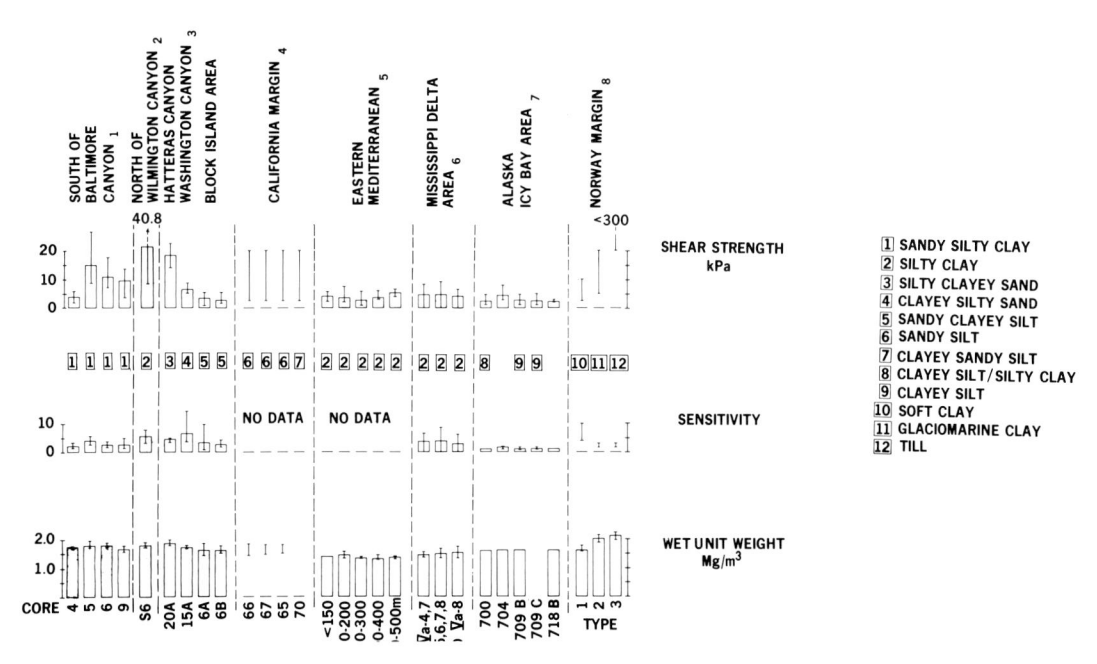

FIG. 7.—Average values and ranges observed in shear strength, sensitivity, wet unit weight, and sediment texture for selected shelfbreaks from: 1. McGregor et al., 1979; 2. Bennett et al., 1977; 3. Keller et al., 1979; 4. Edwards et al., 1980; 5. Almagor, 1978; 6. Booth and Garrison, 1978; 7. Carlson et al., 1978; 8. Rokoengen et al., 1979.

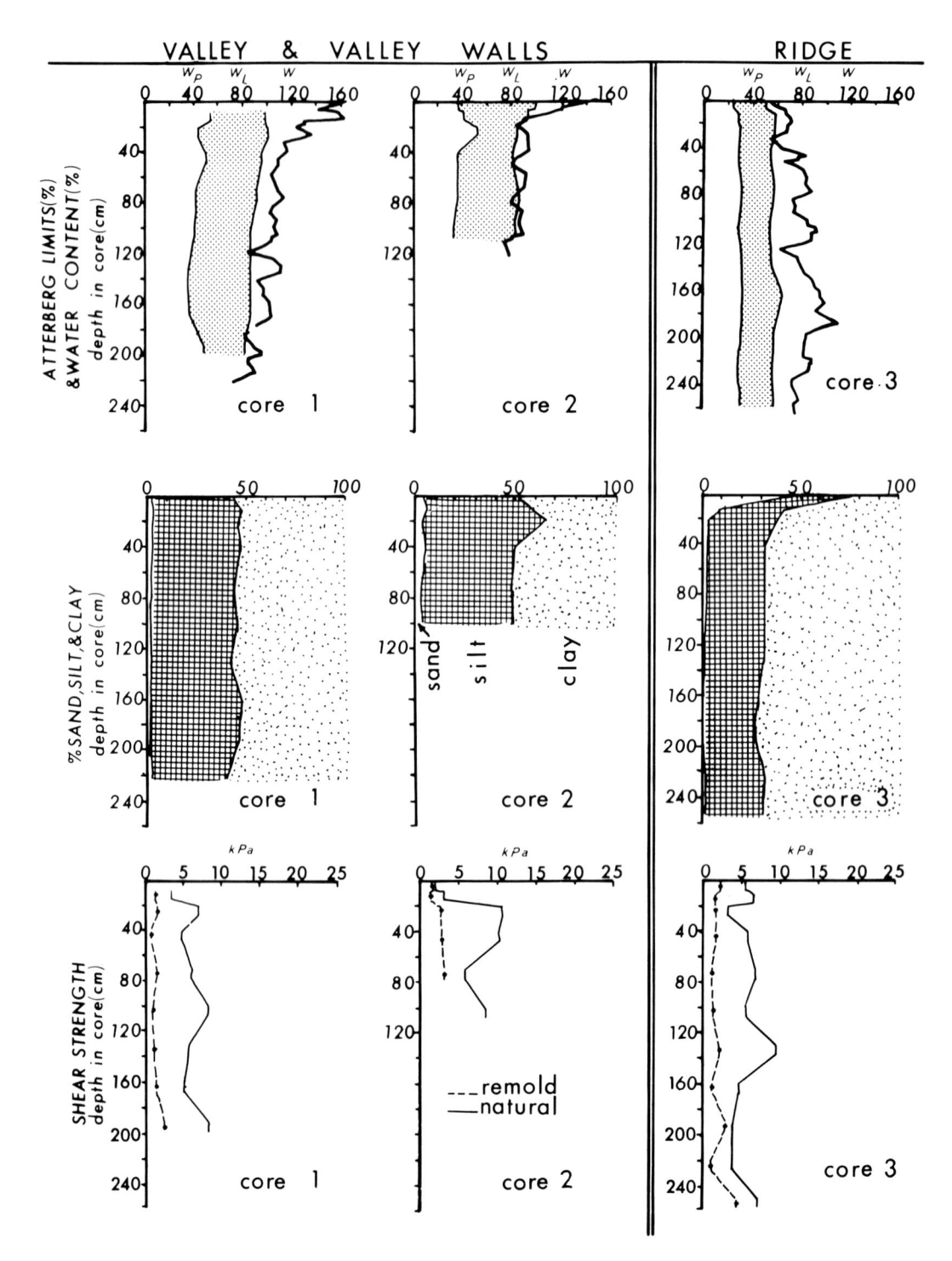

FIG. 8.—Selected geotechnical properties of sediment cores from a local study area south of Baltimore Canyon-U.S. Atlantic Continental Margin (from McGregor et al., 1979). Note significant difference in properties of cores recovered from the area depicted in Fig. 9.

GEOTECHNICAL PROPERTIES AT SHELFBREAKS

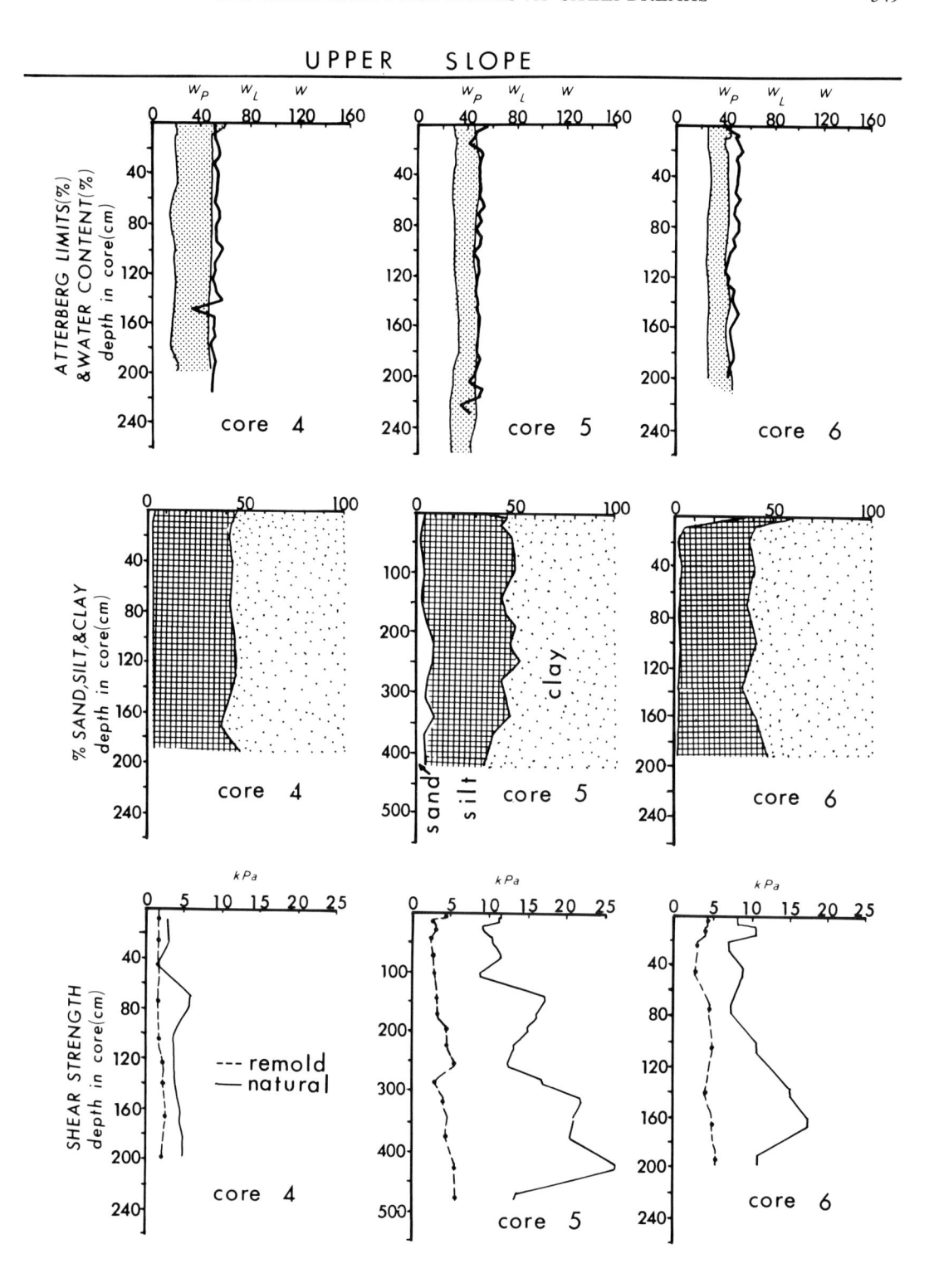

UPPER SLOPE

iterranean and Mississippi Delta sediments are quite similar, both having high percentages of smectite, followed by lesser amounts of kaolinite and illite (Fisk and McClelland, 1959; Almagor, 1978). In contrast, the U.S. Atlantic margin slope sediments north of Hatteras Canyon have high percentages of illite (Doyle et al., 1979). Only minor amounts of smectite have been reported for Wilmington Canyon area sediments (Bennett et al., 1978). The Icy Bay area sediments are reported to have relatively high percentages of kaolinite/chlorite and illite with only small amounts of smectite (Carlson et al., 1978). It is noted that the fine sediment fractions of the California Margin study area are reported to have illite, smectite, and quartz making up approximately 70% with chlorite contributing about 20% with feldspar and amphiboles making up the remaining 10% of the sediment (Edwards et al., 1980). Geotechnical properties (Fig. 6, 7) of these California sediments appear to be approximately "midrange" between the high smectite sediments (eastern Mediterranean and Mississippi Delta) and the high illite sediments (U.S. Atlantic continental slope). Smectite has a high affinity for water (it is

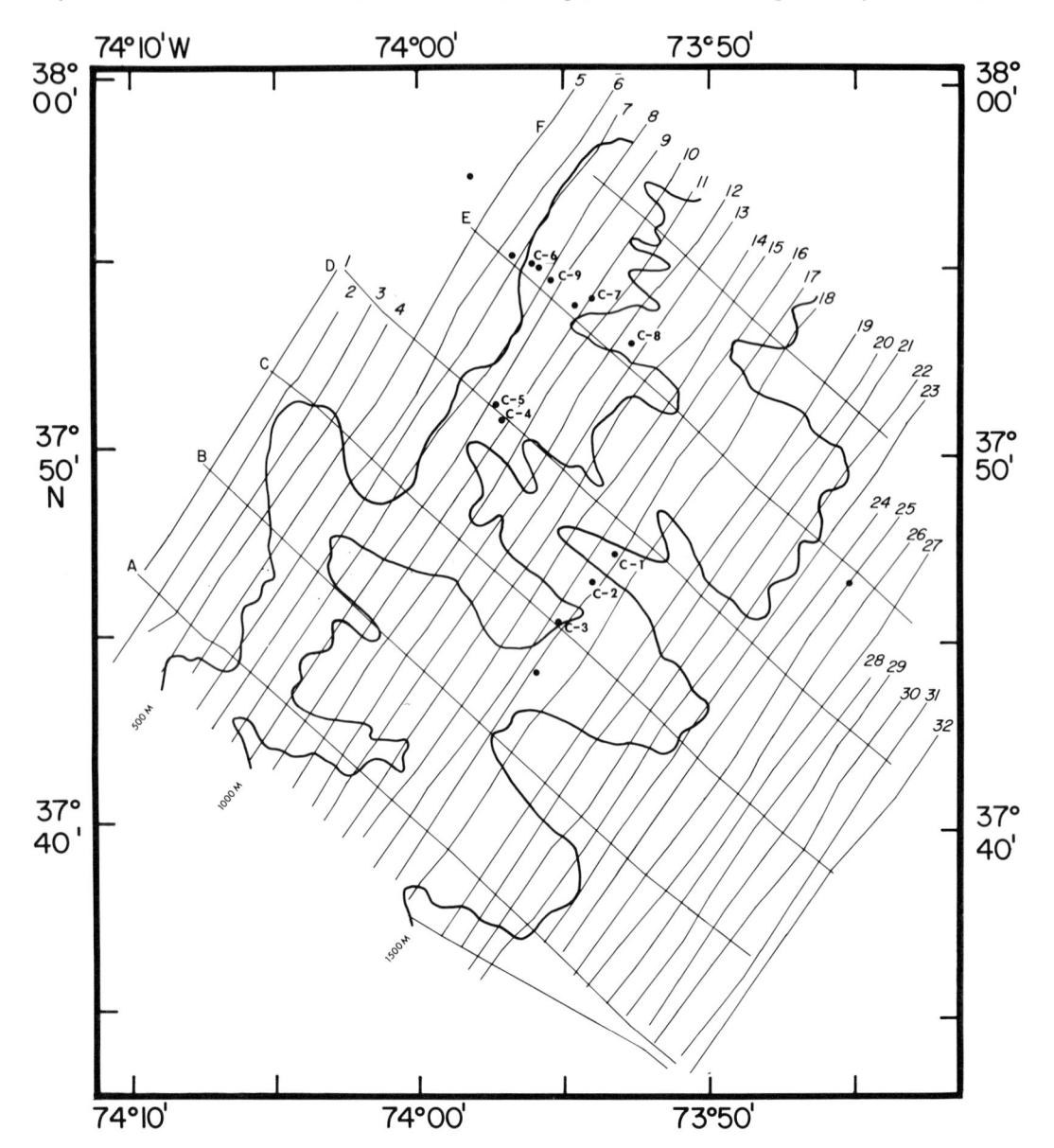

FIG. 9.—Map with index bathymetric contours showing core locations south of Baltimore Canyon. Geotechnical properties of cores are depicted in Fig. 8.

fine-grained compared to illite and kaolinite) and contributes substantially to the plasticity of sediments. Sediments high in smectite generally display high activities (Table 1).

Shear Strength

The shear strength of sediments at shelfbreak areas are highly variable, but on the average, U.S. Atlantic margin sediments are appreciably "stronger" than in the Mississippi Delta, eastern Mediterranean and Icy Bay areas. On the other hand, the overconsolidated tills possess shear strengths (slightly less than 300 kPa) that are magnitudes higher than most of the other margin deposits investigated in this study. The Mediterranean, Mississippi Delta, and Icy Bay area sediments generally average less than 5 kPa, whereas U.S. Atlantic slope sediments summarized here average 4–22 kPa, displaying not only higher strengths than the smectite (Mississippi Delta and eastern Mediterranean) and kaolinite/chlorite and illite (Alaska) rich sediments but also higher overall variability (Fig. 7). High variability in shear strength also is observed for the California margin sediments.

Sensitivity

Much like the variability in shear strength, sensitivity is highly variable in U.S. Atlantic slope sediments. Sensitivity is the ratio of the "natural" to the remolded shear strength and quantifies the potential strength loss upon severe sediment disturbance. The highest sensitivity values and greatest variability observed occur in a core north of Washington Canyon (Fig. 7). The high sensitivity may be related to the relatively high percentage of coarse grained material reported in core 15A (Keller et al., 1979). When remolded, the coarse-grained material would possess minimum cohesion and thus result in relatively high sensitivities despite the overall average to generally low values reported for the "undisturbed" shear strength. Some of the strength values and sensitivities having values of unity and zero, reported for some of the Mississippi Delta sediments, suggest possible core disturbance prior to strength testing (Fig. 7). The lowest shear strengths and sensitivities reported for the Mississippi Delta sediments occur in a mudflow (Booth and Garrison, 1978) which may account for some of these anomalously low values.

Wet Unit Weight

Wet unit weight (Table 1) is not particularly remarkable but values generally follow the water content inversely; the higher the water content, the lower the wet unit weight for sediments having minerals of similar average grain density. The highest wet unit weights ($\gamma_t = 2.1$ Mg/m^3) are found for the overconsolidated till. Lowest values are generally found for the eastern Mediterranean ($\gamma_t = 1.32$ to 1.43 Mg/m^3).

DISCUSSION

Even though similar sediment types often display common geotechnical properties and characteristics, the averages and ranges of values reported here for shelfbreak deposits clearly demonstrates that variability in the mass physical and mechanical properties is the rule rather than the exception. This has been shown elsewhere for other submarine deposits (Keller and Bennett, 1970; Bennett et al., 1970; Bennett et al., 1980). High variability in geotechnical properties holds for both local areas as well as for regional deposits of similar sediment type.

Typical examples of the high variability in geotechnical properties of U.S. East Coast slope sediments are observed in six cores collected from a local study area south of Baltimore Canyon (McGregor et al., 1979). Relatively high variability is evident not only within each core but also among the cores (Fig. 8). Of significance here is the remarkable difference in the properties of cores taken within very short lateral distances (Fig. 9). Upper slope cores are characterized by water contents slightly greater than, or approximately equal to, the liquid limits. Normally consolidated sediments (Fig. 8) display a normal increase in shear strength with depth of burial (core 6) while a sample taken on a small slide scar (core 5) which also shows a somewhat "normal" increase in shear strength with depth, is slightly overconsolidated (McGregor et al., 1979). Immediately downslope from the slide scar, core 4 was taken from the slump block, and although the textures and Atterberg limits are similar, the strength profiles are drastically different (strengths of one third to one fourth and less as compared to the slide scar strength profile). In the same area (Fig. 8) cores taken from a ridge, valley, and valley wall display considerably different properties compared with upper slope sediments (Fig. 9). Water contents are significantly higher than the liquid limits and, although the Atterberg limits of the ridge sediment are similar to core 4 (upper slope), the valley and valley wall sediments (cores 1 and 2) have both higher plastic and liquid limits and a much greater range in plasticity than the upper slope cores (Figs. 8, 9). These sediments show virtually no increase in shear strength but a normal decrease in water content with depth of burial in cores 1 and 2 but no decrease in water content in core 3. Core 3 geotechnical properties are typical of underconsolidated deposits as has been shown earlier (McGregor et al., 1979). The high water contents observed for the valley and valley wall sediments are undoubtedly a function of considerable bioturbation of the sediments as observed during core analysis. Numerous filled and unfilled burrows were observed (ibid). Despite the rather consistent character of the sediment textures (sand-silt-clay percentages), depositional and post-depositional factors have strongly affected the sediments and resulted in high variability in the geotechnical properties over very short lateral and vertical distances. A high degree of variability in geotechnical properties over short lateral distances was found by Inderbitzen (1965) during a detailed study of seafloor stability on the upper slope off Del Mar, California. Thus, the local depositional environment, morphological setting, depositional and post-depositional processes strongly influence the ultimate nature and time dependent changes in the physical and mechanical properties of submarine sedimentary deposits.

CONCLUSIONS

Very limited data are available concerning the geotechnical properties and related processes at shelfbreaks. Within the scope of the published results, however, significant characteristics and differences in the mass physical and mechanical properties of sedimentary deposits have been observed for various continental margin areas. Several physical oceanographic processes, somewhat unique to shelfbreaks in comparison to shallow shelf and deep-sea basin environments, have been identified as important in affecting deposits and in influencing time-dependent changes in the geotechnical properties. Studies of biological and chemical processes at shelfbreaks, as they relate to geotechnical properties of surficial sediments, have received little attention. Likewise, the mechanical processes, although studied in more detail than other processes, have been investigated through descriptive (geophysical) techniques and quantitative (geotechnical) methods.

Shelfbreak textural models were established and verified with published data. These textural models are important in relating sediment types to geotechnical properties and to shelfbreak processes. Soil "state" is strongly dependent upon textural characteristics (particle size and size distribution), mineralogy, and microstructure. Soil "state" also is determined by the primary depositional properties and the post-depositional processes which are active on and within the shelfbreak deposits. The potential for sediment transport and mass movement depends not only upon the soil "state" and physical and mechanical properties of a deposit, but also upon their morphological setting and the sediment's response to active shelfbreak processes.

Sedimentological, geophysical, and geotechnical data from shelfbreak studies have demonstrated clearly that variable sediment textures and the variable mass physical and mechanical properties are the rule rather than the exception. A frontier of scientific and seafloor engineering studies addressing specifically shelfbreak sediment deposits and processes hold promising research endeavors in the ensuing decade, particularly in light of the constant seaward march of seabed exploration and utilization.

ACKNOWLEDGMENTS

This research synthesis was supported by the National Oceanic and Atmospheric Administration (NOAA). The authors appreciate the efforts of Frances Nastav and George Merrill for considerable assistance in the drafting of the figures and help in preparing this manuscript. An earlier version of this manuscript was critically reviewed by Drs. G. Almagor, W. R. Bryant, M. H. Hulbert, G. H. Keller, B. A. McGregor, and Mr. D. N. Lambert. Discussions with Mr. L. Yehle concerning diamicton and mixtum terminology is appreciated.

REFERENCES

ALMAGOR, G., AND WISEMAN, G., 1977, Analysis of submarine slumping in the continental slope off the southern coast of Israel: Marine Geotechnology, v. 2, p. 349–388.

———, 1978, Geotechnical properties of the sediments of the continental margin of Israel: Jour. Sed. Petrology, v. 48, p. 1267–1274.

———, AND GARFUNKEL, Z., 1979, Submarine slumping in continental margin of Israel and northern Sinai: Am. Assoc. Petroleum Geologists Bull., v. 63, p. 324–340.

———, FRYDMAN, S., AND WISEMAN, G., 1980, Stability of continental shelf slopes under earthquake loading effects: Geol. Survey Israel Current Res., p. 96–99.

APEL, J., BYRNE, H., PRONI, J., AND CHARNELL, R., 1975, Observations of oceanic internal and surface waves from the Earth Resources Technology Satellite: Jour. Geophys. Res., v. 80, p. 865–881.

ATTERBERG, A., 1911, Uber die physikalishe Bodenuntersuchung und uber die plastizitat der Torre: Int. Mitt. fur Bodenkunde, v. 1, p. 10–43.

BEARDSLEY, R., BOICOURT, W., AND HANSEN, D., 1976, Physical oceanography of the middle Atlantic Bight, *in* Gross, M., ed., Middle Atlantic Continental Shelf and the New York Bight, Special Symposium: Proc. Amer. Soc. Limn. Oceanogr., v. 2, p. 20–34.

———, AND WINANT, C., 1979, On the mean circulation in the mid-Atlantic Bight: Jour. Phys. Oceanogr., v. 9, p. 612–619.

BENNETT, R. H., KELLER, G. H., AND BUSBY, R. F., 1970, Mass property variability in three closely spaced deep-sea sediment cores: Jour. Sed. Petrology, v. 40, p. 1038–1043.

———, BRYANT, W. R., AND KELLER, G. H., 1977, Clay fabric and geotechnical properties of selected submarine sediment cores from the Mississippi Delta: NOAA Professional Paper 9, 86 p.

———, LAMBERT, D. N., AND HULBERT, M. H., 1978, Geotechnical properties of a submarine slide area on the U.S. Continental Slope northeast of Wilmington Canyon: Marine Geotechnology, v. 2, p. 245–261.

———, AND FARIS, J. R., 1979, Ambient and dynamic pore pressure in fine-grained submarine sediments. Mississippi Delta: Appl. Ocean Res., v. 1, p. 115–123.

———, FREELAND, G. L., LAMBERT, D. N., SAWYER. W. B., AND KELLER, G. H., 1980, Geotechnical properties of surficial sediments in a megacorridor: U.S. Atlantic continental slope, rise and deep-sea basin: Mar. Geology, v. 38, p. 123–140.

———, AND HULBERT, M.H., 1982, Clay microstructure: An historical perspective of clay fabric and physico-chemistry of fine-grained mineral sediments: Oxford, Oxford University Press, in press.

———, BURNS J., CLARKE, T., FARIS, R., FORDE, E., AND RICHARDS, A., 1983, Piezometer probes for assessing effective stress and stability in submarine sediments: NATO Conference on Marine Slides and Other Mass Movements; Conference in 1980, Faro, Portugal: New York, Plenum Press, p. 129–161.

BERGER, W. H., EKDALE, A. A., AND BRYANT, P.P., 1979, Selective preservation of burrows in deep-sea carbonates: Mar. Geology, v. 32, p. 205–230.

BOICOURT W., AND HACKER, P., 1976, Circulation on the Atlantic continental shelf of the United States, Cape May to Cape Hatteras: Mémoires Société Royale des Sciences de Liege, v. 10, p. 187–200.

BOOTH, J. S., AND GARRISON, L. E., 1978, A geological and geotechnical analysis of the upper continental slope adajacent to the Mississippi Delta: Proc. Offshore Tech. Conf. Paper No. 3165, p. 1019–1028.

BRYANT, W. R., BENNETT, R. H., AND KATHERMAN, C. E., 1981, Shear strength, consolidation, porosity, and permeability of oceanic sediments, in Emiliani, C., ed., The Sea, v. 7: New York, Wiley Interscience, p. 1555–1616.

BUTMAN, B., NOBEL, M., AND FOLGER, D., 1979, Long-term observations of bottom current and bottom sediment movement on the mid-Atlantic continental shelf: Jour. Geophys. Res., v. 84, p. 1187–1205.

CACCHIONE, D., AND SOUTHARD, J., 1974, Incipient sediment movement by shoaling internal gravity waves: Jour. Geophys. Res., v. 79, p. 2237–2242.

CADLING, L., AND ODENSTAD, S., 1950, The vane borer: Proc. Royal Swedish Geotech. Inst. No. 2, 87 p.

CARLSON, P. R., LEVY W. P., MOLNIA, B. F., AND HAMPSON, J. C., JR., 1978, Geotechnical properties of sediments from the continental shelf south of Icy Bay, northeastern Gulf of Alaska,: U.S. Geol. Survey Open-File Rept. 78-1071, 28 p.

———, 1978, Holocene slump on continental shelf off Malaspina Glacier, Gulf of Alaska: Am. Assoc. Petroleum Geologists Bull., v. 62, p. 2412–2426.

CASSAGRANDE. A., 1932, Research on the Atterberg limits of soils: Public Roads, v. 13, p. 121–130, p. 136.

———, 1948, Classification and identification of soils: Amer. Soc. Civil Engin. Trans. Pap. 2351, v. 113, p. 901–930.

CHENEY, R., AND RICHARDSON, P., 1976, Observed decay of a cyclonic Gulf Stream ring: Deep-Sea Res., v. 23, p. 143–155.

DEMARS, K. R., NACCI, V. A., KELLEY, W. E., AND WANG, M. C., 1976, Carbonate content: an index property for ocean sediments: Proc. Offshore Tech. Conf., Paper OTC-2627, p. 97–106.

DOYLE, L. J., PILKEY, O. H., AND WOO, C. C., 1979, Sedimentation on the eastern United States' continental slope: Soc. Econ. Paleontologists Mineralogists Spec. Pub. 27, p. 119–129.

EDWARDS, B. D., FIELD, M. E., AND CLUKEY, E. C., 1980, Geological and geotechnical analysis of a submarine slump, California borderland: Proc. Offshore Tech. Conf., Paper 3726, p. 399–410.

EVANS, I., AND SHERRATT, G. G., 1948, A simple and convenient instrument for measuring the shearing resistance of clay soils: Jour. Sci. Instruments and Physics Ind., v. 25, p. 411–414.

FISK H. N., AND MCCLELLAND, B., 1959, Geology of continental shelf, Louisiana: Its influence on offshore foundation design: Geol. Soc. America Bull., v. 70, p. 1369–1394.

FLINT, R. F., SANDERS, J. E., AND ROGERS, J. 1960, Diamictite, a substitute term for smectite: Geol. Soc. America Bull., v. 71, p. 1809.

FRIEDMAN. G., AND SANDERS, J., 1978, Waves and wave-influenced sediments, in Friedman, G., and Sanders, J., eds., Principles of Sedimentology: New York, John Wiley and Sons, p. 464–494.

HADLEY, M., 1964, Wave-induced bottom currents in the Celtic Sea: Mar. Geology, v. 2, p. 164–167.

HAMILTON, E. L., 1971, Elastic properties of marine sediments: Jour. Geophys. Res., v. 76, p. 579–604.

INDERBITZEN, A. L., 1965, An investigation of submarine slope stability, in Ocean Science and Ocean Engineering, v. 2: Mar. Tech. Soc. and Am. Soc. Limnol. Oceanogr., p. 1309–1344.

KELLER, G. H., AND BENNETT, R. H., 1973, Sediment mass physical properties: Panama Basin, northeastern equatorial Pacific, in Van Andel, T. H., Heath, G. R. et al., eds., Initial Reports of the Deep Sea Drilling Project, Volume 16: Washington, D.C., U.S. Govt. Printing Office, p. 499–512.

———, AND ———, 1970, Variations in the mass physical properties of selected submarine sediments: Mar. Geology, v. 9, p. 215–223.

———, AND ———, 1968, Mass physical properties of submarine sediments in the Atlantic and Pacific Basins: Proc XXIII Internat. Geol. Congr., Prague, v. 8, p. 33–50.

———, LAMBERT, R., AND BENNETT, R. H., 1979, Geotechnical properties of continental slope deposits: Cape Hatteras to Hydrographer Canyon: Soc. Econ. Paleontologists Mineralogists Spec. Pub. 27, p. 131–151.

KERR, P. F., 1963, Quick clay: Scientific American, v. 209, p. 132–142.

KOMAR, P., NEUDECK, R., AND KULM, L. 1972, Observations and significance of deep-water oscillatory ripple marks on the Oregon continental margin, *in* Swift, D. J. P., Duane, D., and Pilkey, O. H., eds., Shelf Sediment Transport: Process and Pattern: Stroudsburg, Pennsylvania, Dowden, Hutchinson, and Ross, Inc., p. 601–620.

———, 1976, The transport of cohesionless sediments on continental shelves, *in* Stanley, D. J., and Swift, D. J. P., ed., Marine Sediment Transport and Environmental Management: New York, John Wiley and Sons, Inc., p. 107–126.

KRUMBEIN, W. C., AND PETTIJOHN, F. J., 1938, Manual of Sedimentary Petrography: New York, Appleton-Century-Crofts, Inc., 549 p.

LAMBE, T. W., AND WHITMAN, R. V., 1969, Soil Mechanics: New York, John Wiley and Sons, Inc. 553 p.

LEE, K., AND FOCHT, J., 1976a, Strength of clay subjected to cyclic loading: Marine Geotechnology, v. 1, p. 165–185.

———, AND ———, 1976b, Cyclic testing of soil for ocean wave loading problems: Marine Geotechnology, v. 1, p. 305–335.

LEWIS, K. B., 1970, Slumping on the continental slope inclined at 1°–4°: Sedimentology, v. 16, p. 97–110.

MAYER, D., MOFJELD, H., AND LEAMAN, K., 1983, Near-inertial internal waves observed on the outer shelf in the middle Atlantic Bight in the wake of Hurricane Belle: Jour. Phys. Oceanogr., in press.

MCGREGOR, B. A., BENNETT, R. H., AND LAMBERT, D. N., 1979, Bottom processes morphology and geotechnical properties of the continental slope south of Baltimore Canyon: Appl. Ocean Res., v. 1, p. 177–187.

MOOERS, C., GRAVINE, R., AND MARTIN, W., 1979, Summertime synoptic variability of the middle Atlantic shelf water/slope water front: Jour. Geophys. Res., v. 84, 4837–4854.

MORGAN, C., AND BISHOP, J., 1977, An example of Gulf Stream eddy-induced water exchange in the mid-Atlantic Bight: Jour. Phys. Oceanogr., v. 7, p. 472–479.

MORGENSTERN, N. M., 1967, Submarine slumping and initiation of turbidity currents, *in* Richards, A. F., ed., Marine Geotechnique: Urbana, Illinois, University of Illinois Press, p. 189–220.

NARDIN, T., HEIN, F., GORSLINE, D., AND EDWARDS, B., 1979, A review of mass movement processes, sediment and acoustic characteristics, and contrasts in slope and base-of-slope systems versus canyon-fan-basin floor systems, *in* Doyle, L., and Pilkey, O. H., eds., Geology of Continental Slopes: Soc. Econ. Paleontologists Mineralogists Spec. Pub. 27, p. 61–74.

NEEV, D., ALMAGOR, G., ARAD, A., HALL, J. K., AND GINSBURG, A., 1976, The geology of the southeastern Mediterranean Sea: Geol. Survey Israel Bull. No. 68, 51 p.

NEUMAN, F., BYRNE, H., PRONI, J., AND NELSEN, T., 1977, Acoustic observation of a bottom layer of suspended material on the New England continental shelf: EOS, v. 58, p. 410.

POSTMA, H., 1967, Sediment transport and sedimentation in the estuarine environment, *in* Lauff, G. H., ed., Estuaries: Washington, D.C. Amer. Assoc. Advancement Science, p. 158–179.

READING, H. G., ed., 1978, Sedimentary Environment and Facies: New York, Elsevier, 557 p.

RHOADS, D. C., 1970, Mass properties, stability and ecology of marine muds related to burrowing activity, *in* Crimes, T.P., and Harper, J. C., eds, Trace Fossils: Liverpool, Seel House Press, p. 391–406.

———, 1974, Organism-sediment relations on the muddy seafloor: Oceanogr. Mar. Biol. Ann. Rev., v. 12, p. 263–300.

———, AND BOYER, L., 1983, Animal-sediment relations: The biogenic alteration of sediments, *in* McCall, P.L. and Tevesz, M. J. S., eds., Topics in Geobiology, v. 2: New York, Plenum Press, 336 p.

RICHARDS, A. F., 1962, Investigations of deep-sea sediment cores II. Mass physical properties: U.S. Navy Hydrographic Office Tech. Rept. TR-106, 146 p.

———, 1965, Geotechnical aspects of recent marine sediments, Oslofjord, Norway [abs.]: Am. Assoc. Petroleum Geologists Bull., v. 49, p. 356.

———, 1974, Standardization of marine geotechnics symbols, definitions, units and test procedures, *in* Inderbitzen, A.L., ed., Deep Sea Sediments: New York, Plenum Press, p. 271–292.

RICHARDSON, M. D., AND YOUNG, D. K., 1980, Geoacoustic models and bioturbation: Mar. Geology, v. 38, p. 205–218.

ROKOENGEN K., BUGGE, T., DEKKO, T., GULEIKSRUD, T., LIEN, R. L., AND LOFALDLI, M., 1979, Port and shallow geology of the continental shelf off Norway, *in* Port and Ocean Engineering Under Arctic Conditions: Proc. 5th Intl. Conf., Trondheim, Norway, v. 2, p. 859–875.

RONA, P. A., 1969, Middle Atlantic Continental Slope of United States: deposition and erosion: Am. Assoc. Petroleum Geologists Bull., v. 53, p. 1453–1465.

ROWE, G. T., 1974, The effects of the benthic fauna on the physical properties of deep-sea sediments, in Inderbitzen, A. L., ed., Deep Sea Sediments Physical and Mechanical properties: New York, Plenum Press, p. 381–400.

———, AND HAEDRICH, R. L., 1979, The biota and biological processes of the continental slope, *in* Doyle, L. J., and Pilkey, O. H., ed., Geology of Continental Slopes: Soc. Econ. Paleontologists Mineralogists Spec. Pub. 27, p. 49–59.

SAUNDERS, P., 1971, Anticyclonic eddies formed from shoreward meanders of the Gulf Stream: Deep-Sea Res., v. 18, p. 1207–1219.

SCHERMERHORN, L. J. G., 1966, Terminology of mixed coarse-fine sediments: Jour. Sed. Petrology, v. 36, p. 831–835.

SCHMITZ, J., AND VASTANO, A., 1977, Decay of a shoaling Gulf Stream cyclonic ring: Jour. Phys. Oceanogr., v. 7, p. 479–481.

SEED, H., INDRISS, I., LEE, K., AND MAKDISI, F., 1975, Dynamic analysis of the slide in the lower San Fernando Dam

during the earthquake of February 9, 1971: Jour. Geotech. Eng., v. 101, p. 889–911.
SHEPARD, F., 1973, Submarine Geology: New York, Harper and Row, 577 p.
SKEMPTON, A. W., 1953, Soil mechanics in relation to geology: Proc. Yorkshire Geol. Soc., V. 29, p. 33–62.
——, 1970, The consolidation of clays by gravitational compaction: Geol. Soc. London, v. 125, p. 373–411.
SOUTHARD, J. B., AND CACCHIONE, D., 1972, Experiments on bottom sediment movement by breaking internal waves, *in* Swift, D. J. P., Duane, P., and Pilkey, O. H., eds., Shelf Sediment Transport: Process and Pattern: Stroudsburg, Pennsylvania, Dowden, Hutchinson and Ross, Inc., p. 83–87.
——, AND STANLEY, D., 1976, Shelfbreak processes and sedimentation, *in* Stanley, D. J., and Swift, D. J. P., eds., Marine Sediment Transport and Environmental Management: New York, John Wiley and Sons, p. 351–378.
——, YOUNG, R. A., AND HOLLISTER, C. D., 1971, Experimental erosion of calcareous ooze: Jour. Geophys. Res., v. 76, p. 5903–5909.
SPUDICH, P., AND ORCUTT, J. 1982, Estimation of earthquake ground motions relevant to the triggering of marine mass movements *in* Saxov, S., and Nieuwenhuis, J. K., eds., Slides and Other Mass Movements, NATO Conference Series IV: New York, Plenum Press, p. 219–231.
STANLEY, D. J., ADDY, S., AND BEHRENS, E., 1983, The mudline: Variability of its position relative to shelfbreak, *in* Stanley, D. J., and Moore, G. T., eds., The Shelfbreak: Critical Interface on Continental Margins: Soc. Econ. Paleontologists Mineralogists Spec. Pub. 33, p. 279–298.
instability: interaction of hydrodynamic forces and bottom sediments: Proc. 8th Annual Offshore Tech. Conf., p. 30–33.
——, 1977, Surface waves and bottom sediment response: Marine Geotechnology, v. 2, p. 135–146.
TAYLOR, D. W., 1948, Fundamentals of Soil Mechanics: New York, Wiley and Sons, 700 p.
TERZAGHI, K., AND PECK, R. B., 1967, Soil Mechanics in Engineering Practice: New York, John Wiley and Sons, Inc., 729 p.
VANNEY, J. R. AND STANLEY, D. J., 1983, Shelfbreak physiography: An overview, *in* Stanley, D. J., and Moore, G. T., eds., The Shelfbreak: Critical Interface on Continental Margins: Soc. Econ. Paleontologists Mineralogists Spec. Pub. 33, p. 1–24.
VOORHIS, A., WEBB, D., AND MILLARD, R., 1976, Current structure and mixing in the shelf/slope water front south of New England: Jour. Geophys. Res., v. 81, p. 3695–3708.
WEAR, C. M., STANLEY, D. J., AND BOULA, J., 1974, Shelfbreak physiography between Wilmington and Norfolk Canyons: Mar. Tech. Soc. Jour., v. 8, no. 4, p. 37–48.
YAMAMOTO, T., 1980, Wave-induced stress instabilities in inhomogenous seabeds-stability analysis of seafloor foundations: 17th Ann. Conf. on Coastal Engineering, I.C.E.E. Australia, 2 p.
YOUNG, R. A., AND SOUTHARD, J. B., 1978, Erosion of fine-grained marine sediments: sea-floor and laboratory experiements: Geol. Soc. America Bull., v. 89, p. 663–672.

PART VII
FAUNAL AND ORGANIC MATTER DISTRIBUTION

BENTHIC FORAMINIFERA AT THE SHELFBREAK: NORTH AMERICAN ATLANTIC AND GULF MARGINS

STEPHEN J. CULVER
Department of Geophysical Sciences, Old Dominion University, Norfolk, Virginia 23508,
and Department of Paleobiology, Smithsonian Institution, Washington, D.C. 20560

MARTIN A. BUZAS
Department of Paleobiology, Smithsonian Institution, Washington, D.C. 20560

ABSTRACT

Using all published data from the Atlantic continental margin and the Gulf of Mexico, the distribution of species occurrence was examined in the vicinity of the shelfbreak. Foraminiferal provinces defined by cluster analyses of occurrence data indicate that the termination of inner shelf provinces, in general, coincides with the approximate position of the shelfbreak. The distribution of species occurrence of 30 selected species from the Atlantic continental margin and 30 from the Gulf of Mexico also indicate that the shelfbreak is an important boundary for benthic foraminifera. Most selected species that occur most often at depths of less than 100 m occur occasionally deeper than 100 m. Likewise, most species that usually occur at depths of more than 100 m occur occasionally at shallower depths. Consequently, for these species the upper depth limit cannot be given more weight than the lower depth limit. In general, the frequency of occurrence is a more useful measure than the observed range.

The depth distribution of some species is different in the two areas studied. Consequently, recognition of provincial differences is important when attempting paleoenvironmental analyses. The shelfbreak has been recognized previously as an important boundary for benthic foraminifera through analyses of species abundance and occurrence. In the present study, only species occurrence was used and no distinction was made between live or dead, Pleistocene or Holocene, in place or transported. Evidently, large-scale patterns are recognizable even when possible sources of error are not taken into account. This is encouraging because much of the literature on fossil and modern localities contains only species lists.

INTRODUCTION

Numerous studies indicate that the distribution of benthic foraminifera exhibits distinctive patterns with latitude, longitude and depth. Whenever benthic foraminifera are examined from transects extending across the shelf into deeper water, they invariably show a zonation with depth (see summaries by Phleger, 1960; Murray, 1973; Boltovskoy and Wright, 1976). While disagreement exists about the use of living versus total populations, downslope movement, and dissolution of tests in the placement of boundaries, a depth zonation is universally recognized (Douglas, 1979). In the Gulf of Mexico and south of Cape Cod in the western North Atlantic, the most marked boundary according to Phleger (1960) is at about 100 m, that is, in the vicinity of the shelfbreak.

The criteria used for the depth zonations mentioned above are upper and lower occurrence of species, and species abundance (Phleger, 1960). Many of the early workers listed only species occurrence in their studies. Consequently, if only those papers containing abundance data are considered much distribution information is lost. About 220 papers have been published containing information on the distribution of modern benthic foraminifera from the Atlantic continental margin of North America and the Gulf of Mexico. The distributional data contained in these papers have been extracted (in presence or absence form) and computer-catalogued in Culver and Buzas (1980, 1981a). Distribution maps were drawn (Culver and Buzas, 1980, 1981a) for the most commonly recorded species, defined as those species recorded at 4% or more of the total sample localities. These data serve as the base for the present study.

The purpose of this study is to determine if the boundary in the vicinity of the shelfbreak is recognizable using all published occurrences of benthic foraminifera in the Western North Atlantic and the Gulf of Mexico.

DEPTH ZONATION OF BENTHIC FORAMINIFERA:
MODUS OPERANDI

Benthic foraminifera exhibit discrete patterns of distribution in space and time. Modern forms exhibit distinctive patterns with latitude, longitude, and depth. Unlike other benthic organisms, however, the amount of distributional data available for foraminifera is vast (see Boltovskoy and Wright, 1976; Culver, 1980a).

The first depth zonations of foraminifera from the continental margins of North America were made from analyses of total assemblages (living and dead individuals). The criteria used to distinguish depth zones or biofacies were relative abundance of species (percent of total assemblage) and species occurrences. Often, the shoreward or upper occurrence of a species was given more weight than the seaward or lower occurrence to offset possible downslope transport effects (Phleger, 1960).

When staining techniques became available (Walton, 1952) researchers began to compare the distribution of living and dead or total assemblages. Such comparisons are important because, if the distributions of living species are related to environmental variables and are similar to total distributions, then the distribution of the same or similar species in the fossil record can be used to reconstruct ancient environments. In some cases the transportation or dissolution of tests seriously affects distribution patterns (Murray, 1969; Parker and Athearn, 1959) and some species may change habitat preference with time (Douglas, 1979). Fortunately, however, the distribution of species in the total assemblage is most often very similar to that of the living population (Phleger, 1956; Walton, 1964; Buzas, 1965, 1967; Mello and Buzas, 1968). Indeed, Scott and Medioli (1980) have suggested that the total assemblage is a better representation of overall environmental conditions than the living population. This may be so because during the course of a year, living populations commonly have three or four periods of peak abundances (Boltovskoy and Lena, 1969; Buzas, 1969; Wefer, 1976; Buzas and others, 1977), and the total assemblage is an accumulation of these populations with time. Because only total assemblages are preserved in the fossil record, this viewpoint is comforting to paleoecologists. The overall coincidence of living and dead or total distributions on a large scale, however, indicates that approximately the same distributional pattern will emerge regardless of one's philosophy.

Many of the early workers listed only species occurrences in their studies. For any given number of individuals, data containing abundances of species have more information than presence or absence data, but the amount of information lost by disregarding presence or absence data is considerable. Several studies suggest that the analysis of presence or absence data produces the same patterns as abundance data. Buzas (1967, 1972) analyzed Phleger's (1956) data by canonical variate analysis using mean densities, relative abundance, and presence or absence. The results were nearly identical and were very similar to Phleger's boundaries selected through inspection of the data. Mello and Buzas (1968) analyzed the same data set by cluster analysis using presence or absence data and obtained much the same results. Gill and Tipper (1978) analyzed six multivariate geologic data sets after converting them to binary form. Their results substantially agreed with the original analyses made on the data in metric form. All the above analyses indicate that, given the precision required, the analyses of presence or absence data is quite adequate. This is particularly true when we are analyzing a large data set consisting of hundreds of species distributed over thousands of square kilometers.

The data set used in the present study does not take into account whether or not a recorded occurrence was live or dead, Pleistocene or Holocene, in place or transported. Therefore, the data set analyzed is very large, but unrefined. Two methods are employed: (1) The positions of boundaries of foraminiferal provinces defined by cluster analysis of occurrence data (Buzas and Culver, 1980; Culver and Buzas, 1981b) are examined in relation to the shelfbreak; (2) The distribution of selected species is examined in terms of average depth of occurrence, standard deviation, observed range (upper and lower limits of occurrence), and percent of occurrence above and below 100 m.

BENTHIC FORAMINIFERA AT THE SHELFBREAK:
PROVINCIAL DISTRIBUTIONS

The zoogeography of recent benthic foraminifera has been summarized recently by Buzas and Culver (1980) and Culver and Buzas (1981b, 1982a) who proposed provincial schemes for the North American Atlantic continental margin and the Gulf of Mexico. Foraminiferal provincial boundaries, in general, overlap. For the segments of shelf where most data are available, the northern Gulf of Mexico and the section between Cape Hatteras and Cape Cod off the East Coast of the U.S.A., two provinces overlap in the vicinity of the shelfbreak at approximately 100 m (Figs. 1 and 2). However, a consistent feature appears to be the seaward termination of inner shelf provinces at or very close to the shelfbreak. That is, the seaward boundary of the Gulf Inner Shelf Province and of the East Coast Northern Shelf Province (terminology from Buzas and Culver, 1980, Fig. 1a; Buzas and Culver, 1981b, Fig. 1) is, in general, at the shelfbreak at around 100 m depth (Figs. 1 and 2). [The faunal composition of the East Coast Northern Shelf Province is detailed in Culver and Buzas (1981c); similar compositional data are now available for the Gulf of Mexico (Culver and Buzas, 1983).] This correlates reasonably well with depth zonations proposed by other researchers. Phleger recognized a boundary at approximately 100 m in the northwest Gulf of Mexico (1951a) and off the Texas coast (1956). In the northern Gulf, Lowman (1949) recognized a faunal boundary at 91 m. In the eastern Gulf of Mexico, Bandy (1956) and Parker (1954) reported boundaries at 76 m and 80–100 m respectively. Off the U.S. East Coast previously recorded faunal breaks are a little shallower. For example, Parker (1948) recognized a boundary at 90 m off Maryland, Phle-

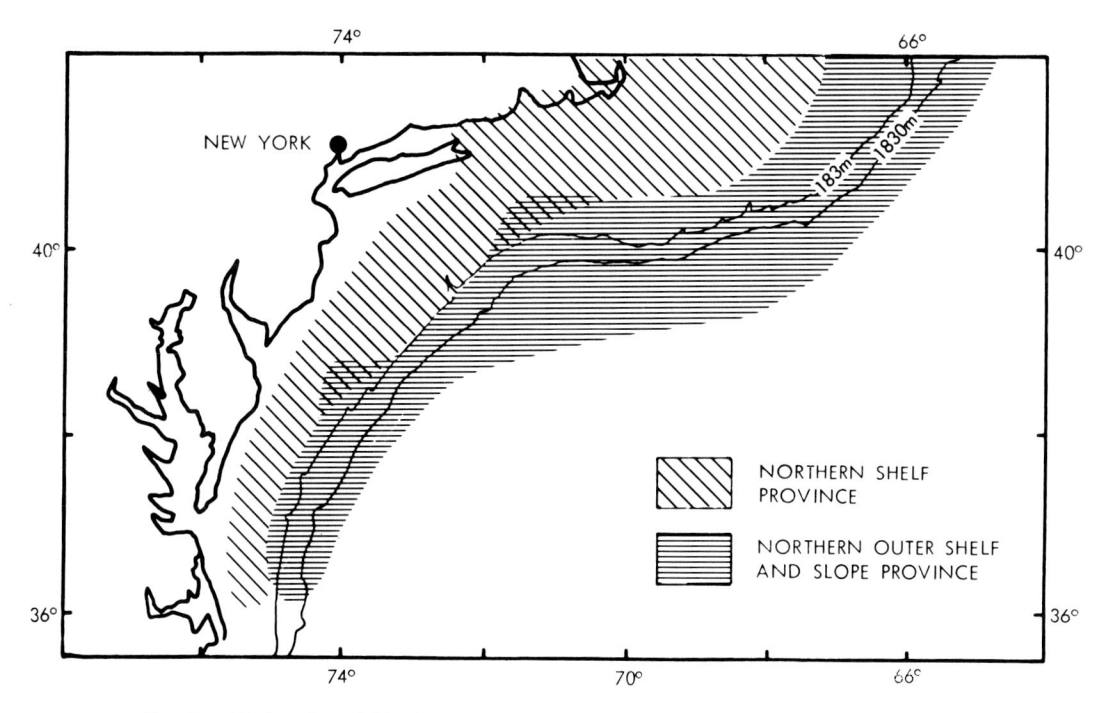

FIG. 1.—Modern foraminiferal provinces near the shelfbreak: Cape Cod to Cape Hatteras.

ger (1952) noted a boundary at 60 to 75 m off New Hampshire, and Murray (1969) reported a boundary at approximately 70 m off Long Island.

South of Cape Hatteras, the seaward boundary of the East Coast Southern Shelf Province is much deeper, at approximately 200 m (Buzas and Culver, 1980, Fig. 1a). Fewer studies, however, have concentrated on this region and this boundary may not be the only one in the shelf and upper slope area. For example, Sen Gupta and Strickert (1982) have

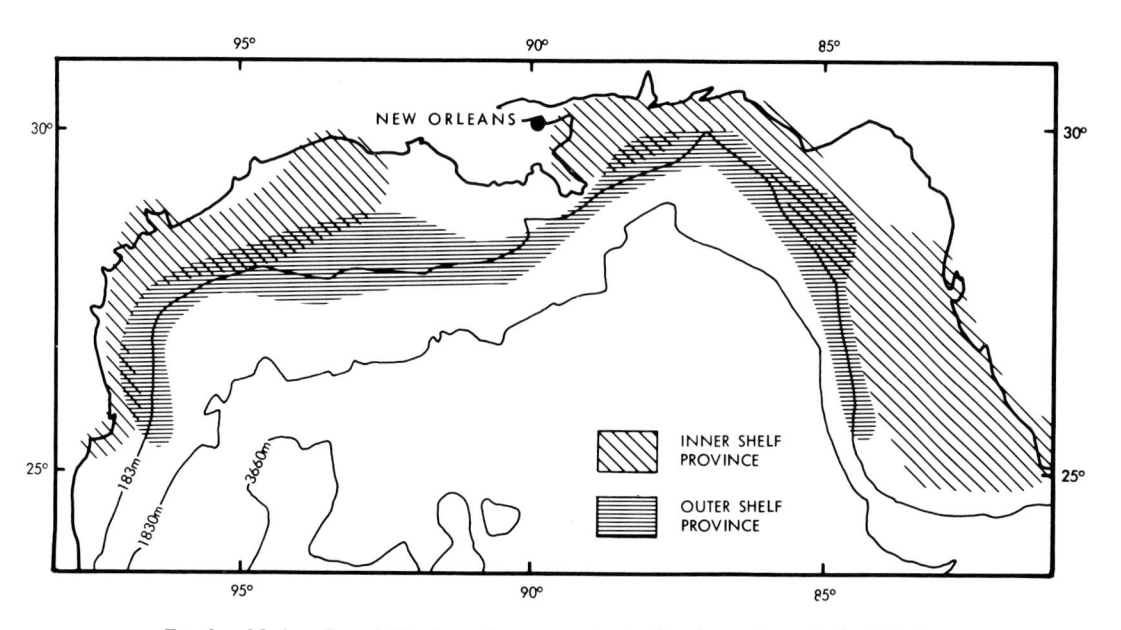

FIG. 2.—Modern foraminiferal provinces near the shelfbreak: northern Gulf of Mexico.

recently recognized an important faunal break at 100 m, just seaward of the shelfbreak at approximately 80 m.

BENTHIC FORAMINIFERA AT THE SHELFBREAK: INDIVIDUAL SPECIES DEPTH DISTRIBUTIONS

One hundred and forty-nine species of recent benthic foraminifera have been recorded at more than 4% of the 542 North American Atlantic continental margin localities (Fig. 3). In the Gulf of Mexico, 295 species have been similarly recorded (at 4%, or more of 426 localities) (Fig. 4). Computer plots of occurrences of these species on contoured charts enabled assignment of the species to particular depth ranges (Culver and Buzas, 1980, Table 1; 1981a, Table 1).

Depth range categories overlap to various degrees: some species are restricted to narrow ranges, others to moderate ranges while still others are recorded from littoral to abyssal depths. For the pres-

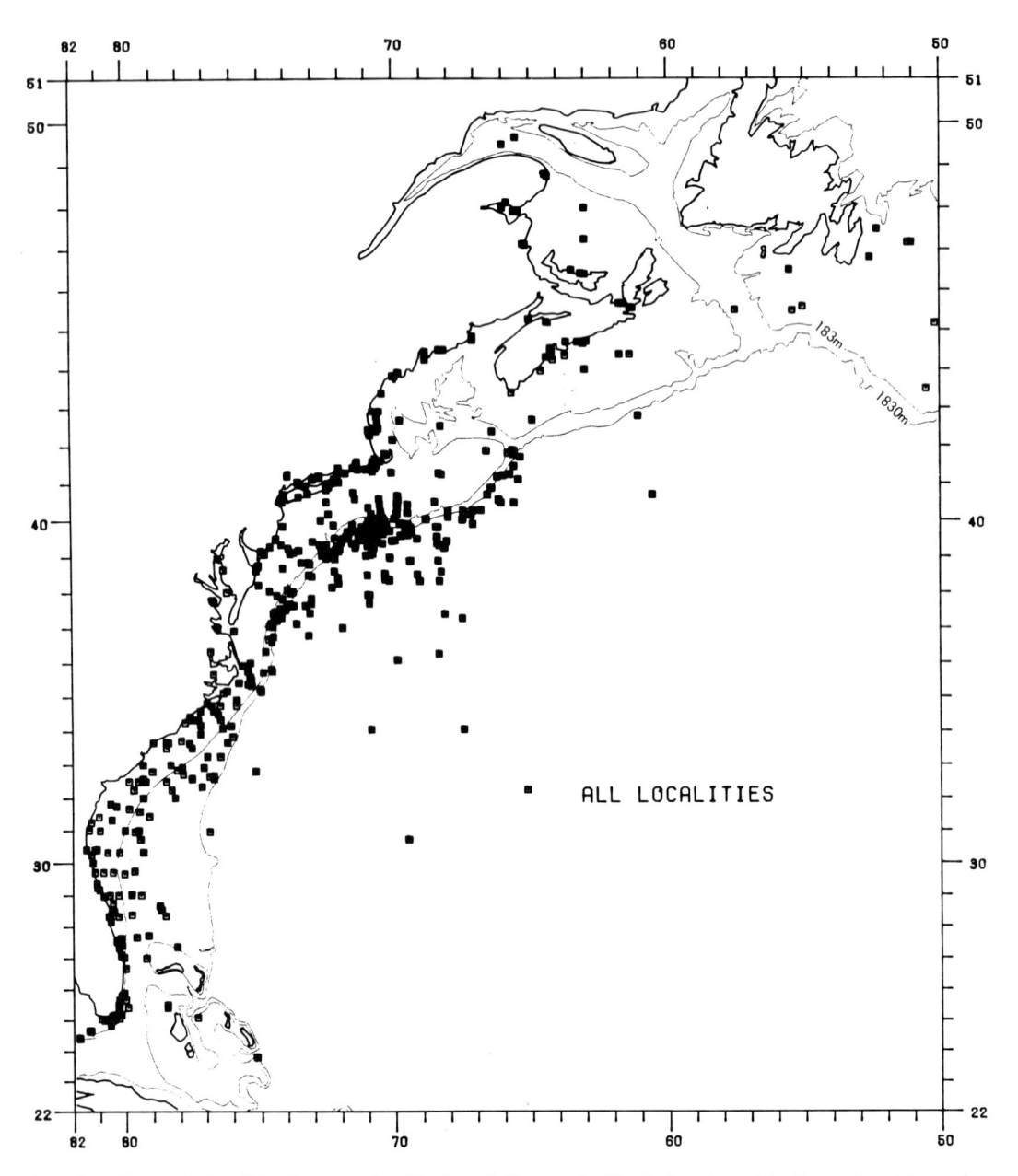

FIG. 3.—All sample localities for recent benthic foraminifera on the North American Atlantic continental margin.

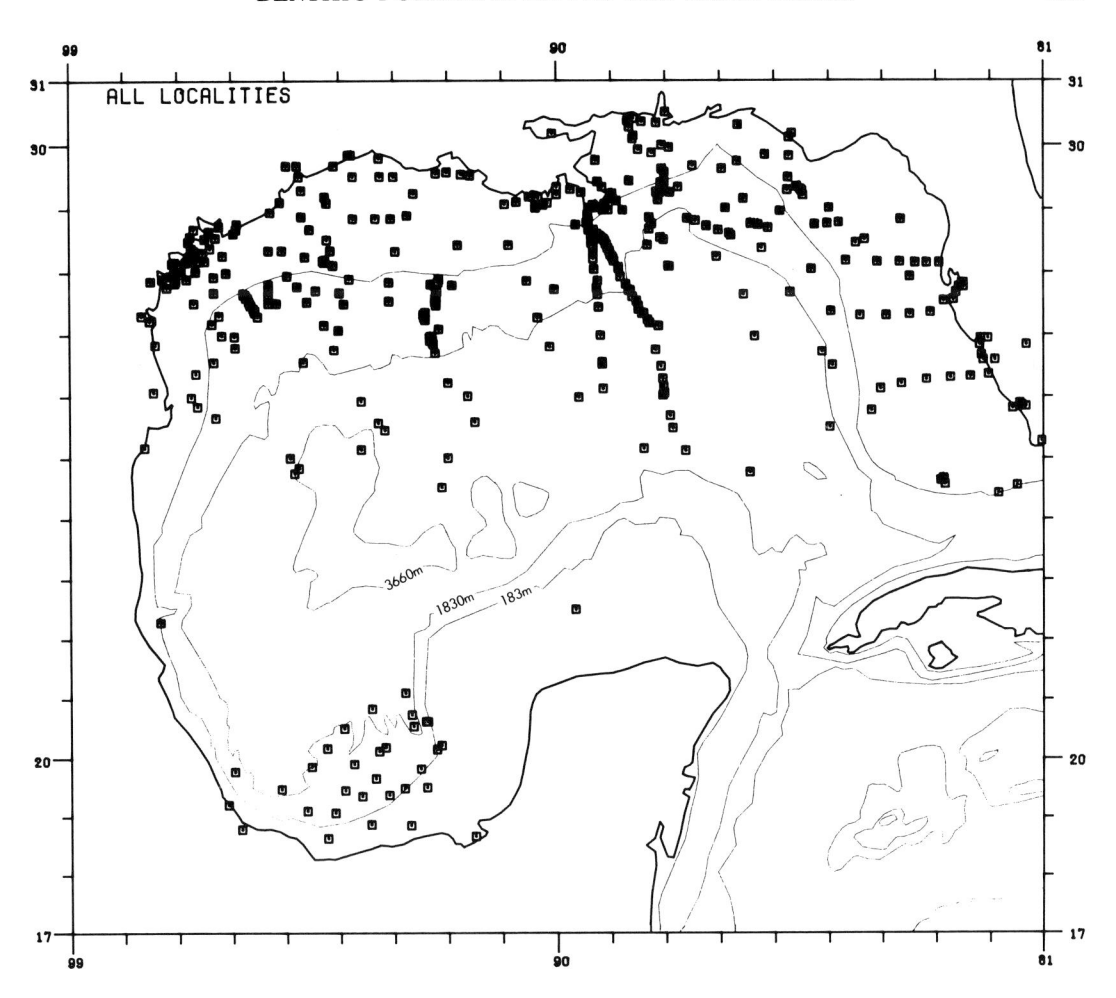

FIG. 4.—All sample localities for recent benthic foraminifera in the Gulf of Mexico.

ent study, those species whose upper and lower depth ranges most closely coincide with the shelfbreak (about 100 m) were selected by examining the distribution maps published by Culver and Buzas (1980, 1981a). Examples of the kinds of patterns selected are shown in Figures 5 through 8. The latitude and longitude of all occurrences of these selected species, obtained from synonymized catalogs (Culver and Buzas 1980, 1981a), were plotted on large scale maps to obtain more precise depth determinations. This information is summarized in Tables 1 through 4.

Examination of Tables 1 through 4 indicates that most selected species have upper and lower limits (observed range) which overlap the 100 m contour to a considerable degree. The mean depth of occurrence is, however, consistently either greater than or less than 100 m. An exception is *Quinqueloculina agglutinans* in the Gulf of Mexico where its mean depth is 285 m. This value is, however, spurious because of one occurrence at 3660 m. The percent of occurrence at less than and more than 100 m is a more reliable depth for judging species usefulness in the present context. On the Atlantic continental margin and in the Gulf of Mexico *Q. agglutinans* occurs most frequently (95% and 93% respectively) at depths of less than 100 m. Similarly, most species whose ranges cross the 100 m contour, are found shallower or deeper than 100 m more than 90% of the time.

DISCUSSION

The position of the provincial boundaries (Figs. 1, 2) and the occurrence of selected species with depth (Tables 1 through 4) indicate that the shelfbreak is an important boundary for benthic foraminifera. The boundary is prominent enough that it is recognizable even though we used all published data without regard for live or dead, fossil or modern, in place or transported occurrences. This con-

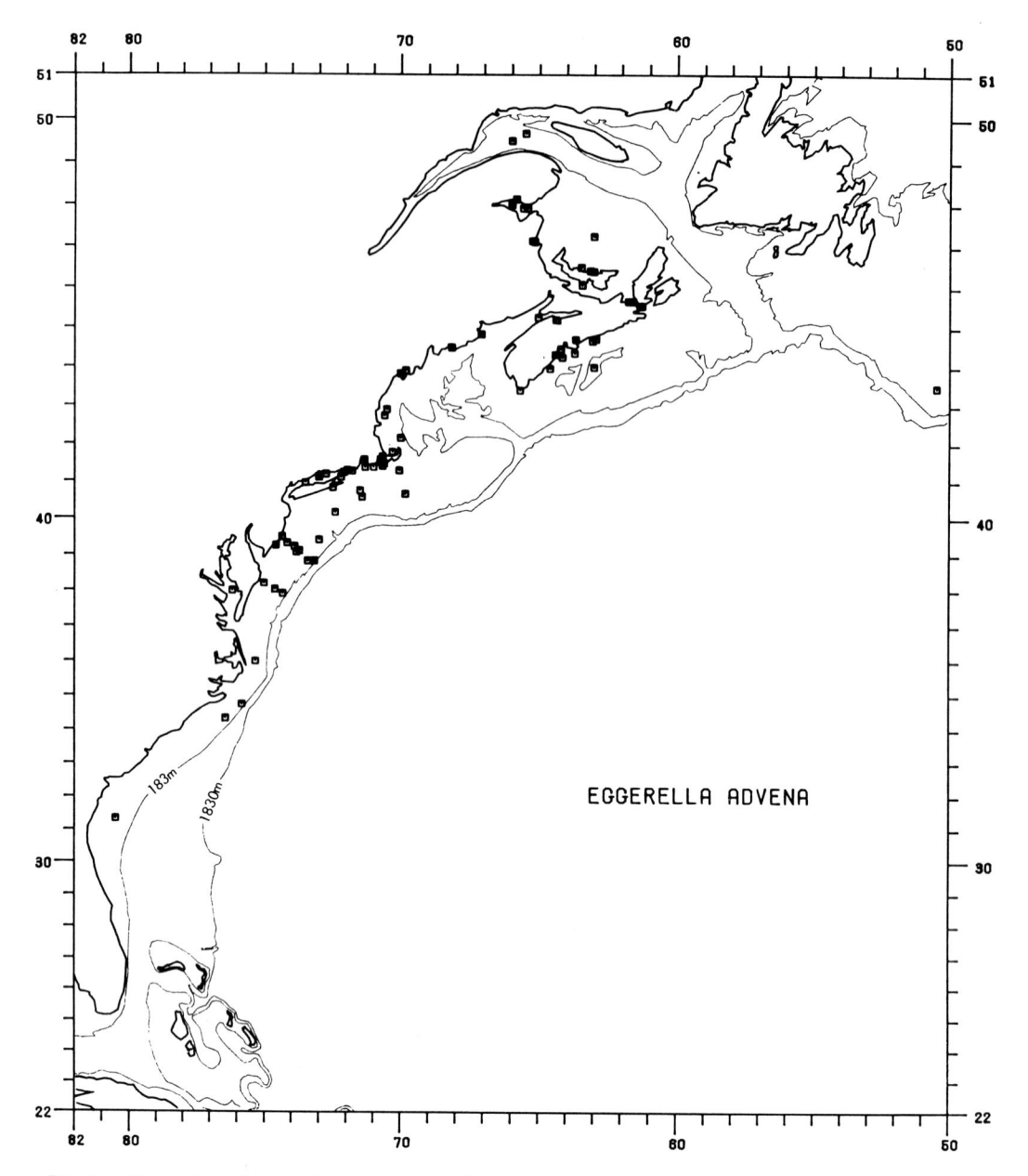

FIG. 5.—Distribution of *Eggerella advena* on the North American Atlantic continental margin: lower depth limit at about 100 m.

clusion corroborates results obtained through consideration of upper and lower occurrences and species abundance (Phleger, 1960).

Some foraminiferal workers regard depth as one of many environmental variables controlling foraminiferal distribution (Murray, 1973; Boltovskoy and Wright, 1976; Douglas, 1979). We believe this is fallacious. As was pointed out by Gibson (1968) and Buzas (1974), depth should not be regarded as an environmental variable at all. The biosphere is three-dimensional and to locate ourselves in this three-dimensional space, Man created latitude, longitude, and distance above and below sea level. Few ecologists would suggest faunal changes are due to longitude, and yet, some are willing to attribute similar changes to depth. Such thinking only causes confusion. The distribution of organisms depends on a multitude of abiotic and biotic vari-

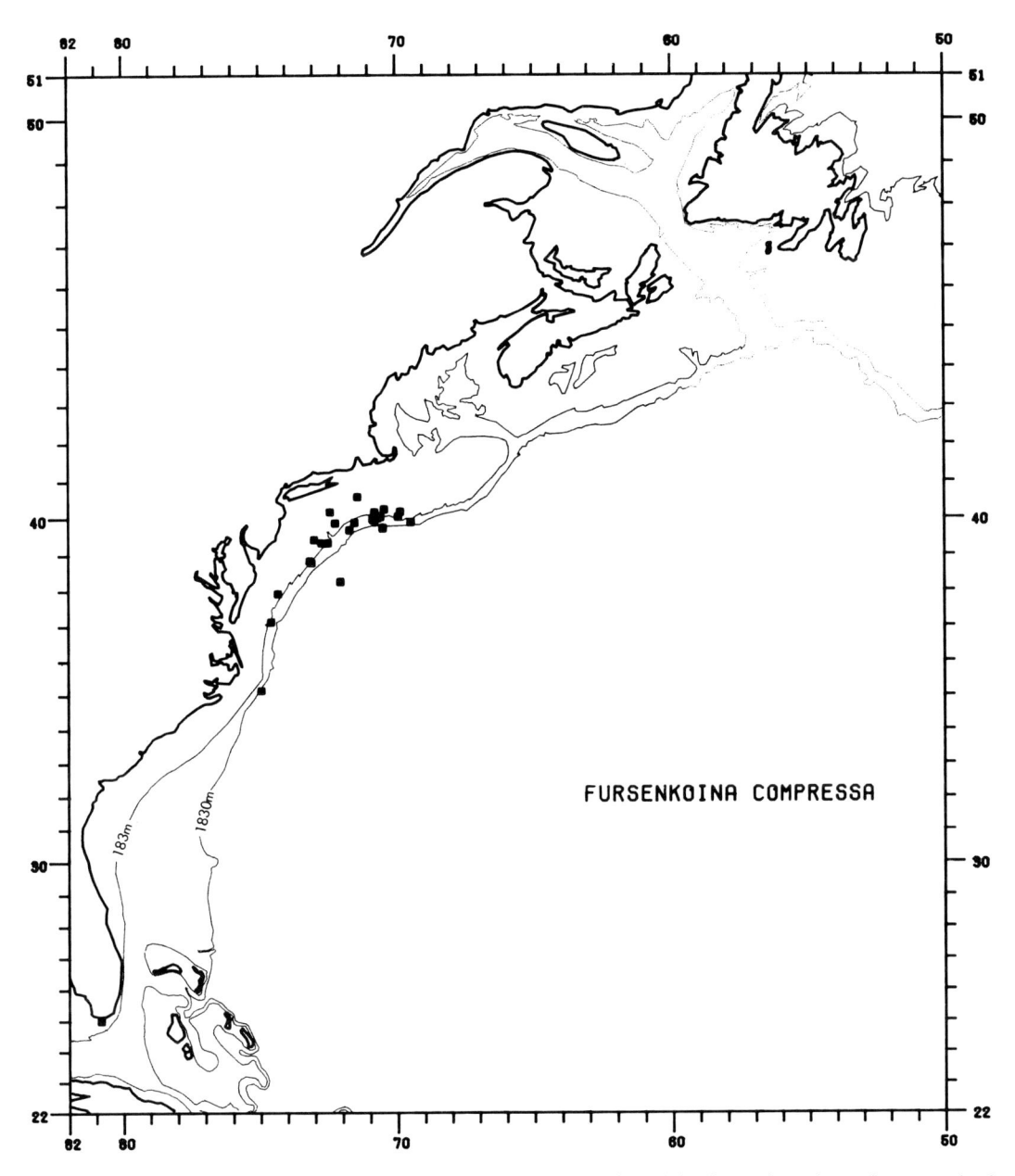

FIG. 6.—Distribution of *Fursenkoina compressa* on the North American Atlantic continental margin: upper depth limit at about 100 m.

ables. We do not believe that the variables defining location, per se, are three of them. At this stage of our ignorance, the best explanation for the boundary observed at the shelfbreak is by Phleger (1960) who concluded the change in distribution is related to the maximum depth of the influence of seasonally variable water masses.

It has been generally considered that downslope transport of shallow water benthic foraminifera is a common phenomenon (e.g. Phleger, 1951b, 1965). While this transport process undoubtedly occurs (e.g. Phleger, 1951b; Elmore and others 1979; Robb and others, 1981), it probably does so along discrete, localized pathways of sediment transport such as submarine chutes and canyons. Associated species normally allow a researcher to correctly interpret the true environment of deposition of a sample containing some transported shal-

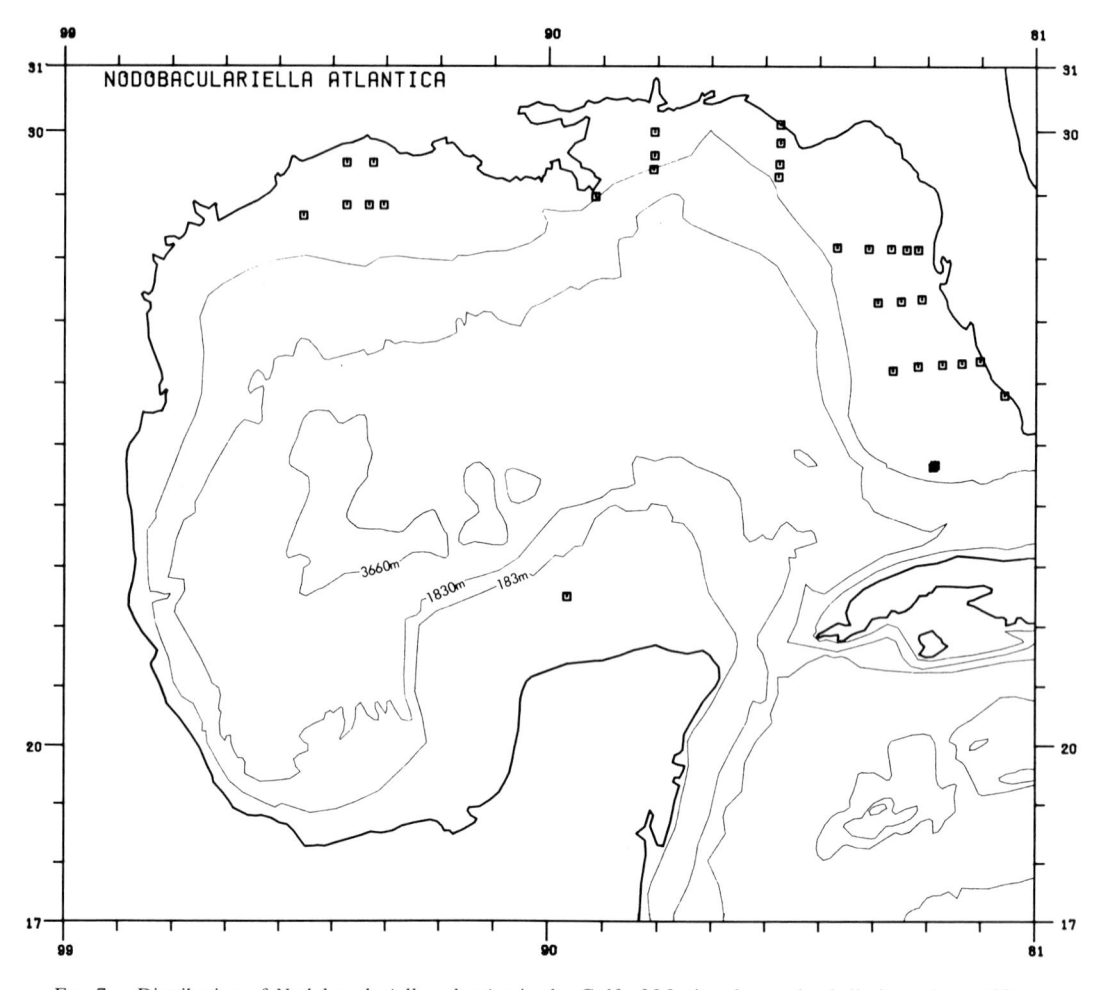

FIG. 7.—Distribution of *Nodobaculariella atlantica* in the Gulf of Mexico: lower depth limit at about 100 m.

low water forms (Phleger, 1960). Deep-water samples containing wholly shallow water faunas, (e.g. Elmore and others, 1979) can be recognized as transported by sedimentological and, indirectly, by seismic evidence (Embley, 1980). On the Atlantic continental margin and in the Gulf of Mexico most of these mass-flow deposits were emplaced during times of lowered sealevel in the Late Pleistocene (Phleger, 1951b; Embley, 1980).

Because of downslope transport, most investigators consider the upper limit of a species depth range to be more reliable than the lower limit (Phleger, 1960). The observed ranges presented in Tables 1 through 4 indicate that for the species we selected this is not so. While species whose mean depth is less than 100 m (Tables 1, 3) occur in much deeper water, species whose mean depth is greater than 100 m (Tables 2, 4) also occur in shallower water. As mentioned previously, we selected these species by examining distributional maps (Culver and Buzas 1980, 1981a) and selecting those whose ranges most closely coincided with the shelf-break. The extreme range of most of these species was not apparent until the data were summarized in the tables. The occurrence of deep water species in shallower water as consistently as the occurrence of shallow water species in deeper water came as a complete surprise. Because the data are from total populations we cannot establish whether this is due to widely dispersed living populations or to transportation landwards, but the latter is probably the case (cf. Murray and Hawkins, 1976; Culver, 1980b).

Tables 1 through 4 also indicate that strict adherence to upper and lower depth limits (observed range) disregards much useful information. Very few of the selected species occur above or below 100 m 100% of the time. Most of them, however, do occur above or below 100 m more than 90% of the time. An occasional occurence in shallower or

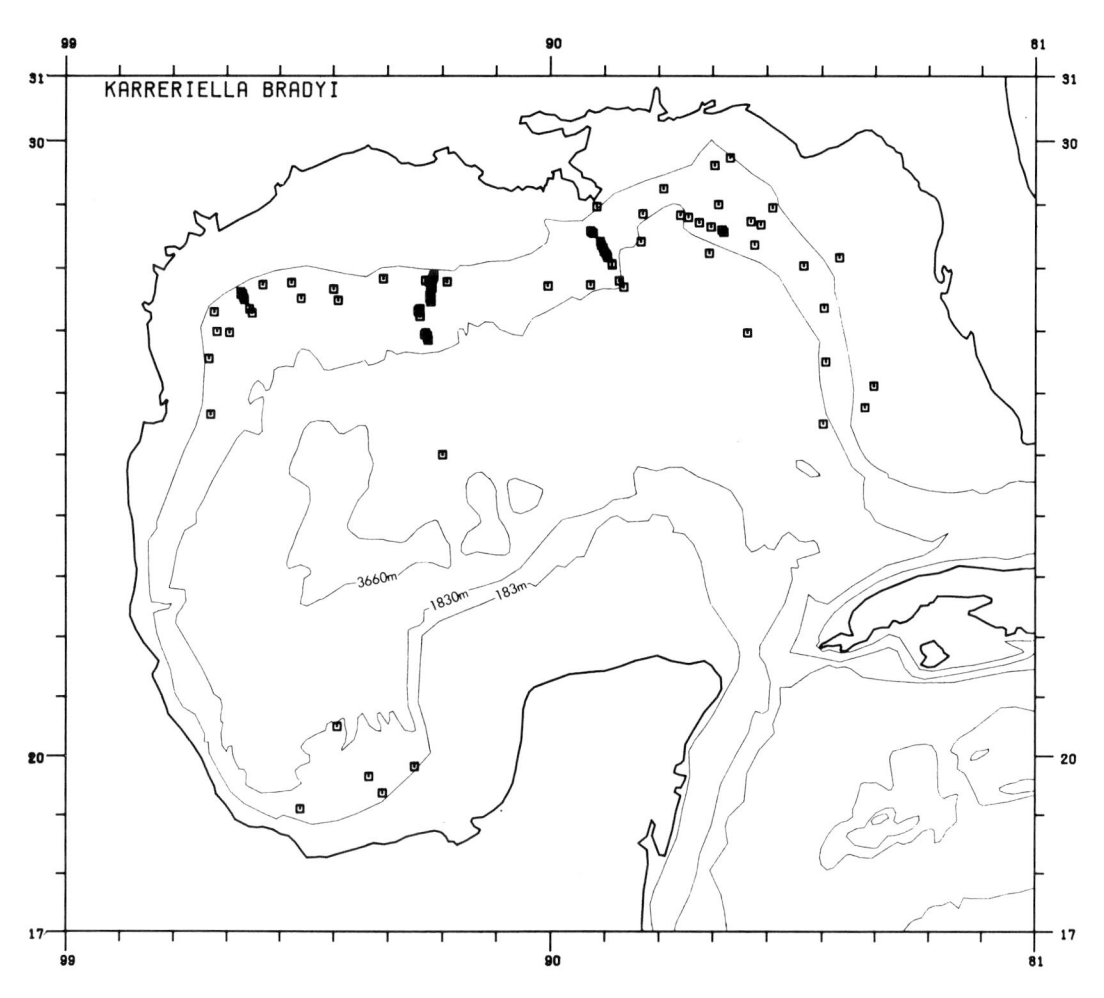

FIG. 8.—Distribution of *Karreriella bradyi* in the Gulf of Mexico: upper depth limit at about 100 m.

deeper water should not deter an investigator from using such species as depth indicators. Considering the ease with which an individual foraminifer can be transported by any number of mechanisms, we should not be surprised to discover a single individual almost anywhere. When using occurrence data this must be kept in mind. Fortunately Figures 1 and 2 and Tables 1 through 4 indicate we can discriminate with a good deal of confidence using occurrence data.

Reference to Tables 1 through 4 illustrates that, in general, different species are utilized in recognizing the shelfbreak on the North American Atlantic continental margin and in the Gulf of Mexico. Only three species, *Elphidium mexicanum*, *Peneroplis proteus* and *Quinqueloculina agglutinans*, have depth limits at approximately 100 m on the Atlantic continental margin and in the Gulf of Mexico also. Thus, it is important to remember that species which are restricted to a particular depth range in one area may have a very different range elsewhere. Two examples will suffice to illustrate this point. *Elphidium excavatum*, previously considered to be a shallow, cold water form, is now known to occur from the Arctic to the Tropics (Culver and Buzas, 1980, 1981a, 1982b) and to live down to abyssal depths on the North American Atlantic continental margin (Culver and Buzas, 1980, their Fig. 54; Schafer and Cole, 1982; Streeter and Lavery, 1982). In the Gulf of Mexico, however, *E. excavatum* is restricted to very shallow water (Culver and Buzas, 1981a, their Fig. 96), and this seems to be the case also in the Caribbean region (Culver and Buzas, 1982b). *Ammobaculites agglutinans* generally occurs between 1000 and 2000 m in the Gulf of Mexico (Culver and Buzas, 1981a, Fig. 5). Off the U.S. Atlantic coast (Culver and Buzas, 1980, Fig. 2), this species occurs on the continental slope at these depths but more often occurs at depths of greater than 2000 m. Therefore, species that are

TABLE 1.—Species with Lower Depth Limit at About 100 m: North American Atlantic Continental Margin

Species	Mean Depth	Standard Deviation	Observed Range	Percent Occurrence Less than 100 m
Ammodiscus catinus	51	31	20–100	89
Asterigerina carinata	56	144	1–800	94
Bolivina pseudoplicata	32	25	5–90	100
Bolivina striatula	45	62	5–255	89
Buliminella elegantissima	30	28	5–100	95
Cancris sagra	60	103	1–450	90
Eggerella advena	33	37	1–230	94
Elphidium advenum	25	21	1–120	98
Elphidium articulatum	46	77	1–350	96
Elphidium discoidale	46	113	1–800	92
Elphidium mexicanum	37	76	1–350	90
Elphidium subarcticum	36	48	5–255	93
Guttulina australis	56	143	10–800	94
Neoconorbina terquemi	51	144	5–800	97
Nonionella atlantica	37	101	1–800	95
Peneroplis proteus	29	58	5–350	97
Quinqueloculina agglutinans	22	24	1–120	95
Quinqueloculina akneriana	42	131	1–800	97
Quinqueloculina compta	50	126	1–800	93
Quinqueloculina poeyana	24	26	1–120	96
Textularia candeiana	82	196	10–915	86
Textularia conica	75	156	10–800	85
Trochammina lobata	34	47	1–350	96
Trochammina ochracea	39	33	5–125	94

TABLE 2.—Species with Upper Depth Limit at About 100 m: North American Atlantic Continental Margin

Species	Mean Depth	Standard Deviation	Observed Range	Percent Occurrence More than 100 m
Bulimina ovata	2087	1104	1–4205	98
Bulimina pyrula	1645	1059	45–4775	97
Eponides umbonatus	1234	1443	20–3500	68
Fursenkoina compressa	409	616	20–2080	67
Pullenia quinqueloba	1006	972	25–3565	78
Uvigerina peregrina	1379	996	10–4205	87

bathymetric indicators for one particular area may not be for other areas.

Just as depth habitats of species may change with geographic location they may also change with time (e.g., Streeter, 1973; Schnitker, 1974; Lohmann, 1978; Blake and Douglas, 1980). It seems that the paleoecologic tool of isobathyal species (Bandy and Echols, 1964; Bandy and Chierici, 1966) may have been over-used in estimating paleobathymetry. Present-day distributions of benthic foraminifera are related to water masses (e.g., Phleger 1960; Streeter, 1973; Schnitker, 1974; Lohmann, 1978; Buzas and Culver, 1980; Culver and Buzas, 1981b; Bock, 1982) and this relationship, if true today, certainly was true also in the past.

CONCLUSIONS

Modern benthic foraminifera invariably show a zonation with depth across continental margins. According to Phleger (1960), who summarized numerous detailed studies based on species abundance and occurrence, the most marked boundary in the northern Gulf of Mexico and in the western North Atlantic south of Cape Cod lies at about 100 m, that is, in the vicinity of the shelfbreak.

Using all published species occurrences in a

TABLE 3.—SPECIES WITH LOWER DEPTH LIMITS AT ABOUT 100 M: GULF OF MEXICO

Species	Mean Depth	Standard Deviation	Observed Range	Percent Occurrence Less than 100 m
Ammonia pauciloculata	61	112	1–550	92
Ammonia rolshauseni	21	18	1–75	100
Elphidium fimbriatulum	30	15	10–55	100
Elphidium galvestonense	16	24	1–150	98
Elphidium mexicanum	25	38	1–205	96
Haynesina germanica	20	29	1–205	98
Nodobaculariella atlantica	33	22	10–20	100
Nouria polymorphinoides	30	19	1–75	100
Peneroplis proteus	32	22	5–80	100
Quinqueloculina agglutinans	285	971	5–3660	93
Quinqueloculina bosciana	38	47	5–230	93
Quinqueloculina subpoeyana	43	37	5–145	93
Sorites orbitolitoides	38	29	5–145	96

TABLE 4.—SPECIES WITH UPPER DEPTH LIMITS AT ABOUT 100 M: GULF OF MEXICO

Species	Mean Depth	Standard Deviation	Observed Range	Percent Occurence More than 100 m
Bulimina aculeata	1182	770	55–3530	97
Cibicides kullenbergi	2090	856	55–3430	98
Cribrostomoides subglobosum	1403	955	40–3475	97
Dentalina communis	1050	817	25–3405	89
Fursenkoina mexicana	903	1027	1–3530	87
Gaudryina atlantica	368	200	95–940	96
Karreriella bradyi	971	746	25–3430	98
Marginulina subaculeata glabrata	352	186	25–730	89
Nodosaria comatula	419	600	25–3345	87
Oridorsalis tener stellatus	741	816	150–3430	100
Reophax hispidulus	1281	1294	15–3345	78
Robertinoides bradyi	1492	1048	25–3345	91
Sigmoilina tenuis	1318	1272	10–3345	82
Textularia mexicana	408	494	25–2485	86
Trochammina tasmanica	1417	884	30–3345	92
Uvigerina auberiana	1427	1195	150–3405	100
Valvulineria laevigata	1039	859	40–3530	96

large, unrefined data set, this major boundary at, or about, the shelfbreak can be recognized in terms of individual species ranges and of foraminiferal provinces.

Although foraminiferal provinces on the American continental margin south of Cape Cod and in the northern Gulf of Mexico overlap in the area of the shelfbreak, the seaward termination of inner shelf provinces is at, or very close to, the shelfbreak. The Gulf of Mexico Inner Shelf Province in the northern Gulf and the Atlantic margin Northern Shelf Province both terminate, in general, at or near 100 m depth in the vicinity of the shelfbreak.

The depth distribution of commonly recorded species (based on all published presence-absence data) indicates that the shelfbreak is also an important boundary for individual species of benthic foraminifera. Most of the sixty selected species have upper and lower limits that overlap the 100 m contour. However, percent of occurrence at less than or more than 100 m shows that species whose ranges cross this contour are found shallower or deeper more than 90% of the time. This indicates that transport downslope does not disrupt foraminiferal distribution patterns any more than transport inshore. Hence, in paleodepth estimate studies, for these sixty species at least, upper depth limit cannot be given more weight than lower depth limit.

The selected species utilized in recognizing the shelfbreak on the North American Atlantic continental margin are generally different from those in the northern Gulf of Mexico. Species that are restricted to a particular depth range in one area may have a different range elsewhere. Recognition of provincial differences, therefore, is very important in paleoenvironmental analyses.

This study shows that large-scale patterns of modern benthic foraminifera can be recognized using unrefined data. The present study, for example, utilized species occurrence and ignored possible sources of error such as live or dead, Pleistocene or Holocene, in place or transported. This helps to explain why benthic foraminifera are so useful in paleoenvironmental and paleobathymetric reconstructions where detailed analyses of all sources of error are not possible.

ACKNOWLEDGMENTS

We thank Laurel Smith for lab assistance and L. B. Isham for drafting. This paper benefited greatly from the reviews of S. W. Petters, R. L. Ellison, G. P. Lohmann and D. Schnitker. The research was supported by two Smithsonian Scholarly Studies awards.

REFERENCES

BANDY, O. L., 1956, Ecology of foraminifera in northeastern Gulf of Mexico: U.S. Geol. Survey, Professional Paper 254F, p. 123–141.

———, AND CHIERICI, M., 1966, Depth-temperature evaluation of selected California and Mediterranean bathyal foraminifera: Mar. Geology, 4, p. 259–271.

———, AND ECHOLS, R., 1964, Antarctic foraminiferal zonation: Antarctic Res. Ser., Am. Geoph. Union, v. 1, p. 73–91.

BLAKE, G. H., AND DOUGLAS, R. G., 1980, Pleistocene occurrence of *Melonis pompilioides* in the California Borderland and its implications for foraminiferal paleoecology: Cushman Found. Foramin. Res., Spec. Publ. 19, p. 59–67.

BOCK, W. D., 1982, Coexistence of deep and shallow water foraminiferal faunas off Panama City, Florida: Geol. Soc. America Bull., v. 93, p. 246–251.

BOLTOVSKOY, E., AND LENA, H., 1969, Seasonal occurrences, standing crop and production in benthic foraminifera of Puerto Desado: Cushman Found. Foramin. Res., Contribs., v. 20, p. 87–95.

———, AND WRIGHT, R., 1976, Recent Foraminifera: Dr. W. Junk b.v., The Hague, 515 p.

BUZAS, M. A., 1965, The distribution and abundance of foraminifera in Long Island Sound: Smithsonian Misc. Coll., v. 149, p. 1–89.

———, 1967, An application of canonical analysis as a method for comparing faunal areas: Jour. Anim. Ecol., v. 36, p. 563–577.

———, 1969, Foraminiferal species densities and environmental variables in an estuary: Limnol. Oceanogr., v. 14, p. 411–422.

———, 1972, Biofacies analysis of presence or absence data through canonical variate analysis: Jour. Paleontology, v. 46, p. 55–57.

———, 1974, Review: Jour. Foramin. Res., v. 4, p. 224.

———, AND CULVER, S. J., 1980, Foraminifera: Distribution of provinces in the western North Atlantic: Science, v. 209, p. 687–689.

———, SMITH R. K., AND BEEM, K. A., 1977, Ecology and systematics of foraminifera in two *Thalassia* habitats, Jamaica, West Indies: Smithsonian Contribs. Paleobiology, No. 31, 139 p.

CULVER, S. J., 1980a, Bibliography of North American recent benthic foraminifera: Jour. Foramin. Res., v. 10, p. 286–302.

———, 1980b, Differential two-way sediment transport in the Bristol Channel and Severn Estuary, U.K.: Mar. Geology, v. 34, p. M39–M43.

———, AND BUZAS, M. A., 1980, Distribution of recent benthic foraminifera off the North American Atlantic coast: Smithsonian Contribs. Marine Sci., No. 6, p. 1–512.

———, AND ———, 1981a, Distribution of recent benthic foraminifera in the Gulf of Mexico: Smithsonian Contribs. Marine Sci., No. 8, p. 1–898.

———, AND ———, 1981b, Foraminifera: distribution of provinces in the Gulf of Mexico: Nature, v. 290, p. 328–329.

———, AND ———, 1981c, Recent benthic foraminiferal provinces on the Atlantic continental margin of North America: Jour. Foramin. Res., v. 11, p. 217–240.

———, AND ———, 1982a, Recent benthic foraminiferal provinces between Newfoundland and Yucatan: Geol. Soc. America, Bull., v. 93, p. 269–277.

———, AND ———, 1982b, Distribution of recent benthic foraminifera in the Caribbean region: Smithsonian Contribs. Marine Sci., No. 14, p. 1–382.

———, AND ———, 1983, Recent benthic foraminiferal provinces in the Gulf of Mexico: Jour. Foramin. Res., v. 13, p. 21–31.

DOUGLAS R. G., 1979, Benthic foraminiferal ecology and paleoecology: Soc. Econ. Paleontologists Mineralogists Short Course No. 6, Houston, Texas, p. 21–53.

ELMORE, R. D., PILKEY, O. H., CLEARY, W. J., AND CURRAN, H. A., 1979, Black shell turbidite, Hatteras Abyssal Plain, western Atlantic Ocean: Geol. Soc. America Bull., v. 90, p. 1165–1176.

EMBLEY, R. W., 1980, The role of mass transport in the distribution and character of deep-ocean sediments with special reference to the North Atlantic: Mar. Geology, v. 38, p. 23–50.

GIBSON, T. G., 1968, Stratigraphy and paleoenvironment of the phosphatic Miocene strata of North Carolina: reply: Geol. Soc. America Bull., v. 79, p. 1437–1448.

GILL, D., AND TIPPER, J. C., 1978, The adequacy of non-metric data in geology: tests using a divisive-omnithetic clustering technique: Jour. Geology, v. 86, p. 241–259.

LOHMANN, G. P., 1978, Abyssal benthonic foraminifera as hydrographic indicators in the western South Atlantic Ocean: Jour. Foramin. Res., v. 8, p. 6–34.

LOWMAN, S. W., 1949, Sedimentary facies in Gulf coast: Am. Assoc. Petroleum Geologists Bull., v. 33, p. 1939–1997.

MELLO, J. F., AND BUZAS, M. A., 1968, An application of cluster analysis as a method of determining biofacies: Jour. Paleontology, v. 42, p. 747–758.

MURRAY, J. W., 1969, Recent foraminifera from the Atlantic continental shelf of the United States: Micropaleontology, v. 15, p. 401–419.

———, 1973, Distribution and Ecology of Living Benthic Foraminiferids: New York, Crane, Russak and Co., 274 p.

———, AND HAWKINS, A. B., 1976, Sediment transport in the Severn Estuary during the past 8,000–9,000 years: Jour. Geol. Soc., v. 132, p. 385–398.

PARKER, F. L., 1948, Foraminifera of the continental shelf from the Gulf of Maine to Maryland: Harvard Mus. Comp. Zool. Bull., v. 100, p. 213–241.

———, 1954, Distribution of the foraminifera in the northeastern Gulf of Mexico: Harvard Mus. Comp. Zool. Bull., v. 111, p. 451–588.

———, AND ATHEARN, W. D., 1959, Ecology of marsh foraminifera in Poponesset Bay, Massachusetts: Jour. Paleontology, v. 33, p. 333–343.

PHLEGER, F. B., 1951a, Ecology of Foraminifera, northwest Gulf of Mexico, Part 1, Foraminifera distribution: Geol. Soc. America Memoir 46, p. 1–88.

———, 1951b, Displaced foraminifera faunas, in Turbidity Currents and the Transportation of Coarse Sediments to Deep Water: A Symposium: Soc. Econ. Paleontologists Mineralogists Spec. Pub. 2, p. 66–75.

———, 1952, Foraminifera ecology off Portsmouth, New Hampshire: Harvard Mus. Comp. Zool. Bull., v. 106, p. 318–390.

———, 1956, Significance of living foraminiferal populations along the central Texas coast: Cushman Found. Foramin. Res., Contribs., v. 7, p. 106–151.

———, 1960, Ecology and Distribution of Recent Foraminifera: Baltimore, The Johns Hopkins Press, 297 p.

———, 1965, Depth patterns of benthonic foraminifera in the eastern Pacific, in Sears, M., ed., Progress in Oceanography, Volume 3: London, Pergamon Press, p. 273–287.

ROBB, J. M., HAMPSON, J. C., JR., AND TWICHELL, D. C., 1981, Geomorphology and sediment stability of a segment of the U.S. continental slope off New Jersey: Science, v. 211, p. 935–937.

SCHAFER, C. T., AND COLE, F. E., 1982, Living benthic foraminifera distributions on the continental slope and rise east of Newfoundland, Canada: Geol. Soc. America Bull., v. 93, p. 207–217.

SCHNITKER, D., 1974, West Atlantic abyssal circulation during the past 120,000 years: Nature, v. 248, p. 385–387.

SCOTT, D. B., AND MEDIOLI, F. S., 1980, Living vs. total foraminiferal populations: their relative usefulness in paleoecology: Jour. Paleontology, v. 54, p. 814–831.

SEN GUPTA, B. K., AND STRICKERT, D. P., 1982, Living benthic foraminifera of the Florida-Hatteras Slope: distribution trends and anomalies: Geol. Soc. America Bull., v. 93, p. 218–224.

STREETER, S. S., 1973, Bottom water and benthonic foraminifera in the North Atlantic: glacial-interglacial contrasts: Quaternary Res., v. 3, p. 131–141.

———, AND LAVERY, S. A., 1982, Holocene and latest glacial benthic foraminifera from the slope and rise off eastern North America: Geol. Soc. America Bull., v. 93, p. 190–199.

WALTON, W. R., 1952, Techniques for recognition of living foraminifera: Cushman Found. Foramin. Res., Contribs., v. 3, p. 56–60.

———, 1964, Recent foraminiferal ecology and paleoecology, in Imbrie, J., and Newell, N., eds., Approaches to Paleoecology: New York, John Wiley and Sons, Inc., p. 151–237.

WEFER, V. G., 1976, Umwelt, produktion and sedimentation benthischer foraminiferen in der westlichen Ostsee: Sonderforschungsbereich, 95, Kiel Univ., 14, 103 p.

DIATOMS IN SEDIMENTS AS INDICATORS OF THE SHELF-SLOPE BREAK

CONSTANCE A. SANCETTA
Lamont-Doherty Geological Observatory of Columbia University, Palisades, New York 10964

ABSTRACT

Diatoms, a group of unicellular algae, are limited to the upper 100 m of the water column, due to their light requirement. Benthic diatoms, consequently, are restricted to the continental shelf, and primarily the inner shelf, coasts, and estuaries. Planktonic diatoms occur on the shelf (neritic) and in deeper waters (pelagic)—certain genera are restricted to one region while others occur in both. Salinity appears to exercise an important control on these distributions, so that a salinity front frequently produces a sharp boundary between populations. The shelf-slope break is usually associated with a sharp salinity gradient, where low-salinity shelf waters encounter a high-salinity oceanic current. The result is that benthic and low-salinity planktonic diatoms characterize shelf sediments, while higher-salinity planktonics and an absence of benthics characterize the slope and ocean basin sediments.

Two techniques for identifying the shelf-slope break are proposed, and examples given from the Bering Sea. The first involves use of the ratio of % benthic specimens to % pelagic-planktonic specimens in a series of sediment samples. The correlation of this ratio with depth is r = .85. A second technique uses quantitative population counts along a shelf-to-basin transect. In this case the shelf-slope break and its associated transition zone are seen to exert an ecologic control on species abundance. Other work in the literature, while not directly addressed to this question, suggests that similar results would be obtained in the Miocene diatomaceous sections of the U.S. East Coast and Japan.

INTRODUCTION

Shelf-Slope Break: Generalities

Physically, the shelf-slope break is a sharp change in gradient of the seafloor, above which the continental shelf has a relatively gentle incline, and below which the continental slope has a steeper gradient, in some cases up to 10°. Essentially, the break marks an inflection point in depth gradient. This change alone affects certain properties of the water column (such as the penetration of light), which in turn affect organisms in the water or on the substrate.

The shelf-slope break is also typically associated with a generally marked gradient in the water column. Commonly it is the region where Shelf Water meets Slope Water. Shelf Water is of lower salinity than Slope Water, due to coastal run-off, and often cooler as a result of winter mixing (e.g. Wright, 1976; Coachman and Charnell, 1979). The resultant density differences produce a shelf-slope front, usually of a sigmoid form. The denser Slope Water intrudes upon the shelf at depth, while the less saline Shelf Water over-rides the Slope Water at the surface. The width of such a frontal zone is variable, depending on the contrast of the two water types as well as upon seasonal variation. The physical properties of these two water types also exert an influence upon the organisms living in Shelf Waters (neritic) or in Slope Waters (pelagic), as certain species are adapted ecologically to lower or higher salinities.

Diatoms

Diatoms are a group of unicellular algae, which secrete an exoskeleton (frustule) of opal, or hydrated silica, $SiO_2 \cdot nH_2O$. The frustule is composed of two valves, one fitting into the other like the lid and bottom of a box, with various ornamentation such as pores, ribs, and nodules. While the frustules of many species are rather weakly-silicified, and consequently dissolve in the water column after death, a small proportion are preserved in the sediments and can be found as fossil assemblages as far back as the Jurassic period.

Most diatoms are obligate autotrophs, which means that they rely upon light energy for photosynthesis and growth. For this reason, they are restricted to the upper 100 to 200 m of the water column—below this depth the light level is too low to support photosynthesis. Benthic (bottom-dwelling) diatoms are thus directly affected by water depth, and are limited to the continental shelves. Planktonic species may be adapted to pelagic (open-ocean) conditions, or to the continental shelves (neritic), but do not show the absolute depth-restriction of the benthics. However, some planktonic groups form spores which spend some time on the substrate before being returned to surface waters by vertical mixing. For these groups, again, the depth of water has a direct effect upon their production.

Both planktonic and benthic diatoms show preferences for certain ranges of temperature, salinity, and nutrient concentrations, although the limits are rather hard to quantify, since in nature the proper-

Copyright © 1983, The Society of
Economic Paleontologists and Mineralogists

ties are never completely independent. The tolerances of certain species have been studied (e.g., Smayda, 1958; Paasche, 1975), but these usually prove to be extremely broad. Nevertheless, diatomists have long recognized that certain species or groups show a strong preference for neritic or pelagic waters (Hustedt, 1930, 1959).

DIATOMS IN THE WATER COLUMN

Little work by marine biologists has been directed toward the relationship of diatom species to the shelf-slope break. Almost all studies of plankton populations are based on short-term sampling (a few weeks to months). Timing of blooms, species succession and dominance, among other parameters, vary greatly from year to year, depending on fortuitous position of seed populations and localized hydrographic events. This makes it very difficult to draw any generalized conclusions as to ecologic preferences of species. Furthermore, most work is limited either to the shelf or to the pelagic realm—remarkably few studies range across the shelf and slope.

A final problem is that most persons studying phytoplankton are interested in bulk productivity and are not taxonomists. Reports usually refer to chlorophyll concentrations or carbon uptake, and may list one or two genera which were particularly common in a sample. The result is that the vast mass of literature on phytoplankton reveals very little information of use to the sedimentologist.

A few biological studies do suggest, if indirectly, that species in the water column occur differentially with respect to the shelf-slope break. Semina (1955), for instance, studied hundreds of plankton samples from the Bering Sea and defined a biologic neritic zone, equivalent to the continental shelf, in which her "neritic" species were >50% of the flora, but she did not draw any conclusions directly correlating this flora with the distinction between Shelf and Slope Waters. Hulbert and Rodman (1963), studying samples off the New York Bight, found that some species occurred only in low-salinity waters

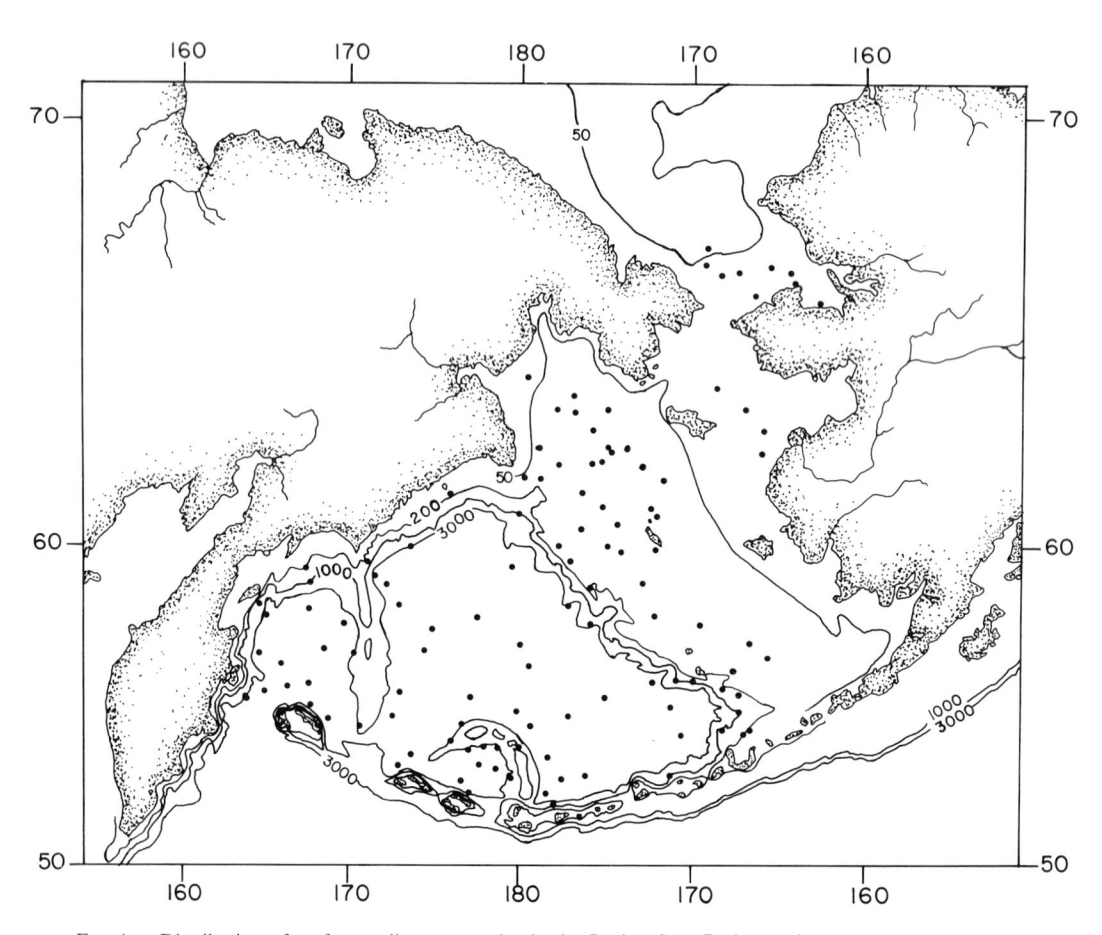

FIG. 1.—Distribution of surface sediment samples in the Bering Sea. Bathymetric contours are in meters.

(<34‰), which were always shallower than the shelf-slope break at 200 m, while other species occurred only in the high-salinity Slope Waters (>34.5‰). Hulbert (1967) and Marshall (1976, 1978) have both noted that there is a sharp change in the ratio of diatoms to calcareous nannoplankton (coccolithophorids) at the shelf-slope break off the southeastern United States, with diatoms far more common in Shelf Waters. Their tables of species occurrence imply that some species of diatoms are restricted to either the Shelf or the Slope waters, but the studies are not quantitative.

DIATOMS IN SEDIMENTS

Ratios of Benthic and Pelagic Diatoms

The most straightforward approach to using diatoms as bathymetric indicators would be to find a consistent correlation between water depth and some distinctive, easily-quantified aspect of the sediment assemblages. The distinction between benthic and planktonic forms would seem a good candidate. This distinction extends to the generic, and even to the family, level, which is convenient for at least two reasons: (1) genera are much longer-ranging in time than species, so that evolutionary changes will not affect the validity of the method; (2) persons who are not specialists can easily learn to identify genera, so that this method is potentially of wide application.

Even on a descriptive, non-quantitative level, it is possible to distinguish shelf, slope, and basin samples by their relative content of benthic and planktonic species. Samples from the continental shelf, especially from the inner shelf, will be dominated by benthic groups, possibly with rare representatives of fresh-water taxa. On the inner shelf, members of littoral groups will be more common among the benthics, while further offshore the freshwater and littoral forms will disappear, as more generalized benthics and neritic planktonics become more important. Below the shelf-slope break, on the continental slope and in the basin, the assemblage will be almost completely dominated

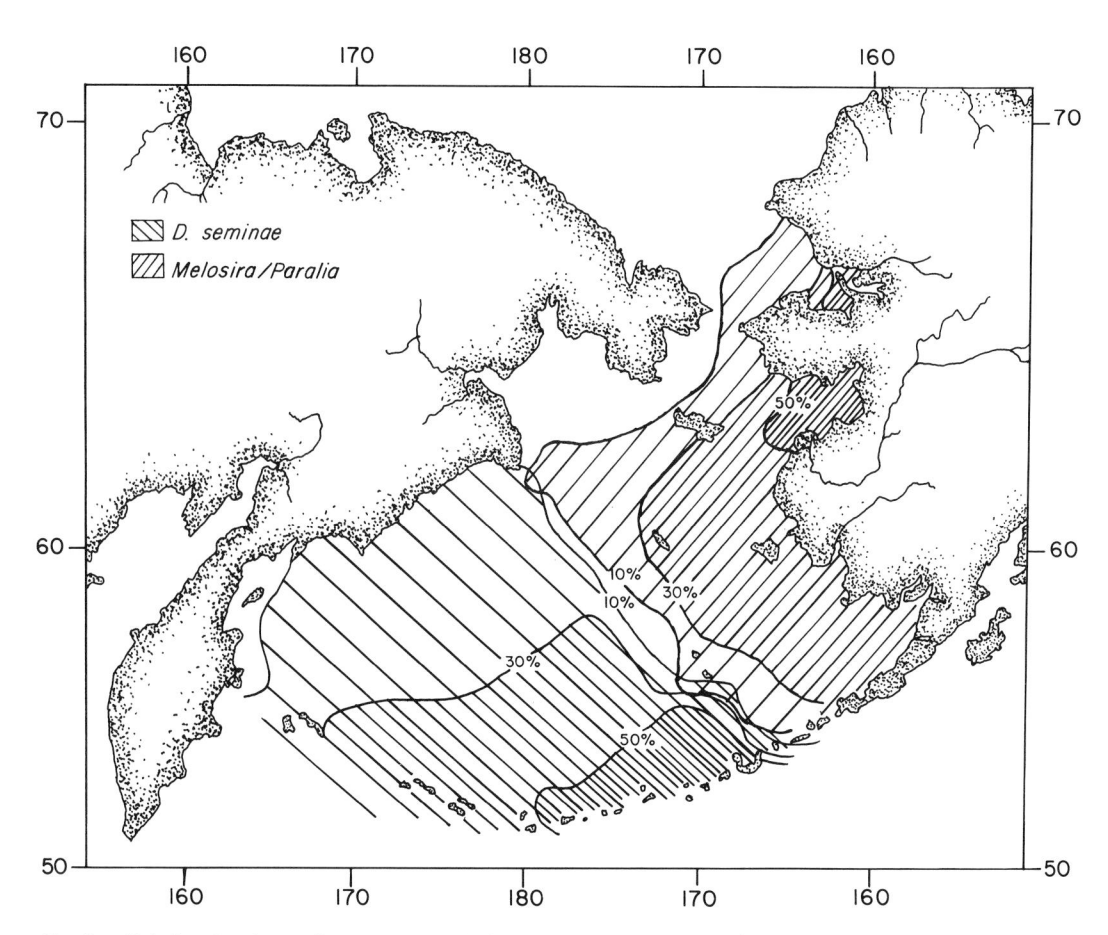

FIG. 2.—Relative abundance of *D. seminae* (pelagic planktonic) and *Melosira/Paralia* (shelf benthics) in sediments.

by fully-marine planktonic forms—benthics will occur only in material displaced from the shelf by currents or sediment flows. Gravitative deposits such as turbidites and slumps can be identified by numerous structural and sedimentologic features, and by an abnormal combination of benthic (shelf) and pelagic diatoms. Several paleontologists working with diatoms have made interpretations as to depth of deposition based on such non-quantitative observations (e.g. Abbott, 1980, on Miocene diatoms of the U.S. East Coast continental shelf; Hanna et al., 1976, on Eocene diatoms of the Falkland Plateau).

If the abundance of the various species has been determined quantitatively, by counting several hundred individuals in a series of samples, it is possible to use some ratio of planktonic-to-benthic individuals and to correlate this ratio with water depth. Unfortunately, a simple ratio using all of the planktonic species will not give good results, because this will include both neritic and pelagic forms, thus blurring the distinction between shelf waters and basin waters. A ratio of pelagic-planktonics to benthics, however, gives quite good results.

In connection with a regional study (Sancetta, 1981b) I counted the total flora in surface sediment samples from the Bering Sea (Fig. 1). The most common pelagic species, *Denticulopsis seminae*, comprises more than 60% of the flora in samples from the basin and decreases sharply in abundance upslope, averaging less than 10% in the outer shelf zone (Fig. 2). The most common benthic species (*Melosira sol* and *Paralia sulcata*) together reach abundances of over 50% in the inner shelf, and decrease to zero below the shelf-slope break (Fig. 2). The ratio:

$$\frac{\%D.\ seminae}{\%M.\ sol\ +\ \%P.\ sulcata}$$

shows a correlation with depth of r = 0.74 (Fig. 3). This correlation is based on all samples from the shelf, some of which are complicated by unusual numbers of freshwater and epontic (ice-dwelling) species. If these samples are eliminated, the correlation rises to r = 0.85 (Fig. 3)—impressively high, considering that this simple ratio ignores the possible effects of current action and reworking, as well as implying that water depth alone is the single controlling variable, when in fact salinity, location of upwelling fronts, and differential feeding by herbivores, among others, must also have an effect.

Shelf-Slope Break as an Ecologic Barrier

As indicated above, the shelf-slope break is associated with a distinct hydrographic front, primarily defined by salinity differences. In reality, this front is not a single sharp line, but a narrow zone where the transition from Shelf Water to Slope Water occurs. While the width of the zone fluctuates seasonally, its average position may be reflected in the sediment assemblages.

The Bering Sea, here again, serves as a good test region for such a study. The eastern shelf is extremely broad (over 500 km) and flat (slope <2 minutes), which provides a wide area over which gradients may be distributed. Kinder and Coachman (1978) identified a persistent shelf-break front, largely defined by salinity gradients separating Basin waters (with no horizontal salinity gradient, and geostrophic flow) and Shelf Waters (with shoreward salinity gradients and tidal flow). Coachman and Charnell (1979) later determined that the transition from Shelf to Basin Water actually occupies a zone about 100 km wide—the outer front of this zone is the shelf-break front, while the inner front (at about the 100 m isobath) marks the boundary between Shelf Water and the transition zone. There are thus, in effect, three zones, separated by two fronts.

A quantitative study was made of the diatoms in surface sediments along a transect in this region (Sancetta, 1981a). The effect of the shelf-break front and the transition zone can be seen in the changes in species abundance (Fig. 4). The pelagic

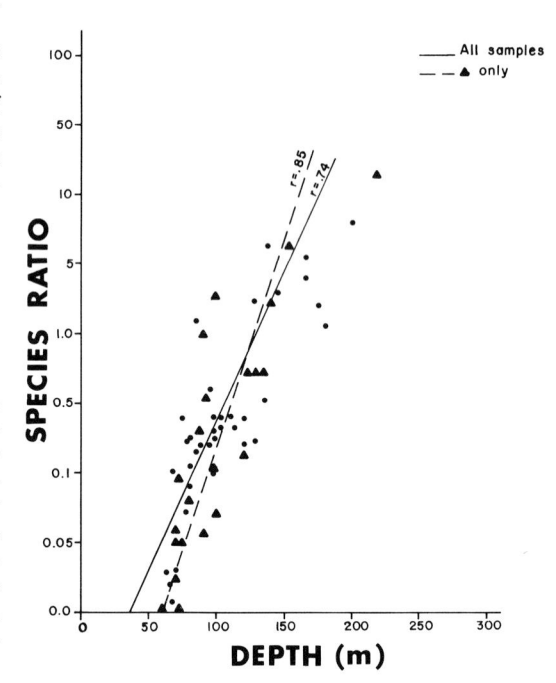

FIG. 3.—Correlation of the *D. seminae*/*Melosira* + *Paralia* ratio with water depth. Triangles represent samples which do not include freshwater or epontic (ice-dwelling) species. Circles represent samples which do contain such species; these latter result in a somewhat lower correlation.

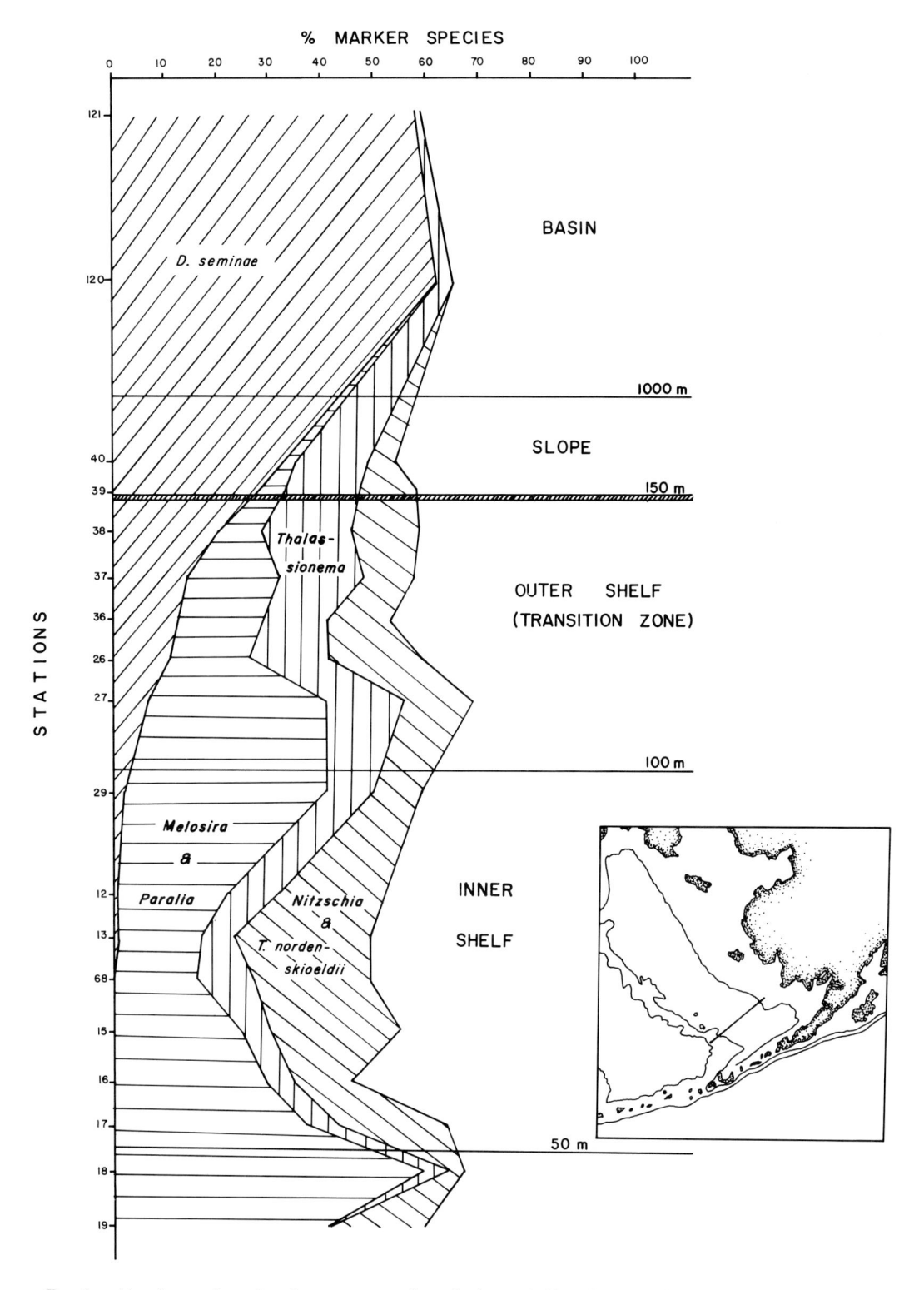

FIG. 4.—Abundance of species along a transect from the inner shelf to the base of the continental slope (location shown in inset). The shelf-slope break is about 150 m.

species *D. seminae*, mentioned above, shows an exponential decrease from the basin through the transition zone, with the sharpest decrease occurring across the shelf-slope break. For this species, the inner front of the transition zone appears to be the limit of penetration on the shelf. The benthic species (*M. sol* and *P. sulcata*), on the other hand, show a very sharp change near the break and its associated front. A second group also shows a distinct decline below the shelfbreak front. These species (*Nitzschia grunowii*, *N. cylindra*, and *Thalassiosira nordenskioeldii*) are closely associated with early spring productivity blooms following initial melting of sea-ice (Sancetta, 1981b)—the sea-ice itself is restricted to the continental shelf and does not occur to any degree in waters over the continental slope. A fourth species (*Thalassionema nitzschioides*), which characterizes the transition zone, shows a decrease slightly seaward of the shelf-slope break. This species appears to prefer the transition zone, its abundance decreasing both landward and seaward of the zone. Numerous other species, which occur at low abundances (2–5%) also show the effect of the shelfbreak front, being consistently present to one side and consistently absent to the other (e.g. *Thalassiosira trifulta* and *Actinocyclus curvatulus* occur in basin sediments; *Delphineis* species and *Thalassiosira decipiens* occur only on the shelf).

Thus, the shelf-slope break, with its associated front, does affect the distribution of species between the shelf and basin, but the effect is generally gradual, due to the transition zone. On a narrower shelf, where the width of the zone is narrower, the effect of the break may be enhanced.

The overall diversity of diatoms, as represented by the Shannon index, also shows a marked relationship to the shelf-slope break (Fig. 5). Diversity is highest near the break on the outer continental shelf, and on the inner shelf. The high values near the break reflect the mixing of Basin and Shelf waters, while the inner shelf is the region of maximal spring productivity (Iverson et al., 1979). The slope and basin sediments, in which *D. seminae* dominates the flora, show much lower values.

FOSSIL DIATOMS AND DEFINITION OF
ANCIENT SHELF-SLOPE BREAKS

Previous work on fossil diatoms usually has concentrated on biostratigraphy, with relatively little interpretation of the depositional environment. Commonly, workers remark that "the environment was probably one of shallow water, as benthic diatoms are common." In principle, however, it should be possible to locate the shelf-slope break, provided that a transect of samples could be collected, and that the ecologic significance of modern counterparts of the flora is known. One possible application of this method would be Miocene diatomaceous sediments of the U.S. East Coast. Abbott (1978, 1980) has derived a biostratigraphic zonation for this region, and the sites he has examined range from 100 km onshore to 400 km offshore on the upper continental slope (modern location). A quantitative analysis of samples from all sites within a biostratigraphic zone might reveal an onshore-offshore gradient, either in the pelagic-to-benthic ratio or in certain indicator species. A sharp change in either gradient might represent the shelf-slope break.

Koizumi (1968) examined Pliocene and Miocene diatoms from sections on the Oga Peninsula of Japan, graphically showing a planktonic-to-benthic ratio and the relative abundance of marine, brack-

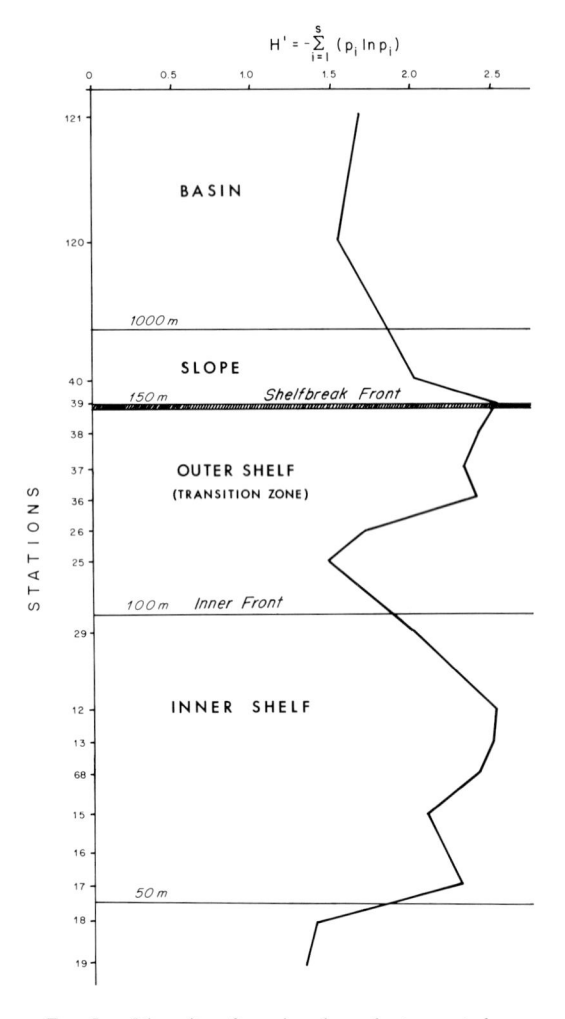

FIG. 5.—Diversity of species along the transect shown in Fig. 4. H' is the Shannon diversity index, where S = the number of species and p_i is the proportion of the *i*th species. $H' = -\sum_{i=1}^{S} (p_i \ln p_i)$

ish, and freshwater diatoms (as species number and as number of specimens). Most such sections are vertical (in time) rather than horizontal (in space), so that it is not possible to examine a "paleo-transect" which could locate the shelf-slope break. However, species abundances or the ratio could at least give some indication as to the distance from the break of that particular locality. It is interesting that Koizumi did find an upward increase in freshwater and brackish diatoms, along with a decrease in the planktonic-to-benthic ratio. Such a change might represent uplift and shoaling of the basin—the appearance of benthics and brackish species might represent the time when the sites were raised above the break. However, Koizumi points out that these assemblages could also be the result of turbidite transport. Further work on the sections is needed to determine the correct interpretation.

These examples, both modern and fossil studies, strongly suggest that diatoms can be used to determine the direction, if not the actual position, of the shelf-slope break, and also the paleobathymetry of the continental shelf in general. Further work should concentrate on collecting samples along cross-shelf transects to test these conclusions.

SUMMARY

1. Studies available to date indicate that some diatoms in the water column show preferences for shelf-only or slope-only waters, perhaps due to salinity sensitivity.

2. It is possible, on a non-quantitative level, to distinguish shelf sediments from slope sediments by their diatom content: shelf sediments have common benthic forms while slope sediments contain mostly pelagic species.

3. In a test case from modern Bering Sea sediments, the pelagic-to-benthic abundance ratio shows a strong positive correlation with depth.

4. Several diatom species in a shelf-to-basin transect of the Bering Sea sediments show sharp changes in abundance across the shelf-slope break, probably in response to a strong salinity front.

5. It is suggested that techniques outlined in this study, applied to ancient sediments, could indicate the relative position of deposits with respect to the shelf-slope break.

ACKNOWLEDGMENTS

I thank W. Reeburg, who provided some samples from the Bering Sea. Other samples were contributed by the Lamont-Doherty and the University of Washington Core Repositories. J. Marra and H. Marshall contributed suggestions concerning biological studies. M. Braun was responsible for photography. The manuscript was reviewed by J. Marra and L. Burckle. Research was supported by NSF Grant OCE79-06368. This is LDGO Contribution No. 3445.

REFERENCES

ABBOTT, W. H., 1978, Correlation and zonation of Miocene strata along the Atlantic margin of North America using diatoms and silicoflagellates: Marine Micropaleontology, v. 3, p. 15–34.
———, 1980, Diatoms and stratigraphically significant silicoflagellates from the Atlantic Margin Coring Project and other Atlantic margin sites: Micropaleontology, v. 26, p. 49–80.
COACHMAN, L. K., AND CHARNELL, R. L., 1979, On lateral water mass interaction—A case study, Bristol Bay, Alaska: Jour. Phys. Oceanogr., v. 9, p. 278–297.
HANNA, G. D., HENDEY, N. I., AND BRIGGER, A. L., 1976, Some Eocene diatoms from South Atlantic cores: California Acad. Sci. Occasional Papers, v. 126, p. 1–26.
HULBERT, E. M., 1967, Some notes on the phytoplankton off the southeastern coast of the United States: Bull. Mar. Sci. v. 17, p. 330–337.
———, AND RODMAN, J., 1963, Distribution of phytoplankton species with respect to salinity between the coast of southern New England and Bermuda: Limnol. and Oceanogr., v. 8, p. 263–269.
HUSTEDT, F., 1930, Die Kieselalgen Deutschlands, Österreichs und der Schweiz, in Rabenhorst, L., Kryptogamen-Flora von Deutschland, Österreich und der Schweiz: Leipzig, Akademische Verlagsgesellschaft m.b.H., v. VII, pt. 1, 920 p.
———, 1959, Die Kieselalgen Duetschlands, Österrichs und der Schweiz, in Rabenhorst, L., Kryptogamen-Flora von Deutschland, Österreich und der Schweiz: Leipzig, Akademische Verlagsgesellschaft m.b.H., v. VII, pt. 2, 845 p.
IVERSON, R. L., COACHMAN, L. K., COONEY, R. T., ENGLISH, T. S., GOERING, J. J., HUNT, G. L., MACAULEY, M. C., MCROY, C. P., REEBURG, W. S., AND WHITLEDGE, T. E., 1979, Ecological significance of fronts in the southeastern Bering Sea, in Livingston, R. L., ed., Ecological Processes in Coastal and Marine Systems: New York, Plenum, p. 437–466.
KINDER, T. H., AND COACHMAN, L. K., 1978, The front overlaying the continental slope in the eastern Bering Sea: Jour. Geophys. Res., v. 83, p. 4551–4559.
KOIZUMI, I., 1968, Tertiary diatom flora of Oga Peninsula, Akita Prefecture, northeast Japan: Scientific Repts. Tohoku Univ., Sendai, Ser. 2, v. 40, p. 171–240.
MARSHALL, H. G., 1976, Phytoplankton distribution along the eastern coast of the U.S.A. I. Phytoplankton composition: Mar. Biology, v. 38, p. 81–89.
———, 1978, Phytoplankton distribution along the eastern coast of the U.S.A. II. Seasonal assemblages north of

Cape Hatteras, North Carolina: Mar. Biology, v. 45, p. 203–208.

PAASCHE, E., 1975, The influence of salinity on the growth of some plankton diatoms from brackish water: Norwegian Jour. Botany, v. 22, p. 209–215.

SANCETTA, C. A., 1981, Diatoms as hydrographic tracers: Example from Bering Sea sediments: Science, v. 211, p. 279–281.

———, 1981b, Oceanographic and ecologic significance of diatoms in surface sediments of the Bering and Okhotsk Seas: Deep-Sea Res., v. 28A, p. 789–817.

SEMINA, G. I., 1955, On two zonal groupings of phytoplankton in Bering Sea sediments [in Russian]: Akad. Nauk SSSR Dokl, 101 p.

SMAYDA, T. J., 1958, Biogeographical studies of marine phytoplankton: Oikos, v. 9, p. 158–191.

WRIGHT, W. R., 1976, The limits of shelf water south of Cape Cod, 1941–1972: Jour. Mar. Res., v. 34, p. 1–14.

INFAUNAL-SEDIMENT RELATIONSHIPS AT THE SHELF-SLOPE BREAK

NORMAN J. BLAKE AND LARRY J. DOYLE
Department of Marine Science, University of South Florida, St. Petersburg, Florida 33701

ABSTRACT

Infauna changes dramatically across the shelf-slope break, along with the physical and chemical parameters of the sediments and overlying water column. Grain size across the transition first increases slightly, then rapidly changes from sand to mud with concomitant increase in clay mineral and organic matter content. Light penetration decreases and there occurs a damping of seasonal temperature fluctuations. Infaunal assemblages change from those characterized by filter feeding organisms to those dominated by deposit feeders. Of the animals with hard parts likely to be preserved in the fossil record, the molluscan order nuculoida, composed of deposit feeders, is heavily represented seaward of the mudline. Biomass and density of organisms first decrease as grain size gets larger near the shelfedge, then increase as the mudline is crossed, then decrease again in the mud downslope. Winnowing recycles fecal material from the shelf infaunal assemblages back into the water column, this contributes to the generally high productivity of shelf waters. Much of the feces seaward of the mudline is incorporated as part of the sediment, contributing to the relatively high organic content. Deposit feeders downslope of the mudline are the primary source of sediment reworking, while physical winnowing processes are more important at and adjacent to the shelfedge.

In the sedimentary record, a sudden change in fossils from groups dominated by filter feeders to groups dominated by deposit feeders may indicate proximity to the shelf-slope break. Such a diagnostic change is associated with a decrease in fossil content of a sand layer and concurrent increase in grain size, followed by a facies change from sand to mud with rapid increase in fossil content, and finally followed by a decrease in fossil content in mud away from the zone of facies change.

INTRODUCTION

A voluminous body of literature exists on the benthic infauna-sediment relationships of continental shelves. These have been summarized by Thorson (1956, 1957) and Emery and Uchupi (1972) for the western North Atlantic. A smaller, but still substantial, body of literature for continental slopes has been summarized by Rowe and Haedrich (1979). Scant attention has been paid to the boundary and transitional zone between the two major parts of the continental margin, the shelf and slope. In this paper we focus on the infauna and sediment at the shelf-slope boundary using as examples those of the western North Atlantic and Gulf of Mexico margins. In addition to our own data, we have utilized material from Day et al. (1971) and from Rowe et al. (1974), significant studies of continental margins that have not specifically focused on the shelf-slope transition.

SHELF-SLOPE TRANSITIONAL ENVIRONMENT

The distribution and speciation of infauna are controlled by interaction of the fauna with the physical and chemical properties of the sediment such as texture, mineralogy, and organic content and the physical and chemical properties of the overlying water column, specifically temperature, salinity, light, waves and currents, and dissolved and particulate organics. Generally, the number of water column physical parameters which potentially can exert control over the infauna decreases with depth across the transition zone which usually begins on the outer edge of the shelf at between about 75 and 100 m and extends to about 150 to 200 m. In addition, variation in seasonal amplitude of any one water column parameter usually decreases and becomes more predictable (Sanders, 1968, 1969; Slobodkin and Sanders, 1969).

Figure 1 diagrammatically depicts a profile of grain size across a non-deltaic continental margin like that of the eastern United States or eastern Gulf of Mexico. Figure 2 shows the nature of sediments along a profile across the shelf-slope transition in the eastern Gulf of Mexico. Emery (1960), Gorsline (1963), Emery and Uchupi (1972), and Stanley and Wear (1978) have noted that sediments in the immediate vicinity of the shelfbreak are often coarser than those of the adjacent continental shelf (see Fig. 1). Presence of coarse sediments in this position is the result of relatively high energy caused by the winnowing action of major shelfedge currents such as the Loop Current and the Gulf Stream, upwelling, and possibly by breaking internal waves (see Southard and Cacchione, 1972, and chapters in parts V and VI, this volume), combined with a relative increase in the sand-sized planktonic foraminifera and authigenic glauconite components of the sediment. On most margins, seaward of the coarse grain-size spike at the outer aspect of the shelf, there occurs a marked transition from shelf sands to slope muds. This abrupt change in texture and mineralogy, referred to as the "mudline" by Stanley and Wear (1978) and Stanley et al. (1983), is not necessarily coincidental with the bathymetric expression of the

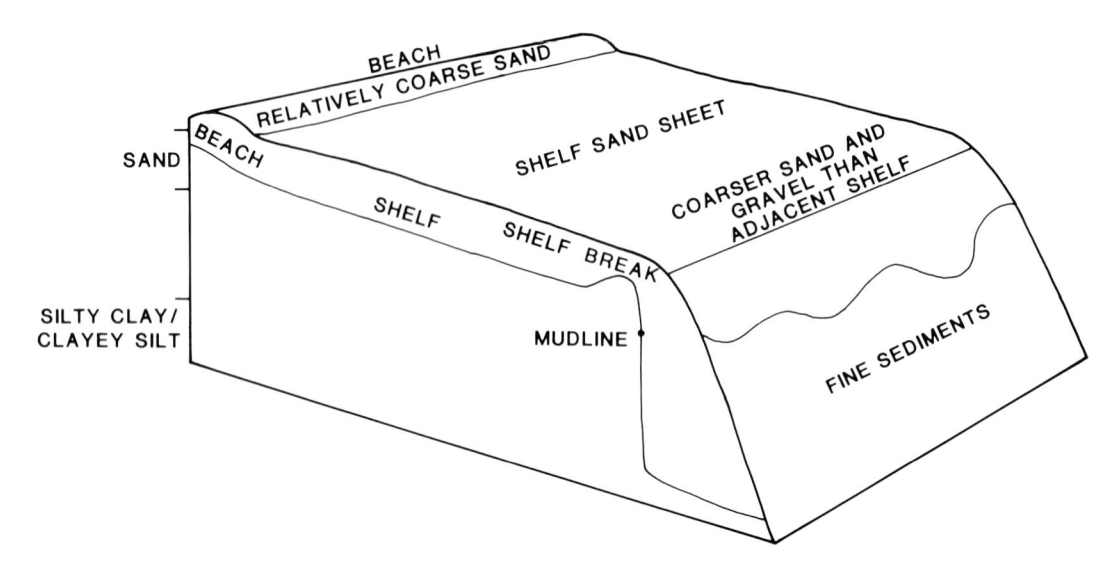

FIG. 1.—Change in sediment grain size across a typical continental shelf and slope.

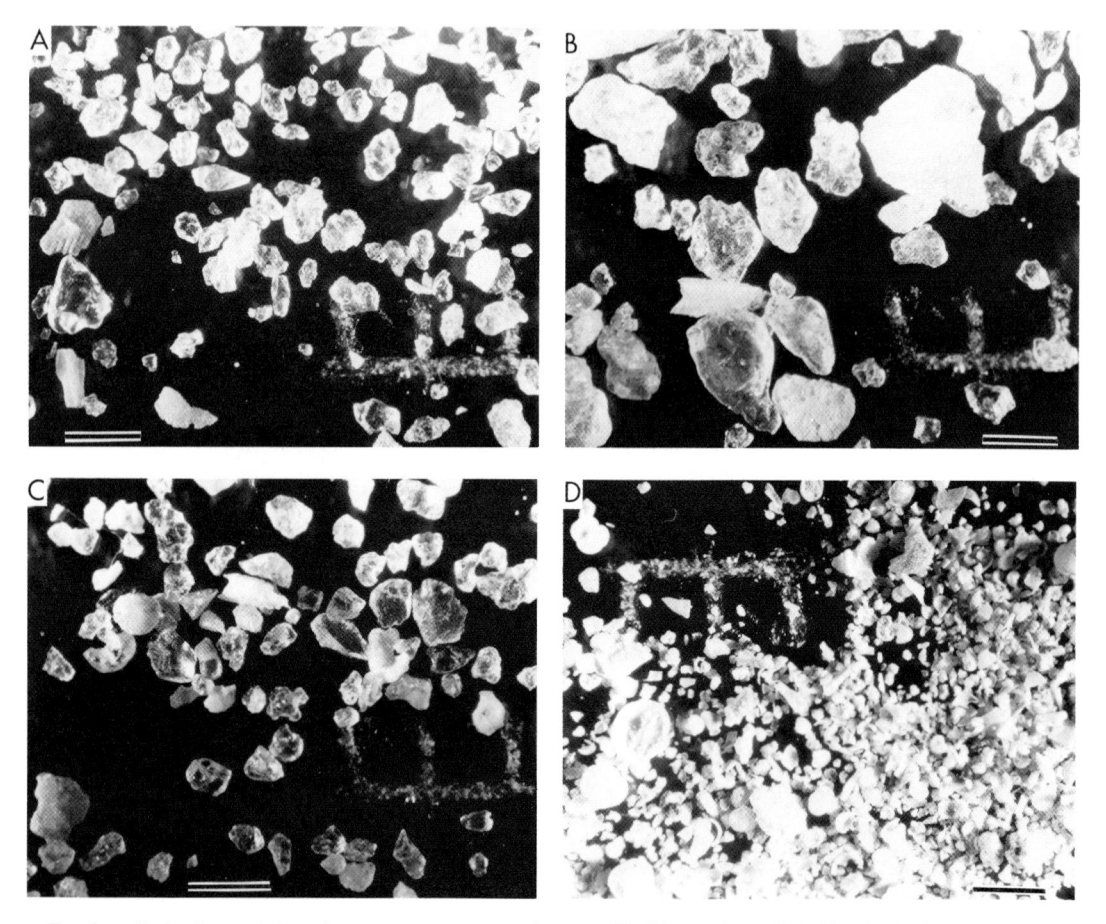

FIG. 2.—Grain size variation along a transect across the west Florida continental shelf and shelf-slope transition. A, 40 m; B, 80 m; C, 95 m; and D, 150 m. Horizontal bar scale = 1 mm.

break. Therefore, the sedimentologic and biologic expressions of the shelf-slope transition may be separate from the physical expression. The mudline itself usually lies downslope from the morphologic break and, as shown diagrammatically in Figure 3, may be quite irregular due to spillover of shelf sands (Doyle et al., 1979; Stanley et al., 1983). On most margins the mudline lies close to the morphologic break, but on those which are swept by major ocean currents, like the southeastern United States (Doyle et al., 1979), the mudline may be displaced several hundred meters downslope or be entirely replaced by a gradual facies change. Above the mudline, infauna are subjected to a relatively high energy winnowing environment; the sector below the mudline occupies a relatively quiet, low energy zone of fine sediment accumulation.

Concomitant with the decrease in grain size across the mudline is a marked change in sediment composition. Total organic carbon ranges from less than 0.5 percent on the Atlantic shelf (Emery and Uchupi, 1972), to about 2 percent on the Atlantic slope (Doyle et al., 1979). Values range from 0.1 percent to 0.4 percent on the carbonate shelf of the eastern Gulf of Mexico and increase less dramatically to above 0.5 percent on the carbonate West Florida slope (Emery and Uchupi, 1972; Doyle and Feldhausen, 1981). On clastic margins like those of the western Atlantic, mineralogy changes from a suite dominated by quartz sand to one rich in clay minerals, with large amounts of mica and planktonic foraminifera in the silt and sand fractions. On the carbonate margin of the eastern Gulf (Fig. 2), carbonate constituents change from sand-sized molluscan fragments on the shelf to planktonic foraminifera in a matrix dominated by coccolith plates and undefined carbonate fragments (Doyle and Sparks, 1980).

Along with the decrease in energy, light penetration decreases with depth across the transition and there is a damping in the seasonal variation of temperature. For example, in water depths of 60 to 70 m off southern New England, temperature shows a 12.5°C seasonal variation, while at a depth of 100 m seasonal variation is reduced to 10°C (Sanders et al., 1965; Sanders, 1969). The continental margin of the eastern Gulf of Mexico shows temperature variations of 8°C at 50 m, 5°C at 100 m and 1° to 2°C at 150 m (Fausak, 1979). On the eastern Gulf of Mexico margin, the permanent thermocline may be superimposed over muted seasonal temperature fluctuations.

SHELF-SLOPE TRANSITION AND INFAUNA

Suspension Feeders versus Deposit Feeders

It is not surprising that there is a major change in the infauna associated with major changes in the substrate and in parameters of the water column across the shelf-slope transition. In general, the infauna changes from one which is dominated by suspension feeders on the shelf side of the shelfbreak to one dominated by deposit feeders on the slope. The molluscan order nuculoida, for instance, which is composed mostly of deposit feeders, often dominates the molluscan fauna seaward of the shelf-

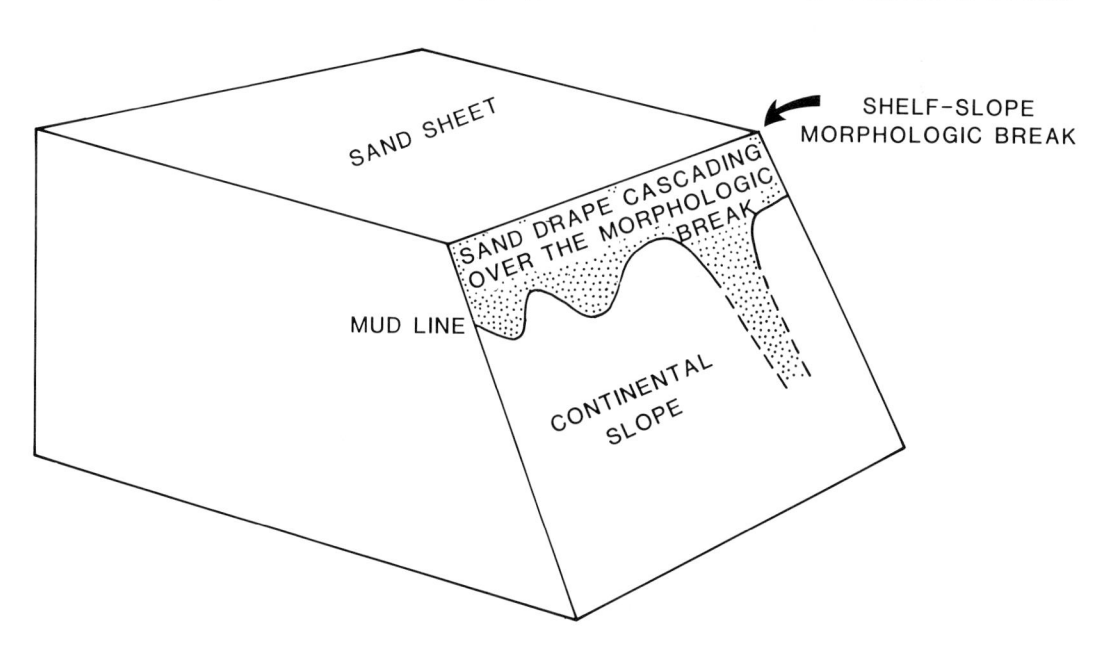

FIG. 3.—Position and typical shape of the mudline along the shelf-slope transition.

edge. In addition to this basic difference in faunal type, other measures used by benthic ecologists to describe infaunal populations or assemblages are also useful in characterizing the shelf-slope transition.

Biomass

Biomass is the weight of animals per unit area of substrate. Representative biomass values of stations across the shelf-slope transition in the northwestern Atlantic and Gulf of Mexico are summarized from the continental margin literature and presented in Table 1. Shelf-slope transition stations of Rowe et al. (1974) in the Gulf average about 1 $g \cdot m^{-2}$. The values obtained for the 14 shelf-slope stations in the eastern Gulf of Mexico off Florida and Alabama average about 3 $g \cdot m^{-2}$ for only polychaetes, molluscs with shells, and crustaceans.

The biomass values from two studies of the northwestern Atlantic (Table 1) vary by an order of magnitude. Those taken from Rowe et al. (1974) give an average biomass of about 4 $g \cdot m^{-2}$ for stations in the shelf-slope transition, while data from Wigley and McIntyre (1964) give an average infaunal biomass of about 63 $g \cdot m^{-2}$ at the shelf-slope transition off Martha's Vineyard.

The biomass of the infauna in the northwestern Atlantic and Gulf of Mexico probably lies somewhere between the values shown in Table 1. Thus, the northwestern Atlantic shelf-to-slope sector has an infaunal biomass of about 4 to 80 $g \cdot m^{-2}$, while the Gulf of Mexico shelf-slope transition has a biomass of less than 1 to 7 $g \cdot m^{-2}$. The biomass of the Gulf of Mexico shelf-slope transition, therefore, may be as much as an order of magnitude less than that of the northwestern Atlantic shelfbreak.

The data presented in Table 1 for the shelf-slope transition reveals several trends for the biomass in the northwestern Atlantic and eastern Gulf of Mexico. The stations discussed by Wigley and McIntyre (1964) and Rowe et al. (1974) in the northwestern Atlantic were positioned at 40°N latitude, while those in the eastern Gulf of Mexico off Florida and Alabama ranged from about 26°N to 30°N latitude. At any particular latitude, biomass tends to be lower at the upper end of the shelf-slope transition, at about 75 m, than it is at 100 m. Biomass usually decreases rapidly below approximately 100 m.

There are limited data to suggest a temperate to tropical decrease in infaunal biomass along a particular isobath of shelf-slope transition. Although not emphasized in their review, Emery and Uchupi (1972) present a map which shows this trend along the western Atlantic and Gulf shelves. Boesch et al. (1977) also observed that the infauna of the south Atlantic shelf are sparser than the infauna on the middle Atlantic shelf. In the eastern Gulf of Mexico, off Florida and Alabama, infaunal biomass of the shelf-slope transition is generally higher for

TABLE 1.—DENSITY AND BIOMASS OF MACROINFAUNA FROM THE WESTERN ATLANTIC AND GULF OF MEXICO SHELF-SLOPE TRANSITIONS

Northwestern Atlantic (Wigley and McIntyre, 1964)

Sta. No.	Depth (m)	Nos. Ind. $\cdot m^{-2}$	Weight (g $\cdot m^{-2}$)
49	69	2145	53.13
50	84	700	78.70
51	99	1920	80.92
52	146	2085	51.46
53	179	1280	53.32
	$\bar{x} = 115$	$\bar{x} = 1626$	$\bar{x} = 63.51$

Northwestern Atlantic (Rowe et al., 1974)

3	77	5490	4.150
4	90	9010	4.725
5	120	4585	3.667
	$\bar{x} = 96$	$\bar{x} = 6362$	$\bar{x} = 4.181$

Middle Atlantic of North Carolina (Day et al., 1971)

7	80	284	—
8	124	494	—
9	160	1018	—
10	205	971	—
	$\bar{x} = 142$	$\bar{x} = 692$	

Eastern Gulf of Mexico (Florida-Alabama)[1]

31	75	1946	7.231
43	80	856	3.390
47	76	581	6.352
26	85	506	4.464
5	99	452	3.153
45	106	756	4.656
58	110	674	4.310
46	122	389	2.562
27	154	174	4.193
57	164	380	2.782
6	171	289	2.820
13	176	133	0.721
36	181	294	0.982
12	189	472	1.922
	$\bar{x} = 128$	$\bar{x} = 754$	$\bar{x} = 3.538$

Eastern Gulf of Mexico-Mississippi Cone (Rowe et al., 1974)

10	190	1547	1.082
11	200	2430	.428
	$\bar{x} = 195$	$\bar{x} = 1988$	$\bar{x} = 0.755$

Western Gulf of Mexico-Texas (Rowe et al., 1974)

| 3 | 90 | 880 | 0.983 |

[1]Polychaetes, molluscs, and crustaceans combined. Polychaete densities and biomass from Vittor (1979) and crustacean densities from Heard (1979). One to nine box cores (20 × 30 cm) were taken at each station which was sampled seasonally 3–7 times.

any particular depth north of Cape San Blas than south of the Cape.

Infaunal Densities

Measurements in numbers of individuals per unit area, infaunal density, are also used to characterize infauna. Like biomass, total density values across the shelf-slope transition of the northwestern Atlantic are 2 to 10 times higher than for the Gulf of Mexico and the middle Atlantic off North Carolina (Table 1). In all three areas, however, faunal densities increase sharply at around 75 m and then decrease somewhere below 100 m. The exact depth of decline seems to vary with the area and the location of the mudline. Off North Carolina, for example, where the mudline is not part of the transition, densities may remain elevated at least down to 205 m (Day et al., 1971).

Most studies which include infaunal biomass densities have not sufficiently sampled the shelf-to-slope break to determine the depth of rapid sediment change, or the precise location of the mudline. In the study of the eastern Gulf of Mexico off Florida and Alabama (Table 1), however, the textural transition as well as the infaunal change is well defined. The 75 to 100 m contour of each transect marks the area where fines have been winnowed away and grain size is higher than on either the adjacent outer shelf or in deeper water on the slope (Fig. 2). It is in this 75 to 100 m zone that infaunal biomass is usually the lowest of the shelf values. Upon crossing the 100 m contour of each transect, grain size rapidly decreases seaward as the mudline is reached (Fig. 2), while biomass and infaunal densities again show an increase. The distance over which this increase occurs is relatively small because biomass and densities decline rapidly before reaching the 200 m contour, even though fines continue to dominate.

DEPOSITIONAL AND INFAUNAL FACIES

Relatively few studies have attempted to map in detail both sediment facies and infauna facies in shelf environments. Most investigators, however, have recognized that infaunal facies are oriented in bands across the shelf and that each region along a transect supports its own characteristic infaunal assemblage (Wigley and McIntyre, 1964; Sanders et al., 1965; Cerame-Vivas and Gray, 1966; Day et al., 1971; Boesch et al., 1977; Blake, 1979; Vittor, 1979).

Cerame-Vivas and Gray (1966) described three assemblages of benthic species on the continental shelf off North Carolina. Two of these assemblages occupy the inner shelf. The assemblage south of Cape Hatteras, on the inner shelf, constitutes what is referred to by biologists as the Carolinian Province, while another assemblage north of the Cape has been described as the Virginia Province. Offshore and extending at least to the shelf-slope break is the Tropical Province which Cerame-Vivas and Gray (1966) projected from 34°N latitude to 36°N latitude. Some overlap in species occurs between the inshore Carolinian Province and the outer shelf Tropical Province, but no species exchange has been observed between the Virginia Province and the Tropical Province. The north-to-south limits of these Provinces are not well defined.

Cerame-Vivas and Gray (1966) attribute these biogeographic provinces to sharp discontinuities in temperature regimes and to prevailing currents. Over a broad scale temperature and circulation factors are of prime importance, but sediment regimes also vary markedly in this area. It thus seems likely that sub-provincial infaunal assemblages can be defined. For instance, Day et al. (1971) described two infaunal assemblages on the shelf-slope transition off North Carolina. The first assemblage occurs in 40 to 120 m where the substrate is fine-to-medium sand. The second assemblage extends to 205 m where the substrate is fine muddy sand and contains more organic matter than the substrate on the shelf. A 37% overlap in species is found between the assemblages. Although the two assemblages can be distinguished, the strong Gulf Stream current in the area (which prevents deposition of fines and development of a well-defined mudline present on other shelf-slope transitions) precludes sharp and distinct assemblage boundaries of the type mapped elsewhere.

To date, some of the most detailed studies of infaunal-sediment relationships on an outer continental shelf are those of Blake (1979) and Vittor (1979). These workers have mapped the molluscan and polychaete facies of the eastern Gulf of Mexico off Florida and Alabama; Doyle and Sparks (1980) have mapped the sediment facies in this region. Figure 4 shows sediment facies across the shelf and Figure 5 shows the molluscan facies based upon Bray-Curtis similarity index (Bray and Curtis, 1957), followed by R-mode and Q-mode cluster analysis (Sokal and Sneath, 1963; Sneath and Sokal, 1973).

Three molluscan assemblages (Blake, 1979) have been defined on the shelf-slope transition of the eastern Gulf of Mexico. These three (Fig. 5) agree closely with those occupied by polychaetes (Vittor, 1979); they also fall within the various sediment zones described by Doyle and Sparks (1980) and Doyle and Feldhausen (1981) and shown in Figure 4. A specific assemblage is largely restricted to one sedimentary regime. Other physical factors such as temperature and depth also are important in determining faunal distribution.

Molluscan assemblage I (Fig. 5), which extends from approximately 28°N latitude to at least 26.5°N latitude at a depth of about 75 to 100 m, is located within the carbonate sand sheet (Fig. 4). The outer part of the sheet marks the beginning of the shelf-slope transition and is characterized by an increase in mean grain size. Fines, and consequently organics, are winnowed away in this relatively high-energy zone. In consequence, the infaunal molluscs present are primarily suspension feeders which rely on the organics in the overlying water column as a food source. Boundaries of this most tropical assemblage vary seasonally as the constituent populations undergo pronounced changes in numbers and composition.

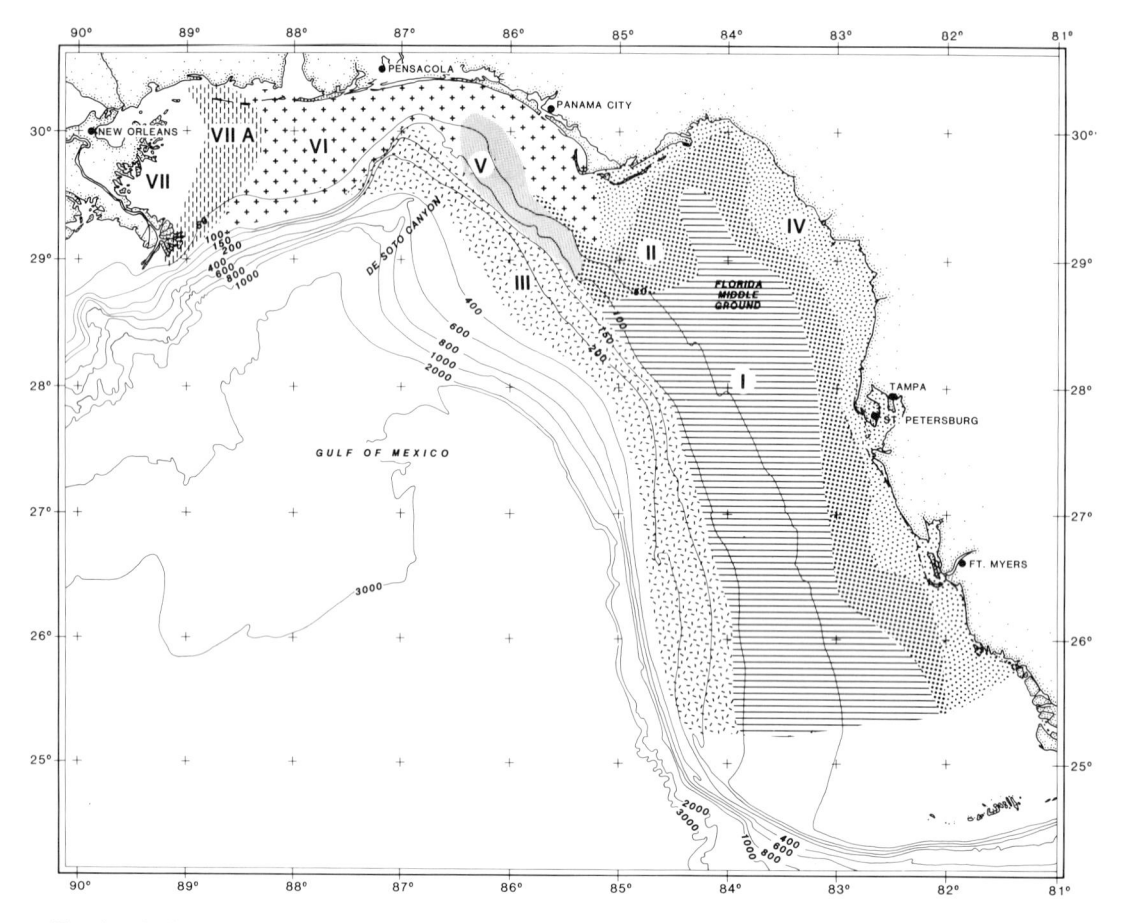

FIG. 4.—Sediment facies map of the eastern Gulf of Mexico continental margin (modified after Doyle and Sparks, 1980). I, MAFLA Carbonate Sand Sheet; II, Carbonate-Quartz Transition Zone; III, West Florida Lime Mud; IV, West Florida Quartz Sand Band; V, Destin Carbonate Facies; VI, MAFLA Quartz Sand Sheet; VII, Mississippi Pro-delta; VIIA, Mississippi Pro-delta/MAFLA Quartz Sand Sheet Transition Zone.

Assemblage II (Fig. 5) occurs in the vicinity of Cape San Blas, also at depths of 75 to 100 m. The molluscs of this area are different from those of assemblage I. Although both assemblages contain mostly suspension-feeding molluscs at the same water depth, the sediment in this northern zone is a mixture of carbonate and quartz (Fig. 4). Also, the molluscs in assemblage II are more Carolinian, molluscs of assemblage I are more Tropical.

Assemblages I and II are separated by the Florida Middle Ground (Fig. 5), an area of carbonate outcrop containing largely epifauna characteristic of tropical carbonate margins. The Florida Middle Ground extends onto the shelf-slope transition and contains many of the hard corals and other fauna associated with more shallow coral reefs. In the area at depths between 75 and 100 m, winnowing by the Florida Loop Current and occasional hurricanes have resulted in the accumulation of shell hash and other carbonates originating from the biological activity of the Middle Ground. Although this area undoubtedly contains some infaunal suspension feeders, it is difficult to quantitatively sample and has not been characterized.

Assemblage III occurs at depths of 100 m to at least 200 m extending along much of the West Florida shelf-slope transition. Ludwig (1964) and Doyle and Sparks (1980) have designated the sediment facies in this area as the West Florida Lime Mud (Fig. 4). This facies is bounded by the mudline of the West Florida shelf-slope transition and contains higher organic carbon than on the shelf. Molluscs, which are rare in comparison to the polychaetes in this assemblage, are primarily deposit feeders and show very little change in seasonal composition or abundance. The environment from 100 to 200 m

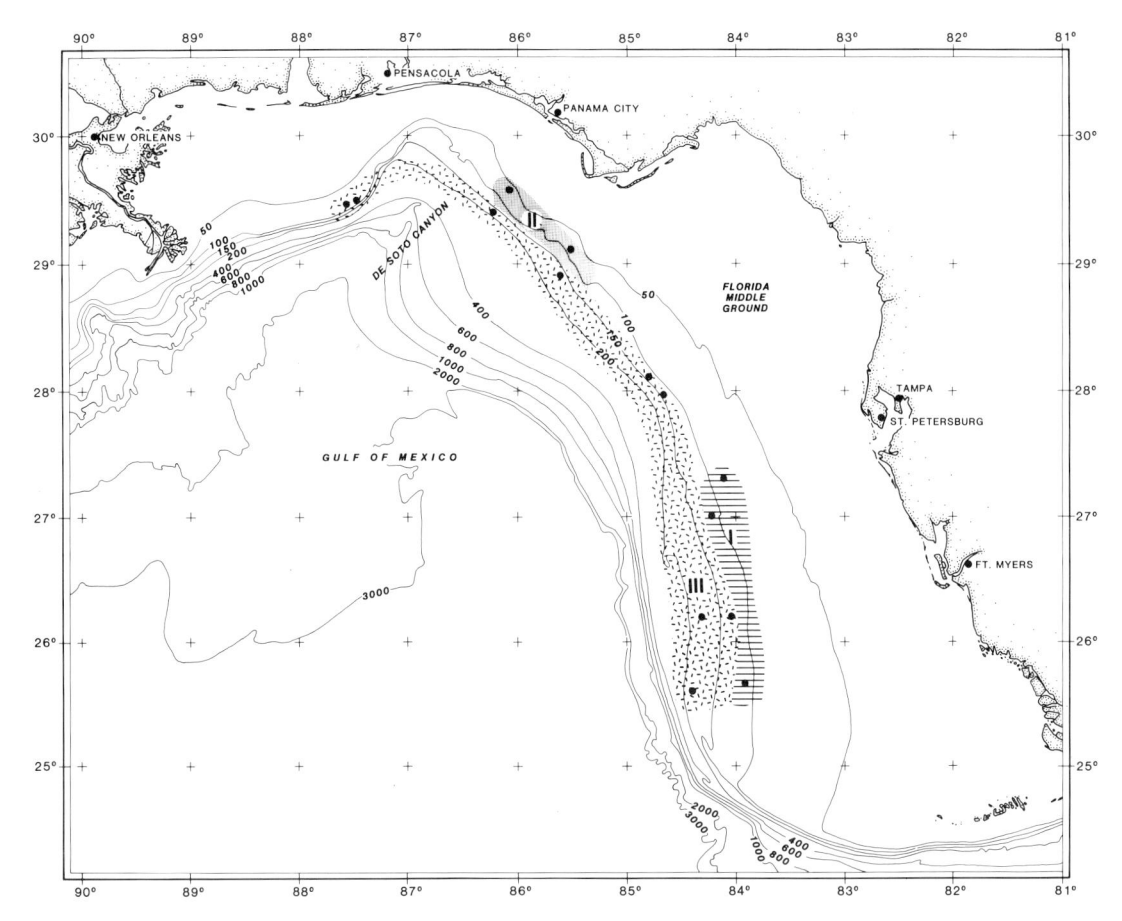

FIG. 5.—Molluscan facies map of the shelf-slope transition of the eastern Gulf of Mexico. I, Tropical suspension feeding assemblage; II, Carolinian suspension feeding assemblage; and III, deposit feeding assemblage.

changes very little unlike the environment between 75 and 100 m which changes latitudinally as well as seasonally.

SYNTHESIS OF INFAUNA-SEDIMENT INTERACTIONS

The percentage of organic carbon normally increases across the shelf and down the slope. An exception occurs in the narrow 75 to 100 m zone of some shelf-to-slope transitions where mean sediment grain size is increased through the winnowing of fines. The infauna of the shelf-slope transition undoubtedly plays a major role in regulating this trend.

On the shelf, particularly at the shallower part (75 to 100 m) of the shelf-slope break, the infauna are essentially suspension feeders which actively remove organic particles from the water column. A portion of the ingested organics is incorporated as part of the infaunal biomass while some is excreted as feces. On the shelf, feces are recycled many times through the benthic food chain and help maintain high productivity. In the higher energy area of the shelf-slope transition, however, the feces are easily suspended and transported downslope. Rapid loss of organic carbon from this zone might partially explain the relatively low infaunal densities and biomass in this sector.

Below 100 m there is a general decrease in winnowing action and an accumulation of fines. Organics generally increase and deposit feeders predominate. At the mudline, faunal densities and biomass are often greater than either in shallower or deeper water. This increase may result largely from an increase in deposited organic carbon and decreased environmental perturbations.

Most of the hundreds of cores collected in the lower part of the shelf-slope transition show little structure, and reworking by the infauna is probably considerable. Sediment creep in which soft, unconsolidated sediment gradually moves downslope has been observed on shelf-slope transitions of as little as 1% grade (Kraft et al., 1979). Sediment creep is particularly common in the northern Gulf of Mexico, where there is rapid sedimentation of fine-grained material near the Mississippi Delta. The primary cause of sediment creep is gravity acting on unconsolidated sediments (Kraft et al., 1979). The reworking of upper sediment layers by the infaunal deposit feeders also probably acts to maintain a high level of porosity and permeability within the sediment, lessen the degree of consolidation, and enhance liquefication and the probability of slope failure (see Bennett and Nelsen, this volume).

CONCLUSIONS

It is clear that the substrate and water column properties play a major role in determination of the infauna across the shelfbreak. Conversely, the infauna reinforce substrate facies changes, acting to further the distinction. These relationships could prove helpful in recognizing the shelf-slope transition in the ancient record. Some criteria which, *along with sedimentary parameters*, may prove important are: (1) Rapid transition to deposit feeding molluscan genera from a mixture of molluscan genera in which filter feeders are heavily represented; and (2) Rapid decrease in numbers of fossils, along with some increase in grain size, followed by rapid increase in fossil content concomitant with a facies change from sand to mud, and finally a sharp decrease in the fossil content of the mud away from the facies change. The geometry of rocks in which such relationships occur could also be useful in interpreting progradation of ancient shelf-to-slope breaks.

REFERENCES

BENNETT, R. H., AND NELSEN, T. A., 1983, Seafloor characteristics and dynamics affecting geotechnical properties at shelfbreaks, *in* The Shelfbreak: Critical Interface on Continental Margins: Soc. Econ. Paleontologists Mineralogists Spec. Pub. 33, p. 333–355.

BLAKE, N. J., 1979, Macroinfaunal molluscs, *in* The Mississippi, Alabama, and Florida Outer Continental Shelf, MAFLA, 1977/1978: Volume IIA, Compendium of Work Element Reports, Washington, Dames and Moore, Bur. Land Management Contract No. AA550-CT7-34, p. 667–698.

BOESCH, D. F., KRAEUTER, J. N., AND SERAFY, D. K., 1977, Distribution and structure of communities of macrobenthos on the outer continental shelf of the Middle Atlantic Bight: 1975–1976 investigations: Virginia Inst. Mar. Sci., Spec. Rept. Appl. Mar. Sci. Oceanogr. Eng. 175., 111 p.

BRAY, I. R., AND CURTIS, J. T., 1957, An ordination of upland forest communities of southern Wisconsin: Ecol. Mono., v. 27, p. 325–349.

CERAME-VIVAS, M. J., AND GRAY, I. E., 1966, The distributional pattern of benthic invertebrates of the continental shelf off North Carolina: Ecology, v. 47, p. 260–270.

DAY, J. H., FIELD, J. G., AND MONTGOMERY, M. P., 1971, The use of numerical methods to determine the distribution of benthic fauna across the continental shelf of North Carolina: Jour. Anim. Ecol., v. 40, p. 93–125.

DOYLE, L. J., 1979, Surficial sediment characters and clay mineralogy, *in* The Mississippi, Alabama, and Florida Outer Continental Shelf; MAFLA, 1977/1978: Volume IIA, Compendium of Work Element Reports, Washington, Dames and Moore, Bur. Land Management Contract No. AA550-CT7-34, p. 311–344.

———, AND FELDHAUSEN, P. H., 1981, Bottom sediments of the eastern Gulf of Mexico examined by traditional and multivariate statistical methods: Math. Geol., v. 13, . 93–117.

———, PILKEY, O. H., AND WOO, C. C., 1979, Sedimentation of the Eastern United States continental slope, *in* Doyle, L. J., and Pilkey, O. H., eds., Geology of Continental Slopes: Soc. Econ. Paleontologists Mineralogists Spec. Pub. 27, p. 119–130.

———, AND SPARKS, T. N., 1980, Sediments of Mississippi, Alabama, and Florida (MAFLA) continental shelf: Jour. Sed. Petrology, v. 50, p. 905–916.

EMERY, K. O., 1960, The Sea Off Southern California: New York, John Wiley & Sons, 366 p.

———, AND UCHUPI, E., 1972, Western North Atlantic Ocean: Topography, Rocks, Structure, Water Life, and Sediments: Tulsa, Oklahoma, Am. Assoc. Petroleum Geologists, 532 p.

FAUSAK, L., 1979, Physical oceanography, *in* The Mississippi, Alabama, Florida, Outer Continental Shelf Environmental Survey, MAFLA, 1977/1978: Volume IIA, Compendium of Work Element Reports, Washington, Dames and Moore, Bur. Land Management Contract No. AA550-CT7-34, p. 888–929.

GORSLINE, D. S., 1963, Bottom sediments of the Atlantic shelf and slope, off the southern United States: Jour. Geology, v. 71, p. 422–440.

HEARD, R., 1979, Macroinfaunal crustaceans, *in* The Mississippi, Alabama, Florida Outer Continental Shelf, MAFLA, 1977/1978: Volume IIA, Compendium of Work Element Reports, Washington, Dames and Moore, Bur. Land Management Contract No. AA550-CT7-34, p. 748–788.

KRAFT, L. M., CAMPBELL, K. J., AND PLOESSEL, M. R., 1979, Some geotechnical engineering problems of upper slope in the northern Gulf of Mexico, *in* Doyle, L. J., and Pilkey, O. H., eds., Geology of Continental Slopes: Soc. Econ. Paleontologists Mineralogists Spec. Pub. 27, p. 25–42.

LUDWIG, J. C., 1964, Sediments in the northeastern Gulf of Mexico, *in* Miller, R. L., ed., Papers in Marine Geology: New York, MacMillan Co., p. 204–238.

ROWE, G. T., AND HAEDRICH, R. L., 1979, The biota and biological processes of the continental slope, *in* Doyle, L.

J., and Pilkey, O. H., eds., Geology of Continental Slopes: Soc. Econ. Paleontologists Mineralogists Spec. Pub. 27, p. 49–60.

———, POLLONI, P. T., AND HORNER, S. G., 1974, Benthic biomass estimates from the northwestern Atlantic Ocean and the northern Gulf of Mexico: Deep-Sea Res., v. 21, p. 641–650.

SANDERS, H. L., 1968, Marine benthic diversity: A comparative study: Amer. Naturalist, v. 102, p. 243–282.

———, 1969, Benthic marine diversity and the stability-time hypothesis, in Woodwell, G. M., and Smith, H. H., eds., Diversity and Stability in Ecological Systems: Brookhaven Symposia in Biology No. 22, p. 71–81.

SANDERS, H. C., HESSLER, R. R., AND HAMPSON, G., 1965, An introduction to the study of deep-sea faunal assemblages along the Gay Head-Bermuda transect: Deep-Sea Res., v. 12, p. 845–867.

SLOBODKIN, L. B., AND SANDERS, H. L., 1969, On the contribution of environmental predictability to species diversity, in Woodwell, G. M., and Smith, H. H., eds., Diversity and Stability in Ecological Systems: Brookhaven Symposia in Biology No. 22, p. 82–95.

SNEATH, P. H., AND SOKAL, R. R., 1973, Numerical Taxonomy: San Francisco, W. H. Freeman and Co., 573 p.

SOKAL, R. R., AND SNEATH, P. H., 1963, Principals of Numerical Taxonomy: San Francisco, W. H. Freeman and Co., 359 p.

SOUTHARD, J. B., AND CACCHIONE, D. A., 1972, Experiments on bottom sediment movement by breaking internal waves, in Swift, D. J., P., Duane, D. B., and Pilkey, O. H., eds., Shelf Sediment Transport: Process and Pattern: Stroudsburg, Pa., Dowden, Hutchinson, and Ross, p. 83–97.

STANLEY, D. J., ADDY, S. K., AND BEHRENS, E. W., 1983, The mudline: variability of its position relative to shelfbreak, in Stanley, D. J., and Moore, G. T., eds., The Shelfbreak: Critical Interface on Continental Margins: Soc. Econ. Paleontologists Mineralogits Spec. Pub. 33, p. 279–298.

STANLEY, D. J., AND WEAR, C. M., 1978, The "mud-line": An erosion-deposition boundary on the upper continental slope: Mar. Geology, v. 28, p. M19–M29.

THORSON, G., 1956, Marine level-bottom communities of recent seas, their temperature adaptation and their "balance" between predators and food animals: Trans. New York Acad. Sci., sec. 2, v. 18, p. 693–700.

———, 1957, Bottom communities (sublittoral or shallow shelf), in Hedgpeth, J. W., ed., Treatise on Marine Ecology and Paleoecology, Volume 1: Geol. Soc. America Memoir No. 67, p. 461–534.

VITTOR, B., 1979, Macrofaunal polychaetes, in, The Mississippi, Alabama, Florida, Outer Continental Shelf, MAFLA, 1977/1978: Volume IIA, Compendium of Work Element Reports, Washington, Dames and Moore, Bur. Land Management Contract No. AA550-CT7-34, p. 699–747.

WIGLEY, R. L., AND MCINTYRE, A. D., 1964, Some quantitative comparisons of offshore meiobenthos and macrobenthos south of Martha's Vineyard: Limnol. and Oceanogr. v. 9, p. 485–493.

ORGANIC MATTER CHARACTERISTICS NEAR THE SHELF-SLOPE BOUNDARY[1]

R. W. JONES
Chevron Oil Field Research Company, P.O. Box 446, La Habra, California 90631

ABSTRACT

The organic facies of sediments deposited near the shelf-slope break depend upon the types and amounts of organic matter available at the depositional site and the early diagenetic history of the organic matter. On a global scale, the organic facies at the shelf-slope break record major tectonic events, eustatic sealevel changes, water circulation, and climate. These overall controls affect a specific sedimentation site through such factors as mean grain size, sedimentation rate, input of terrestrial organic matter, organic productivity in the photic zone, and the dissolved oxygen content of the water column, particularly the water associated with the water-sediment interface.

Oxygen deficient water masses in the oxygen minimum layer (OML) of the World's Ocean intersect the continental margins most often on the upper slope, and it is there that the best potential source rocks now being deposited preferentially exist. Oxygen deficient water locally reaches onto the shelf, most noticeably in areas of shallow upwelling of nutrient-rich water and resulting high productivity, as in offshore southwest Africa. In the past, due to such factors as climate change, different current patterns and eustatic sealevel changes, oxygen deficient water has transgressed well up onto the shelves on a regional basis. Such events have resulted in the deposition of source rocks for much of the world's oil in transgressive shelf deposits.

If the bottom water across the shelf-slope break is oxic, the primary controls on the organic matter distribution are the grain size, sedimentation rate, and the input of terrestrial organic matter. None of these factors can completely offset the negative effects of oxic bottom water on the deposition and preservation of an oil-prone organic facies.

INTRODUCTION

The organic facies of a sediment is determined by the amounts and types of organic matter that arrive at a given depositional site and the chemistry of the early diagenetic environment at that site (Jones and Demaison, 1980, 1982). On a global scale, the organic facies reflect major tectonic events (North, 1979), eustatic sealevel changes (Vail et al., 1977; Vail and Mitchum, 1979), and climate (Lisitsin, 1978). However, to understand the origin of a specific organic facies, we need to evaluate such factors as water circulation, mean grain size, sedimentation rate, terrestrial organic matter input, organic productivity in the photic zone, and the oxygen content of the water column, particularly the bottom water.

The shelf-slope break is a logical place to expect substantial variations in organic facies because many of the controlling parameters listed above exhibit high rates of change in this region. Unfortunately, in studying recent sediments at the shelf-slope break, we are only recording a snapshot of a rather unusual time in geologic history. However, in Holocene marine sediments, the various physical and chemical parameters affecting the organic matter distribution can be accurately measured and the shelf-slope break accurately located. In studying ancient shelf-slope breaks, we cannot measure the dynamic parameters that controlled the organic matter deposition, nor always precisely determine the position of the break in space and time. Fortunately, lithologic and paleontological analyses, well and outcrop data, and seismic stratigraphy can help resolve this problem. Another problem exists. This review suffers from the fact that no systematic study has ever been made that quantifies (most of) the parameters that control the distribution of organic matter at the shelf-slope break. It was necessary to rely on studies made with different objectives in mind. Consequently, some of the tentative interpretations offered are based on meager data.

This paper will examine: (1) the expected effects of physical, chemical, and biological controls on organic matter distribution in marine sediments; (2) the significance of these controls as revealed by an analysis of recent sedimentation near the shelf-slope break in a variety of settings; and (3) shelf-slope breaks in the historical record. The emphasis throughout is on the combination of chemical and physical parameters that control deposition of potential source rocks[2] in association with the shelf-edge.

---

[1] Published with permission of Chevron Oil Field Research Company.

Copyright © 1983, The Society of Economic Paleontologists and Mineralogists

[2] Potential source rocks are rocks whose organic matter is capable of generating commercial volumes of hydrocarbons upon reaching thermal maturity.

PHYSICAL AND CHEMICAL PARAMETERS

Summary

In Quaternary and Upper Tertiary marine sediments the highest concentrations of total organic carbon (TOC) are preferentially found in slope sediments, usually on the upper slope. In the modern oceans, the increase of TOC in sediments on the slope is generally modest and usually correlates with a higher clay content as compared to the shelf sediments and a higher sedimentation rate than on the continental rise and basin floor (Gross et al., 1972; Diester-Haas and Müller, 1979; Exon et al., 1981).

The largest TOC values on either side of the shelf-slope break that are controlled by grain size and sedimentation rate occur in pro-delta muds. Here, the high input of terrestrial organic matter is added to the positive effect of increased clay content and high sedimentation rate (Weber and Daukoru, 1975; Combaz and de Matharel, 1978).

The organic content of the rocks with the best source potential deposited near the shelf-slope break is dominated by the remains of marine plankton deposited under oxygen deficient[3] water. The oxygen deficiency can rise from the basin floor in certain silled basins (Degens and Ross, 1974; Byers, 1977, 1979; Cluff, 1980) or may develop locally due to excessive oxygen demand created by very high productivity (Demaison and Moore, 1980). In the modern oceans, the latter situation is most common on the upper slope, but can occur on the shelf where upwelling and associated planktonic blooms are intense (Calvert and Price, 1971a, b; Burnett et al., 1980).

Effects of Water Circulation

This chapter does not deal with water circulation directly (discussed in Part V, this volume). Rather, the focus is on some of the effects of water circulation at and near a given depositional site that can be measured and correlated with the organic content of the sediment. These include mean grain size, sedimentation rates, organic productivity and the oxygen content of the water.

---

[3]In this chapter the term "oxygen deficient" water refers to water in which the oxygen content is ≤0.5 milliliters of oxygen per liter of water (≤0.5 ml/L). The lower limit (0.0 ml/L) is anoxic water *sensu stricto* (Berner, 1981), whereas the upper limit (0.5 ml/L) is the approximate oxygen content below which bioturbation by deposit feeders becomes significantly depressed and which separates good qualitative and quantitative preservation of organic matter in sediments from poor preservation (Demaison and Moore, 1980).

Mean Grain Size

An inverse correlation between grain size and organic content of sediments has long been recognized (Trask, 1932). Oil-prone source rocks are almost always very fine grained. Hydrology is the dominant control for this observation. Oil-prone organic matter has a density only slightly greater than water (van Krevelen, 1961). It will normally not be deposited with sands and silts in significant quantities.

In addition, the very existence of sands and silts usually implies current activity, water mixing and an oxygenated environment. The high rate of bioturbation and diffusivity of oxygen in the coarse-grained sediments will destroy most of the oil-prone organic matter. The "mudline" (Stanley et al., 1983) must be on the shelf for potential source rocks of significant volume to be deposited above the shelf-slope break. Biogenous sediments can be an exception to this generalization where the grain size is often not determined by bottom currents, but rather by the size of biogenic skeletal remains settling from above (Calvert and Price, 1971b).

Sedimentation Rate

Oxic Bottom Water.—The complex effects of sedimentation rate on the preservation of organic matter are not thoroughly understood, primarily because preservation is also related to the mean grain size of the sediments, productivity in the photic zone, and the oxygen content of the bottom water. Much evidence supports a strong positive correlation between sedimentation rates and organic content for fine-grained sediments (clay-silt) under oxic conditions (Heath et al., 1977; Toth and Lerman, 1977; Froelich et al., 1979; Müller and Suess, 1979). This relationship is created by the shorter residence time of organic matter in oxidizing conditions when sedimentation rates are high. The positive correlation between organic content and sedimentation rate is best defined at high TOC values for highly biogenic sediments.

For high sedimentation rates of nonbiogenic clastics, the correlation breaks down. For example, the highest sedimentation rates of fine-grained sediments are probably found in prodelta muds with TOC contents that rarely exceed 2%. In particular, calculations based on cross sections and TOC values of the Upper Tertiary of the Gulf Coast (in Dow and Pearson, 1978; and Dow, 1978), clearly demonstrate that sedimentation rates of fine-grained slope sediments exceeding 200 m per 10^6 yr have accompanying TOC values <1.0 wt.%. However rapid the sedimentation, the organic content is obviously limited by the organic matter available. In addition, as sedimentation rates increase, the mean grain size of the nonbiogenic sediments is likely to increase. This will usually mean both less organic matter originally and a higher oxygen accessibility to the sediments that increases destruction of organic matter.

The correlation between organic matter content and sedimentation rate also breaks down as the sedimentation rate becomes very low, because a minimum (≤0.1 TOC) amount of refractory organic carbon remains attached to clay minerals independently of the sedimentation rate (Müller and Mangini, 1980).

Oxygen Deficient Bottom Water.—Published data appear to favor the same positive correlation between sedimentation rates and organic content as under oxic water (Richards, 1970; Toth and Lerman, 1977; Müller and Suess, 1979). Deposition under oxygen deficient water, however, clearly results in higher TOC values given comparable depositional conditions (Degens and Ross, 1974; Morris, 1980). In part, this simply reflects the slower and more incomplete breakdown of the organic matter under oxygen deficient conditions (Deuser, 1975) that occurs primarily due to the lack of bioturbation (Demaison and Moore, 1980). As most of the organic matter deposited in an oxygen deficient environment is preserved, the relationship between sedimentation rate and TOC is likely to be controlled by the relative amounts of organic and nonorganic matter input rather than by the overall sedimentation rate. Thus, an inverse rather than a direct correlation between organic content and the sedimentation rate should be expected under oxygen deficient water. Such a relationship was found by Müller and Suess (1979) for shelf and upper slope samples under oxygen deficient water in offshore Peru. Anoxic Holocene sediments in the Black Sea show a well-defined inverse relationship between sedimentation rate and TOC (Degens, 1974; Degens et al., 1978). A higher sedimentation rate (30 cm/10^3 yr) existed for a coccolith ooze with an average TOC of 3 to 5 wt.% than for the underlying carbonate poor unit (sedimentation rate ≅10 cm/10^3 yr) with an average TOC ≅ 15 wt.%. In addition, the underlying Pleistocene unit that was deposited in oxic water with a substantially higher sedimentation rate had greatly diminished TOC values (Degens, 1974). The inverse correlation primarily reflects dilution of a near constant organic input by a varying nonorganic input. Conspicuous examples of dilution are planktonic skeletons arriving from above (Degens, 1974; Morris, 1980), turbidity currents (Emery and Hülsemann, 1962; Degens and Ross, 1974), and the seasonal fine-grained terrestrial input arriving amidst a less variable diatomaceous rain (Calvert, 1966).

In summary, a positive correlation usually exists between the sedimentation rate and the organic content of marine sediments under oxic bottom water near the shelf-slope break—and elsewhere—due to the lowered residence time of organic matter in oxic pore water. An inverse correlation is to be expected under oxygen deficient water due to the variable dilution by clastics of a more constant input of organic matter.

Productivity

No systematic correlation exists between high productivity in the photic zone and the organic content of the underlying sediments (Gross et al., 1972; Demaison and Moore, 1980). High productivity is a necessary, but not a sufficient, condition for the deposition and preservation of an organic-rich, oil-prone organic facies in general, and near the shelf-slope break in particular. High productivity can create the oxygen deficiency at the water-sediment interface that permits preservation of large amounts of organic matter. Somewhat paradoxically, a large amount of organic matter must be oxidized to create the chemical environment that permits the preservation of large amounts of organic matter.

Nutrient-rich upwellings greatly enhance productivity. They often occur inshore of the shelf-slope break and are a primary cause of both the deposition and preservation of the highest organic contents on the upper slopes. The highest organic contents on the outer shelf and upper slope occur where the upwelling waters are also oxygen deficient (Calvert and Price, 1971a, b; Gershanovich et al., 1976). In effect, this requires that the upwelling waters come from the oxygen minimum layer (OML). The classic example of this combination of events is offshore southwest Africa, where the upwelling water comes from the shelf, but the long-term nutrient replacement comes from the north flowing Benguela Current (Calvert and Price, 1971a, b). Organic carbon contents reach 20 wt.% on the middle shelf but remain >5 wt.% well onto the upper slope (Calvert and Price, 1971b).

Intense upwelling and high productivity are not enough to permit the deposition and preservation of abundant organic matter on either side of the shelf-slope break. If the upwelling waters are oxygen-rich as in offshore Brazil (Summerhays et al., 1976) or the oxygen content of the bottom waters is continuously maintained by currents from the polar regions, as in the Grand Banks area of eastern Canada (Worthington and Wright, 1970), the preservation of organic matter will be minimal.

Dissolved Oxygen

The deposition of organic-rich sediments and suppression of bioturbation near the shelf-slope break—and elsewhere—requires the dissolved oxygen content of the bottom water to be <0.5 ml/L (Demaison and Moore, 1980). There are two basic models in which regionally significant oxygen deficiency can be developed that will permit the deposition of potential oil source rocks near the shelf-slope break. The models are: (1) expansion of oxygen deficient bottom water up to and over the shelf-slope break (Fig. 1); and (2) development of an oxygen deficient OML that intersects the water-sediment interface near the shelf-slope break (Fig. 2).

Oxygen Deficient Basin.—If the water in a basin becomes density stratified, due to the formation and preservation of either a thermal or salinity layer, oxygen deficiency will develop from the bottom up. Oxygen is consumed in the water column and on the sea bottom by the destruction of organic matter and not replaced. If stratification is maintained, the oxygen deficiency will rise to a level that is controlled by the depth of mixing of the surface water. The depth of mixing often nearly coincides with the location of the shelf-slope break, but can be both shallower or deeper. The recent history of the Black Sea illustrates these phenomena and that the scenario can be enacted in only a few thousand years (Degens and Ross, 1974). Most of the source rocks deposited on the slopes and outer shelves of interior basins undoubtedly had a similar origin (see later discussion). In the modern oceans the lack of permanent water stratification precludes oxygen deficiency forming from this model.

Oxygen Deficient OML.—For significant source rock deposition without oxygen deficiency on the basin bottom, an oxygen deficient OML ($O_2 < 0.5$ ml/L) is required. An OML exists in the modern oceans between the oxygen-rich surface waters and deep cold, oxygen-rich polar waters. The OML is primarily caused by the consumption of existing oxygen during the oxidation of dead plankton as they sink through the water column. Where plankton production is low to moderate, the OML is oxic, poorly-defined and usually intersects the continental margin well down the continental slope (Wyrtki, 1971). In areas of high productivity near the shelf-slope break, the OML rises and becomes better defined (Calvert and Price, 1971a; Wyrtki, 1971). If the upwelling, nutrient-rich water that literally feeds the plankton originates in the OML, a self-perpetuating cycle may develop that will create oxygen deficient bottom water on the upper slope and occasionally carry it onto the shelf. Offshore Peru and southwest Africa are the type examples in the modern ocean (Calvert and Price, 1971a, b; Gershanovich et al., 1976). The Permian Phosphoria Formation of western Wyoming and environs is the type example from the geologic record (Fig. 3; Claypool et al., 1978).

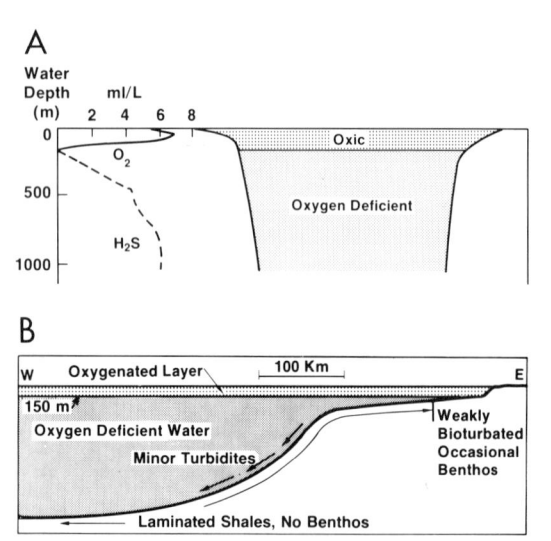

FIG. 1.—Oxygen deficient basin model for development of oxygen deficiency and source rock deposition in association with shelf-slope break. *A*, Black Sea (Thiede and van Andel, 1977) where water circulation generally precludes oxygen deficient water reaching over the shelf-slope break. *B*, Devonian of New York State (Byers, 1977) where oxygen deficient water covered much of the shelf during major transgressions.

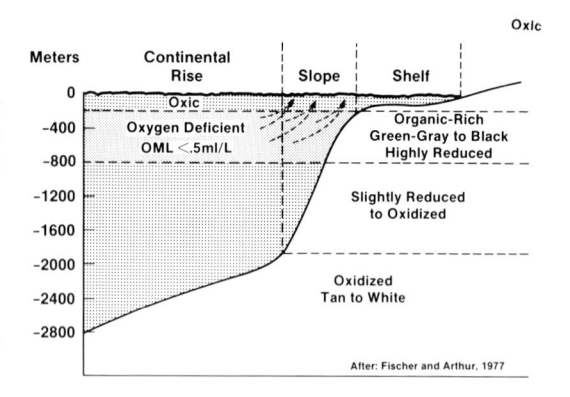

FIG. 2.—Oxygen minimum layer (OML) model for development of oxygen deficiency and source rock deposition near the shelf-slope break. When upwelling of nutrient-rich, but oxygen-poor, water occurs near the shelfedge, the resulting high productivity can lead to the transgression of an oxygen deficient OML onto the shelf and cause the deposition of potential oil-prone source rocks. Offshore Peru, Namibia, and southwest India are modern examples.

RECENT SHELF-SLOPE BREAKS

It is useful to evaluate the distribution of organic matter across the shelf-slope break in six different Holocene geographic settings, from highly oxygen deficient to moderately oxic (Fig. 4).

Peru

The distribution of organic carbon in the shelf and slope sediments of offshore Peru and the factors responsible have been extensively studied (Fair-

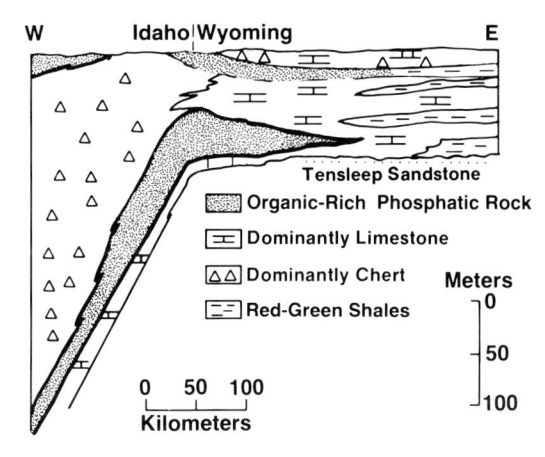

FIG. 3.—Simplified stratigraphic cross-section of the Permian Phosphoria Formation and associated sediments in western Wyoming and environs (after Claypool et al., 1978).

TABLE 1.—Sedimentation Under Oxygen Deficient Water, Outer Shelf and Upper Slope, Offshore Peru (Müller and Suess, 1979)

Water Depth (m)	Productivity (g/m^2/yr)	Sedimentation Rate (cm/10^{3}yr)	TOC Wt. %	% of Productivity
186	330	140	12.5	12
370	330	66	21.2	12

bridge, 1966; Gershanovich et al., 1976; Burnett et al., 1980). Of interest here is the high TOC values created by the oxygen deficiency developed over the shelf-slope break due to the complex interplay of the north flowing Peru Current, offshore winds, high productivity, and the upwelling of oxygen-poor, nutrient-rich waters from the outer shelf (Fig. 5). Two data points from Müller and Suess (1979) show the effects of the highly oxygen deficient bottom waters: high TOC values, on both outer shelf and upper slope, high percentage of organic carbon preservation, and the inverse relationship between TOC and sedimentation rate (Table 1). Although two data points hardly prove a hypothesis, the preservation of an identical percentage of the organic productivity in the two oxygen deficient environments clearly demonstrates the overriding effect of dilution on the TOC.

Northern Indian Ocean

Bottom sediments, shelf-slope topography, and water characteristics of the southwest margin of India have been extensively studied. The shelf-slope break is at a water depth of 100 to 120 m. The inner shelf is floored with Recent terrigenous silts and clays whose TOC values reach 3 wt.% (Kidwai and Nair, 1972), but average 1 to 2% (Paropakari et al., 1978). The outer shelf beyond 60 m is composed of relict Pleistocene algal and oolitic limestones with TOC values <1.0 wt.%. Topographic lows may contain silts and clays with up to 3% organic matter. An excellent correlation exists on the shelf between mean grain size and the organic content of the sediments (Paropakari et al., 1978).

Recent sedimentation rates reach their maximum in the olive-gray argillaceous oozes deposited on the upper continental slope where the bottom water is oxygen deficient (Fig. 6; von Stackelberg, 1972). The oozes contain a high percentage of organic matter that sometimes exceeds 10% but is usually lower (von Stackelberg, 1972; Konjukhov, 1976).

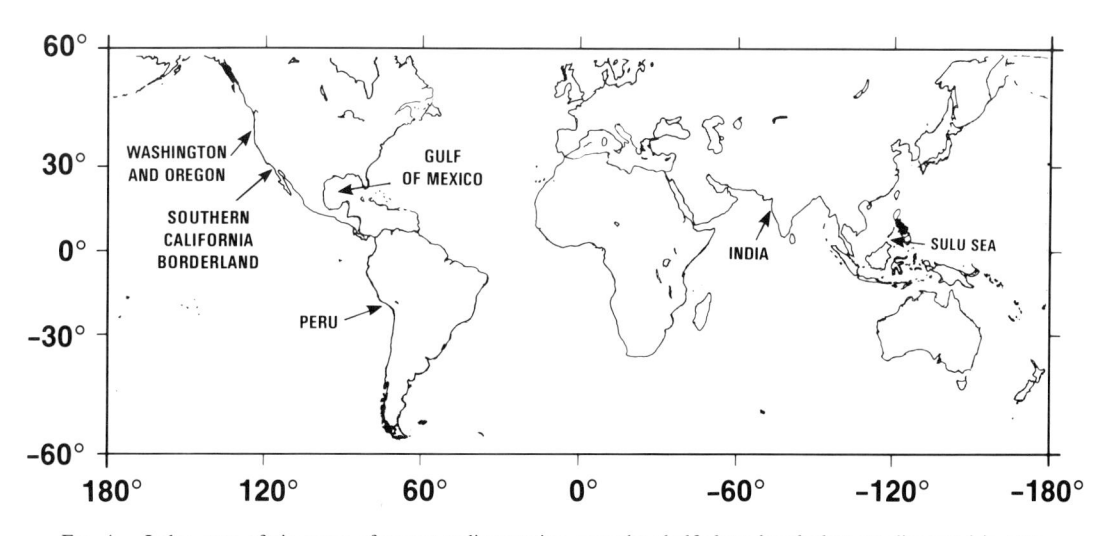

FIG. 4.—Index map of six areas of recent sedimentation near the shelf-slope break that are discussed in text.

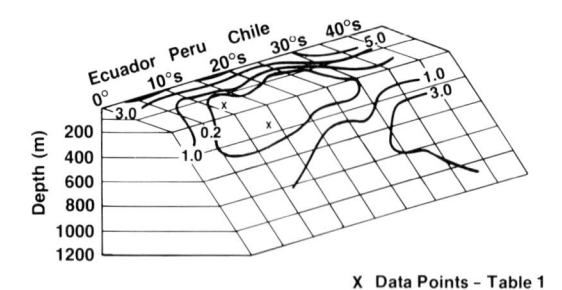

FIG. 5.—Distribution of dissolved oxygen (ml/L) in the bottom water, offshore Peru (after Burnett et al., 1980). Note position of data points from Müller and Suess (1979).

The high sedimentation rates may have placed an upper limit on the TOC values. Rough calculations based on productivity data from Ryther and Menzel (1965) and rate of sedimentation data from von Stackelberg (1972) suggest from 1 to 4% of the organic matter produced in the overlying waters is preserved in the upper slope sediments. In addition, the preservation and distribution of organic matter are affected by seasonal monsoons that create a more complicated water circulation pattern than for offshore Peru, and thereby lead to a less steady state position of oxygen deficient water masses.

Offshore Washington and Oregon

The organic carbon distribution is complex in the surficial sediments of offshore Washington and Oregon (Fig. 7). Interactions exist between an oxygen deficient OML that intersects the slope at about 650 m (Fig. 7), input from the Columbia River, sand-clay distribution, sedimentation rates, and currents preferentially semi-parallel to the shore at all depths (Gross et al., 1972). Productivity is not areally related to TOC values in the underlying sediments (Gross et al., 1972).

Despite the large variability within each gross depositional environment, the expected statistical relationships prevail regarding TOC distribution (Fig. 8). In general, the organic carbon content of the shelf sediments reflects the grain size. Thus, the lowest TOC values are found in the well-oxygenated, coarse-grained sediments of the inner shelf, and the highest values near the shelfedge south of Astoria Canyon, where the mean grain size is fine silt. The highest TOC values are generally on the middle and upper slope where the oxygen deficient OML and generally fine-grained deposits co-exist (Fig. 8).

A profile across the shelf and shelf-slope break west of Grays Harbor, Washington (Table 2; Hedges and Mann, 1979) confirms the more regional observations of Gross et al. (1972). Samples 1 to 7 show a direct correlation between clay content and TOC. In addition, a modest direct correlation of TOC with sedimentation rate in oxic water is shown by comparing samples 4 and 5 that have a nearly identical clay content, and an inverse correlation by sample 7, deposited under oxygen deficient water (Table 2; Fig. 7).

The Columbia River is the major influence on sample 8 as demonstrated by the high sedimentation rate of coarse clastics. As expected, the moderate TOC value is composed almost entirely of terrestrial organic matter and is only half the average TOC of Columbia River discharge (Gross et al., 1972).

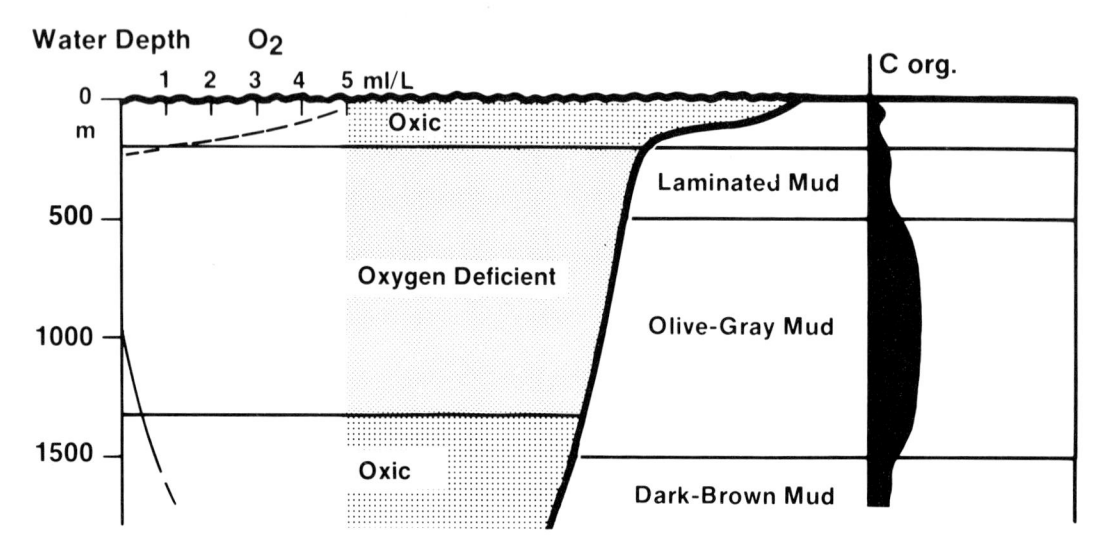

FIG. 6.—Schematic vertical section across the continental margin of southwest India (after von Stackelberg, 1972).

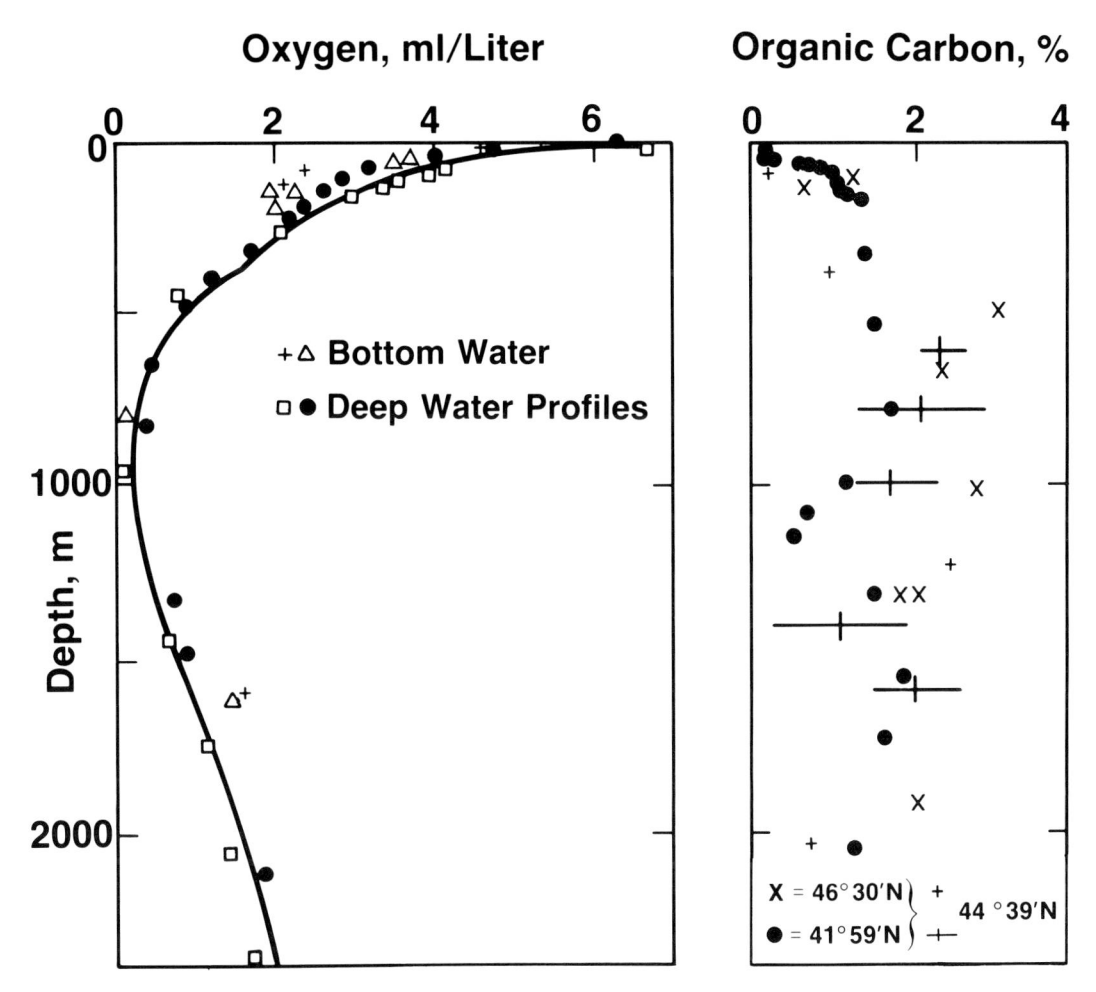

FIG. 7.—Position and intensity of the OML and the distribution of organic carbon in surface sediments, offshore Washington and Oregon (after Gross et al., 1972).

Sulu Sea

Surrounded by islands, the Sulu Sea is part of a tectonically complex area in southeast Asia (Fig. 9; Hamilton, 1977). The Sulu Sea margins are dominated by different sedimentological regimes; the shelf-slope sequence exists in a carbonate-dominated environment off Palawan Island, and in a volcano-clastic environment off Mindanao Island (Exon et al., 1981).

Although the maximum sill depth is about 400 m and the Sulu Sea is nearly 5000 m deep, the Sulu Sea waters are not oxygen deficient, at least above 4000 m (Exon et al., 1981). The dissolved oxygen values drop from ≅4 ml/L near the surface to a poorly defined minimum of 1 ml/L near 1000 m (Fig. 10). Despite the relatively low oxygen values and a high productivity, TOC values are modest (0.1–2.7 wt.%; Fig. 10). They appear to be controlled in both the carbonate (Fig. 10A) and volcanic (Fig. 10B) environments by mean grain size and sedimentation rate rather than by the oxygen content of the overlying water. Moreover, the

TABLE 2.—Properties of Surficial Sediments, East-West Profile off Grays Harbor, Washington (After Hedges and Mann, 1979). Sample 8 is off the Profile, Near the Shelfbreak, Between the Head of Astoria Canyon and the Mouth of the Columbia River

Sample Number	Water Depth (m)	Organic Carbon (Wt %)	Sedimentation Rate (g/m²/yr)	Clay %	Marine OM %
1	40	.20	400	4	82
2	55	.17	440	2	87
3	80	.87	560	10	39
4	100	1.41	440	14	54
5	145	1.05	130	13	79
6	175	.58	260	8	93
7	620	2.58	130	34	97
8	65	.73	1100	7	5

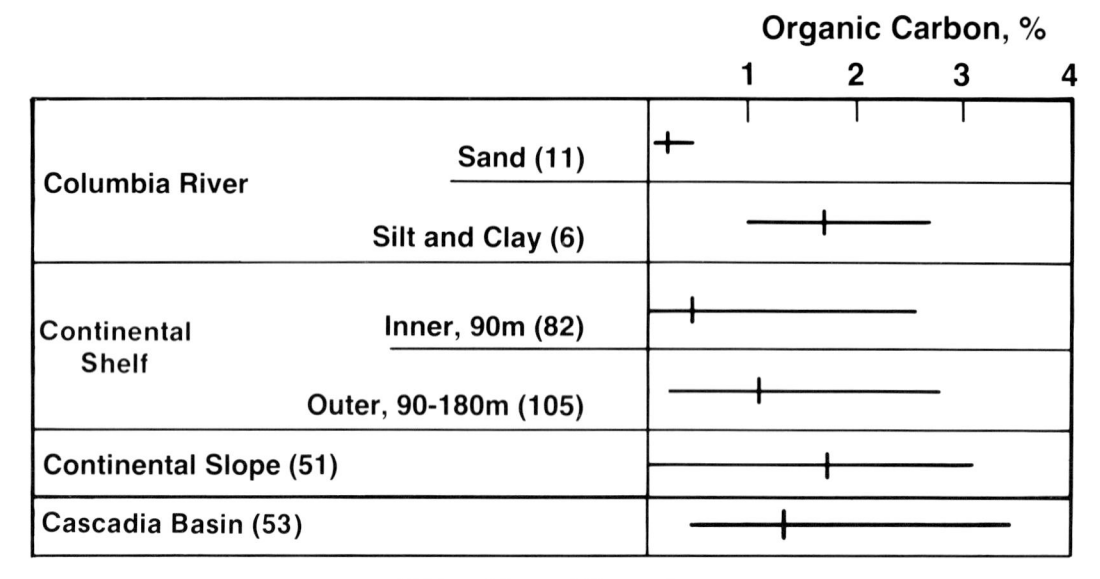

FIG. 8.—Range and median values of organic carbon content of surface sediments offshore Washington and Oregon (after Gross et al., 1972).

amount of organic carbon preserved in the Sulu Sea sediments is <0.5 wt.% of the total productivity of the area (calculations based on data in Exon et al., 1981). This is more than an order of magnitude less than for the anoxic environment of the Black Sea (~5 wt.%, Deuser, 1971) and the oxygen deficient environment of offshore Peru (12 wt.%, Müller and Suess, 1979).

The broad carbonate shelf and upper slope off Palawan Island are dominated by carbonate sands. Finer-grained material, including the organic matter, is winnowed out and TOC values are low (Fig. 10A). Sparse data indicate that the TOC is highest on the slope in an area of rapid deposition and minimum grain size (Exon et al., 1981).

TOC is generally higher on the much narrower shelf and steeper slope off volcanic Mindanao Island than off Palawan (Fig. 10B). However, the shape of the curves is similar. In particular, the largest TOC values are in fine-grained sediments on the upper slope but well beyond the shelf-slope break.

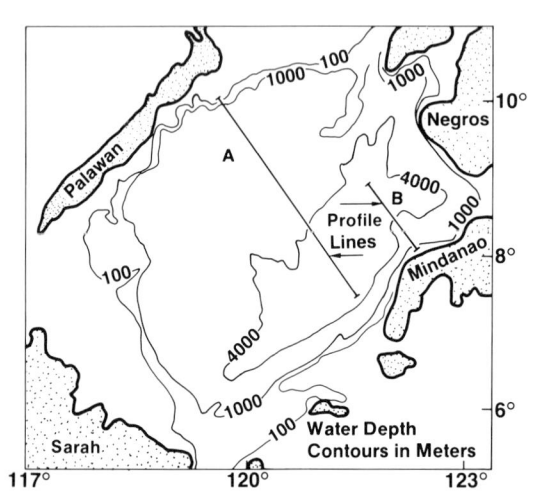

FIG. 9.—Location map, Sulu Sea, southeast Asia (after Exon et al., 1981). Profile data shown in Fig. 10.

Southern California Borderland

Unlike the Sulu Sea, where waters entering the basin at sill depth are well oxygenated, the waters entering the basins of the Southern California Borderland near sill depth are either oxygen deficient, or nearly so (Emery, 1960). This phenomenon is caused by the intersection of a well-defined regional OML ($O_2 \cong 0.5$ ml/L) with the sill depth of many of the nearshore basins (Emery, 1960). The little oxygen available is quickly consumed and TOC values are high on the floor of the basins (Table 3). Because the sills are near the basin bottoms and the regional OML has its oxygen minimum near 700 m, the sediments on most of the basin slopes were deposited under oxic conditions. The substantial TOC that exists in the silts that dominate the basin slope sediments probably reflects rapid deposition under highly productive surface waters.

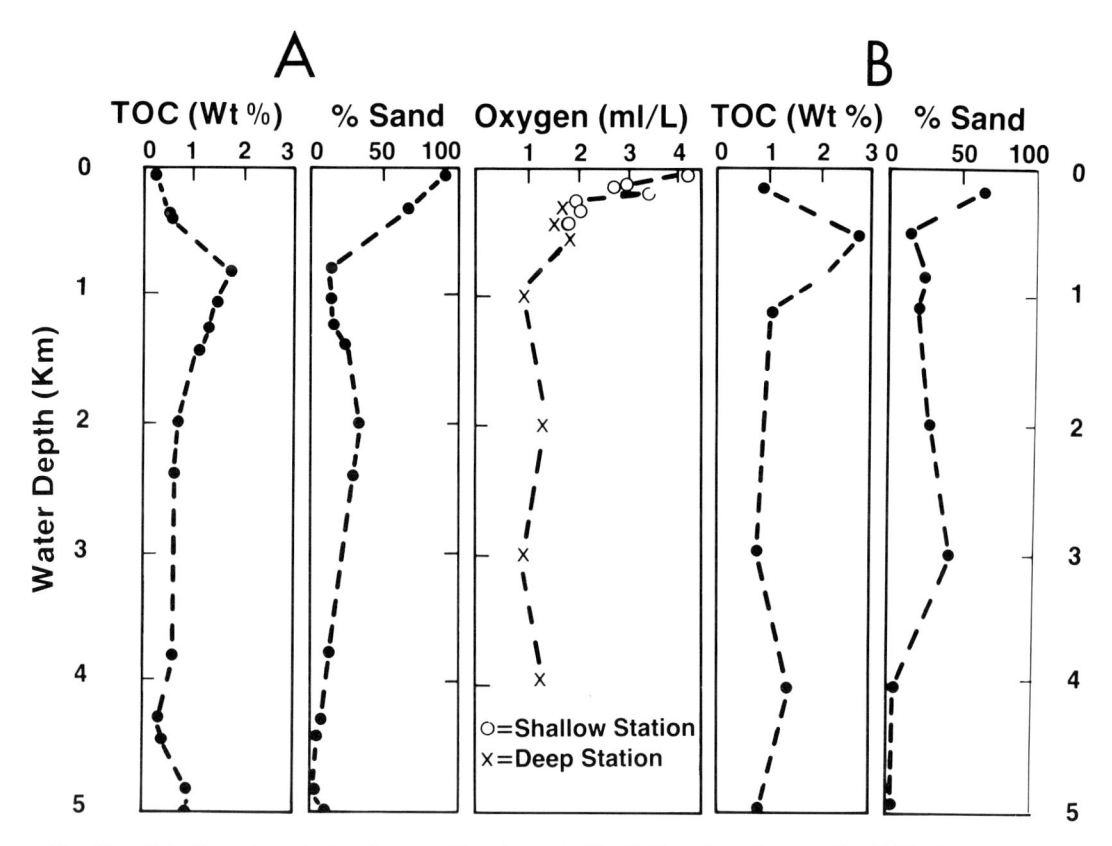

FIG. 10.—Sulu Sea data collected along profiles shown in Fig. 9 (data from Exon et al., 1981). Average oxygen distribution in Sulu Sea (center). A, Profile off carbonate shelf of Palawan Island; B, Profile off volcanic shelf of Mindanao Island.

Shelves of the Southern California Borderland are generally narrow, irregular, and often have an abrupt shelf-slope break controlled by faulting and/or Pleistocene erosion surfaces. They are highly oxygenated and the TOC content usually reflects sedimentary grain size. Along the mainland coast of the Santa Barbara Basin, the highest TOC values (≅2.0 wt.%) are predominantly in the finest sediment, an olive green silt (Emery, 1960). The sediments on the island shelves are coarser than their mainland counterparts and have a lower TOC (Table 3; Emery, 1960).

Northern Gulf of Mexico

Using TOC values of Recent sediments from Trask (1953) and dissolved oxygen data from a 1935 cruise of the R/V ATLANTIS, Richards and Redfield (1953) demonstrated an inverse correlation between organic content in the sediments and the oxygen content of the overlying water (Fig. 11). In particular, the TOC is highest on the outer shelf and upper slope where the oxygen content of the bottom water is lowest. Dow and Pearson (1975) reported a similar location for the TOC maximum in marine shales of Oligocene to Pleistocene age, although the decrease in TOC of the lower slope samples was less pronounced than observed by Richards and Redfield (1953).

In both papers, the reported TOC values are almost all <1.0 wt.% and the contrast in TOC be-

TABLE 3.—VARIATION OF ENVIRONMENT OF DEPOSITION, MEDIAN DIAMETER AND ORGANIC MATTER CONTENT FOR SOME RECENT SEDIMENTS OF THE SOUTHERN CALIFORNIA BORDERLAND (FROM EMERY, 1960)

Environment	Median Diameter (Microns)	Organic Matter (Wt. %)
Mainland Shelves	130 (1773)	.9 (273)
Island Shelves	260 (298)	.6 (168)
Basin Slopes	43 (107)	2.8 (30)
Basin Floors	5.3 (549)	7.0 (80)
Continental Slope	8.0 (20)	4.1 (29)

(20) = Number of Samples

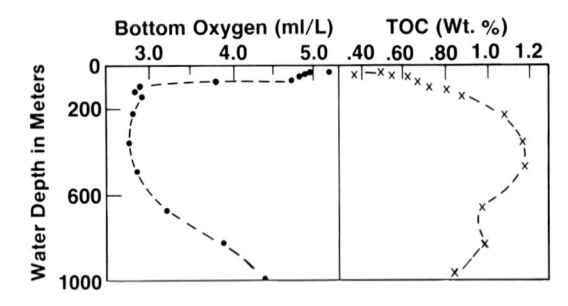

FIG. 11.—Variation of TOC of surficial sediments and the dissolved oxygen content of the water with depth, northern Gulf of Mexico (after Richards and Redfield, 1953).

tween the different depositional environments relatively low. Unpublished Chevron data confirm the TOC distribution with paleo-depth reported in the two papers and also indicate that the organic facies of the slope sediments is more oil-prone than the sediments deposited in both shallower and deeper water.

The OML never drops below 2.5 ml/L of dissolved oxygen, posing the question of whether the co-occurrence of the OML and increased TOC is causal or coincidental. Worldwide data summarized by Demaison and Moore (1980) indicate that ≤0.5 ml/L of dissolved oxygen is necessary to independently and strongly influence the amount of organic carbon preserved in associated sediments. Such low oxygen contents do not exist now in the northern Gulf Coast and there is no reason to believe that they existed in the Tertiary. Thus, alternative explanations must be sought.

Data from the continental shelf of Texas have recorded the increase in organic carbon content toward the shelfedge. Also demonstrated is a correlation with the decreasing mean grain size of the sediments (Berryhill et al., 1977). Using sparker data to locate and core fine-grained, undisturbed Holocene sediments, McKee et al. (1978) found a very good correlation between the thickness of an oxidized zone at the top of the sediments, sedimentation rate, and the TOC (Fig. 12). Their data demonstrate a control on the preservation of organic carbon by the rate of sedimentation regulating the residence time of marine sediments in oxic pore water.

Thus, the conclusion can be made that the slight increase of TOC values in Tertiary sediments of the upper slope in the northern Gulf of Mexico is primarily due to variations in the sedimentation rate and the mean grain size of the sediments. Although well developed, the OML is oxic.

HISTORICAL RECORD

Oxygen Deficient Basins

In cratonic basins the times of maximum areal extent of source rock deposition generally coincided with transgressions of a deep-basin oxygen deficient water body locally over the shelf-slope break.

North America contains many examples. Byers (1977, 1979) proposed a model for the Devonian Sea of New York State in which the top of a 1500± m deep layer of oxygen deficient water and its associated organic-rich sediments extended onto the eastern shelf in times of transgression (Fig. 1B). The oxygen deficient water retreated below the shelf-slope break during the regressions related to deposition of large deltas. A similar control by the varying input of clastic wedges has been proposed for the Cretaceous Pierre and Mowry Shales of the western interior (Nixon, 1973; Byers, 1977, 1979; Byers and Larson, 1979).

Inland seas dominated by shale-carbonate deposition also show the same pattern. Studies by Cluff (1980) on the New Albany Shale (Devonian) of the Illinois Basin demonstrate transgressions of the oxygen deficient waters from the basin deep onto the shelves and the subsequent deposition of lami-

TABLE 4.—Variation of TOC with Paleobathymetry, Oligocene-Pleistocene Age Sediments, Northern Gulf of Mexico (After Dow and Pearson, 1975)

	Plio-Pleistocene	U. Miocene	M. Miocene	L. Miocene	Oligocene	Mean
Inner Shelf 0–100 m	.30 (17)	.19 (27)	.43 (4)	.07 (5)		.23 (53)
Outer Shelf 100–200 m	.53 (54)	.42 (39)	.69 (2)	.36 (7)	.34 (8)	.47 (110)
Upper Slope 200–500 m	.68 (9)	.66 (11)	.67 (26)		.44 (18)	.61 (64)
Lower Slope 500–2000 m	.59 (70)		.63 (9)		.51 (5)	.59 (86)

.53 TOC (Wt. %)
   No. of Samples   Data From: Dow and Pearson, 1975

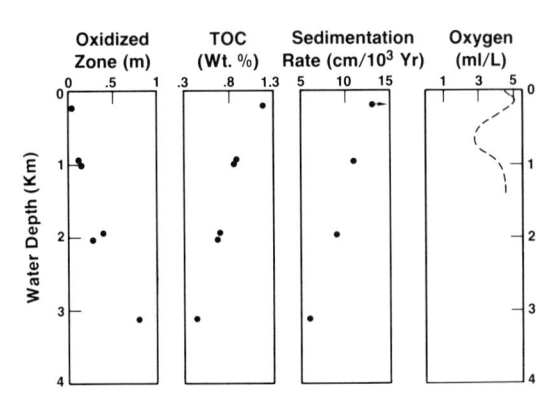

FIG. 12.—Variation of TOC, sedimentation rate, and thickness of oxidized zone at top of sediments with water depth, Holocene sediments, offshore Texas (after McKee et al., 1978).

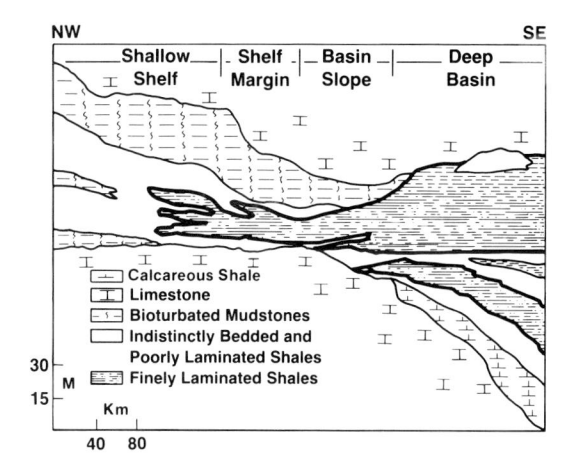

FIG. 13.—Simplified stratigraphic cross-section of the Devonian New Albany Shale, Illinois Basin (after Cluff, 1980).

nated, organic-rich shales across the shelf-slope break (Fig. 13). In the Pennsylvanian Paradox Basin of southeast Utah, the productive carbonate bank complex of the Aneth Field was capped and probably sourced by a black shale that transgressed from the basin deep (Elias, 1963; Hite, 1970). These studies are all well supported by biostratigraphic data that clearly show the enlargement and transgression of basin-wide oxygen deficient water originating in the deepest part of the basin.

Transgressions of oxygen deficient bottom waters into shallow water are not limited to North America. Prior to the main phase of deposition of the Permian Zechstein evaporites in western Europe, the basin was density stratified and highly oxygen deficient. Organic-rich sediments were deposited in the basin deep and within 100 m of the water surface contemporaneously with highly oxygenated reservoir rocks (Fig. 14).

The shallow water Kimmeridgian oil shales described by Hallam (1967) and Morris (1980) had their more massive, thicker counterparts deposited in the deeper portions of the Kimmeridgian Basin in the center of the North Sea. The Toarcian of western Europe, although best known for the bituminous shales that are widespread in the shelf deposits of northern Europe, has bituminous counterparts in the deeper water sediments deposited along the northern margin of the Tethyan Sea in central Europe (Jenkyns, 1980).

In the above examples, a basin-wide stratification was necessary to permit an oxygen deficiency to develop and expand over the shelf-slope break. In some cases, such as the Paradox, Zechstein, and Black Sea Basins, the density stratification was provided by strong salinity gradients, often existing as a prelude to evaporite deposition. However, thermoclines developed during warm and equable climates can also provide the long lasting density stratification that prevents oxygen renewal of bottom waters once they are made oxygen deficient by oxidization of organic matter.

Oxygen Deficient OMLs

Documented examples of OMLs in the geologic record that resulted in deposition of potential source rocks near the shelf-slope break are relatively rare. They should preferentially exist along continental margins. However, many Paleozoic margins have been obscured by later geologic events and the requisite data on later margins are not always available because of proprietary or insufficient data. Nevertheless, some examples exist and others have been suggested.

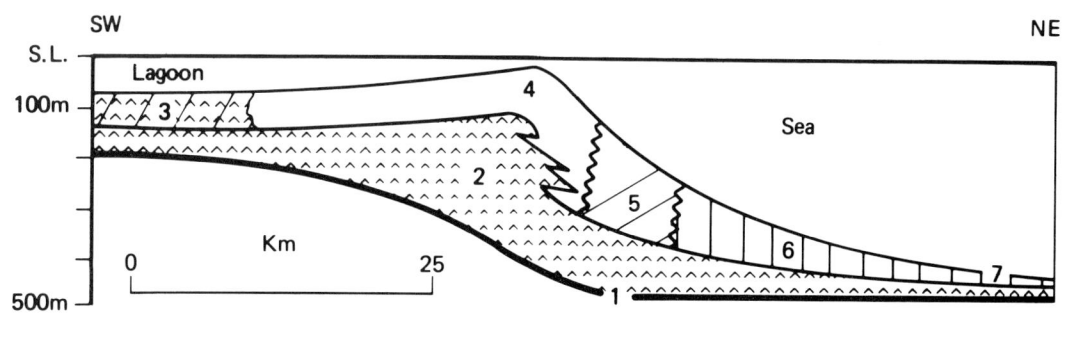

1 Basal Zechstein.
2 Anhydrite.
3 Anhydritic Dolomite.
4 Dolomite Pay, Algal Facies.
5 Stinkdolomit.
6 Stink–Kalk.
7 Stinkschiefer.

FIG. 14.—Diagrammatic cross-section through a marginal part of the Permian Zechstein Basin, Germany, prior to deposition of the main evaporite sequence.

The organic-rich facies of the Phosphoria Formation of western Wyoming and environs was probably deposited under a transgressive oxygen deficient OML caused by intensive shelf-edge upwelling (Fig. 3). Recently, the organic-rich portion of the Monterey Formation of the Miocene of California has been interpreted as being deposited within an oxygen deficient OML (Gilbert and Summerhays, 1981; Govean and Garrison, 1981). This interpretation is given credence by direct analogy with Holocene sedimentation and the oxygen distribution in the Gulf of California (van Andel and Shor, 1964; Gilbert and Summerhays, 1982).

An origin of the organic-rich, phosphate-rich shales of the Pennsylvanian of the Mid-Continent has been ascribed to upwelling on the shelf of oxygen-poor, but oxic, waters during times of maximum transgression (Heckel, 1977). High productivity led to the development of oxygen deficient water at the sediment-water interface and preservation of high TOC values in the sediments. The synchronous, more basinal shales are not as organic-rich as the phosphatic shales on the shelf.

An anoxic OML model has recently been proposed by Comer and Hinch (1981) for the Woodford Formation (Mississippian-Devonian) in and around the Anadarko Basin. They attribute lower organic carbon content and an apparent increase in terrestrial organic matter in the axial portion of the basin to deposition in oxic water under an anoxic OML. However, the organic matter in the Woodford in the axial portion of the Anadarko Basin is thermally post-mature. Both the lower organic carbon contents and the structured appearance of the organic matter could simply reflect a higher thermal maturation. Also, the appropriateness of the basin-wide oxygen deficient model is suggested by the persistence of the pre-Woodford unconformity into the deep portion of the Anadarko Basin (Adler, 1971) and, therefore, an implied moderate water depth for the deposition of the Woodford directly on the unconformity. Another indication of basin-wide oxygen deficiency is the well-established correlation of the Woodford with the organic-rich, oil-prone "Arkansas novaculite" deposited in deeper water of the Ouachita Trough to the southeast of the Anadarko Basin (Briggs, 1974). In any case, the transgression of highly oxygen deficient water over shelf-slope breaks is necessary to explain the distribution of the organic-rich, oil-prone Woodford directly above an unconformity on the surrounding shelves.

CONCLUSIONS

Two distinct models are consistent with the deposition of organic-rich rocks near the shelf-slope break. One, the oxygen deficient basin (ocean) model, involves a basin-wide transgression of oxygen deficient water that originally developed in the basin deep. This model, when accompanied by high productivity on the shelf and an equable climate, is probably the best explanation for most of the source rocks that extend over shelf-slope breaks in the interior of the continents. The second model requires the development of an oxygen deficient OML near the shelf-slope break. This model adequately describes the situation on the margins of Mesozoic to modern oceans and was undoubtedly applicable to similar settings in the geologic past despite the obscuration or obliteration of the geologic record.

In the modern oceans, the organic content of surface sediments reaches a maximum on the slope and usually on the upper slope. To obtain high TOC values, an oxygen deficient OML is necessary. However, the highest TOC values will normally occur on the (upper) slope regardless of how depleted in oxygen the water becomes. This situation occurs because: (1) the slope sediments usually have a substantially smaller mean grain size than shelf sediments, but often a similar sedimentation rate, and (2) they have a comparable grain size to abyssal sediments, but a much faster sedimentation rate.

ACKNOWLEDGMENTS

I thank M. A. Arthur, G. T. Moore, and E. Suess for critically reviewing the manuscript. Special thanks go to G. T. Moore for encouraging me to prepare the paper, unswerving support during its preparation, and a firm editorial hand.

REFERENCES

ADLER, F. J., 1971, Anadarko basin and the central Oklahoma area, *in* Gram, I. H., ed., Future Petroleum Provinces of the United States—Their Geology and Potential: Am. Assoc. Petroleum Geologists Memoir 15, v. 2, p. 1061–1070.

BERNER, R. A., 1981, A new geochemical classification of sedimentary environments: Jour. Sed. Petrology, v. 51, p. 359–365.

BERRYHILL, H. L., ed., 1977, Environmental Studies, South Texas Shelf: U.S. Geol. Survey, Rept. to the U.S. Bur. Land Management, 303 p.

BRIGGS, G., 1974, Carboniferous depositional environments in the Ouachita Mountains—Arkoma Basin area of southeastern Oklahoma, *in* Briggs, G., ed., Carboniferous of the Southeastern United States: Geol. Soc. America, Spec. Paper 148, p. 225–241.

BURNETT, W. C., VEEK, H. H., AND SOUTAR, A., 1980, U-series, oceanographic and sedimentary evidence in support of Recent formation of phosphate nodules off Peru, *in* Bentor, W.Y., ed., Marine Phosphorites: Geochemistry, Occurrence and Genesis: Soc. Econ. Paleontologists Mineralogists Spec. Pub. 29, p. 61–71.

BYERS, C. W., 1977, Biofacies patterns in euxinic basins: A general model, *in* Cook, H. F., and Enos, P., eds., Deep-Water Carbonate Environments: Soc. Econ. Paleontologists Mineralogists Spec. Pub. No. 25, p. 5–17.

―――, 1979, Biogenic structures of black shale environments: Postilla, No. 174, 46 leaves.

―――, AND LARSON, D. W., 1979, Paleoenvironments of Mowry Shale (Lower Cretaceous), western and central Wyoming: Am. Assoc. Petroleum Geologists Bull., v. 63, p. 354–361.

CALVERT, S. E., 1966, Origin of diatom-rich, varved sediments from the Gulf of California: Am. Assoc. Petroleum Geologists Bull., v. 50, p. 546–565.

―――, AND PRICE, N. B., 1971a, Upwelling and nutrient regeneration in the Beneguela Current, October, 1968: Deep-Sea Res., v. 13, p. 505–523.

―――, AND ―――, 1971b, Recent sediments of the South-West African Shelf, *in* Delany, F. M., ed., Atlantic Continental Margins: London Inst. Geol. Sci., p. 175–185.

CLAYPOOL, G. E., LOVE, A. H., AND MAUGHAN, E. K., 1978, Organic geochemistry, incipient metamorphism, and oil generation in black shale members of Phosphoria Formation, western interior, United States: Am. Assoc. Petroleum Geologists Bull., v. 62, p. 98–120.

CLOSS, H., VARIAN, H., AND GARDE, S. C., 1974, Continental margins of India, *in* C. A. Burk and Drake, C. L., eds., The Geology of Continental Margins: New York, Springer-Verlag, 629–639.

CLUFF, R. M., 1980, Paleoenvironment of the New Albany Shale Group (Devonian-Mississippian) of Illinois: Jour. Sed. Petrology, v. 50, p. 767–780.

COMBAZ, A., AND DE MATHAREL, M., 1978, Organic sedimentation and genesis of petroleum in Mahakam delta, Borneo: Am. Assoc. Petroleum Geologists Bull., v. 62, p. 1684–1695.

COMER, J. B., AND HINCH, H. H., 1981, Petrologic factors controlling internal migration and expulsion of petroleum from source rocks: Woodford-Chattanooga of Oklahoma and Arkansas [abs.]: Am. Assoc. Petroleum Geologists Bull., v. 65, p. 912.

DEGENS, E. T., 1974, Cellular processes in Black Sea sediments, *in* Degens, E. T., and Ross, A. A., eds., The Black Sea—Geology, Chemistry and Biology: Am. Assoc. Petroleum Geologists Memoir 20, p. 296–307.

―――, AND ROSS, D. A., eds., 1974, The Black Sea—Geology, Chemistry and Biology: Am. Assoc. Petroleum Geologists Memoir 20, 633 p.

―――, AND OTHERS, 1978, Varve chronology: estimated rates of sedimentation in the Black Sea deep basin, *in* Stoffers, P., Glaubic, S., and Dickman, M., eds., Initial Reports of the Deep Sea Drilling Project, Volume 42, Part 2: Washington, D.C., Govt. Printing Office, p. 499–508.

DEMAISON, G. J., AND MOORE, G. T., 1980, Anoxic environments and oil source bed genesis: Am. Assoc. Petroleum Geologists Bull., v. 64, p. 1179–1209.

DEUSER, W. G., 1971, Organic-carbon budget of the Black Sea: Deep-Sea Res., v. 18, p. 995–1004.

―――, 1975, Reducing environments, *in* Riley, J. P., and Skirrow, G., eds., Chemical Oceanography, Volume 3: New York, Academic Press, p. 1–37.

DIESTER-HASS, L., AND MÜLLER, P. J., 1979, Factors influencing sand fraction components and organic matter in surface sediments of the West African upwelling region, 12°–19°N: "Meteor" Forschungsergeb., Reihe C., v. 31, p. 21–47.

DOW, W. G., 1978, Petroleum source beds on continental slopes and rises: Am. Assoc. Petroleum Geologists Bull., v. 62, p. 1584–1606.

―――, AND PEARSON, D. B., 1975, Organic matter in Gulf Coast sediments: 7th Offshore Tech. Conf. Preprints, Paper OTC 2343, 10 p.

ELIAS, G. K., 1963, Habitat of Pennsylvanian algal bioherms, Four Corners area, *in* Bass, R., and Sharps, S. L., eds., Shelf Carbonates of the Paradox Basin, a Symposium: Four Corners Geol. Soc. 4th Field Conf., p. 185–203.

EMERY, K. O., 1960, The Sea off Southern California: New York, J. Wiley, 366 p.

―――, AND HÜLSEMANN, J., 1962, The relationships of sediments, life and water in a marine basin: Deep-Sea Res., v. 8, p. 165–180.

EXON, N., AND OTHERS, 1981, Morphology, water characteristics and sedimentation in the silled Sulu Sea, southeast Asia: Mar. Geology, v. 39, p. 165–195.

FAIRBRIDGE, R. W., ed., 1966, The Encyclopedia of Oceanography: New York, Van Nostrand-Reinhold, 1021 p.

FISCHER, A. G., AND ARTHUR, M. A., 1977, Secular variations in the pelagic realm, *in* Cook, H. E., and Enos, P., eds., Deep-water Carbonate Environments: Soc. Econ. Paleontologists Mineralogists Spec. Pub. 25, p. 19–50.

FROELICH, P. N., AND OTHERS, 1979, Early oxidation of organic matter in pelagic sediments of the eastern equatorial Atlantic: suboxic diagenesis: Geochim. Cosmochim. Acta, v. 43, p. 1075–1090.

GERSCHANOVICH, D. E., VEBER, V. V., AND KONJUKHOV, A. E., 1976, Organic matter in the bottom sediments of the Peru region of the Pacific Ocean, *in* Vassoevich, N. B., and others, eds., Issledovania Organisheskogo Sovremenyk Iskopaemyk Osadkov: Moscow, Akad Nauk, SSSR, p. 121–128, in Russian.

GILBERT, D., AND SUMMERHAYS, C. P., 1981, Distribution of organic matter in sediments along the California continental margin, *in* Orlofsky, S., ed., Initial Reports of the Deep Sea Drilling Project, Volume 63: Washington, D.C., Govt. Printing Office, p. 757–762.

―――, AND ―――, 1982, Organic facies and hydrocarbon potential in the Gulf of California, *in* Initial Reports of the Deep Sea Drilling Project, Volume 64: Washington, D.C., Govt. Printing Office, in press.

GOVEAN, F. M., AND GARRISON, R. E., 1981, Significance of laminated and massive diatomites in the upper part of the Monterey Formation, California, in Garrison, R. E., and Douglas, R. G., eds., The Monterey Formation and Related Siliceous Rocks of California: Soc. Econ. Paleontologists Mineralogists Res. Symposium, p. 181–198.

GROSS, M. G., AND OTHERS, 1972, Distribution of organic carbon in surface sediment, northeast Pacific Ocean, in Prugher, A. T., and Alberson, A. L., eds., The Columbia River Estuary and Adjacent Ocean Waters: Seattle, Univ. Washington Press, p. 254–264.

HALLAM, A., 1967, The depth significance of shales with bituminous laminae: Mar. Geology, v. 5, p. 481–493.

HAMILTON, W., 1977, Subduction in the Indonesian region, in Talwani, M., and Pitman, W. C., III, eds., Island Arcs, Deep Sea Trenches and Back-arc Basins, Maurice Ewing Series, Volume I: Am. Geoph. Union, p. 15–31.

HEATH, G. R., MOORE, T. C., JR., AND DAUPHIN, J. P., 1977, Organic carbon in deep-sea sediments, in Andersen, R. N., ed., The Fate of Fossil Fuel CO_2 in the Oceans: New York, Plenum Press, p. 627–639.

HECKEL, P. H., 1977, Origin of phosphate black shale facies in Pennsylvanian cyclothems of Mid-Continental North America: Am. Assoc. Petroleum Geologists Bull., v. 61, p. 1045–1068.

HEDGES, J. I., AND MANN, D. C., 1979, The lignin geochemistry of marine sediments from the southern Washington coast: Geochim. et Cosmochim. Acta., v. 43, p. 1809–1818.

HITE, R. J., 1970, Shelf carbonate sedimentation controlled by salinity in the Paradox basin, southeast Utah, in Rau, J. L., and Dellwig, L. F., eds., Third Symposium on Salt, Volume 1: Cleveland, Northern Ohio Geol. Soc., Inc., p. 48–66.

JENKYNS, H. C., 1980, Cretaceous anotic events: from continents to oceans: Jour. Geol. Soc. London, v. 137, p. 171–188.

JONES, R. W., AND DEMAISON, G. J., 1980, Organic facies—stratigraphic concept and exploration tool [abs.]: Am. Assoc. Petroleum Geologists Bull., v. 64, p. 729.

———, 1982, Organic facies-stratigraphic concept and exploration tool: Proc. 1981 ASCOPE '81 Conf., Manila, in press.

KIDWAI, R. M., AND NAIR, R. R., 1972, Distribution of organic matter on the continental shelf off Bombay: a terrigenous-carbonate depositional environment: Indian Jour. Marine Sci., v. 1, p. 116–118.

KONJUKHOV, A. E., 1976, Facies characteristics of contemporary sediments from the western submarine margin of the Indian Peninsula, in Vassoevich, N. B., and others, eds., Issledovania Organisheskogo Veshchestva Sovremenyk Iskopaemyk Osadkov: Moscow, Akad. Nauk SSSR, p. 111–120, in Russian.

LISITSIN, A. P., 1978, Terrigenous sedimentation, climatic zonation and interaction of terrigenous and biogenous material in the ocean: Lith. and Min. Res., v. 12, p. 617–632.

MCKEE, T. R., AND OTHERS, 1978, Holocene sediment and geochemistry of continental slope and intraslope basin areas, northwest Gulf of Mexico, in Bouma, A. H., Moore, G. T., and Coleman, J. M., eds., Framework, Facies and Oil-trapping Characteristics of the Upper Continental Margin, Am. Assoc. Petroleum Geologists, Studies in Geology 7, p. 313–326.

MORRIS, K. A., 1980, Comparison of major sequences of organic-rich mud deposition in the British Jurassic: Jour. Geol. Soc. London, v. 137, p. 157–170.

MÜLLER, P. J., AND MANGINI, A., 1980, Organic carbon deposition rates in sediments of the Pacific manganese nodule belt dated by 230Th and 231Pa: Earth and Planetary Sci. Letters, v. 51, p. 94–114.

———, AND SUESS, E., 1979, Productivity, sedimentation rate, and sedimentary organic matter in the oceans—I. Organic carbon preservation: Deep-Sea Res., v. 22A, p. 1347–1362.

NIXON, R. P., 1973, Oil source beds in Cretaceous Mowry Shale of northwestern interior United States: Am. Assoc. Petroleum Geologists Bull., v. 57, p. 136–161.

NORTH, F. K., 1979, Episodes of source-sediment deposition (1): Jour. Pet. Geol., v. 2, p. 199–218.

PAROPAKARI, A. L., RUO, C. M., AND MURTY, P. S. N., 1978, Geochemical studies on the shelf sediments off Bombay: Indian Jour. Marine Sci., v. 7, p. 8–11.

RICHARDS, F. A., 1970, The enhanced preservation of organic matter in anoxic marine environments, in Hood, D. W., ed., Symposium on Organic Matter in Natural Waters: Univ. Alaska Inst. Marine Sci. Occasional Pub. 1, p. 399–411.

———, AND REDFIELD, A. C., 1953, A correlation between the oxygen content of sea water and the organic content of marine sediments: Deep-Sea Res., v. 1, p. 279–281.

RYTHER, J. H., AND MENZEL, D. W., 1965, On the production, composition and distribution of organic matter in the Western Arabian Sea: Deep-Sea Res., v. 12, p. 199–209.

STANLEY, D. J., ADDY, S. K., AND BEHRENS, E. W., 1983, The mudline: variability of its position relative to shelfbreak, in Stanley, D. J., and Moore, G. T, eds., The Shelfbreak: Critical Interface on Continental Margins: Soc. Econ. Paleontologists Mineralogists Spec. Pub. 33, p. 279–298.

SUMMERHAYS, C. P., DE MELO, U., AND BARETTO, H. T., 1976, The influence of upwelling on suspended matter and shelf sediments off southeastern Brazil: Jour. Sed. Petrology, v. 46, p. 819–828.

THIEDE, J., AND VAN ANDEL, T. H., 1977, The paleoenvironment of anaerobic sediments in the Late Mesozic South Atlantic Ocean: Earth and Planetary Sci. Letters, v. 33, p. 301–309.

TOTH, D. J., AND LERMAN, A., 1977, Organic matter reactivity and sedimentation rates in the ocean: Am. Jour. Sci., v. 277, p. 465–485.

TRASK, P. D., 1932, Origin and environments of source sediments of petroleum: Houston, Gulf Publishing Co., 323 p.
——, 1953, The sediments of the Western Gulf of Mexico: Part II—Chemical studies of sediments of the Western Gulf of Mexico: Mass. Inst. Tech. and Woods Hole Oceanographic Inst., Phys. Oceanog. and Meteor., v. 12, no. 4, p. 47–120.
VAIL, P. R., AND MITCHUM, R. M., JR., 1979, Global cycles of sea level change and their role in exploration: 10th World Petroleum Cong., Preprint, Bucharest Panel Discussion 2, Paper 4, 11 p.
——, ——, AND THOMPSON, S., III, 1977, Seismic stratigraphy and global changes of sea level, pt. 4: Global cycles of relative changes of sea level, in Payton, C. E., ed., Seismic Stratigraphy Applications to Hydrocarbon Exploration: Am. Assoc. Petroleum Geologists Memoir 26, p. 83–97.
VAN ANDEL, T. H., AND SHOR, G. G., 1964, Marine Geology of the Gulf of California: Am. Assoc. Petroleum Geologists Memoir 3, 408 p.
VAN KREVELEN, D. W., 1961, Coal: Amsterdam, Elsevier, 514 p.
VON STACKELBERG, U., 1972, Faziesverteilung in Sedimenten des indisch-pakistanischen Kontinentalrandes (Arabisches Meer): "Meteor" Forschungsergebnisse, Reihe C, no. 9, p. 1–173.
WEBER, K. J., AND DAUKORU, E., 1975, Petroleum geology of the Niger Delta: Proc. 9th World Pet. Congr., v. 2, p. 209–221.
WORTHINGTON, L. V., AND WRIGHT, W. R., 1970, North Atlantic Ocean Atlas: Woods Hole, Massachusetts, Woods Hole Oceanogr. Inst., 24 p., 58 plates.
WYRTKI, K., 1971, Oceanographic Atlas of the International Indian Ocean Expedition: Washington, D.C., National Science Foundation, 531 p.

PART VIII
ECONOMIC PROSPECTS
AND LEGAL ASPECTS

OCCURRENCES OF OIL AND GAS IN ASSOCIATION WITH THE PALEO-SHELFBREAK

WILLIAM C. KRUEGER, JR.
Amoco Production Co., Tulsa, Oklahoma 74102

F. K. NORTH
Consultant, La Habra, California 90631

ABSTRACT

Hinge belts between shelves and basins are sought by petroleum explorationists as preferred sites of hydrocarbon accumulation. Shelf-slope breaks between continents and oceans are the largest and most durable of such belts. They may or may not be associated with plate boundaries. Ancient shelf-slope breaks of this order are not, however, especially favorable sites for hydrocarbon accumulation. Hinge belts having all the structural and lithological hallmarks of true shelf-slope breaks may, however, be developed within restricted or interior basins, or they may face such basins rather than facing open oceans. These inward-facing breaks are highly favorable sites for hydrocarbon accumulation, although it is commonly impossible to demonstrate more than general coincidence between the accumulations and the breaks.

Shelf-slope breaks providing control for hydrocarbon accumulations in interior basins may survive with little or no deformation, as in the Permian Basin. Those forming parts of the margins of extensional ocean basins may similarly survive the stabilization of their basins, as along the Atlantic Ocean margin. Those within convergent margins become deformed or concealed by incorporation into orogenic belts like those along the Tethyan Sea margins. Three principal source-reservoir lithological associations are considered: carbonate-evaporite, mixed carbonate and clastic, and wholly clastic. The second is the most likely to result in large hydrocarbon accumulations in immediate proximity to the breaks; the last is least likely to do so. Notable examples described herein are: (1) the inner foothills fields of Iran and Iraq and the Golden Lane-Reforma-Campeche fields of Mexico, representing the carbonate-evaporite association; (2) the Permian basin, the Cretaceous of the U.S. outer Gulf Coast, and the fields of the Brazilian Campos and Southeast China Sea basins, representing mixed lithological associations; (3) the Arkoma and Upper Assam Valley basins as deformed representatives; and (4) the northwest Australian shelf and the Texas Gulf Coast Eocene breakover as undeformed representatives of the clastic facies association.

INTRODUCTION

Postulates axiomatic to many petroleum scientists are that hydrocarbons are generated most successfully in the sediments of basins, which are regions undergoing sufficiently continuous subsidence to acquire thicker sediments than their surroundings; that the hydrocarbons then migrate toward the margins of the basins in order to reduce lithostatic load pressure; and that they finally become pooled where sedimentary rocks of a different type interfinger with the basinal sediments. These "different" sediments have been deposited in areas less continuously subsiding, hence more commonly agitated by currents, waves, and organisms, and they are consequently endowed with the porosity that is the fundamental requisite of a fluid accumulation.

The zone that separates a basin from its shallower surroundings is commonly called the hinge zone or hinge belt. Hinge belts have been accorded a preferred position for the development of large oil fields (and for some large gas fields) by nearly all petroleum geologists (Weeks, 1952; Knebel and Rodriguez-Eraso, 1956; Halbouty et al., 1970; Klemme, 1980). The most continuous and long-lived hinge belts are those that separate the continents from the ocean basins; they are the shelf-slope breaks at the continental margins. The shelf-slope break is a dynamic boundary. It represents profound changes in physiography, energy, lithology, fauna, and the chemistry of the marine system.

Exploration for petroleum in many areas of the world has been governed by the belief of geologists and geophysicists that stratigraphic mapping and/or seismic profiling has identified such a change of slope of the strata, and that this change of slope can be interpreted as having been original and depositional, not secondary and structural. The hope, then, is that it represents an increase in the gradient of the seafloor across the area on which a prospective sedimentary succession was being deposited.

The break between shelf and slope is commonly envisaged as that marking the oceanward edge of a simple, idealized continental shelf, leading downwards to a simple continental slope and rise and, thence, into the abyssal plain of a true ocean. The rather clear evidence of thousands of known oil and gas accumulations, from some 125 basins, is that this "type" shelfbreak is unlikely to become the locus of important oil or gas deposits. Modern shelf-edges and upper slopes of this type may be the sites of rich accumulations of organic matter, but only under exceptional circumstances have these become the sources of rich accumulations of hydrocarbons. Shelfedges facing open oceans lack access to effective source sediments, from which hydrocarbons are matured, expelled, and preserved.

Famous alignments of large oil fields ascribed to hinge-belt or hinge-line controls (eastern Venezuela, Los Angeles, Sirte, and central Sumatran basins, and the Frio trend of the U.S. Gulf Coast) are assigned to different sub-categories in familiar classifications of petroleum basins. In all these and similar cases, however, the field trends are intracratonic in position; so, in all probability, are the sites of accumulation of their principal source sediments. The petroleum-rich hinge belts were not paleoshelf-slope breaks.

Most of the trends of oil and gas fields deduced here to have developed along, or close to, paleo-shelf-slope breaks did so within one of three settings. Some were associated with breaks facing relatively narrow or enclosed seaways having some structural/topographic isolation from an open ocean. A newly-opened, extensional ocean basin must present such a setting early in its existence; the circum-Atlantic basins did so during Early Cretaceous time. The setting may become long-lived if complex or non-uniform rifting creates outer ridges or platforms along the new margins, isolating two-sided basins behind them. The Avalon basin off Newfoundland and the Lewis trough behind the Rankin platform off northwestern Australia are examples.

The second setting is within an interior seaway undergoing closure during subduction or collision. An excellent example is provided by the shelf-slope break which extends across the northern Black Sea today and continued across the southern Caspian Sea during Neogene time. It was formed by late Mesozoic rifting between the northern margin of the Tethyan plate and its flanking volcanic arc, and now lies closely behind the boundary between the Alpine and Hercynian systems of southeastern Europe. The Caucasus-Caspian sector of the break is highly petroliferous. The two-sided character of the basin arises during a critical phase of convergence at which maximum basinward slopes are acquired, as in the Iran-Iraq and Arkoma-Ouachita oilfield belts discussed in this chapter.

The third setting is within a wholly interior, intracratonic seaway, devoid of immediate association with any plate boundary and undergoing neither extension nor closure. Notable examples in prolifically petroliferous regions are the flexure crossing the San Joaquin Basin of California during the Miocene, and the western side of the Central Basin Platform of early and middle Permian time in the Permian Basin.

The terms shelf and slope in reference to petroleum deposits refer to features that existed at the time the rocks now forming the commercial reservoirs were deposited. The position of the present shelf-slope break is important to the exploration process for operational and economic reasons; it is not the present break that is discussed in this paper. To avoid confusion, we use here the more unwieldy, but less ambiguous, terms paleoshelf and paleoslope.

TECTONIC CONDITIONS AT SHELF-SLOPE BREAKS

Because they represent lengthy discontinuities in the crust's geometric and geologic pattern, shelf-slope breaks are prone to tectonic disruption. Breaks in immediate association with plate boundaries, like those west of the Americas, are episodically the loci of subduction- or transform-related deformation. The normal growth faults which developed as consequences of the break are likely to be replaced by continent-verging thrust faults, and the contents of the slope (and the rise) thrust over the shelf and the continent. The fronts of the resulting orogenic belts (or of the flysch belts that they commonly contain) must bear some direct spatial relationship, possibly close, to the original shelf-slope break, and oil or gas fields are common parallel to such fronts. Obvious examples are the fields in front of the Zagros, Ouachita, Appalachian, and coastal Venezuelan mountains and of the Naga Hills in Upper Assam. The fields are more likely to lie some distance shelfward or slopeward of the paleobreak than along it, but the actual relationship may defy convincing detection.

Shelf-slope breaks marking extensional plate boundaries are not at continental margins except during the earliest rifting phase, as on the two sides of the present Red Sea. They may, however, become richly petroliferous if they are aborted and left behind by the development of more successful breaks which eventually become the plate boundaries. The Avalon basin off Newfoundland, the Gabon basin of West Africa, and the Rankin platform off northwestern Australia are examples.

The oil and gas accumulations most obviously identifiable with shelf-slope breaks are those on trailing, Atlantic-type margins now far removed from plate boundaries. Except off the mouths of large rivers, these may have remained close to the shelf-slope break to the present day; the fields of the Campos basin off Brazil will serve as an illustration of this type.

FACIES CONSIDERATIONS

Carbonate-Evaporite Facies

The most spectacular oil field developments at, or close to, paleo-shelfedges happen to occur in successions dominantly of carbonate-evaporite facies. Both the Iran-Iraq belt and the Tampico-Campeche region of Mexico were parts of the carbonate rich subequatorial seaway of Cretaceous time, when the coincidence of all factors favorable for the generation and preservation of oil seems to have occurred (Irving et al., 1974). Where the edge of a carbonate bank or ramp (Ahr, 1973; Murris, 1980) coincides with a paleoshelf-slope break—as on the Blake-Bahamas plateau during the Mesozoic and Cenozoic—shelfedge reefs may be highly suitable reservoir rocks. They can be richly petroliferous, however, only where they had access to organically-rich, noncarbonate, nonevaporitic source sediments. In such favorable settings, the spectacular nature of the oil fields may deceive one into thinking of them as grandiose examples of a fundamentally common phenomenon. This is not the case. With relatively trivial exceptions, pre-Jurassic carbonate banks and reefs contain no very large fields near true continental paleo-shelfedges.

Mixed Lithologies

A far commoner setting is that in which lithofacies are markedly mixed. It is also the setting most easily recognized for its control by the paleobreak, especially in the change from reef carbonate to off-reef noncarbonate rock. Outside the petroleum context, the basin-margin facies changes most frequently cited in the literature are of this type: Middle Devonian of the Rhineland, Lower Permian of the downwarp west of the Ural Mountains, Permian of the Delaware and Midland basins, and Halstatt facies of the Triassic of the Carnic (Austrian) Alps.

Although the Permian basin is commonly cited as a classic case of carbonate-reservoired oil fields with evaporite seals, it contains hundreds of oil and gas accumulations in sandstone reservoirs, and the principal oil source sediments are assuredly dark shales with little carbonate content. The lower Paleozoic fields in front of the Ouachita and Arbuckle mountains, the Cretaceous of the U.S. Gulf Coast and of the circum-Atlantic margins, and the Neogene of southeast Asia provide numerous comparable examples, some of which are described below.

Clastic Facies

The third principal facies association, that essentially of clastic rocks without important carbonates or evaporites, is much less certainly ascribed to shelfbreak control. It is clear that the rare petroliferous regions in fore-arc basins, like the Sacramento Valley gas fields within the Great Valley sequence of California and the slightly younger oil fields in southwestern Ecuador, are likely to cross paleoshelf-slope breaks. It is also clear that compressional basins containing strata of the same age across the geosyncline-foreland boundary must cross paleoshelf-slope breaks. Individual sandstone bodies of types suitable for petroleum reservoir rock are nonetheless unlikely to cross uninterruptedly over such breaks. They are more likely to stop abruptly at them. Shelfedge sands are common features of the stratigraphic record. An association consisting of well-sorted sandstones, wedging out laterally into mudstones and overlying pelites of the upper slope and then massive sandstones and pelites of the lower slope is the hallmark of a progradational paleo-shelfedge. There are hundreds of petroliferous examples, but if the original dip is not reversed the hydrocarbons are likely to migrate toward the paleo-shoreline and will not now be in the position of the paleobreak. Recently described examples are the small fields in the Upper Cretaceous Sussex and Shannon sandstones of the western flank of the Powder River basin in Wyoming (Brenner, 1978). Both sand bodies terminate in abrupt depositional slopes at the paleobreak.

Conversely, sand bodies formed by transport across the break are likely to accumulate downslope from it, possible at the mouths of channels at the foot of the paleoslope, on the rise, or on the basin floor. Notable reservoir rocks interpreted as submarine fan deposits (the Spraberry sands and silts in the Permian basin of West Texas, the Paleocene and Eocene sandstones of Forties, Frigg, and other giant North Sea fields, the Retrench and Herrera sands in the Miocene of Trinidad, the Pliocene sands of the Ventura and Los Angeles basins) owe their existence to suitable paleoslopes. They are not to be identified, however, with the breaks at the tops of those paleoslopes now. Exceptions must be made for productive sands of turbidite fan origin with apices coinciding with the paleo-shelfbreaks. A number of relatively small, stratigraphically-trapped fields in the Miocene Stevens sandstone of California appear to share this characteristic, and are described hereafter.

The actual lithofacies characterizing any stretch of shelf or paleo-shelfbreak depends on an array of controls—geometry of the break, source of sediment, climate, whether the slope is constructional or erosional, and numerous other factors—that are beyond consideration in this chapter but are discussed elsewhere in this volume.

PALEO-SHELFBREAK OIL FIELDS
OF CARBONATE ASSOCIATION

Iran-Iraq Oil Field Belt

The productive Persian Gulf basin is occupied by a northeastward-thickening wedge of sedimentary rocks, largely carbonates, lying atop the Arabian sub-plate. This sub-plate, part of the African plate, became a downgoing slab which was partly consumed below the Iranian sub-plate in mid-Tertiary time. The suture now lies within the Zagros Mountains.

Throughout Jurassic and Early Cretaceous time, the continental margin of the Arabian sub-plate separated a carbonate-dominated platform, on the southwest, from the Tethyan Ocean to the northeast. The continental slope fluctuated in position, receding westward during episodes of rising sealevel, with pelagic deposits resting on shelf carbonates, and advancing seaward during falling sealevel, when the carbonate platform built out over slope and basinal marls. The position of the slope remained, however, consistently to the northeast of the present Iranian mountain front (Murris, 1980). Reconstructions by Falcon (1958) suggest that the deepest part of the trough then lay as much as 65 km northeast of the Iranian oilfield belt (Fig. 1). Contemporary depositional gradients derived from isopachs were 20 to 30 m · km^{-1}. This time interval saw the deposition of the principal source sediments of the region (Dunnington, 1967; Murris, 1980).

By late Turonian-early Senonian time, consumption of the leading edge of the downgoing platform had begun, initiating a collision orogeny and the creation of an open marine foredeep through what is now the Iran-Iraq oilfield belt. This seaway lasted through latest Cretaceous and earliest Tertiary time. Northeast of it, synorogenic flysch deposits accumulated; southwest of it, a steep, narrow paleobreak separated the seaway from the shallow evaporitic platform on the craton (Fig. 2). The axis of the flysch trough lay more than 100 km northeast of the oilfield belt (Falcon, 1958), but it migrated progressively southwestward during Paleocene and early Eocene time.

Oil from Cretaceous sources accumulated in Cretaceous limestone reservoirs during this latest Cretaceous-earliest Tertiary time interval (Dunnington, 1958; Murris, 1980). The late Tertiary folding phase then created the belt of giant anticlinal traps in which the Oligo-Miocene Asmari limestone now contains the bulk of the re-migrated oil. The innermost line of anticlinal fields, closest to the mountain front, includes, at its two ends, two of the largest of the foothills fields, Kirkuk in Iraq and Gach Saran in Iran (Fig. 1). The Kirkuk anticline is nearly 100 km long, that at Gach Saran nearly 60 km; structural closures are nearly 800 m at Kirkuk (Fig. 3) and almost 2500 m at Gach Saran. The two fields between them contain ultimately recoverable reserves of oil approaching 5×10^9 m^3.

From the Turonian to the Middle Eocene, the Kirkuk-Gach Saran trend was approximately coincident with the hinge zone between the carbonate-evaporite shelf, to the southwest, and the rapidly sinking flysch trough to the northeast. Both the oilfield belt and the present mountain front are essentially parallel, therefore, to the fluctuating margin of an unstable shelf, undergoing the early stages of subduction. In pre-Senonian time, the shelfedge lay well back from the present mountain front, and faced northeast. From the Senonian to the mid-Tertiary, the paleobreak faced an interior seaway between the rising orogenic belt on the northeast and the shallow shelf on the southwest.

The association of the fields with the fluctuating paleo-shelfedge has been studied in fascinating detail (Daniel, 1954; Thomas, 1948; Dunnington, 1958; Falcon, 1958; Slinger and Crichton, 1959; Murris, 1980). At Kirkuk, the flexure that had separated the southwestern platform from the northeastern basin during the Cretaceous had become reversed by early Tertiary time, and a Paleogene reef developed along it with a fore-reef, intraplatformal basin to its southwest (Fig. 2). The strike of this reef became obliquely transected by the strike of the subsequent folding, so that the exploitation of the field has provided an oblique cross-section through the back-reef, reef, fore-reef, and basinal facies (Fig. 3).

By the time the Kirkuk reef was deposited, the flysch trough had been replaced by a lagoonal phase leading to emergence. During deposition of the Miocene evaporitic caprocks, the axis of this lagoonal regime lay essentially in the position of the oilfield belt, nearly 100 km southwestward of its position during the late Cretaceous and early Tertiary. There was no longer any true basin. The southwest flank of the lagoonal depositional area had become exceedingly gentle, its gradient less

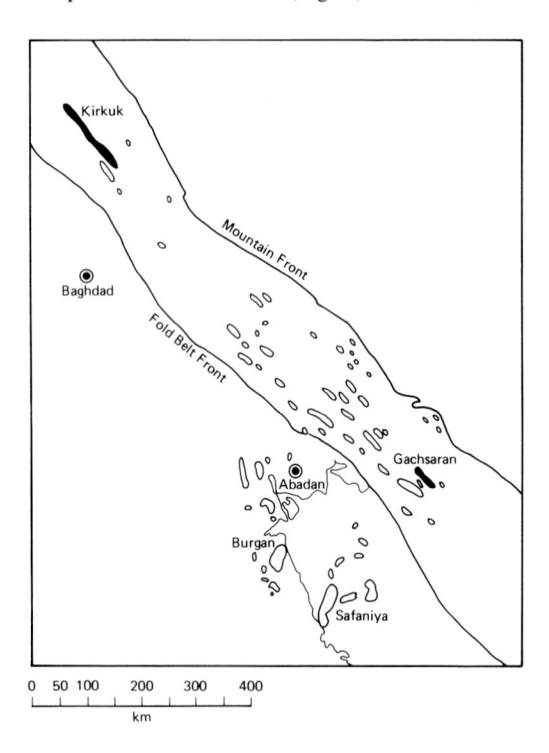

FIG. 1.—Oil fields in the Iran-Iraq foothills belt lying between the front of the fold belt and the "mountain front" of more complex structure. Fields of northern Persian Gulf shown for geographic reference.

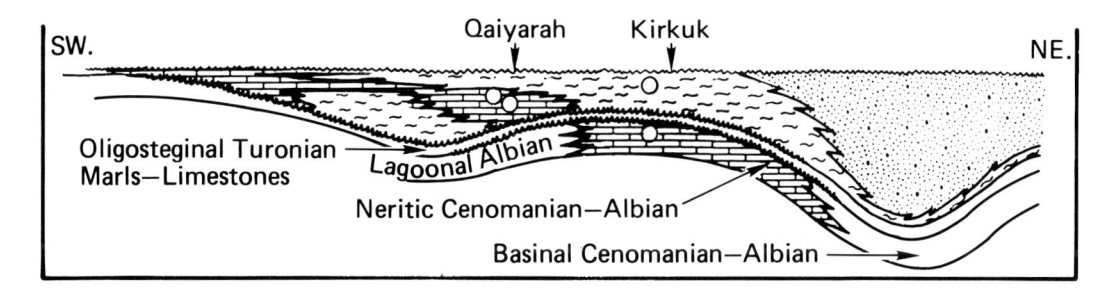

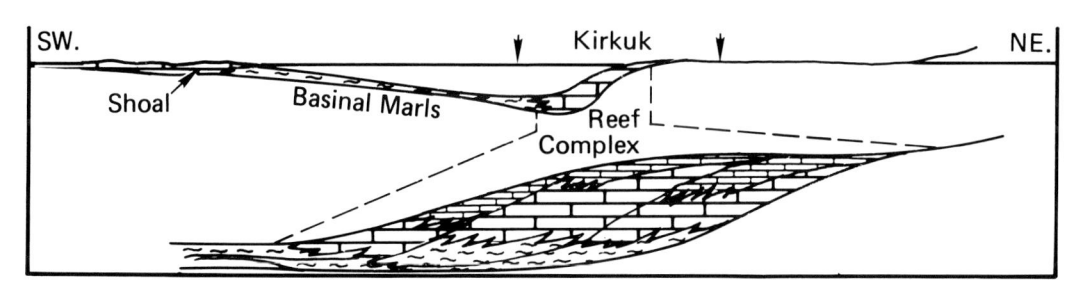

FIG. 2.—Diagrammatic cross-sections, not to scale, showing position of Kirkuk structure during Late Cretaceous (top) and Oligocene (bottom). Cretaceous flysch trough lay to northeast, and Oligocene intraplatformal basin to southwest, of present field position (from Dunnington, 1958).

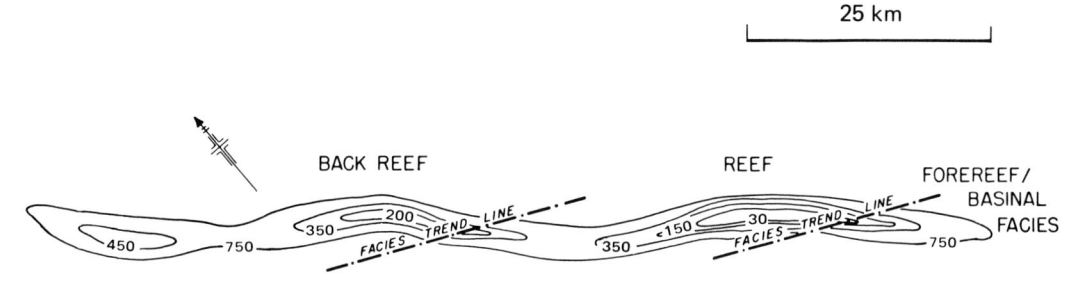

FIG. 3.—Generalized structural map on top of Main Limestone in Kirkuk field, Iraq; contours in meters below sealevel (modified from Daniel, 1954).

than 10 m·km^{-1}; its northeast flank, bounded by a rising orogene, was much steeper, possibly 50 m·km^{-1} or more (Falcon, 1958).

Tampico Region of Mexico

One of the largest reef masses in the geological record is that encircling what is now the Gulf of Mexico during Albian time. In the United States, it is referred to as the Edwards or Stuart City trend (Beebe, 1968), on the opposite side of the Gulf as the Great Carbonate Bank of Yucatan (Viniegra, 1981). The reef effectively rimmed the older Jurassic salt basin. In its western arc, the reef was discontinuous because of interruptions by channels in which deeper-water sediments were deposited. In this interrupted stretch, a giant elliptical atoll grew on an uplifted platform of pre-Jurassic rocks (Figs. 4, 5).

The Cretaceous limestone of the Old Golden Lane oil fields was a shelf deposit containing patches of miliolid-rudistid reef material, now provided with remarkable porosity because of solution of the fossils and extensive fracturing. Northeast of this shelf deposit (toward the older, Jurassic salt basin) lay a shallow lagoon (Fig. 6). Southwest of the shelf an apron of reef debris accumulated, and

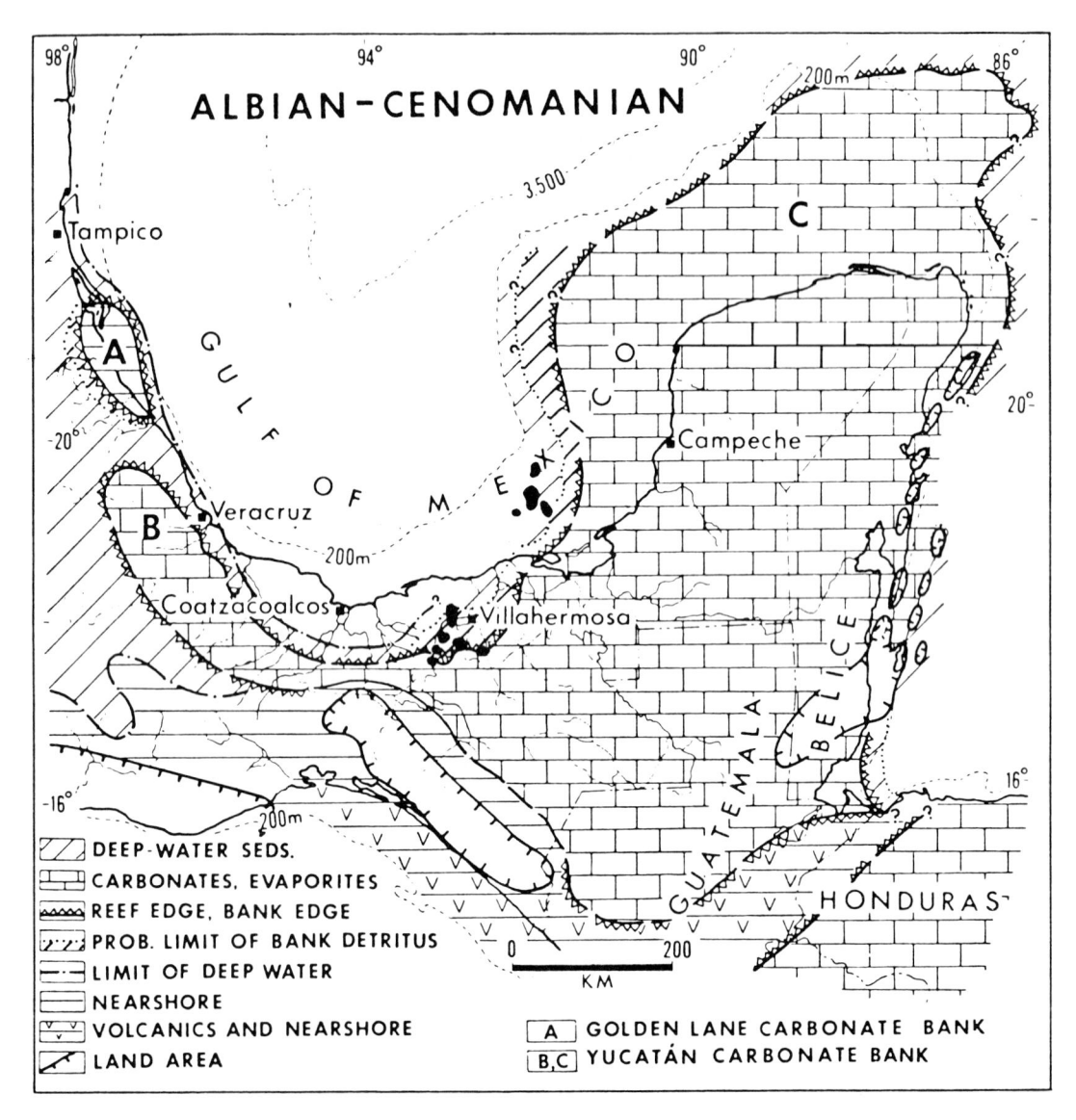

FIG. 4.—Distribution of oil fields in Reforma and Campeche regions of Mexico, on base showing Albian-Cenomanian paleogeography (from Viniegra, 1981).

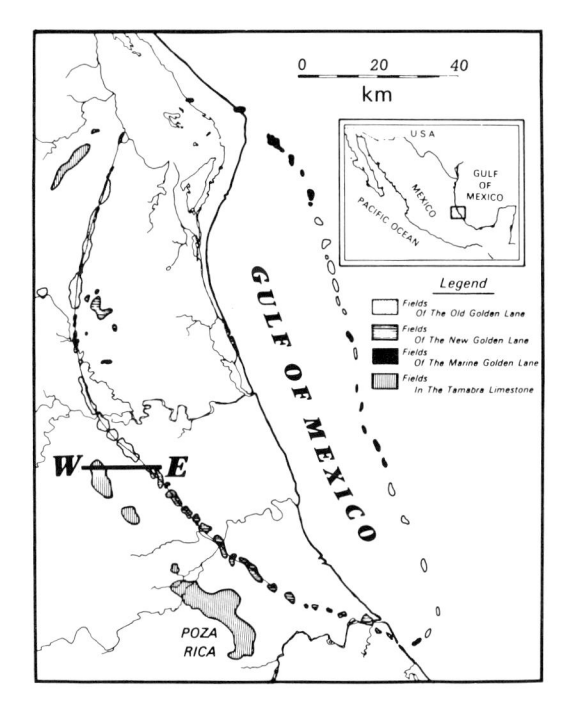

FIG. 5.—Golden Lane and Poza Rica fields, eastern Mexico (from Viniegra and Castillo-Tejero, 1970).

on the west-facing shelfedge, controlled by a high rim in the underlying platform, another, wider reef trend developed with a much more various organic foundation (including corals). This second reef trend shed abundant fore-reef talus downslope into the basin to the west, in which micritic limestones containing planktonic microfossils were deposited. The eastern shelf limestone of the Old Golden Lane represents the El Abra facies; the basinal limestones represent the Tamaulipas facies; the intermediate facies of mixed reef and fore-reef talus is the Tamabra facies (Fig. 6). This last alone is dolomitized;

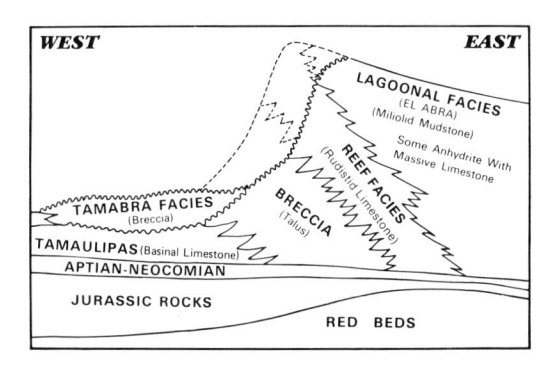

FIG. 6.—Diagrammatic west-east cross-section of Golden Lane reef, showing interpreted facies relations at paleo-shelfbreak (from Viniegra and Castillo-Tejero, 1970).

it contains the Poza Rica oil field. Where reef growth was least interrupted, up to 1600 m of reef rock developed. Terrigenous material is essentially absent, the reservoir rocks being of reef, back-reef, lagoonal, and downslope reef debris facies (Barnetche and Illing, 1956; Coogan et al., 1972; Viniegra and Castillo-Tejero, 1970).

Before the end of Eocene time, some 1000 m of relief had been developed between the higher Golden Lane trend and the lower Poza Rica trend, through a combination of down-to-the-west tilting and faulting along the edge of the Golden Lane paleoshelf. Parts of the El Abra limestone were exposed in the latest Cretaceous and earliest Tertiary. The Upper Cretaceous and Eocene, however, are much thicker over the highest parts of the structure than they are off its flanks, much of the Eocene being in fact restricted to the top of the structure. Post-Eocene strata, in contrast, are present only around the flanks of the structure, because the westward tilting was reversed during the Oligocene. This reversal resulted in a regional southeast dip of about 1.5°, allowing much of the oil to accumulate in the Tamabra facies along the updip, western margin. With the rise of the thrust belt to the west, the atoll complex which formerly faced a depositional basin to the west now faces the open Gulf of Mexico to the east. The eastern arc of the closed atoll lies offshore and is called the Marine Golden Lane (Fig. 5). The crustal instability controlling the paleoshelf-slope zone is here strikingly reminiscent of that northeast of the Persian Gulf.

An equally productive region of Mexico that follows a carbonate paleo-shelfedge but has not undergone structural reversal is the Campeche-Reforma

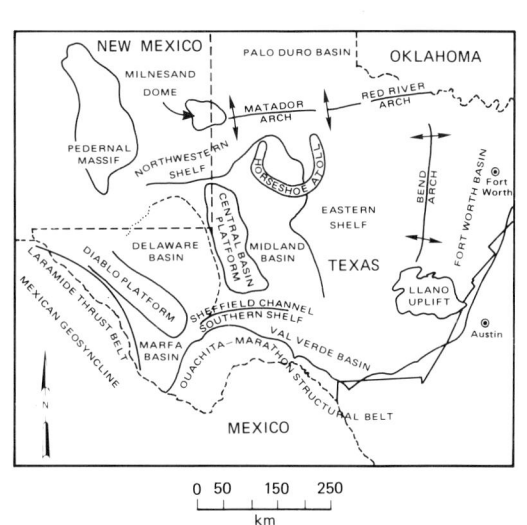

FIG. 7.—Index map, Permian Basin (from Hartman and Woodard, 1971).

province of the southeast (Santiago, 1980; Viniegra, 1981). The carbonate bank itself is not significantly productive; the giant accumulations occur in dolomitic and limestone breccias and calcarenites derived from the bank edge to the east and south (Fig. 4). The bank edge was so persistent that reservoir rocks of talus from it retained the same trend from Albian time through the Paleocene.

PALEO-SHELFBREAK PETROLEUM ACCUMULATIONS IN ENVIRONMENTS OF MIXED LITHOLOGIES

Permian Basin of West Texas-New Mexico

The intracratonic Permian basin of late Paleozoic time was occupied by an interior sea lying between newly elevated fold mountains to the south and the Precambrian Transcontinental arch to the north. It contained no continental shelf or slope in the strict sense. The mid-Pennsylvanian orogenies, however, rejuvenated a north-trending horst block across the basin, separating an eastern (Midland) from a western (Delaware) basin by the Central Basin Platform (Fig. 7; see also Galley, 1958). The margins of this central platform became remarkably steep and abrupt; the outer rims of the whole basin also at times acquired abrupt inner slopes. The central platform and the basin's rims acted as carbonate-hoarding shelves (Adams et al., 1951), starving the two deep basins of sediment supply and permitting the accumulation there of only thin, dark mudstones.

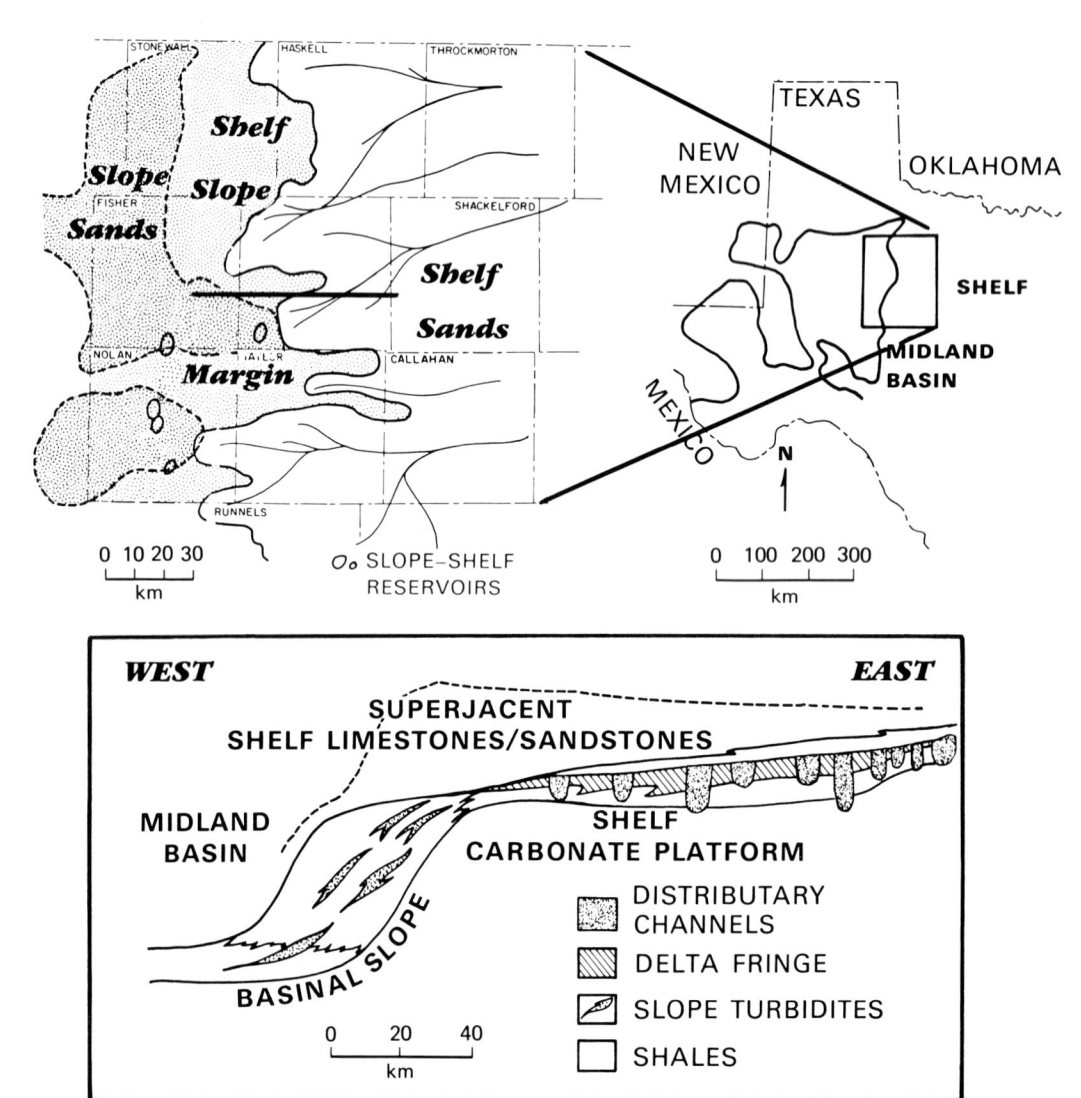

FIG. 8.—Distribution of shelf-slope sands, Eastern Shelf, Midland Basin (from Galloway and Brown, 1972).

During Pennsylvanian time, the principal reef growth took place along the eastern rim of the Midland basin (Fig. 8), building the rim upwards and outwards as the basin itself deepened. The shelf behind the rim accumulated fluvio-deltaic sediments derived from highlands farther to the east. These clastic sediments interfingered westward with limestones and shales which formed the paleo-shelfedge bank system. This system, in turn, overran the edge and interfingered abruptly with the slope system of thick mudstones interspersed with sandstone fans (Fig. 9). Thus, the eastern shelf of the Midland basin prograded into the basin by deposition on both paleo-shelfedge and paleoslope. The preserved relief between the paleoshelf margin and the floor of the starved basin ranges from 180 to 330 m, with dips up to 5° (Galloway and Brown, 1972). The most common petroleum accumulations are in sandstones of the paleo-shelfedge, lower paleoslope, and distal paleoslope, and in stratigraphic traps.

The Palo Duro basin, north of the Midland basin and separated from it by an east-west uplift, was very similar to it but shallower (Handford and Dutton, 1980). It shares the shelfedge characteristics of the Midland basin, with the added advantage of thick arkosic sandstone reservoir rocks derived from Precambrian highlands to the north, but with leaner or less mature source rocks it has no remotely comparable production.

Shelfedge reef production of Pennsylvanian age is minor. The largest single reef, the Horseshoe Atoll, grew not on the edge of the Midland basin, but in the middle of it (Figs. 7, 10). This reef was extraordinarily long-lived and eventually built up about 750 m of reef rock despite the general absence of organisms capable of constructing a rigid framework (Vest, 1970) (Fig. 10B).

At the close of the Pennsylvanian, the central platform was uplifted greatly and the western (Delaware) basin deeply depressed. The margins received almost continuous carbonate buildups during Permian time, individual reefs coalescing into wave-resistant barriers with exceedingly abrupt depositional fronts facing the deep basin. Westward tilting without significant tectonism raised Pennsylvanian reef mounds along the eastern slope of the Horseshoe Atoll to elevations 400 m higher than younger, Permian mounds along the atoll's southwestern side (Vest, 1970).

No subsequent important reef development took place in the basins, which received successions much thicker than those on the platforms and the rims and almost entirely clastic. The sinking basins were, nonetheless, never filled with sediment during the platform reef phase, which reached its acme

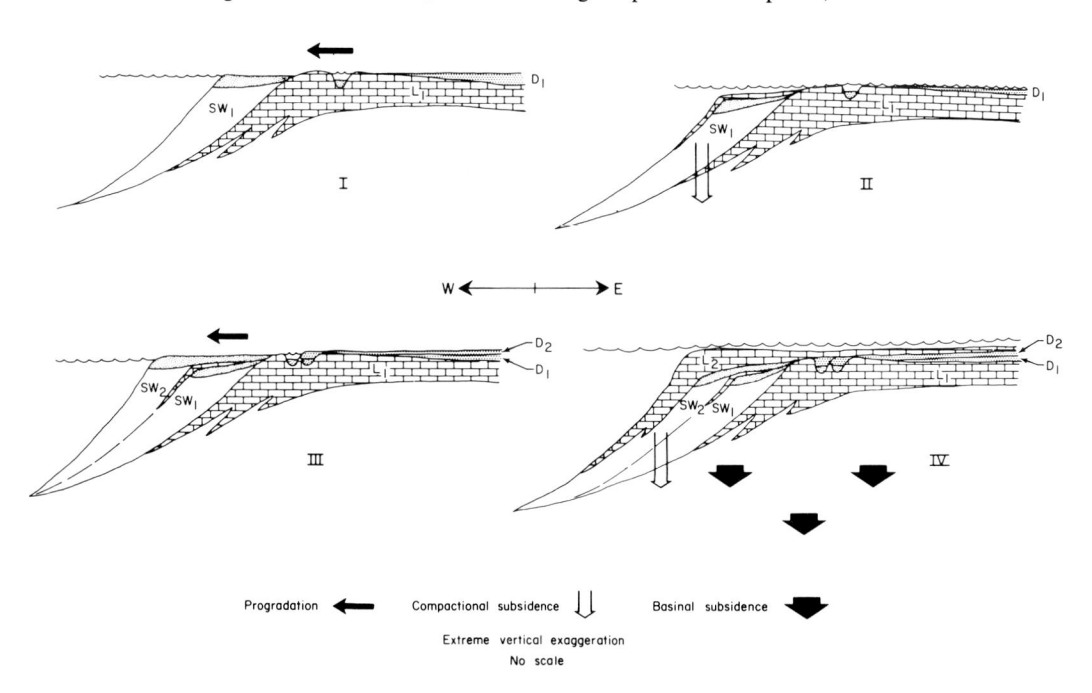

FIG. 9.—Development of shelf margin, Eastern Shelf of Permian Basin, West Texas (from Galloway and Brown, 1972). *I*, Advancing delta D_1 prograades across shelfedge bank and constructs a slope wedge (SW_1), with slope sandstone lenses. *II*, Temporary abandonment of delta and slope wedge, and deposition of carbonate cap. *III*, Second delta D_2 prograades new slope wedge (SW_2), with more slope sandstone lenses. *IV*, Abandonment of delta complex allows growth of shelfedge bank complex (L_2) at the new shelfedge.

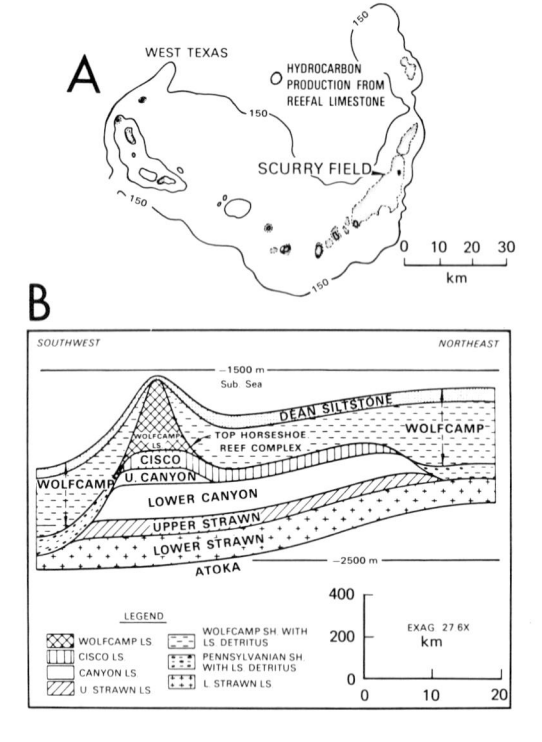

FIG. 10.—*A*, Isopachous map of Horseshoe Reef Complex, West Texas, showing location of significant production from reef limestone along crest of atoll (from Vest, 1970). *B*, Southwest-northeast schematic cross-section through thickest known part of Horseshoe Atoll. Off-reef talus is included with basinal facies (from Vest, 1970).

in mid-Permian (Leonardian and Guadalupian) time. From the Late Pennsylvanian to the Guadalupian, fine quartz sandstones were carried down the slopes and into marginal channels, becoming slope and basinal sandstone reservoirs for oil and both associated and nonassociated gas (Fig. 8) (Galley, 1958; Hartman and Woodard, 1971).

The topographic relief across the basin during most of Permian time measured in excess of 1000 m. The Delaware basin became a small, deep, stagnant sea in which the combined Wolfcampian (Lower Permian) and Leonardian section is three times as thick as on the Central Basin Platform to the east. Around much of the southwestern rim of the basin (Diablo Platform), the two series are absent. The final phase for the basin as a whole came at the close of Permian time with its infilling by evaporites.

Paleo-shelfbreaks have been richly productive both around and within the Permian basin from both Late Pennsylvanian and Permian reservoir rocks. A marked trend of shelfbreak reefs, the Abo reefs, is aligned along the edge of the northwestern shelf, north of the Delaware basin (LeMay, 1972). The reefs, of Leonardian age, are positioned exactly along the paleobreak, with shelf sediments immediately north of them and basinal sediments immediately to the south. The most spectacular alignments of fields with the breaks, however, occur along the western and eastern edges of the Central Basin Platform. Leonardian oil production is notable from reefs along the eastern edge and from siltstones and sandstones around the slopes and in the Midland basin. Gas production dominates in Leonardian rocks along the western edge of the platform. Lower Guadalupian oil was also concentrated along the eastern edge of the platform, sealed by contemporaneous evaporites from the east. In the upper Guadalupian, very prolific oil and gas have been produced from almost continuous reefs along the western edge, from Eunice-Monument in the north to Estes in the south, and from sandstone reservoirs around the platform slopes and in the basins and channels beyond them.

Of the ultimate recoverable reserves from known accumulations in the Permian basin, some 85% of the 5×10^9 m^3 of oil will come from reservoir rocks of the ages described here, and a high proportion of this from fields along the abrupt platform edges and basin rims. Of total known recoverable gas reserves of the order of 2×10^{12} m^3, perhaps 90% of the associated portion will come from Pennsylvanian and Permian strata; a much lower proportion (perhaps one-third) of the nonassociated gas will be similarly derived, more coming from the deep Delaware basin (Galley, 1971). Future exploration potential along the paleo-shelfbreaks cannot be great, because so much earlier exploration has been concentrated along them, but nonreef discoveries (from reef debris accumulations like those in Mexico's Tamabra facies, and from lagoonal reservoirs in stratigraphic traps sealed by evaporites or muds) certainly may be expected.

Cretaceous Sandstone Fields of United States Gulf Coast

The most famous sandstone reservoir rock in the Mesozoic of the U.S. Gulf Coast basin is the Woodbine Formation of the East Texas field, Cenomanian in age. To the east, in Louisiana, it is called the Tuscaloosa Formation. In its oil-productive area, the Woodbine-Tuscaloosa sandstone was a shallow-water deposit associated with a regional unconformity. Toward the south and southeast, however, the downdip equivalents of the formation crossed the break between shelf and slope, remaining a sandstone in lithology. The seaward edge of the continental slope of this age lies in the position of a flexure immediately updip from (north of) the gulf-encircling Albian reef, the Edwards "trend" (Stehli et al., 1972). Deep sea lay to the south of the flexure and of the reef trend; an epicontinental sea lay to the north.

On the oceanward side of the reef trend, sand carried over and through the reef was deposited by

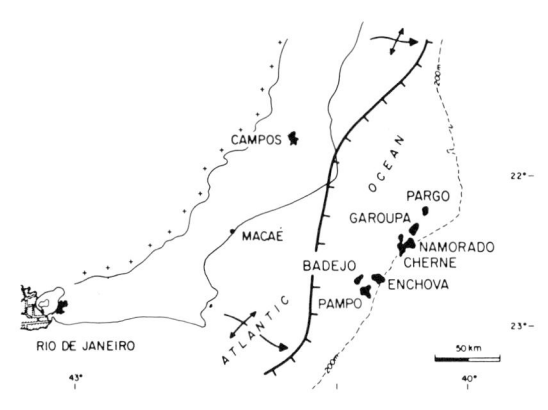

FIG. 11.—Location map of Campos Basin, offshore Brazil, showing oil fields discovered to end of 1978. Major normal fault shown by hachured line (from Bacoccoli et al., 1980).

turbidity currents on the deep slope, in offlapping, shingle-like wedges prograding obliquely toward the south. These sands are gas-bearing in numerous individually modest fields in both southeastern Texas and southern Louisiana (Siemers, 1978; Funkhouser et al., 1980).

Oil Fields of Campos Basin, Offshore Brazil

The South Atlantic basin opened by continental separation during late Mesozoic time, creating a new ocean within an ancient shield. It does not occupy, as the North Atlantic basin does, the site of a complex of deformed belts developed within an earlier seaway. Its present continental margins are therefore much simpler than those around the North Atlantic, and not far removed from the margins caused by the original separation.

The closeness of the present margin to the Cretaceous margin is most clearly seen in the Campos basin, which lies wholly offshore to the east of Rio de Janeiro (Fig. 11; Bacoccoli et al., 1980). The Lower Cretaceous rests directly on basalt flows dated at 120 Ma. The characteristic Neocomian lacustrine strata and Aptian salt represent the rift stage of the separation; it contains some oil. The principal reservoirs, however, are provided by an Albian shelf limestone (Macae Formation, as in the Garoupa field) and by deeper-water turbidite sandstones which were interbedded with basinal shales during a transgression across the Albian limestone shelf. Channel and lobe sandstones appear to have coalesced into a fan-like deposit occupying a top-

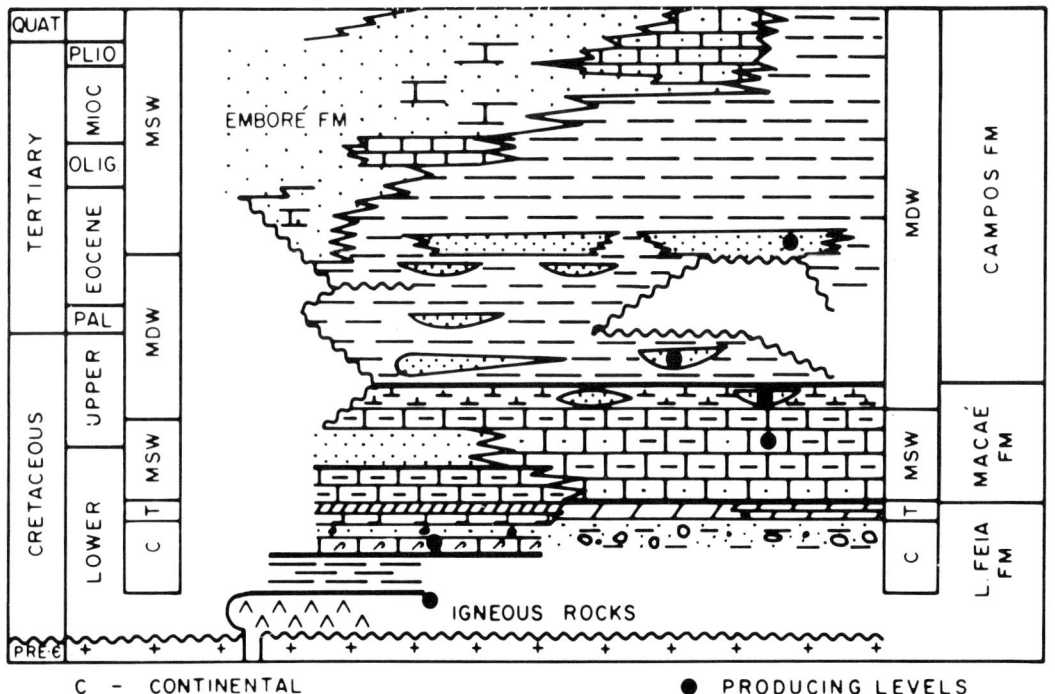

FIG. 12.—Simplified lithostratigraphic column for the Campos Basin, distinguishing shallow- and deep-water deposits (from Bacoccoli et al., 1980).

ographic low below the paleo-shelfedge; this negative structure was reversed during late Cretaceous halokinesis. The reservoir sandstones are, in turn, overlain by deep-water shales containing more lenses of turbidite sandstone and then by prograding slope shales with algal reefs along the paleo-shelfedge (Fig. 12). All known Campos basin oil accumulations are concentrated along a northeast-southwest trend close to the present shelfedge (Fig. 11).

Oil Fields Along Neogene Margin of South China Sea

Between the outer Banda and Philippine arcs, on the east, and nuclear southeast Asia, on the west, is a complex of deformed rocks of late Mesozoic and earliest Tertiary age, including abundant ophiolites. The northwestern margin of this knot of orogenesis forms the narrow continental shelf off East Malaysia (Sarawak-Brunei-Sabah), Palawan Island, and western Luzon.

A late Eocene subduction event severely deformed a pile of clastic sediments some 10 km thick, creating a new marginal trough beneath the Sunda shelf. This trough and its paleo-shelfedge have been progressively displaced toward the northwest since Eocene time (Haile, 1969). In East Malaysia (northwestern Borneo), a mid-Tertiary succession nearly 5 km thick consists mostly of shales (including source sediments), but with episodic reef development along the edge of the Eocene fold belt. Paleo-shelfedge sands of late Miocene and Pliocene ages built outwards over the bathyal shales, reaching a thickness of some 6 km as growth faults developed below them and clay diapirs were intruded. Oil accumulations are concentrated near the junction between the "paralic" and the "neritic" facies in the paleo-shelfedge sands (Schaub and Jackson, 1958). Nearly all pre-Pliocene production is now offshore.

Offshore from Palawan Island, similarly, the deformed Eocene and older rocks are unconformably overlain by seaward-dipping Oligo-Miocene beds (Fig. 13). Oil occurs in limestones of this age, including true reefs (Saldivar-Sali et al., 1981). The limestone was overlapped by bathyal claystones and calcareous siltstones, followed by shelfedge block faulting, uplift, and erosion. The oil occurrences still display a close association with the 100-fathom (180 m) isobath today (Fig. 14). Plio-Pleistocene reef limestones now cover all older rocks and are themselves apparently unfaulted.

PALEO-SHELFBREAK PETROLEUM ACCUMULATIONS
IN CLASTIC SUCCESSIONS

Arkoma Basin, Southern United States

The Arkoma basin is the narrow foredeep between the folded and thrust-faulted Ouachita mountains and the massif of the Ozark uplift to the north (Fig. 15). In its present form, the basin did not begin to form until the end of Mississippian time. Throughout the earlier Paleozoic, however, sedimentary lithofacies in the Ouachita trough reflect a much deeper-water environment there than on the shelf to the north. The Lower Paleozoic Ouachita facies of graptolitic shales, cherts, limestones, and boulder conglomerates is overlain by the late Devonian novaculites and then by Mississippian flysch.

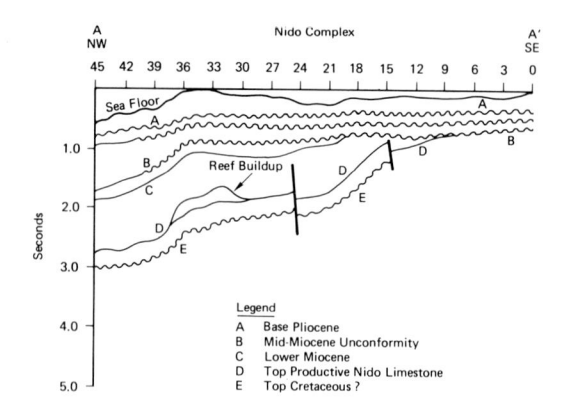

FIG. 13.—West-east seismic time-section across the Nido complex, offshore Palawan Island. See Fig. 14 for line of section (after Saldivar-Sali et al., 1981).

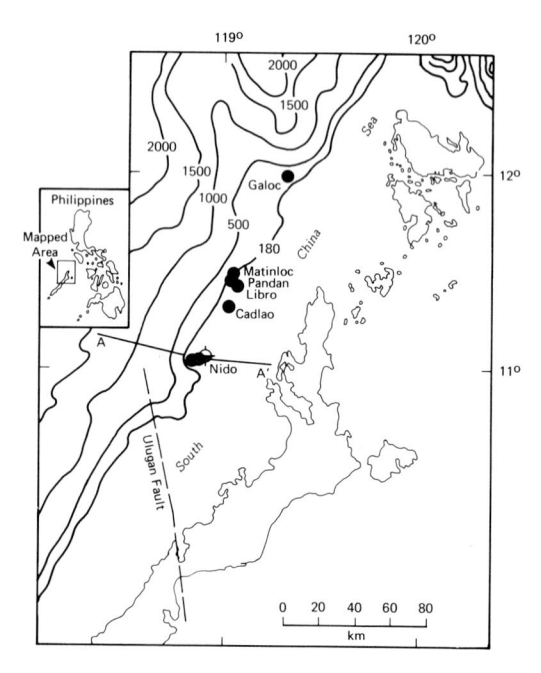

FIG. 14.—Northwest Palawan Island, showing positions of offshore oil discoveries along present shelf-slope break. Bathymetric contours in meters (after Saldivar-Sali et al., 1981).

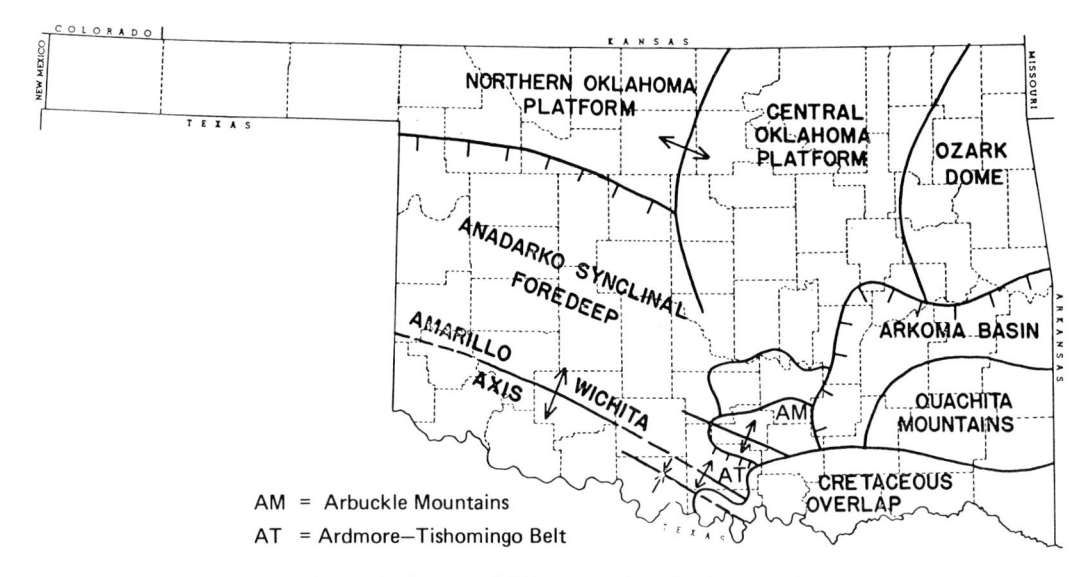

FIG. 15.—Principal tectonic features of Oklahoma referred to in text (after Mairs, 1966).

Within the productive area of the Arkoma basin, the identification of a paleoslope facies is clearest in the Viola Formation of mid- to late-Ordovician age.

In the north, the relatively thin "platform" facies of the Viola Formation (Mairs, 1966) consists essentially of light-colored limestone, extensively dolomitized and with numerous sandstone interbeds. The "slope" facies, to the southwest, is somewhat thicker and more easily divisible into mappable subunits; several of these subunits fail to extend across the platform to the north. The slope facies is typified by abundant chert and by layers of greenish shale instead of sandstone; the limestones are darker-colored than those of the platform and of pelletoid texture. Farther to the southwest, in the Ardmore-Tishomingo belt (Fig. 15), a true basinal or Ouachita facies is encountered, almost wholly clastic (and graptolitic). The boundaries between the different facies belts are rather abrupt. The regional southward dip of the Viola Formation is about 27 m·km^{-1} in the northern part of the basin, but more than 60 m·km^{-1} in the basin; the effective dip is further increased by down-to-the-basin faults. The most important oil fields producing from the "slope" facies of the Viola Formation are the Fitts and Jessie fields.

The zone of distinction between the "shelf" and the "basin" had migrated farther south by the end of Ordovician time, and is now within the deformed belt. Southward tilting was rejuvenated in the late Devonian, as renewed uplift took place in the Ozarks to the north. The Arkoma basin then became the area of maximum deposition in Middle Pennsylvanian (Atokan) time, coincident with the first uplift of the Ouachita mountains. Intense subsequent deformation created a great variety of structural traps (and fractured reservoir rocks). The front of the Ouachita mountains as a geological province is conventionally placed at the Choctaw thrust fault (Fig. 16). Numerous minor thrusts, however, extend for at least 15 km northward of the Choctaw fault, and one of the prominent anticlines raised by these smaller thrust faults contains the Red Oak-Norris gas field (Fig. 16) (Six, 1968). Production is from sandstones in the lower half of the Atoka Formation, immediately northwest of the deep axis of the Arkoma basin; the Atoka Formation is here about 1800 m thicker than it is over the northern shelf of the basin 15 km farther north. The structure of the field itself, in fact, is characterized by normal faults along its northern flank facing the thrust faults along its southern flank (Fig. 17).

Immediately north of the Red Oak-Norris field is the Kinta gas field (Fig. 16) (Woncik, 1968). Although it lies within the northern "subshelf" area of the Arkoma basin, it is on a pronounced surface anticline. The faulted northern limit of the commercial accumulation (in Atokan and older sandstones) approximately coincides with the hinge-belt of the basin.

Although the whole pre-Pennsylvanian section (excluding the Arbuckle Formation at the bottom) is less than 300 m thick, the Pennsylvanian alone increases in thickness from 1200 m to 3000 m in a distance of 15 km from north to south across the field. Nearly all of this increase occurs in the Atoka part of the section, and takes place across large growth faults (one having a downthrow of 1200 m in its own right; Fig. 18). The regional southward dip is as much as 60 to 100 m·km^{-1}, and increases markedly only a short distance to the south.

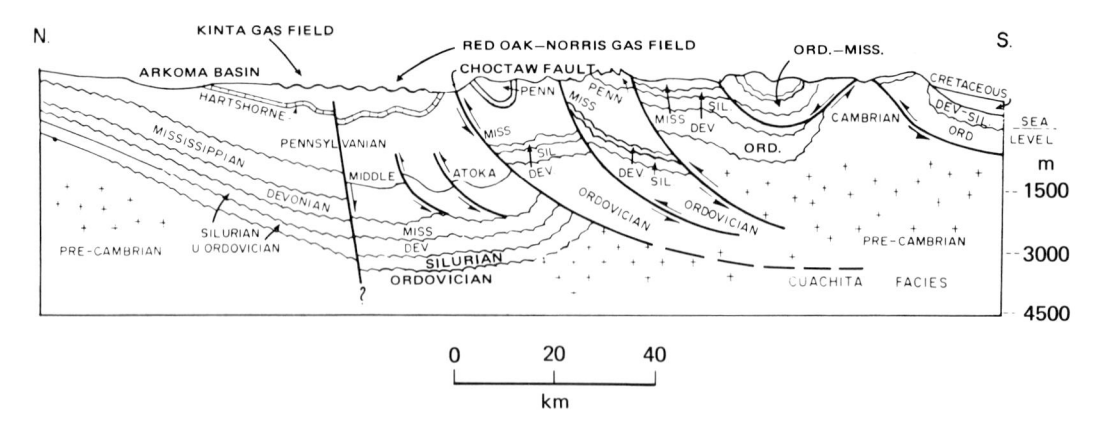

FIG. 16.—Generalized north-south cross section, southern Arkoma Basin and Ouachita Mountains, Oklahoma, showing positions of Kinta and Red Oak-Norris gas fields (after Six, 1968).

Oil Fields of Upper Assam Valley, India

The Arkoma basin has illustrated the control of minor petroleum accumulation by a paleo-shelf-break now impinged upon by a thrust belt which has overridden the paleoslope and rise. In the Upper Assam Valley of northeastern India (Raju, 1968), most of the oil fields (Nahorkatiya, Moran) are in fault-block traps immediately in front of the thrust belt. However, the field found first (Digboi) is in the hanging wall of the principal thrust fault (Fig. 19).

The area was land during the Cretaceous; the Tertiary succession rests on Precambrian basement. The depositional trough developed in the early Eocene as a collisional foreland basin, southeast of a strongly demarcated hinge line extending northeastward from the region of present Calcutta. At the end of the Oligocene, the whole area again became emergent, and succeeding sediments are largely a nonmarine molasse facies. Late Miocene uplift and erosion created another major unconformity. The Tertiary succession below the late Miocene unconformity is about 4 km thick on the Assam shelf, nearly 15 km in the Surma valley to the southwest, where the lower part of the succession is in flysch facies. The Pliocene Himalayan orogeny resulted

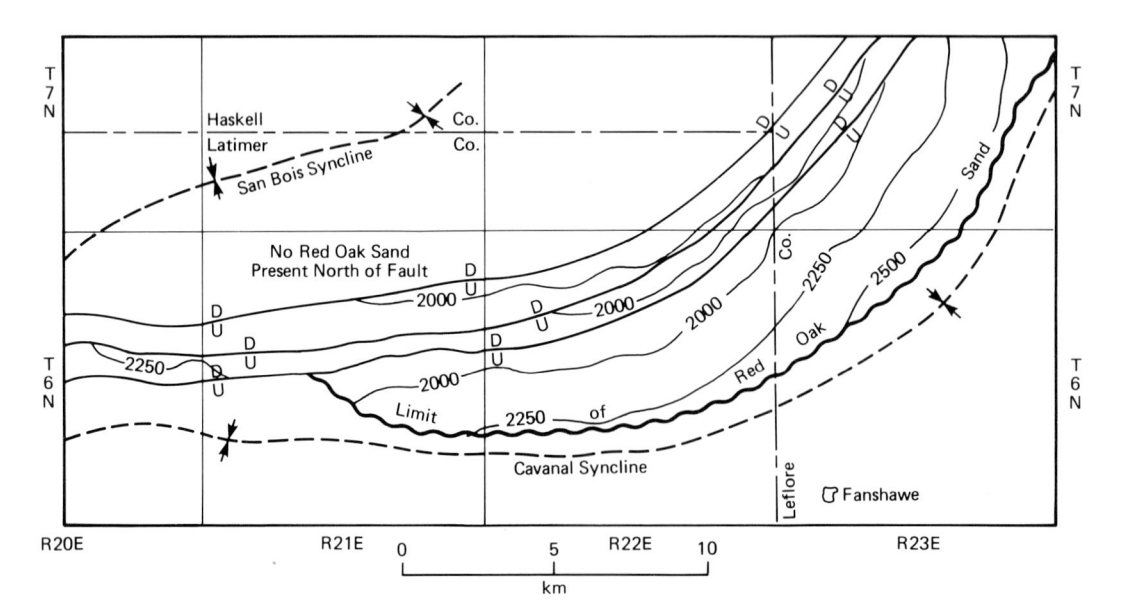

FIG. 17.—Structure map of Red Oak-Norris gas field, Arkoma Basin, Oklahoma. Contour interval 250 m (modified from Six, 1968).

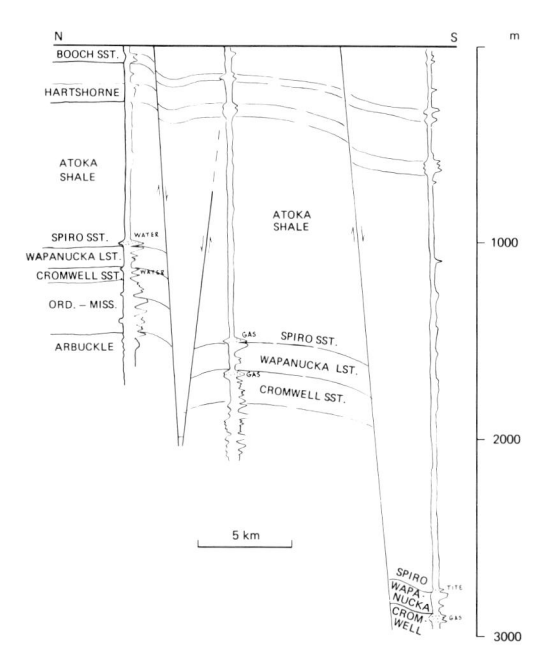

FIG. 18.—Cross-section through part of the Kinta gas field, Arkoma Basin, Oklahoma. Spiro and Cromwell sandstones are predominantly water-bearing in the shelf block of the growth-fault system. Downthrown blocks contain gas if porosity is high enough (from Woncik, 1968).

in the creation of the present valley by violent overthrusting from both sides. Earlier, down-to-the-basin normal faults along the paleo-shelfedge were converted into overthrust nappes causing great shortening of the southeastern side of the basin (Fig. 19). The thrusts are almost entirely in Oligocene sedimentary rocks referred to as "subflysch." Most of the oil occurrences are close to the post-Oligocene unconformity.

Oil and Gas Fields of Northwest Australian Shelf

It was observed in the introduction that early-formed horst blocks, isolated behind newly rifted continental margins during breakup, may create their own shelf-slope breaks on areally local scales. If such a break faces an interior, pull-apart basin, both the horst block itself and its paleo-shelfbreak may become the sites of oil or gas accumulations. The Rankin platform, on the northwestern Australian shelf, illustrates both conditions (Fig. 20).

The platform was isolated by early rifting in late Triassic time, and rejuvenated in the mid-Jurassic, between the Pilbara block of the Australian shield and the Exmouth plateau (Crostella and Chaney, 1978). The main paleobreak of those ages lies at the edge of the Exmouth plateau, facing the early Indian Ocean. Between the platform and the Pilbara block, however, is the Dampier sub-basin; its principal element is the linear Lewis trough, which represents the rejuvenation of an older rift valley or failed arm (Figs. 20 and 21).

On the Rankin platform, large gas reserves occur in Triassic fluvial sandstones unconnected with any shelfbreak. Much or all of the Jurassic is absent over the platform, but in the Lewis trough to the southeast, it is nearly 5 km thick and no doubt contains the principal source sediments. The platform was re-transgressed at the beginning of Cretaceous time. Lower Cretaceous strata are less than 50 m

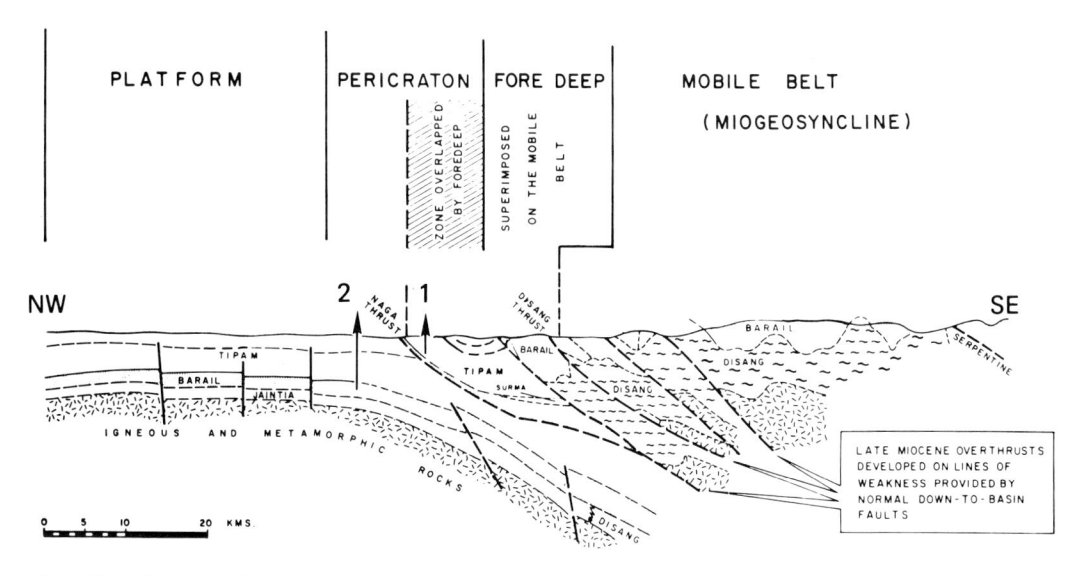

FIG. 19.—Diagrammatic section across Upper Assam valley oil field area. Locations 1 and 2 are Digboi and Nahorkatiya fields, projected (from Raju, 1968).

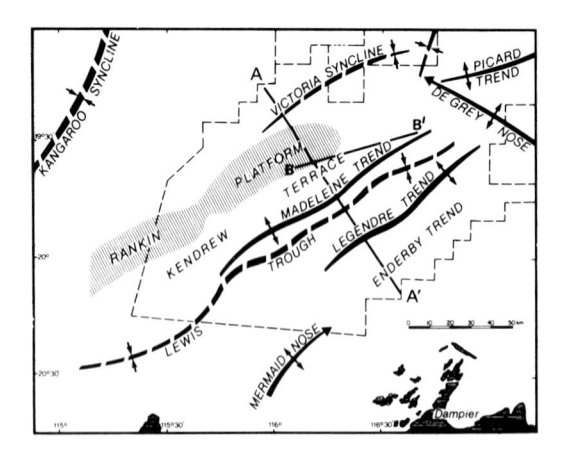

FIG. 20.—Tectonic elements of the Dampier sub-basin, northwest shelf of Australia (from Crostella and Chaney, 1978).

thick on the platform whereas more than 1000 m of a predominantly shale section were deposited in the trough. Post-rifting, shallow marine sandstones, of both Late Jurassic and Early Cretaceous ages, lie at the eastern and southern margins of the platform and extend beyond them. They form reservoirs for minor oil accumulations in several discoveries on or close to the paleobreak (Eaglehawk, Egret, and Lambert discoveries).

Gas Fields of Eocene Paleo-Shelfedge, South Texas

The Lower Wilcox (Eocene) deltaic sandstone reservoirs of the Texas Gulf Coast are famous. Less familiar are the Upper Wilcox deltaic sands, which lie further south (extending into Mexico) and, although stratigraphically younger, also deeper than the Lower Wilcox sands because they prograded over the paleo-shelfedge (Edwards, 1981). As progradation proceeded across the muds of the unstable prodelta slope, growth faults were activated, greatly enhancing the abruptness of the paleobreak. As they cross the growth fault zone, the delta complexes thicken from an average of about 180 m to nearly 1000 m; in one transection a single delta complex thickens tenfold over a distance of 24 km. The complexes are gas-bearing in the Rosita and other fields.

An Unusual Example from California

We earlier suggested that sands carried beyond the shelfbreak are likely to end up at the mouths of channels far down the slope. Fields trapped in such sand bodies are not now spatially related to the paleo-shelfbreaks. The West Thornton gas field, in the Sacramento Valley of California, has been interpreted (Dickas and Payne, 1967) to be trapped at the head of a channel cut into the paleo-shelfbreak (Fig. 22A).

According to these authors, the incision of the channel (called the Meganos Channel; e.g. May et al., 1983) took place on the seafloor starting from the shelf-slope break. It was cut into a Paleocene sandstone and filled with clay (Fig. 22B). Paleobathymetric analyses indicate that the channel extended from neritic to upper bathyal depths, over a distance of at least 80 km with an average gradient of about 2°. Along its course across the paleoshelf and slope, the channel acted as a conduit for sediment moving from the shelf to the basin. The plugging of its upper reaches by clay, however, created submarine paleogeomorphic traps by forming lateral seals to the Paleocene sandstone reservoir below (Fig. 22B). One must wonder whether this apparently freakish trapping mechanism is as rare as our present knowledge makes it appear.

California also provides convincing examples of fields in turbidite fan reservoirs actually at the paleo-shelfbreaks. In the San Joaquin Valley, a number of stratigraphic trap fields in the Miocene Stevens sandstone occur at the updip, proximal edges

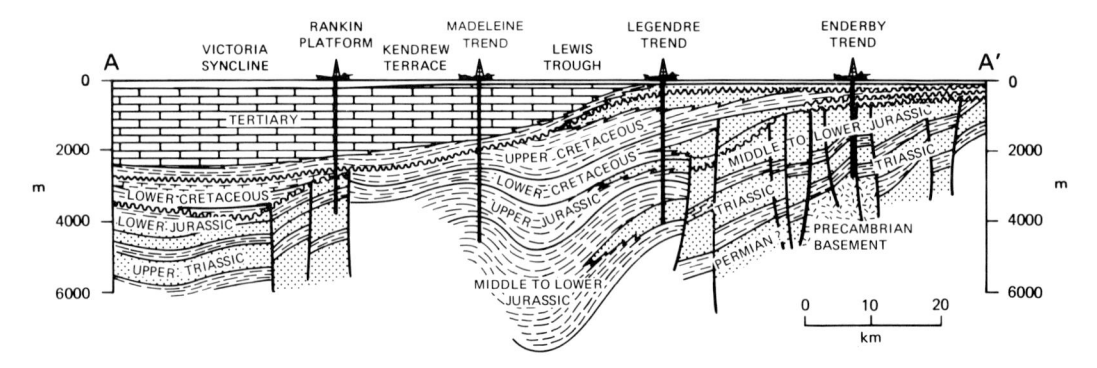

FIG. 21.—Structural section across the Dampier sub-basin, northwest shelf of Australia, showing relationship between Rankin platform and intraplatformal Lewis trough (from Crostella and Chaney, 1978).

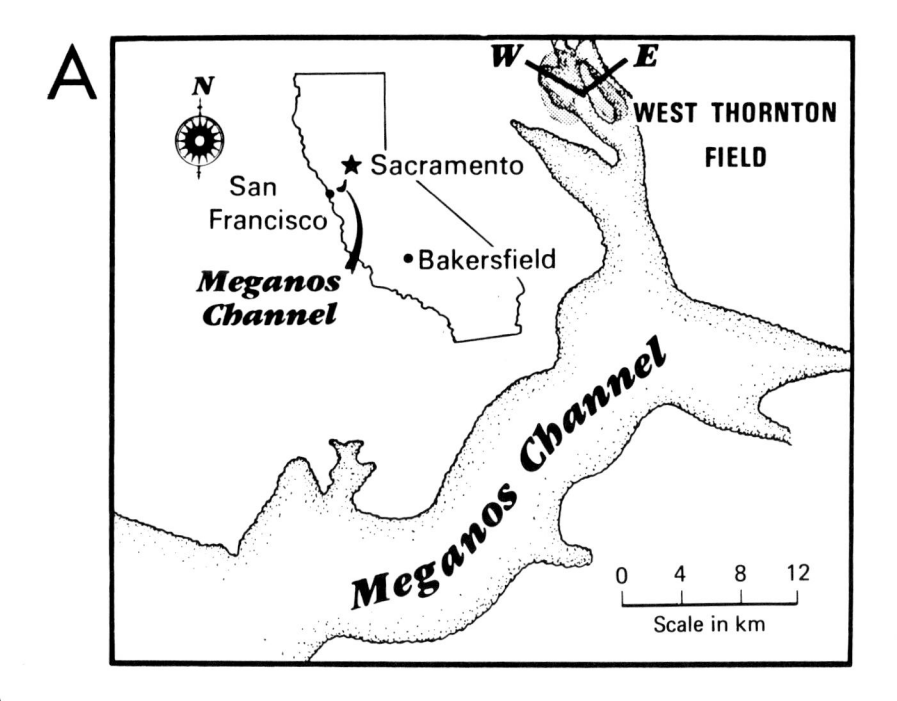

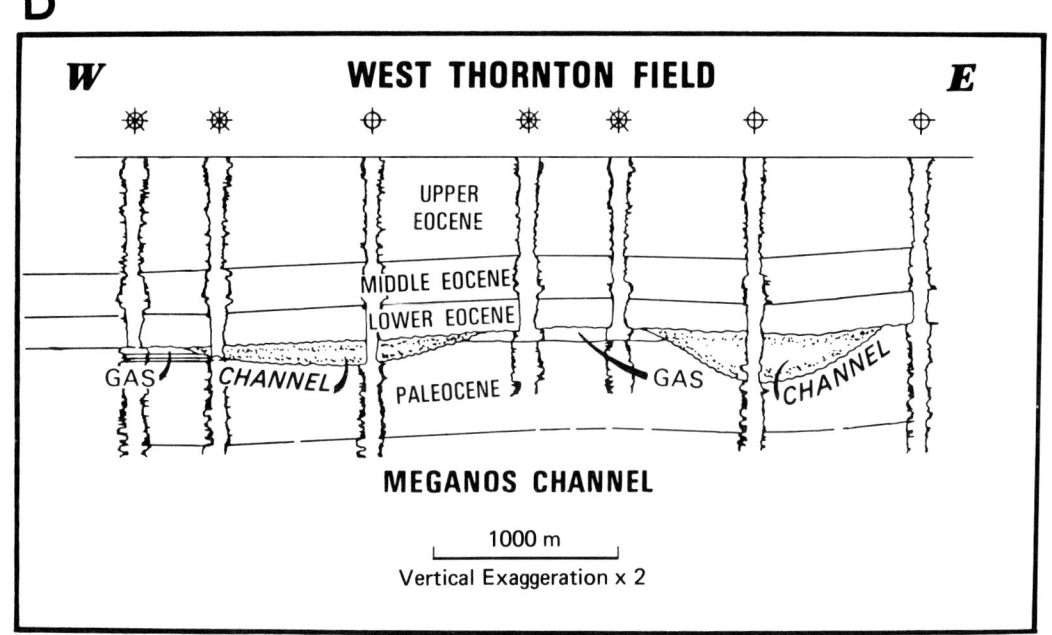

FIG. 22.—Location map and west-east cross-section of West Thornton gas field and Meganos Channel, Sacramento Valley, California (modified from Dickas and Payne, 1967).

of their individual fans, along a "mobile zone of basin-margin wedging" (Macpherson, 1978). A succession of fans was fed directly into the Miocene source sediment downdip, and was overlapped by porcellanitic shales.

CONCLUSIONS

With only a few, albeit spectacular, exceptions, oil and gas fields closely assignable to paleo-shelf-edge locales are far below the first rank. Among the greatest concentrations of hydrocarbons known, those in the Mesozoic of the Persian Gulf basin, west of the foothills fields, and in the Maracaibo, upper U.S. Gulf Coast, West Siberian, Ural-Volga, Sirte, Alberta, and North Sea basins lie wholly within cratonic interiors, devoid of any association with paleo-shelfedges. The "hinge belts" so sought after by petroleum explorationists are seldom true paleo-shelfedges facing open oceans.

Many petroliferous basins must lie across paleo-breaks which are older than the principal reservoir rocks, so that the association of the fields with the breaks is tenuous and not causative. The Niger delta and the South Caspian Tertiary basin are obvious examples. Within both extensional and compressional regimes, gas appears to dominate over oil where clastic productive sections are closely assignable to paleo-shelfedge positions. Far more oil is found in such positions in carbonate-evaporite regimes, especially if these also contain important sandstone reservoir rocks. Of the examples cited, only the Reforma-Campeche area of Mexico appears to represent a paleo-shelfedge in the commonly accepted connotation of that term, as a breakover between continent and ocean. The Iran-Iraq foothills belt is aligned parallel to a paleobreak which fluctuated rather widely between the Early Cretaceous and the mid-Tertiary (the time span encompassing the source and reservoir rocks). The break most closely identifiable with the lines of the fields faced an intraplatformal seaway and was backed by a newly rising orogene.

The closest coincidences between field trends and paleo-shelfedges occur where true shelfedges are simulated by fault-bounded uplifts (extensional or compressional) within wholly intracratonic seaways. The geometric and lithologic characteristics of such local breaks are often similar to those considered typical of true shelfedges, save only for the absence of oceanic or transitional crust. The margins of the Permian basin and of the Central Basin Platform within it exemplify this condition.

ACKNOWLEDGMENTS

We acknowledge critical and invaluable review and editorial comments from William K. Gealey, Robert W. Jones, John S. Kelley, George T. Moore, Russell W. Pfeil, Jr., and Parke D. Snavely III. The choice and interpretation of the examples remain the responsibility of the authors. One of us (WCK) is indebted to Amoco Production Company for permission to prepare and publish this paper.

REFERENCES

ADAMS, J. E., FRANZEL, H. N., RHODES, M. L., AND JOHNSON, D. P., 1951, Starved Pennsylvanian Midland Basin: Am. Assoc. Petroleum Geologists Bull., v. 35, p. 2600–2607.

AHR, W. M., 1973, The carbonate ramp—an alternative to the shelf model: Trans. Gulf Coast Assoc. Geol. Socs., v. 23, p. 222–225.

BACOCCOLI, G., MORALES, R. G., AND CAMPOS, O. A. J., 1980, The Namorado oil field: a major discovery in the Campos Basin, Brazil, in Halbouty, M. T., ed., Giant Oil and Gas Fields of the Decade 1968–1978: Am. Assoc. Petroleum Geologists Memoir 30, p. 329–338.

BARNETCHE, A., AND ILLING, L. V., 1956, The Tamabra Limestone of the Poza Rica Oil Field, Veracruz, Mexico: 20th Internat. Geol. Congr., Mexico, 30 p.

BEEBE, B. W., 1968, Deep Edwards Trend of South Texas, in Beebe, B. W., ed., Natural Gases of North America: Am. Assoc. Petroleum Geologists Memoir 9, v. 1, p. 961–975.

BRENNER, R. L., 1978, Sussex sandstone of Wyoming—example of Cretaceous offshore sedimentation: Am. Assoc. Petroleum Geologists Bull., v. 62, p. 181–200.

COOGAN, A. H., BEBOUT, D. G., AND MAGGIO, C., 1972, Depositional environments and geologic history of Golden Lane and Poza Rica Trend, Mexico, an alternative view: Am. Assoc. Petroleum Geologists Bull., v. 56, p. 1419–1447.

CROSTELLA, A., AND CHANEY, M. A., 1978, Petroleum geology of Australia's Outer Dampier Sub-basin: Oil & Gas Journal, 25 Sept., 1978, p. 162–178.

DANIEL, E. J., 1954, Fractured reservoirs of Middle East: Am. Assoc. Petroleum Geologists Bull., v. 38, p. 774–815.

DICKAS, A. B., AND PAYNE, J. L., 1967, Upper Paleocene buried channel in Sacramento Valley, California: Am. Assoc. Petroleum Geologists Bull., v. 51, p. 873–882.

DUNNINGTON, H. V., 1958, Generation, migration, accumulation, and dissipation of oil in northern Iraq, in Weeks, L. G., ed., Habitat of Oil: Tulsa, Am. Assoc. Petroleum Geologists, p. 1194–1251.

———, 1967, Stratigraphical distribution of oilfields in the Iraq-Iran-Arabia Basin: Jour. Inst. Petroleum, v. 53, p. 129–161.

EDWARDS, M. B., 1981, Upper Wilcox Rosita Delta System of South Texas: growth-faulted shelfedge deltas: Am. Assoc. Petroleum Geologists Bull., v. 65, p. 54–73.

FALCON, N. L., 1958, Position of oil fields of southwest Iran with respect to relevant sedimentary basins, in Weeks, L. G., ed., Habitat of Oil: Tulsa, Am. Assoc. Petroleum Geologists, p. 1279–1293.

FUNKHOUSER, L. W., BLAND, F. X., AND HUMPHRIS, C. C., JR., 1980, The Deep Tuscaloosa Gas Trend of S. Louisiana: Oil & Gas Journal, 8 September, p. 96–101.

GALLEY, J. E., 1958, Oil and geology in the Permian Basin of Texas and New Mexico, in Weeks, L. G., ed., Habitat of Oil: Tulsa, Am. Assoc. Petroleum Geologists, p. 395–446.

———, 1971, Summary of petroleum resources in Paleozoic rocks of Region 5—north, central, and west Texas and eastern New Mexico, in Cram, I. H., ed., Future Petroleum Provinces of the United States—Their Geology and Potential: Am. Assoc. Petroleum Geologists Memoir 15, v. 1, p. 726–737.

GALLOWAY, W. E., AND BROWN, L. F., JR., 1972, Depositional systems and shelf-slope relationships in Upper Pennsylvanian rocks, north-central Texas: Univ. Texas, Bur. Econ. Geol. Rept. Investigations 75, 62 p.

HAILE, N. S., 1969, Geosynclinal theory and the organizational pattern of the north-west Borneo Geosyncline: Quart. Jour. Geol. Soc., v. 124, p. 171–194.

HALBOUTY, M. T., KING, R. E., KLEMME, H. D., DOTT, R. H., SR., AND MEYERHOFF, A. A., 1970, Factors affecting formation of giant oil and gas fields, and basin classification, in Halbouty, M. T., ed., Geology of Giant Petroleum Fields: Am. Assoc. Petroleum Geologists Memoir 14, p. 528–555.

HANDFORD, C. R., AND DUTTON, S. P., 1980, Pennsylvanian-early Permian depositional systems and shelf-margin evolution, Palo Duro Basin, Texas: Am. Assoc. Petroleum Geologists Bull., v. 64, p. 88–106.

HARTMAN, J. K., AND WOODARD, L. R., 1971, Future petroleum resources in post-Mississippian strata of north, central, and west Texas and eastern New Mexico, in Cram, I. H., ed., Future Petroleum Provinces of the United States—Their Geology and Potential: Am. Assoc. Petroleum Geologists Memoir 15, v. 1, p. 752–803.

IRVING, E., NORTH, F. K., AND COUILLARD, R., 1974, Oil, climate, and tectonics: Canadian Jour. Earth Sciences, v. 11, p. 1–17.

KLEMME, H. D., 1980, Petroleum basins—classifications and characteristics: Jour. Petroleum Geology, v. 3, p. 187–207.

KNEBEL, G. M., AND RODRIGUEZ-ERASO, G., 1956, Habitat of some oil: Am. Assoc. Petroleum Geologists Bull., v. 40, p. 547–561.

LEMAY, W. J., 1972, Empire Abo Field, southeast New Mexico, in King, R. E., ed., Stratigraphic Oil and Gas Fields—Classification, Exploration Methods, and Case Histories: Am. Assoc. Petroleum Geologists Memoir 16, p. 472–480.

MACPHERSON, B. A., 1978, Sedimentation and trapping mechanism in Upper Miocene Stevens and older turbidite fans of southeastern San Joaquin Valley, California: Am. Assoc. Petroleum Geologists Bull., v. 62, p. 2243–2274.

MAIRS, T., 1966, A subsurface study of the Fernvale and Viola Formations in the Oklahoma portion of the Arkoma Basin: Tulsa Geol. Soc. Digest, v. 34, p. 60–81.

MAY, J. A., WARME, J. E., AND SLATER, R. A., 1983, Role of submarine canyons on shelfbreak erosion and sedimentation: Modern and ancient examples, in Stanley, D. J., and Moore, G. T., eds., The Shelfbreak: Critical Interface on Continental Margins: Soc. Econ. Paleontologists Mineralogists Spec. Pub. 33, p. 315–332.

MURRIS, R. J., 1980, Middle East: stratigraphic evolution and oil habitat: Am. Assoc. Petroleum Geologists Bull., v. 64, p. 597–618.

RAJU, A. T. R., 1968, Geological evolution of Assam and Cambay Tertiary Basins of India: Am. Assoc. Petroleum Geologists Bull., v. 52, p. 2422–2437.

SALDIVAR-SALI, A., OESTERLE, H. G., AND BROWNLES, D. N., 1981, Geology of offshore northwest Palawan, Philippines, I: Oil & Gas Journal, 30 November, p. 119–128.

SANTIAGO, A. J., 1980, Giant fields of the Southern Zone—Mexico, in Halbouty, M. T., ed., Giant Oil and Gas Fields of the Decade 1968–1978: Am. Assoc. Petroleum Geologists Memoir 30, p. 339–385.

SCHAUB, H. P., AND JACKSON, A., 1958, The Northwestern Oil Basin of Borneo, in Weeks, L. G., ed., Habitat of Oil: Tulsa, Am. Assoc. Petroleum Geologists, p. 1330–1336.

SIEMERS, C. T., 1978, Submarine fan deposition of the Woodbine-Eagle Ford Interval (Upper Cretaceous), Tyler County, Texas: Trans., Gulf Coast Assoc. Geol. Socs., v. 28., p. 493–533.

SIX, D. A., 1968, Red Oak-Norris Gas Field, Brazil Anticline, Latimer and LeFlore Counties, Oklahoma, in Beebe, B. W., ed., Natural Gases of North America: Am. Assoc. Petroleum Geologists Memoir 9, v. 2, p. 1644–1657.

SLINGER, F. C. P., AND CRICHTON, J. G., 1959, The geology and development of the Gachsaran Field, southwest Iran: Proc. 5th World Petroleum Congr., Sec. 1, p. 349–375.

STEHLI, F. G., CREATH, W. B., UPSHAW, C. F., AND FORGOTSON, J. M., JR., 1972, Depositional history of Gulfian Cretaceous of East Texas Embayment: Am. Assoc. Petroleum Geologists Bull., v. 56, p. 38–67.

THOMAS, A. N., 1948, The Asmari limestone of southwest Iran: 18th Internat. Geol. Congr., Pt. 6, Sec. E, p. 35–44.

VEST, E. L., 1970, Oil fields of Pennsylvanian-Permian Horseshoe Atoll, West Texas, in Halbouty, M. T., ed., Geology of Giant Petroleum Fields: Am. Assoc. Petroleum Geologists Memoir 14, p. 185–203.

VINIEGRA, F. O., 1981, Great Carbonate Bank of Yucatan, southern Mexico: Jour. Petroleum Geology, v. 3, p. 247–278.

———, AND CASTILLO-TEJERO, C., 1970, Golden Lane fields, Veracruz, Mexico, in Halbouty, M. T., ed., Geology of Giant Petroleum Fields: Am. Assoc. Petroleum Geologists Memoir 14, p 309–325.

WEEKS, L. G., 1952, Factors of sedimentary basin development that control oil occurrence: Am. Assoc. Petroleum Geologists Bull., v. 36, p. 2071–2124.

WONCIK, J., 1968, Kinta Gas Field, Haskell County, Oklahoma, in Beebe, B. W., ed., Natural Gases of North America: Am. Assoc. Petroleum Geologists Memoir 9, v. 2, p. 1636–1643.

MINERAL DEPOSITS AT THE SHELFBREAK[1]

MICHAEL J. CRUICKSHANK
Henry Krumb School of Mines, Columbia University, New York, New York 10027

T. JOHN ROWLAND JR.
Minerals Management Service, Department of the Interior, Reston, Virginia 22092

ABSTRACT

The shelfbreak, the transitional zone between the outer continental shelf and slope, is a unique environment with respect to some mineral deposits. Physical and chemical processes exert considerable influence over the occurrence of minerals formed in this environment. Most prominent are geochemical deposition of phosphorus, uranium, and other metals, resulting from upwelling and areas of oxygen minima; physical concentrations of heavy minerals such as tin oxides or gold, resulting from eustatic changes, surface and internal waves and geostrophic currents; biological deposits such as limestones, and precious corals associated with coral bioherms, reefs and atolls; and strata-bound deposits of sulfur, salt, or coal, associated with bedded or diapiric structures. While almost any economic mineral deposit of the type found in continental land masses may be located at the shelfbreak, only phosphorite, some heavy mineral placers, and certain precious corals appear to be products of the zone's unique environment. Because of the water depth and distance from land, prospecting might usefully be limited at the present time to these types of deposit. As the zone overlaps the various legal boundaries of national and international jurisdiction, any prospecting activity should include a complete investigation of the legal factors involved.

INTRODUCTION

The typical continental margin comprises a shelf, slope, and rise seaward of the continental land mass. While both the shelf and slope have been described with respect to form and variety, the shelfbreak (zone where the continental shelf drops off toward abyssal depths) has proved more elusive to define precisely. It has been compared to the coastal zone on land (Southard and Stanley, 1976). Occurring worldwide at an average depth of 130 m, the shelfbreak has been measured at depths ranging from 24 m to 500 m. Differences in depth are ascribed in part to structural differences rather than simply to irregular eustatic changes in sealevel affecting the world margins. The International Commission on Nomenclature of Ocean Bottom Features (ICNOFB), while recognizing the great variability of the shelfbreak, and its 130 m worldwide average depth, defined it by convention at a depth of 200 m and where a marked increase in slope is apparent.

For the discussion of mineral deposits to follow, the provinces of the continental margin (Fig. 1A) are summarized in simple fashion:

Continental shelf—marine zone between the coast and the shelfbreak.
Shelfbreak—zone of rapid steepening of the seafloor slope, often occurring at water depths between 50 and 200 m.
Continental slope—zone seaward of the shelfbreak where the seafloor slopes toward the deep-sea, with an average gradient of 4°.
Continental rise—more gradual slope toward deep-sea bed, with depths commonly ranging between about 2500 and 4000 m.
Continental terrace—step-like features at depths intermediate between the continental shelf and slope extending beyond the continental shelf, often formed by extension of shelf strata beneath the continental shelf.

Dietz and Menard (1951), Curray (1965, 1966), and Shepard (1973) have reviewed the variety of modern continental margins. The remarkable diversity of shelfbreak configuration is highlighted in this volume by Vanney and Stanley (1983).

MINERAL DEPOSITION AND OCCURRENCES AT THE SHELFBREAK

Physical and chemical processes and conditions at the seafloor exert considerable influence on the occurrence of mineral deposits at the shelfbreak. Among the more prominent are: (1) geochemical deposition resulting from upwelling and areas of oxygen minima; (2) physical concentrates resulting from eustatic changes and oceanographic properties (surface and internal waves, geostrophic currents); (3) biological deposits, associated with coral bioherms, reefs, and atolls; and (4) bedded or diapiric structures.

Coastal upwelling and the associated geochemical environments that result in mineral deposition tend to occur on the western margins of continents

---

[1] Publication authorized by the Director, Minerals Management Service, Department of the Interior.

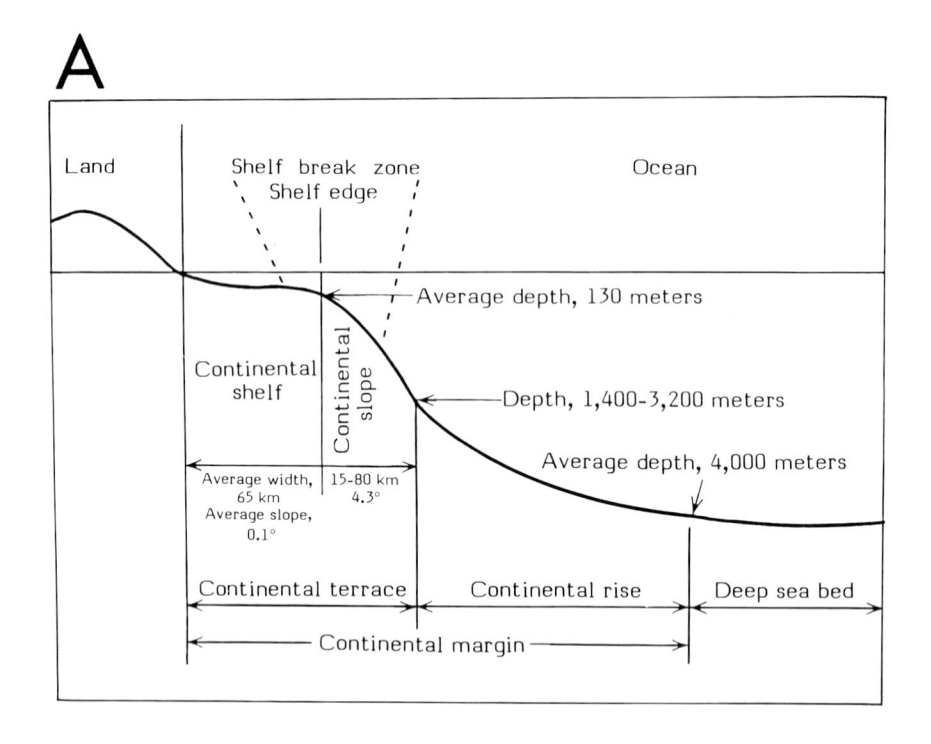

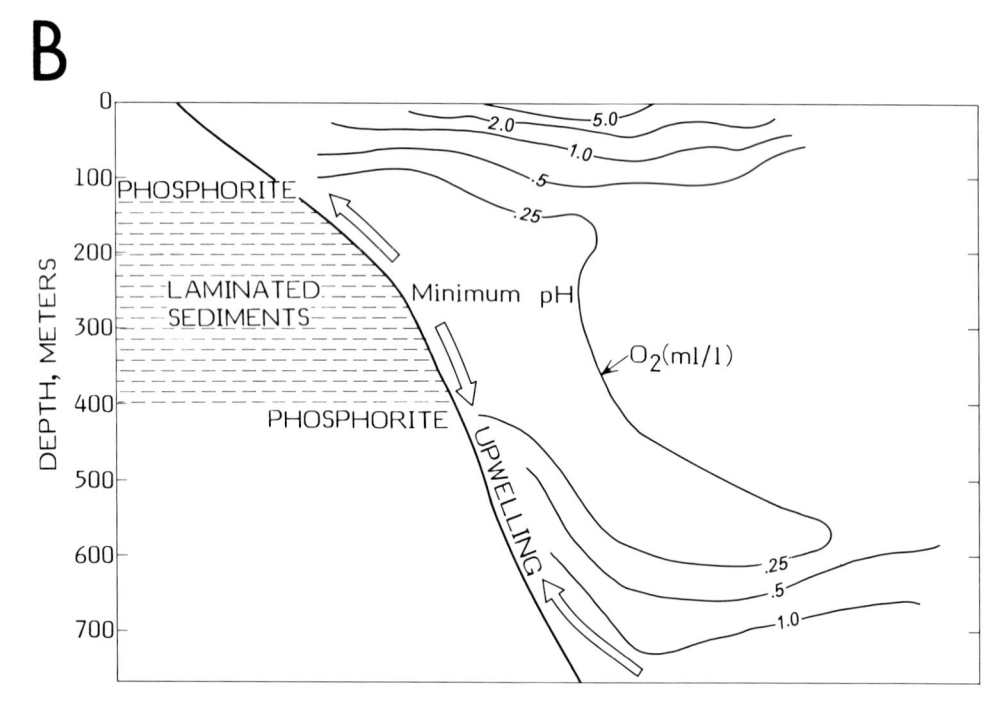

FIG. 1.—*A*, Schematic diagram of an idealized continental margin profile including shelfbreak zone (adapted from Cronan, 1980). *B*, Profile of the oxygen minimum zone, upwelling, pH, and phosphorite distribution in some areas at shelfbreak depths (100 to 400 m). Adapted from Burnett and Sheldon (1979).

and are associated with the deposition of uranium-enriched phosphorite, and other minerals. Cold, low-salinity, bottom water, rich in oxygen and nutrients, flows upward to the surface, increasing biotic productivity and inducing high biogenic precipitation, while changes in temperature and pressure of the upwelling watermass and the saturation capacities for certain elements result in chemical precipitation of minerals. This process commonly occurs in the depth range of the shelfbreak (100 to 400 m), and is a principal factor in the formation of marine phosphorite in various geographic regions (Fig. 1B).

The southwest African shelf and shelfbreak is a region where recent phosphorite formation has been associated with the upwelling process according to Bezrukov and Senin (1971), Baturin et al. (1972) and Birch (1973). The relation between coastal upwelling and mineral deposition has been studied off Chile, Peru, and northwestern and southwestern Africa by, for example, Veeh et al. (1973) and Manheim et al. (1975). On the basis of this relationship, the shelfbreak in regions of upwelling in both modern and the ancient settings should be considered as potential targets for economic mineral exploration. Many of the known marine phosphorite deposits (Fig. 2) encompass areas of the shelfbreak throughout the world and many of these deposits are associated with modern or ancient coastal upwelling.

In certain areas, such as the India-Pakistan margin, oceanic water masses impinge on the continental slope and shelfbreak coincident with the oxygen-minimum zone. The conditions that develop as a consequence are similar in some ways to the conditions occurring in closed anoxic basins such as the Black Sea. In the absence of adequate chemical oxidation and consumption by heterotrophic organisms, organic material accumulates and the geochemical nature of sediments is different from the more typical oxidized areas owing to differences in such factors as pH, CO_2, and redox potential which influence the nature of minerals concentrated in the vicinity.

Eustatic fluctuations during the Pleistocene and the effects of subsequent hydrodynamic processes have provided the potential for relict placer and aggregate deposits at and adjacent to the shelfbreak. At least four Pleistocene stages of glacial advance coupled with interglacial stages of melting produced pulsations of sealevel estimated to be from 100 m to 150 m. During this period, present continental margins were repeatedly exposed and the shorelines were temporarily located near the shelfedge. Continental glaciers produced a supply of fine-grained material, and the continental margins experienced periods of rapid deposition. During the periods of glaciation, the sea was about 130 m below its present level, a depth which in many regions of the world is coincident with that of modern shelfbreaks. Sedimentary processes associated with modern inner shelf areas were active but transposed geographically to the region of the shelfbreak. At such times, longshore currents may have been unimportant because the seafloor at the shelfedge sloped steeply.

Since the late Pleistocene rise in sealevel, erosion of the seafloor by currents has become more im-

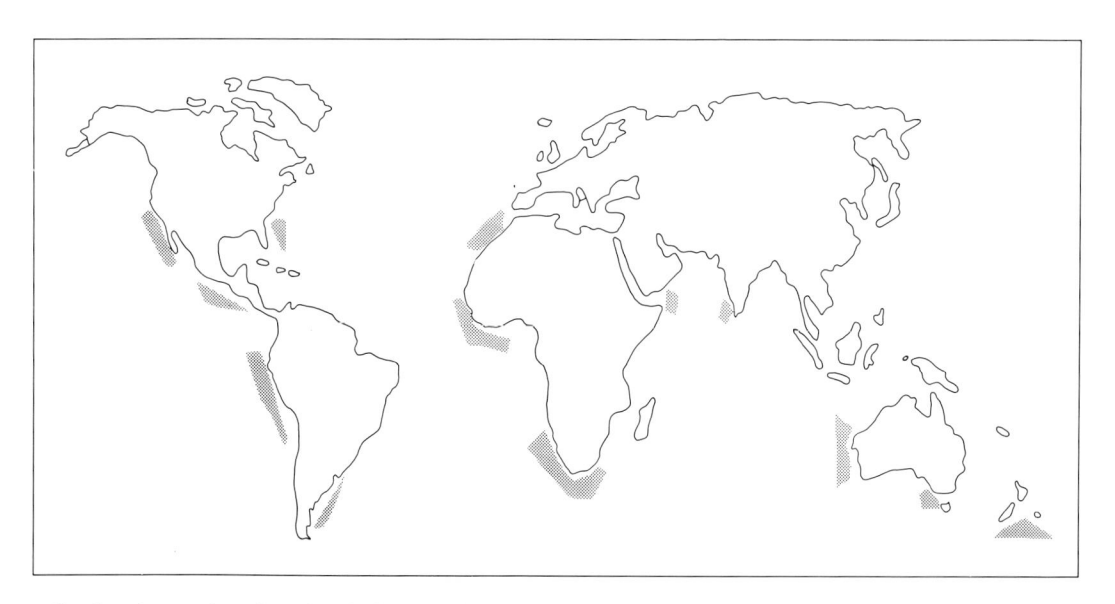

FIG. 2.—Areas of marine phosphorite deposits (shaded) on continental margins, including the shelfbreak zone (adapted from Emery and Noakes, 1968).

portant at the shelfbreak and outer shelf. Reworking of surface sediments results in the transport of some fractions over the continental margin. Studies of the continental margin off New England by MacIlvaine and Ross (1979) highlight the effects of the interaction of Pleistocene relict material and modern environmental processes. Bottom currents, density flows, geostrophic currents, friction from tidal forces, and possibly the impingement of internal waves exert influence over the shelfbreak zone (see chapters in Part V of this volume). Consideration of the hydrodynamic processes along with such factors as sediment source and geological history is crucial for establishing targets for exploration for placer and aggregate deposits.

Areas of the continental margins, including the shelfbreak zone where significant placer deposits may occur, are illustrated on a world map (Fig. 3). Environments of possible placer mineral occurrences near the shelfbreak include submerged beaches (Fig. 4A), depressions on the seafloor (Fig. 4B), or buried river valleys (Fig. 4C). The submerged terraces and relict beaches may be sites of placer deposits of Pleistocene origin. Shelfbreak sectors off North America, western Europe, southeast Asia, and southern Africa provide high potential for such deposits. Most placer deposits on the shelves have been found along shelf extensions of river valleys and submerged beaches (Emery and Noakes 1968).

Enormous volumes of calcareous material including limestone, dolomite, precious corals, and calcareous sands compose the framework and detritus of bioherm structures. Large areas such as those off west Florida, northeast Brazil, the Campeche Reef off the Yucatan, as well as the numerous atoll structures of the Pacific are classical bio-calcareous locales. The most common types of reefs are: fringing reefs, barrier reefs, atolls, table reefs, faros, and coral knolls. Most coral reefs, regardless of the size or kind, have similar environments and associated organisms and hence have similiar biologically-derived mineral deposits. Atolls are the most common type of coral reef. The oval shaped atolls, most abundant in the Pacific, rise from a deep basaltic base. They are surrounded by narrow, steep terraces which represent a potential source of coral, precious coral, coral sands, and other material including limestone and dolomite. Large accumulations of limestone, corals, some rare and precious corals, and calcareous debris are present in areas associated with the shelfbreak on the atolls and other bioherm structures (Newell, 1971; Emery et al., 1954).

Bedded deposits and diapiric structures occur on continental margins of many regions of the world. Many bedded deposits under the continental margins are extensions of land deposits. In some instances, the extensions conformably extend into the subsurface of the shelfbreak zone. Others may occur in such regions as a result of faulting or specific depositional event. Diapiric structures containing sulfur, evaporites, or other mobile materials have the potential to migrate into the vicinity of the shelfbreak from more landward or deeper sources. Diapiric minerals may outcrop or rise to the prox-

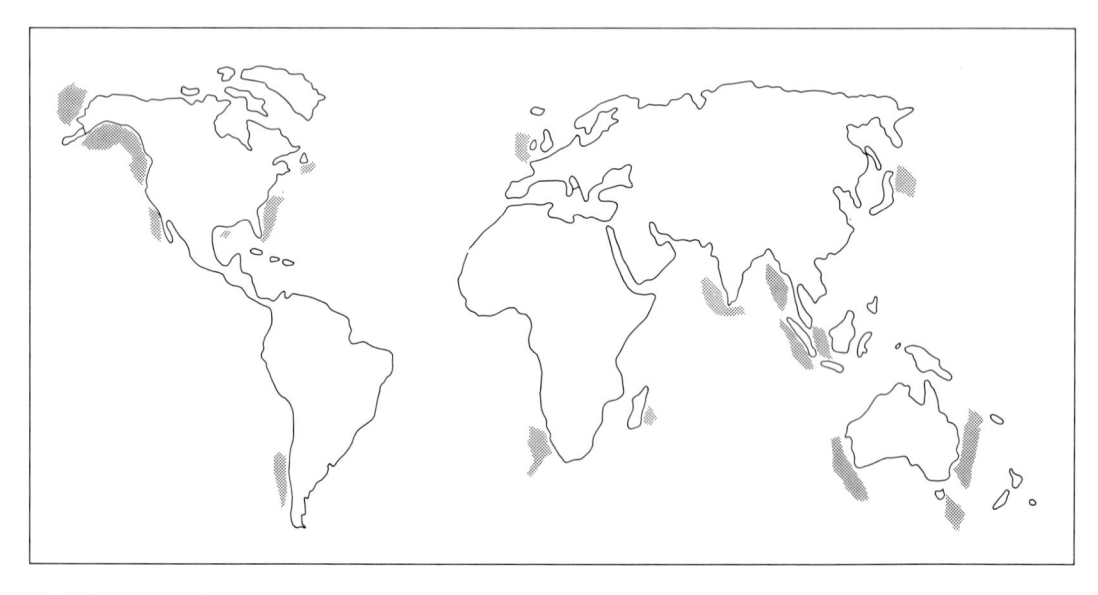

FIG. 3.—Areas of potential marine placer deposits (shaded). Adapted from Emery and Noakes (1968) and Cronan (1980).

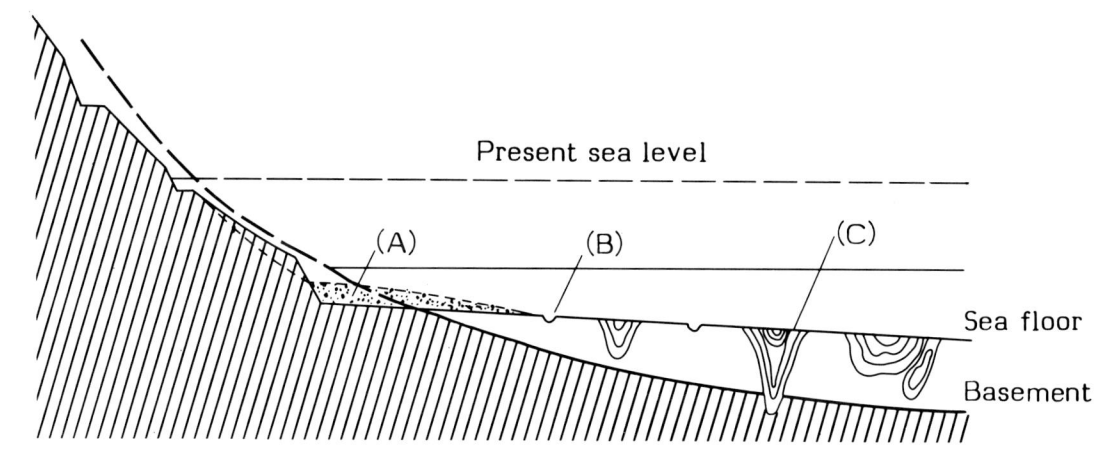

FIG. 4.—Schematic diagram of environments comprising potential placer mineral deposits (adapted from Cronan, 1980). Placers on submerged beaches (A); trapped in surface depressions on the seafloor (B); and in buried river valleys (C).

imity of the surface at the shelfbreak. Exposure of bedded and diapiric structures may be further enhanced by faulting and mass wasting.

Sulfur, potash, oilshale, coal and evaporite minerals may occur at and beneath the shelfbreak zone. In addition, minerals associated with spreading, rift and fracture zones, such as surficial polymetallic sulfides, plutonic intrusions, and disseminated and vein deposits are locally recorded. Their occurrence is related to geological factors only, and their formation in most cases is analogous to that of similar deposits on land rather than to any factors unique to the shelf-to-slope transition zone.

MINERAL DEPOSITS AT THE SHELFBREAK

Off the United States, mineral deposits of phosphorite, metalliferous nodules, gravels, and precious corals have been identified at the shelfbreak zone. Phosphorites occur on bathymetric highs in the Southern California Borderland in areas of upwelling affected by bottom currents. The deposits are mostly replacements of exposed Miocene car-

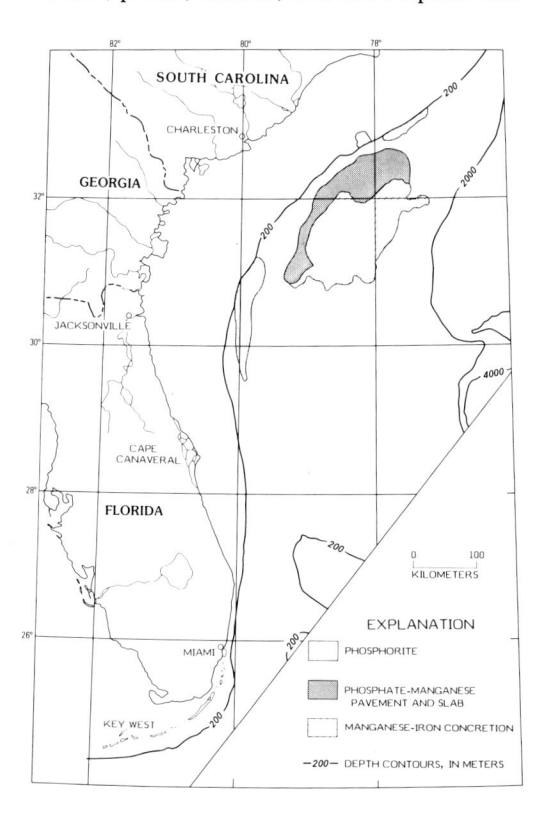

FIG. 5.—Phosphorite and manganese concretion deposits on the Blake Plateau off southeastern United States (adapted from Manheim, Pratt and Mcfarlin, 1980).

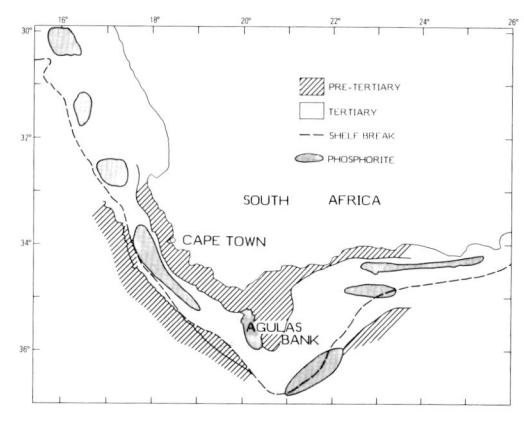

FIG. 6.—Shelfbreak zone off South Africa and associated phosphorite deposits (adapted from Fuller, 1974).

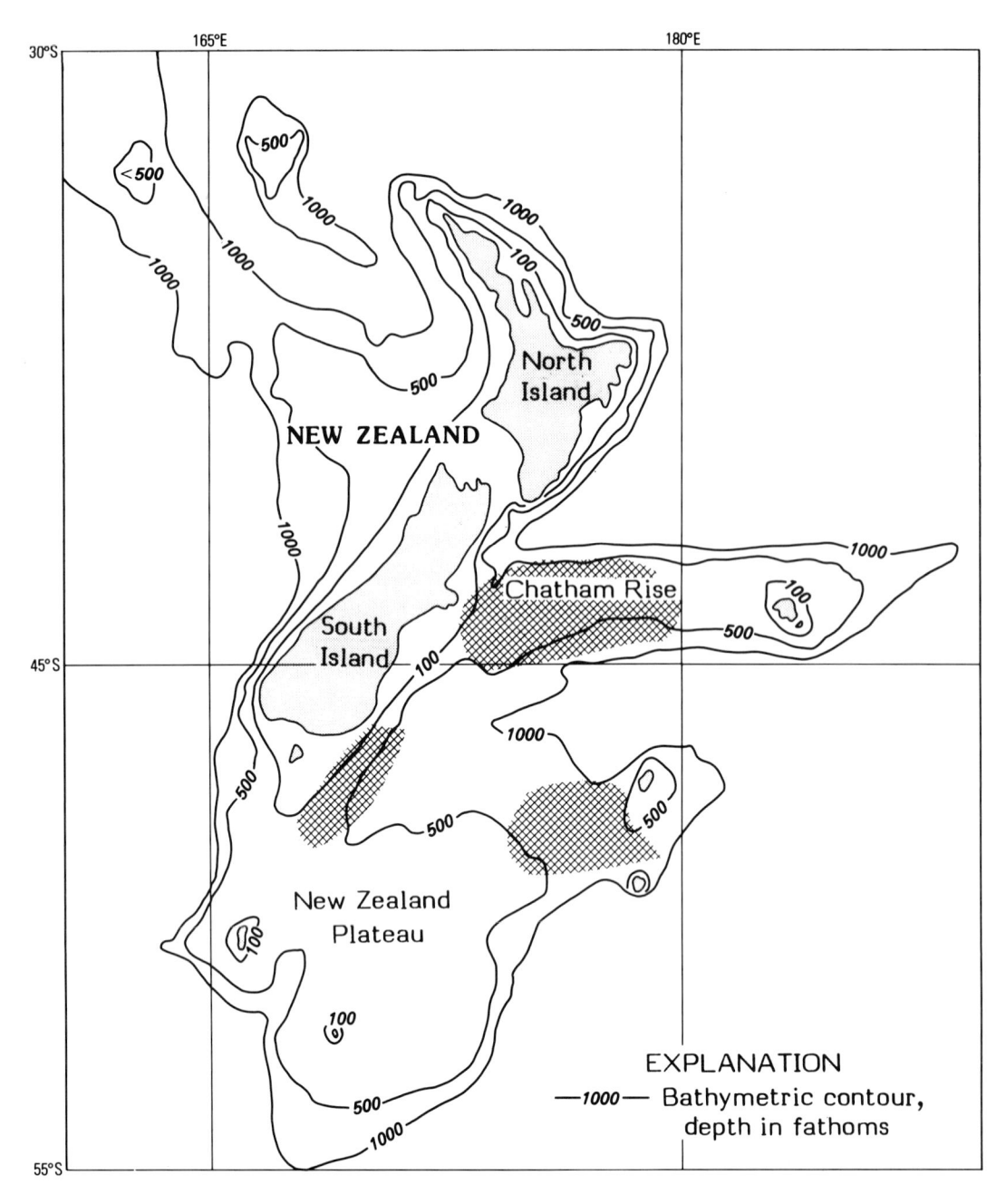

FIG. 7.—Chart showing Chatham Rise region and locations of marine phosphorites (adapted from Burnett and Sheldon, 1979).

bonate rocks and cobbles, but some fine-grained sands and mud are also phosphatized. Economic studies indicate the deposits, reportedly containing up to 117 million tons of 28 percent P_2O_5 materials, to be marginally economic at this time (Department of the Interior, 1979).

On the Blake Plateau off the southeast Atlantic States similar phosphorite deposits (Fig. 5) occur in phosphatized carbonate debris, cobbles, and slabs, much of it along a pseudo-shelfbreak at about 200 m depth. Unlike the California coast, no phosphatization is taking place at this time, and much of the phosphorite is now coated or masked by ferromanganese crustations (Manheim et al., 1980) that are perhaps associated with the contemporary Gulf Stream. Estimates of 2 billion metric tons of phosphorite and about 1.2 billion metric tons of mixed ferro-manganese/phosphorite pavement have been given (Manheim, 1980). Shelfbreak gravels are indicated off the northeast coastal states and Alaska, where Pleistocene gravel beds have been reworked (Molnia, 1979; Manheim, 1979). Off Hawaii, and potentially off Alaska, precious corals have been identified and collected in the deep waters at the shelfbreak (Grigg, 1979).

Other areas of the world where shelfbreak phosphorite deposits are known or probably occur include the Agulhas Plateau off South Africa (Fig. 6), the Atlantic coast of Morocco, the Chatham Rise off New Zealand (Fig. 7), and the coastal area off Peru and Chile. In the last case, deposits are confined to two narrow bands at about 100 m and 400 m in depth which coincide with the oxygen minimum zones in an area of strong upwelling current. Here, the deposits appear to be formed by direct precipitation within the openings of skeletal material (Cronan, 1980).

In New Zealand, gold deposits recently discovered off the southwest coast of South Island appear to be concentrated in increasingly coarser form and matrix of unconsolidated sands as one approaches the shelfbreak (Robin Falconer, 1981, pers. comm.). This suggests an offshore source for the gold and its reworking during lower sealevel stands. Similar indications occur near the edge of the Sunda Shelf, in southeast Asia, where coarse sediments increase toward the shelfbreak in areas of tinbearing alluvium. In certain areas, such as the head of the Gulf of California, fracture zones may extend across the shelfbreak resulting in the deposition of hydrothermal sulfides and other minerals. Such occurrences would be in the nature of fortuitous deposits, however, and would not necessarily be specifically restricted to the shelfbreak zone.

There is a possibility of finding relict deposits of metalliferous muds formed under anoxic conditions during the early stages of continental rifting. The metal content of such hydrothermal precipitates was influenced by the composition of continental rocks from both sides of the spreading center. Such deposits might usefully be sought in areas underlain by evaporites, as in the Red Sea (Emery and Skinner, 1977).

LEGAL ASPECTS

We recall that the shelfbreak occurs at diverse depths, and its distance from the coast may vary from only a few kilometers (off Chile, for example) to as much as 300 km (off the coast of Alaska). Shelfbreak zones off islands or submerged continental remnants may be thousands of miles from continental land masses. Thus, the location of the zone may lie within: (1) the territorial seas of adjacent states; (2) the 200-mile Exclusive Economic Zone (EEZ) already claimed by some nations; or (3) within the legally-defined "deep seabed beyond the limits of national jurisdiction." International law does not fully or satisfactorily cover the exploitation of minerals in these areas. Clearly, considerations applied to their possible economic recovery should also include a thorough investigation of the relevant international legal factors (see Ross and Emery, 1983, this volume).

CONCLUSIONS

Although almost any mineral deposit of the type found in continental land masses may be fortuitously located in the shelf-to-slope transition zone, only phosphorites, some heavy mineral placers, and certain precious corals appear to be products of the zone's unique geological and oceanographic history. In view of the relatively greater depth of the shelfbreak zone and the consequent technical problems of working in the open marine environment above this zone, mineral prospecting preferably should be directed towards these types of deposit. As for other deposits, there is greater probability of finding them in more favorable environments than the narrow, limited shelfbreak zone. Since such a small part of the outer continental margins have been adequately mapped for exploration at this time, the possibilities are good for serendipitous discoveries of minerals other than those identified to date.

ACKNOWLEDGMENTS

We thank the reviewers and editors for providing many suggestions used in the preparation of this chapter.

REFERENCES

BATURIN, G. N., MERKULOUA, K. I., AND CHALOU, P. I., 1972, Radiometric evidence for recent formation of phosphatic nodules in marine shelf sediments: Mar. Geology, v. 13, p. M37–M41.

BIRCH, G. F., 1973, Unconsolidated sediments off the Cape West coast: Univ. Cape Town, Mar. Geol. Progr., Joint Geol. Survey, Tech. Rept. 5, p. 48–66.

BEZRUKOV, P. L., AND SENIN, K. M., 1971, Sedimentation on the West African Shelf: Internat. Council Scientific Unions/Spec. Committee Oceanogr. Res. Working Party 31, Symposium, Cambridge, England, 1970, Inst. Geol. Rept. No. 70/16, p. 3–7.

BURNETT, W. C., AND SHELDON, R. P., 1979, Report on the marine phosphatic sediments workshop: Honolulu, Hawaii, East-West Center, 65 p.

CRONAN, D. S., 1980, Underwater Minerals: London, Academic Press, 360 p.

CURRAY, J. R., 1965, Late Quaternary history, continental shelves of the United States, in Wright, H. E., Jr., and Frey, D. G., eds., The Quaternary of the United States: Princeton, Princeton Univ. Press, p. 723–735.

———, 1966, Continental terrace, in Fairbridge, R. W., ed., The Encyclopedia of Oceanography: New York, Reinhold, p. 1021.

DIETZ, R. S., AND MENARD, H. W., 1951, Origin of abrupt change in slope at continental shelf margin: Am. Assoc. Petroleum Geologists Bull., v. 35, p. 1994–2016.

DEPARTMENT OF THE INTERIOR, 1979, Economic feasibility study of OCS mining of phosphorites, offshore southern California: U.S. Dept. Interior, Program Feasibility Document, OCS Hard Minerals Leasing, NTIS No. P601-192670, Appendix 15, 51 p.

EMERY, K. O., TRACEY, J. I., AND LADD, H. S., 1954, Geology of Bikini and nearby atolls. Part 1. Geology: U.S. Geol. Survey Professional Paper 260-A, 265 p.

———, 1968, Relict sediments on the continental shelves of the world: Am. Assoc. Petroleum Geologists Bull., v. 52, p. 445–464.

———, AND NOAKES, L. C., 1968, Economic placer deposits on the continental shelf: Econ. Commission for Asia and the Far East Tech. Bull., v. 1, p. 95–111.

———, AND SKINNER, B. J., 1977, Mineral deposits of the deep ocean floor: Marine Mining, v. 1, p. 1–71.

FULLER, A. O., 1974, Phosphate occurrences on the western and southern areas and continental shelves of Southern Africa: Econ. Geology, v. 74, p. 221–231.

GRIGG, R. W., 1979, Hawaii's Precious Corals: Honolulu, Island Heritage Ltd., 64 p.

HEEZEN, B. C., AND MENARD, H. W., 1963, Topography of the deep sea floor, in Hill, M. N., ed., The Sea, Volume 3: New York, Wiley, p. 233–280.

KUENEN, P. H., 1933, Geology of coral reefs, in The Snellius Expedition, 1929–1930, Geological Results, Volume 5, Part 2: Leiden, Netherlands, E. J. Brill, p. 124.

MACILVAINE, J. C., AND ROSS, D. A., 1979, Sedimentary processes on the continental slope of New England: Jour. Sed. Petrology, v. 49, p. 563–574.

MANHEIM, F., ROWE, G. T., AND JIPA, D., 1975, Marine phosphorite formation off Peru: Jour. Sed. Petrology, v. 45, p. 243–251.

———, 1979, Potential hard minerals and associated resources on the Atlantic and Gulf continental margins: U.S. Dept. Interior, Program Feasibility Document, OCS Hard Minerals Leasing, NTIS No. PB 81-192643, Appendix 12, p. 42.

———, PRATT, R. M., AND MCFARLIN, P. F., 1980, Composition and origin of phosphorite deposits of the Blake Plateau, in Bentor, Y. K., ed., Marine Phosphorites: Geochemistry, Occurrences, and Genesis: Soc. Econ. Paleontologists Mineralogists Spec. Pub. 29, p. 117–137.

MOLNIA, G. F., 1979, Sand and gravel resources of the continental shelf of Alaska: U.S. Dept. Interior, Program Feasibility Document, OCS Hard Minerals Leasing, NTIS No. PB 81-192577, Appendix 5, p. 11.

NEWELL, N. D., 1971, An outline history of tropical organic reefs: American Museum Novitates, No. 2465, p. 37.

ROSS, D. A., AND EMERY, K. O., 1983, The shelfbreak: some legal aspects, in Stanley, D. J., and Moore, G. T., eds., The Shelfbreak: Critical Interface on Continental Margins: Soc. Econ. Paleontologists Mineralogists Spec. Pub. 33, p. 437–441.

SHEPARD, F. P., 1973, Submarine Geology: New York, Harper and Row, 517 p.

SOUTHARD, J. B., AND STANLEY, D. J., 1976, Shelf-break processes and sedimentation, in Stanley, D. J., and Swift, D. J. P., eds., Marine Sediment Transport and Environmental Management: New York, Wiley-Interscience, p. 351–377.

VANNEY, J.-R., AND STANLEY, D. J., 1983, Shelfbreak physiography: an overview, in Stanley, D. J., and Moore, G. T., eds., The Shelfbreak: Critical Interface on Continental Margins: Soc. Econ. Paleontologists Mineralogists Spec. Pub. 33, p. 1–24.

VEEH, H. W., BURNETT, W. C., AND SONTAR, A., 1973, Contemporary phosphorites on the continental margin of Peru: Science, v. 181, p. 844–845.

THE SHELFBREAK: SOME LEGAL ASPECTS[1]

DAVID A. ROSS AND K. O. EMERY
Department of Geology and Geophysics, Woods Hole Oceanographic Institution,
Woods Hole, Massachusetts 02543

ABSTRACT

Conferences on Law of the Sea have had the objective of increasing the area of ocean floor subject to control by adjacent coastal countries. These extensions of jurisdiction have paid little attention to carefully defined and relatively easily identified geological boundaries such as the shelfbreak. Indeed, a geological term often is used in a legal sense that far exceeds the geological meaning, resulting in unnecessary confusion. The recently concluded Third United Nations Conference on the Law of the Sea adds an area of the ocean subject to national control equal to that of the land area of the world. Certain aspects of the remaining area of deep-ocean floor, such as mining, will also be controlled and taxed by an international authority. It is possible future oceanographers may have little opportunity for research without permission and regulation by governments of either coastal nations or the United Nations. One result could be increased research and knowledge of the ocean floor that is under the jurisdiction of industrialized countries and decreased effort in the rest of the ocean.

INTRODUCTION

Geologists have taken considerable care to define the major physiographic features of the earth, including those of the seafloor (see, for example, Heezen et al., 1959). Attention to detail has helped in the understanding of such features. Unfortunately, at least for the ocean, those in the legal and political arena have shown little concern for the terminology that has been carefully evolved. This is especially true for the continental shelf and outer continental margin (Emery, 1981). Such actions have led to confusion as well as a failure to develop a logical scenario for defining the legal boundaries of the continental margin and the deep ocean. This chapter discusses the above as it relates to the shelfbreak or shelfedge, and whether the shelfbreak, especially for marine scientific research, has any legal status in the Draft Convention from the Third United Nations Conference on the Law of the Sea (UNCLOS III).

BACKGROUND

According to Shepard (1973, p. 277) the continental shelf averages about 75 km (40 nautical miles) in width. The greatest change in slope occurs at a depth of 130 m (71 fathoms). As is well-known both to geologists, and more recently to lawyers, these values have a considerable range.

The shelfbreak (or shelfedge) marks the boundary between the continental shelf and continental slope and is the most evident physiographic feature of the ocean floor. It has been known for more than two centuries. Following the *Challenger* Expedition, it was stated that the 100-fathom line represents the outer limit of the continental shelf (Murray and Renard, 1891, p. 185). This 100-fathom depth, and its translation to about 200 meters, often have been used as general depths for the shelfbreak, even though the actual shelfbreak depths range between less than 50 and more than 400 m.

A common, although sometimes hard to apply, definition of the shelfbreak is the first major increase in gradient at the outermost part of the continental shelf (Dietz and Menard, 1951; Wear et al., 1974; Vanney and Stanley, 1983). As a physiographic feature, the shelfbreak is essentially ubiquitous, being found on all continental margins in all oceans. However, its depth, shape, and distance from shore can vary depending on the structural characteristics of a given continental margin and the region's geological and sedimentary history.

Echo-sounding and continuous seismic profiling throughout the world show that the shelfbreak region may (or may not) mark the position of a change in the deep underlying structure, but essentially it separates the very flat continental shelf from the slightly steeper continental slope of less than 5 degrees. It should be appreciated that most echo-sounding and seismic profiles have large vertical exaggerations (usually more than 10 times) that accentuate the slope differences between the continental shelf and slope. Without such a vertical exaggeration, a shelfbreak occasionally can be difficult to clearly identify. Various methods have helped to identify the position of the shelfbreak, and several are shown in Figure 1. The choice of shelfbreak can be especially unclear if a region has a series of terraces or step-like breaks in slope (see for example, profiles C, D, E and F on Fig. 1). In plan view, the shelfbreak can vary from being sinuous, cren-

---

[1]Contribution No. 5198 of the Woods Hole Oceanographic Institution.

Copyright © 1983, The Society of
Economic Paleontologists and Mineralogists

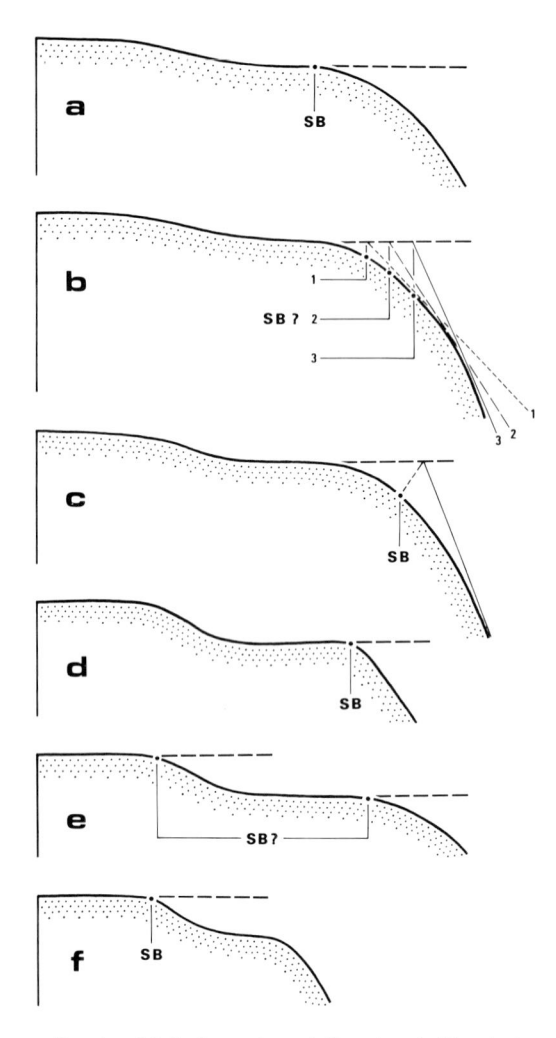

FIG. 1.—Methods used to define the shelfbreak (*a*, method used in this volume; *b-c* two other possible methods; *d-f*, more complex situations due to terraces at or near the shelfbreak = SB). Note that vertical scale is highly exaggerated. From Vanney and Stanley (this volume) who modified it from Wear et al. (1974) and Southard and Stanley (1976).

ulated or even discontinuous (see Farre et al., 1983; Vanney and Stanley, 1983).

STATUS OF THE SHELFBREAK PRIOR TO UNCLOS III

The concept of a shelfbreak in legal or political usage is relatively modern but an infrequent one. Since the 17th century, a basic feature used in legal aspects of the ocean is the territorial sea which was long held to extend to 3 miles off a country's coast; within this territorial sea the coastal state had almost complete sovereignty. This concept, however, was never officially ratified by any international agreement. The area beyond the territorial sea, called the high seas, was considered as *res nullius* or belonging to no one. The width of the shelf or depth of the shelfbreak was irrelevant.

After World War II, interest in the ocean increased, mainly because of the discovery of marine resources by private companies of industrial nations. One of the first major challenges to the 3-mile territorial sea came from United States' President H. S. Truman in 1945. Truman made two policy statements (called the Truman Proclamations). The first concerned the natural resources of the seabed and subsoil (legal jargon for the ocean floor and underlying sediments and rock), and said that "the Government of the United States regards the natural resources of the subsoil and seabed of the continental shelf beneath the high seas but contiguous to the coasts of the United States as appertaining to the United States, subject to its jurisdiction and control." The second proclamation concerned conservation zones for fishing in the water column.

Truman's Proclamations did not actually define the continental shelf, although a White House press statement released at the same time said that the continental shelf is generally considered as that submerged land which is contiguous to the continent and which is covered by no more than 100 fathoms of water. So, by inference, the shelfbreak was considered to be 100 fathoms or less and the shelf was given a depth definition. Defining the shelf by depth is a good approach for a nation having a wide, shallow shelf. However, other countries not so well endowed also were anxious to extend their jurisdiction (Truman's Proclamation actually never really extended U.S. jurisdiction over an area but just focused on resources; this subtlety was lost or not considered by others). In particular, Peru, Chile, and Ecuador in 1947 extended their jurisdiction and sovereignty over the water, seabed and subsoil out to 200 nautical miles from their coasts; they essentially declared a 200-mile territorial sea (Chile did not go quite as far as the others). This claim far exceeds the width of their continental shelves but takes into account that these countries border the Peru-Chile trench and have a very narrow shelf. It would not have been in their interest to make a claim based on depth; likewise, there is nothing special about 200 nautical miles perhaps other than that it is a nice round number like 200 meters.

This extension and several others led to the First (1958) and Second (1960) United Nations Conferences on the Law of the Sea. These were legal procedures, and physiography and other fields of geology were not seriously considered. Among the goals of these meetings were the definitions of widths of the territorial sea and of the continental shelf; they failed on both counts. Concerning the territorial sea, it stated "the outer limit of the territorial sea is the line every point of which is at a distance from the nearest point of the baseline equal to the breadth of the territorial sea" (Article 6, United Nations 1958b Convention on the Territorial Sea and Contiguous Zone). A contiguous zone (which included the territorial sea) was defined as not extending beyond 12 nautical miles from the coast. Thus, the territorial sea cannot be more than 12 nautical miles wide. The question of the width or depth of the continental shelf was considered in the United Nations (1958a) Convention of the Continental Shelf and the results were even less satisfying. In this latter document, the shelf was defined as "the seabed and subsoil of the submarine areas to a depth of 200 meters, or beyond that limit, to where the superadjacent waters, admits of the exploitation of the natural resources of the said areas." Thus, a combination of depth and distance was applied—200 meters—and as far beyond as exploitation can occur. Coastal states were given sovereign rights over the shelf for natural resources and consent was needed for scientific research. These poorly defined terms, combined with further extensions of claims and increased interest in revenue from marine resources (manganese nodules, in particular), led to the Third Conference—UNCLOS III—that began in the early 1970s and ended in 1982. As an aside, one cannot help but compare the over-enthusiasm for the economics of nodules that existed in the 1960s with that for ocean-ridge sulfides in the early 1980s. The result of the former has been an increase in coastal state jurisdiction over the ocean; the sulfide enthusiasm (often exaggerated in the press) could lead to similar consequences.

Following the first two Conferences, the question for United States marine scientists was, where can or cannot one work and under what regulations? The conditions for marine research are strongly influenced by the official position of the United States Government. For example, the United States Government presently (May, 1982) recognizes a 3-mile wide territorial sea, unless the research concerns living resources. For research on the continental shelf, the Department of State for many years required foreign permission for work in depths less than 200 m; more recently, this has been extended (actually restricted is a better word) to 600 m (perhaps recognizing the ability to exploit deeper). These shelf restrictions do not apply to research in the water column, as for most types of physical oceanography where territorial-sea limits apply. Seismic profiling and echo-sounding is considered water column research (by the U.S. Department of State); bottom sampling isn't and permission of the coastal country is needed for the territorial sea and beyond. For fisheries research using commercial gear or taking resources in commercial quantities, permission is required from the coastal country out to 200 nautical miles from its coast. Note, however, that the United States' position might be quite different from that of foreign coastal countries—putting a researcher in a dilemma. In any case, the lack of satisfactory resolution of several issues, including those previously mentioned, led to UNCLOS III.

UNCLOS III

In the Draft Convention resulting from UNCLOS III the continental shelf, as stated in Article 76, "comprises the sea-bed and subsoil of the submarine areas that extend beyond its (a coastal states') territorial sea throughout the natural prolongation of its land territory to the outer edge of the continental margin or to a distance of 200 nautical miles from the baselines from which the breadth of the territorial sea is measured where the outer edge of the continental margin does not extend up to that distance." This definition clearly does not make any geological sense, since the so-called "legal continental shelf" can contain the geological shelf, slope, rise, ridges, plateaus, and even trenches and portions of the abyssal plain. Likewise, the "natural prolongation of its land" should not extend to the continental rise or trenches whose geology and structure is typical of the oceans and not of land (see, in particular, Emery, 1981 for a detailed discussion of this point for the Atlantic Ocean). It becomes even more confusing! In subsequent parts of Article 76 the continental margin is defined as comprising "the submerged prolongation of the land mass of the coastal state and consists of the sea-bed and subsoil of the shelf, the slope and the rise. It does not include the deep-ocean floor with its oceanic ridges . . ." If the continental margin extends beyond 200 nautical miles, its outer edge is established by a complex (and very difficult to establish) series of criteria, but it may not be more than 350 nautical miles (some exceptions for particular countries permit an even further extension) from the baseline from which the territorial sea is measured. This is referred to as the outer limits of the continental shelf, which we agree doesn't make any sense; these points are summarized in Figure 2.

We have found no reference to the shelfbreak or shelfedge in the Draft Convention, although a rather confusing term—the foot of the continental slope—plays an important role in the above criteria. It is defined as "in the absence of evidence to the contrary, the foot of the continental slope shall be determined as the point of maximum change in the gradient at its base." Again, the definition is poor and open to considerable interpretation (see, in particular, Hedberg, 1979 and references therein). These above points, although perhaps somewhat humorous on first reading, really are quite important. Within the legal continental shelf (which can

LEGAL DEFINITION OF OUTER LIMIT OF "CONTINENTAL SHELF"

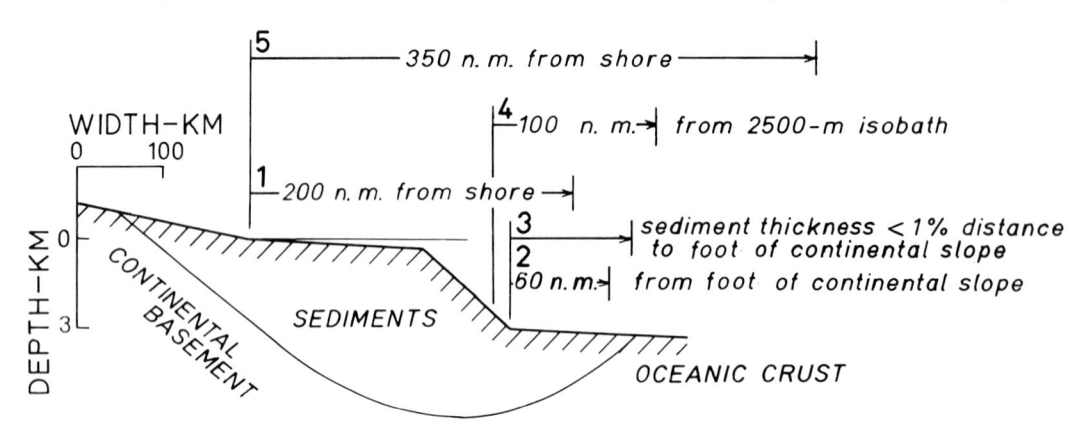

FIG. 2.—Five components of the legal definition of the outer limit of the "legal continental shelf" as presented in Article 76 of the Draft Convention on the Law of the Sea (modified from Emery, 1981).

cover about 42 percent of the ocean and thus slightly more than the world's land area) there are considerable constraints on marine scientific research. These restraints have been detailed elsewhere (Ross, 1982) and will not be repeated here other than to say that a consent regime for marine scientific research, with certain obligations, exists in most of this region.

Certainly, in the scientific sense, it would have been valuable to define the legal boundaries of the ocean following basic geological concepts. Being realistic, one must appreciate the difficulty since some countries would get less, or more, than others—a point that is politically unacceptable. Considering this, it is unfortunate that the well-established physiographic terms had to be "re-defined" (and very poorly so) to justify in part new enclosure of the ocean by the countries of the world.

As marine scientists, we may suffer restraints in our activities in the name of world economics, if oceanographic activities are restricted in the 42 percent of the ocean comprising the "legal continental shelf." Ironically, this situation is a partial result of our successful research which has led to many of the marine mineral discoveries. The finding of such deposits then has led to restriction or enclosure by nations that have made no ocean-floor studies and whose politicians have little knowledge or interest in the ocean other than as a possible source of unearned income. A likely result is increased concentration of study within regions of ocean floor bordering industrial nations and focus on the shelf-to-slope sector as defined by marine scientists.

Probable new and unexpected findings of possible economic value will be made in these regions.

SUMMARY

Although geologists have taken considerable care in defining the major physiographic features of the earth, including the shelf and shelfbreak, lawyers have often severely misused these definitions. This is particularly true for the continental shelf, which in the legal sense (of UNCLOS III) can include the continental shelf, continental slope, continental rise and indeed even parts of the deep sea such as the abyssal plain. The shelfbreak, or shelfedge, has unfortunately not been used in such discussions—usually a combination of depth and distance were applied to define the edge of the shelf.

In the recently concluded UNCLOS III the legal continental shelf (which for marine science comes under coastal state control) can include as much as 42 percent of the ocean. The consent regime for marine science could restrict research opportunities to areas off industrial or developed nations that have their own marine science capability.

ACKNOWLEDGMENTS

We would like to thank Robert W. Knecht, Jack W. Pierce, Elazar Uchupi and the editors for reviewing this article. Support for David A. Ross came from the Department of Commerce, NOAA, Office of Sea Grant under Grant No. NA80-AA-D-00077 (E/L-1) and the Pew Memorial Trust, but the views are those of the authors.

REFERENCES

DIETZ, R. S., AND MENARD, H., 1951, Origin of abrupt change in slope at continental shelf margin: Am. Assoc. Petroleum Geologists Bull., v. 35, p. 1194–1216.
EMERY, K. O., 1981, Geological limits of the "continental shelf": Ocean Development and International Law Journal, v. 10, p. 1–11.
FARRE, J. A., McGREGOR, B. A., RYAN, W. B. F., AND ROBB, J. M., 1983, Breaching the shelfbreak: passage from youthful to mature phase in submarine canyon evolution, *in* Stanley, D. J., and Moore, G. T., eds., The Shelfbreak: Critical Interface on Continental Margins: Soc. Econ. Paleontologists Mineralogists Spec. Pub. 33, p. 25–39.
HEDBERG, H. B., 1979, Ocean floor boundaries: Science, v. 204, p. 135–144.
HEEZEN, B. C., THARP, M., AND EWING, M., 1959, The Floors of the Oceans, I. The North Atlantic: Geol. Soc. America Spec. Paper 65, 122 p.
MURRAY, J., AND RENARD, A. F., 1891, Deep Sea Deposits: Scientific Results of the Voyage of H.M.S. *Challenger*, 1872–1876: London, Her Majesty's Stationery Office, 525 p.
ROSS, D. A., 1982, Marine science and the law of the sea: EOS, v. 62, p. 650–652.
SHEPARD, F. P., 1973, Submarine Geology, 3rd Edition: New York, Harper and Row, 517 p.
SOUTHARD, J. B., AND STANLEY, D. J., 1976, Shelf-break processes and sedimentation, *in* Stanley, D. J., and Swift, D. J. P., eds., Marine Sediment Transport and Environmental Management: New York, Wiley-Interscience, p. 351–377.
UNITED NATIONS, 1958a, Convention on the Continental Shelf—Convention on the Law of the Sea Adopted by the United Nations Conference at Geneva, 1958: U.N. Doc.A/Conf. 13/55.
———, 1958b, Convention on the Territorial Sea and the Contiguous Zone—Convention on the Law of the Sea Adopted by United Nations Conference in Geneva, 1958: U.N. Doc.A/Conf. 13/L. 52.
VANNEY, J.-R., AND STANLEY, D. J., 1983, Shelfbreak physiography: an overview, *in* Stanley, D. J., and Moore, G. T., eds., The Shelfbreak: Critical Interface on Continental Margins: Soc. Econ. Paleontologists Mineralogists Spec. Pub. 33, p. 1–24.
WEAR, C. M., STANLEY, D. J., AND BOULA, J. E., 1974, Shelfbreak physiography between Wilmington and Norfolk Canyons: Marine Tech. Soc. Jour., v. 8, p. 37–48.

SUBJECT INDEX

A

Acadian orogeny, 81–82, 90–91
　plate collision, 80
Accretion unit, 129
Accretionary basin, *See* Trench-slope basin
Accretionary prism, 97, 100–101, 104, 109, 113, 115–116, 306–307
Active margin, *See also* Active shelfedge, 79, 83, 90–91, 97–104, 293
　ancient, 107–116
　basin development, 107–116
　convergent setting, 107, 109, 110–113
　geotechnical studies on, 333–334, 346
　protoceanic setting, 107, 109, 113
　transform setting, 107, 111
　U.S. West Coast, 222
— shelfedge, 90–91, *See also* Active margin
　Borden Formation, 90
　Fort Payne ramp, 91
Aegean Sea, islands in
　unstable rifted margin, SB on, 19
Africa
　Mesozoic carbonate SB, 191
　plate collision, 79, 99
　Proterozoic carbonate SB, 190
— margin, northwest, 51
　phosphorite, 435
　upwelling-mineral deposits, 431
— —, southwest, *See also* Namibia
　high productivity off, 391
　organic carbon off, 393
　phosphorite, 431, 435
　upwelling off, 393–394
　upwelling, mineral deposits, 431
　shelfbreak placer deposits, 432
— —, west, 51, 57, 140, 145, 191
African plate, 411
Aggregate
　Alaska, southern, 435
　placer, 431–432
Alabama margin
　infauna biomass of, 384–385
　infauna density on, 385
Alaska margin, Arctic
　polar non-glaciated region, SB on, 10
— —, southern, *See also* Gulf of Alaska, 299
　aggregate, 435
　Cook Inlet, 233
　Copper River, 233, 302
　paleomagnetic studies, 103
　sediment loads of rivers, 302
　shelfedge, 300, 304, 310
Aleutian Islands
　trench off, 222
Algae, 83, 85, 89, 198, 395
　blue-green, 211
　calcareous, 198
　coccolith plates, 383
　codiacian, 172
　coralline, 172, 174, 176, 179
　dasyclad acean, 210
　diatoms, 373–379
　green, 84, 201, 211
　red, 84, 211
Algal
　mound, 284, 288
　mud, 196
　reef, 135, 169, 290, 420
　ridge, 176, 181, 186
　sand, 282
Alpine orogeny
　cause of sealevel fall, 47
— tectonics, 207, 212, 214
　major events, 212
—-Himalayan belt
　Cenozoic carbonate SB, 191
　Mesozoic carbonate SB, 191–192, 194, 197, 203
Alps, 207, 212
　source of sediments, 72
　uplift of, 69
—, French
　Grès d'Annot, 325, 327
　submarine canyons in, 315
Amazon
　canyon, 130
　delta, 136
— cone, 139
—-Guiana-Orinoco shelf
　tropical fluvial dominated region, 14
American plates
　western margins of, 410
Anadarko Basin
　Woodford Formation, 402
　starved basin, 80, 88, 91
Anaerobic conditions, *See* Anoxic conditions
Anchoralis-latus Zone (Mississippian), *See also* Biostratigraphic zone, 79–83, 85–86, 88, 90–91
Andaman Sea
　tropical fluvial dominated region, 14
Andes-type convergence, 97
Aneth field
　source rocks for, 401
Anoxic environment, 160, 203, 341
　cause of benthonic segregation, 160
　Eastern Interior trough, 80
　mineral enrichment in sediments, 431, 435
　Monterey Formation (Miocene), 111
　off Namibia, 342
　off Peru, 342
　restricts bioturbation, 342
— model
　Woodford Formation, 402
— water, *See also* Bottom water, Oxygen deficient water, 392–393, 402
　Black Sea, 393, 398
Antarctic margin
　polar glacial region, SB on, 9
Antarctica, glaciation in, 46
Antler foreland trough, trough, 80, 82–83, 90
　submarine fan deposits, 83
　orogeny, 80–82
　turbidite deposits, 83

— highlands, 82, 90–91
plate collision, 80
Apennine Mountains
  source of sediments, 72
  uplift of, 69
Appalachian
  basin, 81
  plate collision, 80
— foldbelt
  Africa-North America collision, 99
— mountains, 80, 83, 90–91, 410–411
  denudation rate of, 54
  Paleozoic carbonate SB, 190, 195
  sediment origin, 287
— region, 159–164
Arabia
  Mesozoic carbonate SB, 195, 201
Aragonite, 201
Arizona, 90
  Redwall-Escabrosa shelf, 89
Arkansas
  Burlington shelfedge, 88, 90
— novaculite
  source rock, 402
Arkoma basin, 420–421
  Atoka Formation, 421
  clastic facies, 409
  flysch, 420
  novaculite, 420
  oil fields, 410, 421
  paleo-shelfbreak, 422
  Red Oak-Norris gas field, 421
  reservoir rocks, 420
  Viola Formation, 420
Asia, southeast
  collision, 422
  ophiolites, 420
  orogenesis, 420
  paleo-shelfbreak, 420
  shelfbreak placer deposits, 432
Asmari Limestone, (Oligo-Miocene)
  re-migrated oil, 412
Assam valley, *See* Upper Assam Valley
Astoria Canyon, 243, 396
Atlantic basin
  continental separation, 419
— margin, 293
  contrast with Gulf, 292
  subsidence curve for, 197
— Ocean, 17, 419
  Cretaceous marginal basins, 411, 419
  Early Cretaceous basins of, 410, 419
  Gulf of Maine, 233–234
  legal aspects, 439
  margins of, 223
  Middle Atlantic Bight, 233–235, 238–239, 241, 243, 247
  New York Bight, 233–234, 241
  South Atlantic Bight, 233–235, 237–240, 243, 247–248,
  western, 327
— —, North, 344
— —, northwest, 381–384
  infauna biomass of, 384
  infauna density of, 384

organic carbon on, 383
— —, South
  desiccation of, 318
  salt basin, 419
—-type margin, *See* Passive margin
Atoka Formation (Pennsylvanian)
  Arkoma basin, 421
  Red Oak-Norris gas field, 421
Atoll, 169–170, 176, 180, 191, 414–415, 429, 432
  Golden Lane oil fields, 414–415
  Midland basin, Horseshoe atoll, 417
Atoll margin, 176
  Bikini, 173, 176
  carbonate margin, classification of, 169, 172
  Eniwetok, 173, 176
Atterberg limits, 337–338, 347, 351
Aulacogens, 190
Australia
  Cenozoic carbonate SB, 191
  Great Barrier Reef, 172, 174, 176, 180
  Paleozoic carbonate SB, 190, 196
  Proterozoic carbonate SB, 190
Australian shelf, northwest, *See* Northwest Australian shelf.
Avalon basin
  rift basin, aborted, 410

B

Back-reef, *See* Reef
Backarc basin, margin, 107
Backstripping, 41, 45
Bahamas
  Mesozoic carbonate SB, 191
  sand-shoal-dominated margin, 173–174, 176, 178–179
  Tongue of the Ocean, 179, 319
  tropical non-fluvial region, SB in, 12
Baja California, 242
  canyons off, 318–319
— California, eastern
  rifted substrate, SB on, 12
Balleny Group (Oligocene)
  submarine canyons in, 327
Baltimore Canyon, 318, 337, 351
  hanging valleys of, 28, 36
  shelf indenting, 28, 36
— Canyon trough, 48
Barbados, 100, 102
  tropical non-fluvial region, SB in, 12
Barnett Formation (Mississippian), 89
Basalt, 172, 419
Base level, 53–54
Basin margin facies
  Chemung, 161
  deposited by bottom currents, 162
Bay of Biscay
  temperate region, SB in, 18
Beaufort Sea
  polar non-glaciated region, SB on, 10
Bedforms, 275
  large near canyon heads, 300
  ripples, 272, 282, 320–321, 325
  sand waves, 184, 272, 282
—, giant
  Bering Sea, 225
  Cook Inlet, 225
  formation by, 225

SUBJECT INDEX

North Sea, 225
Bedload, 223, 227–228, 245, 265, 270–271, 273, 323, 328
Belize, 180
  reef dominated margin, 172, 174
Bengal coast
  tropical fluvial dominated region, 14
Benguela Current, 393
Benioff zone, 99
Benthic boundary layer (BBL), 275
  Flower Garden Bank, 255–256
  shelfedge, 219, 222, 225–228
  studies, 251
Bering Sea, 373, 376, 379
  fault related, canyons in, 318
  giant bedforms, 225
  polar non-glaciated region, SB on, 10
  shelf-slope break, 373
Bikini atoll
  atoll margin, 173, 176
Bioerosion, 315, 319, 328
Biofacies model, 85–88
Bioherms, 10, 15, 83, 90, 186, 198, 429, 432
  Arkansas, 88
  Burlington shelfedge, 88
  Chappel shelf, 89
  Fort Payne Formation, 91
  Lake Valley Limestone, 89
  Missouri, 88
  Oklahoma, 88
Biological productivity, 222
— zonation, 199
  reef dominated margin, 169, 174, 176, 181, 186
  vertical, 198
Biomass, 381
Biostratigraphic zone, See also Anchoralis-latus Zone
  compression of, 107, 110–111, 116
  conodont, compression of, 88
Bioturbation, 30, 91, 159, 162, 164, 176, 222, 227, 293–294, 325, 327, 340–342, 351, 387, 409
  lack of, 393
  limited by anoxic condition, 342
  organic matter destruction, 393
Black Sea, 318, 393–394, 410
  anoxic water, 393, 398
Bloom
  diatom, 374, 378
  plankton, 392
  radiolaria plankton, 82
Borden Formation (Mississippian), 81, 83, 90–91
  active shelfedge, 90–91
  delta, 79, 81, 83, 90–91
  foreset beds of, 91
  Illinois basin, 83, 90
  Indiana, 91
  Kentucky, 91
  Michigan basin, 90
  prodelta part, 90
  topset beds of, 91
  trace fossils, 88
Bottom current, 224, 226, 228, 284, 288, 309, 316, 319, 340, 342, 344, 392, 432–433
  sediment resuspension by, 287, 290
— sediment studies, 27, 279
  geotechnical studies of, 333–352

Hudson Canyon head, 288
  in New York Bight, 280
  near Desoto Canyon, 281, 287
  off Cape Hatteras, 280
  off Mid-Atlantic States, 280, 285
  off Mississippi delta, 281
  Scotia shelf, 266–267
  submarine canyons, 316
— water, See also Anoxic water, Oxygen deficient water
  anoxic, 393, 398
  oxic, 391–393, 397, 400, 402
  oxygen deficient, 393, 395, 398, 400–402
Bottomset beds
  New Providence Shale, 90
  Springville Shale, 90
Boundary current, 219, 223–224, 228, 328
Brachiopods, 88–89, 91, 162, 164
Brazer Dolomite (Mississippian), 84
  Utah, 84
  Wyoming, 84
Brazil
  Campos basin, 409–410, 419–420
  Garoupa field, 419
—, margin
  upwelling on, 393
Breccia, 327–328
  as reservoirs, 416
—, internal, 207–214
  carbonate, 207–214
  cause of, 207
  comparison with conglomerates, 208
  composition, 207
  correlation with Alpine tectonics, 207
  fitting of, 207–208
  Hydra, Greece, 207, 209, 212–214
  Italy, 207
  mass flow, 207–208, 210–212
  origin of, on Hydra, 213–214
  other terms for, 207
  types of, 208
  United States, 207
—, tectonic, 100
Brecciation, 207, 210–214
Bryozoans, 83–84, 88–89, 91, 164, 198, 211
Burlington Limestone (Mississippian), shelf, shelfedge, 88, 90–91
  Arkansas, 88
  carbonate bank, 88
  extension of Chappel shelf, 88
  Illinois, 88
  Iowa, 88
  Missouri, 88
  Oklahoma, 88
Burma-Indonesia
  tropical fluvial dominated region, 14
Bypass, sediment, 14, 26, 36–37, 61, 74, 104, 107, 110, 116, 191, 201, 223, 288, 290, 292, 295, 307, 325, 327

C

Caballos-Arkansas Island Chain, 83, 91
  welt, 82
Calabrian Arc
  temperate region, SB in, 18
Caledonian belt

Paleozoic carbonate SB, 190
California, See also Pacific Coast, 103, 115, 321–322, 434
  Carmelo Formation, 315, 327
  central, 300, 302
  Delmar Formation, 323
  Eel River, 222, 305
  forearc basin, 110
  Great Valley sequence, 103, 113, 411
  Meganos Canyon, 318, 327
  Monterey Formation, 111, 213, 402
  northern, 300, 302
  Pigeon Point Formation, 315, 327
  San Andreas fault, 107
  San Joaquin valley, 113, 410–411, 424–425
  San Pedro shelf survey, 226–227
  sediment loads of rivers, 302
  shelfedge, 300, 304
  southern, 224–225, 300, 308, 315, 318
  Southern California Borderland, 300, 305, 433
  Stevens Sandstone, 411, 424
  Torrey Submarine Canyon, 324–325, 327–328
  unstable rifted margin, SB on, 19
  West Thornton gas field, 424
— Current, 223–224
— margin, See also Pacific Coast
  central, 236
  currents on, 242, 248
  geotechnical studies on, 346–352
  southern, 234, 243
  tides, 238
  upwelling, 236, 239, 242–243
Campeche bank
  non-rimmed margin, 173, 179–180
— fields, 409, 415–416
  carbonate-evaporite facies, 409
  Mexico, 409
Campos basin, 409–410, 419–420
  Brazil, 409–410
  carbonate and clastic facies, 409, 419–420
  Garoupa field, 419
  oil fields, 410
  paleo-shelfbreak, 420
  reef, algal, 420
  turbidites, 419
Canada
  Nova Scotia shelf, 234
  Paleozoic carbonate SB, 192–194, 196, 203
  Proterozoic carbonate SB, 190, 192
  Scotia shelf, 197, 234, 265–275
  western, 79, 83
Canyon, See Submarine canyon
Canyon fill creep, 35
Cape Hatteras, 248, 280, 285, 344, 360–361, 385
  bottom sample studies, off, 280
  circulation off, 233–235, 239
  narrow shelf off, 281
  Type I mudline, 281, 292, 295
Capitan Limestone (Permian), 193
Carbonate and clastic facies, 409, 411, 426
  Campos basin, 409, 419–420
  Delaware basin, 411, 416–418
  Gulf Coast, 409, 411
  Midland basin, 411, 416–418
  Permian basin, 409, 411, 416–418

South China Sea basin, 409
— bank, 88, 411
  Burlington shelf, 88
  Chappel shelf, 88–89
  Yucatan, 12, 179–180, 415–416
— breccia, See Breccia, internal
—-evaporite facies, 409–411, 426
  Campeche fields, 409–410, 426
  Golden Lane fields, 409–410
  Iran-Iraq oil fields, 409–414, 426
  Mexico, 409–410, 418, 426
—-evaporite-clastic facies, 113
— fixation, 189, 191–192, 195–196, 201, 203
— margin, 169–186, See also Cenozoic, Mesozoic, Paleozoic, and Proterozoic carbonate SB,
  ancient, 189–203
  sediment production, 169
— margins, classification of, 169, 170, 172
  atoll margin, 169, 176, 186
  non-rimmed margins, 169, 179–180, 186
  reef-dominated margin, 169, 172–176, 186
    biological zonation, 169
  sand-shoal-dominated margin, 169, 176–179, 186
— margins, processes controlling, 180, 186
  antecedent topography, 169, 180–185
  sealevel history, 169, 184–185
  tectonism, 169, 180
  wave energy, 169–170, 178, 180–181, 185, 192, 198–199
— platform, 7, 82–84, 88, 90–91, 190, 210, 212, 412
  Canada, western, 79, 83
  United States, 79, 82, 83
— sand shoal, See also Sand-shoal-dominated margins, 169, 172–173, 176–179, 185–186, 189, 191–192, 194, 197–199, 201, 203
Caribbean
  Cenozoic carbonate SB, 191
Carmelo Formation (Paleocene)
  California, 315, 327
Carolinian Province, 385–386
Carteret Canyon, 25, 28, 31–33
  slope failure, 32–33
Castle Reef Dolomite (Mississippian)
  Montana, 90
Catskill delta, Formation (Devonian), 80, 159–160
— sea, 159, 164
  near paleoequator, 159
Celtic Sea, 18
  temperate region, SB in, 18
Cenozoic carbonate SB, 189, 191, 203
  Alpine-Himalayan belt, 191
  Australia, 191
  Caribbean, 191
  Gulf of Mexico, 191
Central America
  Mesozoic carbonate SB, 191
  tropical fluvial dominated region, 14
— Basin Platform, See also Permian basin, 426
  oil fields, 410, 417–418
  reservoirs, 418
Cephalopods, 210–211
Chappel Limestone (Mississippian), shelf, shelfedge, 88–89
  bioherms, 88
  conodonts, 89
  Llano uplift, 88–89

SUBJECT INDEX

upwelling, 89
Texas, 89
Chattanooga Shale (Mississippian), 91
Chemung (Devonian), 161
Chile rise, 97
— margin, 431, 435
   upwelling-mineral deposits, 431
   phosphorite, 435
Chlorite, 350–351
Cincinnatti arch, 90
Clastic dike, 325
— facies. *See also* Flysch, 409, 411, 420
   Arkoma basin, 409, 420–421
   Great Valley Sequence, 411
   Gulf Coast, 409, 418, 424
   Northwest Australian shelf, 409, 423–424
   Permian basin, 411
   Upper Assam Valley, 409, 422–423
   West Thornton gas field, 424
— shelfedge, classification of, 139–140
— wedge, 400
Clinoform facies
   deposited by turbidity currents, 162
   Portage Shale, 161
Clinoforms, 82, 129, 141, 144, 148, 159–164
   deposited by turbidity currents, 162
   limit by pycnocline, 159–160
Coal, 113
Coarsening upward sequence, 149
Coccolith plates, 383
Coccolithophorid, calcareous nannoplankton, 200, 375
Coelenterates, 162
Collision, *See* Plate collision
Color, 83–84
   oxidation-reduction state, 84
Colorado
   Redwall-Escabrosa Limestone, shelf, 89
Columbia River, 235, 237, 305, 396
Compression regime, *See* Convergent margin
Computer model
   coastal region, 219
— plots
   foraminifera data bank, 362
—-catalog, *See* Data bank
Cone, *See* Submarine fan
Conglomerate, 162–163, 193–194, 196, 199, 208, 327, 420
   comparison with internal breccias, 208
Congo canyon, 130, 318
Conodonts, 83, 210
   biofacies model, 79, 85
   biostratigraphic compression of, 88
   Chappel Limestone, 89
   Deseret basin, 85
   Springville Shale, 88
   upwelling, 85
   zonation, 79
Contemporaneous fault, *See* Growth fault
Continent-ocean boundary, *See* Ocean-continent boundary
Continental crust, 98, 107, 110, 113, 196
— drift, 191, 197
   Atlantic basin, 419
— margin
   definition of, 437

Continental shelf, *See* Shelf
Convergent margin, *See also* Plate collision, Subduction, 4–5, 9, 12, 14, 18, 97–104, 107, 110–111, 113, 293, 426
   Andes-type, 97
   forearc basin, 9, 14, 97, 102–103, 107, 109–110, 113, 115
   Himalaya-type, 97
   intramassif basin, 107, 109
   Japan-type, 97
   Pacific, 299
   South America, 97
   trench-slope basin, 97, 107, 109, 115
Cook Inlet
   Alaska, 225
   giant bedforms, 225
Cooling, *See* Thermal effects
Copper River
   Alaska, 223
   discharge of, 302
Coral, precious, 429, 435
   at shelfbreak, 435
   Hawaii, 435
   location of, 432
Corals, 83–84, 135, 164, 173–174, 176, 186, 198, 211, 290, 386, 415
   branching, 174
   colonial rugose, 84
   head corals, 174
   mound, 284
   sand, 282
   scleractinian, 172, 185, 201
   solitary, 84, 88
Coriolis force, 258–259
   Eastern Interior Sea, 83
   Illinois basin, 83, 91
   Madison Sea, 83
Corsica
   canyons off, 318
   unstable rifted margin, SB on, 19
COST B-2 well, 45
Counter-current, 224
Cratonic crust, *See* Continental crust
Creep, 19, 121, 139, 144, 315, 320, 325, 327–328, 333, 345, 387
   lack of, 320
Cretaceous delta, 400
Crinoid, 84, 89
Cross stratification, 164, 325
Crustaceans, 384
Current(s), *See also* Wave energy, Wave(s), 1, 25, 184, 241, 300, 322, 328, 376, 381, 385, 391, 396, 409, 431, *See also* Gulf Stream, Loop Current
   at Scotia shelfbreak, 265, 268
   at shelfedge, 219
   Benguela, 393
   bottom, 18, 72, 160–162, 224, 226, 228, 284–287, 309, 316, 319, 340, 342, 344, 392, 432–433
   boundary, 219, 223–224, 228, 328
   counter-, 224
   detection by satellite, 224
   first observed in canyons, 321–322
   geostrophic, 237, 241, 263, 281, 300, 376, 429, 432
   Gulf of Alaska, 242
   importance at shelfbreak, 431–432

in submarine canyons, 225
longshore, 225, 227, 241–243, 284, 287, 291, 315, 318–319, 322, 431
offshore, 89, 290
on California margin, 242, 248
on Scotia shelf, 265, 270, 272
paleo-surface, 83
paleo-tidal, 83, 91
Peru, 395
rip, 315, 322
seasonal variation of, 267
surface, 429
tides, 1, 18, 61, 64, 159, 162, 169–170, 176, 179–180, 219–220, 222, 224–225, 227, 233, 238, 257–258, 265, 268, 279, 281, 287, 300, 315, 320, 322, 328, 376, 432
turbidity, 25, 37, 135, 152, 159, 161–162, 220, 245, 265, 281, 300, 315–316, 319, 322, 324–325, 327–328, 376, 379, 393, 419, 432
Western Boundary, 235
— meter, 224, 226, 237–238, 251–253, 257–258, 266–267, 274
— meter data, 220, 224, 226, 251, 258, 262, 322
lack of on Scotia margin, 274
Scotia shelf, 265–268, 270, 272–273
Scotia shelfbreak, 265–266, 273
Scotia slope, 265–266, 273
— ripples, *See* Ripples
Czechoslovakia
Tethyan canyons, 315, 318, 327

D

Danube delta, 121
Data bank
computer plots from, 362
foraminifera, 359–360, 368
Debris flow, 121, 195, 199, 201, 208, 315, 319–320, 324–325, 327–328
in submarine canyons, 37, 281
— lobe, 124
— sheet, 200
—-flow deposits
calcareous foraminifera in, 85
Decollement, 144, 148
Deep Sea Drilling Project, 333
—-sea fan, *See* Submarine fan
Degradation curve, 56
Degradation rate, *See* Denudation rate
Delaware basin, 411, 416–418
gas fields, 418
starved, 416
turbidite rservoirs, 418
Delmar Formation (Eocene), 323
Delta, *See also* Progradation by deltas, Mississippi delta
abandoned, St. Bernard, 125
Borden, 79, 81, 83, 90–91
Catskill, 80, 159–160
Cretaceous, 400
Danube, 121
Devonian, 400
Ganges-Brahmaputra, 121
late Pleistocene, 134
late Wisconsin, 136
MacKenzie, 139
Niger, 110, 139–141, 426

Nile, 136, 139
Pocono, 80
progradation by, 14, 121–122
submarine, 15
Tertiary, 147
— outbuilding, 110, 113
—-front instability, 121, 124–125
—, shelfedge, 36
Deltaic, complex, 79, 91
Borden delta, delta front, 79, 81, 83, 90
Catskill delta, 80
Midland basin, margin of, 417
Pocono delta, 80
Rosita field, 424
Wilcox, 424
Density current, *See* Current(s), Turbidity current
— stratification, 160
Denticulopsis seminae, 376, 378
Denudation rate, 53, 55
Appalachian Mountains, 54
Himalayan Mountains, 54
Depocenter, 140, 148, 152, 290, 319
Great Valley sequence, California, 103, 113
multiple, as Gulf of Mexico, 144
single, as Niger, 144
Depositional equilibrium, 64, 75
compared with H, 61
Depth zonation
foraminifera, 359–360, 362–370
Deseret basin, starved basin, 79, 80, 83, 90–91
carbonate platform, 82
conodont zonation, 85
foraminifera, 85
radiolaria, 86
Utah, Idaho, and Nevada, 79, 83, 91
— Limestone (Mississippian)
Utah, 89
Desiccation
Mediterranean Sea, 318
South Atlantic, 318
Desoto Canyon
bottom sediment studies, 281, 287
mudline near, 279–280, 287, 293–295
Devonian delta, 400
Dewatering features, 325, 327
Diagensis, 153, 201, 333,
carbonate, 169, 176, 181, 186, 189, 193, 201, 203
Diapirism, 15, 121, 130, 139, 144–145, 147, 152–153, 299–300, 429, *See also* Structural deformation, Tectonics
clay, 420
salt, 45, 139, 145, 147, 251, 293, 295, 420
shale, 139, 145, 147, 293
Diastem, 193
Diatomaceous sections (Miocene)
Japan, 373, 378
U.S. East Coast, 373, 378
Diatom(s)
benthonic, 373, 375–376, 378
planktonic-to-benthonic ratio, 373, 375–376, 378–379
blooms, 374, 378
Denticulopsis seminae, 376, 378
depth restricted, 373–379
displaced benthic, 376
distribution controlled by salinity, 373–379

SUBJECT INDEX

diversity of, 378
factors controlling range of, 373–374
Melosira sol, 376, 378
New York Bight study, 373–374
occurrence back to Jurassic, 373
Paralia sulcata, 376, 378
planktonic, 373, 375–376, 378
Thalassiosira, decipiens, 378
to calcareous nannoplankton ratio, 375
U.S., southeast, 375
Differential compaction, 144
Digboi field
  oil field, 422
  Upper Assam Valley, 422
Displaced fauna, 125, 135
  benthonic diatoms, 376
  foraminifera, benthonic, 359–360, 363, 365, 367, 369
Distribution, foraminifera
  change, cause of, 365
  maps of, 359, 363
Dolomitization, 201, 203, 210, 213
Downgoing slab, *See* Subduction
Downwarping, 207, 214
Downwelling, 233, 238, 245, 290, 328,
  lack of, 243
Driving subsidence, 41–42, 45, 50–51, 54, 56

E

Eastern Interior sea, foreland trough, trough, 80, 83, 88, 90
  anoxic conditions in, 80
  paleoccurrents, 83
Echinoderms, 83, 211
  crinoid, 84, 89
  crinoid material, 88, 91
Echinoids, 174
Ecological zonation, *See* Biological zonation
Ecuador
  tropical fluvial dominated region, 14
Eel River
  California, 222, 305
Ekman flow, 241–242, 263
Electric log, *See* Well log
Elphidium mexicanum
  shelfbreak indicator, 367
Eniwetok atoll
  atoll margin, 173, 176
Epeiric sea, 83
Epicontinental sea, 159–164
Equilibrium depth H, *See also* Shelfbreak depth, 72–74
Eros Limestone (Triassic), 210
Erosion, 9, 12, 15, 18, 25–26, 28, 35–36, 48, 61, 64, 69, 74, 101, 116, 135, 279–280, 282, 288, 291, 311, 315–316, 318–319, 322–323, 328, 333, 342, 344–345, 422
Escabrosa Limestone (Mississippian), *See* Redwall-Escabrosa Limestone
Espirto Santo shelf
  tropical non-fluvial region, SB in, 12
Europe
  Paleozoic carbonate SB, 190, 196
  Plate collision, 79
  shelfbreak placer deposits, 432
  Zechstein evaporites, 401

Eustatic sealevel change, *See* Sealevel change
Evaporite, 111, 193–194, 401, 411, 418, 433
  in Red Sea, 435
Exclusive Economic Zone (EEZ)
  legal aspects, 435
Extensional basin, *See* Passive Margin

F

Facies boundary, 161–162
Fan, *See* Submarine fan
Fault
  control of submarine canyons, 318
  shelf-slope break control of, 399
Faults, *See* Diapirism, Seismicity, Structural deformation, Tectonics
Feeders, infauna
  deposit, 381, 383, 386–388
  filter, 381, 383, 385–388
Fining upward sequence, 315, 325, 327
Flexure, 207, 214
Florida margin
  infauna biomass on, 384–385
  infauna density on, 385
  reef-dominated margin, 172
  tropical non-fluvial region, SB in, 12
— shelf
  east
    sediment transport on, 248
    upwelling on, 243
  northeast, 233, 243
  northwest, 287
  south, 174
  west, 172, 179–180, 279, 282, 284, 295
—, slope
  northeast, 248
Flower Garden Bank
  benthic boundary layer studies, 255–256
  East, 251–253
  West, 251–253, 258, 261
  surveys of, 252
Fluid pressure, excess, *See* Geopressure
Fluxoturbidites, 325, 327
Flysch, 79, 82–83, 90, 103, 116, 410, 412
  Arkoma basin, 420
  Upper Assam Valley, 422–423
Foot of slope
  legal aspects, 439
Foraminifera, 83, 172, 179, 186, 200, 210–211, 287
  agglutinate, 85–86
  calcareous, benthic, 85–86, 89
  Deseret basin, 85
  in debris-flow deposits, 85
  planktonic, absence in Paleozoic, 85
  Redwall-Escabrosa shelf, 89
  use in shelfedge recognition, 86
—, benthonic, 342
  at shelfbreak, 359–360, 362–363, 366–370
  computer plot of, 362
  data bank of, 359–360, 362, 368
  depth zonation, 359–360, 362–370
  displacement of, 359–360, 363, 365, 367
  distribution change, cause of, 365
  distribution maps of, 359, 363
  Elphidium mexicanum, 367

Gulf of Mexico, northern, 359–360, 362–363, 366–370
  Maryland shelf, 360
  methods of study, 359–360, 363
  New Hampshire, 361
  Peneroplis proteus, 367
  provinces of, 359–365, 367–370
  Quinqueloculina agglutinans, 363, 367
  range variation, 359, 367–370
  species used, 359, 363, 367–370
  test dissolution, 359–360
  Texas shelf, 360
  U.S. East Coast, 359–363, 366–370
  zoogeography of, 360–362
—, planktonic, 381, 383
Forearc basin, 9, 97, 102–103, 109, 115
  California, 110
  convergent margin, 107, 113
  Mentawei, 102
  progradation of, 110
  unstable convergent margins, 9, 14
— setting, 113, 115
Foreland basin, 422
— trough, 79, 91
  Antler, 80, 82
  Eastern Interior, 80
  Ouachita, 80, 91
Fore-reef, *See* Reef
Foreset beds, 19, 121, 134, 139–140
  Borden Formation, 91
— slope, 130
Fort Payne Formation (Mississippian), ramp, 83, 91
  Active shelfedge, 91
  bioherms, 91
Fracture zone, 432, 435
Frio Formation (Oligocene), 145, 149
  Texas, 147

G

Gabon basin
  rift basin, aborted, 410
Gach-Saran oil field, *See* Kirkuk-Gach Saran oil fields
Ganges-Brahmaputra delta, 121
Garoupa field
  Brazil, 419
  Campos basin, 419
  oil field, 419
Gas field(s), *See also* Petroleum accumulation
  Delaware basin, 418
  location of, 409–410
  Midland basin, 418
  Rankin platform, 410, 423
  Red Oak-Norris, 421
  Rosita field, 424
  West Thornton, 424
Gas generation, *See also* Methane production, 340
Gas seep, 293–294
Gastropods, 91
Genesee Group (Devonian)
  New York, 162–164
Geopressure, 139–140, 152–153
  dynamic model, 152–153
  source of water, 153
  static model, 152–153
  updip limit of, 141

Geoprobe, 226–228
Geostrophic current, 237, 241, 263, 281, 300, 376, 429, 432
Geotechnical properties, 333–335, 337–346, 352
  Atterberg limits, 337–338, 347, 351
  processes affecting, 338–346
  sediment types, 335–337, 346–351
  shelfbreak areas, 346–352
— studies
  active martin, 333–334, 346
  at shelfbreak, 333–334, 340–341, 343–347, 350–352
  California margin, 346–352
  Gulf of Alaska, 334–335, 346–347, 351
  Israel margin, 345
  Massachusetts coast, 342
  Mediterranean, eastern, 334, 346–351
  Mississippi delta, 334, 343–344, 346–351
  Norway margin, 334–335, 344, 346–347
  passive margin, 333–334, 346
  purpose of, 333
  type of, 333
  U.S. East Coast, 334–335, 342, 344, 346–351
  U.S. West Coast, 334, 346–351
  value of, 334
Geothermal gradient, 97, 104, 203
Gevaram Canyon (Cretaceous)
  Israel, 327
Glacial agents
  effect on shelfbreak, 7–9, 19
Glaciation, 7–8, 15, 19, 35, 41, 46, 51, 57, 72, 222, 431
  cause of sealevel changes, 41, 46–48, 57
  effect on mudline, 293–294
Glauconite, 65, 84, 110, 287, 381
Glomar Challenger, 102
GLORIA
  side-scan sonar, long range, 28, 31
Gold
  off New Zealand, 435
  offshore source of, 435
  placers, 429, 431–432
Golden Lane oil fields, 409–410, 415–416
  atoll, 414–415
  carbonate-evaporite facies, 409–410
  Mexico, 409–410
  paleo-shelfbreak, 415
  Poza Rica oil field, 415
Graded sands, 309, 321, 325, 327
Grain size, mean, 396–398, 402
  correlations, 392, 395–396, 399–400
Grand Banks
  polar glacial regions, SB on, 9
  upwelling on, 393
Gravitative processes, 14, 19, 121–136, 144, 199, 308, 316, 319, 324, 327–328, 333–334, 338, 345, 352, 366, 376
  chutes, 124
  creep, 19, 35, 121, 139, 144, 315, 320, 325, 327–328, 333, 345, 387
  debris flow, 37, 121, 195, 199, 201, 208, 281, 315, 319–320, 324–325, 327–328
  debris lobe, 124
  debris sheets, 200
  failures, propagation of, 136
  gullies, 124
  lack of, 160, 162–163

liquefaction, 333, 345
mass flow, 207–208, 210–212, 214
massive failures, 129–130, 135–136
mudflow, 121, 124–125, 152, 351
progradation of shelfedge by, 125, 136
reactivitation of, 121
rock fall, 337
rotational slide, 121, 124
sand flow, 25, 225, 315, 320, 325, 327–328
sediment flow, 334, 337
slide, 121, 139, 144, 207–208, 299, 319–320, 334, 337, 345, 351
slope failure, 4–5, 15, 19
slump, 5, 9, 121, 124, 130, 133–136, 139, 144, 150, 194, 201, 210–211, 213–214, 224, 288, 290, 293, 299, 307, 319–320, 327–328, 333, 345, 351, 376
spillover, of sand, 315, 320, 328
Gravity deposits, 113
Great Barrier Reef, Australia, 180
reef dominated margin, 172, 174, 176
— Valley sequence
California, 411
clastic facies, 411
depocenter, 103, 113
Greece
Hydra, 207, 209, 212–214
Greenland
polar glacial region, SB on, 9
Grès d'Annot Canyon
Alps, French, 325, 327
Growth fault, 14, 45, 124–125, 129, 139–140, 144, 147–150, 153, 293, 295, 410, 420–421, 424
growth ratio, 147, 149
log character of, 149
petroleum accumulation, 129
Guinea
tropical fluvial dominated region, 14
Gulf Coast, 149, 152, 410, See also Gulf of Mexico
carbonate-clastic facies, 409, 411, 418–419
clastic facies, 409, 418, 424
petroleum accumulation, 426
Texas, 409, 419
— of Aden
rifted substrate, SB on, 12
— of Alaska, See also Alaska, 103, 222–223, 302, 305, 309
currents in, 242
geotechnical study, 334–335, 346, 351
shelf valleys in, 300
— of Asinara
unstable rifted margin, SB on, 19
— of California, 402, 435
— of California, eastern margin
tropical fluvial dominated region, 14–15
— of Genova
progradation of shelf, 68–69, 74
subsiding margin, 69, 72
— of Honduras
tropical fluvial dominated region, 14–15
— of Lion, 72
unstable rifted margin, SB on, 19
— of Maine, 233–234
— of Mexico, See also Gulf Coast, Alabama, Florida, Louisiana, Texas, 110, 136, 139–153, 279
Cenozoic carbonate SB, 191

contrast with Atlantic, 292
Desoto Canyon, 280, 287
Flower Garden Bank, 251–253, 255–256, 258, 261
Gulf Coast Tertiary Basin, 149, 152
hurricanes in, 257, 262
Loop Current, 263, 285, 288, 381, 386
Mesozoic carbonate SB, 191
multiple depocenters, 144
reef, 414, 418
salt basin, in, 414
tropical fluvial dominated region, 14
waves in, 256–263
winter fronts over, 257, 260–262
— — —, northeast, 280, 287–288, 381, 383
infauna biomass of, 384–385
infauna density of, 381, 384–385
organic carbon in, 383
— — —, northern, 279–280, 293, 315, 327, 359–360, 362–363, 366–370, 392, 399–400
organic carbon, 399–400
mudline types in, 281, 290, 293, 295
— — —, northwest, 251–263, 270
— stream, 219, 223–224, 233, 235, 237, 239, 241, 248, 279, 282, 284, 381, 385, 435, See also Current(s)
eddies, 265, 274, 345
filaments, 239–241, 243, 263
meanders, 240–241, 243
ring, 345
sediment reworking by, 282
variation in, 241
Gullies
gravitative process, 124

H

H, See Shelfbreak depth
Halokinesis, See Diapirism
Hawaii
precious coral, 435
Heavy mineral placers, 429, 435
Heavy minerals, 273
Hellanic Arc
temperate region, SB in, 18
Hemipelagic sediments, 32, 325, 327–328
Hercynian system, 190, 410
High seas, definition, 438
Himalaya-type convergence, 97
Himalayan Mountains
denudation rate of, 54
— orogeny, 422
cause of sealevel fall, 47
Horseshoe atoll, See Midland basin
Horst, 12, 19, 416, 423
Hot spot, 45, 47
Hudson Canyon, 28, 279, 321–322
bottom samping in, 288
flow in, 245
mudline near, 288, 293, 295, 320
submersible dives in, 288
— River, 235
ancestral, 288
— shelf valley, 318
currents in, 288
Human activity
effects on mudline, 294–295

Humid region
  sedimentation rates in, 299
Hurricanes, 285, 386
  Gulf of Mexico, 257, 262
Hydra, island of
  Greece, 207, 209, 212, 214
  internal breccias, origin of, 213–214
  melange on, 214
Hydrocarbon accumulation, *See* Petroleum accumulation, Oil or Gas field
—, source rocks, *See* Source rocks
Hydrocarbons
  maturation of, 97, 104
  migration of, 409
  thermal maturity of, 153

I

Iberian margin, western, 72
  progradation of shelfbreak, 61, 74
Idaho
  Deseret basin, 79, 91
  Madison shelfedge, 90
  -Montana border, 90
Illinois basin, starved basin, 80, 83, 88, 91, 400
  Borden Formation, 83, 90
  Coriolis effects, 83
  sedimentation rate, 90
  upwelling, 83
—, 91
  Burlington Shelf, 88
  Springville Shale, 88
Illite, 350–351
India
  Upper Assam Valley, 409, 422–423
Indian Ocean, northern, 395–396
  Northwest Australian shelf, 423
  organic matter, 395–396
  oxygen minimum layer, 431
Indiana
  Borden Formation, 91
Inertial waves, 257–260, 262–263
Infauna assemblage, 385–388
  affected by wave energy, 383
  at shelf-slope break, 381, 383–385, 387–388
  Carolinian Province, 385–386
  controlled by, 381
  deposit feeder, 381, 383, 386–388
  effect on organic carbon, 387
  filter feeder, 381, 383, 385–388
  Tropical Province, 385–386
  Virginia Province, 385
— biomass, 384, 387
  Alabama margin, 384–385
  Atlantic, northwest, 384
  depth effect on, 384
  Florida margin, 384–385
  Gulf of Mexico, northeastern, 384–385
  latitude effect on, 384
  temperature affect on, 384
— density, 381, 384, 387
  Alabama margin, 385
  Atlantic, northwest, 384
  Florida margin, 385
  Gulf of Mexico, 384
  increase at mudline, 385

North Carolina margin, 384–385
Internal breccia, *See* Breccia, internal
— waves, 5, 61, 72, 161, 219–220, 222, 224, 227–228, 237–238, 242, 251, 253, 257–258, 265, 279, 281, 287–288, 300, 315, 322, 328, 340, 343, 381, 429, 432
  effects in submarine canyons, 225
  on San Pedro shelf, 227
International Court of Justice, 20
Intramassif basin
  convergent margin, 107, 109
Iowa
  Burlington shelf, 88
Iran-Iraq foothills fields, 409–414, 426
  Kirkuk-Gach Saran oil fields, 412–414
  orogenesis in, 412–414
  source rocks of, 412
Iraq-Iran, *See* Iran-Iraq foothills fields
Isopach of sediment, 133–134, 139, 145, 152
Isostatic adjustment
  in polar glacial regions, 7, 9
— equilibrium, 43
Israel, margin
  geotechnical study, 345
  Gevaram Canyon, 327
Italy
  breccia, internal, 207

J

Jamaica
  reef-dominated margin, 172, 174, 176
Japan, 373
  canyons off, 318
  planktonic-to-benthonic ratio, 378
  subduction zone, 102
  temperate region, SB in, 18
  trench, 104
  -type, convergence, 97
— Sea, 113
Java ridge, 97, 102

K

Kaolinite, 350–351
Karst, 10, 181, 203
Kentucky, 91
  Borden Formation, 88, 91
Kicking Horse Rim
  North America, 196
Kimmeridgian shale
  North Sea source rocks, 401
Kirkuk reef, 412
—-Gach Saran oil field, 412–414

L

La Jolla Canyon, 319, 322
  fault related, 318
Lake Valley Limestone (Mississippian), shelf, shelfedge, 89
  bioherms, 89
  New Mexico, 89
Landes Marginal Plateau
  progradation on, 64
Landsat image, *See* Satellite imagery
Law of the Sea, *See* United Nations

SUBJECT INDEX

Le Danois Bank
  progradation on, 64
Leadville Limestone (Mississippian), 89
Legal aspects
  Atlantic Ocean, 439
  Chile, 438
  Exclusive Economic Zone, 435
  foot of slope, 439
  high seas, 438
  Law of the Sea, 437–440
  marine resources, 438–439
  mineral prospecting, 429, 435
  Peru, 438
  physiography, 437
  shelfbreak, 437–440
  shelf outer limit, 438
  territorial sea, 435, 438
  Truman Proclamations, 438
  UNCLOS-III, 437, 439–440
  United Nations, 437–440
  United States, 438
Lesser Antilles
  tropical non-fluvial region, SB in, 12
Lewis trough
  Northwest Australian shelf, 410
  rift valley, 423
  source rocks, 423
Ligurian Sea
  unstable rifted margin, SB on, 19
Lindenkohl Canyon, 26, 30–31
Liquefaction, 333, 345
Lisbon, 65
  slope failure off, 74
Lithofacies model, 83–84
Lithologic associations, *See* Carbonate-evaporite facies, Carbonate-evaporite-clastic facies, Carbonate and clastic facies, and Clastic facies
Llano uplift
  Chappel Limestone, 88, 89
  Texas, 89
Longshore currents, 225, 227, 241–243, 284, 287, 291, 315, 318–319, 322
  lack of, 431
Loop Current, *See also* Current(s)
  Gulf of Mexico, 263, 285, 288, 381, 386
Los Angeles basin, 410
  fields in submarine fans, 411
Louisiana
  Tuscaloosa Formation, 418–419
  — shelf, 251, 288, 290, 292, 295
  Type III mudline, 288
  Type IV mudline, 290
Luzon subduction zone, 99

M

MacKenzie delta, 139
Madison sea, shelf, shelfedge, 83, 89–90
  Coriolis effect, 83
  paleoequator, 83
  upwelling, 83
Magdalena
  canyon, 130
  delta, 136
Magmatic arc, *See also* Volcanic arc, 107, 113

Manganese nodules, 439
  legal aspects, 439
Mantle, 97, 99–100, 104
Marathon basin, starved basin, 80, 89, 91
Marine resources, 438–439
  legal aspects, 439
  manganese nodules, 439
Maryland, shelf
  foraminifera province, 360
Mass movement, *See* Gravitative processes
Massachusetts, coast
  geotechnical study, 342
Mature rifted margin, *See* Passive margin
McGowan Creek Formation (Mississippian), 90
Mediterranean, eastern
  geotechnical study, 334, 346–351
— margin, western, 72
  progradation of shelfbreak, 61
— region, 212
  unstable rifted margin, SB on, 19
— Sea
  canyons of, 318
  desiccation of, 318
Meganos Channel (Tertiary)
  California, 318, 327
  West Thornton gas field, 424
Melange, 100–103, 212
  on Hydra, 214
  in Yugoslavia, 212
Melosira sol, 376, 378
Mesozoic carbonate SB, 189–191, 203
  Africa, 191
  Alpine-Himalayan belt, 191–192, 194, 197, 203
  Arabia, 195, 201
  Bahamas, 191
  Central America, 191
  Gulf of Mexico, 191
  North America, 191
Messinian erosion surface, 68–69, 72, 74
Metalliferous deposits, 189, 203
  potential, 432
Meteoric water, 176, 201
Methane production, biochemical, *See also* Gas generation, 121
Mexico, 90
  Campeche Bank, 173, 179–180
  Campeche fields, 409–410
  carbonate-evaporite facies, 409–410, 418
  Golden Lane fields, 409–410, 415–416
  northeast, 144
  Poza Rica field, 415
  Reforma field, 409, 426
  Sonora, 89
  Wilcox, 424
Michigan basin, 80, 83
  Borden Formation, 90
Mid-Atlantic states, 17, 25–37, 225, 228, 295
  bottom sample studies off, 280, 285
  major lithofacies off, 279
  narrow shelf off, 282
  submersible dives off, 287
  Type II mudline, 285, 288
  Type IV mudline, 290
Mid-ocean ridge system, 47
  change in length, 46

change in volume, 46
thermal origin of, 46
Middle Atlantic Bight (MAB), 233, 238, 243, 344–345
  character, of, 234
  fronts, 235, 238, 241
  sediment dispersal in, 247
  submarine canyons in, 234, 243
  tides, 238
  upwelling, 235, 239, 243
Midland basin, 411, 416–418
  deltaic sediments, 417
  gas fields, 418
  Horseshoe atoll, 417
  paleo-shelfbreak, 417
  reef, 417–418
  reservoir rocks, 417
  source rocks, 417
  starved basin, 416–417
Midway Formation (Paleocene), 144
Mindanao Island, 396
— thrust, 99
Mineral deposits
  at shelfbreak, 429
  related to upwelling, 429–431
— enrichment
  in anoxic environment, 435
— prospecting
  legal aspects, 429, 435
Mineralization
  submarine canyon related, 316
Mission Canyon Limestone (Mississippian), 90
Mississippi Canyon, 121, 130–136
  formation of, 130–136
  infilling of, 133–136
  petroleum industry leasing near, 130
  radiometric dating, 132
— delta, 12, 14, 49, 121–136, *See also* Delta, Progradation by deltas, 147, 251, 263, 281, 288
  bottom sample studies, 281
  geotechnical study, 334, 343–344, 346–351
  sedimentation rate off, 388
  shelfedge of, 122, 124, 136
  subsidence of, 121, 293
  Type III mudline, 288
  Type IV mudline, 290, 292
— fan, 130
  volume of, 136
— River, 12, 14, 251, 288, 290
  sediment input to Gulf, 279, 285, 290, 295
— Trough, 288
  formation of, 290
  Type III mudline, 288
Missouri
  Burlington shelfedge, 88
Mixed facies, *See* Carbonate and clastic facies
Molasse, 81, 90, 116
  Upper Assam Valley, 422
Molluscs, 169, 172, 186, 207, 385–386, 388
  deposit feeder, 381, 383, 386–388
  filter feeder, 381, 383, 385–388
  nuculoida, 381, 383, 386–388
Montana
  Castle Reef Dolomite, 90
  -Idaho border shelfedge, 90
  Madison shelfedge, 83, 89–90

Monte Cristo Limestone (Mississippian), 89
Monterey Canyon, 318
— Formation (Miocene)
  anoxic conditions, 111
  California, 111, 213, 402
  oil in, 213
Moroccan shelf
  temperate region, SB in, 18
Mudflow
  gravitative process, 121, 124–125, 152, 351
Mudline, 320, 344, 381, 383, 386–388, 392
  controlling factors, 280–295
  definition of, 279, 295
  early use of term, 279
  erosion-deposition boundary, 279
  glaciation effects on, 293–294
  human activity, effects of, 294
  in Hudson Canyon, 320
  in submarine canyons, 279–281, 287–288, 293–295
  increase of infauna density at, 385
  location of, 279–295
  relative to source rocks, 392
  seismicity effects on, 280
—, types of
  I-at depth on slope, 279, 281–285, 292, 294–295
    off Cape Hatteras, 281, 292, 295
    province boundary, 285
    west Florida margin, 282, 284, 295
  II-on uppermost slope, 279–281, 285–288, 292, 294–295
    Florida shelf, northwest, 287
    northeastern Gulf of Mexico, 287, 295
    off Mid-Atlantic states, 285, 295
  III-near shelfbreak, 279, 281, 288–290, 292, 294–295
    head of Hudson Canyon, 288, 293, 295
    head of Mississippi trough, 288
    Louisiana shelf, 288
    off Mississippi delta, 288, 295
    off Rio Grande, 288
    Texas shelf, 288
  IV-shoreward of shelfbreak, 279, 281, 290–292, 295
    Mid-Atlantic states, 290
    Mississippi delta, 290, 292, 295
    province boundary, 290

N

Namibia, *See also* Africa, southwest
  anoxic environment off, 342
Nappe, 423
Nazca plate, 97, 101
Nematode, 342
Nepheloid layer, 251, 290, 315, 321, 343
Nevada, 90
  Deseret starved basin, 79, 83, 91
New Albany Shale (Devonian-Mississippian), 80, 400
— Hampshire shelf
  foraminifera province, 361
— Mexico, 196
  Lake Valley shelfedge, 89
— Providence Shale (Mississippian)
  bottomset beds, 90
— York, 400
  Genesee Group, 162–164
  West Falls Group, 164
— — Bight, 233–234, 241, 285

bottom sample studies on, 280
diatom salinity-depth study, 374–375
— Zealand, 104
  Balleny Group, canyons in, 327
  gold deposits off, 435
  phosphorite off, 435
  submarine canyons in, 315
  temperate region, SB in, 18
Newport Canyon, 319
Niger delta, 110, 139–141, 426
Nile canyon, 130
— delta, 136, 139
  single depocenter, 144
— River, 294
Ninas Island, Sumatra
  tropical fluvial dominated region, 14
Non-rimmed margin, 179–180
  Campeche Bank, 173, 179–180
  carbonate margin, classification of, 169
  Florida, west, 173, 179
Normal fault, 61, 109, 148, 421–423
North America, 91, 103, 222, 295, 400–401
  eastern, 191
  in Devonian time, 159
  margin, 359, 367
    Mesozoic carbonate SB, 191
    Paleozoic carbonate SB, 190, 193, 195–196
    shelfbreak placer deposits, 432
  western, 299, 306, 327
— America plate, 79, 223
  collision, 79, 82, 99
  craton, 79–80
— Carolina margin
  infauna density on, 384–385
— Sea, 57
  fields in submarine fans, 411
  giant bedforms, 225
  petroleum accumulation, 426
  rift basin, 214
  source rock for, 401
  tide dominated, 268
  transgression, 214
Northwest Australian shelf
  clastic facies, 409
  Indian Ocean, 423
  Lewis trough, 410, 423
  passive margin, 423
  Rankin platform, 410
  rift basin, aborted, 410, 423–424
  rifting, 423–424
Norway, margin
  geotechnical study, 334–335, 337, 344, 346–347
  polar glaciated region, SB on, 9
Nova Scotia shelf, SB, slope, See Scotia shelf, SB, slope
Novaculite, 82–83
  Arkoma basin, 420
Nuculoida, 381
  deposit feeder, 381, 383, 386–388

O

^{18}O, 46
Oblique configuration, 64–66, 69, 72, 74
  on mature slowly subsiding margins, 61, 75
— unit, 125
  seismic facies, 124, 129

Ocean basin, margins of
  Atlantic, 223
  currents off, 223
  Pacific, 223
—-continent boundary, 41, 45, 48, 51, 54, 56–57
Oceanic crust, 101–102, 107, 197, 426
  age of, 97
  subduction of, 97
  subsidence of, 45
Oceanographer Canyon, 28, 322
  shelf identing, 28
Oceanographic instrumentation, 228, 251
  bottom moored, 226
  current meter, 224, 226, 237–238, 251–253, 257–258, 266–267, 274
  Geoprobe, 226–228
  moored system, 252–255
  profiling system (PHISH), 251–252, 256, 261
    data acquisition, 252
    measurement types, 252
  tripod system, 220, 228
Offlap margin, 189, 192–196, 203
Offshore construction
  effects on sediments, 337
— design and engineering, 333–334, 343, 352
Oil
  in Monterey shale, 213
  in transgressive shelf deposits, 391
— field(s), See also Petroleum accumulation
  Arkoma-Quachita fields, 410, 421
  Campos basin, 409–410, 419–420
  Central Basin Platform fields, 410, 418
  Digboi field, 422
  Garoupa field, 419
  Golden Lane fields, 409–410, 415–416
  Iran-Iraq foothills fields, 409–414
  Kirkuk-Gach Saran oil fields, 412–414
  location of, 409–410
  Palawan Island, 420
  Poza Rica field, 415
  Rankin platform, 423
  South China Sea, 409, 420
  Upper Assam Valley, 410, 422–423
  Viola Formation, production from, 421
—-prone Organic facies, See Organic facies
Oklahoma
  Burlington shelfedge, 88
Old Red (Sandstone) Continent
  reconstruction of, 159
Olistostromes, 33
Onlap, 305
  margin, 189, 194, 196
  of slope, 41
Oolite, 179, 186, 282, 395
Ooze, 169, 176, 179, 186, 199–200, 203
Ophiolites, 101, 212
  Asia, southeast, 420
Oregon, See also Pacific Coast, 18
  sediment loads of rivers, 302
  shelfedge, 300, 304
  temperate region, SB in, 18
— margin, 236, 238, 396
  organic carbon on, 396
  sediment dispersal on, 245, 247
  tides, 238

upwelling, 236–237, 239, 242–243
Organic carbon, *See also* Organic matter, 83–84, 387, 402
   affected by infauna, 387
   at shelf-slope break, 392
   correlations, 393, 396, 399–400, 402
   Gulf of Mexico, northern, 383, 399–400
   in oxygen deficient water, 393
   occurrences of, 392, 402
   off Oregon, 396
   off Peru, 395
   off southwest Africa, 393
   off Washington, 396
   on Atlantic margin, 383
   Southern California Borderland, 398–399
   variation of, 387
   West Florida slope, 383
— facies
   at shelf-slope break, 391, 400, 402
   definition of, 391
   factors affecting, 391–394
   oil-prone, 391, 393, 400
— matter, *See also* Organic carbon, 121, 308, 319–320, 328, 342, 381, 385, 387
   affected by sedimentation rate, 392–393, 396, 400, 402
   correlations, 392–393, 395
   destruction of, 393–394
   dilution of, 393–394
   dissolved, 381
   distribution of, 391–393, 402
   factors affecting distribution of, 391–394
   Indian Ocean, northern 395–396
   recycled, 387
   terrestrial, 391, 396, 402
— productivity, *See* Productivity
Orogenesis, 190–191
   Asia, southeast, 420
   in Iran-Iraq, 412–414
   Permian basin, 416
Orogeny, 81–83, 90–91
   Acadian, 80–82, 90–91
   Alpine, 47
   Antler, 80, 82
   Himalayan, 47, 422
   proto-Ouachita, 80, 82
Ouachita foreland trough, trough, 80, 82–83, 88, 90–91, 402, 420
   upwelling, 83, 88, 91
— mountains, 410–411, 420
— oil fields, *See* Oil field(s), Arkoma-Ouachita
Oxidation-reduction state, 84
Oxygen deficiency models, 393–394, 402
— deficient water, *See also* Anoxic, Bottom water, 160, 391, 393–396, 401–402
   organic carbon in, 393
   transgression of 400–402
— minimum layer (OML), 111, 391, 393–394, 396, 398, 400–402, 429, 435
   in northern Indian Ocean, 431
Ozark Island, 88

P

Pacific Coast, *See also* United States, West Coast, California margin, Oregon margin, Washington margin, 233, 242, 248

character of, 234
fronts, 235
sediment dispersal on, 247
submarine canyons on, 243
— Ocean, 242
   margins of, 223
— plate, 102
— shelfedge, *See* Shelfedge
—-type margin, *See* Plate collision, Convergent margin, Subduction
Palawan Island, 396, 420
   oil, 420
   reef, 420
Paleoceanographic map
   United States, 80, 83
Paleoequator, 159
   Madison sea, 83
Paleogeographic-lithofacies map
   United States, 80, 82–83
Paleomagnetic studies
   Alaska, 103
Paleozoic carbonate SB, 189–190, 203
   Australia, 190, 196
   Appalachian mountains, 190, 195
   Caledonian belt, 190
   Canada, 192–194, 196, 203
   Europe, western, 190, 196
   North America, 190, 193, 195–196
   Russia, 190
   United States, 195–196
Paradox Basin
   Aneth field, 401
Paralia sulcata, 376, 378
Passive margin, 25, 41, 45, 56, 76, 111, 116, 197, 203, 292–293, 409–410, 426
   Atlantic, 409–410
   carbonate, 190–191
   formation, 41
   formation by rifting, 3
   geotechnical studies on, 333–334, 346
   Gulf of Mexico, 110
   Niger delta, 110
   Northwest Australian shelf, 423
   petroleum accumulation on, 410
   prograding shelfbreaks, 61
   Rankin platform, 423
   shelfbreak types, 3–4, 9
   subsidence of, 41, 43, 45
   U.S. East Coast, 222
— shelfedges, 79–80, 83–84, 88–91
   Burlington, 88
   Chappel, 88–89
   Lake Valley, 89
   Madison, 90
   Redwall-Escabrosa, 89–90
Patagonia
   polar non-glaciated region, SB on, 10
Patch reef, *See* Reef
Pelagonian Platform, 212
Pelecypods, 164, 208
Peneroplis proteus
   shelfbreak indicator, 367
Permian basin, 411, 416–418, 426
   carbonate and clastic facies, 409, 411, 416–418
   Central Basin Platform, 410, 417–418, 426

clastic facies, 411
  Delaware basin, 411, 416–418
  fields in submarine fans, 411
  Midland basin, 411, 416–418
  orogenesis in, 416
  paleo-shelfbreak, 418
  reserves, 418
  starved basin, 416
  Texas, 411
— reef complex, 193–194, 201
  Capitan Limestone, 193
Persian Gulf
  paleo-shelfbreak, 415
  petroleum accumulation, 426
Peru-Chile trench, 438
— Current, 395
—, margin, 393–395, 398
  anoxic condition, 342
  phosphorite, 435
  upwelling-mineral deposits, 431
Petroleum accumulation, See also Oil or Gas field, 189, 191, 203, 409–410, 412, 415, 422
  growth faults, importance to, 129
  Gulf Coast, 426
  North Sea, 426
  on passive margins, 410
  Persian Gulf, 426
— association
  with paleo-shelfbreak, 409–412, 417
— exploration, 294, 409
  environmental aspects on slope, 26
  importance of shelfbreak, 410
— industry
  Mississippi Trough, leasing near, 130
— maturation, See Hydrocarbons
— reserves, 425–426
  Permian basin, 418
— traps
  in submarine canyons, 316
Phosphoria Formation (Permian)
  Wyoming, 394, 402
Phosphorite, 84, 111, 429, 431, 433, 435
  lack of present formation, 434–435
  off Chile, 435
  off New Zealand, 435
  off northwest Africa, 435
  off Peru, 435
  off southeast U.S., 434
  off southwest Africa, 431, 435
  Southern California Borderland, 433
  uranium enriched, 431
Physiography, 1–20, 437
  legal aspects, 437–438
  "redefinition" for legal terminology, 440
Pigeon Point Formation (Cretaceous)
  California, 315, 327
Piracy
  of canyons, 318–319
Placers, 429, 431–432
  at shelfbreak, 429, 431–432
  gold, 429
  heavy minerals, 429, 435
  location of on shelf, 432
  off North America, 432
  off southeast Asia, 432

off southwest Africa, 432
off western Europe, 432
offshore source for, 435
relict, 431
tin oxide minerals, 429, 435
Plankton
  radiolaria, blooms of, 82
Planktonic fraction, 1
—-to-benthonic ratio
  as environmental indication, 378–379
  as shelf-slope indicator, 373, 375–376, 378–379
Plate boundary, 409–410
— collision, See also Convergent margin, Seismicity, Subduction, Tectonics, 47, 80, 97, 115, 191, 222, 426
  Acadian orogeny, 80
  Africa-North America, 99
  Africa, 79
  Antler orogeny, 80
  Appalachian highlands, 80, 91
  Europe, 79
  North America, 79, 82
  Proto-Ouachita orogeny, 80, 82
  South America, 79, 82
  Upper Assam Valley, 422
— reconstruction, See Reconstruction
— tectonics, 222
Pocono delta, 80
Polar glacial regions, SB on, 7–9
  glacial agents, 7
  isostatic adjustment, 7, 9
  relatively stable margin, 9
    Antarctic margin, 9
    chamber SB variant, 9
    embanked SB variant, 9
    Grand Banks, 9
    Ross Sea, 9
    Weddell Sea, 9
  SB on unstable convergent margins, 9
    Alaska, southern, 9
    Aleutians, 9
    Antarctic peninsula, 9
    Antarctica, east, 9
    Chile southern margin, 9
    fault controlled, 9
    forearc basin, 9
    Greenland, 9
    isostatic adjustment, 9
    Norway, 9
    sill, 9
    submarine valley, 9
— non-glaciated regions, SB on, 9–10
  Alaska margin, Arctic, 9
  Beaufort Sea, 9
  Bering Sea, 9
  Patagonia, 9
  Siberia, 9
Pollution, sources of, 316
Polychaetes, 342, 384–386
Pore pressure, 144, 343–344
  excess, 213–214, 293
  very high, 345
Portage Shale (Devonian)
  clinoform facies, 161
Portuguese shelf, western, 72–73

Poza Rica field
  Golden Lane, 415
  reef, 415
Preparis Island
  tropical fluvial dominated region, 14
Process-boundary, 159, 161–164
Prodelta, 12, 14, 110, 113, 116, 288, 290
  Borden Formation, part of, 90
  deposits, 164, 392
Productivity, 191, 391, 394, 402
  demand on oxygen, 392, 402
  high, 342, 381, 387
  in Sulu Sea, 397
  off Peru, 395
  off southwest Africa, 391
  Southern California Borderland, 398
Progradation, 5, 14–15, 19, 51, 54, 57, 65, 69, 71–72, 113, 159, 164, 308, 324, 328, 388, 411, 419, 424
  "accretion unit", 129
  ancient unstable margins, 140–144
  carbonate, 79, 82, 88–89, 91, 192, 194, 201
  clastic, 80–81, 90–91
  clastic margin, unstable, 139–153
  complex sigmoid-oblique, 72
  deltaic, 292
  effect on shelfbreak, 1, 5
  evidence from electric logs, 148
  evolution of structural style, 144–152
  forearc basin, 110
  Gulf of Genova shelf, 68–69, 71–72, 74
  Iberian margin, western, 61, 72
  infilling of canyons, 124
  Landes Marginal Plateau, 64
  Le Danois Bank, 64
  Mediterranean margin, western, 61, 72
  oblique configuration, 64–66, 69, 72, 74–75
  of delta, 14
  off U.S. East Coast, 125
  seismic units, 125, 129
  shelf-slope break, active, 107
  shelfbreaks, 61, 64, 72, 74–75
  sigmoid configuration, 66, 69, 72, 74–75
  — by deltas, *See also* Delta, Mississippi delta
  rapid off certain rivers, 121–122
  sealevel change effect, 122, 125
  shelfedge, 121, 125
Prograding margins, *See* Progradation
— shelfbreaks, *See* Progradation
Proterozoic carbonate, SB, 189–190, 198
  Africa, 190
  Australia, 190
  Canada, 190, 192
  Russia, 190
Proto-Ouachita orogeny, 80, 82
  plate collision, 80
Protoceanic setting, 107, 109, 113
  Gulf of California, 109
  Red Sea, 109
Provenance, 201, 287
Provence
  unstable rifted margin, SB on, 19
Provinces
  foraminifera, 359–365, 367–370
Pycnocline, 1, 159–162, 224, 237, 257, 343, 401
  basin margin-clinoform junction, 159
  sedimentary process separation, 159–162

Q
Quinqueloculina agglutinans, 363
  shelfbreak indicator, 367

R
Radiolaria, 83, 87, 207–208, 211
  blooms, 82
  Deseret basin, 85–86
  guide to shelfedge recognition, 87
Radiolarian chert, 98
Radiolarites, 211–212
Radiometric dating, 122, 125, 129
  lack of data off deltas, 136
  Mississippi Canyon formation, 132, 134
Ramp, 91, 179
  carbonate, 195, 199, 411
Rankin platform
  gas fields, 410, 423
  Northwest Australian shelf, 423
  paleo-shelfbreak, 423
  reservoir rocks, 424
  rift basin, aborted, 410, 423–424
Ratio
  planktonic-to-benthonic, 373, 375–376, 378–379
  diatom to calcareous nannoplankton, 375
Reciprocal sedimentation, 201
Reconstruction, 98
  basin, 107
  Old Red (Sandstone) Continent, 159
  paleoenvironments, 107
— of margins
  in middle Osage (Mississippian) time, 79
  North America, 91
  United States, 79
Red Oak-Norris gas field
  Arkoma basin, 421
  Atoka Formation, 421
— Sea, 410
  evaporites in, 435
  rifted substrate, SB on, 12
Redwall-Escabrosa Limestone (Mississippian), shelf, shelfedge, 89–90
  Arizona, 89
  Colorado, 89
  foraminifera, 89
Reef complex, 111
—, 11, 15, 61, 135, 169–186, 191, 194, 197–198, 201, 203, 211–212, 222, 290, 386, 412, 415, 417, 420, 429, 432
  algal, 135, 169, 290, 420
  atoll, 169–170, 176, 180, 191, 414–415, 417, 429, 432
  back-reef, 174, 185–186, 412, 415
  barrier, 170, 189, 198–199, 203, 432
  Campos basin, 420
  Central Basin Platform, 410, 418
  complex, 111
  correlation in size with fauna, 198
  distribution of, 170–173
  drowning of, 12
  effects on shelves, 3, 7, 10, 19–20
  foreslope, 189
  fore-reef, 174, 176, 186, 211, 412, 415
  fringing, 170, 178, 432
  Great Barrier, 172, 174, 176, 180
  Gulf of Mexico, 414, 418

Kirkuk, 412
knoll, 199, 203, 423
lack of, due to upwelling off
   Chile, 10
   Namibia, 10
   Peru, 10
   Western Sahara, 10
lack, due to abundant clastics, 12
Midland basin, 417–418
mounds, 198
Palawan Island, 420
patch reef, 89, 174
Poza Rica field, 415
precious coral, 429, 432, 435
reef-front, 174
reservoirs, 411, 415
sediment influx, effect on, 181
talus, 414–415
Reefbuilding organisms, 189
Reef-dominated margin, 172–176
   Belize, 172, 174
   biological zonation in, 169
   carbonate margin, classification of, 169
   Florida, south, 172, 174
   Florida, west, 180
   Great Barrier Reef, Australia, 172, 174, 176
   Jamaica, 172, 174
Reforma fields
   carbonate-evaporite facies, 409, 426
   Mexico, 409
Regradation, 61
   effect on shelfbreak, 1, 5
Regression, 50–51, 56, 64, 69, 71–72, 74, 150, 185, 194, 302
Relict feature, 64–65, 71
   channels, 18
   erosion surfaces, 322
   estuaries and lagoons, 18
   lake, 10
   late Pleistocene shelfbreak, 69
   offshore bar, 18–19
   paleodelta, 14–15
   permafrost, below seafloor, 10
   sands, 247
   sediments, 7, 18
   shelf sediment, 320
   shelfbreak, 5, 19
   surface, 1
Reservoir rocks, 409, 412, 426
   Arkoma basin, 420–421
   breccias, 416
   Central Basin Platform, 418
   Midland basin, 417
   Rankin platform, 424
   reefs, and related facies, 411, 414–415, 418
   San Joaquin Valley, 411, 424–425
   submarine canyon related, 316
   submarine fan reservoirs, 411, 419–420, 424–425
   turbidites, 411, 418, 424–425
   Wilcox, 424
Retrogression, 121
   by gullies, 124
   by large failures, 124
Rift, 212, 432, 435
— basin, 207
   North Sea, 214

— basins, aborted
   Avalon basin, 410
   Gabon basin, 410
   Rankin platform, 410, 423–424
— valley
   Lewis trough, 423
Rifted margin, *See* Passive margin
Rifting, 5, 41–42, 45–46, 109–110, 410, 419, 423–424
Northwest Australian shelf, 423–424
   passive margin formation, 3
   phase of crustal thinning, 43
   sediment accumulation, 43
   SB on rifted substrate, 12
   SB on unstable rifted margins, 9, 12, 14–15, 18–20
Rio Grande River, 291
   mouth of, 290
   Type III mudline off, 288
Riony River
   canyon off, 318
Rip currents, 315, 322
Ripple marks, 162–163, 343
Ripples, 282, 320–321, 325
   Scotia shelf, 272
River
   Columbia, 235, 237, 305, 396
   Copper, 223, 302
   Eel, 222, 305
   Mississippi, 12, 14, 251, 279, 285, 288, 290, 295
   Nile, 294
   Rio Grande, 288, 290–291
   Riony, 318
   Santa Clara, 305
   sediment source to ocean, 1, 7, 14, 19, 47, 62, 72, 121, 136, 222–223, 279, 282, 295, 299, 302
   sediment trapped in estuaries, 287
   sediment loads of, 302
   source of sediment to canyons, 315
— discharge
   related to turbidity currents, 322
—-submarine canyon association, 130
Rock color, *See* color
— fall, 337
Rollover structure, 121, 129, 144, 147
Rosita field
   delta complex, 424
   gas field, 424
   Wilcox, 424
Ross shelf
   polar glacial region, SB on, 9
Rotational slide, 121, 124
Roundness, 207–208
Rudist, 198, 201
Russia
   Paleozoic carbonate SB, 190
   Proterozoic carbonate SB, 190
Ryukyu subduction zone, thrust, 99

S

Sable Island, 265, 267
Sahul shelf
   tropical non-fluvial region, SB in, 12
St. Bernard delta, 290
   abandoned, 125
Salinity
   diatom distribution control, 373–379
   New York Bight study, 374–375

Salt basin
  in Gulf of Mexico, 414
  in South Atlantic, 419
— flow, See Diapirism
— tectonics, See Diapirism
San Andreas fault
  California, 107, 234
San Joaquin Valley, 113, 410
  California, 410
  reservoir rocks, 411
  source rocks, 425
  Stevens Sandstone, 411, 424–425
  submarine fan reservoirs, 424–425
  turbidites, 424
San Pedro shelf survey
  California, 226–227
  effects of waves on, 227
  upwelling, 227
Sand flow, 25, 225, 315, 320, 325, 327–328
Sand waves, 184, 282
  Scotia shelf, 272
Sand-shoal-dominated margin, 176–179, See also Carbonate sand shoal
  Bahamas, 173–174, 178
  carbonate margin, classification of, 169
Sangihe subduction zone, 99
Santa Clara River, 305
Satellite imagery, 224, 239, 241, 243
  current detection by, 224
  infrared, 237, 239
  Landsat, 225
SB, Shelfbreak abbreviation
Scotia shelf, 197, 234, 265–275
  bottom sediment studies on, 266–267
  current meter studies on, 265–268, 270, 272–273
  ripples, 272
  sand waves on, 272
  sediment studies on, 265, 268–275
  sediment transport on, 265, 272–273
  source of water, on, 265
— shelfbreak, 265–275
  current meter studies at, 265, 268
  sediment studies at, 265, 268–275
  sediment transport at, over, 265, 272–273, 275
— slope, 265–275
  current meter data on, 265–266, 268
  sediment transport on, 273, 275
Scripps Canyon, 319
  fault related, 318
Sea MARC I, See Side-scan sonar, mid-range
Sealevel
  high stand, 5, 61, 64, 71, 74, 79, 88, 91, 176, 327
  low stand, 5, 10, 15, 17–19, 25, 37, 61, 69, 74, 125, 129, 147, 176, 288, 290, 294, 308, 311, 318–319, 431, 435
  standstill, 57, 64, 194
  surface depression, 241
Sealevel change, 15, 45, 48, 50–51, 56–57, 61, 69, 72, 75, 173, 179–180, 184–185, 189, 192, 194–196, 294, 299, 308–309, 316, 327, 391, 429
  defined, 192
  effect on delta, 122, 125
  effect on mudline, 279–280, 288, 293–295
  fall, 19, 41, 45–48, 50–51, 54, 56–57, 80, 125, 135, 174, 192, 195, 201, 214, 294, 315, 323, 328, 366, 412
  fall due to orogenies, 47
  glaciation related, 41, 48, 51
  history of, 310
  not related to glaciation, 41, 49, 51, 54
  rate of, 315
  rise, 5, 10, 12, 18, 44–48, 69, 71, 79, 80–81, 83, 88–89, 91, 122, 129, 135, 174, 178–179, 192, 194–195, 197, 199, 201, 203, 315, 318, 322, 324, 328, 431
Seamount, 212
Sediment
  at shelfbreak, 381, 388
  coarsening upward sequence, 149
  fining-upward sequence, 315, 325, 327
  mineral enrichment in, 435
— accumulation
  rifting related, 43
— barriers, 300, 303, 309–310
— bypassing, See Bypassing
— color, See Color
— dynamics
  at shelfbreak, 219, 222–225, 227–228, 251
— flow, 334, 337
— influx, 139, 159
  effect on reefs, 181
  rate of, 144
— loading, 148, 197, 293, 340
  cause of subsidence, 41, 45
— loads of rivers
  California, 302
  Oregon, 302
  southern Alaska, 302
  Washington, 302
— origin
  Appalachians, 287
— properties, 333–335, 337–346, 352
  post-depositional processes, 333, 340–346, 352
  primary depositional properties, 333, 338–340, 352
— relict, 272, 287, 290–291, 309
— resuspension, 223–224, 228, 258–260, 279–280, 288, 294–295, 304, 343–344
  by bottom currents, 287
  by storm currents, 225
  effects of water density on, 261
  outer Texas shelf, 257
  transport over shelfbreak, 260–262
— textural model, 333–338, 352
— textures, 265
  grain size distribution, 270–272
  grain size studies, 268–270
— transport, 150, 180, 208, 265, 272–273, 294–295, 299–300, 315–316, 321–322, 345, 352, 366
  at shelfbreak, 219–228, 316
  direction of, 265
  fauna displacement by, 365, 369
  geologic and oceanic factors, 219, 223–225, 228
  models for, 265
  on east Florida shelf, 248
  over shelfbreak, 186, 228, 265, 272–273, 275, 279, 282, 284, 290, 292–293, 299, 307, 333, 343, 345, 383
  resuspension, 223–225
  sand spillover, 319
  seen from submersibles, 320
Sedimentary processes
  at shelfbreak, 219–222, 227–228

SUBJECT INDEX

boundary between, 159, 161–162, 164
  on shelfedge, 251, 256–263
  separation of, 159–164
    by pycnocline, 159–160
Sedimentation rate, 12, 69, 72, 110, 121, 129, 136, 152, 293–295, 305, 309–311, 340–341, 345, 391, 396–398
  correlations, 392–393, 396, 400–402
  effect on organic matter, 392, 396, 400–402
  high, 299, 307
  high off deltas, 121
  Illinois basin, 90
  in humid region, 299
  in semiarid region, 299
  low, 305–307
  low off U.S. East Coast, 121, 125
  off late Pleistocene deltas, 134
  off Mississippi delta, 388
  oxic bottom water, 392–393, 397, 400, 402
  oxygen deficient bottom water, 393, 395, 398, 400 (Devonian), 401–402
Seismic facies, 122, 124–125, 299–311, *See also* Shelfedge, types of
  "accretion unit", 129
  oblique unit, 124–125, 129
  progradation of, 125, 129
  sigmoidal unit, 124–125
Seismicity, *See also* Plate collision, Structural deformation, Subduction, Tectonics, 213–214, 319, 337, 340, 343, 345
  Benioff zone, 99
  cause of slumping, 224
  deep focus, 97
  effect on mudline position, 280, 293, 295
  focal depth, 97
  related to subduction rate, 97
  U.S. West Coast, 222, 299–300, 308
Semiarid region
  sedimentation rates in, 299
Shear strength, 121, 328, 333, 337, 342, 345, 347, 351
Shear velocity, 270–271, 273–275
Shelf
  comparison of, U.S., 299
  Florida, 273, 279, 282, 284, 287, 295
  Florida, east, 243, 248
  legal definition of, 437–440
  Louisiana, 251, 288, 290, 292, 295
  narrow off Cape Hatteras, 281
  Texas, 281, 288, 290–291, 294
— outer limit, definition, 20, 438–439
— valleys, 299
  buried, 318
  Gulf of Alaska, 300
  Hudson, 288, 318
—-slope boundary, 97
—-— break, 394
  Bering Sea, 373, 376, 379
  definition of, 373
  fault-controlled, 399
  organic carbon distribution, 392, 396, 400
  organic facies at, 391, 400, 402
  transgression across, 400–402
—-— —, active, 110, 113, 115–116, *See also* Trench-slope basin
  clastic-starved, 107, 111, 113
  progradational, deltaic, 107

—-— —, carbonate, 169–186, 189–203, *See also* Carbonate platform, Shelfedge, carbonate
  ancient, 185
  Cenozoic, 189, 191, 203
  deformation of, 169, 185
  evolution of, 189
  growth rate, 200
  location determined by, 189
  Mesozoic, 189–191, 203
  occurrence of, 169
  offlap margin, 189, 192–195
  onlap margin, 189, 194
  Paleozoic, 189–190, 203
  Proterozoic 189–190, 198
  types of, 189, 192–197
—-— — indicator, *See* Shelfbreak indicator
—-— front, 373, 376, 378
Shelfbreak
  as benthic foraminifera boundary, 359–360, 362–363, 366–370
  average distance from shore, 2
  benthic boundary layer, 219, 222, 225–228
  canyon evolution at, 25–37
  circulation, 233–248
  classification basis, 1–3
    climatic events, 1, 5, 7
    eustatic events, 1, 5, 10, 12, 15, 17–18
    sediment processes, 1, 3, 7, 15
    sediment supply, 1, 3, 7, 14, 18–19
    structural framework, 1, 3, 7, 14–15, 18–19
    substrate motion, rate of, 1, 3–4, 19
  classification of, 7–19
    polar glacial regions, 7–9
    polar non-glaciated regions, 9–10
    temperate regions, 15–19
    tropical fluvial dominated regions, 12–15
    tropical non-fluvial regions, 10–12
  definition of, 1, 19, 280, 334, 429, 437–438, 440
  effects of canyons, 315
  fronts, 233–238, 241
  glacial agents, effect on, 7, 19
  geotechnical properties at, 346–352
  geotechnical studies at, 333–334, 340–341, 343–347, 350–352
  importance of currents at, 431–432
  importance in petroleum exploration, 410
  infauna at, 381, 383–385, 387–388
  influence of sedimentary processes, 5
  legal definition of, 438, 440
  mineral deposits at, 429–435
  mudline position relative to, 279–295
  off Nova Scotia, 265–275
  on active margins, 97–104
  outer limit, 438–439
  placers at, 429, 431–432
  physiography, 1–20
  reef complex, 111
  relatively stable margins, 3–4, 9, 12–15
  sediment at, 381, 388
  sediment dispersal at, 233, 245–248
  sediment dynamics, at, 219, 222–225, 227–228
  sediment processes, at, 219, 220–222, 227–228
  sediment textural model, 333–338, 352
  sediment transport at, 219–228, 316
  sediment transport over, 279, 282, 284
  tectonism at, 410

SUBJECT INDEX

unstable convergent margin, 4–5, 9, 12–14, 18
unstable rifted margin, 5, 9, 12, 14–15, 18–19
upwelling at, 238, 243
— configuration
  convex-up, 2
  series of terraces, 2
— depth (H), 61, 65–66, 69, 72, 74
— indicator
  Elphidium mexicanum, 367
  Peneroplis proteus, 367
  planktonic-to-benthonic ratio, 373, 375–376, 378–379
  quantitative population count, 373
  Quinqueloculina agglutinans, 367
—, paleo-
  Arkoma basin, 422
  Asia, southeast, 420
  Campos basin, 420
  Golden Lane, 415
  Midland basin, 417
  Permian basin, 418
  Persian Gulf, 415
  petroleum association, and lack of, 409–412, 417, 425–426
  Rankin platform, 423
  West Thornton gas field, 424
Shelfedge, 213–214
  active, 90–91
  Alaska, southern, 300, 304, 310
  ancient, 139–144
  California, 300, 304
  carbonate, 83–84, *See also* Carbonate platform, Shelf-slope break, carbonate
  delta, 36
  factors controlling sedimentation at, 299, 302, 309–310
  flow convergence at, 251
  interface of two realms, 139
  off Mississippi delta, 122, 124
  Oregon, 300, 304
  passive, 88–90
  progradation by deltas, 121–122
  progradation by gravitative processes, 125, 136
  recognition by foraminifera, 86
  recognition by radiolaria, 87
  sediment dynamics, at, 251
  sedimentary processes on, 251, 256–263
  variation in flow at, 251
  Washington, 300, 304
  West Coast, 299–311
—, types of, *See also* Seismic facies,
  U.S. West Coast, 300, 308
  composite, 299, 305, 308–309, 311
  cut-fill, 299–305, 308, 311
  draped, 299, 304, 306, 310
  prograded, 299, 304, 306–307, 310
  starved, 299, 304–306, 310
  upbuilt-outbuilt, 299, 304, 307–308, 310
Siberia
  Polar non-glaciated region, SB on, 10
Side-scan sonar, 121, 124–125, 130
  mud volcanoes, detection of, 129
—-scan sonar, long-range
  GLORIA, 28, 31
—-scan sonar, mid-range, 25–37
  Carteret Canyon, 31
  Sea MARC I, 27, 31

Wilmington Canyon, 28
Sigmoid configuration, 64, 69, 72, 74
  on young subsiding margins, 61, 75
— unit
  seismic facies, 124–125
Sill, *See also,* Welt, 9, 82, 91, 111
Skewness, 287, 325
Slide, 121, 139, 144, 207–208, 299, 319, 334, 337, 345, 351
  lack of, 320
Slope failure
  Carteret Canyon, 32–33
  in canyons, 25, 32–33, 36
  on steep slope off Lisbon, 74
Slump, 5, 9, 121, 124, 130, 133–136, 139, 144, 150, 194, 201, 210–211, 213–214, 224, 288, 290, 293, 299, 307, 319–320, 327–328, 333, 345, 351, 376
  seismically induced, 224
Smectite, 350–351
Soil "state", *See* Sediment properties
Sonar image interpretation, 28
  submersibles, use in, 28
Sorting, 208, 271–272
Source rocks, 84, 203, 316, 392, 402, 410–411, 420, 426
  Arkansas novaculite, 402
  best type, 392
  deep water shales, 139
  definition of, 391
  deposition of, 400
  for Aneth field, 401
  for North Sea, 401
  Kimmeridgian shale, 401
  Lewis trough, 423
  Midland basin, 417
  of Iran-Iraq oil fields, 412
  relative to mudline, 392
  San Joaquin Valley, 425
— sediments, *See* Source rocks
South America, 101
  convergence, 97
  plate collision, 79, 82
— Atlantic Bight (SAB), 237, 243
  character of, 234
  fronts, 235, 238
  sediment dispersal in, 247–248
  tides, 238
  upwelling, 235, 239–240
— China Sea basin
  carbonate and clastic facies, 409, 420
  oil fields on, 420
— Toms Canyon, 28, 30–31
Southern California Borderland, 300, 305
  organic carbon, 398–399
  phosphorite, 433
  productivity, 398
Spanish shelf, north, 64–65, 72
Spillover, of sand, 315, 320
Sponge, 198, 211
Spreading rate
  volume change of mid-ocean ridge, 46
Springville Shale (Mississippian)
  bottomset beds, 90
  compressed conodont fauna, 88
  Illinois, 88

Starved basin, *See also* Anadarko, Deseret, Illinois, Marathon basins, 79, 90–91, 194–195, 203
  Delaware basin, 416
  Midland basin, 416–417
  Permian basin, 416
— margin, 41
Stevens Sandstone (Miocene)
  California, 411, 424
  San Joaquin Valley, 411, 424–425
  submarine fan reservoir, 411, 424–425
Storm waves, 5, 61, 220, 223, 225, 228, 287–288, 290, 299–300, 320, 322, 343
  Mid-Atlantic coast, 228
  resuspension of sediments by, 225
  winter, 287–288
Storms, major, 164, 180, 251, 257, 260–262, 265, 284–285, 307
Strike-slip fault, 110, 213
Stromatoporoids, 198, 201
Structural deformation, 3–5, 9, 18–19, *See also* Diapirism, Seismicity, Subduction, Tectonics, 65, 110–111, 121, 197, 212–214, 415, 420
  block-faulted crust, 109
  compressive, 4, 110
  down-to-basin fault, 144, 147
  downwarping, 207, 214
  fault block, 148, 150, 299
  growth fault, 14, 45, 124–125, 129, 139–140, 144, 147–150, 153, 293, 295, 410, 420–421, 424
  horst, 12, 19, 416, 423
  imbrication, 4, 100, 104, 109, 116
  Mindanao thrust, 99
  nappes, 423
  normal fault, 61, 109, 148, 421–423
  orogenesis, 189
  Peru megathrust, 101
  regional tectonics, 192, 196–197, 199
  rift basin, 207, 214, 410, 423–424
  rifting, 3, 5, 9, 12, 14–15, 18–20, 41–43, 45–46, 109–110, 410, 419, 423
  Ryukyu thrust, 99
  strike-slip fault, 110, 213
  tectonic uplift, 97
  thrust fault, 99, 101, 145, 410, 415, 420–423
  thrust sheets, exposures in, 196
  tilting, 5, 9, 14, 18, 65
  transform tectonics, 15, 107, 110
  up-to-basin fault, 145
  wrench tectonics, 111, 113, 116
Structural evolution, prograding margin
  local scale, 145, 147
  regional scale, 144–145
  shelf margin deltas, 147–152
Subduction, *See also* Plate collision, Convergent margin, Seismicity, Structural deformation, Tectonics, 46–47, 115–116, 293, 410–412, 420, 426
  allochthonous slab, 4
  imbrication, 4
  oceanic crust, 97
  one cycle, 102–103
  rate affecting seismicity, 97
  rate of, 97, 99
  three cycle, 102–103
  two cycle, 102
— complex, *See* Accretionary prism

— zone, *See also* Plate collision, Convergent margin, Seismicity, Structural deformation, Tectonics, 97–104
  active, 97
  incipient, 98–100
  steady state, 97–98, 100–101, 103–104
  Aleutian, 99, 103
  Bering shelf, south edge, 99
  collision type, 98
  inactive, 97–98, 103–104
  Japan, 102
  Lesser Antilles, 102
  Luzon, 99
  Ryukyu, 99
  Sunda, 102
  Sangihe, 99
  Taiwan, 99
Submarine canyon, 5, 9, 14–15, 18, 25–37, 74, 201, 233, 248, 293, 299, 334
  Alps, French, 315
  Amazon, 130
  association with rivers, 130
  Astoria, 243, 396
  Balleny Group, 327
  Baltimore, 28, 36, 318, 337, 351
  Bering Sea, 225, 318
  bypassing in, 26, 36–37
  canyon fill creep, 35
  Carteret Canyon, 25, 28, 31–33
  circulation in, 243–245
  Congo, 130, 318
  currents first observed in, 321
  currents in, 225
  debris flows in, 37, 281
  definition used, 315
  Delmar Formation, 323
  Desoto, 279–281, 287, 293–295
  effects on shelfbreak, 315
  erosion of, 25–37
  evolution of, 25–37
  fauna displacement in, 365
  flow in, 322
  formation of, 130–136, 316–319
  Gevaram, 327
  Grès d'Annot, 325, 327
  hanging valleys of, 28, 36
  Hudson, 28, 245, 279, 288, 293, 295, 320–322
  Hudson shelf valley, 288, 318
  in Middle Atlantic Bight, 234, 243
  internal waves in, 225
  La Jolla, 318–319, 322
  Lindenkohl, 26, 30–31
  Magdalena, 130
  mature phase, 25, 37
  meandering in, 25, 29, 37
  Mediterranean Sea, 318
  Meganos Channel, 318, 327, 424
  mineralization related, 316
  Mississippi, 121, 130–136
  Mississippi Trough, 288, 290
  Monterey, 318
  mudline in, 281
  New Zealand, 315
  Newport, 319
  Nile, 130
  Oceanographer, 28, 322

off Baja California, 318–319
off California, 322
off Corsica, 318
off eastern U.S., 318
off Japan, 318
piracy of, 318–319
problems in study of, 316
related to low sea level stands, 25
reservoir rocks, 316
Riony River, canyon off, 318
sampling by submersibles, 28
Scripps, 318–319
sediment trapped by, 227
shelf-indenting, 25, 28, 36–37
slope failure, formation by, 25, 32–33, 36
South Atlantic, 318
South Toms, 28, 31
straight, 25, 29–30
Tethyan, 315, 318, 327
Torrey, 324–325, 327–328
traps for petroleum, 316
turbidity currents in, 245, 281
U.S. East Coast, 25–37, 225
U.S. West Coast, 225, 243, 318, 322
upwelling in, 245
Wilmington, 25, 28–29, 36, 279, 295, 318, 320–321, 343, 350
youthful phase, 25, 37
— — fill, 324–327
  coarse grained, 315
  distribution of, 316
  fine grained, 315
— fan, 74, 104, 111, 121, 129, 136
  Antler trough, 83
  Campos basin, 419–420
  Los Angeles basin fields in, 411
  Mississippi, 130
  North Sea fields in, 411
  off U.S. West Coast, 222
  Permian basin fields in, 411
— fan reservoirs
  Campos basin, 419–420
  Permian basin, 411
  San Joaquin Valley, 410, 424–425
  Stevens Sandstone, 411, 424–425
— rise, See Sill
Submersible dives
  Hudson Canyon head, 288
  observation of current flow, 321
  observation of sediment transport, 320
  off East Coast, 282
  off Mid-Atlantic states, 287
Submersibles, 28, 256, 258, 316
  sampling in
    Baltimore Canyon, 28
    Oceanographer Canyon, 28
    South Toms Canyon, 28
    Wilmington Canyon, 28
  use in sonar image interpretation, 28
Subsidence, 4, 12, 18, 41, 43, 45, 55, 81, 110, 148, 159, 195, 203, 210, 212, 302, 306, 308, 315, 327–328
  backstripping, 41
  curve, 45, 197
  driving subsidence, 43
  due to salt withdrawal, 148

due to sediment loading, 41, 45
due to thermal cooling, 43, 45, 51, 54, 69, 74, 197
Mississippi delta, 293
oceanic lithosphere, 45
passive margins, 41, 45
rate of, 192
tectonic, cause of, 197
Subsidence rate, 43, 45, 48, 50–51, 56–57, 121, 139, 148–150, 153, 180
  Mississippi delta, 121
  passive margin, 41, 45
  starved margin, 41
Subsiding margins, 66, 74–75
  Gulf of Genova, 69, 72
Sulu Sea, 397–398
  productivity, 397
Surface waves, 219–220, 222–224, 237, 256–257, 260, 300, 328, 340, 343, 429
  on San Pedro shelf, 227
Suspended load, 245, 265, 270–271, 273, 290, 316, 323

T

Taconic Orogeny, 195
Taiwan subduction zone, 99
Tectonics, See also Diapirism, Plate collision, Seismicity, Subduction, Structural deformation, 72, 97, 101, 107–116, 139, 159, 180, 190, 195–197, 207, 212, 214, 219, 222, 292, 299, 308, 315–316, 327–328, 334, 391
  at shelfbreak, 410
  lack of, 417
Temperate regions, SB in, 15–19
  relatively stable margins, 15–18
    Bay of Biscay, 18
    Celtic Sea, 18
    Moroccan shelf, 18
    U.S. East Coast, 15
  unstable convergent margins, 18
    Calabrian Arc, 18
    Hellenic Arc, 18
    Japan, 18
    New Zealand, 18
    Oregon, 18
    Washington, 18
  unstable rifted margins, 18–19
    Aegean Sea, islands in, 19
    California, 18
    Corsica, 19
    Gulf of Asinara, 19
    Gulf of Lion, 19
    Ligurian shelf, 19
    Mediterranean, 18
    Provence, 19
    Tyrrhenian Sea, 19
Tennessee, 91
Terrace
  definition of, 429
Territorial sea, 438
  Chile, 438
  Ecuador, 200 nautical miles, 438
  legal aspects, 435
  legal definition of, 438
  Peru, 200 nautical miles, 438
  United States, 3-mile wide, 438–439
Test dissolution

Foraminifera, 359–360
Tethyan canyons (Eocene-Oligocene)
  Czechoslovakia, 315, 318, 327
— Sea, 190, 207, 401, 412
Tethys plate, 409–410
Texas, 90, 400
  Chappel Limestone, shelf, 89
  Frio Formation, 145, 149
  Gulf Coast, 409, 419
  Llano uplift, 89
  Permian basin, 411
  Wilcox, 424
— margin, 251, 258
  Flower Garden Bank, 251–253, 255–256, 258, 261
  resuspension of sediment, 257
— shelf, 290
  foraminifera province, 360
  south, 281, 294
  Type III mudline, 288, 291
Thalassiosira decipiens, 378
Thermal effects, 47, 51
  cooling as cause of subsidence, 43, 45, 51, 54, 69, 74, 197
  mid-ocean ridge system, 46–47
— maturation, *See* Hydrocarbons, maturation of
Thermocline, 224, 342, 383, 401
Thixotropism, 342
Threshold velocity, 287–288
Thrust faults, 99, 101, 145, 410, 415, 420–423
Tides, 1, 18, 61, 64, 159, 162, 169–170, 176, 179–180, 219–220, 222, 224–225, 228, 257–258, 265, 268, 279, 281, 287, 300, 315, 320, 322, 328, 376, 432
  North Sea, in 268
  on San Pedro shelf, 227
  on U.S. East Coast, 233, 238
  on U.S. West Coast, 233, 238
Tin oxide minerals
  offshore source of, 435
  placers, 429
Tongue of the Ocean, 179
  axial valley in, 319
Topset beds, 19, 110, 113, 116, 139
  Borden Formation, 91
Torrey Submarine Canyon (Eocene), 324, 327–328
  tripartite fill, 324–325
Total organic carbon, (TOC), *See* Organic carbon
Trace fossils, 83, 87–88
  Borden Formation, 88
Traction deposits, 328
— flow, 315–316, 321, 327
Transcontinental arch, 82, 88–90, 416
Transform fault, 15
Transform margin, 107, 111
Transgression, 37, 41, 50–51, 55–57, 69, 71–72, 152, 185, 214, 265, 294, 302, 308–309, 319, 402
  across shelf-slope break, 400–402
  North Sea, 214
  oxygen deficient water, 400–402
  source rock deposition, 400
Transgressive shelf deposits
  oil in, 391
— unit, time, 135
Transitional crust, 107, 426
Transmissivity, 252, 259
  extent of, 262

Trench, 97–104
  off Aleutian Islands, 222
—-slope basin, 97, *See also* Shelf-slope break, active
  convergent margin, 107, 109, 115
  formation of, 109
Trilobites, 91
Tropical fluvial dominated regions, SB on, 12–15
  relatively stable margins, SB on, 12–14
    Amazon-Guiana-Orinoco shelf, 12
    Bengal coast, 12
    Guinea, 12
    Gulf of Mexico, 12
  unstable convergent margins, SB on, 14
    Burma-Indonesia, 14
    Central America, 14
    Ecuador, 14
    Ninas Island, Sumatra, 14
    Preparis Island, 14
    Venezuela, islands off, 14
  unstable rifted margins, SB on, 14–15
    Andaman Sea, 14
    Gulf of California, eastern margin, 14–15
    Gulf of Honduras, 14–15
    Venezuela, northeast, 14
— non-fluvial regions, SB in, 10–12
  reefs, 10
  relatively stable margins
    Bahamas, 12
    Espirito Santo shelf, 12
    Florida, 12
    Sahul shelf, 12
    Yucatan, 12
  rifted substrate
    Baja California, eastern, 12
    Gulf of Aden, 12
    horsts, 12
    Red Sea, 12
    rifts, 12
  unstable convergent margins
    Barbados, 12
    Lesser Antilles, 12
— Province, 385–386
Truman Proclamations
  fishing zones, 438
  natural resources, 438
Tsunami, 220, 300–301
Turbid plume, *See* Sediment resuspension
Turbidite fan, *See* Submarine fan
Turbidites, 14, 113, 159–162, 164, 195, 199–200, 203, 211, 325, 327
  Antler trough, 83
  Campos basin, 419–420
  Delaware basin, 418
  reservoirs, 410–411, 418–420, 424–425
  San Joaquin Valley, 411, 424
Turbidity currents, 135, 152, 159, 161–162, 220, 265, 300, 315–316, 319, 324–325, 327–328, 376, 379, 393, 419, 432
  canyon formation, 25, 37
  in submarine canyons, 245, 281
  related to river discharge, 322
Tuscaloosa Formation (Cretaceous), 144–145
  Louisiana, 418–419
Tyrrhenian Sea
  unstable rifted margin, SB on, 19

U

U.S., See United States
Unconformity, 68–69, 72, 100, 102, 107, 110, 116, 129, 176, 308, 323, 399, 402, 418, 422–423
United Nations Law of Sea
  1958 Conference, 438–439
  1960 Conference, 438–439
  UNCLOS-III, 437, 439–440
—— 1958 Convention
  Continental Shelf, 439
  Territorial Sea and Contiguous Zone, 439
United States, 85
  breccia, internal, 207
  Gulf Coast, 409–411, 414–415, 418
  Law of the Sea, 437
  orogeny in, 79–83
  paleoceanographic map of, 80
  paleoceanography of, 83
  paleogeographic-lithofacies maps of, 80
  paleogeography of, 82–83
  Paleozoic carbonate SB, 195–196
  reconstruction of margins, 79
  territorial sea, 439
  —— margin, 233–248
    East Coast, 25–37, 41, 57, 125, 222–225, 233–234, 237–238, 243, 248, 280–282, 288, 290, 293, 318, 320, 322, 334–335, 342, 344, 346–351, 359–363, 366–370, 373, 378, 381, 383–384
      low sedimentation rate on, 121
      submarine canyons on, 25–37
      temperate region, SB in, 18
    east-west coast comparison, 299
    Gulf of Mexico, northern, 359–360, 362–363, 366–370, 381, 383–385
    Mid-Atlantic states, 17, 25–37, 225, 228, 279–280, 285, 287–288, 290, 295
    northeast, 80
    northwest, 224, 299, 301
    southeast, 375, 434
    West Coast, 222–225, 233–234, 237–238, 242, 245, 248, 310, 318, 322, 334, 346–350, See also Pacific Coast
      seismicity on, 299–300, 308
      shelfedge, 299–311
      shelfedge types, 299, 304–311
Upper Assam Valley
  clastic facies, 409, 422–423
  collision, 422
  Digboi field, 422
  flysch, 422–423
  molasse, 422
  oil fields, 410, 422–423
Upwelling, 223, 233, 237–239, 328, 342, 376, 381, 391–392, 395, 402
  at shelfbreak, 238, 243
  California margin, 236, 239, 242
  cause of off-shelf transport, 228
  Chappel shelfedge, 89
  conodont development, 85
  Illinois basin, 83
  in submarine canyons, 245
  lack of, 243
  Madison sea, 83
  Middle Atlantic Bight, 235, 239
  off Brazil, 393
  off Peru, 393–395
  off southwest Africa, 393–394
  off U.S. West Coast, 224
  on east Florida shelf, 243
  on Grand Banks, 393
  Oregon margin, 236–237, 239, 242
  Ouachita trough, 83, 88, 91
  prevention of reef development
    Chile, 10
    Namibia, 10
    Peru, 10
    Western Sahara, 10
  related to mineral deposition, 429–431, 433, 435
  San Pedro shelf, 227
  South Atlantic Bight, 235, 239–240
  Washington margin, 236–237, 239, 242
Uranium enriched phosphorite, 431
USSR, See Russia
Utah, 89–90
  Brazer Dolomite, 84
  Deseret Limestone, 89
  Deseret starved basin, 79, 83, 91

V

Var River, 12
Venezuela, islands off
  tropical fluvial dominated region, 14
—, northeast
  tropical fluvial dominated region, 14
Viola Formation (Ordovician)
  Arkoma basin, 421
  oil fields, producing from, 421
Virginia Province, 385
Volcanic arc, 97, 99–100, 102, 109
  Granada, 102
  Trinidad-Tobago, 102

W

Washington margin, See also Pacific Coast, 236, 238, 396
  organic carbon on, 396
  sediment dispersal on, 245, 247
  sediment loads of rivers, 302
  shelfedge, 300, 304
  submarine canyons on, 243
  temperate region, SB in, 18
  tides, 238
  upwelling, 236–237, 239, 242–243
Water mass fronts, 219, 222, 228
Waulsortian mound, 83, 88–91
Wave energy, See also Current(s), Wave(s), 5, 15, 61–62, 169, 176, 178, 181, 194–195, 208, 291–292, 295, 299, 305–309, 311, 343–345
  correlation with latitude, 223
  effect on infauna, 383
  high, 307–308, 381, 383, 385, 387
  low, 306, 308, 383
Wave-base, 159
Wave(s), 233, 239, 241–242, 257–263, 409, See also Current(s), Wave energy
  in Gulf of Mexico, 256–263
  inertial, 257–260, 262–263
  internal, 5, 61, 72, 161, 219–220, 222, 224, 227–228, 237–238, 242, 251, 253, 257–258, 265, 279, 281,

287–288, 300, 315, 322, 328, 340, 343, 381, 429, 432
  storm, 5, 161, 220, 223, 225, 228, 287–288, 290, 299–300, 320, 322, 343
  surface, 219–220, 222–224, 227, 237, 256–257, 260, 287, 290–291, 300, 328, 340, 343
  tides, 219–220, 222, 224–225, 227–228, 300
  tsunami, 220, 300–301
Weddell Sea
  polar glacial region, SB on, 9
Well log
  correlation of, 140, 148
  electric, 121, 148–149
  growth fault character on, 149
Welt, *See also* Sill, 79, 82, 91
  Caballos-Arkansas Island Chain, 82–83, 91
West Falls Group (Devonian)
  New York, 164
— Florida slope
  organic carbon on, 383
— Thornton gas field
  California, 424
  Meganos Channel, 424
  paleo-shelfbreak, 424
Western Boundary Current, 235
Wilcox Group (Paleocene-Eocene), 145, 149
  deltaic sediments, 424
  Mexico, 424
  reservoirs, 424
  Texas, 424

Wilmington Canyon, 25, 28–29, 36, 279, 295, 318, 320–321, 343, 350
  meandering in, 25, 29
  shelf-indenting, 28
  straight, 29
Woodbine Formation (Cretaceous), 144
Woodford Formation (Devonian-Mississippian)
  Anadarko Basin, 402
  anoxic model for, 402
Wrench-tectonics, 111, 113
Wyoming, 411
  Brazer Dolomite, 84
  Phosphoria Formation, 394, 402

Y

Yucatan, 179–180
  carbonate bank, 415–416
  tropical non-fluvial region, SB in, 12
Yugoslavia
  melange in, 212

Z

Zechstein evaporites (Permian)
  Europe, western, 401
Zoogeographic province
  Carolinian, 385–386
  Tropical, 385–386
  Virginia, 385
Zoogeography
  foraminifera, 360–362

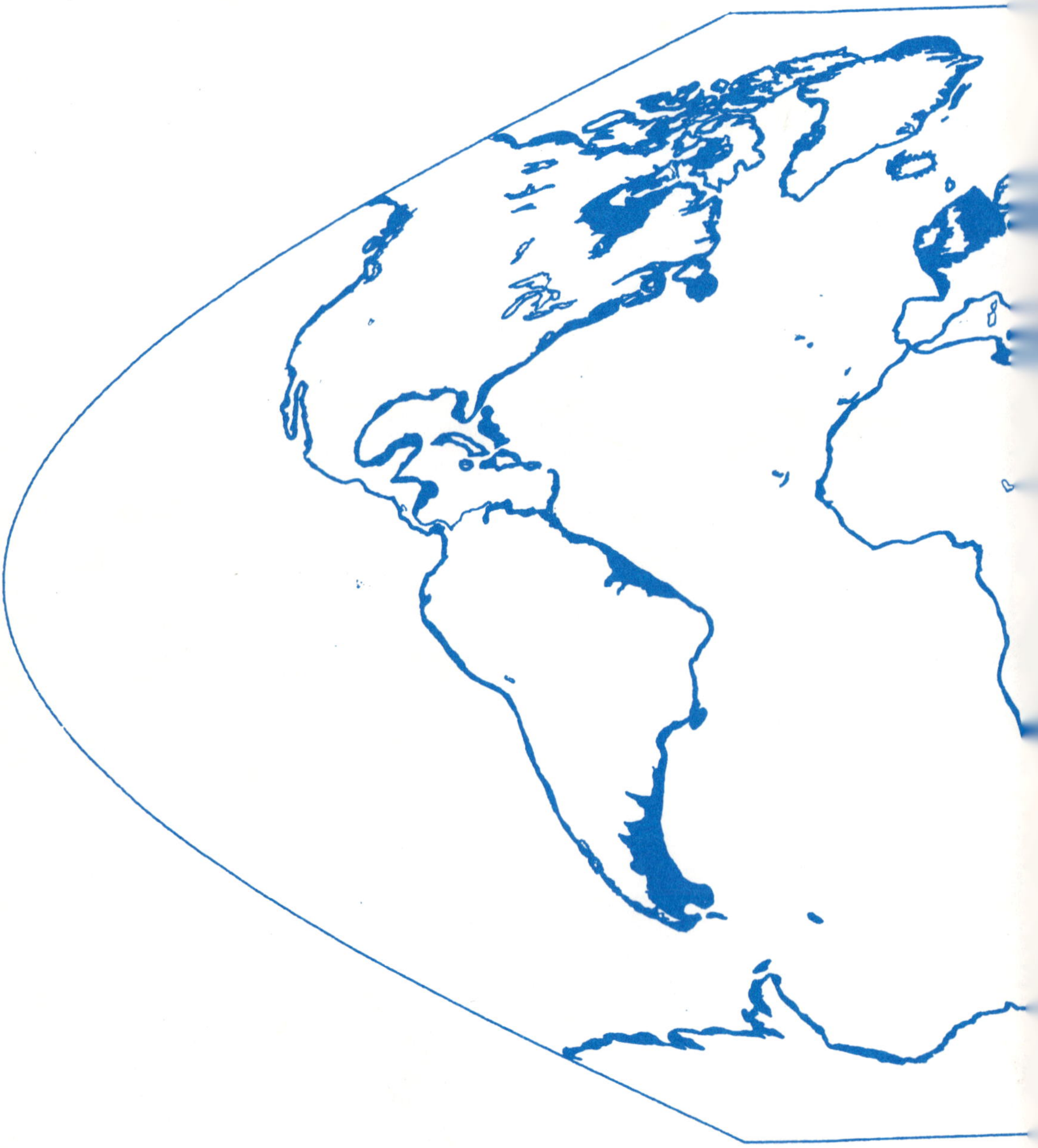